The TCP/IP Protocol Suite

D1298925

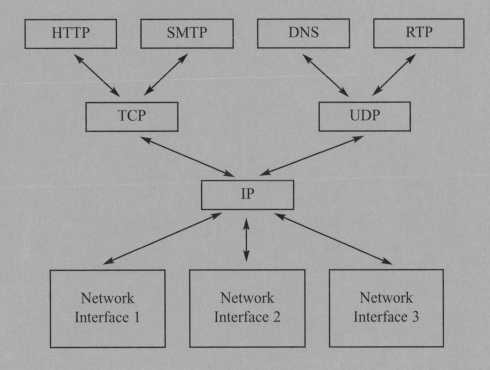

The hourglass shape of the TCP/IP protocol suite underscores the features that make TCP/IP so powerful. The operation of the single IP protocol over various networks provides independence from the underlying network technologies. The communication services of TCP and UDP provide a network-independent platform on which applications can be developed. By allowing multiple network technologies to coexist, the Internet is able to provide ubiquitous connectivity and to achieve enormous economies of scale.

Communication Networks

McGraw-Hill Series in Computer Science

SENIOR CONSULTING EDITOR

C.L. Liu, University of Illinois at Urbana-Champaign

CONSULTING EDITOR

Allen B. Tucker, Bowdoin College

Fundamentals of Computing and Programming
Computer Organization and Architecture
Computers in Society/Ethics
Systems and Languages
Theoretical Foundations
Software Engineering and Database
Artificial Intelligence
Networks, Parallel and Distributed Computing
Graphics and Visualization
The MIT Electrical and Computer Science Series

McGraw-Hill Series in Electrical and Computer Engineering

SENIOR CONSULTING EDITOR

Stephen W. Director, University of Michigan, Ann Arbor

Circuits and Systems
Communications and Signal Processing
Computer Engineering
Control Theory and Robotics
Electromagnetics
Electronics and VLSI Circuits
Introductory
Power
Antennas, Microwaves, and Radar

PREVIOUS CONSULTING EDITORS

Ronald N. Bracewell, Colin Cherry, James F. Gibbons, Willis W. Harman, Hubert Heffner, Edward W. Herold, John G. Linvill, Simon Ramo, Ronald A. Rohrer, Anthony E. Siegman, Charles Susskind, Frederick E. Terman, John G. Truxal, Ernst Weber, and John R. Whinnery

Communication Networks

Fundamental Concepts and Key Architectures

Alberto Leon-Garcia
University of Toronto

Indra Widjaja

Boston   Burr Ridge, IL   Dubuque, IA   Madison, WI
New York   San Francisco   St. Louis
Bangkok   Bogotá   Caracas   Lisbon   London   Madrid   Mexico City
Milan   New Delhi   Seoul   Singapore   Sydney   Taipei   Toronto

McGraw-Hill Higher Education

A Division of The McGraw-Hill Companies

COMMUNICATION NETWORKS
Fundamental Concepts and Key Architectures

Copyright © 2000 by The McGraw-Hill Companies, Inc. All rights reserved. Printed in the United States of America. Except as permitted under the United States Copyright Act of 1976, no part of this publication may be reproduced or distributed in any form or by any means, or stored in a database or retrieval system, without the prior written permission of the publisher.

This book is printed on acid-free paper.

3 4 5 6 7 8 9 0 DOC/DOC 9 0 9 8 7 6 5 4 3 2 1

ISBN 0-07-242349-8

Vice president/Editor-in-chief: *Kevin T. Kane*
Publisher: *Thomas Casson*
Executive editor: *Elizabeth A. Jones*
Senior developmental editor: *Kelley Butcher*
Senior marketing manager: *John T. Wannemacher*
Project manager: *Jim Labeots*
Senior production supervisor: *Heather D. Burbridge*
Freelance design coordinator: *Gino Cieslik*
Cover illustration: *Henk Dawson*
Supplement coordinator: *Rose M. Range*
Compositor: *Keyword Publishing Services*
Typeface: *10/12 Times Roman*
Printer: *R. R. Donnelley & Sons Company*

Library of Congress Cataloging-in-Publication Data
Leon-Garcia, Alberto.
    Communication networks / Alberto Leon-Garcia, Indra Widjaja.
      p. cm.
    Includes bibliographical references and index.
    ISBN 0-07-242349-8 (alk. paper)
    1. Telecommunication systems. 2. Computer networks. I. Widjaja, Indra, 1960- II.
Title.
TK5101.L46 2000
621.382′1–dc21
                                             99-052271

http://www.mhhe.com

To my teachers
  who taught me how to learn for the sake of learning,
and to my students
  who taught me how to learn for the sake of teaching.
  —alg

To my parents: with gratitude for your love and support.
  —iw

PREFACE

OBJECTIVE

Communication networks have entered an era of fundamental change where market and regulatory forces have finally caught up with the relentless advance of technology, as evidenced by the following:

- The explosive growth of multimedia personal computing and the World Wide Web, demonstrating the value of network-based services.
- The deregulation of the telecommunications industry opening the door to new access network technologies (digital cellular systems, cable modems, high-speed DSL modems, direct broadcast satellite systems, satellite constellation networks, broadband wireless cable) that will cause telecommunications infrastructure to migrate towards a flexible packet-based backbone network technology.
- The explosion in available bandwidth due to optical transmission technology and the entry of new national and global backbone service providers.
- The emergence of the Internet suite of protocols as the primary means for providing ubiquitous connectivity across the emerging network of networks.
- The predominance of data traffic over voice traffic dictating that future networks will be designed for data, and that telephone voice service must eventually operate—possibly solely—over the Internet.

Thus, the main architectural elements of the network of networks that will emerge in the next ten years are becoming more evident. The purpose of this book is to introduce electrical engineering, computer engineering, and computer science students to fundamental network architecture concepts and to their application in these emerging networks.

TARGET COURSES

The book is designed for introductory one-semester or one-year courses in communication networks in the upper-level undergraduate and first-year graduate programs. The second half of the book can be used in more advanced courses that deal with the details of current network architectures. The book can also be used by engineering and computer professionals seeking an introduction to networking.

As prerequisites the book assumes a general knowledge of computer systems and programming, and elementary calculus. In certain parts of the text, knowledge of elementary probability is useful but not essential.

APPROACH AND CONTENT

Networks are extremely complex systems consisting of many components whose operation depends on many processes. To understand networks it is essential that students be exposed to *the big picture of networks* that allows them to see how the various parts of the network fit into one whole. We have designed the book so that students are presented with this big picture at the beginning of the book. The students then have a context in which to place the various topics as they progress through the book.

The book attempts to provide a *balanced view of all important elements of networking*. This is a very big challenge in the typical one-semester introductory course which has very limited time available. We have organized the book so that all the relevant topics can be covered at some minimum essential level of detail. Additional material is provided that allows the instructor to cover certain topics in greater depth.

The book is organized into four sections: the first section provides the big picture; the second section develops fundamental concepts; the third section deals with advanced topics and detailed network architectures; and in the fourth section two appendices provide important supporting material.

Big Picture First: Networks, Services, and Layered Architectures

This section begins in Chapter 1 with a discussion of network-based applications that the student is familiar with (World Wide Web, e-mail, telephone call, and home video entertainment). These examples are used to emphasize that modern networks must be designed to support a wide range of applications. We then discuss the evolution of telegraph, telephone, and computer networks, up to the present Internet. This historical discussion is used to identify the essential functions that are common to all networks. We show how there is usually more than one way to carry out a function, for example, connectionless versus circuit-switched transfer of information, and that the specific structure of a network is determined by a combination of technological, market, and regulatory factors at a given point in time.

The view of the network as a provider of services to applications is developed in Chapter 2. We consider the e-mail and Web browsing applications, and we explain the application layer protocols that support these, namely HTTP, SMTP, and DNS. We also explain how these protocols in turn make use of the communication services provided by TCP and UDP. Together these examples motivate the notion of layering, leading naturally to a discussion of the OSI reference model. A detailed example is used to show how Ethernet, PPP, IP, TCP, and UDP work together to support the application layer protocols. The key notions of addressing and encapsulation are developed in this example. Chapter 2 concludes with two optional sections: an introduction to sockets and an introduction to additional application layer protocols and to several TCP/IP utilities. We believe that the student will be familiar with some of the application layer topics, and so Chapter 2 can serve as a bridge to the less visible topics relating to the internal operation of a network. Sockets and TCP/IP utilities provide the basis for very useful and practical exercises and experiments that provide students with some "hands on" networking experience.

Fundamental Network Architecture Concepts

The second section develops the fundamental concepts of network architecture, proceeding from the physical layer to the network layer. We complement the discussion of fundamental concepts with sections that explore trends in network architecture.

Chapter 3 deals with digital transmission including error detetection. We identify the bit rate requirements that applications impose on the network, and then we examine the transmission capabilities of existing and emerging networks. We introduce the relationship between bandwidth, bit rate, and signal-to-noise ratio, and then develop the basic digital transmission techniques, using modem standards as examples. The properties of various media (copper wires, coaxial cable, radio, optical fiber) and their possible role in emerging access networks are then discussed. This chapter contains more material than can be covered in the introductory course, so it is written to allow the instructor to pick and choose what sections to cover.

Chapter 4 discusses digital transmission systems and the telephone network. The first few sections deal with properties of current and emerging optical networks. The digital multiplexing hierarchy and the SONET standard are introduced. We develop the fault recovery features of SONET rings and we emphasize the capability of SONET optical networks to create arbitrary logical topologies under software control. We then introduce wavelength division multiplexing and explain how WDM optical networks share the flexible network configuration features of SONET. The design of circuit switches for traditional telephone networks and for future optical networks is discussed next. The latter sections deal with telephone networks, with a focus on the signaling system that enables telephone service and associated enhanced services, e.g., caller ID, 800-call. We consider the telephone network and the layered architecture of its signaling system. We discuss the frequency reuse concept and its application in telephone and satellite cellular networks.

Chapter 5 is the usual place to discuss data link controls. Instead of dealing immediately with this topic, we first introduce the notions of peer-to-peer protocols and service models. ARQ protocols that provide reliable transfer service are developed in detail as specific examples of peer-to-peer protocols. The detailed discussion gives the student an appreciation of what is involved in implementing a protocol. The end-to-end and hop-by-hop approaches to deploying peer-to-peer protocols are compared, and additional examples of peer-to-peer protocols are introduced for flow control and for timing recovery. We also preview the reliable stream service provided by TCP. The details of HDLC and

PPP data link standards are then presented. Finally we discuss the sharing of a data link by multiple packet flows and introduce the notion of multiplexing gain.

Chapter 6 deals with the transfer information across shared media, using LANs and wireless networks as specific examples. We begin with an introduction to broadcast networks and to approaches to sharing a medium. We explain the function of LANs and their placement in the OSI reference model. We consider random access as well as scheduling approaches to transferring packets across a shared medium. We examine the impact of delay-bandwidth product on performance, and we show why this dictates the evolution of Ethernet from a shared medium access technique to a switched technique. In addition to token ring and FDDI LANs, we also present a full discussion of the IEEE 802.11 wireless LAN standard. We also discuss FDMA, TDMA, and CDMA channelization approaches to sharing media and we show their application in various existing cellular radio networks. We have taken great care to make the difficult topic of CDMA accessible to the student.

Chapter 7 deals with packet switching networks. To provide a context for the chapter we begin by presenting an end-to-end view of packet transfer across the Internet. We then develop the notions of datagram and virtual-circuit packet switching, using IP and ATM as examples. We introduce basic design approaches to packet switches and routers. Shortest-path algorithms and the link state and distance vector approaches to selecting routes in a network are presented next. ATM and the concept of label switching are introduced, and the relationship between Quality-of-Service and traffic shaping, scheduling and call admission control is developed. The chapter includes a discussion of TCP and ATM congestion control.

Key Architectures and Advanced Topics

The third section shows how the fundamental networking concepts are embodied in two key network architectures, ATM and TCP/IP. The section also deals with the interworking of ATM and TCP/IP, as well as with enhancements to TCP/IP to provide secure and more responsive communications.

Chapter 8 presents a detailed discussion of TCP/IP protocols. We examine the structure of the IP layer and the details of IP addressing, routing, and fragmentation and reassembly. We discuss the motivation and present the features of IPv6. We introduce UDP, and examine in detail how TCP provides reliable stream service and flow control end-to-end across a connectionless packet network. RIP, OSPF, and BGP are introduced as protocols for synthesizing routing tables in the Internet. Multicast routing is also introduced.

Chapter 9 deals with the architecture of ATM networks. The ATM layer is explained, and Quality-of-Service and the ATM network service categories are presented. The various types of ATM adaptation layer protocols are discussed next. ATM signaling and PNNI routing are introduced.

Chapter 10 deals with the interworking of IP and ATM and with proposed enhancements to IP. We consider the various approaches for operating IP over

ATM networks. We then introduce Multiprotocol Label Switching which is the most promising example for operating IP over ATM and other link layer protocols. Finally we introduce RSVP, Integrated Services IP, and Differentiated Services IP which together provide mechanisms for providing Quality-of-Service over IP.

Chapter 11 provides an introduction to network security protocols. The various categories of threats that can arise in a network are used to identify various types of security requirements. Secret key and public key cryptography are introduced and their application to providing security is discussed. We develop protocols that provide security across insecure networks and we introduce protocols for establishing security associations and for managing keys. These general protocols are then related to the IP security protocols and to transport layer security protocols.

Chapter 12 deals with multimedia information and networking. We begin with an introduction to the properties of image, audio, and video signals. We discuss the various compression schemes that are applied to obtain efficient digital representations, and we describe the relevant compression standards. We then introduce the RTP protocol for transmitting real-time information across the Internet. Finally, we close the loop in the discussion of "plain old telephone service" by reviewing the various signaling protocols that are being developed to support multimedia communications in general, and IP telephony in particular, over the Internet.

The book ends with an Epilogue that discusses trends in network architecture and identifies several areas that are likely to influence the development of future networks.

Appendices

Appendix A deals with network performance models. Network performance is an integral part of network design and operation. In the text we use quantitative examples to illustrate the tradeoffs involved in various situations. We believe that an intuition for performance issues can be developed without delving into the underlying mathematics. Delay and loss performance results are introduced in the sections that deal with multiplexing, trunking, and medium access control. In these sections, the dynamics of the given problem are described and the key performance results are presented. The purpose of Appendix A is to develop the analysis of the performance models that are cited in the text. These analyses may be incorporated into more advanced courses on communication networks.

Appendix B provides an introduction to network management. The basic functions and structure of a network management system are introduced as well as the Simple Network Management Protocol (SNMP). We present the rules for describing management information, as well as the collection of objects, called Management Information Base, that are managed by SNMP. We also introduce remote monitoring (RMON) which offers extensive network diagnostic, planning, and performance information.

HOW TO USE THIS BOOK

The book was designed to support a variety of introductory courses on computer and communication networks. By appropriate choice of sections, the instructor can make adjustments to provide a desired focus or to account for the background of the students. Chapter 1 to Chapter 8 contain the core material (and more) that is covered in the typical introductory course on computer networks. For example, at the University of Toronto a 40 lecture-hour introductory undergraduate course in computer networks covers the following: Chapter 1 (all); Chapter 2 (all) including a series of lab exercises using sockets; Chapter 3 (sections 3.1, 3.2, 3.5, 3.6, 3.8.1 to 3.8.5); Chapter 4 (sections 4.1 to 4.3); Chapter 5 (all); Chapter 6 (sections 6.1 to 6.4, 6.6.1, 6.6.2); Chapter 7 (all); and Chapter 8 (sections 8.1 to 8.5). For courses that spend more time on the material in Chapter 8 or later, the material from Chapters 3 and 4 can be dropped altogether. The book contains enough material for a two-semester course sequence that provides an introductory course on computer networks followed by a course on emerging network protocols.

PEDAGOGICAL ELEMENTS

The book contains the following pedagogical elements:

- *Numerous Figures*. Network diagrams, time diagrams, performance graphs, state transition diagrams are essential to effectively convey concepts in networking. The 574 figures in the book are based on a set of Microsoft PowerPoint® course presentations that depend heavily on visual representation of concepts. A set of these presentation charts is available to instructors.
- *Numerous Examples*. The discussion of fundamental concepts is accompanied with examples illustrating the use of the concept in practice. Numerical examples are included in the text wherever possible.
- *Text Boxes*. Commentaries in text boxes are used to discuss network trends and interesting developments, to speculate about future developments, and to motivate new topics.
- *Problems*. The authors firmly believe that learning must involve problem solving. The book contains 589 problems. Each chapter includes problems with a range of difficulties from simple application of concepts to exploring, developing or elaborating various concepts and issues. Quantitative problems range from simple calculations to brief case studies exploring various aspects of certain algorithms, techniques, or networks. Simple programming exercises involving sockets and TCP/IP utilities are included where appropriate.
- An *Instructor's Solutions Manual* is available from McGraw-Hill.
- *Chapter Introductions*. Each chapter includes an introduction previewing the material covered in the chapter and in the context of the "big picture".

- *Chapter Summaries and Checklist of Important Terms*. Each chapter includes a summary that reiterates the most important concepts. A checklist of important terms will aid the student in reviewing the material.
- *References*. Each chapter includes a list of references. Given the introductory nature of the text, references concentrate on pointing to more advanced materials. Reference to appropriate Internet Engineering Taskforce (IETF) RFCs and research papers is made where appropriate, especially with more recent topics.
- *A web site*. The following Web site contains links to the on-line version of the solutions manual, the Powerpoint slides*, author information, and other related information: www.mhhe.com/leon-garcia.

ACKNOWLEDGMENTS

The material in the book was developed over many years in introductory as well as advanced courses in networking, both in regular undergraduate and graduate programs as well as in programs with an orientation towards professional practice. We acknowledge the feedback from the many students who participated in these courses and who used various versions of the manuscript. In particular we thank the students from CETYS University. We also acknowledge the input of the graduate students who served as teaching assistants in these courses, especially Dennis Chan, Yasser Rasheed, Mohamed Arad, Massoud Hashemi, Hasan Naser, and Andrew Jun.

We thank Anindo Banerjea, Raouf Boutaba, Michael Kaplan, and Gillian Woodruff for many exciting conversations on networking. Anindo and Raouf graciously provided some of the material that is presented in Chapter 2. We would also like to thank Anwar Elwalid and Debasis Mitra for their continued encouragement and interest in the book. We thank Yau-Ren Jenq for reviewing the fair queueing discussions in detail.

We are especially grateful to Irene Katzela for testing the manuscript in her courses. We also thank Ray Pickholtz for testing various versions of the text, including the beta version, and for his many valuable suggestions and his continued encouragement.

We thank the reviewers for their many useful comments on the various versions of the manuscript: Subrata Banerjee (Stevens Institute of Technology), John A. Copeland (Georgia Institute of Technology), Mario Gerla (UCLA), Rohit Goyal (Ohio State University), Gary Harkin (Montana State University), Melody Moh (San Jose State University), Kihong Park (Purdue University–West Lafayette), Raymond L. Pickholtz (The George Washington University), Chunming Qiao (SUNY Buffalo), Arunabha Sen

---

*The *Instructor's Solutions Manual* and the Powerpoint slides are password protected. See the website for information on how to obtain one.

(Arizona State University), Stuart Tewksbury (West Virginia University), and Zhi-li Zhang (University of Minnesota).

We would also like to acknowledge the many friends from Nortel Networks for showing us the many facets of networking. We thank Sidney Yip for opening the door to many years of interaction. We also thank Richard Vickers, Marek Wernik, and Jim Yan for many illuminating conversations over the years. We especially thank Tony Yuen for sharing his vast knowledge of the networking industry and for continuously showing how the big picture is actually bigger!

We thank Eric Munson from McGraw-Hill for persuading us to take the plunge with this project, and Betsy Jones, Executive Editor, for providing decisive support at key times. In addition we thank the production team at McGraw-Hill for their patience, ideas, and continued support, especially Kelley Butcher and Jim Labeots.

IW would like to thank to his wife Liesye for the constant encouragement and for putting up with him during the many nights and weekends spent writing the book, especially during the final stages.

Finally, ALG would like to thank his soulmate, Karen Carlyle, who went beyond the usual putting up with an author's neglect, to assuming the role of project manager, designer, transcriber and real-time editor for the book.

With the help of the many reviewers, professors, and students who have used early versions of this book we have tried to make the complex and fluid topic of network architecture as approachable, up-to-date and error-free as possible. We welcome all comments and suggestions on how to improve the text. Please contact us via the text's website with any ideas you may have.

Alberto Leon-Garcia
Indra Widjaja

CONTENTS

CHAPTER 1

Communication Networks and Services

The operation of modern communication networks is a very complex process that involves the interaction of many systems. In the study of networks, it is easy to get lost in the intricacy of the details of the various component systems and to lose track of their role in the overall network. The purpose of this and the next chapter is to present students with the "big picture" so that they can place the various components in the context of the overall network.

We begin with a discussion of how the design of networks has traditionally been driven by the services they provide. Some of these services, such as mail, are so basic that they outlive the underlying technology and even the underlying network design. We present several examples of services that are revisited in the course of the book, namely, electronic mail (e-mail), Web browsing, and telephony.

We next consider the problem of designing networks to provide these services. First we present a general discussion on the structure of networks, and we introduce essential functions that all networks must provide. We then present three design approaches to providing these functions: message switching, circuit switching, and packet switching. Each design approach is presented in the context of a sample network, namely, the telegraph network, telephone network, and Internet, respectively. This discussion serves two purposes: to show how the essential functions are incorporated into the design of each network and to provide a historical perspective of networks. We also discuss how the architectures (overall design) of the networks have changed with changes in technology and the prevailing regulatory and business environment.

The context provided by this and the next chapter is intended to prepare students to deal with not only existing networks but also future network technologies and architectures. Finally, at the end of the chapter we give an overview of the book that relates the remaining chapters to the context introduced here.

1.1   NETWORKS AND SERVICES

A **communication network**, in its simplest form, is a set of equipment and facilities that provides a service: the transfer of information between users located at various geographical points. In this textbook, we focus on networks that use electronic or optical technologies. Examples of such networks include telephone networks, computer networks, television broadcast networks, cellular telephone networks, and the Internet.

Communication networks provide a service much like other ubiquitous utilities, and many analogies can be drawn between communication networks and other utility systems. For example, communication networks, such as cable or broadcast television, provide for the distribution of information much like the water supply or electricity power systems that distribute these commodities to users. Communication networks also provide access for gathering information much like sewer or garbage collection systems, which gather various materials from users. On the other hand, communication networks exhibit tremendous flexibility in their use and in this respect they are closest to transporation networks.

Communication networks, along with transportation networks, have become essential infrastructure in every society. Both types of networks provide flexible interconnectivity that allows the flow of people and goods in the case of transportation and the flow of information in the case of communications. Both transportation and communication networks are "enabling" in that they allow the development of a multiplicity of new services. For example, the development of a postal service presupposes the availability of a good transportation system. Similarly, the development of an e-mail service requires the availability of a communication network.

The ability of communication networks to transfer communication at extremely high speeds allows users to *gather information* in large volumes nearly instantaneously and, with the aid of computers, to almost immediately exercise *action at a distance*. These two unique capabilities form the basis for many existing services and an unlimited number of future network-based services.

We will now discuss several services that are supported by current networks. The services are examined from the point of view of user requirements, that is, quality of service, features, and capabilities. The viewpoint here is that networks should ultimately be designed to meet the requirements of the user applications. We refer to these services and their requirements when discussing various issues throughout the book.

Radio and television broadcasting are probably the most common communication services. Various "stations" ("programs") transmit an ensemble of signals simultaneously over radio or cable distribution networks. Aside from selecting the station of interest, the role of the user in these services is passive. Relatively high audio and video quality is expected, but a significant amount of delay (in the order of seconds or more) can be tolerated even in "live" broadcasts.

Telephone service is the most common real-time service provided by a network. Two people are able to communicate by transmitting their voices across the network. The service is "connection-oriented" in the sense that the users must first interact with the network to set up a connection, as shown in Figure 1.1.

The telephone service has a real-time requirement in that users cannot interact as in normal face-to-face conversation if the delays are greater than a fraction of a second (approximately 250 milliseconds). The service must also be reliable in the sense that once the connection is established it must not be interrupted because of failures in the network. At a minimum the delivered voice signal must be intelligible, but in most situations the users expect a much higher quality that enables the listener not only to recognize the speaker but also to discern subtleties in intonation, mood, and so on. A high degree of availability is another requirement: Telephone users expect the network to be capable of completing the desired connection almost all the time. Security and privacy of the conversation is a consideration in some situations.

The telephone service can be enhanced in a number of ways. For example, the 800 service provides toll-free (and possibly long distance) service to the caller where the costs of the call are automatically billed to the subscriber of the service.

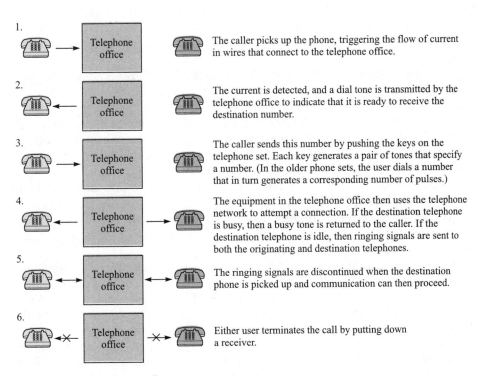

1.
The caller picks up the phone, triggering the flow of current in wires that connect to the telephone office.

2.
The current is detected, and a dial tone is transmitted by the telephone office to indicate that it is ready to receive the destination number.

3.
The caller sends this number by pushing the keys on the telephone set. Each key generates a pair of tones that specify a number. (In the older phone sets, the user dials a number that in turn generates a corresponding number of pulses.)

4.
The equipment in the telephone office then uses the telephone network to attempt a connection. If the destination telephone is busy, then a busy tone is returned to the caller. If the destination telephone is idle, then ringing signals are sent to both the originating and destination telephones.

5.
The ringing signals are discontinued when the destination phone is picked up and communication can then proceed.

6.
Either user terminates the call by putting down a receiver.

FIGURE 1.1 Telephone call setup

Similarly, in credit-card or calling-card services the cost of a call is automatically billed to the holder of the card. Clearly, security and fraud are issues here.

Telephone networks provide a broad class of call management services that use the originating number or the destination number to determine the handling of a call. For example, in *call return* the last originating number is retained to allow it to be automatically called by the destination user at a later point in time. *Caller ID* allows the originating number, and sometimes name, of the originating call to be displayed to the destination user when the receiving device is display capable. *Voice mail* allows a destination user to have calls forwarded to a message-receiving device when the destination user is not available.

Cellular telephone service extends the normal telephone service to mobile users who are free to move within a regional area covered by an interconnected array of smaller geographical areas called cells. Each cell has a radio transmission system that allows it to communicate with users in its area. The use of radio transmission implies design compromises that may result in lower voice quality, lower availability, and greater exposure to eavesdropping. In addition, the cellular system must handle the "handing off" of users as they move from one cell to another so that an ongoing conversation is not terminated abruptly. Some cellular providers also support a roaming service where a subscriber is able to place calls while visiting regional areas other than the subscriber's home base. Note that the mobility aspect to the roaming service is not limited to cellular (or wireless) communications. Indeed, the need for mobility arises whenever a subscriber wishes to access a service from anywhere in the world.

Electronic mail (e-mail) is another common network service. The user typically provides a text message and a name and/or address to a mail application. The application interacts with a local mail server, which in turn transmits the message to a destination server across a computer network. The destination user retrieves the message by using a mail application, such as shown in Figure 1.2. E-mail is not a real-time service in that fairly large delays can be tolerated. It is also not necessarily connection-oriented in that a network connection does not need to be set up expressly for each individual message. The service does require reliability in terms of the likelihood of delivering the message without errors and to the correct destination. In some instances the user may be able to request delivery confirmation. Again security and privacy may be a concern.

Many applications that involve an interaction between processes running in two computers may be characterized by **client/server** interaction. For example, a client may initiate a process to access a given file on some server. The **World Wide Web (WWW) application** typifies this interaction. The WWW consists of a framework for accessing documents that are located in computers connected to the Internet. These documents consist of text, graphics, and other media and are interconnected by links that appear within the documents. The WWW is accessed through a browser program that displays the documents and allows the user to access other documents by clicking one of these links. Each link

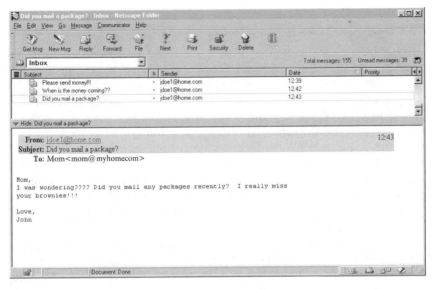

FIGURE 1.2 Retrieving e-mail (*Netscape Communicator screenshots © 1999 Netscape Communications Corporation. Used with permission.*)

provides the browser with a **uniform resource locator (URL)** that specifies the name of the machine where the document is located as well as the name of the file that contains the requested document.

For example, in Figure 1.3 a user is looking at the WorldWide News home page to find out the latest events. The URL of the home page is identified in the pull-down window labeled 'Go to' toward the top of the page. The user wants to find out more about some recent weather events, and so she puts the cursor over the text about the snowfall in LA. The field at the bottom then identifies the URL of the linked document, in this case:

http://www.wwnews.com/ltnews/lasnow

The first term, http, specifies the retrieval mechanism to be used, in this case, the **HyperText Transfer Protocol (HTTP)**. Next the URL specifies the name of the host machine, namely, www.wwnews.com. The remaining data gives the path component, that is, the URL identifies the file on that server containing the desired article. By clicking on a highlighted item in a browser page, the user begins an interaction to obtain the desired file from the server where it is stored, as shown in Figure 1.4.

In addition to text the files in the WWW may contain audio and images that can involve large amounts of information. While the user does not require real-time response, excessive delay in retrieving files reduces the degree of interactivity of the overall application where the user seeks information, reads it, and again seeks additional information by clicking on other items or on links to other Web sites. The overall delay is determined by the delays in accessing the servers as well as the time required to transmit the files through the network.

WHAT IS A PROTOCOL?

In dealing with networks we run into a multiplicity of protocols, with acronyms such as HTTP, FTP, TCP, IP, DNS, and so on. What is a protocol, and why are there so many? A **protocol** is a set of rules that governs how two communicating parties are to interact. In the Web browsing example, the HTTP protocol specifies how the Web client and server are to interact.

The purpose of a protocol is to provide some type of service. HTTP enables the retrieval of Web pages. Other examples of protocols are: File Transfer Protocol (FTP) for the transfer of files, Simple Mail Transfer Protocol (SMTP) for e-mail, Internet Protocol (IP) for the transfer of packets, and Domain Name System (DNS) for IP address lookup.

Chapter 2 shows how the overall communications process can be organized into a stack of layers. Each layer carries out a specific set of communication functions using its own protocol, and each layer builds on the services of the layer below it. In the Web example HTTP uses the connection service provided by the Transmission Control Protocol (TCP). TCP uses the packet transfer service provided by IP which in turn uses the services provided by various types of networks.

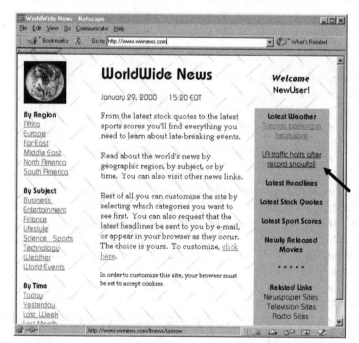

FIGURE 1.3 World Wide Web example (*Netscape Communicator screenshots © 1999 Netscape Communications Corporation. Used with permission.*)

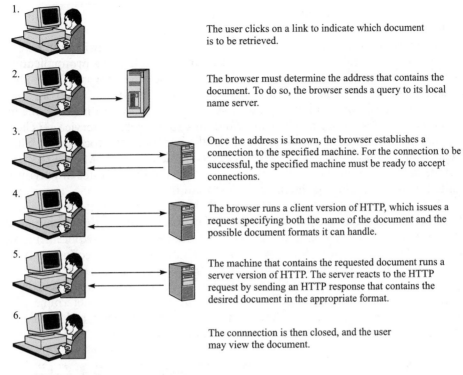

1. The user clicks on a link to indicate which document is to be retrieved.

2. The browser must determine the address that contains the document. To do so, the browser sends a query to its local name server.

3. Once the address is known, the browser establishes a connection to the specified machine. For the connection to be successful, the specified machine must be ready to accept connections.

4. The browser runs a client version of HTTP, which issues a request specifying both the name of the document and the possible document formats it can handle.

5. The machine that contains the requested document runs a server version of HTTP. The server reacts to the HTTP request by sending an HTTP response that contains the desired document in the appropriate format.

6. The connnection is then closed, and the user may view the document.

FIGURE 1.4 Retrieving a Web page

Video on demand characterizes another type of interactive service. The objective of the service is to provide access to a video library, that is, a kind of "video jukebox" located at some remote site, and to provide the type of controls available in a video cassette recorder (VCR), such as slow motion, fast forward, reverse, freeze frame, and pause. The user initiates the service by accessing a menu from which a selection is made. A number of transactions may follow, for example, to provide payment for the service. When these transactions have been completed, a server that contains the selection begins to transmit the video information across the network to the user. Video involves enormous amounts of information, so it is not feasible to transmit and store the entire movie as a single file. Instead the video information is sent as a stream of "frames" that contain individual pictures that constitute the video.

The video-on-demand application is not real-time and can tolerate delay as long as the responsiveness expected in using the VCR-type controls is not affected. However the stream of frames must flow through the network in a steady fashion in the sense that the frame jitter, or the delays between consecutive frames, does not vary too much. Excessive jitter can result in a frame not being available at the receiver after the previous frame has been played out, which would then result in impairments in the video quality. The service requires the delivery of relatively high audio and video quality. Security and privacy are a concern mostly in the initial transactions where the selection and payment are made.

Streamed audiovisual services over the Internet provide an example of a service with some of the features of video on demand. Here an application such as RealPlayer^{TM} can be used to access a "channel" that provides an audiovisual stream to the client machine. By selecting a channel from a control panel, such as the one shown in Figure 1.5, a process is initiated by which a multimedia server begins to send a stream of information to the client process. The client application processes the stream to display audio and a moving picture much like a television displays a program. The service is on demand in the sense that the user determines when the display is initiated, but the degree of interactivity is limited, and the quality of the image is less than that of commercial television.

A more complex class of interactive services results when more than two users are involved. For example, **audio conferencing** involves the exchange of voice signals among a group of speakers. In addition to the requirements of normal telephone service, the network must be able to provide connectivity to the participants and somehow combine the various voice signals to emulate the way voice signals are combined in a normal group discussion. The limitation to voice information means that the visual cues that normally help determine who speaks next are absent, so the group interaction is somewhat awkward and unnatural. The addition of real-time video information can help provide visual

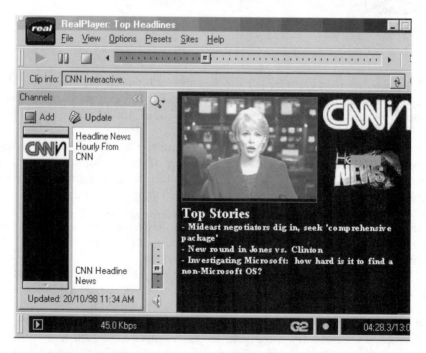

FIGURE 1.5 A RealPlayer segment over the Internet (*Copyright © 1995–2000 RealNetworks, Inc. All rights reserved. RealPlayer is a trademark of RealNetworks, Inc.*)

cues and help produce a more natural conferencing situation. However, the addition of video brings additional requirements in terms of the volume of information that needs to be transmitted, as well as the quality, delay, and jitter requirements of video. The problem remains of displaying the visual information in a way conducive to normal human interaction. Various means of superimposing the images of the participants in a natural setting and of using large screen displays have been explored.

An even more demanding class of services arises when the **audio-visual conferencing** requirements are combined with a demanding real-time response requirement. This type of service might arise in a number of possible settings. It may be required in a situation where a group of people and a variety of machines interact with real phenomena, for example, a complex surgery by a team of surgeons where some of the participants are located remotely. It may also arise in various real-time distributed simulations or even in distributed interactive video games involving geographically separated players. Many recent films, usually of dubious quality, have explored the potential of various immersive, virtual reality technologies that would make use of this type of network service. As you work through the book, you will learn how to specify the requirements that must be met by a communication network to provide this class of challenging services.

1.2   APPROACHES TO NETWORK DESIGN

In the previous section we examined networks from the point of view of the services they provide to the user. We saw that different user applications impose different requirements on the services provided by the network in terms of transfer delay, reliability of service, accuracy of transmission, volume of information that can be transferred, and, of course, cost and convenience. In this section we examine networks from the point of view of the network designer.

The task of the designer is to develop an overall network design that meets the requirements of the users in a cost-effective manner. In the next section, we present a number of essential functions that must be provided by any network. We also discuss how the topology of a network develops as the network grows in scale. Subsequent sections present three approaches to providing the essential functions that are required by a network: message switching, circuit switching, and packet switching. We use the telegraph network, the telephone network, and the Internet as examples of how these three approaches have been applied in practice.

1.2.1   Network Functions and Network Topology

The essential function of a network is to transfer information between a source and a destination. The source and destination typically comprise **terminal** equipment that attaches to the network, for example, a telephone or a computer. This process may involve the transfer of a single block of information or the transfer of a stream of information as shown in Figure 1.6. The network must be able to provide **connectivity** in the sense of providing a means for information to flow among users. This basic capability is provided by **transmission systems** that transmit information by using various media such as wires, cables, radio, and optical fiber. Networks are typically designed to carry specific types of **information representation**, for example, analog voice signals, bits, or characters.

A strong analogy can be made between communication networks and transportation networks. Roads and highways are analogous to transmission lines. Minor roads provide access to higher-speed highways. A specific segment of road corresponds to a point-to-point transmission line. The intersections and highway interchanges that allow the transfer of traffic between highways correspond to **switches**, which transfer the information flow from one transmission line to another. Two actions are required by a driver as his or her vehicle enters an interchange. First the driver must *decide* which exit corresponds to his or her destination; when dealing with information, we refer to this action as **routing**. Once the exit has been determined, the driver must actually *move* the vehicle through that exit; when dealing with information, we call this action **forwarding**.

Figure 1.7a shows how a switch can be used to provide the connectivity between a community of users that are within close distance of each other. The pairwise interconnection of users would require N x (N − 1) lines, which is not sustainable as N grows large. By introducing a switch, the number of lines is reduced to N. In effect, the access transmission lines are extended to a central location, and the "network" that provides the connectivity is reduced to equipment in a room or even a single box or chip. We refer to the access transmission lines and the first switch as an **access network**. These access networks concentrate the information flows prior to entering a backbone network. For example, this approach is used to connect users to the nearest telephone central office.

Typically, users are interested in communicating with other users in another community, just as they might want a high-speed highway to connect the two communities. The communication network provides transmission lines, or "trunks," that interconnect the switches of the two communities. The longer distance of these lines implies higher cost, so **multiplexers** are used to concentrate the traffic between the communities into the trunks that connect the switches, as

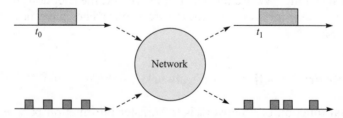

FIGURE 1.6 A network transfers information among users

(a) A switch provides the network to a cluster of users.

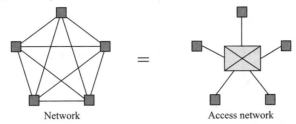

Network   Access network

(b) A multiplexer connects two access networks.

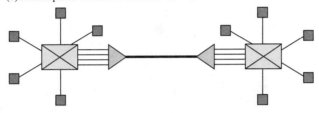

FIGURE 1.7 Role of switches and multiplexers in the network

shown in Figure 1.7b. The associated demultiplexer and switch then direct each information flow to the appropriate destination.

The need to communicate extends to more than two communities, so multiplexers and trunks are used to interconnect a number of access networks to form a *metropolitan network* as shown in Figure 1.8a. Figure 1.8a also shows how the metropolitan network can be viewed as a network that interconnects access subnetworks. A metropolitan network might correspond to a transportation system associated with a county or large city. Similarly, metropolitan subnetworks (e.g., large cities) are interconnected into a *regional network* (e.g., state or province). Figure 1.8b, in turn, shows how a *national network* can be viewed as a network of regional subnetworks that interconnect metropolitan networks.

The above **hierarchical network topology** arises because of geography, cost considerations, and communities of interest among the users. A *community of interest* is defined as a set of users who have a need to communicate with each other. Switches are placed to interconnect a cluster of users where it makes economic sense. Typically, users are more likely to communicate with users who are nearby so most of the traffic tends to stay within the switch. Traffic to more distant users is aggregated into multiplexers that connect to more distant switches through a **backbone network**. A similar process of providing interconnections through multiplexers and switches takes place at higher levels of aggregation, that is, metropolitan to regional and then regional to national networks. In the case of roads and railroads, in particular, it is possible for different operators to compete in providing connectivity between regional centers. This situation also applies to networks, where different long-distance operators or Internet service providers can compete in providing intereconnection between regional networks.

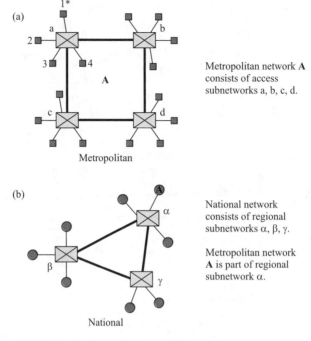

Metropolitan network **A** consists of access subnetworks a, b, c, d.

National network consists of regional subnetworks α, β, γ.

Metropolitan network **A** is part of regional subnetwork α.

FIGURE 1.8 Hierarchical network topology

Traditionally, a larger portion of the traffic has tended to stay at the lower layers in the hierarchy. However the pattern of traffic flow has started to change as the cost of communications has tended to become independent of distance. For example, when you click on a browser link, you do not worry about distance or cost. Consequently, the portion of traffic that is flowing into the backbone of networks is increasing. In addition, as cost becomes less of a factor, community of interest becomes a stronger determinant of traffic flows. These two trends, distance-independent communities of interest and lower communications costs, will in time lead to very different network topologies.

Addressing is required to identify which network input is to be connected to which network output. Here we again see that it is natural to specify addresses according to a hierarchy: number, street, city, state, country. Thus in Figure 1.8 the user identified by an asterisk in part (a) has an address α.A.a.1. Just as there is more than one way to go from New York to Los Angeles, there is more than one possible way to interconnect users in a communication network. We saw above that routing involves the selecting of a path for the transfer of information among users. The use of hierarchical addresses facilitates the task of routing. For example, in routing a letter through the postal system we are first concerned about getting to the right country, then the right state, then the city, and so on. We will see that the hierarchical approach to addressing is in wide use because it simplifies the task of routing information across large networks.

In networks we will find two types of addressing: hierarchical addressing in **wide area networks** and flat addressing in **local area networks**. Hierarchical addressing is required in the wide area networks because they facilitate routing as indicated above. The addressing problem in local area networks is analogous to the problem of finding a building in a university campus. Buildings are identified by a name or a number that does not provide any location information. Flat addresses are acceptable in local area networks because the number of nodes is relatively small.

The network operation must also ensure that network resources are used effectively under normal as well as under problem conditions. **Traffic controls** are necessary to ensure the smooth flow of information through the network, just as stop signs and traffic lights help prevent car collisions. In addition, when congestion occurs inside the network as a result of a surge in traffic or a fault in equipment, the network should react by applying **congestion** or **overload control** mechanisms to ensure a degree of continued operation in the network. In the case of roads, drivers are instructed to take a different route or even to stay home! Similarly, during congestion, information traffic may be rerouted or prevented from entering the network.

Finally, we note one additional necessary network function. Just as highways and roads require crews of workers to maintain them in proper condition, networks require extensive support systems to operate. In networks these functions fall under the category of **network management** and include monitoring the performance of the network, detecting and recovering from faults, configuring the network resources, maintaining accounting information for cost and billing purposes, and providing security by controlling access to the information flows in the network.

In this section we have introduced the following essential network functions: terminal; transmission; information representation; switching, including routing and forwarding; addressing; traffic control; congestion control; and network management. The following list summarizes the functions that a network must provide.

- Basic user service—the primary service or services that the network provides to its users.
- Switching approach—the means of transferring information flows between communication lines.
- Terminal—the end system that connects to the network.
- Information representation—the format of the information handled by the network.
- Transmission system—the means for transmitting information across a physical medium.
- Addressing—the means for identifying points of connection to the network.
- Routing—the means for determining the path across the network.
- Multiplexing—the means for connecting multiple information flows into shared connection lines.

In the remainder of this section, we consider specific approaches to providing these functions. In Chapter 2 we show how these functions are organized into an overall layered network architecture.

1.2.2 Message, Packet, and Circuit Switching

Several approaches can provide the essential network functions identified in the previous sections. These approaches are characterized by how information is organized for transmission, multiplexing, routing, and switching in a network. Each approach defines the internal operation of a network and specifies a basic information transfer capability. The service or services that are provided to the user build on this basic transfer capability.

First we consider telegraph networks. Here the network operation was based on message switching, which provides for the transfer of text messages called telegrams. Next we consider the telephone network, which uses circuit switching as its mode of operation to provide its basic service, the telephone call. Finally, we consider packet switching, which forms the basis for the Internet Protocol (IP). You will see that the Internet provides two basic types of services to its users and that these services are built on top of a packet transfer capability. For each of these three networks, we explain how the essential network functions are provided. Table 1.1 summarizes some of the features of these three networks. We also provide a few historical notes on the evolution of these networks. Figure 1.9 indicates the transmission capabilities of these networks over the past 150 years.

Function	Telegraph network	Telephone network	Internet
Basic user	Transmission of telegrams	Bidirectional, real-time transfer of voice signals	Datagram and reliable stream service between computers
Switching approach	Message switching	Circuit switching	Connectionless packet switching
Terminal	Telegraph, teletype	Telephone, modem	Computer
Information representation	Morse, Baudot, ASCII	Analog voice or PCM digital voice	Any binary information
Transmission system	Digital over various media	Analog and digital over various media	Digital over various media
Addressing	Geographical addresses	Hierarchical numbering plan	Hierarchical address space
Routing	Manual routing	Route selected during call setup	Each packet routed independently
Multiplexing	Character multiplexing, message multiplexing	Circuit multiplexing	Packet multiplexing, shared media access networks

TABLE 1.1 Essential functions of network operation

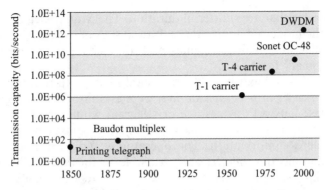

FIGURE 1.9 Evolution of telecommunications capacity

1.2.3 Telegraph Networks and Message Switching

In 1837 Samuel B. Morse demonstrated a practical telegraph that provided the basis for **telegram** service, the transmission of text messages over long distances. The text was encoded using the Morse code into sequences of dots and dashes. Each dot or dash was communicated by transmitting short and long pulses of electrical current over a copper wire. By relying on two signals, telegraphy made use of a **digital transmission system**. The Morse code shown in Table 1.2 is an example of an efficient **binary representation** for text information. The time required to transmit a message is minimized by having more frequent letters

	Morse code	Probability of occurrence		Morse code	Probability of occurrence
A	. —	0.08149	S	. . .	0.06099
B	— . . .	0.01439	T	—	0.10465
C	— . — .	0.02757	U	. . —	0.02458
D	— . .	0.03787	V	. . . —	0.00919
E	.	0.13101	W	. — —	0.01538
F	. . — .	0.02923	X	— . . —	0.00166
G	— — .	0.01993	Y	— . — —	0.01982
H	. . . .	0.05257	Z	— — . .	0.00077
I	. .	0.06344	1	. — — — —	
J	. — — —	0.00132	2	. . — — —	
K	— . —	0.00420	3	. . . — —	
L	. — . .	0.03388	4	. . . . —	
M	— —	0.02535	5	. . . . .	
N	— .	0.07096	6	—	
O	— — —	0.07993	7	— — . . .	
P	. — — .	0.01981	8	— — — . .	
Q	— — . —	0.00121	9	— — — — .	
R	. — .	0.06880	0	— — — — —	

TABLE 1.2 International Morse code

assigned to strings that are shorter in duration. The Morse telegraph system is a precursor of the modern digital communication system in which all transmission takes place in terms of binary signals and all user information must first be converted to binary form.

In 1851 the first submarine cable was established between London and Paris. Eventually, *networks of telegraph stations* were established, covering entire continents. In these networks a message or telegram would arrive at a telegraph-switching station, and an operator would make a **routing** decision based on destination **address** information. The operator would then store the message until the communication line became available to **forward** the message to the next appropriate station. This process would be repeated until the message arrived at the destination station. **Message switching** is used to describe this approach to operating a network. Addressing, routing, and forwarding are elements of modern computer networks.

The **information transmission rate** (in letters per second or words per minute) at which information could be transmitted over a telegraph circuit was initially limited to the rate at which a single human operator would enter a sequence of symbols. An experienced operator could transmit at a speed of 25 to 30 words per minute, which, assuming five characters per word and 8 bits per character, corresponds to 20 bits per second (bps) in Figure 1.9.

A subsequent series of inventions attempted to increase the rate at which information could be transmitted over a single telegraph circuit by **multiplexing** the symbols from several operators onto the same communication line. One multiplexing system, the Baudot system, used **characters**, groups of five binary symbols, to represent each letter in the alphabet. The Baudot multiplexing system could interleave characters from several telegraph operators into a single transmission line.

The Baudot system eventually led to the modern practice of representing alphanumeric characters by groups of binary digits as in the **ASCII** code (short for American Standard Code for Information Interchange). The Baudot system also eventually led to the development of the **teletype terminal**, which could be used to transmit and receive digital information and was later used as one of the early input/output devices for digital computer systems. A Baudot multiplexer telegraph with six operators achieved a speed of 120 bits/sec.

Another approach to multiplexing involves **modulation**, which uses a number of sinusoidal pulses to carry multiple telegraphy signals. For example, each of the binary symbols could be transmitted by sending a sinusoidal pulse of a given frequency for a given period of time, say, frequency f_0 to transmit a "0" and f_1 to transmit a "1." Multiple sequences of binary symbols could be transmitted simultaneously by using multiple pairs of frequencies for the various telegraphy signals. These modulation techniques formed the basis for today's modems.

Prior to the invention of telegraphy, long-distance communication depended primarily on messengers who traveled by foot, horse, or other means. In such systems a message might propagate at a rate of tens of kilometers per day. In the late 1700s, visual telegraph networks using line-of-sight semaphore systems reduced the time required to deliver a message. In 1795 it was reported that a signal in one such system in England took only 3 minutes to traverse a distance of 800 kilometers [Stumpers 1984]. The invention of the electric telegraph in the early 1800s extended the range of communications beyond the line of sight and made long-distance communication almost instantaneous; for example, an electric signal would take less than 3 milliseconds to cover the 800 kilometers.[1] Clearly, electrical communications had marked advantages over all other forms of communications. Indeed, the telegraph gave birth to the "news" industry; to this day some newspapers have the name *The Daily Telegraph*.

1.2.4   Telephone Networks and Circuit Switching

In 1875, while working on the use of sinusoidal signals for multiplexing in telegraphy, Alexander Graham Bell recognized that direct transmission of a voice signal over wires was possible. In 1876 Bell developed a system that could transmit the entire voice signal and could form the basis for voice communication, which we now know as the **telephone**. The modern telephone network was developed to provide basic telephone service, which involves the two-way, real-time transmission of voice signals across a network.

The telephone and telegraph provided services that were fundamentally different. The telegraph required an expert operator with knowledge of Morse code, while the telephone terminal was very simple and did not require any expertise. Consequently the telephone was targeted as a direct service to end users, first in the business and later in residential markets. The deployment of telephones grew quickly, from 1000 phones in 1877 to 50,000 in 1880 and 250,000 in 1890.

Connectivity in the original telephone system was provided by an **analog transmission system**. The transmitted electrical signal is analogous to the original voice signal; that is, the signal is proportional to the sound pressure in speech. It was quickly recognized in the early days of telephony that providing dedicated lines between each pair of users is very costly. Switches were introduced shortly after the invention of the telephone to minimize the cost of providing connectivity between a community of users.

In its simplest form, the switch consists of a **patch cord panel** and a human operator as shown in Figure 1.10a. The originating user picks up the telephone and in the process activates a signal in the circuit that connects it to the telephone office. The signal alerts the operator that a connection is requested. The operator takes the requested name and checks to see whether the desired user is available. If so, the operator establishes a connection by inserting the two ends of a cord into the sockets that terminate the lines of the two users as shown in Figure 1.10b. This connection allows electrical current, and the associated voice signal, to flow between the two users. This end-to-end connection is maintained for the

---

[1]The speed of light in a vacuum is 3×10^8 meters/second; in cable, it is 2.3×10^8 meters/second; in optical fiber, it is 2×10^8 meters/second.

(a) A switch in the form of an operator with a patch cord panel (not shown).

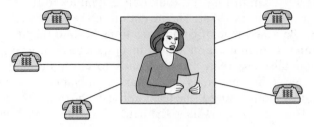

(b) Cords interconnect user sockets providing end-to-end connection.

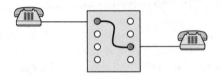

FIGURE 1.10 Switching

duration of the call. When the users are done with their conversation, they "hang up" their telephones, which generates a signal indicating that the call is complete. The two telephone lines are then available to make new connections. We say that telephone networks are **connection-oriented** because they require the setting up of a connection before the actual transfer of information can take place. The transfer mode of a network that involves setting up a dedicated end-to-end connection is called **circuit switching**.

Note that in circuit switching the **routing** decision is made when the path is set up across the network. After the call has been set up, information is "forwarded" continuously across each switch in the path. No additional address information is required after the call is set up.

The telephone network has undergone a gradual transition to its present state, where it is almost completely based on digital transmission and computer technology. This transition began with the invention of the transistor in 1948 and accelerated with the invention of integrated circuits in the 1960s, leading to the development of **digital transmission systems** that could carry voice in a more cost-effective manner. Digital transmission systems were designed to carry binary information in the form of 0s and 1s. Thus these systems required that the analog voice signal be converted into a binary sequence, using a technique called **pulse code modulation (PCM)**. The T-1 digital transmission system was first deployed in 1962 to carry voice traffic between telephone offices. This **multiplexing** system could handle 24 voice calls for a total transmission rate of 1.5 Mbps in Figure 1.9.

Even as digital transmission systems were being deployed, the new digital transmission segments had to interface to existing analog switches. Upon arrival at an analog switch, the digital signal would be reconverted into analog voice signals for switching and then reconverted to digital form for transmission in the next hop. This situation eventually led to the invention of **digital switches** that could switch the voice signals in digital form. Thus a voice call would need to be digitized only once upon entering the network; then it would be transmitted and switched in digital form until it reached the other end of the backbone network. The call would then be converted to analog form for transmission over the pair of wires that connects the user to the network.

As the number of users and geographical extent of the user population increased, telephone networks grew along the lines shown in Figure 1.11. Users are served by an **access network** that connects them to a *local* **central office** (CO) switch. The switches themselves are interconnected with higher-speed communication lines through *tandem* switches. **Multiplexing** is used to combine many calls in these high-speed lines. Tandem switches, in turn, connect to toll switches that are used to provide long-distance connections. The result is **hierarchical network topology**, such as the one shown in Figure 1.11. Figure 1.9 shows the steady increase in the transmission rates that have been deployed in these backbone transmission lines over the years. The telephone digital multiplexing systems have advanced tremendously with the introduction of optical fiber transmission. In 2000 we are seeing the introduction of 1600 gigabit/second dense wavelength-division multiplexing (DWDM) systems.

The hierarchical network topology of the telephone network is complemented by a hierarchical decimal **numbering system** for dialing connections in the telephone network. For example in the 10-number system used in North America, the area code specifies a subarea that has been assigned a three-digit number. The next three numbers are the exchange code, which identifies specific switching facilities in a central office within the subarea. The final four digits specify a specific line that connects the user to the central office.

Another advance in telephone networks involved the introduction of **computer control** for the setting up of connections in a switch. These computers would examine a request for a call as it came in, check to see whether the

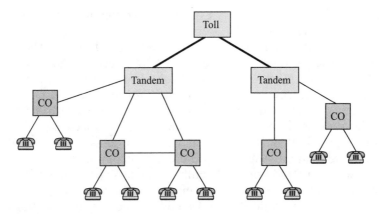

FIGURE 1.11 Hierarchical telephone network structure

destination was available, and if so, make the appropriate connection. The use of computers to control the switch provided great flexibility in modifying the control and in introducing new features. It also led to the introduction of a separate **signaling** network to carry the messages between the switch computers.

In the early 1970s telephone companies realized that the signaling network (and its computer control) could be used to introduce **enhanced telephone services**. Credit-card calls, long-distance calls, 800 calls, and other services could all be implemented using this more capable signaling network. In the case of credit-card calls, a recorded message could request the credit-card number. The digits would be collected, and a message would be sent to a database to check the credit-card number; if authorized, the call would then be set up. The signaling network also enabled **mobility**, which is the capability of a network to direct calls to users as they roam away from their home network.

1.2.5   The Internet and Packet Switching

The Internet Protocol (IP) provides a means for transferring information across multiple, possibly dissimilar, networks. Before discussing packet switching and the organization of the Internet, we need to present some of the types of computer networks that the Internet operates on.

COMPUTER NETWORKS

The first computer network was the Semi-Automatic Ground Environment (SAGE) system developed between 1950 and 1956 for air defense systems [Green 1984]. The system consisted of 23 computer networks, each network connecting radar sites, ground-to-air data links, and other locations to a central computer. The SABRE airlines reservations system, which was introduced in 1964, is cited as the first large successful commercial computer network, and it incorporated many of the innovations of the SAGE system.

Early computers were extremely expensive, so techniques were developed to allow them to be shared by many users. In time-sharing systems the processor would visit each job awaiting service in round-robin fashion. The computer would spend a fixed duration of time, called a time slice, executing instructions from each job before moving to the next job. The need to access time-shared computers led to the development of *terminal-oriented networks*. We will see below that this development led to tree-topology networks with the host computer at the root node. As the cost of computers dropped and their use in organizations proliferated, it became necessary to connect to more than one computer. This situation required a different topology than that of terminal-oriented networks. In addition, as "dumb" terminals were replaced by "intelligent" terminals and later by personal computers, it became necessary to develop networks that were more flexible and could provide communications among many computers.

The ARPANET was the first major effort at developing a network to inter-connect *computers* over a wide geographical area.[2] We emphasize the fact that the "users" of this network were full-fledged computers, not terminals. As such, the users of this network had processing and storage resources not available in previous terminal equipment. It therefore became possible to develop powerful networking protocols that made use of this processing capability at the edge of the network and to simplify the operation of the equipment inside the network. This approach is in marked contrast to the telephone network, where the intelligence resides inside the network, not in the telephone set. The TCP/IP protocols that emerged out of the ARPANET project form the basis for today's Internet.

The trend towards a proliferation of inexpensive computers, and hence toward a paradigm that supports the Internet approach to networking, was recognized in the mid-1980s. Paul Green, in a 1985 article, notes that a SAGE system computer occupied a four-story building, could execute 150,000 instructions per second, and had 32 kilobytes of memory (RAM) and 600 kilobytes of magnetic storage (disk). By the early 1980s these capabilities could be provided by a desktop computer (see problem 15). Paul Green also quotes Bob Metcalf as making the following statement in 1983, "In the 1960's there was one mainframe per company, in the 1970's [there was added] one minicomputer per branch office, in the 1980's one microcomputer per office or home." Metcalf then conjectured that "in the 1990's there will be one nanocomputer in each appliance." In the year 2000 we can say that Metcalf was close to the mark, for we are indeed seeing the proliferation of computers in appliances. As these computers are required to communicate, we will see a demand for computer networks of a scale much larger than that of the global telephone network.

Terminal-oriented networks

Figure 1.12a shows an arrangement that allows a number of terminals to time-share a computer. Each terminal, initially a **teletype printer** and later a **video terminal**, is connected by a set of wires to the computer, enabling the terminal to input instructions and to obtain results from the computer. Initially all terminals were located in a room adjacent to the host computer. Access from terminals located farther from the host computer became possible as communication lines with greater geographical reach became available. Eventually, **modem** devices for transmitting digitial information were introduced so that terminals could access the host computer via the public switched telephone network (PSTN), as shown in Figure 1.12b.

Certain applications required a large number of geographically distributed terminals to be connected to a host computer. For example, a set of terminals at various travel agencies in a city might need to access the same computer. In most of these applications, the terminals would generate messages in a **bursty** manner;

---

[2]The Advanced Research Projects Agency (ARPA) of the U.S. Department of Defence funded the development of the ARPANET.

(a) Time-shared computers & cables for input devices

(b) Dial in

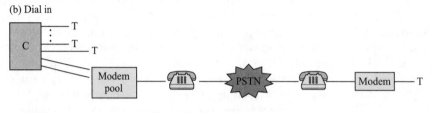

T = terminal

FIGURE 1.12 Terminal-oriented networks

that is, the message transmissions would be separated by long idle times. The cost of providing individual lines to each terminal could be prohibitive. **Line-sharing techniques** were developed to allow a number of terminals to share a communication line. Figure 1.13 shows a multidrop line arrangement, allowing several terminals to share one line to and from the computer. This system uses a master/slave polling arrangement whereby the host computer sends a poll message to a specific terminal on the outgoing line. All terminals listen to the outgoing line, but only the terminal that is polled replies by sending any information that is ready for transmission on the incoming line.

Another means for sharing a communication line among terminals generating bursty traffic involves the use of **statistical multiplexers**, as shown in Figure 1.14. Messages from a terminal are **encapsulated** in a frame that contains a header and the user information. The header provides the **address** information necessary to identify the terminal. Additional **framing** information is usually required to allow the demultiplexer to delineate the beginning and end of each message. The messages from the various terminals are collected by the concentrator, ordered into a queue, and transmitted one at a time over the communication line to the host computer. Note that *messages may be lost* if a surge of message arrivals occurs without sufficient buffers to store the arrivals in the multiplexer. The host computer sorts out the messages from each terminal and carries out the necessary processing.

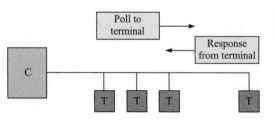

FIGURE 1.13 Sharing a multidrop line

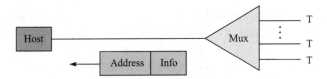

FIGURE 1.14 Multiplexer systems

Early data transmission systems that made use of telephone lines had to deal with errors in transmission arising from a variety of sources: interference from spurious external signals, impulses generated by analog switching equipment, and thermal noise inherent in electronic equipment. **Error-control techniques** were developed to ensure virtually error-free communication of data information. In addition to the header, each block of information would have a number of "check bits" appended prior to transmission. These check bits would enable the receiver to detect whether the received blocks contained errors and, if so, to request transmissions. In this manner very high levels of reliability in data transmission could be achieved.

Figure 1.15 shows a typical terminal-oriented network circa 1970. Remote concentrators/multiplexers at regional centers connect to a host computer, using high-speed lines. Each remote concentrator gathers messages using lower-speed lines from various sites in its region. Multipoint lines such as those discussed above could be used in such lines, for example, Atlanta in the figure. Note that routing and forwarding are straightforward in these tree-topology networks as all information flows from terminals to the central host and back.

Computer-to-computer networks

The basic service required of computer networks is the transfer of messages from any computer connected to the network to any other computer connected to the network. This function is similar to the message switching service provided by

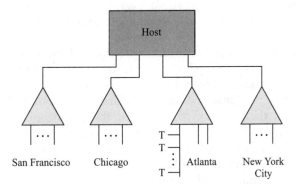

FIGURE 1.15 Typical terminal-oriented network circa 1970

telegraph systems. An additional requirement of a computer network is that it provide short transit times for interactive messages. This requirement suggests that a limit be imposed on the size of messages that are allowed to enter the network, since long messages can result in long waiting times for interactive traffic. **Packet switching** addresses this problem. The network is designed to transfer variable-length blocks of information up to some specified maximum size. User messages that do not fit inside a single packet are segmented and transmitted using multiple packets.

Two types of packet transfer mode result from the two basic approaches to routing and forwarding. The first approach uses **connectionless packet transfer**, or **datagrams**, where each packet is routed independently of all other packets. This approach was used in the ARPANET and later the Internet and is discussed in the next section. The second approach involves first setting up a **virtual circuit (VC)** across switches and links in the network and then forwarding all subsequent packets along the VC. This approach generalizes the procedures followed in the telephone network and is used by Asynchronous Transfer Mode (ATM) networks, which are discussed later in the book.

The ARPANET

The **ARPANET** was developed in the late 1960s to provide a test bed for **wide area network** packet switching research. **Packet switching** was viewed as a promising approach to enable the sharing of computer resources. A **packet** is a block that consists of a header with a destination address attached to user information and is transmitted as a single unit across a network, much as a telegram would be transmitted in a telegraph network. The ARPANET consisted of packet switches interconnected by communication lines that provided multiple paths for interconnecting host computers over wide geographical distances. The packet switches were implemented by dedicated minicomputers, and each packet switch was connected to at least two other packet switches to provide alternative paths in case of failures. The communications lines were leased from public carriers and initially had a speed of 56-kbps lines. The resulting network had a topology, such as shown in Figure 1.16. The ARPANET was designed to transmit packets of information no longer than a given maximum length, about 1000 bits. Therefore, messages from a host might require segmentation into several packets.

ARPANET packet communications were **connectionless** in the sense that no connection setup was required prior to the transmission of a packet. Thus packets could be transmitted without incurring the connection setup delay. Each packet or **datagram** contained **destination address information** that enabled the packet switches in the network to carry out the **routing** of the packet to the destination. Each packet switch maintained a routing table that specified the output line that was to be used for each given destination. Packets were then buffered to await transmission on the appropriate link. Packets from different users would thus be multiplexed into the links between packet switches. Because no connection setup was necessary, no prior allocation of bandwidth or buffer resources was made for each user.

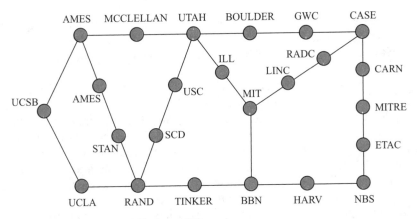

FIGURE 1.16 ARPANET circa 1972

Each packet switching node implemented a **distributed route synthesis** algorithm to maintain its routing tables. The algorithm involved the exchange of information between neighboring nodes. This arrangement enabled the routing tables to be updated in response to changes in traffic or topology and gave ARPANET the capability to adapt to faults in the network. The packets would simply be routed around the points of failure after the failures were detected and the routing tables updated. Because routes could change, it was also possible for packets to arrive out of order at the destination.

Each packet switch in ARPANET contained a limited amount of buffering for holding packets. To prevent packet switches from becoming congested, an end-to-end **flow control** was developed to limit the number of packets that a host can have in transit.

In addition to addressing many fundamental packet networking problems, the ARPANET also developed several lasting applications. These included e-mail, remote login, and file transfer.

Ethernet local area networks

The emergence of workstations that provided computing at lower cost led to a proliferation of individual computers within a department or building. To minimize the overall system cost, it was desirable to share (then) expensive devices such as printers and disk drives. This practice gave rise to the need for **local area networks (LANs)**. The requirements of a LAN are quite different from those of a wide area network. In a LAN the small distances between computers implied that low-cost, very high speed and relatively error-free communication was possible. Complex error-control procedures were largely unnecessary. In addition, in the local environment machines were constantly being moved between labs and offices, which created the administrative problem of keeping track of the location of a computer at any given time. This problem is easily overcome by giving the **network interface card (NIC)** for each machine a **unique address** from a flat address space and by **broadcasting** all messages to all machines in the LAN. A

medium access control protocol becomes essential to coordinate access to the transmission medium in order to prevent collisions. A variety of topologies can provide the broadcasting feature required by LANs, including ring, bus, and tree networks.

The most successful LAN, the **Ethernet** shown in Figure 1.17a, involved transmission over a **bus topology** coaxial cable. Stations with messages to transmit would first sense the cable (carrier sensing) for the presence of ongoing transmissions. If no transmissions were found, the station would proceed to transmit its message. The station would continue to monitor the cable in an attempt to detect collision. If collisions were detected, the station would abort its transmission. The introduction of sensing and collision detection significantly improved the efficiency of the transmission cable.

The bus topology of the original Ethernet has a disadvantage in terms of the cost of coaxial wiring relative to the cost of telephone cable wires as well as in terms of fault handling. Twisted-pair Ethernet was developed to provide lower-cost through the use of conventional unshielded copper wires such as those used for telephones. As shown in Figure 1.17b, the computers are now connected by copper wires in a **star topology** to a hub that can be located in a closet in the same way that telephones are connected in a building. The computers transmit packets using the same random access procedure as in the original Ethernet, except that collisions now occur at the hub, where the wires converge.

THE INTERNET

In the mid-1970s after the ARPANET packet-switching network had been established, ARPA began exploring data communications using satellite and mobile packet radio networks. The need to develop protocols to provide packet communications across multiple, possibly dissimilar, networks soon became apparent. An **internetwork** or **internet** involves the interconnection of multiple networks into a single large network, as shown in Figure 1.18. The component networks may differ in terms of their underlying technology and operation. For example, these networks could consist of various types of LANs, ATM net-

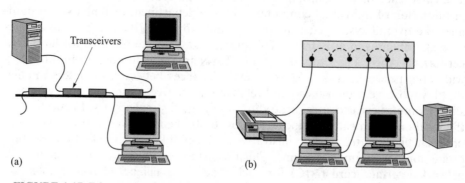

(a)   (b)

FIGURE 1.17 Ethernet local area network

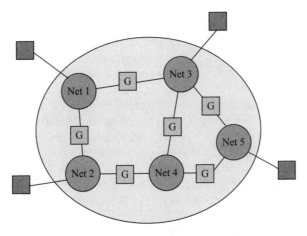

G = gateway

FIGURE 1.18 An internetwork

works, or even a single point-to-point link. The power in the internet concept is that it allows different networks to coexist and interwork effectively.

The **Internet Protocol (IP)** was developed to provide for the connectionless **transfer of packets** across an internetwork. In IP the component networks are interconnected by special packet switches called **gateways** or **routers**. IP routers direct the transfer of IP packets across an internet. After a routing decision is made, the packets are placed in a buffer to await transmission over the next network. In effect, packets from different users are **statistically multiplexed** in these buffers. The underlying networks are responsible for transferring the packets among routers.

IP currently provides **best-effort service**. That is, IP makes every effort to deliver the packets but takes no additional actions when packets are lost, corrupted, delivered out of order, or even misdelivered. In this sense the service provided by IP is unreliable. The student may wonder why one would want to build an internetwork to provide unreliable service. The reason is that providing reliability inside the internetwork introduces a great deal of complexity in the routers. The requirement that IP operates over any network places a premium on simplicity.

The design of IP attempts to keep the operation within an internet simple by relegating complex functions to the edge of the network. The connectionless orientation means that the routers do not need to keep any state information about specific users or their packet flows. This situation allows IP to scale to very large networks. Similarly, when congestion occurs inside an internet, packets are discarded. The end-to-end mechanisms at the edges of the network are responsible for recovery of packet losses and for adapting to the congestion.

IP uses a limited **hierarchical address space** that has location information embedded in the structure. IP addresses consist of four bytes usually expressed in decimal-dotted notation, for example 128.100.11.56. IP addresses consist of two parts: a network ID and a host ID. Machines in the same geographical location share common portions of the address, which allows routers to handle addresses with the same prefix in the same manner. The Internet also provides a **name space** to refer to machines connected to the Internet, for example tesla. comm.toronto.edu. The name space also has a hierarchical structure, but it is administrative and not used in the routing operation of the network. Automatic translation of names to addresses is provided by the **Domain Name System (DNS)**.

The transfer of individual blocks of information using datagrams can support many applications. However, many applications also require the reliable transfer of a stream of information in the correct sequence or order. The **Transmission Control Protocol (TCP)** was developed to provide reliable transfer of stream information over the connectionless IP. TCP operates in a pair of end hosts across an IP internet. TCP provides for error and flow control on an end-to-end basis that can deal with the problems that can arise because of lost, delayed, or mis-delivered IP packets. TCP also includes a mechanism for reducing the rate at which information is transmitted into a internet when congestion is detected. TCP exemplifies the IP design principle, which is that complexity is relegated to the edge of the network, where it can be implemented in the host computers.

Throughout the text we use the term *Internet* as defined by the Federal Networking Council [FNC 1995]. *Internet* refers to the global information system that

a. *Is logically linked together by a global unique address space based on the Internet Protocol (IP) or its subsequent extensions/follow-ons.*

b. *Is able to support communications using the Transmission Control Protocol/Internet Protocol (TCP/IP) suite or its subsequent extensions/follow-ons, and/or other IP-compatible protocols.*

c. *Provides, uses, or makes accessible, either publicly or privately, high-level services layered on the communications and related infrastructure described herein.*

WHAT IS A DOMAIN NAME?

While each machine on a network has an IP address, this address usually provides little information on the use of the machine or who owns it. To make the addresses more meaningful to people, host names were introduced. For example, the IP address 128.100.132.30 is not as informative as the corresponding domain name toronto.edu, which, of course, identifies a host belonging to the University of Toronto.

Today host names are determined according to the *DNS*. The system uses a hierarchical tree topology to reflect different administrative levels. Below the root level are the familiar terms such as com, org, and edu as well as country identifiers such as jp and ca for Japan and Canada, respectively. When written linearly, the lowest level is the far-left term. Thus the domain name of the Network Architecture Lab at the University of Toronto is nal.toronto.edu, where nal is a subdomain of toronto.edu.

DNS is discussed further in Chapter 2.

1.2.6   Discussion on Switching Approaches

We have seen three approaches to designing networks. Telegraph networks were introduced primarily to show the continuity between them and modern computer networks. We conclude the section on network design with a brief comparison of circuit and packet switching. We address both of these topics in greater detail in the remainder of the book, but it is worthwhile to make a few comparisons at this point.

The telephone network was designed to provide real-time voice communications. Circuit switching provides this by setting up end-to-end paths across the network that allow information to flow across the network with very low delay. Circuit switching is well matched to the steady information flow of voice signals. However, circuit switching is not suitable for the transfer of many types of data traffic. The call setup introduces too much delay before information can be transmitted. The bursty nature of data traffic also results in poor utilization of the transmission capability provided by the dedicated connection. An additional disadvantage of circuit switching is that it must maintain state information on all its connections. The failure of a switching node or transmission line requires new circuits to be completely set up.

The Internet uses connectionless packet switching to provide robustness with respect to failures through its capability to reroute traffic around points of failure. Connectionless packet switching requires that transmission capability be expended in the transfer of header information in every packet. This overhead reduces the amount of transmission capacity available to transfer user information. On the one hand, because connectionless packet switching avoids the call setup delay of circuit switching, the former does not need to keep any state information on connections. On the other hand, additional processing is required to carry out the routing decision that has to be made for every packet. Connectionless packet switching also requires the computers at the edge of the network to deal with the effects of the unreliable nature of packet transfer. Although TCP provides an effective mechanism for handling reliable and ordered transfer, achieving low delay for real-time applications still poses a challenge in the Internet. We will see how IP is developing to meet these **Quality-of-Service** challenges in later chapters.

This section has described how networks emerge, evolve, and disappear. In the late 1700s the visual telegraph network extended over thousands of kilometers in Europe. This network vanished completely after the introduction of the electric telegraph. The telegraph network was gradually displaced by the telephone network over a period of many years. What does the future hold for telephone networks and the Internet?

In Section 1.1 we described a broad range of services that networks can provide. Some of these services are already provided, whereas others require capabilities that cannot be met by existing networks. Since the mid-1970s an effort has been underway to develop an "integrated services" network design that can provide traditional telephone services as well as data communication services. The ATM network architecture discussed later in the book was developed to meet this requirement. The TCP/IP architecture of the Internet is also evolving to meet this requirement. Both ATM and TCP/IP networks are based on packet switching. However, ATM is connection-oriented, whereas IP is connectionless. As we discuss the various essential functions of the network, one of the recurring themes will be the comparison of these two approaches to network design.

1.3 KEY FACTORS IN COMMUNICATION NETWORK EVOLUTION

In the previous section we traced the evolution of communication networks from telegraphy to the emerging integrated services networks. Before proceeding with the technical details of networking, however, we pause to discuss factors that influence the evolution of communication networks. Figure 1.19 shows the three traditional factors: technology, regulation, and market. To these we add standards, a set of technical specifications followed by manufacturers or service providers, as a fourth factor.

A traditional axiom of telecommunications was that a new telecommunications service could succeed only if three conditions were satisfied. First of all the technology must be available to implement the service in a cost-effective manner. Second, government regulations must permit such a service to be offered. Third, the market for the service must exist. These three conditions were applicable in the monopoly environment where a single provider made all the decisions regarding design and implementation of the network. The move away from single providers of network services and manufacturers of equipment made compliance with recognized standards essential.

The availability of the technology to implement a service in and of itself does not guarantee its success. Numerous failures in new service offerings can be traced back to the nontechnology factors. Frequently new services fall in gray areas where the regulatory constraints are not clear. For example, most regula-

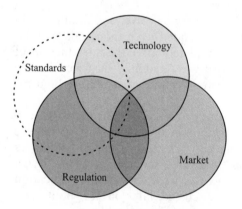

FIGURE 1.19 Factors determining success of a new service

tory policies regarding television broadcasting are intended for radio broadcast and cable systems; however, it is not clear that these regulations apply to television over the Internet. Also, it is seldom clear ahead of time that a market exists for a given new service. For example, the deployment of videotelephony has met with failure several times in the past few decades due to lack of market.

1.3.1   Role of Technology

Technology always plays a role in determining what can be built. The capabilities of various technologies have improved dramatically over the past two centuries. These improvements in capabilities have been accompanied by reductions in cost. As a result, many systems that were simply impossible two decades ago have become not only feasible but also cost-effective.

Of course, fundamental physical considerations place limits on what technology can ultimately achieve. For example, no signal can propagate faster than the speed of light, and hence there is a minimum delay or latency in the transfer of a message between two points a certain distance apart. However, while bounded by physical laws, substantial opportunities for further improvement in enabling technologies remain.

The capabilities of a given technology can be traced over a period of time and found to form an S-shaped curve, as shown in Figure 1.20a. During the initial phase the capabilities of the technology improve dramatically, but eventually the capabilities saturate as they approach fundamental limitations. An example of this situation is the capability of copper wires to carry information measured in bits per second. As the capabilities of a given technology approach saturation, innovations that provide the same capabilities but within a new technology class arise. For example, as copper wire transmission approached its fundamental limitation, the class of coaxial cable transmission emerged, which in turn was replaced by the class of optical fiber transmission. The optical

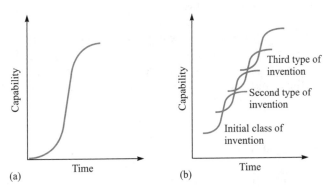

FIGURE 1.20 Capability of a technology as an S curve (based on Martin 1977)

fiber class has much higher fundamental limits in terms of achievable transmission rates and its S curve is now only in the early phase. When the S curves for different classes of technologies are superimposed, they form a smooth S curve themselves, as shown in Figure 1.20b.

In discussing the evolution of network architecture in section 1.2, we referred to the capability curve for information transmission shown in Figure 1.9. The figure traces the evolution from telegraphy to analog telephony, computer networks, digital telephony, and the currently emerging integrated services networks. In the figure we also note various milestones in the evolution of networking concepts. Early telegraphy systems operated at a speed equivalent to tens of bits per second. Early digital telephone systems handled 24 voice channels per wire, equivalent to about 1,500,000 bits per second. In 1997 optical transmissions systems could handle about 500,000 simultaneous voice channels, equivalent to about 10^{10} bits per second (10 gigabits per second)! In the year 2000 systems can operate at rates of 10^{12} bits per second (1 terabit per second) and higher! These dramatic improvements in transmission capability have driven the evolution of networks from telegraphy messaging to voice telephony and currently to image and video communications.

In addition to information transmission capacity, a number of other key technologies have participated in the development of communication networks. These include signal processing technology and digital computer technology. In particular, computer memory capacity and computer processing capacity play a key role in the operation of network switches and the implementation of network protocols. These two technologies have thus greatly influenced the development of networks. For more than three decades now, computer technology has improved at a rate that every 18 to 24 months the same dollar buys twice the performance in computer processing, and computer storage. These improvements have resulted in networks that not only can handle greater volumes of information and greater data rates but also can carry out more sophisticated processing and hence support a wider range of services.

It should be noted that the advances in capabilities in technology do not just happen. They are due to the creativity of numerous engineers and scientists that work relentlessly in pursuing these advances. It should also be noted that advances in the core technologies are not the only factor in providing advances in technology. The development of new algorithms to design, control, and manage large systems are key to handling the complexity associated with modern systems. Indeed, most of this book is dedicated to presenting the key concepts that are used in the design, control, and management of modern communication networks.

1.3.2 Role of Regulation

Traditional communication services in the form of telephony and telegraphy have been government regulated. Because of the high cost in deploying the requisite infrastructure and the importance of controlling communications, governments often chose to operate communications networks as monopolies. The planning of communication networks was done over time horizons spanning several decades. This planning accounted for providing a very small set of well-defined communication services, for example, telegraph and "plain-old telephone service" (POTS). These organizations were consequently not very well prepared to introduce new services at a fast rate.

The last three decades have seen a marked move away from monopoly environment for communications. The Carterfone decision by the U.S. Federal Communications Commission (FCC) in 1968 opened the door for the connection of non-telephone-company telephone equipment to the telephone network. The breakup in 1984 of the AT&T system into an independent long-distance carrier and a number of independent regional telephone operating companies opened the way for further competition. Initially, customers had a choice of long-distance carriers. More recently, with the development of wireless radio technologies and advances in cable TV systems, competition is now possible in the access portion of the network, which connects the customer to the main telephone backbone network. Indeed the Telecommunications Act of 1996 opens the way for a far wider participation of established and new companies in the development and deployment of new services and new technologies.

The trend toward deregulation of telecommunications services has also taken place on an international basis. Countries such as the United Kingdom, New Zealand, and Australia in particular have experimented with novel approaches to providing telecommunications services in a competitive environment. As indicated above, the development of new technologies, wireless radio in particular, has enabled businesses to provide complete telecommunications services using entirely new technology infrastructures. In the third world and in countries with emerging industries, these wireless technologies have a cost advantage over the traditional wire-based systems and hence facilitate the introduction of competition by enabling the deployment of new alternative service providers.

In spite of the trend towards deregulation, telecommunications will probably never be entirely free of government regulation. For example, telephone service is now considered an essential "lifeline" service in many countries, and regulation plays a role in ensuring that access to a minimal level of service is available to everybody. Regulation can also play a role in addressing the issue of which information should be available to people over a communications network. For example, many people agree that some measures should be available to prevent children from accessing pornography over the Internet. However, there is less agreement on the degree to which information should be kept private when transmitted over a network. Should encryption be so secure that no one, not even the government in matters of national security, can decipher transmitted information? These questions are not easily answered. The point here is that regulation on these matters will provide a framework that determines what types of services and networks can be implemented.

1.3.3   Role of the Market

The existence of a market for a new service is the third factor involved in determining the success of a new service. This success is ultimately determined by a customer's willingness to pay, which, of course, depends on the cost, usefulness, and appeal of the service. *For a network-based service, the usefulness of the service frequently depends on there being a critical mass of subscribers.* For example, telephone or e-mail service is of limited use if the number of reachable destinations is small. In addition, the cost of a service generally decreases with the size of the subscriber base due *to economies of scale*, for example, the cost of terminal devices and their components. The challenge then is how to manage the deployment of a service to first address a critical mass and then to grow to large scale.

As examples, we will cite one instance where the deployment to large scale failed and another where it succeeded. In the early 1970s a great amount of investment was made in the United States in developing the Picturephone service, which would provide audio-visual communications. The market for such a service did not materialize. Subsequent attempts have also failed, and only recently are we starting to see the availability of such a service piggybacking on the wide availability of personal computers.

As a second example, we consider the deployment of cellular radio telephony. The service, first introduced in the late 1970s, was initially deployed as a high-end service that would appeal to a relatively narrow segment of people who had to communicate while on the move. This deployment successfully established the initial market. The utility of being able to communicate while on the move had such broad appeal that the service mushroomed over a very short period of time. The explosive growth in the number of cellular telephone subscribers prompted the deployment of new wireless technologies.

1.3.4   Role of Standards

Standards are basically agreements, with industrywide, national, and possibly international scope, that allow equipment manufactured by different vendors to be interoperable. Standards focus on interfaces that specify how equipment is physically interconnected and what procedures are used to operate across different equipment. Standards applying to data communications between computers specify the hardware and software procedures through which computers can correctly and reliably "talk to one another." Standards are extremely important in communications where the value of a network is to a large extent determined by the size of the community that can be reached. In addition, the investment required in telecommunications networks is very high, and so network operators are particularly interested in having the choice of buying equipment from multiple, competing suppliers, rather than being committed to buying equipment from a single supplier.

Standards can arise in a number of ways. In the strict sense, de jure standards result from a consultative process that occurs on a national and possibly international basis. For example, many communication standards, especially for telephony, are developed by the **International Telecommunications Union (ITU)**, which is an organization that operates under the auspices of the United Nations. Almost every country has its own corresponding organization that is charged with the task of setting national communication standards. In addition, some standards are set by nongovernmental organizations. The TCP/IP protocol suite and the Ethernet local area network are two examples of this type of standard.[3] De facto standards arise when a certain product, or class of products, becomes dominant in a market. For example, personal computers based on Intel microprocessors and the Microsoft Windows operating system are a standard in this sense.

The existence of standards enables smaller companies to enter large markets such as communication networks. These companies can focus on the development of limited but key products that are guaranteed to operate within the overall network. This environment results in an increased rate of innovation and evolution of both the technology and the standards.

On a more fundamental level, standards provide a framework that can guide the decentralized activities of the various commercial, industrial, and governmental organizations involved in the development and evolution of networks.

1.4 BOOK OVERVIEW

This chapter provides the "big picture" of the communication network. We have identified fundamental concepts that form the basis for the operation of these networks. These fundamental concepts appear in various forms in different network architectures. The particular form at any given time depends on the available technology and the prevailing commercial and regulatory environment. Concepts appear, disappear, and reappear as demonstrated by the example of the digital network in telegraphy that disappeared, only to reappear in the modern Internet. In the remainder of the text we systematically develop these fundamental concepts. We hope that the student will then be well prepared to deal with future network architectures.

Chapter 2 begins with several examples that demonstrate how two familiar applications—Web browsing and e-mail—make use of underlying communication services. These examples illustrate the fundamental concept of layering. We then present *a layered reference model for network design* and we introduce the notion of a *network protocol*. We use the Internet as an example of how this layered design is implemented. The remainder of the book elaborates on the

---

[3]The Internet Engineering Task Force (IETF) is responsible for the development of Internet standards. The IEEE.802 committee deals with LAN standards.

details of these concepts and on how they are applied in present and emerging network architectures. The student is encouraged to keep the big picture in mind when delving into the detailed working of networks and their protocols. We also include an optional section on UNIX sockets to show how applications can access the TCP/IP communications services and to provide a basis for the development of lab exercises. An optional section on useful TCP/IP applications is also included.

Chapter 3 deals with the transmission systems required to transfer digital information. The chapter begins with a brief discussion of the properties of text, voice, images, and video in order to identify the basic transmission requirements that a network handling these information flows must meet. The chapter then considers the transmission capability of media and the fundamental limitations and trade-offs inherent in digital transmission. The basic operation of modems is investigated, and examples from various "physical layer" standards are described. The chapter concludes with an introduction to some error-detection and -correction techniques that are used to provide error-free communications over unreliable transmission links.

Chapter 4 introduces the transmission systems that form the basis for existing telephone and computer networks. The chapter discusses the operation of the telephone network and the types of digital transmission capabilities that it can provide. This chapter is placed here because from the point of view of computer communications, the role of the telephone network is to provide physical connections between computers. We believe that an understanding of the basics of telephone networks is important as the Internet evolves to provide services tranditionally offered by the telephone network, including telephony itself. The chapter also introduces mobile cellular networks.

Chapter 5 begins with a discussion of peer-to-peer protocols that two end stations can implement to provide reliable communication across an unreliable link or an unreliable network. We consider other adaptation functions that can be provided by peer-to-peer protocols, including flow control and timing recovery. We then consider standards for the data link layer, which deals with the transfer of frames of information between machines that are directly connected by a physical link. We conclude with a discussion of how statistical multiplexing can be used to share a physical link among several users.

Chapter 6 discusses the various classes of medium access control techniques for sharing broadcast transmission media. The application of medium access control techniques to LANs, satellite networks, and wireless networks and relevant standards are discussed. We also discuss the interconnection of LANs by bridges.

In Chapter 7 we discuss the operation of packet-switching networks. We discuss addressing and its interplay with routing table design. We also discuss the design and operation of the network routing algorithms and protocols. Shortest path and associated techniques are developed and applied to the routing problem. We compare the connection-oriented and connectionless approaches to packet switching, and we discuss the design of packet switches and routers. ATM and IP are introduced as examples. We introduce the notion of packet scheduling and its role in providing Quality-of-Service. We also discuss the basic approaches to providing congestion control in a packet network.

Chapter 8 builds on the previous chapter to provide a full discussion of TCP/IP internets. The structure of the IP packet and the operation of the IP protocol are described. Various protocols associated with IP are also discussed. Approaches to providing sufficient IP addresses in the face of explosive growth in the Internet are introduced. The structure of IPv6, the next version of this protocol, is also discussed. We explain how the two transport layer protocols, TCP and User Datagram Protocol (UDP), build on the connectionless packet transfer service of IP to provide the communication services on which user applications are built. Protocols for the dynamic assignment of IP addresses and support for mobile users are also discussed. Finally, the principal IP routing algorithms are introduced, and the extension to multicast routing is also discussed.

Chapter 9 also builds on Chapter 7 to provide a full discussion of ATM networks. The ATM header and the operation of ATM networks is described. The notion of a traffic connection contract to provide Quality-of-Service guarantees and the associated traffic control mechanisms are introduced. The role of the adaptation layer in providing a variety of communication services to the user is also discussed. Addressing and the signaling procedures to set up ATM connections are also introduced, as well as routing protocols in ATM networks.

Chapter 10 considers advanced network architecture issues. Approaches for operating IP over ATM networks are discussed. Approaches to achieving low delay packet transfer and of providing Quality-of-Service guarantees in the Internet are also discussed.

Chapter 11 addresses the issue of providing security in a network. Various application scenarios are introduced, and several types of security services are discussed. The recent set of standards for providing security services as part of IP are also introduced. Finally, we introduce cryptographic algorithms and explain how they are incorporated into network security protocols.

Chapter 12 deals with the representation of multimedia information and the requirements multimedia services impose on the underlying network. The chapter discusses the techniques used in converting analog information into digital form. Methods for achieving efficient representations for all forms of digital information are also discussed. The applications of these techniques to fax, digital voice, image, and video and the relevant standards are discussed. The Real-Time Transport Protocol (RTP) for supporting real-time multimedia services over IP is introduced. Associated session control protocols for establishing and controlling real-time sessions are introduced and related to the issue of providing telephone service across the Internet and conventional telephone networks.

The book ends with an epilogue that cites several active areas of development that are likely to profoundly affect future network architectures. The student is encouraged to track these developments.

Performance evaluation is a key aspect of network design. Throughout the text we discuss performance issues. In Appendix A we carry out a development of network models that parallels the discussion of performance issues in the text. Courses that include performance modeling can readily incorporate these sections into the syllabus. Network management is the set of activities that deal with the monitoring and maintenance of the many processes and resources that are required for a network to function correctly. In Appendix B we give an overview of network management protocols that are used to describe and exchange management information.

CHECKLIST OF IMPORTANT TERMS

access network
addressing
backbone network
best-effort service
bursty
circuit switching
communication network
congestion control
connection-oriented
connectionless packet transfer
datagram
Domain Name System (DNS)
encapsulation
error-control techniques
Ethernet
forwarding
framing
gateway
hierarchical address space
hierarchical network topology
Hypertext Transfer Protocol (HTTP)
information representation
information transmission rate
International Telecommunications Union (ITU)
Internet
internetwork or internet
Internet Protocol (IP)
local area network (LAN)
message switching
multiplexing
numbering system
packet switching
Quality-of-Service (QoS)
routers
routing
services
signaling
statistical multiplexing
switch
Transmission Control Protocol (TCP)
transmission system
uniform resource locator (URL)
virtual circuit (VC)
wide area network (WAN)
World Wide Web (WWW)

FURTHER READING

"100 Years of Communications Progress," *IEEE Communications Magazine*, Vol. 22, No. 5, May 1984. Contains many excellent articles on the history of communications and predictions on the future of networks.

Bylanski, P. and D. G. W. Ingram, *Digital Transmission Systems*, Peter Peregrinus Ltd., England, 1980. Interesting discussion on Baudot telegraph system.

Carne, E. B., *Telecommunications Primer: Signals, Building Blocks, and Networks*, Prentice-Hall, Englewood Cliffs, New Jersey, 1995. Excellent introduction to telephone networks.

Davies, D. W., D. L. A. Barber, W. L. Price, and C. M. Solomonides, *Computer Networks and Their Protocols*, John Wiley & Sons, New York, 1979.

Federal Networking Council Resolution, http://www.fnc.gov/Internet_res.html, October 1995.

Goralski, W. J., *Introduction to ATM Networking*, McGraw-Hill, New York, 1996. Presents an interesting discussion on the history of telegraphy and telephony.

Green, P. E., "Computer Communications: Milestones and Prophecies," *IEEE Communications Magazine*, May 1984, pp. 49–63.

Hunter, C. D., "The Real History of the Internet," http://www.asc.upenn.edu/usr/chunter/agora_uses/chapter_2.html.

Kahn, R. E., "Resource-Sharing Computer Communication Networks," *Computer Networks: A Tutorial*, M. Abrams, R. P. Blanc, and I. W. Cotton, eds., IEEE Press, 1978, pp. 5-8–5-18.

Leiner, B. M., V. G. Cerf, D. D. Clark, R. E. Kahn, L. Kleinrock, D. C. Lynch, J. Postel, L. G. Roberts, and S. Wolff, "A Brief History of the Internet," http://www.isoc.org/internet-history/brief.html, May 1997.

Martin, J., *Future Developments in Telecommunications*, Prentice-Hall, Englewood Cliffs, New Jersey, 1977. Details Martin's vision of the future of networking; interesting to look back and see how often his predictions were on target.

Perlman, R., *Interconnections: Bridges and Routers*, Addison-Wesley, Reading, Massachusetts, 1992. Excellent discussion of fundamental networking principles.

Schwartz, M., R. R. Boorstyn, and R. L. Pickholtz, "Terminal-Oriented Computer-Communication Networks," *Computer Networks: A Tutorial*, M. Abrams, R. P. Blanc, and I. W. Cotton, eds., IEEE Press, 1978, pp. 2-18–2-33.

Stevens, W. R., *TCP/IP Illustrated, Volume 1: The Protocols*, Addison-Wesley, Reading, Massachusetts, 1994.

Stumpers, F. L. H. M., "The History, Development, and Future of Telecommunications in Europe," *IEEE Communications Magazine*, May 1984, pp. 84–95.

Yeager, N. J. and R. E. McGrath, *Web Server Technology: The Advanced Guide for World Wide Web Information Providers*, Morgan Kaufmann, San Francisco, 1996.

RFC1160, V. Cerf, "The Internet Activities Board," May 1990.

PROBLEMS

1. a. Describe the step-by-step procedure that is involved from the time you deposit a letter in a mailbox to the time the letter is delivered to its destination. What role do names, addresses, and mail codes (such as ZIP codes or postal codes) play? How might the letter be routed to its destination? To what extent can the process be automated?

   b. Repeat part (a) for an e-mail message. At this point you may have to conjecture different approaches about what goes on inside the computer network.

   c. Are the procedures in parts (a) and (b) connection-oriented or connectionless?

2. a. Describe what step-by-step procedure might be involved inside the network in making a telephone connection.

   b. Now consider a personal communication service that provides a user with a personal telephone number. When the number is dialed, the network establishes a connection to wherever the user is located at the given time. What functions must the network now perform to implement this service?

3. Explain how the telephone network might modify the way calls are handled to provide the following services:
   a. Call display: the number and/or name of the calling party is listed on a screen before the call is answered.
   b. Call waiting: a special sound is heard when the called party is on the line and another user is trying to reach the called party.
   c. Call answer: if the called party is busy or after the phone rings a prescribed number of times, the network gives the caller the option of leaving a voice message.
   d. Three-way calling: allows a user to talk with two other people at the same time.

4. a. Suppose that the letter in problem 1 is sent by fax. Is this mode of communications connectionless or connection-oriented? real-time or non-real-time?
   b. Repeat part (a) for a voice-mail message left at a given telephone.

5. Suppose that network addresses are scarce and are assigned so that they are not globally unique; in particular, suppose that the same block of addresses may be assigned to multiple organizations. How can the organizations use these addresses? Can users from two such organizations communicate with each other?

6. Explain the similarity between the domain name system and the telephone directory service.

7. Consider the North American telephone-numbering plan discussed in the chapter. Could this numbering plan be used to route packets among users connected to the telephone network?

8. a. Describe the similarities and differences in the services provided by (1) a music program delivered over broadcast radio and (2) music delivered by a dedicated CD player.
   b. Describe how these services might be provided and enhanced by providing them through a communications network.

9. a. Use the World Wide Web to visit the sites of several major newspapers. How are these newspapers changing the manner in which they deliver news over the Internet?
   b. Now visit the Web sites of several major television networks. How are they changing the manner in which they deliver news over the Internet? What differences, if any, exist between the approaches taken by television networks and newspapers?

10. Discuss the advantages and disadvantages of transmitting fax messages over the Internet instead of the telephone network.

11. a. Suppose that an interactive video game is accessed over a communications network. What requirements are imposed on the network if the network is connection-oriented? connectionless?
    b. Repeat part (a) if the game involves several players located at different sites.
    c. Repeat part (b) if one or more of the players is in motion, for example, kids in the back of the van during a summer trip.

12. Discuss the similarities between the following national transportation networks and a communications network. Is the transportation system more similar to a telephone network or to a packet network?

a. Railroad network.

b. Airline network.

c. Highway system.

d. Combination of (a), (b), and (c).

13. In the 1950s standard containers were developed for the transportation of goods. These standard containers could fit on a train car, on a truck, or in specially designed container ships. The standard size of the containers makes it possible to load and unload them much more quickly than non-standard containers of different sizes. Draw an analogy to packet-switching communications networks. In your answer identify what might constitute a container and speculate on the advantages that may come from standard-size information containers.

14. The requirements of world commerce led to the building of the Suez and Panama Canals. What analogous situations might arise in communication networks?

15. Two musicians located in different cities want to have a jam session over a communication network. Find the maximum possible distance between the musicians if they are to interact in real-time, in the sense of experiencing the same delay in hearing each other as if they were 10 meters apart. The speed of sound is approximately 330 meters/second, and assume that the network transmits the sound at the speed of light in cable, 2.3×10^8 meters/second.

16. The propagation delay is the time required for the energy of a signal to propagate from one point to another.

a. Find the propagation delay for a signal traversing the following networks at the speed of light in cable (2.3×10^8 meters/second):

a circuit board	10 cm
a room	10 m
a building	100 m
a metropolitan area	100 km
a continent	5000 km
up and down to a geostationary satellite	$2 \times 36,000$ km

b. How many bits are in transit during the propagation delay in the above cases if bits are entering the above networks at the following transmission speeds: 10,000 bits/second; 1 megabit/second; 100 megabits/second; 10 gigabits/second.

17. In problem 16, how long does it take to send an L-byte file and to receive a 1-byte acknowledgment back? Let $L = 1$, 10^3, 10^6, and 10^9 bytes.

18. *BYTE*, April 1995, gives the following performance and complexity of Intel ×86 processors:

1978	8086	0.33 Dhrystone MIPs	29,000 transistors
1982	286	1.2 Dhrystone MIPs	134,000 transistors
1985	386	5 Dhrystone MIPs	275,000 transistors
1989	486	20 Dhrystone MIPs	1.2 million transistors

1993	Pentium	112 Dhrystone MIPs	3.1 million transistors
1995	P6	250+ Dhrystone MIPs	5.5 million transistors

Plot performance and complexity versus time in a log-linear graph and compare it to the growth rate discussed in the text. Access the Intel Web site, http://www.intel.com, to get updated figures on their processors.

19. Use your Web browser to access a search engine and retrieve the article "A Brief History of the Internet" by Leiner, Cerf, Clark, Kahn, Kleinrock, Lynch, Postel, Roberts, and Wolff. Answer the following questions:
   a. Who was J. Licklider, and what was his "galactic network" concept?
   b. Who coined the term *packet*?
   c. What (who?) is an IMP?
   d. Did the ARPANET use NCP or TCP/IP?
   e. Was packet voice proposed as an early application for Internet?
   f. How many networks did the initial IP address provide for?

20. Use your Web browser to access a search engine and retrieve the following presentation from the ACM 97 conference: "The Folly Laws of Predictions 1.0" by Gordon Bell. Answer the following questions:
   a. At what rate have processing, storage, and backbone technologies improved since 1950? How does this rate compare to advances in telephony?
   b. What is Moore's Law?
   c. What's the point of making predictions?
   d. What is the difficulty in anticipating trends that have exponential growth?
   e. Who was Vannevar Bush, and why is he famous?
   f. What is the size in bytes of each frame in this presentation? What is the size in bytes of the audio clip for a typical frame? What is the size of the video clip for a typical scene?

21. Use your Web browser to access cnn.com and play a news video clip. Speculate about how the information is being transported over the Internet. How does the quality of the audio and video compare to that of broadcast or cable television?

22. The official standards of the Internet community are published as a Request for Comment, or RFC. Use your Web browser to access the IETF Web page, http://www.ietf.org.
   a. Find and retrieve the RFC titled "Internet Official Protocol Standards." This RFC had number 2400 at the time of writing. This RFC gives the state of standardization of the various Internet protocols. What is the state and standard number of the following protocols: IP, UDP, TCP, TELNET, FTP, DNS, ARP?
   b. Find and retrieve the RFC titled "Assigned Numbers." This RFC, number 1700 at the time of writing, contains all the numbers and constants that are used in Internet protocols. What are the port numbers for Telnet, ftp, and http?

CHAPTER 2

Applications and Layered Architectures

architecture, n. Any design or orderly arrangement perceived by man.
design, n. The invention and disposition of the forms, parts, or details of something according to a plan.[1]

Communication networks can be called upon to support an extremely wide range of services. We routinely use networks to talk to people, to send e-mail, to transfer files, and to retrieve information. Business and industry use networks to carry out critical functions, such as the transfer of funds and the automated processing of transactions, and to query or update database information. Increasingly, the Internet is also being used to provide "broadcast" services along the lines of traditional radio and television. It is clear then that the network must be designed so that it has the flexibility to provide support for current services and to accommodate future services. To achieve this flexibility, *an overall network architecture or plan is necessary.*

The overall process of enabling two devices to communicate effectively across a network is extremely complex. In Chapter 1 we identified the many functions that are required to enable effective communication. Early network designers recognized the need to develop architectures that would organize these functions into a coherent form. As a result, in the early 1970s various computer companies developed proprietary network architectures. A common feature to all of these was the grouping of the communication functions into related and manageable sets called **layers**. We use the term **network architecture** to refer to a set of protocols that specify how every layer is to function.

---

[1]Definitions are from *The American Heritage Dictionary of the English Language*, Houghton Mifflin Co., 1978.

The decomposition of the overall communications problem into a set of layers resulted in a number of benefits. First of all, the design process was simplified once the functions of the layers and their interactions were defined clearly. Second, the layered approach led to flexibility in modifying and developing the network. In contrast, a monolithic approach to network design in which a single large body of software met all the network requirements at a given point in time would quickly become obsolete and also would be extremely difficult and expensive to modify. The layered approach accommodates incremental changes much more readily.

In this chapter we develop the notion of a layered architecture, and we provide examples from TCP/IP, the most important current network architecture. The discussion is organized as follows:

- Web browsing and e-mail applications are used to demonstrate the operation of a protocol within a layer and how it makes use of underlying communication services.
- The Open Systems Interconnection (OSI) reference model is discussed to show how the overall communication process can be organized into functions that are carried out in seven layers.
- The TCP/IP architecture is introduced and compared to the OSI reference model.
- The big picture is completed by presenting a detailed end-to-end example that shows how the various layers work in a typical TCP/IP internet.

Two optional sections present material that is useful in developing lab exercises and experiments involving TCP/IP:

- Berkeley sockets, which allow the student to write applications that use the services provided by the TCP/IP protocols.
- Additional TCP/IP applications and utilities.

2.1 EXAMPLES OF LAYERING

This section uses concrete examples to illustrate what is meant by a protocol and to show how layers interact. Together the examples also show the advantages of layering. The examples use two familiar applications, namely, e-mail and Web browsing. We present a simplified discussion of the associated protocols. Our purpose here is to relate familiar applications to the underlying network services that are the focus of this textbook.

All the examples discussed in this section involve a **client/server** relationship. A server process waits for incoming requests by listening to a **port**. Client processes make *requests* as required. The servers provide *responses* to those requests. The server software usually runs in the background and is referred to as a **daemon**. For example, `httpd` refers to the server daemon for HTTP.

Example—HTTP and Web Browsing

Let us consider the example of browsing through the World Wide Web introduced earlier in Chapter 1. The **HyperText Transfer Protocol (HTTP)** specifies rules by which the client and server interact so as to retrieve a document. The rules also specify how the request and response are phrased. The protocol assumes that the client and server can exchange messages directly. In general, the client software needs to set up a two-way connection prior to the HTTP request.

Table 2.1 shows the sequence of events that are involved in retrieving a document. In step 1 a user selects a document by clicking on its corresponding link. For example, the browser may extract the URL associated with the following link:

http://www.comm.utoronto.ca/infocom/index.html

The client software must usually carry out a Domain Name System (DNS) query to determine the IP address corresponding to the host name, www.comm. utoronto.ca. (We discuss how this query is done in the next example.) The client software then sets up a TCP connection with the WWW server (identified by port 80) at the given IP address (step 2). The client end identifies itself by an "ephemeral" port number that is used only for the duration of the connection.

After the connection is established, the client uses HTTP to request a document (step 3). The request message specifies the method or command (GET), the

Event	Message content
1. User selects document.	
2. Network software of client locates the server host and establishes a two-way connection.	
3. HTTP client sends message requesting document.	GET/infocom/index.html HTTP/1.0
4. HTTP daemon listening on TCP port 80 interprets message.	
5. HTTP daemon sends a result code and a description of the information that the client will receive.	HTTP/1.1 200 OK Server: Apache/1.2.5 FrontPage 3.0.4 Content-Length: 414 Content-Type: text/html
6. HTTP daemon reads the file and sends the requested file through the TCP port.	\<html\> \<head\> \<title\> IEEE Infocom '99 - The Future is Now...
7. HTTP daemon disconnects the connection.	
8. Text is displayed by client browser, which interprets the HTML format.	

TABLE 2.1 Retrieving a document from the Web

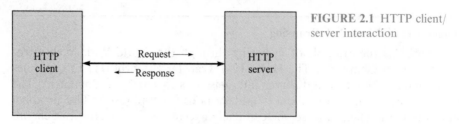

FIGURE 2.1 HTTP client/server interaction

document (infocom/index.html), and the protocol version that the browser is using (HTTP/1.0). The server daemon identifies the three components of the message and attempts to locate the file (step 4).

In step 5 the daemon sends a status line and a description of the information that it will send. Result code 200 indicates that the client request was successful and that the document is to follow. The message also contains information about the server software, the length of the document (414 bytes), and the content type of the document (text/html). If the request was for an image, the type might be image/gif. If the request is not successful, the server sends a different result code, which usually indicates the type of failure, for example, 404 when a document is not found.

In step 6 the HTTP daemon sends the file over the TCP connection and then proceeds to close the connection (step 7). In the meantime, the client receives the file and displays it (step 8). To fetch images that may follow the above document, the browser must initiate additional TCP connections followed by GET interactions.[2]

The HTTP example clearly indicates that a protocol is solely concerned with the interaction between the two peer processes, that is, the client and the server. The protocol assumes that the message exchange between peer processes occurs directly as shown in Figure 2.1. Because the client and server machines are not usually connected directly, a connection needs to be set up between them. In the case of HTTP, we require a two-way connection that transfers a stream of bytes in correct sequential order and without errors. The TCP protocol provides this type of *communication service* between two machines connected to a network. Each HTTP process inserts its messages into a buffer, and TCP transmits the contents of the buffer to the other TCP process in blocks of information called segments, as shown in Figure 2.2. Each segment contains port number information in addition to the HTTP message information. *HTTP is said to use the service provided by TCP in an underlying layer.* Thus the transfer of messages between HTTP client and server in fact is *virtual* and occurs *indirectly* via the TCP connection as shown in Figure 2.2. Later you will see that TCP, in turn, uses the service provided by IP.

---

[2]The student may verify the above message exchange by using a Telnet program. See problem 49.

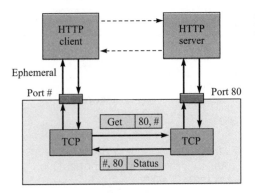

FIGURE 2.2 TCP provides a pipe between the HTTP client and HTTP server

---

Example—DNS Query

The HTTP example notes that the client needs to first perform a DNS query to obtain the IP address corresponding to the domain name. This step is done by sending a message to a DNS server. The DNS is a distributed database that resides in multiple machines on the Internet and is used to convert between names and addresses and to provide e-mail routing information. Each machine maintains its own database and acts as a DNS server that other systems can query. A protocol has been defined whereby a DNS query can initiate a series of interactions in the DNS system to resolve a name or address. We now consider a simple case where the resolution takes place in the first server. Table 2.2 shows the basic steps required for this example.

After receiving the address request, a process in the host, called the resolver, composes the message shown in step 2 and sends it to the local server using the User Datagram Protocol (UDP) communication service. The OPCODE value in the header indicates that the message is a standard query. The question portion of the query contains the following information: QNAME identifies the domain name that is to be translated. The DNS server can handle a variety of queries, and the type is specified by QTYPE. In the example, QTYPE = A requests a translation of a name to an IP address. QCLASS requests an Internet address (some name servers handle non-IP addresses).

The message returned by the server has the Response and Authoritative Answer bits set in the header. This setting indicates that the response comes from an authority that manages the domain name. The question portion is identical to that of the query. The answer portion contains the domain name for which the address is provided. This portion is followed by the Time-to-Live field, which specifies the time in units of seconds that this information is to be cached by the client. Next are the two values for QCLASS and QTYPE. IN again indicates that it is an Internet address. Finally, the address of the domain name is given.

In this example the DNS query and response messages are transmitted by using the communication service provided by the **User Datagram Protocol**. The UDP client attaches a header to the user information to provide port

Event	Message content
1. Application requests name to address translation.	

2. Resolver composes query message.	`Header: OPCODE=SQUERY`
	`Question:`
	`QNAME=tesla.comm.toronto.edu.,`
	`    QCLASS=IN, QTYPE=A`

3. Resolver sends UDP datagram encapsulating the query message.	

4. DNS server looks up address and prepares response.	`Header: OPCODE=SQUERY, RESPONSE,`
	`    AA`
	`Question: QNAME=`
	`tesla.comm.toronto.edu.,`
	`    QCLASS=IN, QTYPE=A`
	`Answer: tesla.comm.toronto.edu.`
	`    86400 IN A 128.100.11.56`

5. DNS sends UDP datagram encapsulating the response message.	

TABLE 2.2 DNS query and response

information (port 53 for DNS) and encapsulates the resulting block in an IP packet. The UDP service is connectionless; no connection setup is required, and the datagram can be sent immediately. Because DNS queries and responses consist of short messages, UDP is ideally suited for conveying them.

The DNS example shows again how a protocol, in this case the DNS query protocol, is solely concerned with the interaction between the client and server processes. The example also shows how the transfer of messages between client and server, in fact, is virtual and occurs indirectly via UDP datagrams.

Example—SMTP and E-mail

Finally, we consider an e-mail example, using the **Simple Mail Transfer Protocol (SMTP)**. Here a mail client application interacts with a local SMTP server to initiate the delivery of an e-mail message. The user prepares an e-mail message that includes the recipient's e-mail address, a subject line, and a body. When the user clicks Send, the mail application prepares a file with the above information and additional information specifying format, for example, plain ASCII or MIME extensions to encode non-ASCII information. The mail application has the name of the local SMTP server and may issue a DNS query for the IP address. Table 2.3 shows the remaining steps involved in completing the transfer of the e-mail message to the local SMTP server.

Before the e-mail message can be transferred, the application process must set up a TCP connection to the local SMTP server (step 1). Thereafter, the SMTP protocol is used in a series of exchanges in which the client identifies itself, the sender of the e-mail, and the recipient (steps 2–8). The client then transfers the message that the SMTP server accepts for delivery (steps 9–12) and ends the mail session. The local SMTP server then repeats this process with the destination

Event	Message content
1. The mail application establishes a TCP connection (port 25) to its local SMTP server.	

2. SMTP daemon issues the following message to the client, indicating that it is ready to receive mail.	`220 tesla.comm.toronto.edu ESMTP`
	`    Sendmail 8.9.0/8.9.0; Thu,`
	`    2 Jul 1998 05:07:59 -0400 (EDT)`

3. Client sends a HELO message and identifies itself.	`HELO bhaskara.comm.utoronto.ca`

4. SMTP daemon issues a 250 message, indicating the client may proceed.	`250 tesla.comm.toronto.edu Hello`
	`    bhaskara.comm [128.100.10.9],`
	`    pleased to meet you`

5. Client sends sender's address.	`MAIL FROM:` `<banerjea@comm.utoronto.ca>`
6. If successful, SMTP daemon replies with a 250 message.	`250 <banerjea@comm.utoronto.ca>...` `    Sender ok`
7. Client sends recipient's address.	`RCPT TO: <alg@nal.utoronto.ca>`
8. A 250 message is returned.	`250 <alg@nal.utoronto.ca>...` `    Recipient ok`
9. Client sends a DATA message requesting permission to send the mail message.	`DATA`
10. The daemon sends a message giving the client permission to send.	`354 Enter mail, end with ''.'' on` `    a line by itself`

11. Client sends the actual text.	`Hi Al,`
	`This section on email sure needs`
	`    a lot of work...`

12. Daemon indicates that the message is accepted for delivery. A message ID is returned.	`250 FAA00803 Message accepted for` `    delivery`
13. Client indicates that the mail session is over.	`QUIT`
14. Daemon confirms the end of the session.	`221 tesla.comm.toronto.edu` `    closing connection`

TABLE 2.3 Sending e-mail

SMTP server. To locate the destination SMTP server, the local server may have to perform a DNS query of type **MX** (mail exchange). SMTP works best when the destination machine is always available. For this reason, users in a PC environment usually retrieve their e-mail from a mail server using the **Post Office Protocol (POP)** instead.

The e-mail, DNS query, and HTTP examples show how multiple protocols can operate by using the communication services provided by the TCP and UDP protocols. In fact, both TCP and UDP operate by using the connectionless packet network service provided by IP. Indeed, an entire suite of protocols has been developed to operate on top of IP, thereby demonstrating the usefulness of the layering concept. New services can be quickly developed by building on the services provided by existing layer protocols.

2.2   THE OSI REFERENCE MODEL

The early network architectures developed by various computer vendors were not compatible with each other. This situation had the effect of locking in customers with a single vendor. As a result, there was pressure in the 1970s for an open systems architecture that would eventually lead to the design of computer network equipment that could communicate with each other. This desire led to an effort in the **International Organization for Standardization (ISO)** first to develop a reference model for **open systems interconnection (OSI)** and later to develop associated standard protocols. *The OSI reference model provided a framework for talking about the overall communications process* and hence was intended to facilitate the development of standards. The reference model incorporated much of the available knowledge from the research community and has served a useful role in network design for more than two decades.

2.2.1   Unified View of Layers, Protocols, and Services

The **OSI reference model** partitions the overall communication process into functions that are carried out by various layers. In each layer a process on one machine carries out a conversation with a **peer process** on the other machine, as shown in Figure 2.3. In OSI terminology the processes at layer n are referred to as **layer n entities**. Layer n entities communicate by exchanging **protocol data units (PDUs)**. Each PDU contains a **header**, which contains protocol control information, and usually user information in the form of a **service data unit (SDU)**. The behavior of the layer n entities is governed by a set of rules or conventions called the **layer n protocol**. In the HTTP example the HTTP client and server applications acted as peer processes. The processes that carry out the transmitter and receiver functions of TCP also constitute peer processes at the layer below.

 The communication between peer processes is virtual in the sense that no direct communication link exists between them. *For communication to take place, the layer n + 1 entities make use of the services provided by layer n.* The transmission of the layer n + 1 PDU is accomplished by passing a block of information from layer n + 1 to layer n through a software port called the layer n **service access point (SAP)**, as shown in Figure 2.4. Each SAP is identified by a unique identifier. The block of information consists of control information and a layer n

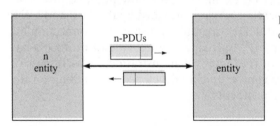

FIGURE 2.3 Peer-to-peer communication

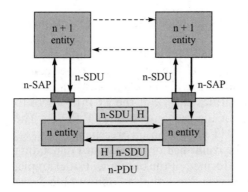

FIGURE 2.4 Layer services

SDU, which is the layer $n+1$ PDU itself. The layer n entity uses the control information to form a header that is attached to the SDU to produce the layer n PDU. Upon receiving the layer n PDU, the layer n peer process uses the header to execute the layer n protocol and, if appropriate, to deliver the SDU to the corresponding layer $n+1$ entity. The communication process is completed when the SDU (layer $n+1$ PDU) is passed to the layer $n+1$ peer process.[3]

In principle, the layer n protocol does not interpret or make use of the information contained in the SDU.[4] We say that the layer n SDU, which is the layer $n+1$ PDU, is *encapsulated* in the layer n PDU. This process of **encapsulation** narrows the scope of the dependencies between adjacent layers to the service definition only. In other words, *layer $n+1$, as a user of the service provided by layer n, is only interested in the correct execution of the service required to transfer its PDUs. The details of the implementation of the layers below layer $n+1$ are irrelevant.*

The service provided by layer n typically involves accepting a block of information from layer $n+1$, transferring the information to its peer processes, which in turn delivers the block to the user at layer $n+1$. The service provided by a layer can be connection oriented or connectionless. A **connection-oriented service** has three phases.

1. Establishing a connection between two layer n SAPs. The setup involves negotiating connection parameters as well as initializing "state information" such as the sequence numbers, flow control variables, and buffer allocations.
2. Transferring n-SDUs using the layer n protocol.
3. Tearing down the connection and releasing the various resources allocated to the connection.

Connectionless service does not require a connection setup, and each SDU is transmitted directly from SAP to SAP. In this case the control information

---

[3]It may be instructive to reread this paragraph where a DNS query message constitutes the layer $n+1$ PDU and a UDP datagram constitutes the layer n PDU.
[4]On the other hand, accessing some of the information "hidden" inside the SDU can sometimes be useful.

that is passed from layer n + 1 to layer n must contain all the address information required to transfer the SDU.

In the HTTP example in Section 2.1 the HTTP client process uses the services provided by TCP to transfer the HTTP PDU, which consists of the request message. A TCP connection is set up between the HTTP client and server processes, and the TCP transmitter/receiver entities carry out the TCP protocol to provide a reliable message stream service for the exchange of HTTP PDUs. The TCP connection is then released after the HTTP response is received.

The services provided by a layer can be **confirmed** or **unconfirmed** depending on whether the sender must eventually be informed of the outcome. For example, connection setup is usually a confirmed service. Note that a connectionless service can be confirmed or unconfirmed depending on whether the sending entity needs to receive an acknowledgment.

Information exchanged between entities can range from a few bytes to multimegabyte blocks or continuous byte streams. Many transmission systems impose a limit on the maximum number of bytes that can be transmitted as a unit. For example, Ethernet LANs have a maximum transmission size of approximately 1500 bytes. Consequently, when the number of bytes that needs to be transmitted exceeds the maximum transmission size of a given layer, it is necessary to divide the bytes into appropriate-sized blocks.

In Figure 2.5a a layer n SDU is too large to be handled by the layer n−1, and so **segmentation** and **reassembly** are applied. The layer n SDU is *segmented* into multiple layer n PDUs that are then transmitted using the services of layer n−1. The layer n entity at the other side must *reassemble* the original message from the sequence of layer n PDUs it receives.

On the other hand, it is also possible that the layer n SDUs are so small as to result in inefficient use of the layer n−1 services, and so **blocking** and **unblocking** may be applied. In this case, the layer n entity may *block* several layer n SDUs into a single layer n PDU as shown in Figure 2.5b. The layer n entity on the other side must then *unblock* the received PDU into the individual SDUs.

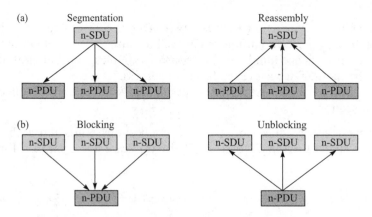

FIGURE 2.5 Segmentation/reassembly and blocking/unblocking

2.2.2 The Seven-Layer Model

In addition to clarifying the notions of protocols, layers, and the interlayer interactions, the OSI activity produced a reference model. The OSI reference model divides the basic communication functions into the seven layers shown in Figure 2.6. Here it is assumed that application A in one computer is communicating with application B in another computer. Applications A and B communicate using an application layer (layer 7) protocol. The purpose of the **application layer** is to provide services that are frequently required by applications that involve communications. In the WWW example the browser application uses the HTTP application layer protocol to access a WWW document. Application layer protocols have been developed for file transfer, virtual terminal (remote login), e-mail, name service, network management, and other applications.

The **presentation layer** is intended to provide the application layer with independence from differences in the representation of data. In principle, the presentation layer should first convert the machine-dependent information provided by application A into a machine-independent form and later convert the machine-independent form into a machine-dependent form suitable for application B. For example, different computers use different codes for representing characters and integers and also different conventions as to whether the first or last bit is the most significant bit.

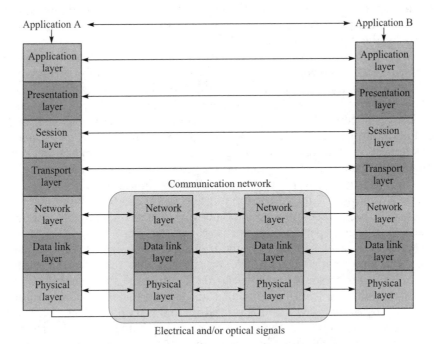

FIGURE 2.6 The OSI reference model

The **session layer** enhances a reliable transfer service provided by the transport layer by providing dialogue control. The session layer can be used to control the manner in which data is exchanged. For example, certain applications require a half-duplex dialogue where the two parties take turns transmitting information. Other applications require the introduction of synchronization points that can be used to mark the progress of an interaction and can serve as points from which error recovery can be initiated. For example, this type of service may be useful in the transfer of very long files over connections that have short times between failures.

The **transport layer** is responsible for the end-to-end transfer of messages from a session entity in the source machine to a session entity in the destination machine. The transport layer protocol involves the transfer of transport layer PDUs called **segments** and is carried out only in the end computer systems. The transport layer uses the services offered by the underlying network to provide the session layer with a transfer of messages that meets a certain quality of service. The transport layer can provide a variety of services. At one extreme a connection-oriented service can provide error-free transfer of a sequence of bytes or messages. The associated protocol carries out error detection and recovery, and sequence and flow control. At the other extreme an unconfirmed connectionless service provides for the transfer of individual messages. In this case the role of the transport layer is to provide the appropriate address information so that the messages can be delivered to the appropriate destination session layer entity. Transport layers may be called upon to do segmentation and reassembly or blocking and unblocking to match the size of the messages produced by the session layer to the packet sizes that can be handled by the network layer.

Processes typically access the transport layer through **socket** interfaces. We discuss the socket interface in the Berkeley UNIX application programming interface (API) in an optional section at the end of this chapter.

The transport layer can be responsible for setting up and releasing connections across the network. To optimize the use of network services, the transport layer may multiplex several transport layer connections onto a single network layer connection. On the other hand, to meet the requirements of a high-throughput transport layer connection, the transport layer may use splitting to support its connection over several network layer connections.

Note from Figure 2.6 that the top four layers are end to end and involve the interaction of peer processes across the network. The lower three layers involve interaction of peer-to-peer processes across a single hop.

The **network layer** provides for the transfer of data in the form of **packets** across the communication network. A key aspect of this transfer is the routing of the packets from the source machine to the destination machine, typically traversing a number of transmission links and network nodes where routing is carried out. By *routing* we mean the procedure that is used to select a path across the network. The nodes in the network must work together to perform the routing effectively. This function makes the network layer the most complex layer in the reference model. The network layer is also responsible for dealing with the *congestion* that occurs from time to time because of temporary surges in packet traffic.

When the two machines are connected to the same packet-switching network, as in Figure 2.7, a single routing procedure is used. However, when the two machines are connected to different networks, the transfer of data must traverse two or more networks that possibly differ in their internal routing. In this case **internetworking** procedures are necessary to route the data between gateways that connect the intermediate networks, as shown below in Figure 2.8. The internetworking procedures must also deal with differences in addressing and differences in the size of the packets that are handled within each network. This **internet sublayer** of the network layer assumes the responsibility for hiding the details of the underlying network(s) from the upper layers. This function is particularly important given the large and increasing number of available network technologies for accomplishing packet transfer.

As shown in Figure 2.6, each intermediate node in the network must implement the lower three layers. Thus one pair of network layer entities exists for each hop of the path required through the network.

The **data link layer** provides for the transfer of **frames** (blocks of information) across a transmission link that *directly* connects two nodes. The data link layer inserts *framing* information to indicate the boundaries of the frames. It also inserts control and address information in the header and check bits to enable recovery from transmission errors, as well as flow control. The data link control is particularly important when the transmission link is prone to transmission errors. Historically, the data link layer included the case where multiple terminals are connected to a host computer in point-to-multipoint fashion. (In Chapter 5 we discuss High-level Data Link Control (HDLC) and Point-to-Point Protocol (PPP), which are two standard data link controls that are in wide use.)

The OSI data link layer was defined so that it included the functions of **LANs**. The notion of a link, then, includes the case where multiple nodes are

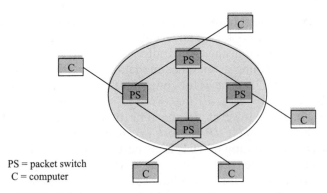

FIGURE 2.7 A packet-switching network using a uniform routing procedure

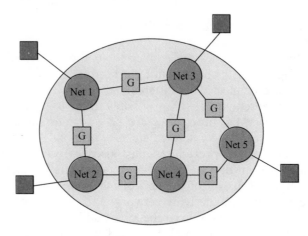

G = gateway/router

FIGURE 2.8 An internetwork

connected to a broadcast medium. As before, frames flow directly between nodes. A *medium access control* procedure is required to coordinate the transmissions from the machines into the medium. Later in this chapter we discuss how the Ethernet LAN standard addresses this case.

The **physical layer** deals with the transfer of **bits** over a communication channel, for example, copper wire pairs, coaxial cable, radio, or optical fiber. The layer is concerned with the particular choice of system parameters such as voltage levels and signal durations. The layer is also concerned with the set up and release of the physical connection as well as with mechanical aspects such as socket type and number of pins.

Earlier in this section we noted that each layer adds a header, and possibly a trailer, to the SDU it accepts from the layer above. Figure 2.9 shows the headers and trailers that are added as a block of application data works its way down the seven layers. At the destination each layer reads its corresponding header to determine what action to take and eventually passes the SDU to the layer above after removing the header and trailer.

In addition to defining a reference model, an objective of the ISO activity was the development of *standards* for computer networks. This objective entailed *specifying the particular protocols* that were to be used in various layers of the OSI reference model. However, in the time that it took to develop the OSI protocol standards, the *TCP/IP network architecture* emerged as an alternative for OSI. The free distribution of TCP/IP as part of the Berkeley UNIX® ensured the development of numerous applications at various academic institutions and the emergence of a market for networking software. This situation eventually led to the development of the global Internet and to the dominance of the TCP/IP network architecture.

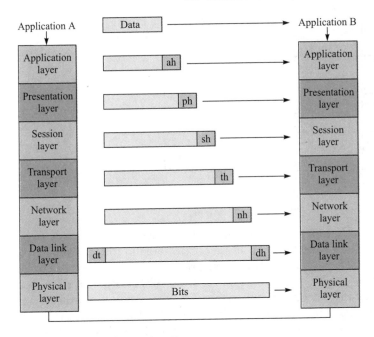

FIGURE 2.9 Headers and trailers

2.3 OVERVIEW OF TCP/IP ARCHITECTURE

The TCP/IP network architecture is a set of protocols that allow communication across multiple diverse networks. The architecture evolved out of research that had the original objective of transferring packets across three different packet networks: the ARPANET packet-switching network, a packet radio network, and a packet satellite network. The military orientation of the research placed a premium on robustness with regard to failures in the network and on flexibility in operating over diverse networks. This environment led to a set of protocols that are highly effective in enabling communications among the many different types of computer systems and networks. Indeed, the Internet has become the primary fabric for interconnecting the world's computers. In this section we introduce the TCP/IP network architecture. The details of specific protocols that constitute the TCP/IP network architecture are discussed in later chapters.

Figure 2.10a shows the **TCP/IP network architecture**, which consists of four layers. The application layer provides services that can be used by other applications. For example, protocols have been developed for remote login, for e-mail, for file transfer, and for network management. The TCP/IP application layer incorporates the functions of the top three OSI layers. The HTTP protocol discussed in Section 2.2 is actually a TCP/IP application layer protocol. Recall that the HTTP request message included format information and the HTTP protocol defined the dialogue between the client and server.

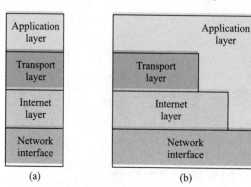

FIGURE 2.10 TCP/IP network architecture

The application layer programs are intended to run directly over the transport layer. Two basic types of services are offered in the transport layer. The first service consists of reliable connection-oriented transfer of a byte stream, which is provided by the **Transmission Control Protocol (TCP)**. The second service consists of best-effort connectionless transfer of individual messages, which is provided by the **User Datagram Protocol (UDP)**. This service provides no mechanisms for error recovery or flow control. UDP is used for applications that require quick but necessarily reliable delivery.

The TCP/IP model does not require strict layering, as shown in Figure 2.10b. In other words, the application layer has the option of bypassing intermediate layers. For example, an application layer may run dirctly over the internet layer.

The **internet layer** handles the transfer of information across multiple networks through the use of gateways or routers, as shown in Figure 2.11. The internet layer corresponds to the part of the OSI network layer that is concerned with the transfer of packets between machines that are connected to different networks. It must therefore deal with the routing of packets across these networks as well as with the control of congestion. A key aspect of the internet layer is the definition of *globally unique addresses* for machines that are attached to the

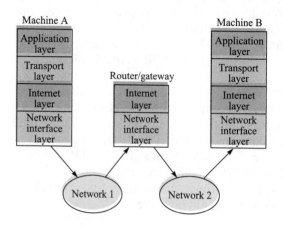

FIGURE 2.11 The internet layer and network interface layers

Internet. The internet layer provides a single service, namely, *best-effort connectionless packet transfer*. IP packets are exchanged between routers without a connection setup; the packets are routed independently, and so they may traverse different paths. For this reason, IP packets are also called **datagrams**. The connectionless approach makes the system robust; that is, if failures occur in the network, the packets are routed around the points of failure; there is no need to set up the connections. The gateways that interconnect the intermediate networks may discard packets when congestion occurs. The responsibility for recovery from these losses is passed on to the transport layer.

Finally, the **network interface layer** is concerned with the network-specific aspects of the transfer of packets. As such, it must deal with part of the OSI network layer and data link layer. Various interfaces are available for connecting end computer systems to specific networks such as X.25, ATM, frame relay, Ethernet, and token ring. These networks are described in later chapters.

The network interface layer is particularly concerned with the protocols that access the intermediate networks. At each gateway the network access protocol encapsulates the IP packet into a packet or frame of the underlying network or link. The IP packet is recovered at the exit gateway of the given network. This gateway must then encapsulate the IP packet into a packet or frame of the type of the next network or link. This approach provides a clear separation of the internet layer from the technology-dependent network interface layer. This approach also allows the internet layer to provide a data transfer service that is transparent in the sense of not depending on the details of the underlying networks. The next section provides a detailed example of how IP operates over the underlying networks.

Figure 2.12 shows some of the protocols of the TCP/IP protocol suite. The figure shows two of the many protocols that operate over TCP, namely, HTTP and SMTP. The figure also shows DNS and Real-Time Protocol (RTP), which operate over UDP. The transport layer protocols TCP and UDP, on the other hand, operate over IP. Many network interfaces are defined to support IP. The salient part of Figure 2.12 is that all higher-layer protocols access the network

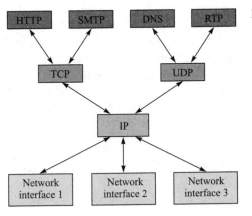

FIGURE 2.12 TCP/IP protocol graph

interfaces through IP. This feature provides the capability to operate over multiple networks. The IP protocol is complemented by additional protocols (ICMP, IGMP, ARP, RARP) that are required to operate an internet. These protocols are discussed in Chapter 8.

The hourglass shape of the TCP/IP protocol graph underscores the features that make TCP/IP so powerful. The operation of the single IP protocol over various networks provides independence from the underlying network technologies. The communication services of TCP and UDP provide a network-independent platform on which applications can be developed. By allowing multiple network technologies to coexist, the Internet is able to provide ubiquitous connectivity and to achieve enormous economies of scale.

2.3.1 TCP/IP Protocol: How the Layers Work Together

In this section we provide a detailed example of how the layering concepts discussed in the previous sections are put into practice in a typical TCP/IP network scenario. We show

- Examples of each of the layers.
- How the layers interact across the interfaces between them.
- How the PDUs of a layer are built and what key information is in the header.
- The relationship between physical addresses and IP addresses.
- How an IP packet or datagram is routed across several networks.

This example will complete our goal of providing the big picture of networking. In the remainder of the book we systematically examine the details of the various components and aspects of networks.

Consider the network configuration shown in Figure 2.13a. A server, a workstation, and a router are connected to an Ethernet LAN, and a remote PC is connected to the router through a point-to-point link. From the point of view of IP, the Ethernet LAN and the point-to-point link constitute two different networks as shown in Figure 2.13b.

Each host in the Internet is identified by a **globally unique IP address**. Strictly speaking, the IP address identifies the host's network interface rather than the host itself. A node that is attached to two or more physical networks is called a router. In this example the router attaches to two networks with each network interface assigned to a unique IP address. An IP address is divided into two parts: a *network id* and a *host id*. The network id must be obtained from an organization authorized to issue IP addresses. In this example we use simplified notation and assume that the Ethernet has net id 1 and that the point-to-point link has a net id 2. In the Ethernet we suppose that the server has IP address (1,1), the workstation has IP address (1,2), and the router has address (1,3). In the point-to-point link, the PC has address (2,2), and the router has address (2,1).

On a physical network the attachment of a device to the network is often identified by a *physical address*. The format of the physical address depends on the particular type of network. For example, Ethernet LANs use 48-bit

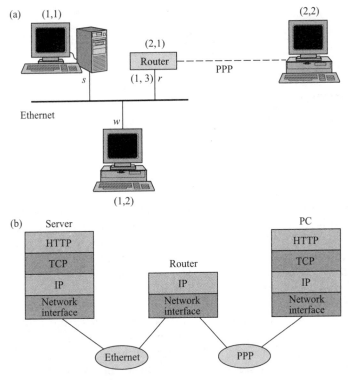

FIGURE 2.13 An example of an internet consisting of an Ethernet LAN and a point-to-point link

addresses. Each Ethernet network interface card (NIC) is issued a globally unique medium access control (MAC) or physical address. When a NIC is used to connect a machine to any Ethernet LAN, all machines in the LAN are automatically guaranteed to have unique addresses. Thus the router, server, and workstation also have physical addresses designated by r, s, and w, respectively.

First, let us consider the case in which the workstation wants to send an IP datagram to the server. The IP datagram has the workstation's IP address and the server's IP address in the IP packet header. We suppose that the IP address of the server is known. The IP entity in the workstation looks at its routing table to see whether it has an entry for the complete IP address. It finds that the server is directly connected to the same network and that the sender has physical address s.[5] The IP datagram is passed to the Ethernet device driver, which prepares an Ethernet frame as shown in Figure 2.14. The header in the frame contains the source physical address, w, and the destination physical address, s. The header

---

[5]If the IP entity does not know the physical address corresponding to the IP address of the sender, the entity uses the Address Resolution Protocol (ARP) to find it. ARP is discussed in Chapter 8.

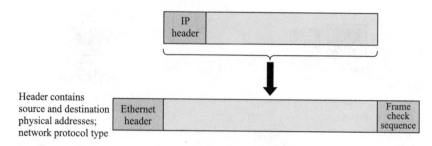

FIGURE 2.14 IP datagram is encapsulated in an Ethernet frame

also contains a protocol type field that is set to the value that corresponds to IP. The type field is required because the Ethernet may be carrying packets for other non-IP protocols. The Ethernet frame is then broadcast over the LAN. The server's NIC recognizes that the frame is intended for its host, so the card captures the frame and examines it. The NIC finds that the protocol type field is set to IP and therefore passes the IP datagram up to the IP entity.

Next let us consider the case in which the server wants to send an IP datagram to the personal computer. The PC is connected to the router through a point-to-point link that we assume is running PPP as the data link control.[6] We suppose that the server knows the IP address of the PC and that the IP addresses on either side of the link were negotiated when the link was set up. The IP entity in the server looks at its routing table to see whether it has an entry for the complete IP address of the PC. We suppose that it doesn't. The IP entity then checks to see whether it has a routing table entry that matches the network id portion of the IP portion of the IP address of the PC. Again we suppose that the IP entity does not find such an entry. The IP entity then checks to see whether it has an entry that specifies a default router that is to be used when no other entries are found. We suppose that such an entry exists and that it specifies the router with address (1,3).

The IP datagram is passed on the Ethernet device driver, which prepares an Ethernet frame. The header in the frame contains the source physical address, s, and the destination physical address, r. However, the IP datagram in the frame contains the destination IP address of the PC, (2,2), not the destination IP address of the router. The Ethernet frame is then broadcast over the LAN. The router's NIC captures the frame and examines it. The card passes the IP datagram up to its IP entity, which discovers that the IP datagram is not for itself but is to be routed on.

The routing tables at the router show that the machine with address (2,2) is connected directly on the other side of the point-to-point link. The router encapsulates the IP datagram in a PPP frame that is similar to that of the Ethernet frame shown in Figure 2.14. However, the frame does not require physical

---

[6]PPP is discussed in Chapter 5.

address information, since there is only one "other side" of the link. The PPP receiver at the PC receives the frame, checks the protocol type field, and passes the IP datagram to its IP entity.

The preceding discussion shows how IP datagrams are sent across an internet. Next let's complete the picture by seeing how things work at the higher layers. Consider the browser application discussed in the beginning of the chapter. We suppose that the user at the PC has clicked on a Web link of a document contained in the server and that a TCP connection has already been established between the PC and the server.[7] Consider what happens when the TCP connection is confirmed at the PC. The HTTP request message GET is passed to the TCP layer, which encapsulates the message into a TCP segment as shown in Figure 2.15. The TCP segment contains an ephemeral port number for the client process, say, c, and a well-known port number for the server process, 80 for HTTP.

The TCP segment is passed to the IP layer, which in turn encapsulates the segment into an Internet packet. The IP packet header contains the IP addresses of the sender, (2,2), and the destination, (1,1). The header also contains a protocol field, which designates the layer that is operating above IP, in this case TCP. The IP datagram is then encapsulated using PPP and sent to the router, which routes the datagram to the server using the procedures discussed above.

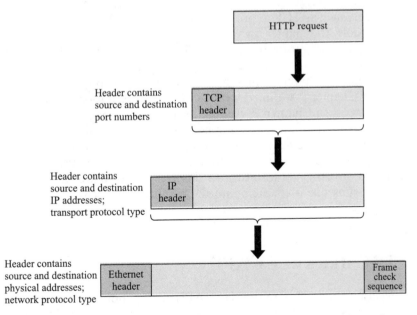

FIGURE 2.15 Encapsulation of PDUs in TCP/IP and addressing information in the headers

---

[7]The details of how a TCP connection is set up are described in Chapter 8.

Eventually the server NIC captures the Ethernet frame and extracts the IP datagram and passes it to the IP entity. The protocol field in the IP header indicates that a TCP segment is to be extracted and passed on to the TCP layer. The TCP layer, in turn, uses the port number to find out that the message is to be passed to the HTTP server process. A problem arises at this point: The server process is likely to be simultaneously handling multiple connections to multiple clients. All these connections have the same destination IP address; the same destination port number, 80; and the same protocol type, TCP. How does the server know which connection the message corresponds to? The answer is in how an end-to-end process-to-process connection is specified.

The source port number, the source IP address, and the protocol type are said to define the sender's *socket address*. Similarly, the destination port number, the destination IP address, and the protocol type define the destination's socket address. Together the source socket address and the destination socket address uniquely specify the connection between the HTTP client process and the HTTP server process. For example, in the earlier HTTP example the sender's socket is (TCP, (2,2), ephemeral #), and the destination's socket is (TCP, (1,1), 80). The combination of these five parameters (TCP, (2,2), ephemeral #, (1 1), 80) uniquely specify the process-to-process connection.

This completes the discussion on the transfer of messages from process to process over a TCP/IP internet. In the remaining chapters we examine the details of the operation of the various layers. In Chapters 3 and 4 we consider various aspects of physical layers. In Chapter 5 we discuss peer-to-peer protocols that allow protocols such as TCP to provide reliable service. We also discuss data link control protocols. In Chapter 6 we discuss LANs and their medium access controls. In Chapter 7 we return to the network layer and examine the operation of routers and packet switches as well as issues relating to addressing, routing, and congestion control. Chapter 8 presents a detailed discussion of the TCP and IP protocols. In Chapter 9 we introduce ATM, a connection-oriented packet network architecture. In Chapter 10 we discuss advanced topics, such as the operation of IP over ATM and new developments in TCP/IP architecture. In Chapter 11 we introduce enhancements to IP that provide security. Finally, in Chapter 12 we discuss the support of real-time multimedia services over IP. From time to time it may be worthwhile to return to this example to place the discussion of details in the subsequent chapters into the big picture presented here.

2.4 THE BERKELEY API[8]

An Application Programming Interface (API) allows application programs (such as Telnet, Web browsers, etc.) to access certain resources through a predefined and preferably consistent interface. One of the most popular of the APIs that

---

[8]This section is optional and is not required for later sections. A knowledge of C programming is assumed.

provide access to network resources is the Berkeley socket interface, which was developed by a group at the University of California at Berkeley in the early 1980s. The socket interface is now widely available on many UNIX machines. Another popular socket interface, which was derived from the Berkeley socket interface, is called the Windows sockets or Winsock and was designed to operate in a Microsoft® Windows environment.

By hiding the details of the underlying communication technologies as much as possible, the socket mechanism allows programmers to write application programs easily without worrying about the underlying networking details. Figure 2.16 shows how two applications talk to each other across a communication network through the socket interface. In a typical communication session, one application operates as a server and the other as a client. The server provides services upon request by the client.

This section explains how the socket mechanism can provide services to the application programs. Two modes of services are available: *connection-oriented* and *connectionless*. With the connection-oriented mode, an application program must first establish a connection to the other end before the actual communication (i.e., data transfer) can take place. The connection is established if the other end agrees to accept the connection. Once the connection is established, data will be delivered through the connection to the destination in sequence. Typically, a connection-oriented model provides a reliable delivery service. With the connectionless mode an application program sends its data immediately without waiting for the connection to get established at all. This mode avoids the setup overhead found in the connection-oriented mode. However, the price to pay is that an application program may waste its time sending data when the other end is not ready to accept it. Moreover, data may not arrive at the other end if the network decides to discard it. Worse yet, even if data arrives at the destination, it may not

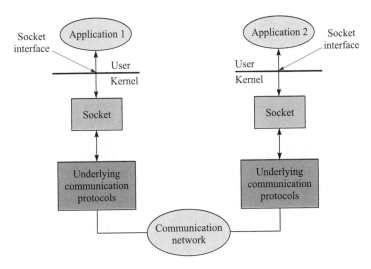

FIGURE 2.16 Communications through the socket interface

arrive in the same order as it was transmitted. The connectionless mode is often said to provide *best-effort service*, since the network would try its best to deliver the information but cannot guarantee delivery.

Figure 2.17 shows a typical diagram of the sequence of socket calls for connection-oriented communication. The server begins by carrying out a *passive open* as follows. The socket call creates a TCP socket. The bind call then binds the well-known port number of the server to the socket. The listen call turns the socket into a listening socket that can accept incoming connections from clients. Finally, the accept call puts the server process to sleep until the arrival of a client connection. The client does an *active open*. The socket call creates a socket on the client side, and the connect call establishes the TCP connection to the server with the specified destination socket address. When the TCP connection is completed, the accept function at the server wakes up and returns the descriptor for the given connection, namely, the source IP address, source port number, destination IP address, and destination port number. The client and server are now ready to exchange information.

Figure 2.18 shows the sequence of socket calls for a connectionless communication. Note that no connection is established prior to data transfer. The recvfrom call returns when a complete UDP datagram has been received. For both types of communication, the data transfer phase may occur in an arbitrary number of exchanges.

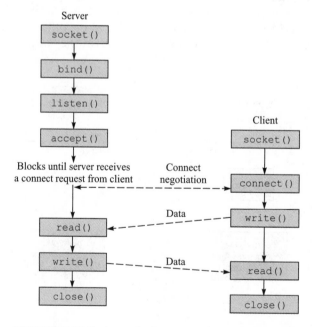

FIGURE 2.17 Socket calls for connection-oriented communication

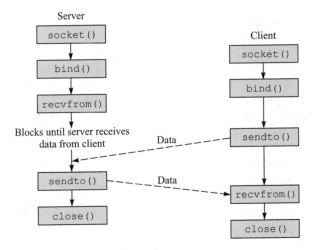

FIGURE 2.18 Socket calls for connectionless communication

2.4.1 Socket System Calls

Socket facilities are provided to programmers through C system calls that are similar to function calls except that control is transferred to the operating system kernel once a call is entered. To use these facilities, the header files <sys/types.h> and <sys/socket.h> must be included in the program.

Before an application program (client or server) can transfer any data, it must first create an endpoint for communication by calling socket. Its prototype is

```
int socket(int family, int type, int protocol);
```

where family identifies the family by address or protocol. The address family identifies a collection of protocols with the same address format, while the protocol family identifies a collection of protocols having the same architecture. Although it may be possible to classify the family based on addresses or protocols, these two families are currently equivalent. Some examples of the address family that are defined in <sys/socket.h> include AF_UNIX, which is used for communication on the local UNIX machine, and AF_INET, which is used for Internet communication using TCP/IP protocols. The protocol family is identified by the prefix PF_. The value of PF_XXX is equal to that of AF_XXX, indicating that the two families are equivalent. We are concerned only with AF_INET in this book.

The type identifies the semantics of communication. Some of the types include SOCK_STREAM, SOCK_DGRAM, and SOCK_RAW. A SOCK_STREAM type provides data delivery service as a sequence of bytes and does not preserve message boundaries. A SOCK_DGRAM type provides data delivery service in blocks of bytes called datagrams. A SOCK_RAW type provides access to internal network interfaces and is available only to superuser.

The protocol identifies the specific protocol to be used. Normally, only one protocol is available for each family and type, so the value for the protocol argument is usually set to 0 to indicate the default protocol. The default protocol of SOCK_STREAM type with AF_INET family is TCP, which is a connection-oriented protocol providing a reliable service with in-sequence data delivery. The default protocol of SOCK_DGRAM type with AF_INET family is UDP, which is a connectionless protocol with unreliable service.

The socket call returns a nonnegative integer value called the socket descriptor or handle (just like a file descriptor) on success. On failure, socket returns −1.

After a socket is created, the bind system call can be used to assign an address to the socket. Its prototype is

```
int bind(int sd, struct sockaddr *name, int namelen);
```

where sd is the socket descriptor returned by the socket call, name is a pointer to an address structure that contains the local IP address and port number, and namelen is the size of the address structure in bytes. The bind system call returns 0 on success and −1 on failure. The sockaddr structure is a generic address structure and has the following definition:

```
struct sockaddr {
        u_short    sa_family;           /* address family  */
        char       sa_data[14];         /* address  */
};
```

where sa_family holds the address family and sa_data holds up to 14 bytes of address information that varies from one family to another. For the Internet family the address information consists of the port number that is two bytes long and an IP address that is four bytes long. The appropriate structures to use for the Internet family are defined in <netinet/in.h>:

```
struct in addr {
        u_long    s_addr;               /* 32-bit IP address */
};
struct sockaddr_in {
        u_short    sin_family;          /* AF_INET */
        u_short    sin_port;            /* TCP or UDP port */
        struct     in_addr  sin_addr;   /* 32-bit IP address */
        char       sin_zero[8];         /* unused */
};
```

An application program using the Internet family should use the sockaddr_in structure to assign member values and should use the sockaddr structure only for casting purposes in function arguments. For this family sin_family holds the value of the identifier AF_INET. The structure member sin_port holds the local port number. Port numbers 1 to 1023 are normally reserved for system use. For a server, sin_port contains a well-known port number that clients must know in advance to establish a connection. Specifying a port number 0 to `bind` asks the system to assign an available port number. The structure member `sin_addr` holds the local IP address. For a host with multiple IP addresses, `sin_addr` is typically set to `INADDR_ANY` to indicate that the server is willing to accept communication through any of its IP addresses. This setting is useful for a host with multiple IP addresses. The structure member `sin_zero` is used to fill out `struct sockaddr_in` to 16 bytes.

Different computers may store a multibyte word in different orders. If the least significant byte is stored first, it is known as **little endian**. If the most significant byte is stored first, it is known as **big endian**. For any two computers to be able to communicate, they must use a common data format while transferring the multibyte words. The Internet adopts the big-endian format. This representation is known as **network byte order** in contrast to the representation adopted by the host, which is called **host byte order**. It is important to remember that the values of `sin_port` and `sin_addr` must be in the network byte order, since these values are communicated across the network. Four functions are available to convert between the host and network byte order conveniently. Functions `htons` and `htonl` convert an unsigned short and an unsigned long, respectively, from the host to network byte order. Functions `ntohs` and `ntohl` convert an unsigned short and an unsigned long, respectively, from the network to host byte order. We need to use these functions so that programs will be portable to any machine. To use these functions, we should include the header files `<sys/types.h>` and `<netinet/in.h>`. The appropriate prototypes are

```
u_long htonl(u_long hostlong);
u_short htons(u_short hostshort);
u_long ntohl(u_long netlong);
u_short ntohs(u_short netshort);
```

A client establishes a connection on a socket by calling `connect`. The prototype is

```
int connect(int sd, struct sockaddr *name, int namelen);
```

where `sd` is the socket descriptor returned by the `socket` call, `name` points to the server address structure, and `namelen` specifies the amount of space in bytes pointed to by `name`. For connection-oriented communication, `connect` attempts to establish a virtual circuit between a client and a server. For connectionless communication, `connect` stores the server's address so that the client can use a socket descriptor when sending datagrams, instead of specifying the server's address each time a datagram is sent. The `connect` system call returns 0 on success and −1 on failure.

A connection-oriented server indicates its willingness to receive connection requests by calling `listen`. The prototype is

```
int listen(int sd, int backlog);
```

where `sd` is the socket descriptor returned by the `socket` call and `backlog` specifies the maximum number of connection requests that the system should queue while it waits for the server to accept them (the maximum value is usually 5). This mechanism allows pending connection requests to be saved while the server is busy processing other tasks. The listen system call returns 0 on success and −1 on failure.

After a server calls listen, it can accept the connection request by calling accept with the prototype

```
int accept(int sd, struct sockaddr *addr, int *addrlen);
```

where sd is the socket descriptor returned by the socket call, addr is a pointer to an address structure that accept fills in with the client's IP address and port number, and addrlen is a pointer to an integer specifying the amount of space pointed to by addr before the call. On return, the value pointed to by addrlen specifies the number of bytes of the client address information.

If no connection requests are pending, accept will block the caller until a connection request arrives. The accept system call returns a new socket descriptor having nonnegative value on success and −1 on failure. The new socket descriptor inherits the properties of sd. The server uses the new socket descriptor to perform data transfer for the new connection. A concurrent server can accept further connection requests using the original socket descriptor sd.

Clients and servers may transmit data using write or sendto. The write call is usually used for a connection-oriented communication. However, a connectionless client may also call write if it has a connected socket (that is, the client has executed connect). On the other hand, the sendto call is usually used for a connectionless communication. Their prototypes are

```
int write(int sd, char *buf, int buflen);
int sendto(int sd, char *buf, int buflen, int flags,
    struct sockaddr *addrp, int addrlen);
```

where sd is the socket descriptor, buf is a pointer to a buffer containing the data to transmitted, buflen is the length of the data in bytes, flags can be used to control transmission behavior such as handling out-of-band (high priority) data but is usually set to 0 for normal operation, addrp is a pointer to the sockaddr structure containing the address information of the remote hosts, and addrlen is the length of the address information. Both write and sendto return the number of bytes transmitted on success or −1 on failure.

The corresponding system calls to receive data read and recvfrom. Their prototypes are

```
int read(int sd, char *buf, int buflen);
int recvfrom(int sd, char * buf, int buflen, int flags,
    struct sockaddr *addrp, int *addrlen);
```

The parameters are similar to the ones discussed above except buf is now a pointer to a buffer that is used to store the received data and buflen is the length of the buffer in bytes. Both read and recvfrom return the number of bytes received on success or −1 on failure. Both calls will block if no data arrives at the local host.

If a socket is no longer in use, the application can call `close` to terminate a connection and return system resources to the operating system. The prototype is

```
int close(int sd);
```

where `sd` is the socket descriptor to be closed. The `close` call returns 0 on success and −1 on failure.

2.4.2   Network Utility Functions

Library routines are available to convert a human-friendly domain name such as markov.ece.arizona.edu into a 32-bit machine-friendly IP as 10000000 11000100 00011100 10111101 and vice versa. To perform the conversion we should include the header files `<sys/socket.h>`, `<sys/types.h>`, and `<netdb.h>`. The appropriate structure that stores the host information defined in the `<netdb.h>` file is

```
struct hostent {
    char *h_name;            /* official name of host */
    char **h_aliases;        /* alias name this host uses */
    int h_addrtype;          /* address type */
    int h_length;            /* length of address */
    char **h_addr_list;      /* list of addresses from name
                                server */
};
```

The `h_name` element points to the official name of the host. If the host has name aliases, these aliases are pointed to by `h_aliases`, which is terminated by a NULL. Thus `h_aliases[0]` points to the first alias, `h_aliases[1]` points to the second alias, and so on. Currently, the `h_addrtype` element always takes on the value of `AF_INET`, and the `h_length` element always contains a value of 4. The `h_addr_list` points to the list of network addresses in network byte order and is terminated by a NULL.

NAME-TO-ADDRESS CONVERSION FUNCTIONS

Two functions are used for routines performing a name-to-address-conversion: `gethostbyname` and `gethostbyaddr`.

```
struct hostent *gethostbyname (char *name);
```

The function `gethostbyname` takes a domain name at the input and returns the host information as a pointer to `struct hostent`. The function returns a NULL on error. The parameter `name` is a pointer to a domain name of a host whose information we would like to obtain. The function `gethostbyname` obtains the host information either from the file `/etc/hosts` or from a name server. Recall that the host information includes the desired address.

```
struct hostent *gethostbyaddr (char *addr, int len, int type);
```

The function `gethostbyaddr` takes a host address at the input in network byte order, its length in bytes, and type, which should be `AF_INET`. The function returns the same information as `gethostbyname`. This information includes the desired host name.

The IP address is usually communicated by people using a notation called the *dotted-decimal notation*. As an example, the dotted-decimal notation of the IP address 10000000 11000100 00011100 10111101 is 128.196.28.189. To do conversions between these two formats, we could use the functions `inet_addr` and `inet_ntoa`. The header files that must be included are `<sys/types.h>`, `<sys/socket.h>`, `<netinet/in.h>`, and `<arpa/inet.h>`.

IP ADDRESS MANIPULATION FUNCTIONS

Two functions are used for routines converting addresses between a 32-bit format and the dotted-decimal notation: `inet_nota and inet_addr`.

```
char *inet_ntoa(struct in_addr in);
```

The function `inet_ntoa` takes a 32-bit IP address in network byte order and returns the corresponding address in dotted-decimal notation.

```
unsigned long inet_addr(char *cp);
```

The function `inet_addr` takes a host address in dotted-decimal notation and returns the corresponding 32-bit IP address in network byte order.

Example—Communicating with TCP

As an illustration of the use of the system calls and functions described above, let us show two application programs that communicate via TCP. The client prompts a user to type a line of text, sends it to the server, reads the data back from the server, and prints it out. The server acts as a simple echo server. After responding to a client, the server closes the connection and then waits for the next new connection. In this example each application (client and server) expects a fixed number of bytes from the other end, specified by BUFLEN. Because TCP is stream oriented, the received data may come in multiple pieces of byte streams independent of how the data was sent at the other end. For example, when a transmitter sends 100 bytes of data in a single `write` call, the receiver may receive the data in two pieces—80 bytes and 20 bytes—or in three pieces—10 bytes, 50 bytes, and 40 bytes—or in any other combination. Thus the program has to make repeated calls to `read` until all the data has been received. The following program is the server.

```
/* A simple echo server using TCP */
#include <stdio.h>
#include <sys/types.h>
#include <sys/socket.h>
#include <netinet/in.h>

#define SERVER_TCP_PORT    3000   /* well-known port */
```

```
#define BUFLEN          256          /* buffer length */

int main(int argc, char **argv)
{
    int     n, bytes_to_read;
    int     sd, new_sd, client_len, port;
    struct  sockaddr_in server, client;
    char    *bp, buf[BUFLEN];

    switch(argc) {
    case 1:
        port = SERVER_TCP_PORT;
        break;
    case 2:
        port = atoi(argv[1]);
        break;
    default:
        fprintf(stderr, "Usage: %s [port]\n", argv[0]);
        exit(1);
    }

    /* Create a stream socket */
    if ((sd = socket(AF_INET, SOCK_STREAM, 0)) == -1) {
        fprintf(stderr, "Can't create a socket\n");
        exit(1);
    }

    /* Bind an address to the socket */
    bzero((char *)&server, sizeof(struct sockaddr_in));
    server.sin_family = AF_INET;
    server.sin_port = htons(port);
    server.sin_addr.s_addr = htonl(INADDR_ANY);
    if (bind(sd, (struct sockaddr *)&server,
    sizeof(server)) == -1) {
      fprintf(stderr, "Can't bind name to socket\n");
      exit(1);
    }

    /* queue up to 5 connect requests */
    listen(sd, 5);

    while (1) {
      client_len = sizeof(client);
      if ((new_sd = accept(sd, (struct sockaddr *)
      &client, &client_len)) == -1) {
        fprintf(stderr, "Can't accept client\n");
        exit(1);
      }

      bp = buf;
```

```
        bytes_to_read = BUFLEN;
        while ((n = read(new_sd, bp, bytes_to_read)) > 0) {
          bp += n;
          bytes_to_read -= n;
        }

        write(new_sd, buf, BUFLEN);
        close(new_sd);
      }
    close(sd);
    return(0);
}
```

The client program allows the user to identify the server by its domain name.
Conversion to the IP address is done by the gethostbyname function. Again, the
client makes repeated calls to read until no more data is expected to arrive. The
following program is the client.

```
/* A simple TCP client */
#include <stdio.h>
#include <netdb.h>
#include <sys/types.h>
#include <sys/socket.h>
#include <netinet/in.h>

#define SERVER_TCP_PORT     3000
#define BUFLEN             256          /* buffer length */

int main(int argc, char **argv)
{
    int     n, bytes_to_read;
    int     sd, port;
    struct  hostent     *hp;
    struct  sockaddr_in  server;
    char    *host, *bp, rbuf[BUFLEN], sbuf[BUFLEN];

    switch(argc) {
    case 2:
        host = argv[1];
        port = SERVER_TCP_PORT;
        break;
    case 3:
        host = argv[1];
        port = atoi(argv[2]);
        break;
    default:
        fprintf(stderr, "Usage: %s host[port]\n", argv[0]);
        exit(1);
    }
```

```
    /* Create a stream socket */
    if ((sd = socket(AF_INET, SOCK_STREAM, 0)) == -1) {
        fprintf(stderr, "Can't create a socket\n");
        exit(1);
    }

    bzero((char *)&server, sizeof(struct sockaddr_in));
    server.sin_family = AF_INET;
    server.sin_port = htons(port);
    if ((hp = gethostbyname(host)) == NULL) {
        fprintf(stderr, "Can't get server's address\n");
        exit(1);
    }
    bcopy(hp->h_addr, (char *)&server.sin_addr,
        hp->_length);

    /* Connecting to the server */
    if (connect(sd, (struct sockaddr *)&server,
    sizeof(server)) == -1) {
        fprintf(stderr, "Can't connect\n");
        exit(1);
    }
    printf("Connected: server's address is %s\n",
        hp->_name);

    printf("Transmit:\n");
    gets(sbuf);            /* get user's text */
    write(sd, sbuf, BUFLEN); /* send it out */

    printf("Receive:\n");
    bp = rbuf;
    bytes_to_read = BUFLEN;
    while ((n = read (sd, bp, bytes_to_read)) > 0) {
        bp += n;
        bytes_to_read -= n;
    }
    printf("%s\n", rbuf);

    close(sd);
    return(0);
}
```

Example—Using the UDP Protocol

Let us now take a look at client/server programs using the UDP protocol. The following source code is a program that uses the UDP server as an echo server as before. Note that data receipt can be done in a single call with recvfrom, since UDP is blocked oriented.

```c
/* Echo server using UDP */
#include <stdio.h>
#include <sys/types.h>
#include <sys/socket.h>
#include <netinet/in.h>

#define SERVER_UDP_PORT   5000    /* well-known port */
#define MAXLEN         4096       /* maximum data length */

int main(int argc, char **argv)
{
    int      sd, client_len, port, n;
    char     buf[MAXLEN];
    struct   sockaddr_in server, client;

    switch(argc) {
    case 1:
        port = SERVER_UDP_PORT;
        break;
    case 2:
        port = atoi(argv[1]);
        break;
    default:
        fprintf(stderr, "Usage: %s [port]\n", argv[0]);
        exit(1);
    }

    /* Create a datagram socket */
    if ((sd = socket(AF_INET, SOCK_DGRAM, 0)) == -1) {
        fprintf(stderr, "Can't create a socket\n");
        exit(1);
    }

    /* Bind an address to the socket */
    bzero((char *)&server, sizeof(server));
    server.sin_family = AF_INET;
    server.sin_port = htons(port);
    server.sin_addr.s_addr = hton1(INADDR_ANY);
    if (bind(sd, (struct sockaddr *)&server,
    sizeof(server)) == -1) {
        fprintf(stderr, "Can't bind name to socket\n");
        exit(1);
    }

    while (1) {
        client_len = sizeof(client);
        if ((n = recvfrom(sd, buf, MAXLEN, 0,
        (struct sockaddr *)&client, &client_len)) < 0) {
            fprintf(stderr, "Can't receive datagram\n");
            exit(1);
```

```
        }

        if (sendto(sd, buf, n, 0,
        (struct sockaddr *)&client, client_len) != n) {
            fprintf(stderr, "Can't send datagram\n");
            exit(1);
        }
    }
    close(sd);
    return(0);
}
```

The following client program first constructs a simple message of a predetermined length containing a string of characters a, b, c, ..., z, a, b, c, ..., z, ... The client then gets the start time from the system using gettimeofday and sends the message to the echo server. After the message travels back, the client records the end time and measures the difference that represents the round-trip latency between the client and the server. The unit of time is recorded in milliseconds. This simple example shows how we can use sockets to gather important network statistics such as latencies and jitters.

```
/* A simple UDP client which measures round trip delay */
#include <stdio.h>
#include <string.h>
#include <sys/time.h>
#include <netdb.h>
#include <sys/types.h>
#include <sys/socket.h>
#include <netinet/in.h>

#define SERVER_UDP_PORT   5000
#define MAXLEN        4096    /* maximum data length */
#define DEFLEN        64      /* default length */

long delay(struct timeval t1, struct timeval t2);

int main(int argc, char **argv)
{
    int     data_size = DEFLEN, port = SERVER_UDP_PORT;
    int     i, j, sd, server_len;
    char    *pname, *host, rbuf[MAXLEN], sbuf[MAXLEN];
    struct  hostent    *hp;
    struct  sockaddr_in    server, client;
    struct  timeval    start, end;

    pname = argv[0];
    argc--;
    argv++;
    if (argc > 0 && (strcmp(*argv, "-s") == 0)) {
        if (--argc > 0 && (data_size = atoi(*++argv))) {
            argc--;
```

```
            argv++;
        }
        else {
            fprintf (stderr,
            "Usage: %s [-s data_size] host [port]\n",
            pname);
            exit(1);
        }
    }
    if (argc > 0) {
        host = *argv;
        if (--argc > 0)
            port = atoi(*++argv);
    }
    else {
        fprintf(stderr,
        "Usage: %s [-s data_size] host [port]\n", pname);
        exit(1);
    }

    /* Create a datagram socket */
    if ((sd = socket(AF_INET, SOCK_DGRAM, 0)) == -1) {
        fprintf(stderr, "Can't create a socket\n");
        exit(1);
    }

    /* Store server's information */
    bzero((char *)&server, sizeof(server));
    server.sin_family = AF_INET;
    server.sin_port = htons(port);
    if ((hp = gethostbyname(host)) == NULL) {
        fprintf(stderr,
    "Can't get server's IP address\n");
        exit(1);
    }
    bcopy(hp->h_addr, (char *)&server.sin_addr,
        hp->_length);

    /* Bind local address to the socket */
    bzero((char *)&client, sizeof(client));
    client.sin_family = AF_INET;
    client.sin_port = htons(0);
    client.sin_addr.s_addr = hton1(INADDR_ANY);
    if (bind(sd, (struct sockaddr *)&client,
    sizeof(client)) == -1) {
        fprintf(stderr, "Can't bind name to socket\n");
        exit(1);
    }
    if (data_size > MAXLEN) {
        fprintf(stderr, "Data is too big\n");
```

```
            exit(1);
        }
        /* data is a, b, c, ..., z, a, b, ... */
        for (i = 0; i < data_size; i++) {
            j = (i < 26) ? i : i % 26;
            sbuf[i] = "a" + j;
        }

        gettimeofday(&start, NULL); /* start delay measure */

        /* transmit data */
        server_len = sizeof(server);
        if (sendto(sd, sbuf, data_size, 0, (struct sockaddr *)
            &server, server_len) == -1) {
            fprintf(stderr, "sendto error\n");
            exit(1);
        }

        /* receive data */
        if (recvfrom(sd, rbuf, MAXLEN, 0, (struct sockaddr *)
            &server, &server_lens) < 0) {
            fprintf(stderr, "recvfrom error\n");
            exit(1);
        }

        gettimeofday(&end, NULL); /* end delay measure */

        printf ("Round-trip delay = %ld ms.\n",
            delay(start, end));

        if (strncmp(sbuf, rbuf, data_size) != 0)
            printf("Data is corrupted\n");

        close(sd);
        return(0);
}

/*
 * Compute the delay between t1 and t2 in milliseconds
 */
long delay (struct timeval t1, struct timeval t2)
{
        long d;

        d = (t2.tv_sec - t1.tv_sec) * 1000;
        d += ((t2.tv_usec - t1.tv_usec + 500) / 1000);
        return(d);
}
```

It is important to remember that datagram communication using UDP is unreliable. If the communication is restricted to a local area network environment, say within a building, then datagram losses are extremely rare in practice, and the above client program should work well. However, in a wide area network environment, datagrams may be frequently discarded by the network. If the reply from the server does not reach the client, the client will wait forever! In this situation, the client must provide a timeout mechanism and retransmit the message. Also, further reliability may be provided to reorder the datagram at the receiver and to ensure that duplicated datagrams are discarded.

2.5 APPLICATION PROTOCOLS AND TCP/IP UTILITIES

Application protocols are high-level protocols that provide services to user applications. These protocols tend to be more visible to the user than other types of protocols. Furthermore, application protocols may be user written, or they may be standardized applications. Several standard application protocols form part of the TCP/IP protocol suite, the more common ones being Telnet, File Transfer Protocol (FTP), and SMTP. Coverage of the various TCP/IP application layer protocols is beyond the scope of this textbook. The student is referred to "Internet Official Protocol Standards," which provides a list of Internet protocols and standards [RFC 2400]. In this section the focus is on applications and utilities that can be used as tools to study the operation of the Internet.

2.5.1   Telnet

Telnet is a TCP/IP protocol that provides a standardized means of accessing resources on a remote machine where the initiating machine is treated as local to the remote host. In many implementations Telnet can be used to connect to the port number of other servers and to interact with them using a command line. For example, the HTTP and SMTP examples in section 2.1 were generated this way.

The Telnet protocol is based on the concept of a *network virtual terminal (NVT)*, which is an imaginary device that represents a lowest common denominator terminal. By basing the protocol on this interface, the client and server machines do not have to obtain information about each other's terminal characteristics. Instead, each machine initially maps its characteristics to that of an NVT and *negotiates options* for changes to the NVT or other enhancements, such as changing the character set.

The NVT acts as a character-based terminal with a keyboard and printer. Data input by the client through the keyboard is sent to the server through the Telnet connection. This data is echoed back by the server to the client's printer. Other incoming data from the server is also printed.

Telnet commands use the seven-bit U.S. variant of the ASCII character set. A command consists minimally of a two-byte sequence: the Interpret as Command (IAC) escape character followed by the command code. If the command pertains to option negotiation, that is, one of WILL, WONT, DO, or DONT, then a third byte contains the option code. Table 2.4 lists the Telnet command names, their corresponding ASCII code, and their meaning.

A substantial number of Telnet options can be negotiated. Option negotiations begin once the connection is established and may occur at any time while connected. Negotiation is symmetric in the sense that either side can initiate a negotiation. A negotiation syntax is defined in RFC 854 to prevent acknowledgment loops from occurring.

Telnet uses one TCP connection. Because this type of connection is full duplex and identified by a pair of ports, a server may have more than one Telnet connections simultaneously. Once the connection is established, the default is for the user, that is, the initiator of the connection, to enter a login name and password. By default the password is sent as clear text, although more recent versions of Telnet offer an authentication option.

Name	Code	Meaning
EOF	236	End of file.
SUSP	237	Suspend cursor process.
ABORT	238	Abort process.
EOR	239	End of record.
SE	240	End of subnegotiation parameters.
NOP	241	No operation.
Data mark	242	The data stream portion of a synch signal. This code should always be accompanied by a TCP urgent notification.
Break	243	NVT character BRK.
Interrupt process	244	The function IP.
Abort output	245	The function AO.
Are you there	246	The function AYT.
Erase character	247	The function EC.
Erase line	248	The function EL.
Go ahead	249	The GA signal.
SB	250	Indicates that what follows is subnegotiation of the indicated option.
WILL (option code)	251	Option negotiation.
WONT (option code)	252	Option negotiation.
DO (option code)	253	Option negotiation.
DONT (option code)	254	Option negotiation.
IAC	255	Data byte 255.

TABLE 2.4 Telnet commands

2.5.2   File Transfer Protocol

File Transfer Protocol (FTP) is another commonly used application protocol. FTP provides for the transfer of a file from one machine to another. Like Telnet, FTP is intended to operate across different hosts, even when they are running different operating systems or have different file structures.

FTP requires two TCP connections to transfer a file. One is the *control connection* that is established on port 21 at the server. The second TCP connection is a *data connection* used to perform a file transfer. A data connection must be established for each file transferred. Data connections are used for transferring a file in either direction or for obtaining lists of files or directories from the server to the client. Figure 2.19 shows the role of the two connections in FTP.

A control connection is established following the Telnet protocol from the user to the server port. FTP commands and replies are exchanged via the control connection. The user protocol interpreter (PI) is responsible for sending FTP commands and interpreting the replies. The server PI is responsible for interpreting commands, sending replies, and directing the server data transfer process (DTP) to establish a data connection and transfer. The commands are used to specify information about the data connection and about the particular file system operation being requested.

A data connection is established usually upon request from the user for some sort of file operation. The user PI usually chooses an ephemeral port number for its end of the operation and then issues a passive open from this port. The port number is then sent to the server PI using a PORT command. Upon receipt of the port number via the control connection, the server issues an active open to that same port. The server always uses port 20 for its end of the data connection. The user DTP then waits for the server to initiate and perform the file operation.

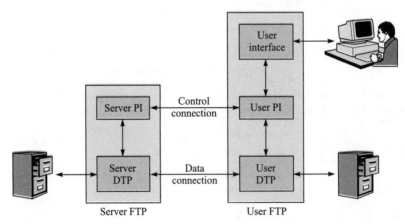

PI = Protocol interpreter
DTP = Data transfer process

FIGURE 2.19 Transferring files using FTP

Note that the data connection may be used to send and receive simultaneously. Note also that the user may initiate a file transfer between two nonlocal machines, for example, between two servers. In this case there would be a control connection between the user and both servers but only one data connection, namely, the one between the two servers.

The user is responsible for requesting a close of the control connection, although the server performs the action. If the control connection is closed while the data connection is still open, then the server may terminate the data transfer. The data connection is usually closed by the server. The main exception is when the user DTP closes the data connection to indicate an end of file for a stream transmission. Note that FTP is not designed to detect lost or scrambled bits; the responsibility for error detection is left to TCP.

The Telnet protocol works across different systems because it specifies a common starting point for terminal emulation. FTP works across different systems because it can accommodate several different file types and structures. FTP commands are used to specify information about the file and how it will be transmitted. In general, three types of information must be specified. Note that the default specifications must be supported by every FTP implementation.

1. File type. FTP supports ASCII, EBCDIC, image (binary), or local. Local specifies that the data is to be transferred in logical bytes, where the size is specified in a separate parameter. *ASCII is the default type*. If the file is ASCII or EBCDIC, then a vertical format control may also be specified.
2. Data structure. FTP supports file structure (a continuous stream of bytes with no internal structure), record structure (used with text files), and page structure (file consists of independent indexed pages). *File structure is the default specification*.
3. Transmission mode. FTP supports stream, block, or compressed mode. When transmission is in stream mode, the user DTP closes the connection to indicate the end of file for data with file structure. If the data has block structure, then a special two-byte sequence indicates end of record and end of file. *The default is stream mode*.

An *FTP command* consists of three or four bytes of uppercase ASCII characters followed by a space if parameters follow, or by a Telnet end of option list (EOL) otherwise. FTP commands fall into one of the following categories: access control identification, data transfer parameters, and FTP service requests. Table 2.5 lists some of the common FTP commands encountered.

Every command must produce at least one *FTP reply*. The replies are used to synchronize requests and actions and to keep the client informed of the state of the server. A reply consists of a three-digit number (in alphanumeric representation) followed by some text. The numeric code is intended for the user PI; the text, if processed, is intended for the user. For example, the reply issued following a successful connection termination request is "221 Goodbye". The first digit indicates whether and to what extent the specified request has been completed. The second digit indicates the category of the reply, and the third digit provides

Command	Meaning
ABOR	Abort the previous FTP command and any data transfer.
LIST	List files or directories.
QUIT	Log off from server.
RETR filename	Retrieve the specified file.
STOR filename	Store the specified file.

TABLE 2.5 Some common FTP commands

additional information about the particular category. Table 2.6 lists the possible values of the first two digits and their meanings.

In this case of the goodbye message, the first 2 indicates a successful completion. The second digit is also 2 to indicate that the reply pertains to a connection request.

2.5.3   IP Utilities

A number of utilities are available to help in finding out about IP hosts and domains and to measure Internet performance. In this section we discuss PING, which can be used to determine whether a host is reachable; Traceroute, a utility to determine the route that a packet will take to another host; Netstat, which provides information about the network status of a local host; and tcpdump, which captures and observers packet exchanges in a link. We also discuss the use of Telnet with standard TCP/IP services as a troubleshooting and monitoring tool.

Reply	Meaning
1yz	Positive preliminary reply (action has begun, but wait for another reply before sending a new command).
2yz	Positive completion reply (action completed successfully; new command may be sent).
3yz	Positive intermediary reply (command accepted, but action cannot be performed without additional information; user should send a command with the necessary information).
4yz	Transient negative completion reply (action currently cannot be performed; resend command later).
5yz	Permanent negative completion reply (action cannot be performed; do not resend it).
x0z	Syntax errors.
x1z	Information (replies to requests for status or help).
x2z	Connections (replies referring to the control and data connections).
x3z	Authentication and accounting (replies for the login process and accounting procedures).
x4z	Unspecified.
x5z	File system status.

TABLE 2.6 FTP replies—the first and second digits

PING

PING is a fairly simple application used to determine whether a host is online and available. The name is said to derive from its analogous use in sonar operations to detect underwater objects.[9] PING makes use of Internet Control Message Protocol (ICMP) messages. The purpose of ICMP is to inform sending hosts about errors encountered in IP datagram processing or other control information by destination hosts or by routers. ICMP is discussed in Chapter 8. PING sends one or more ICMP Echo messages to a specified host requesting a reply. PING is often used to measure the round-trip delay between two hosts. The sender sends a datagram with a type 8 Echo message and a sequence number to detect a lost, reordered, or duplicated message. The receiver changes the type to Echo Reply (type 0) and returns the datagram. Because the TCP/IP suite incorporates ICMP, any machine with TCP/IP installed can reply to PING. However, because of the increased presence of security measures such as firewalls, the tool is not always successful. Nonetheless, it is still the first test used to determine accessibility of a host.

In Figure 2.20 PING is used to determine whether the NAL machine is available. In this example, the utility was run in an MS-DOS session under Windows 95. The command in its simplest form is `ping <hostname>`. The round-trip delay is indicated, as well as the time-to-live (TTL) value. The TTL is the maximum number of hops an IP packet is allowed to remain in the network. Each time an IP packet passes through a router, the TTL is decreased by 1. When the TTL reaches 0, the packet is discarded.

TELNET AND STANDARD SERVICES

Because ICMP operates at the IP level, PING tests the reachability of the IP layer only in the destination machine. PING does not test the layers above IP. A number of standard TCP/IP application layer services can be used to test the layers above IP. Telnet can be used to access these services for testing purposes. Examples of these services include Echo (port number 7), which echoes a character back to the sender, and Daytime (port number 13), which returns the time

```
C:\WINDOWS>ping nal.toronto.edu

Pinging nal.toronto.edu [128.100.244.3] with 32 bytes of data:

Reply from 128.100.244.3: bytes=32 time=118ms TTL=243
Reply from 128.100.244.3: bytes=32 time=118ms TTL=243
Reply from 128.100.244.3: bytes=32 time=118ms TTL=243
Reply from 128.100.244.3: bytes=32 time=118ms TTL=243

C:\WINDOWS>
```

FIGURE 2.20 Using PING to determine host accessibility

---

[9]PING is also reported to represent the acronym Packet Internet Groper [Murhammer 1998].

and date. A variety of utilities are becoming available for testing reachability and performance of HTTP and Web servers. The student is referred to [CAIDA 98].

TRACEROUTE

A second TCP/IP utility that is commonly used is Traceroute. This tool allows users to determine the route that a packet takes from the local host to a remote host, as well as latency and reachability from the source to each hop. Traceroute is generally used as a debugging tool by network managers.

Traceroute makes use of both ICMP and UDP. The sender first sends a UDP datagram with TTL = 1 as well as an invalid port number to the specified destination host. The first router to see the datagram sets the TTL field to zero, discards the datagram, and sends an ICMP Time Exceeded message to the sender. This information allows the sender to identify the first machine in the route. Traceroute continues to identify the remaining machines between the source and destination machines by sending datagrams with successively larger TTL fields. When the datagram finally reaches its destination, that host machine returns an ICMP Port Unreachable message to the sender because of the invalid port number deliberately set in the datagram.

NETSTAT

The netstat utility queries a host about its TCP/IP network status. For example, netstat can find the status of the network drivers and their interface cards, such as the number of in packets, out packets, errored packets, and so on. It can also find the state of the routing table in a host, which TCP/IP server processes are active in the host, and which TCP connections are active.

TCPDUMP

The tcpdump program can capture and observe IP packet exchanges on a network interface. The program usually involves setting an Ethernet network interface card into a "promiscuous" mode so that the card listens and captures every frame that traverses the network. A packet filter is used to select the IP packets that are of interest in a given situation. These IP packets and their higher-layer contents can then be observed and analyzed. Because of security concern, normal users typically cannot run the tcpdump program. The book by Stevens provides numerous examples of the operation of the TCP/IP protocols using this tool [Stevens 1994].

SUMMARY

This chapter describes how network architectures are based on the notion of layering. Layering involves combining network functions into groups that can be implemented together. Each layer provides a set of services to the layer above it; each layer builds its services using the services of the layer below. Thus applications are developed using application layer protocols, and application layer protocols are built on top of the communication services provided by TCP and UDP. These transport protocols in turn build on the datagram service provided by IP, which is designed to operate over various network technologies. IP allows the applications above it to be developed independently of specific underlying network technologies. The network technologies below IP range from full-fledged packet-switching networks, such as ATM, to LANs, and individual point-to-point links.

The Berkeley socket API allows the programmer to develop applications using the services provided by TCP and UDP. Various TCP/IP utilities and tools allow the programmer to determine the state and configuration of a TCP/IP network. Beginning students can use these tools to get some hands on experience with the operation of TCP/IP.

CHECKLIST   OF IMPORTANT TERMS

application layer
♦ big endian
client/server
confirmed/unconfirmed service
connectionless service
connection-oriented service
daemon
data link layer
encapsulation
frame
header
♦ host byte order
hypertext transfer protocol (HTTP)
International Organization for
   Standardization (ISO)
internet layer
internetworking
layer
layer n entity
layer n protocol
♦ little endian
network architecture

♦ network byte order
network interface layer
network layer
open systems interconnection (OSI)
OSI reference model
packet
peer process
physical layer
Point-to-Point Protocol (PPP)
port
presentation layer
protocol data unit (PDU)
segment
service access point (SAP)
service data unit (SDU)
session layer
Simple Mail Transfer Protocol (SMTP)
socket
TCP/IP network architecture
Transport Control Protocol (TCP)
transport layer
User Datagram Protocol (UDP)

FURTHER READING

CAIDA, "CAIDA Measurement Tool Taxonomy," http://www.caida.org/Tools/taxono-my.html, October 1998.

Comer, D. E. and Stevens, D. L., *Internetworking with TCP/IP, Vol. III: Client-Server Programming and Applications*, Prentice Hall, Englewood Cliffs, New Jersey, 1993.

Davies, D. W., D. L. A. Barger, W. L. Price, and C. M. Solomonides, *Computer Networks and Their Protocols*, John Wiley & Sons, New York, 1979.

Murhammer, M. W., O. Atakan, S. Bretz, L. R. Pugh, K. Suzuki, and D. H. Wood, *TCP/IP Tutorial and Technical Overview*, Prentice Hall PTR, Upper Saddle River, New Jersey, 1998.

Perlman, R., *Interconnections: Bridges and Routers*, Addison-Wesley, Reading, Massachusetts, 1992.

Peterson, L. L. and B. S. Davie, *Computer Networks: A Systems Approach*, Morgan Kaufmann, San Francisco, 1996.

Piscitello, D. M. and A. L. Chapin, *Open Systems Networking: TCP/IP and OSI*, Addison-Wesley, Reading, Massachusetts, 1993.

Sechrest, S., "An Introductory 4.4 BSD Interprocess Communication Tutorial," *Computer Science Network Group*, UC Berkeley.

Stevens, W. R., *TCP/IP Illustrated, Volume 1: The Protocols*, Addison-Wesley, Reading, Massachusetts, 1994.

Stevens, W. R., *UNIX Network Programming*, 2nd edition, Prentice Hall, Englewood Cliffs, New Jersey, 1998.

Yeager, N. J. and R. E. McGrath, *Web Server Technology: The Advanced Guide for World Wide Web Information Providers*, Morgan Kaufmann, San Francisco, 1996.

RFC 821, J. Postel, "Simple Mail Transfer Protocol," August 1982.

RFC 854, J. Postel and J. Reynolds, "Telnet Protocol Specification," May 1983.

RFC 959, J. Postel and J. Reynolds, "File Transfer Protocol," October 1985.

RFC 1034, Mockapetris, "Domain Names—Concepts and Facilities," November 1987.

RFC 1035, Mockapetris, "Domain Names—Implementation and Specification," November 1987.

RFC 2068, R. Fielding, J. Geetys, J. Mogul, H. Frystyk, T. Berners-Lee, "Hypertext Transfer Protocol," January 1997.

RFC 2151, G. Kessler and S. Shepard, "Internet & TCP/IP Tools and Utilities," June 1997.

RFC 2400, J. Postel and J. Reynolds, eds., "Internet Official Protocol Standards," September 1998.

PROBLEMS

1. Explain how the notion of layering and internetworking make the rapid growth of applications such as the World Wide Web possible.

2. a. What universal set of communication services is provided by TCP/IP?
   b. How is independence from underlying network technologies achieved?
   c. What economies of scale result from (a) and (b)?

3. What difference does it make to the network layer if the underlying data link layer provides a connection-oriented service versus a connectionless service?

4. Suppose transmission channels become virtually error free. Is the data link layer still needed?

5. Why is the transport layer not present inside the network?

6. Which OSI layer is responsible for the following?
   a. Determining the best path to route packets.
   b. Providing end-to-end communications with reliable service.
   c. Providing node-to-node communications with reliable service.

7. Should connection establishment be a confirmed service or an unconfirmed service? what about data transfer in a connection-oriented service? connection release?

8. Does it make sense for a network to provide a confirmed, connectionless packet transfer service?

9. Explain how the notion of multiplexing can be applied at the data link, network, and transport layers. Draw a figure that shows the flow of PDUs in each multiplexing scheme.

10. Give two features that the data link layer and transport layer have in common. Give two features in which they differ. Hint: Compare what can go wrong to the PDUs that are handled by these layers.

11. a. Can a connection-oriented, reliable message transfer service be provided across a connectionless packet network? Explain.
    b. Can a connectionless datagram transfer service be provided across a connection-oriented network?

12. An internet path between two hosts involves a hop across network A, a packet-switching network, to a router and then another hop across packet-switching network B. Suppose that packet-switching network A carries the packet between the first host and the router over a two-hop path involving one intermediate packet switch. Suppose also that the second network is an Ethernet LAN. Sketch the sequence of IP and non-IP packets and frames that are generated as an IP packet goes from host 1 to host 2.

13. Does Ethernet provide connection-oriented or connectionless service?

14. Ethernet is a LAN so it is placed in the data link layer of the OSI reference model.
    a. How is the transfer of frames in Ethernet similar to the transfer of frames across a wire? How is it different?
    b. How is the transfer of frames in Ethernet similar to the transfer of frames in a packet-switching network? How is it different?

15. Suppose that a group of workstations is connected to an Ethernet LAN. If the workstations communicate only with each other, does it make sense to use IP in the workstations? Should the workstations run TCP directly over Ethernet? How is addressing handled?

16. Suppose two Ethernet LANs are interconnected by a box that operates as follows. The box has a table that tells it the physical addresses of the machines in each LAN. The box listens to frame transmissions on each LAN. If a frame is destined to a station at the other LAN, the box retransmits the frame onto the other LAN; otherwise, the box does nothing.

   a. Is the resulting network still a LAN? Does it belong in the data link layer or the network layer?

   b. Can the approach be extended to connect more than two LANs? If so, what problems arise as the number of LANs becomes large?

17. Suppose all laptops in a large city are to communicate using radio transmissions from a high antenna tower. Is the data link layer or network layer more appropriate for this situation? Now suppose the city is covered by a large number of small antennas covering smaller areas. Which layer is more appropriate?

18. Suppose that a host is connected to a connection-oriented packet-switching network and that it transmits a packet to a server along a path that traverses two packet switches. Suppose that each hop in the path involves a point-to-point link, that is, a wire. Show the sequence of network layer and data link layer PDUs that is generated as the packet travels from the host to the server.

19. Suppose an application layer entity wants to send an L-byte message to its peer process, using an existing TCP connection. The TCP segment consists of the message plus 20 bytes of header. The segment is encapsulated into an IP packet that has an additional 20 bytes of header. The IP packet in turn goes inside an Ethernet frame that has 18 bytes of header and trailer. What percentage of the transmitted bits in the physical layer corresponds to message information if $L = 100$ bytes? 500 bytes? 1000 bytes?

20. Suppose that the TCP entity receives a 1.5-megabyte file from the application layer and that the IP layer is willing to carry blocks of maximum size 1500 bytes. Calculate the amount of overhead incurred from segmenting the file into packet-sized units.

21. Suppose a TCP entity receives a digital voice stream from the application layer. The voice stream arrives at a rate of 8000 bytes/second. Suppose that TCP arranges bytes into block sizes that result in a total TCP and IP header overhead of 50 percent. How much delay is incurred by the first byte in each block?

22. How does the network layer in a connection-oriented packet-switching network differ from the network layer in a connectionless packet-switching network?

23. Identify session layer and presentation layer functions in the HTTP protocol.

24. Suppose we need a communication service to transmit real-time voice over the Internet. What features of TCP and what features of UDP are appropriate?

25. Consider the end-to-end IP packet transfer examples in Figure 2.13. Sketch the sequences of IP packets and Ethernet and PPP frames that are generated by the three examples of packet transfers: from the workstation to the server, from the server to the PC, and from the PC to the server. Include all relevant header information in the sketch.

26. Suppose a user has two browser applications active at the same time and suppose that the two applications are accessing the same server to retrieve HTTP documents at the same time. How does the server tell the difference between the two applications?

27. What is the difference between a physical address, a network address, and a domain name?

28. The Domain Name System has a hierarchical structure, for example, comm.toronto.edu. Explain how a DNS query might proceed if the local name server does not have the IP address for a given host.

29. What is wrong with the following methods of assigning host id addresses?
    a. Copy the address from the machine in the next office.
    b. Modify the address from the machine in the next office.
    c. Use an example from the vendor's brochure.

30. Suppose a machine is attached to several physical networks. Why does it need a different IP address for each attainment?

31. Suppose a computer is moved from one department to another. Does the physical address need to change? Does the IP address need to change? Does it make a difference if the computer is a laptop?

32. Suppose the population of the world is 4 billion people and that there is an average of 1000 communicating devices per person. How many bits are required to assign a unique host address to each communicating device? Suppose that each device attaches to a single network and that each network on average has 10,000 devices. How many bits are required to provide unique network ids to each network?

33. Can the Internet protocol be used to run a homogeneous packet-switching network, that is, a network with identical packet switches interconnected with point-to-point links?

34. Is it possible to build a homogeneous packet-switching network with Ethernet LANs interconnecting the packet switches? If so, can connection-oriented service be provided over such a network?

35. In telephone networks one basic network is used to provide worldwide communications. In the Internet a multiplicity of networks are interconnected to provide global connectivity. Compare these two approaches, namely, a single network versus an internetwork, in terms of the range of services that can be provided and the cost of establishing a world-wide network.

36. Consider an internetwork architecture that is defined using gateways/routers to communicate across networks but that uses a connection-oriented approach to packet switching? What functionality is required in the routers? Are any additional constraints imposed on the underlying networks?

37. The internet below consists of three LANs-interconnected by two routers. Assume that the hosts and routers have the IP addresses as shown.
    a. Suppose that all traffic from network 3 that is destined to H1 is to be routed directly through router R2 and that all other traffic from network 3 is to go to network 2. What routing table entries should be present in the network 3 hosts and in R2?

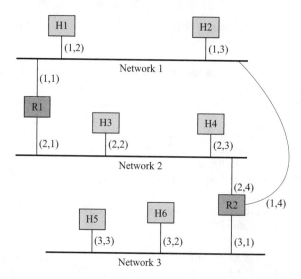

b. Suppose that all traffic from network 2 to network 3 is to be routed directly through R2. What routing table entries should be present in the network 1 hosts and in R2?

38. Explain why it is useful for application layer programs to have a "well-known" TCP port number?

39. Use a Web browser to connect to cnn.com. Explain what layers in the protocol stack are involved in the delivery of the video newscast.

40. Use a Web browser to connect to an audio program, say, www.rsradio.com (Rolling Stone Radio) or www.cbc.com (CBC Radio). Explain what layers in the protocol stack are involved here. How does this situation differ from the delivery of video in problem 39?

41. Which of the TCP/IP transport protocol (UDP or TCP) would you select for the following applications: packet voice, file transfers, remote login, multicast communication (i.e., multiple destinations).

42. Use the Telnet program to send an e-mail by directly interacting with your local mail server.

43. The nslookup program can be used to query the Internet domain name servers. Use this program to look up the IP address of www.utoronto.ca.

44. Use PING to find the round-trip time to the home page of your university and to the home page of your department.

45. Use netstat to find out the routing table for a host in your network.

46. Suppose regularly spaced PING packets are sent to a remote host. What can you conclude from the following results?

   a. No replies arrive back.

   b. Some replies are lost.

   c. All replies arrive but with variable delays.

   d. What kind of statistics would be useful to calculate for the round-trip delays?

47. Suppose you want to test the response time of a specific Web server. What attributes would such a measurement tool have? How would such a tool be designed?

48. A denial-of-service attack involves loading a network resource to the point where it becomes nonfunctional. Explain how PING can be used to carry out a denial-of-service attack.

49. HTTP relies on ASCII characters. To verify the sequence of messages shown in Table 2.1, use the Telnet program to connect to a local Web site.

50. Discuss the similarities and differences between the control connection in FTP and the remote control used to control a television. Can the FTP approach be used to provide VCR-type functionality to control the video from a video-on-demand service?

51. Use a Web browser to access the Cooperative Association for Internet Data Analysis (CAIDA) Web page (http://www.caida.org/Tools/taxonomy.html) to retrieve the CAIDA measurement tool taxonomy document. You will find links there to many free Internet measurement tools and utilities.

52. Run the UDP client and server programs from the Berkeley API section on different machines, record the round-trip latencies with respect to the size of the data, and plot the results.

53. In the TCP example from the Berkeley API section, the message size communicated is fixed regardless of how many characters of actual information a user types. Even if the user wants to send only one character, the programs still sends 256 bytes of messages— clearly an inefficient method. One possible way to allow variable-length messages to be communicated is to indicate the end of a message by a unique character, called the sentinel. The receiver calls `read` for every character (or byte), compares each character with the sentinel value, and terminates after this special value is encountered. Modify the TCP client and server programs to handle variable-length messages using a sentinel value.

54. Another possible way to allow variable-length messages to be communicated is to precede the data to be transmitted by a header indicating the length of the data. After the header is decoded, the receiver knows how many more bytes it should read. Assuming the length of the header is two bytes, modify the TCP client and server programs to handle variable-length messages.

55. The UDP client program in the example from the Berkeley API section may wait forever if the datagram from the server never arrives. Modify the client program so that if the response from the server does not arrive after a certain timeout (say, 5 seconds), the `read` call is interrupted. The client then retransmits a datagram to the server and waits for a new response. If the client does not receive a response after a fixed number of trials (say, 10 trials), the client should print an error message and abandon the program. Hint: Use the `sigaction` and `alarm` functions.

56. Modify the UDP client to access a date-and-time server in a host local to your network. A date-and-time server provides client programs with the current day and time on demand. The system internal clock keeps the current day and time as a 32-bit integer. The time is incremented by the system (every second). When an application program (the server in this case) asks for the date or time, the system consults the internal clock and formats the date and time of day in human-readable format. Sending any datagram to a date-and-time server is equivalent to making a request for the current date and time; the server responds by returning a UDP message containing the current date and time. The date-and-time server can be accessed in UDP port 13.

CHAPTER 3

Digital Transmission Fundamentals

In Chapter 1 we saw that the development of the early network architectures was motivated by very specific applications. Telegraphy was developed specifically for the transfer of text messages. We saw that Morse and Baudot pioneered the use of binary representations for the transfer of text and that digital transmission systems were developed for the transfer of the resulting binary information streams. Later telephony was developed for the transfer of analog voice information. The invention of **Pulse Code Modulation (PCM)** paved the way for voice to be transmitted over digital transmission networks. In the same way that the Morse and Baudot codes standardized the transfer of text, PCM standardized the transfer of voice in terms of 0s and 1s. We are currently undergoing another major transition from analog to digital transmission technology, namely, the transition from analog television systems to entirely digital television systems. When this transition is complete, all major forms of information will be represented in digital form. This change will open the way for the deployment of digital networks that can transfer the information for all the major types of information services.

Digital transmission is the core technology that enables the integration of services in a network. In this chapter we present the fundamental concepts concerning digital transmission. The chapter is organized into the following sections:

1. *Digital representation of information.* The basic properties of text, image, voice, audio and video information are presented. These information types can be viewed as flows that traverse the network. We examine the types of information flows that are generated by the various types of information, and we consider the requirements that they place on the networks that carry them.
2. *Why digital transmission?* We explain the advantages of digital transmission over analog transmission. We develop the basic abstraction of a transmission link as a pipe that transfers bits for the data link layer. We present the key

parameters that determine the transmission capacity of a physical medium. We also indicate where various media are used in digital transmission systems. This section is a summary of the following two sections.[1]

3. *Characterization of communication channels.* We discuss communication channels in terms of their ability to transmit pulse and sinusoidal signals. We introduce the concept of bandwidth as a measure of a channel's ability to transmit pulse information.

4. *Fundamental limits of digital transmission.* We discuss binary and multilevel digital transmission systems, and we develop fundamental limits on the bit rate that can be obtained over a channel.

5. *Line coding.* We introduce various formats for transmission binary information and discuss the criteria for selecting an appropriate line code.

6. *Modems and digital modulation.* We discuss digital transmission systems that use sinusoidal signals, and we explain existing telephone modem standards.

7. *Properties of transmission media.* We discuss copper wire, radio, and optical fiber systems and their role in access and backbone digital networks. Examples from various physical layer standards are provided.

8. *Error detection and correction.* We present coding techniques that can be used to detect and correct errors that may occur during digital transmission. These coding techniques form the basis for protocols that provide reliable transfer of information.

3.1 DIGITAL REPRESENTATION OF INFORMATION

Networks are driven by the applications they support and must therefore be designed to accommodate the requirements imposed by the information types used in the applications. These information types include text, speech, audio, data, images, and video. In this section we are concerned with (1) the properties of these information types and (2) the requirements that the digital representation of these information types impose on the network. We are concerned in particular with the volume of binary information that needs to be transferred for each information type. The remainder of the chapter is concerned with techniques for carrying out the transfer of such information over various types of communications systems.

It is useful at this point to identify which layers in the OSI reference model we are dealing with. This section examines the information in its original form in the application. The application generates blocks or streams of information that all the layers in the protocol stack must handle. The remainder of the chapter

---

[1]This arrangement allows sections 3.3 and 3.4 to be skipped in courses that are under tight time constraints. Nevertheless, we believe that the remaining sections make the topic of digital transmission quite accessible, so we encourage all students to read the remainder of the chapter independently.

deals with the physical layer, which handles the transfer of raw bits. A useful analogy is to view the information generated by the application as a flow that needs to be carried by the network. The digital transmission systems at the physical layer are the pipes that carry the information flows.

3.1.1 Binary Representations of Different Information Types

Information can be categorized into two broad categories: information that occurs naturally in the form of a single *block* and *stream* information that is produced continuously by a source of information. Table 3.1 gives examples of **block-oriented information**, which include data files, black-and-white documents, and pictures. Table 3.2 gives examples of stream information such as audio and video.

The most common examples of block information are files that contain text, numerical, or graphical information. We routinely deal with these types of information when we send e-mail and when we retrieve documents. These blocks of information can range from a few bytes to several hundred kilobytes and occasionally several megabytes. The normal form in which these files occur can contain a fair amount of statistical redundancy. For example, certain characters and patterns such as *e* and *the*, occur very frequently. **Data compression** utilities such as compress, zip, and other variations exploit these redundancies to encode the original information into files that require less disk storage space.[2] Some modem standards also apply these data compression schemes to the information prior to transmission. The **compression ratio** is defined as the ratio of the number of bits in the original file to the number of bits in the compressed file. Typically

Information type	Data compression technique	Format	Uncompressed	Compressed (compression ratio)	Applications
Text files	Compress, zip, and variations	ASCII	Kbytes to Mbytes	(2–6)	Disk storage, modem transmission
Scanned black-and-white documents	CCITT Group 3 facsimile standard	A4 page @ 200 × 200 pixels/inch and options	256 Kbytes	15–54 Kbytes (1-D) 5–35 Kbytes (2-D) (5–50)	Facsimile transmission, document storage
Color images	JPEG	8-×-10 inch photo scanned @ 400 pixels/inch	38.4 Mbytes	1.2–8 Mbytes (5–30)	Image storage or transmission

TABLE 3.1 Block-oriented information

---

[2]The details of the data compression techniques discussed in this section are found in Chapter 12.

the compression ratio for these types of information is two or more, thus providing an apparent doubling or more of the transmission speed or storage capacity.

A facsimile system scans a black-and-white document into an array of dots that are either white or black. The CCITT Group 3 facsimile standards provide for resolutions of 200, 300, or 400 dots per horizontal inch and 100, 200, or 400 vertical lines per inch. For example, a standard A4 page at 200×100 pixels/inch (slightly bigger than 8.5×11 inches) produces 256 kilobytes prior to compression. At a speed of 28.8 kbps, such an uncompressed page would require more than 1 minute to transmit. Existing fax compression algorithms can reduce this transmission time typically by a factor of 8 to 16.

An individual color image produces a huge number of bits. A *pixel* is defined as a single dot in a digitized image. For example, an 8-×-10-inch picture scanned at a resolution of 400×400 pixels per square inch yields $400 \times 400 \times 8 \times 10 = 12.8$ million pixels; see Table 3.1. A color image is decomposed into red, green and blue subimages as shown in Figure 3.1. Normally eight bits are used to represent each of the red, green, and blue color components, resulting in a total of 12.8 megapixels $\times$ 3 bytes/pixel $=$ 38.4 megabytes. At a speed of 28.8 kbps, this image would require about 3 hours to transmit! Clearly, data compression methods are required to reduce these transmission times.

The *Graphics Interchange Format (GIF)* takes image data, in binary form, and applies lossless data compression. **Lossless data compression** schemes produce a compressed file from which the original data can be recovered *exactly*. (Facsimile and file compression utilities also use lossless data compression.) However, lossless data compression schemes are limited in the compression rates they can achieve. For this reason, GIF is used mainly for simple images such as line drawings and images containing simple geometrical shapes. On the other hand, **lossy data compression** schems produce a compressed file from which only an *approximation* to the original information can be recovered. Much higher compression schemes are possible. In the case of images, lossy compression is acceptable as long as there is little or no visible degradation in image quality. The *Joint Photographic Experts Group (JPEG)* standard provides a lossy compression algorithm that can be adjusted to balance image quality versus file size. For example, JPEG can typically produce a high-quality reproduction with a compression ratio of about 15. Combined with the fastest telephone modems, say

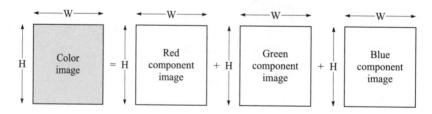

Total bits before compression = 3 × H × W pixels × B bits/pixel = 3 HWB

FIGURE 3.1 Color image had three components

56 kbps, this compression ratio reduces the transmission time of the image in Table 3.1 to several minutes. Clearly in the case of images, we either make do with lower resolution and/or lower quality images or we procure higher speed communications.

Analog information such as voice, music, or video occurs in a steady stream. Table 3.2 lists the properties of this type of information. In the case of a voice or music signal, the sound, which consists of variations in air pressure, is converted into a voltage that varies continuously with time.

The first step in digitizing such a signal is to sample the values of the signal every T seconds as shown in Figure 3.2a. Clearly, the value of T that is used depends on how fast the signal varies with time. For example, for a PCM telephone-quality voice, the signal is sampled at a rate of 8000 samples/second, that is, $T = 1/8000 = 125$ microseconds as shown in Figure 3.2a. The precision of the sample measurements determines the bit rate as well as the quality of the approximation. For example, Figure 3.2b shows how each of the signal samples can be approximated by one of the eight levels. Each level can then be represented by a three-bit number. In the case of telephone systems, the samples are represented by 8 bits in resolution, resulting in a bit rate of 8000 samples/second × 8 bits/sample = 64 kbps. The bit rate can be reduced by reducing the resolution, but doing so results in a less accurate reproduction of the original signal.

Information type	Compression technique	Format	Uncompressed	Compressed	Applications
Voice	PCM	4 kHz voice	64 kbps	64 kbps	Digital telephony
Voice	ADPCM (+ silence detection)	4 kHz voice	64 kbps	32 kbps (16 kbps)	Digital telephony, voice mail
Voice	Residual-excited linear prediction	4 kHz voice	64 kbps	8–16 kbps	Digital cellular telephony
Audio	MPEG audio MP3 compression	16–24 kHz audio	512–748 kbps	32–384 kbps	MPEG audio
Video	H.261 coding	176 × 144 or 352 × 288 frames @ 10–30 frames/second	2–36.5 Mbps	64 kbps–1.544 Mbps	Video conferencing
Video	MPEG-2	720 × 480 frames @ 30 frames/second	249 Mbps	2–6 Mbps	Full-motion broadcast video
Video	MPEG-2	1920 × 1080 frames @ 30 frames/second	1.6 Gbps	19–38 Mbps	High-definition television

TABLE 3.2 Properties of analog and video information

(a) Original waveform and the sample values

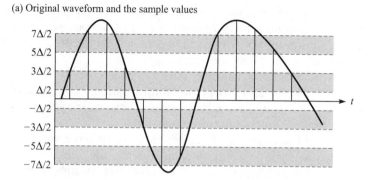

(b) Original waveform and the quantized values

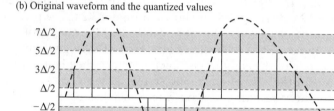

FIGURE 3.2 Sampling of a speech signal

The high cost of transmission in certain situations, for example, undersea cable, satellite, and wireless systems, has led to the development of more complex algorithms for reducing the bit rate while maintaining a telephone-quality voice signal. Differential PCM (DPCM) encodes the difference between successive samples. Adaptive DPCM (ADPCM) adapts to variations in voice-signal level. Linear predictive methods adapt to the type of sound. These systems can reduce the bit rate of telephone-quality voice to the range 8 to 32 kbps. Despite the fact that they are "lossy," these schemes achieve compression and high quality due to the imperceptibility of the approximation errors that are introduced.

Music signals vary much more rapidly than voice signals do. Thus, for example, audio compact disc (CD) systems sample the music signals at 44 kilo-samples/second at a resolution of 16 bits. For a stereo music system the result is a bit rate of 44,000 samples/second × 16 bits/sample × 2 channels = 1.4 megabits/second. One hour of music will then produce 317 Mbytes of information. The subband coding technique used in the MPEG audio standard can reduce this bit rate by a factor of 14 kbps to about 100 kbps.

Video signals ("moving pictures" or "flicks") can be viewed as a succession of pictures that is fast enough to give the human eye the appearance of continuous motion. If there is very little motion, such as a close-up view of a face in a videoconference, then the system needs to transmit only the differences between successive pictures. Typical videoconferencing systems operate with frames of 176 × 144 pixels at 10 to 36 frames/second as shown in Figure 3.3a. The color of each pixel is initially represented by 24 bits, that is, 8 bits per color component. When compressed, these videoconferencing signals produce bit rates in the range of several hundred kilobits/second as shown in Table 3.2.

Broadcast television requires greater resolution (720 × 480 pixels per frame) than video conferencing requires, as shown in Figure 3.3b, and can contain a high degree of motion. The MPEG-2 coding system can achieve a reduction from the uncompressed bit rate of 249 Mbps to the range of 2 to 6 Mbps. The recently approved Advanced Television Systems Committee (ATSC) U.S. standard for high-definition television applies the MPEG-2 coding system in a system that operates with more detailed 1920 × 1080 pixel frames at 30 frames/second as shown in Figure 3.3c. The 16:9 aspect ratio of the frame gives a more theaterlike experience; ordinary television has a 4:3 aspect ratio. The uncompressed bit rate is 1.6 gigabits/second. The MPEG-2 coding can reduce this to 19 to 38 Mbps, which can be supported by digital transmission over terrestrial broadcast and cable television systems.

3.1.2   Network Requirements of Different Information Types

In the preceding discussion we focused primarily on the amount of information required to represent different types of information, that is, bits per file for blocks of information and bits per second for streams of information. To handle a particular type of information, a network or transmission system must be capable of transferring the associated *volume of information*. The various types

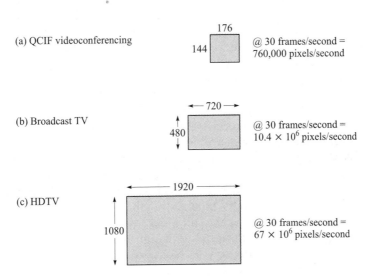

FIGURE 3.3 Video image pixel rates

of information also impose additional requirements on the network or transmission system. The *accuracy* of the information delivered and the *timeliness* of the delivery are of particular interest.

This chapter shows that digital transmission systems have some small, but nonzero, rate at which they introduce errors into the delivered information. Different types of information have different degrees of tolerance to such **transmission errors**. For example, data files in general cannot tolerate any errors. The transfer of data files therefore involves the use of error-detection schemes that identify transmissions that contain errors and initiate error-recovery procedures. Other types of information, such as facsimile, can tolerate some errors. Typically, as the compression ratio of a scheme increases, the importance of the information conveyed by every bit in the compressed sequence increases accordingly. Thus errors introduced in the compressed sequence can lead to high or even catastrophic distortion. For example, PCM speech, which is not highly compressed, can tolerate fairly high bit error rates. On the other hand, highly compressed speech, for example, using residual-excited linear predictive coding, will be quite vulnerable to errors and therefore necessitates the use of error-correction coding. Error-detection and -correction schemes are discussed in the last section of this chapter.

Certain applications have timeliness requirements associated with the delivery of information. For example, a request for a block of information may involve a presentation deadline. The network may therefore be required to deliver the block within a certain maximum **delay**. The time to deliver a block of L bits of information includes the propagation delay as well as the block transmission time: $t_{prop} + L/R$. The propagation delay $t_{prop} = d/c$ where d is that distance that the information has to travel and c is the speed of light in the transmission medium. Clearly, the designer cannot change the speed of light, but the distance that the information has to travel can be controlled through the placement of the file servers. The time to transmit the file can be reduced by increasing the transmission bit rate R.

Stream information can be viewed as a sequence of blocks of information as shown in Figure 3.4a. An application may impose a maximum delay on *every* block of information within the stream. For example, real-time communications between people requires a maximum delay of about 250 ms to ensure interactivity close to that of normal conversation.

As a stream of blocks traverses the network, the spacing between the information blocks can be altered. **Jitter** is defined as the variation in the delay between consecutive blocks. Figure 3.4b shows that variable delays incurred while traversing a network can introduce jitter through "bunching up." In stream applications the decoder that plays back the information must be fed a steady stream of information blocks for the system to operate correctly. Typically, the decoder assumes that all blocks will arrive within some maximum delay, and it plays back the information within such a delay, as shown in Figure 3.4c. The receiver uses a buffer to hold blocks of information until their playback time. The jitter is a measure of how much the blocks of information will tend to bunch up and hence of the buffer size that will be required to hold the blocks.

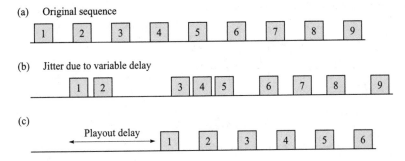

FIGURE 3.4 Temporal impairments for stream information

In the remainder of this chapter we are concerned only with the bit rate and the transmission error properties of digital transmission systems. We also consider the problem of dealing with transmission errors. In Chapter 5 we return to the discussion of the adaptation techniques that can be used to help an application deal with the various types of impairments incurred during transfer through a network.

3.2   WHY DIGITAL COMMUNICATIONS?[3]

A transmission system makes use of a physical **transmission medium** or **channel** that allows the propagation of energy in the form of pulses or variations in voltage, current, or light intensity as shown in Figure 3.5. In analog communications the objective is to transmit a waveform, which is a function that varies continuously with time, as shown in Figure 3.6a. For example, the electrical signal coming out of a microphone corresponds to the variation in air pressure corresponding to sound. This function of time must be reproduced exactly at the output of the analog communication system. In practice, communications channels do not satisfy this condition, so some degree of distortion is unavoidable.

In digital transmission the objective is to transmit a given symbol that is selected from some finite set of possibilities. For example, in binary digital

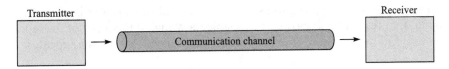

FIGURE 3.5 General transmission system

---

[3]This section summarizes the main results of sections 3.3 and 3.4, allowing these two sections to be skipped if necessary.

(a) Analog transmission: all details must be reproduced
    accurately

FIGURE 3.6 Analog versus digital transmission

• e.g., AM, FM, TV transmission

(b) Digital transmission: only discrete levels need to be
    reproduced

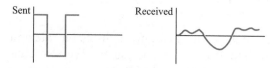

• e.g., digital telephone, CD audio

transmission the objective is to transmit either a 0 or a 1. This can be done, for instance, by transmitting positive voltage for a certain period of time to convey a 1 or a negative voltage to convey a 0, as shown in Figure 3.6b. The task of the receiver is to determine the input symbol with high probability. The positive or negative pulses that were transmitted for the given symbols can undergo a great degree of distortion. Where signaling uses positive or negative voltages, the system will operate correctly as long as the receiver can determine whether the original voltage was positive or negative.

The cost advantages of digital transmission over analog transmission become apparent when transmitting over a long distance. Consider, for example, a system that involves transmission over a pair of copper wires. As the length of the pair of wires increases, the signal at the output is attenuated and the original shape of the signal is increasingly distorted. In addition, interference from extraneous sources, such as radiation from car ignitions and power lines, as well as noise inherent in electronic systems result in the addition of random noise to the transmitted signal. To transmit over long distances, it is necessary to introduce repeaters periodically to regenerate the signal, as shown in Figure 3.7. Such signal regeneration is fundamentally different for analog and digitial transmissions.

In an analog communication system, the task of the repeater is to regenerate a signal that resembles as closely as possible the signal at the input of the repeater segment. Figure 3.8 shows the basic functions carried out by the repeater. The

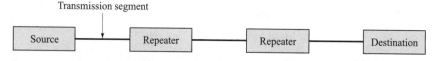

FIGURE 3.7 Typical long-distance link

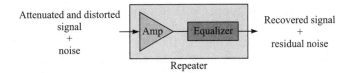

FIGURE 3.8 An analog repeater

input to the repeater is an attenuated and distorted version of the original transmitted signal plus the random noise added in the segment. First the repeater
deals with the attenuation by amplifying the received signal. To do so the repeater multiplies the signal by a factor that is the reciprocal of the attenuation a. The
resulting signal is still distorted by the channel.

The repeater next uses a device called an **equalizer** in an attempt to eliminate
the distortion. You will see in the next section that the source of the distortion in
the signal shape has two primary causes. The first cause is that different frequency components of the signal are attenuated differently.[4] In general, high-
frequency components are attenuated more than low-frequency components.
The equalizer compensates for this situation by amplifying different frequency
components by different amounts. The second cause is that different frequency
components of a signal are delayed by different amounts as they propagate
through the channel. The equalizer attempts to provide differential delays to
realign the frequency components. In practice it is very difficult to carry out
the two functions of the equalizer. For the sake of argument, suppose that the
equalization is perfect. The output of the repeater then consists of the original
signal plus the noise.

In the case of analog signals, the repeater is limited in what it can do to deal
with noise. If it is known that the original signal does not have components
outside a certain frequency band, then the repeater can remove noise components that are outside the signal band. However, the noise within the signal band
cannot be reduced and consequently the signal that is finally recovered by the
repeater will contain some noise. The repeater then proceeds to send the recovered signal over the next transmission segment.

The effect on signal quality after multiple analog repeaters is similar to that
in repeated recordings using analog audiocassette tapes or VCR tapes. The first
time a signal is recorded, a certain amount of noise, which is audible as hiss, is
introduced. Each additional recording adds more noise. After a large number of
recordings, the signal quality degrades considerably.[5] A similar effect occurs in
the transmission of analog signals over multiple repeater segments.

---

[4]Periodic signals can be represented as a sum of sinusoidal signals using Fourier series. Each sinusoidal
signal has a distinct frequency. We refer to the sinusoidal signals as the "frequency components" of the
original signal. (Fourier series are reviewed in Appendix 3B.)
[5]Digital recording techniques will be introduced in consumer products in the near future, and so this
example will become obsolete! Another example involves noting the degradation in image quality as a
photocopy of a photocopy is made.

Next consider the same copper wire transmission system for digital communications. Suppose that a string of 0s and 1s is conveyed by a sequence of positive and negative voltages. As the length of the pair of wires increases, the pulses are increasingly distorted and more noise is added. A digital repeater is required as shown in Figure 3.7. The sole objective of the repeater is to determine with high probability the original binary stream. The repeater also uses an equalizer to compensate for the distortion introduced by the channel. However, the repeater does not need to completely regenerate the original shape of the transmitted signal. It only needs to determine whether the original pulse was positive or negative. To do so, the repeater is organized in the manner shown in Figure 3.9.

A timing recovery circuit keeps track of the intervals that define each pulse. The decision circuit then samples the signal at the midpoint of each interval to determine the polarity of the pulse. In a property designed system, in the absence of noise, the original symbol would be recovered every time, and consequently the binary stream would be regenerated exactly over any number of repeaters and hence over arbitrarily long distances. However, noise is unavoidable, which implies that errors will occur from time to time. An error occurs when the noise signal is sufficiently large to change the polarity of the original signal at the sampling point. Digital transmission systems are designed for very low bit error rates, for example, 10^{-7}, 10^{-9}, or even 10^{-12}, which corresponds to one error in every trillion bits!

The impact on signal quality in multiple digital repeaters is similar to the digital recording of music where the signal is stored as a file of binary information. We can copy the file digitally any number of times with extremely small probabilities of errors being introduced in the process. In effect, the quality of the sound is unaffected by the number of times the file is copied.

The preceding discussion shows that digital transmission has superior performance over analog transmission. Digital repeaters eliminate the accumulation of noise that takes place in analog systems and provide for long-distance transmission that is nearly independent of distance. Digital transmission systems can operate with lower signal levels or with greater distances between repeaters than analog systems can. This factor translates into lower overall system cost and was the original motivation for the introduction of digital transmission.

Over time, other benefits of digital transmission have become more prominent. Networks based on digital transmission are capable of handling *any* type of information that can be represented in digital form. Thus digital networks are

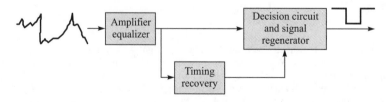

FIGURE 3.9 A digital repeater

suitable for handling many types of services. Digital transmission also allows networks to exploit the advances in digital computer technology to increase not only the volume of information that can be transmitted but also the types of processing that can be carried out within the network, that is, error correction, data encryption, and the various types of network protocol processing that are the subject of this book.

3.2.1 Basic Properties of Digital Transmission Systems

The purpose of a **digital transmission** system is to transfer a sequence of 0s and 1s from a transmitter (on the left end) to a receiver (on the right) as shown in Figure 3.10. We are particularly interested in the **bit rate** or transmission speed as measured in bits/second. The bit rate R can be viewed as the cross-section of the information pipe that connects the transmitter to the receiver.

The transmission system uses pulses or sinusoids to transmit binary information over a physical transmission medium. A fundamental question in digital transmission is *how fast* can bits be transmitted *reliably* over a given medium. This capability is clearly affected by several factors including:

- The amount of *energy* put into transmitting each signal.
- The *distance* that the signal has to traverse (because the energy is dissipated and dispersed as it travels along the medium).
- The amount of *noise* that the receiver needs to contend with.
- The *bandwidth* of the transmission medium, which we explain below.

A transmission channel can be characterized by its effect on tones of various frequencies. A (sinusoidal) tone of a given frequency f is applied at the input, and the tone at the output of the channel is measured. The ability of the channel to transfer a tone of the frequency f is given by the **amplitude-response function** $A(f)$, which is defined as the ratio of the amplitude of the output tone divided by the amplitude of the input tone. Figure 3.11 shows the typical amplitude-response functions of a low-pass channel and its idealized counterpart. The **bandwidth** of a channel is defined as the range of frequencies that is passed by a channel. Our first major result in the following sections will be to show that the rate at which pulses can be transmitted over a channel is proportional to the bandwidth. In essence, we will show that if a channel has bandwidth W, then the narrowest pulse that can be transmitted over the channel has width $\tau = 1/2W$ seconds. *Thus the fastest rate at which pulses can be transmitted into the channel is given by the Nyquist rate: $r_{max} = 2W$ pulses/second.*

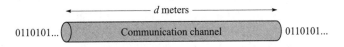

FIGURE 3.10 A digital transmission system

(a) Low-pass and idealized low-pass channel

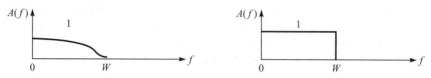

(b) Maximum pulse transmissiom rate is $2W$ pulses/second

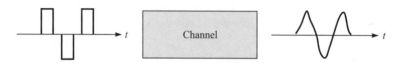

FIGURE 3.11 Typical amplitude-response functions

We can transmit binary information by sending a pulse with amplitude $+A$ to send a 1 bit and $-A$ to send a 0 bit. Each pulse transmits one bit of information, so this system then has a bit rate of $2W$ pulses/second * 1 bit/pulse = $2W$ bps. We can increase the bit rate by sending pulses with more levels. For example, if pulses can take on amplitudes from the set $\{-A, -A/3, +A/3, +A\}$ to transmit the pairs of bits $\{00, 01, 10, 11\}$, then each pulse conveys two bits of information and the bit rate is $4W$ bps. In general, if we use $M = 2^m$ levels, say, $\{-A, -(M-3)A/(M-1), \ldots, +(M-3)A/(M-1), +A\}$, then the bit rate will be $2Wm$ bps. *Thus, if we use **multilevel transmission** pulses that can take on $M = 2^m$ amplitude levels, we can transmit at a bit rate of $2Wm$ bits/second*:

$$R = 2W \text{ pulses/second * } m \text{ bits/pulse} = 2Wm \text{ bits/second}$$

In the absence of noise, the bit rate can be increased without limit by increasing the number of signal levels M. However, noise is an impairment encountered in all communication channels. Noise consists of extraneous signals that are added to the desired signal at the input to the receiver. Figure 3.12 gives two examples where the desired signal is a square wave and where noise is added to the signal. In the first example the amplitude of the noise is less than that of the desired signal, and so the desired signal is discernable even after the noise has been added. In the second example the noise amplitude is greater than that of the desired signal, which is now more difficult to discern. The **signal-to-noise ratio (SNR)**, defined in Figure 3.12, measures the relative amplitudes of the desired signal and the noise. The SNR is usually stated in decibels (dB). Returning to multilevel transmission, suppose we increase the number of levels while keeping the maximum signal levels $\pm A$ fixed. Each increase in number of signal levels requires a reduction in the spacing between levels. At some point these reductions will imply significant increases in the probability of detection errors as the noise will be more likely to convert the transmitted signal level into other signal levels. *Thus the presence of noise limits the reliability with which the receiver can correctly determine the information that was transmitted.*

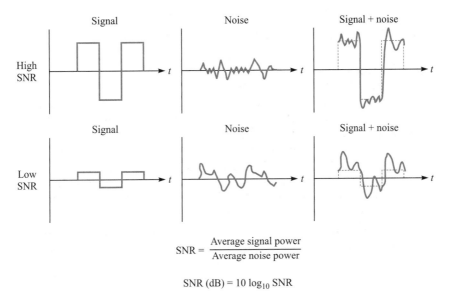

$$SNR = \frac{\text{Average signal power}}{\text{Average noise power}}$$

$$SNR \text{ (dB)} = 10 \log_{10} SNR$$

FIGURE 3.12 Signal-to-noise ratio

The **channel capacity** of a transmission system is the maximum rate at which bits can be transferred reliably. We have seen above that the bit rate and reliability of a transmission system are affected by the channel bandwidth, the signal energy or power, and the noise power. Shannon derived an expression for channel capacity. He also showed that reliable communication is not possible at rates above this capacity. The *Shannon channel capacity* is given by the following formula:

$$C = W \log_2(1 + \text{SNR})\text{bits/second}$$

As an example, consider telephone modem speeds. A modem operating over a telephone line has a maximum useful bandwidth of about 3400 Hz. If we assume an SNR of 40 dB, which is slightly over the maximum possible in a telephone line, Shannon's formula gives a channel capacity of 44.8 kbps. The V.90 modems that were introduced in 1998 operate at a rate of 56 kbps, well in excess of the Shannon bound! How can this be? The explanation is given on the next page.

Table 3.3 shows the bit rates that are provided by current digital transmission systems over various media. Twisted pairs present a wide set of options in telephone access networks and in Ethernet LANs. Cable television modems provide high speed transmission over coaxial cable. Wireless links also provide some options in access networks and LANs. Optical fiber transmission provides the high bandwidths required in backbone networks. Dense wavelength division multiplexing systems will provide huge bandwidths and will profoundly affect network design.

SHANNON CHANNEL CAPACITY AND THE 56KBPS MODEM

The Shannon channel capacity for a telephone channel gives a maximum possible bit rate of 44.8 kbps at 40 dB SNR. How is it that the V.90 modems achieve rates of 56 kbps? In fact, a look at the fine print shows that the bit rate is 33.6 kbps inbound into the network. The modem signal must undergo an analog-to-digital conversion when it is converted to PCM at the entrance to the telephone network. This step introduces the PCM approximation error or noise. At the maximum allowable signal level, we have a maximum possible SNR of 39 dB, so the 56 kbps is not attainable in the inbound direction, and hence the inbound operation is at 33.6 kbps. In the direction from the Internet server provider (ISP) to the user, the signal from the ISP is already digital and so it does not need to undergo analog-to-digital conversion. Hence the quantization noise is not introduced, a higher SNR is possible, and speeds approaching 56 kbps can be achieved from the network to the user.

Digital transmission system	Bit rate	Observations
Telephone twisted pair	33.6 kbps	4 kHz telephone channel
Ethernet over twisted pair	10 Mbps	100 meters over unshielded twisted pair
Fast Ethernet over twisted pair	100 Mbps	100 meters using several arrangements of unshielded twisted pair
Cable modem	500 kbps to 4 Mbps	Shared CATV return channel
ADSL over twisted pair	64–640 kbps inbound 1.536–6.144 Mbps outbound	Uses higher frequency band and coexists with conventional analog telephone signal, which occupies 0–4 kHz band
Radio LAN in 2.4 GHz band	2 Mbps	IEEE 802.11 wireless LAN
Digital radio in 28 GHz band	1.5–45 Mbps	5 km multipoint radio link
Optical fiber transmission system	2.4–9.6 Gbps	Transmission using one wavelength
Optical fiber transmission system	1600 Gbps and higher	Multiple simultaneous wavelengths using wavelength division multiplexing

TABLE 3.3 Bit rates of digital transmission systems

3.3 CHARACTERIZATION OF COMMUNICATION CHANNELS

A communication channel is a system consisting of a physical medium and associated electronic and/or optical equipment that can be used for the transmission of information. Commonly used physical media are copper wires, coaxial cable, radio, and optical fiber. Communications channels can be used for the transmission of either digital or analog information. Digital transmission involves the transmission of a sequence of pulses that is determined by a corresponding digital sequence, typically a series of binary 0s and 1s. Analog transmission involves the transmission of waveforms that correspond to some analog signal, for example, audio from a microphone or video from a television camera. Communication channels can be characterized in two principal ways: frequency domain and time domain.

3.3.1 Frequency Domain Characterization

Figure 3.13 shows the approach used in characterizing a channel in the frequency domain. A sinusoidal signal $x(t) = \cos(2\pi f t)$ that oscillates at a frequency of f cycles/second (Hertz) is applied to a channel. The channel output $y(t)$ usually consists of a sinusoidal signal of the same frequency but of different amplitude and phase:[6]

$$y(t) = A(f)\cos(2\pi f t + \varphi(f)) = A(f)\cos(2\pi f(t - \tau(f))). \tag{1}$$

The channel is characterized by its effects on the input sinusoidal signal. The first effect involves an attenuation of the sinusoidal signal. This effect is characteized by the **amplitude-response** function $A(f)$, which is the ratio of the output amplitude to the input amplitude of the sinusoids at frequency f. The second effect is a **phase shift** $\varphi(f)$ in the output sinusoid relative to the input sinusoid. In general, both the amplitude response and the phase shift depend on the frequency f of the sinusoid. You will see later that for various communication channels the attenuation increases with f. Equation 1 also shows that the output $y(t)$ can be viewed as the input attenuated by $A(f)$ and delayed by $\tau(f)$.

The frequency-domain characterization of a channel involves varying the frequency f of the input sinusoid to evaluate $A(f)$ and $\varphi(f)$. Figure 3.14 shows the amplitude-response and phase-shift functions for a "low pass" channel. In this channel very low frequencies are passed, but very high frequencies are essentially eliminated. In addition, frequency components at low frequencies are not phase shifted, but very high frequencies are shifted by 90 degrees.

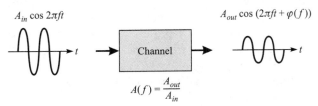

FIGURE 3.13 Channel characterization—frequency domain

---

[6]The statement applies to channels that are "linear." For such channels the output signal corresponding to a sum of input signals, say, $x_1(t) + x_2(t)$, is equal to the sum of the outputs that would have been obtained for each individual input; that is, $y(t) = y_1(t) + y_2(t)$.

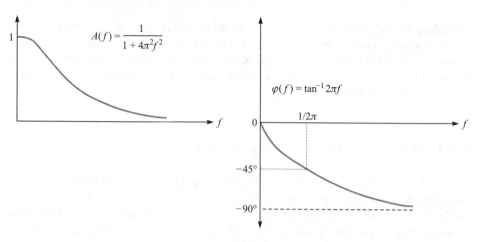

FIGURE 3.14 Example of amplitude-response function and phase-shift function

The **attenuation** of a signal is defined as the reduction or loss in signal power as it is transferred across a system. The attenuation is usually expressed in dB:

$$\text{attenuation} = 10\log_{10}\frac{P_{in}}{P_{out}}$$

The power in a sinusoidal signal of amplitude is $A^2/2$. Therefore, the attenuation in the channel in Figure 3.13 is given by $P_{in}/P_{out} = A_{in}^2/A_{out}^2 = 1/A^2(f)$.

The amplitude-response function $A(f)$ can be viewed as specifying a window of frequencies that the channel will pass. The **bandwidth** W of a channel measures the width of the window of frequencies that are passed by the channel. Figure 3.11 shows two typical amplitude-response functions. The low-pass channel passes low-frequency components and increasingly attenuates higher frequency components. In principle this channel has infinite bandwidth, since every frequency is passed to some degree. However, for practical purposes, signals above the specified frequency are considered negligible. We can define such a frequency as the bandwidth and approximate the amplitude-response function by the idealized low-pass function in Figure 3.11a. Another typical amplitude-response function is a "band pass" channel that passes frequencies in the range f_1 to f_2 instead of low frequencies (see Figure 3.27). The bandwidth for such a channel is $W = f_2 - f_1$.

Communication systems make use of electronic circuits to modify the frequency components of an input signal. When circuits are used in this manner, they are called *filters*. Communication systems usually involve a tandem arrangement of a transmitter filter, a communication channel, and a receiver filter. The overall tandem arrangement can be represented by an overall amplitude-response function $A(f)$ and a phase-shift function $\varphi(f)$. The transmitter and receiver filters are designed to give the overall system the desired amplitude-response and delay properties. For example, devices called loading coils were added to telephone wire pairs to provide a flat amplitude-response function in the frequency range where telephone voice signals occur. Unfortunately, these coils also introduced a much higher attenuation at the higher frequencies, greatly reducing the bandwidth of the overall system.

Let us now consider the impact of communication channels on other signals. The effect of a channel on an arbitrary input signal can also be determined from $A(f)$ and $\varphi(f)$ as follows. Many signals can be represented as the sum of sinusoidal signals:

$$x(t) = \sum a_k \cos(2\pi f_k t).$$

For example, periodic functions have the preceding form with $f_k = kf_0$ where f_0 is the fundamental frequency.

Now suppose that a periodic signal $x(t)$ is applied to a channel with a channel characterized by $A(f)$ and $\varphi(f)$. The channel attenuates the sinusoidal component at frequency kf_0 by $A(kf_0)$, and it also phase shifts the component by $\varphi(kf_0)$. The output signal of the channel, which we assume to be linear, will therefore be

$$y(t) = \sum a_k A(kf_0) \cos(2\pi kf_0 t + \varphi(kf_0)).$$

This expression shows how the channel distorts the input signal. In general, the amplitude-response function varies with frequency, and so the channel alters the relative weighting of the frequency components. In addition, the different frequency components will be delayed by different amounts, altering the relative alignment between the components. Not surprisingly, then, the shape of the output $y(t)$ generally differs from $x(t)$. Note that the output signal will have its frequencies restricted to the range where the amplitude-response function is nonzero. Thus the bandwidth of the output signal is necessarily less than that of the channel. Note also that if $A(f)$ is equal to a constant, say, C, and if $\varphi(f) = 2\pi f t_d$, over the range of frequencies where a signal $x(t)$ has nonnegligible components, then the output $y(t)$ will be equal to the input signal scaled by the factor C and delayed by t_d seconds.

Example—Effect of a channel on the shape of the output signal

Suppose that binary information is transmitted at a rate of 8 kilobits/second. A binary 1 is transmitted by sending a rectangular pulse of amplitude 1 and of duration 0.125 milliseconds, and a 0 by sending a pulse of amplitude -1. Consider the signal $x_3(t)$ in Figure 3.15 that corresponds to the repetition of the pattern 10000001 over and over again. This periodic signal can be expressed as a sum of sinusoids as follows using Fourier series:

$$x_3(t) = -0.5 + \left(\frac{4}{\pi}\right)\left\{ \sin\left(\frac{\pi}{4}\right)\cos(2\pi 1000t) + \sin\left(\frac{2\pi}{4}\right)\right.$$
$$\left. \cos(2\pi 2000t) + \sin\left(\frac{3\pi}{4}\right)\cos(2\pi 3000t) + \ldots \right\}$$

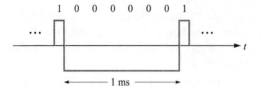

1 0 0 0 0 0 0 1

··· ···

1 ms

FIGURE 3.15 Signals corresponding to repeated octet patterns

Suppose that the signal is passed through a communication channel that has $A(f) = 1$ and $\varphi(f) = 0$ for f in the range 0 to W. Figures 3.16a, b, and c show the output (the solid line) of the communication channel for values of W (1.5 kHz, 2.5 kHz, and 4.5 kHz) that pass the frequencies only to the first, second, and fourth harmonic, respectively. As the bandwidth of the channel increases, more of the harmonics are passed and the output of the channel more closely approximates the input. This example shows how the bandwidth of the channel affects the ability to transmit digital information in the form of pulses. Clearly, *as bandwidth is decreased, the precision with which the pulses can be identified is reduced.*

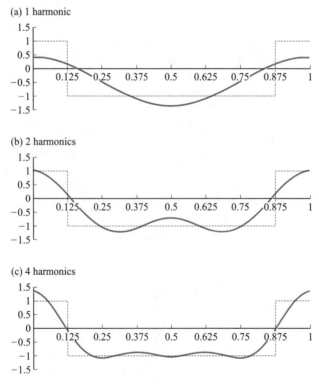

FIGURE 3.16 Output of low-pass communication channels for input signal in Figure 3.15

3.3.2   Time Domain Characterization

Figure 3.17 considers the time domain characterization of a communication channel. A very narrow pulse is applied to the channel at time $t = 0$. The energy associated with the pulse appears at the output of the channel as a signal $h(t)$ some propagation time later. The propagation speed, of course, cannot exceed the speed of light in the given medium. The signal $h(t)$ is called the **impulse response** of the channel. Invariably the output pulse $h(t)$ is spread out in time. The width of the pulse is an indicator of how quickly the output follows the input and hence of how fast pulses can be transmitted over the channel. In digital transmission we are interested in maximizing the number of pulses transmitted per second in order to maximize the rate at which information can be transmitted. Equivalently, we are interested in minimizing the time T between consecutive input pulses. This minimum spacing is determined by the degree of interference between pulses at the output of the channel.

Suppose that we place transmitter and receiver filters around a communication channel and suppose as well that we are interested in using the frequencies in the range 0 to W Hz. Furthermore, suppose that the filters can be selected so that the overall system is an idealized low-pass channel; that is, $A(f) = 1$, and $\varphi(f) = 2\pi ft_d$. It can be shown that the impulse response of the system is given by

$$h(t) = s(t - t_d)$$

which is a delayed version of

$$s(t) = \frac{\sin(2\pi Wt)}{2\pi Wt}$$

Figure 3.18 shows $s(t)$. It can be seen that this function is equal to 1 at $t = 0$ and that it has zero crossings at nonzero integer multiples of $T = 1/2W$. Note that the pulse is mostly confined to the interval from $-T$ to T, so it is approximately $2T = 2/2W = 1/W$ seconds wide. Thus we see that *as the bandwidth* W *increases, the width of the pulse* s(t) *decreases, suggesting that pulses can be input into the system more closely spaced, that is, at a higher rate.* The next section shows that the signal $s(t)$ plays an important role in the design of digital transmission systems.

Recall that $h(t)$ is the response to a narrow pulse at time $t = 0$, so we see that our ideal system has the strange property that its output $h(t) = s(t - t_d)$ anticipates the input that will be applied and begins appearing at the output before time $t = 0$. In practice, this idealized filter cannot be realized; however, delayed and slightly modified versions of $s(t)$ are approximated and implemented in real systems.

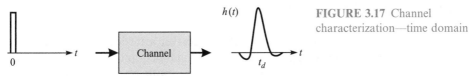

FIGURE 3.17 Channel characterization—time domain

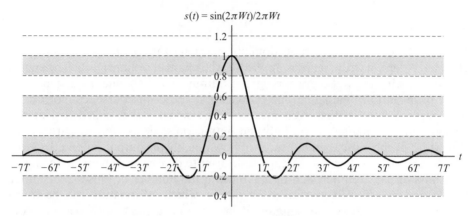

FIGURE 3.18 Signaling pulse with zero intersymbol interference

3.4   FUNDAMENTAL LIMITS IN DIGITAL TRANSMISSION

In this section we consider **baseband transmission**, which is the transmission of digital information over a low-pass communication channel. The quality of a digital transmission system is determined by the transmission rate or bit rate at which information bits can be transmitted reliably. Thus the quality is measured in terms of two parameters: *transmission speed*, or *bit rate*, in bits per second and the *bit error rate*, the fraction of bits that are received in error. We will see that these two parameters are determined by the bandwidth of the communication channel and by the SNR, which we will define formally later in the section.

Figure 3.19 shows the simplest way to transmit a binary information sequence. Every T seconds the transmitter accepts a binary information bit and transmits a pulse with amplitude $+A$ if the information bit is a 1 and with $-A$ if the information bit is a 0. In the examples in section 3.3, we saw how a channel distorts an input signal and limits the ability to correctly detect the polarity of a pulse. In this section we show how the problem of channel distortion is addressed and how the pulse transmission rate is maximized at the same time. In particular, in Figure 3.19 each pulse at the input results in a pulse at the output. We also show how the pulses that arrive at the receiver can be packed as closely as possible if they are shaped appropriately by the transmitter and receiver filters.

3.4.1   The Nyquist Signaling Rate

Let $p(t)$ be the basic pulse that appears at the receiver after it has been sent over the combined transmitter filter, communication channel, and receiver filter. The first pulse is transmitted, centered at $t = 0$. If the input bit was 1, then $+Ap(t)$

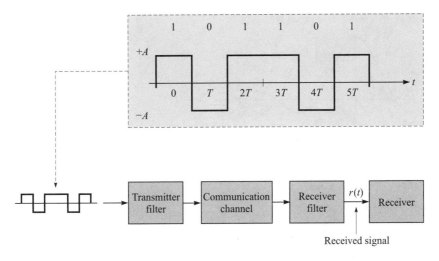

FIGURE 3.19 Digital baseband signal and baseband transmission system

should be received; if the input was 0, then $-Ap(t)$ should be received instead. For simplicity, we assume that the propagation delay is zero. To determine what was sent at the transmitter, the receiver samples the signal it receives at $t = 0$. If the sample is positive, the receiver decides that a 1 was sent; if the sample is negative, the receiver decides that a 0 was sent.

Every T seconds the transmitter sends an additional information bit by transmitting another pulse with the appropriate polarity. For example, the second bit is sent at time $t = T$ and will be either $+Ap(t - T)$ or $-Ap(t - T)$, depending on the information bit. The receiver samples its signal at $t = T$ to determine the corresponding input. However, the pulses are sent as part of a sequence, and so the total signal $r(t)$ that appears at the receiver is the *sum* of all the inputs:

$$r(t) = \sum_k A_k p(t - kT)$$

According to this expression, when the receiver samples the signal at $t = 0$, it measures

$$r(0) = A_0 p(0) + \sum_{k \neq 0} A_k p(-kT)$$

In other words, the receiver must contend with *intersymbol interference* from all the other transmitted pulses. What a mess! Note, however, that all the terms in the summation disappear if we use a pulse that has zero crossings at $t = kT$ for nonzero integer values k. The pulse $s(t)$ introduced in section 3.3.2 and shown in Figure 3.18 satisfies this property. This pulse is an example of the class of *Nyquist pulses* that have the property of providing *zero intersymbol interference* at the times $t = kT$ at the receiver. Figure 3.20a shows the three pulses

(a) Three separate pulses for sequence 110

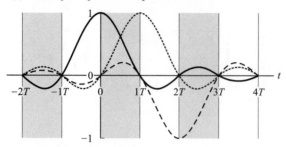

FIGURE 3.20 System response to binary input 110

(b) Combined signal for sequence 110

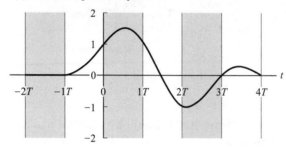

corresponding to the sequence 110, before they are added. Figure 3.20b shows the signal that results when the three pulses are combined. It can be seen that the combined signal has the correct values at $t = 0$, 1, and 2.

The above transmission system sends a bit every T seconds, where $T = 1/2 W$ and W is the bandwidth of the overall system in Figure 3.19. For example, if $W = 4$ kHz, then pulses would be sent every $T = 1/8000 = 125$ microseconds, which corresponds to a rate of 8000 pulses/second. A bit is sent with every pulse, so the bit rate is 8000 bits/second. The **Nyquist Signaling Rate** is defined by

$$r_{max} = 2W \, \text{pulses/second}$$

The Nyquist rate r_{max} is the maximum signaling rate that is achievable through an ideal low-pass channel with no intersymbol interference.

We already noted that the ideal low-pass system in Figure 3.11 cannot be implemented in practice. Nyquist also found other pulses that have zero intersymbol interference but that require some additional bandwidth. Figure 3.21 shows the amplitude-response function for one such pulse. Here a transition region with odd symmetry about $f = W$ is introduced. The more gradual roll-off of these systems makes the appropriate transmitter and receiver filters simpler to attain in practice.

The operation of the baseband transmission systems in this section depends critically on having the receiver synchronized precisely to intervals of duration T. Additional processing of the received signal is carried by the receiver to obtain this synchronization. If the receiver loses synchronization, then it will start sampling the signal at time instants that do contain intersymbol interference. The use

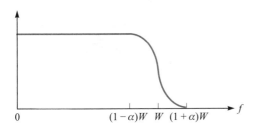

FIGURE 3.21 Raised cosine transfer function

of pulses corresponding to the system in Figure 3.21 provides some tolerance to small errors in sampling time.

3.4.2   The Shannon Channel Capacity

Up to this point we have been assuming that the input pulses can have only two values, 0 or 1. This restriction can be relaxed, and the pulses can be allowed to assume a greater number of values. Consider **multilevel transmission** where binary information is transmitted in a system that uses one of 2^m distinct levels in each input pulse. The binary information sequence can be broken into groups of m bits. Proceeding as before, each T seconds the transmitter accepts a group of m bits. These m bits determine a unique amplitude of the pulse that is to be input into the system. As long as the signaling rate does not exceed the Nyquist rate $2W$, the interference between pulses will still be zero, and by measuring the output at the right time instant we will be able to determine the input. Thus if we suppose that we transmit $2W$ pulses/second over a channel that has bandwidth W and if the number of amplitude values is 2^m, then the **bit rate** R of the system is

$$R = 2W \text{ pulses/second} {}^* m \text{ bits/pulse} = 2Wm \text{ bits/second}$$

In principle we can attain arbitrarily high bit rates by increasing the number of levels 2^m. However, we cannot do so in practice because of the limitations on the accuracy with which measurements can be made and also the presence of random noise. The random noise implies that the value of the overall response at time $t = kT$ will be the sum of the input amplitude plus some random noise. This noise can cause the measurement system to make an incorrect decision. *To keep the probability of decision errors small, we must maintain some minimum spacing between amplitude values* as shown in Figure 3.22. Here four signal levels are shown next to the typical noise. In the case of four levels, the noise is not likely to cause errors. Figure 3.22 also shows a case with eight signal levels. It can be seen that if the spacing between levels becomes too small, then the noise signals can cause the receiver to make the wrong decision.

We can make the discussion more precise by considering the statistics of the noise signal. Figure 3.23 shows the Gaussian probability density function, which is frequently a good model for the noise amplitudes. The density function gives

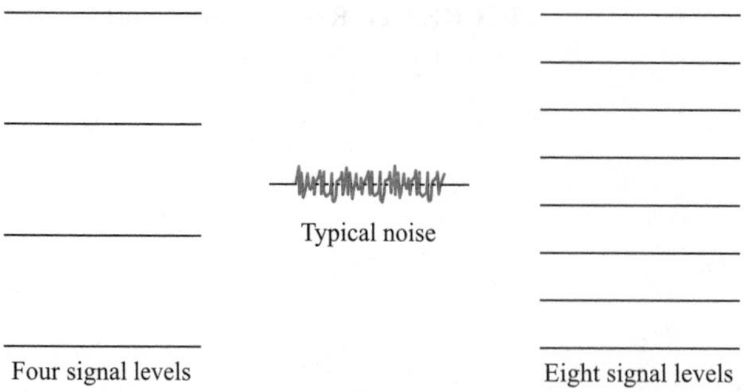

FIGURE 3.22 Effect of noise on transmission errors as number of levels is increased

Typical noise

Four signal levels

Eight signal levels

the relative frequency of occurrence of the noise amplitudes. It can be seen that for the Gaussian density, the amplitudes are centered around zero. The average power of this noise signal is given by σ^2, where σ is the standard deviation.

Consider how errors occur in multilevel transmission. If we have maximum amplitudes $\pm A$ and M levels, then the separation between adjacent levels is $\delta = 2A/(M - 1)$. When an interior signal level is transmitted, an error occurs if the noise causes the received signal to be closer to one of the other signal levels. This situation occurs if the noise amplitude is greater than $\delta/2$ or less than $-\delta/2$. Thus the probability of error for an interior signal level is given by

$$P_e = \int_{-\infty}^{-\delta/2} \frac{1}{\sqrt{2\pi}\sigma} e^{-x^2/2\sigma^2} dx + \int_{\delta/2}^{\infty} \frac{1}{\sqrt{2\pi}\sigma} e^{-x^2/2\sigma^2} dx = 2 \int_{\delta/2\sigma}^{\infty} \frac{1}{\sqrt{2\pi}} e^{-x^2/2} dx$$

$$= 2Q\left(\frac{\delta}{2\sigma}\right)$$

The expression on the right-hand side is evaluated using tables or analytic aproximations [Leon-Garcia, 1994, chapter 3]. Figure 3.24 shows how P_e varies with $\delta/2\sigma = A/(M - 1)\sigma$. For larger values of separation $\delta/2\sigma$, large decreases in the probability of error are possible with small increases in $\delta/2\sigma$. However, as the number of signal levels M is increased, $\delta/2\sigma$ is decreased, leading to large increases in the probability of error. We conclude that the bit rate cannot be increased to arbitrarily high values by increasing M without incurring signifi-cantly higher bit error rates.

We have now seen that two parameters affect the performance of a digital transmission system: bandwidth and SNR. Shannon addressed the question of determining the maximum achievable bit rate at which reliable communication is

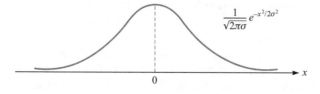

$$\frac{1}{\sqrt{2\pi}\sigma} e^{-x^2/2\sigma^2}$$

FIGURE 3.23 Gaussian probability density function

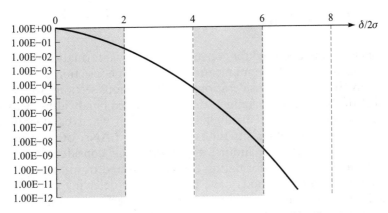

FIGURE 3.24 Probability of error for an interior signal level

possible over an ideal channel of bandwidth W and of a given SNR. The phrase *reliable communication* means that it is possible to achieve arbitrarily small error probabilities by using sufficiently complex coding. Shannon derived the channel capacity for such a channel under the condition that the noise has a Gaussian distribution. This **channel capacity** is given by the following formula:

$$C = W \log_2(1 + \text{SNR}) \text{ bits/second}$$

Shannon showed that the probability of error can be made arbitrarily small only if the transmission rate R is less than channel capacity C. Therefore, the channel capacity is the maximum possible transmission rate over a system with given parameters.

SHANNON CHANNEL CAPACITY OF TELEPHONE CHANNEL

Consider a telephone channel with W = 3.4 kHz and SNR = 10,000. The channel capacity is then

$$C = 3400 \log_2(1 + 10000) = 44,800 \text{ bits/second}$$

The following identities are useful here: $\log_2 x = \ln x/\ln 2 = \log_{10} x/\log_{10} 2$. We note that the SNR is usually stated in dB. Thus if SNR = 10,000, then in dB the SNR is

$$10 \log_{10} \text{SNR dB} = 10 \log_{10} 10000 = 40 \text{ dB}$$

The above result gives a bound to the achievable bit rate over ordinary analog telephone lines.

3.5 LINE CODING

Line coding is the method used for converting a binary information sequence into a digital signal in a digital communications system. The selection of a line coding technique involves several considerations. In the previous sections, we focused on maximizing the bit rate over channels that have limited bandwidths. Maximizing bit rate is the main concern in digital transmission when bandwidth is at a premium. However, in other situations, such as in LANs, other concerns are also of interest. For example, another important design consideration is the ease with which the bit timing information can be recovered from the digital signal. Also, some line coding methods have built-in error detecting capabilities, and other methods have better immunity to noise and interference. Finally, the complexity and the cost of the line code implementations are always factors in the selection for a given application.

Figure 3.25 shows various line codes that are used in practice. The figure shows the digital signals that are produced by the line codes for the binary sequence 101011100. The simplest scheme is the unipolar **nonreturn-to-zero (NRZ) encoding** in which a binary 1 is transmitted by sending a $+A$ voltage level, and a 0 is transmitted by sending a 0 voltage. If binary 0s and 1s both occur with probability 1/2, then the average transmitted power for this line code is $(1/2)A^2 + (1/2)0^2 = A^2/2$. The **polar NRZ encoding** method that maps a binary 1 to $+A/2$ and binary 0 to $-A/2$ is more efficient than unipolar NRZ in

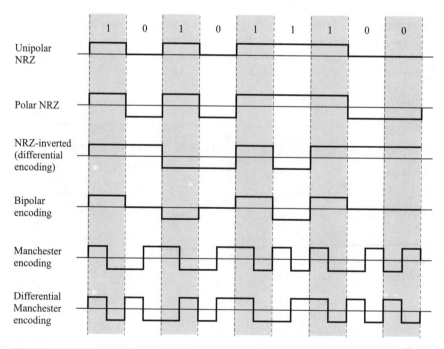

FIGURE 3.25 Line coding methods

terms of average transmitted power. Its average power is given by $(1/2)(+A/2)^2 + (1/2)(-A/2)^2 = A^2/4$.

The spectrum that results from applying a given line code is of interest. We usually assume that the binary information is equally likely to be 0 or 1 and that they are statistically independent of each other, much as if they were produced by a sequence of independent coin flips. The unipolar and the polar NRZ encoding methods have the same frequency components because they produce essentially the same variations in a signal as a function of time. Strings of consecutive 0s and consecutive 1s lead to periods where the signal remains constant. These strings of 0s and 1s occur frequently enough to produce a spectrum that has its components concentrated at the lower frequencies as shown in Figure 3.26.[7] This situation presents a problem when the communications channel does not pass low frequencies. For example, most telephone transmission systems do not pass the frequencies below about 200 Hz.

The **bipolar encoding** method was developed to produce a spectrum that is more amenable to channels that do not pass low frequencies. In this method binary 0s are mapped into 0 voltage, thus making no contribution to the digital signals; consecutive 1s are alternately mapped into $+A/2$ and $-A/2$. Thus a string of consecutive 1s will produce a square wave with the frequency $1/2T$ Hz. As a result, the spectrum for the bipolar code has its frequency content centered around the frequency $1/2T$ Hz and has small content at low frequencies as shown in Figure 3.26.

Timing recovery is an important consideration in the selection of a line code. The timing-recovery circuit in the receiver monitors the transitions at the edge of the bit intervals to determine the boundary between bits. Long strings of 0s and

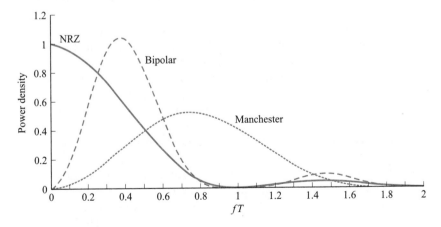

FIGURE 3.26 Spectra for different line codes

---

[7]The formulas for the spectra produced by the line codes in Figure 3.26 can be found in [Smith 1985, pp. 198–203].

1s in the binary and the polar binary encodings can cause the timing circuit to lose synchronization because of the absence of transitions. In the bipolar encoding long strings of 1s result in a square wave that has strong timing content; however, long strings of 0s still pose a problem. To address this problem, the bipolar line codes used in telephone transmission systems place a limit on the maximum number of 0s that may be encoded into the digital signal. Whenever a string of N consecutive 0s occurs, the string is encoded into a special binary sequence that contains 0s and 1s. To alert the receiver that a substitution has been made, the sequence is encoded so that the mapping in the bipolar line code is violated; that is, two consecutive 1s do not alternate in polarity.

A problem with polar coding is that a systematic error in polarity can cause all 0s to be detected as 1s and all 1s as 0s.[8] The problem can be avoided by mapping the binary information into *transitions* at the beginning of each interval. A binary 1 is transmitted by enforcing a transition at the beginning of a bit time, and a 0 by having no transition. The signal level within the actual bit time remains constant. Figure 3.25 shows an example of how **differential encoding**, or **NRZ inverted**, carries out this mapping. Starting at a given level, the sequence of bits determines the subsequent transitions at the beginning of each interval. Note that differential encoding will lead to the same spectrum as binary and polar encoding. However, errors in differential encoding tend to occur in pairs. An error in one bit time will provide the wrong reference for the next time, thus leading to an additional error in the next bit.

Example—Ethernet and Token-Ring Line Coding

Bipolar coding has been used in long-distance transmission where bandwidth efficiency is important. In LANs, where the distances are short, bandwidth efficiency is much less important than cost per station. The Manchester encodings shown in Figure 3.25 are used in Ethernet and token-ring LAN standards. In **Manchester encoding** a binary 1 is denoted by a transition from $A/2$ to $-A/2$ in the middle of the bit time interval, and a binary 0 by a transition from $-A/2$ to $A/2$. The presence of a transition in the middle of every bit interval makes timing recovery particularly easy and also results in small content at low frequencies. However, the pulse rate is essentially double that of binary encoding, and this factor results in a spectrum with significantly larger bandwidth as shown in Figure 3.26. **Differential Manchester encoding**, which is used in token-ring networks, retains the transition in the middle of every bit time, but the binary sequence is mapped into the presence or absence of transitions in the beginning of the bit intervals. In this type of encoding, a binary 0 is marked by a transition at the beginning of an interval, whereas a 1 is marked by the absence of a transition.

---

[8]This polarity inversion occurs when the polar-encoded stream is fed into a phase modulation system such as the one discussed in section 3.6.

Note that the Manchester encoding can be viewed as the transmission of two pulses for each binary bit. A binary 1 is mapped into the binary pair of 10, and the corresponding polar encoding for these two bits is transmitted: A binary 0 is mapped into 01. The Manchester code is an example of a *m*B*n*B code (where *m* is 1 and *n* is 2) in which *m* information bits are mapped into $n > m$ encoded bits. The encoded bits are selected so that they provide enough pulses for timing recovery and limit the number of pulses of the same level. For example, the optical fiber transmission system in the Fiber Distributed Data Interface (FDDI) LAN uses a 4B5B line code.

3.6 MODEMS AND DIGITAL MODULATION

In section 3.4 we considered digital transmission over channels that are low pass in nature. We now consider band-pass channels that do not pass the lower frequencies and instead pass power in some frequency range from f_1 to f_2, as shown in Figure 3.27. We assume that the bandwidth of the channel is $W = f_2 - f_1$ and discuss the use of modulation to transmit digital information over this type of channel. The basic function of the modulation is to produce a signal that contains the information sequence and that occupies frequencies in the range passed by the channel. A **modem** is a device that carries out this basic function. In this section we first consider the principles of digital modulation, and then we show how these principles are applied in telephone modem standards.

Let f_c be the frequency in the center of the band-pass channel in Figure 3.27; that is, $f_c = (f_1 + f_2)/2$. The sinusoidal signal $\cos(2\pi f_c t)$ has all of its power located precisely at frequency f_c. The various types of modulation schemes involve imbedding the binary information sequence into the transmitted signal by varying, or modulating, some attribute of the sinusoidal signal. In **amplitude shift keying (ASK)** the sinusoidal signal is turned on and off according to the information sequence as shown in Figure 3.28a. The demodulator for an ASK system needs only to determine the presence or absence of a sinusoid in a given time interval. In **frequency shift keying (FSK)**, shown in Figure 3.28b, the frequency of the sinusoid is varied according to the information. If the information bit is a 0, the sinusoid has frequency $f_1 = f_c - \varepsilon$, and if it is a 1, the sinusoid has a frequency $f_2 = f_c + \varepsilon$. The demodulator for an FSK system must be able to determine which of two possible frequencies is present at a given time. In

FIGURE 3.27 Digital modulation

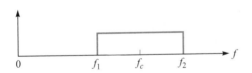

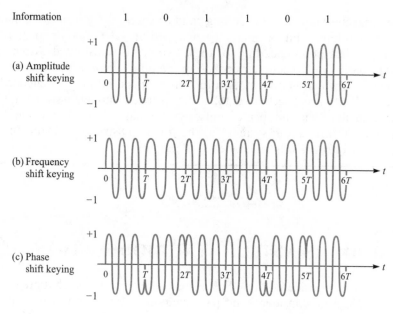

FIGURE 3.28 Amplitude, frequency, and phase modulation techniques

phase shift keying (PSK), the phase of the sinusoid is altered according to the information sequence. In Figure 3.28c a binary 1 is transmitted by $\cos(2\pi f_c t)$, and a binary 0 is transmitted by $\cos(2\pi f_c t + \pi)$. Because $\cos(2\pi f_c t + \pi) = -\cos(2\pi f_c t)$, we note that this PSK scheme is equivalent to multiplying the sinusoidal signal by $+1$ when the information is a 1 and by -1 when the information bit is a 0. Thus the demodulator for a PSK system must be able to determine the phase of the received sinusoid with respect to some reference phase. In the remainder of this section, we concentrate on phase modulation techniques.

Consider the problem of transmitting a binary information sequence over an ideal band-pass channel using PSK. We are interested in developing modulation techniques that can achieve pulse rates that are comparable to that achieved by Nyquist signaling over low-pass channels. In Figure 3.29c we show the waveform $Y_i(t)$ that results when a binary 1 is transmitted using a cosine wave with amplitude $+A$, and a binary 0 is transmitted using a cosine wave with amplitude $-A$. The corresponding modulator is shown in Figure 3.30a. Every T seconds the modulator accepts a new binary information symbol and adjusts the amplitude A_k accordingly. In effect, as shown in Figure 3.29c, the modulator transmits a T-second segment of the signal as follows:

$+A\cos(2\pi f_c t)$ if the information symbol is a 1.

$-A\cos(2\pi f_c t)$ if the information symbol is a 0.

(a) Information
1 0 1 1 0 1
(b) Baseband signal $X_i(t)$
+A
−A
0 T 2T 3T 4T 5T 6T
t
(c) Modulated signal $Y_i(t)$
+A
−A
0 T 2T 3T 4T 5T 6T
t
(d) $2Y_i(t)\cos(2\pi f_c t)$
+2A
−2A
0 T 2T 3T 4T 5T 6T
t

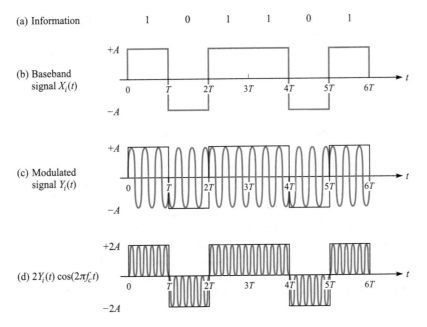

FIGURE 3.29 Modulating a signal

Note that the modulated signal is no longer a pure sinusoid, since the overall transmitted signal contains glitches between the T-second intervals, but its primary oscillations are still around the center frequency f_c; therefore, we expect that the power of the signal wil be centered about f_c and hence located in the range of frequencies that are passed by the band-pass channel.

By monitoring the polarity of the signal over the intervals of T seconds, a receiver can recover the original information sequence. Let us see more precisely how this recovery may be accomplished. As shown in Figure 3.30b suppose we

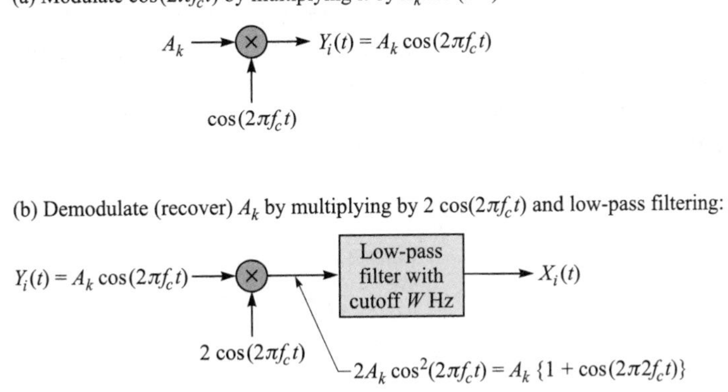

(a) Modulate $\cos(2\pi f_c t)$ by multiplying it by A_k for $(k\text{-}1)T < t < kT$:
A_k
$Y_i(t) = A_k \cos(2\pi f_c t)$
$\cos(2\pi f_c t)$
(b) Demodulate (recover) A_k by multiplying by $2\cos(2\pi f_c t)$ and low-pass filtering:
$Y_i(t) = A_k \cos(2\pi f_c t)$
Low-pass filter with cutoff W Hz
$X_i(t)$
$2\cos(2\pi f_c t)$
$2A_k\cos^2(2\pi f_c t) = A_k\{1 + \cos(2\pi 2 f_c t)\}$

FIGURE 3.30 Modulator and demodulator

multiply the modulated signal $Y_i(t)$ by $2\cos(2\pi f_c t)$. The resulting signal is $+2A\cos^2(2\pi f_c t)$ if the original information symbol is a 1 or $-2A\cos^2(2\pi f_c t)$ if the original information symbol is 0. Because $2\cos^2(2\pi f_c t) = (1 + \cos(4\pi f_c t))$, we see that the resulting signals are as shown in Figure 3.29d. By smoothing out the oscillatory part with a so-called low-pass filter, we can easily determine the original baseband signal $X_i(t)$ and the A_k and subsequently the original binary sequence.

When we developed the Nyquist signal result in section 3.4.1, we found that for a low-pass channel of bandwidth W Hz the maximum signaling rate is $2W$ pulses/second. It can be shown that the system we have just described can transmit only W pulses/seconds over a band-pass channel that has bandwidth W.[9] Consequently, the time per bit is given by $T = 1/W$. Thus this scheme attains only half the signaling rate of the low-pass case. Next we show how we can recover this factor of 2 by using Quadrature Amplitude Modulation.

Suppose we have an original information stream that is generating symbols at a rate of $2W$ symbols per second. In **Quadrature Amplitude Modulation (QAM)** we split the original information stream into two sequences that consist of the odd and even symbols, say, B_k and A_k, respectively, as shown in Figure 3.31. Each sequence now has the rate W symbols per second. Suppose we take the even sequence A_k and produce a modulated signal by multiplying it by $\cos(2\pi f_c t)$; that is, $Y_i(t) = A_k \cos(2\pi f_c t)$ for a T-second interval. As before, this modulated signal will be located within the band of the band-pass channel. Now suppose that we take the odd sequence B_k and produce another modulated signal by multiplying it by $\sin(2\pi f_c t)$; that is, $Y_q(t) = B_k \sin(2\pi f_c t)$ for a T-second interval. This modulated signal also has its power located within the band of the band-pass channel. We finally obtain a composite modulated signal by adding $Y_i(t)$ and $Y_q(t)$, as shown in Figure 3.31.

$$Y(t) = Y_i(t) + Y_q(t) = A_k \cos(2\pi f_c t) + B_k \sin(2\pi f_c t).$$

This equation shows that we have generated what amounts to a two-dimensional modulation scheme. The first component A_k is called the **in-phase component**; the second component B_k is called the **quadrature-phase component**.

We now transmit the sum of these two modulated signals over the band-pass channel. The composite sinusoidal signal $Y(t)$ will be passed without distortion by the linear band-pass channel. We now need to demonstrate how the original information symbols can be recovered from $Y(t)$. We will see that our ability to do so depends on the following properties of cosines and sines:

$$2\cos^2(2\pi f_c t) = 1 + \cos(4\pi f_c t)$$
$$2\sin^2(2\pi f_c t) = 1 - \sin(4\pi f_c t)$$
$$2\cos(2\pi f_c t)\sin(2\pi f_c t) = 0 + \sin(4\pi f_c t)$$

---

[9]In the remainder of this section, we continue to use the term *pulse* for the signal that is transmitted in a T-second interval.

Modulate $\cos(2\pi f_c t)$ and $\sin(2\pi f_c t)$ by multiplying them by A_k and B_k, respectively, for $(k\text{-}1)T < t < kT$:

FIGURE 3.31 QAM modulator

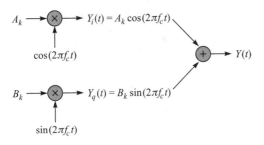

These properties allow us to recover the original symbols as shown in Figure 3.32. By multiplying $Y(t)$ by $2\cos(2\pi f_c t)$ and then low-pass filtering the resulting signal, we obtain the sequence A_k. Note that the cross-product term $B_k(t)\sin(4\pi f_c t)$ is removed by the low-pass filter. Similarly, the sequence B_k is recovered by multiplying $Y(t)$ by $2\sin(2\pi f_c t)$ and low-pass filtering the output. Thus QAM is a two-dimensional system that achieves an effective signaling rate of $2W$ pulses per second over the band-pass channel of W Hz. This result matches the performance of the Nyquist signaling procedure that we developed in section 3.4.

The two-dimensional nature of the above signaling scheme can be used to plot the various combinations of levels that are allowed in a given signaling interval of T seconds. (Note: It is important to keep in mind that $T = 1/W$ in this discussion). In the case considered so far, the term multiplying the cosine function can assume the value $+A$ or $-A$; the term multiplying the sine function can also assume that value $+A$ or $-A$. In total, four combinations of these values can occur. These are shown as the four points in the two-dimensional plane in Figure 3.33. At any given T-second interval, only one of the four points in this **signal constellation** can be in use. It is therefore clear that in every T-second interval we are transmitting two bits of information. As in the case of baseband

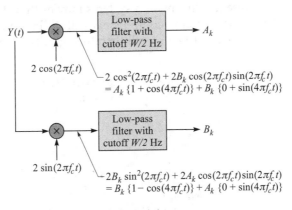

FIGURE 3.32 QAM demodulator

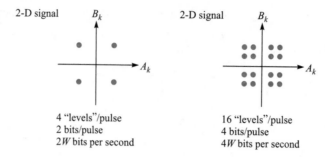

FIGURE 3.33 Signal constellations

signaling, we can increase the number of bits that can be transmitted per T-second interval by increasing the number of levels that are used. Figure 3.33 shows a 16-point constellation that results when the terms multiplying the cosine and sine functions are allowed to assume four possible levels. In this case only one of the 16 points in the constellation is in use in any given T-second interval, and hence four bits of information are transmitted at every such interval.

Another way of viewing QAM is as the simultaneous modulation of the amplitude and phase of a carrier signal, since

$$A_k \cos(2\pi f_c t) + B_k \sin(2\pi f_c t) = (A_k^2 + B_k^2)^{1/2} \cos(2\pi f_c t + \tan^{-1} B_k/A_k)$$

Each signal constellation point can then be seen as determining a specific amplitude and phase.

Many signal constellations that are used in practice have nonrectangular arrays of signal points such as those shown in Figure 3.34. To produce this type of signaling, we need to modify the above encoding scheme only slightly. Suppose that the constellation has 2^m points. Each T-second interval, the transmitter accepts m information bits, identifies the constellation point assigned to these bits, and then transmits cosine and sine signals with the amplitudes that correspond to the constellation point.

The presence of noise in transmission systems implies that the pair of recovered values for the cosine and sine components will differ somewhat from the transmitted values. This pair of values will therefore specify a point in the plane

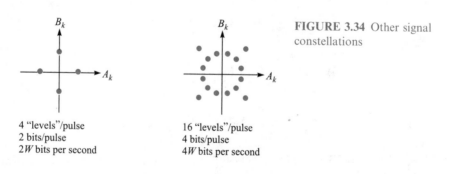

FIGURE 3.34 Other signal constellations

that deviates from the transmitted constellation point. The task of the receiver is to take the received pair of values and identify the closest constellation point.

3.6.1 Signal Constellations and Telephone Modem Standards

Signal constellation diagrams of the type shown in Figure 3.33 and Figure 3.34 are used in the various signaling standards that have been adopted for use over telephone lines. For the purposes of data communications, most telephone channels have a usable bandwidth in the range $f_1 = 500$ Hz to $f_2 = 2900$ Hz. This implies $W = 2400$ and hence a signaling rate of $1/T = W = 2400$ pulses/second. Table 3.4 lists some of the parameters that specify the ITU V.32bis and V.34bis modem standards that are in current use. Each standard can operate at a number of speeds that depend on the quality of the channel available. Both standards operate at a rate of 2400 pulses/second, and the actual bit rate is determined by which constellation is used. The QAM 4 systems uses four constellation points and hence two bits/pulse, giving a bit rate of 4800 bps.

Trellis modulation systems are more complex in that they combine error-correction coding with the modulation. In trellis modulation the number of constellation points is 2^{m+1}. At every T-second interval, the trellis coding algorithm accepts m bits and generates $m + 1$ bits that specify the constellation point that is to be used. In effect, only 2^m out of the 2^{m+1} possible constellation points are valid during any given interval. This extra degree of redundancy improves the robustness of the modulation scheme with respect to errors. In Table 3.4, the trellis 32 system has 2^5 constellation points out of which 16 are valid at any given time; thus the bit rate is $4 \times 2400 = 9600$ bps. Similarly, the trellis 128 system gives a bit rate of $6 \times 2400 = 14,400$ bps.

The V.34bis standard can operate at rates of 2400, 2743, 2800, 3000, 3200, or 3429 pulses/second. The modem precedes communications with an initial phase during which the channel is probed to determine the usable bandwidth in the given telephone connection. The modem then selects a pulse rate. For each of these pulse rates, a number of possible trellis encoding schemes are defined. Each encoding scheme selects a constellation that consists of a subset of points from a superconstellation of 860 points. A range of bit rates are possible, including 2400, 4800, 9600, 14,400, 19,200 and 28,800, 31,200 and 33,600 bps.

V.32bis	Modulation	Pulse rate
14,000 bps	Trellis 128	2400 pulses/second
9600 bps	Trellis 32	2400 pulses/second
4800 bps	QAM 4	2400 pulses/second
V.34bis		
2400–33,600 bps	Trellis 960	2400–3429 pulses/second

TABLE 3.4 Modem standards

It is instructive to consider how close the V.34bis modem comes to the maximum possible transmission rate predicted by Shannon's formula. Suppose we have a maximum useful bandwidth of 3400 Hz and assume a maximum SNR of 40 dB, which is a bit over the maximum possible in a telephone line. Shannon's formula then gives a maximum possible bit rate of 44,880 bits/second. It is clear then that the V.34bis modem is coming close to achieving the Shannon bound. In 1997 a new class of modems, the ITU-T V.90 standard, was introduced that tout a bit rate of 56,000 bits/second, well in excess of the Shannon bound! As indicated at the beginning of this chapter, the 56 kbps speed is attained only under particular conditions that do not correspond to the normal telephone channel.[10]

BEEEEEEP, CHIRP, CHIRP, KTWANG, KTWANG, shhhhh, SHHHHHH

What are the calling and answering V.34 modems up to when they make these noises? They are carrying out the handshaking that is required to set up communication. V.34 handshaking has four phases:

Phase 1: Because the modems don't know anything about each other's characteristics, they begin by communicating using simple, low-speed, 300 bps FSK. They exchange information about the available modulation modes, the standards they support, the type of error correction that is to be used, and whether the connection is cellular.

Phase 2: The modems perform probing of the telephone line by transmitting tones with specified phases and frequencies that are spaced 150 Hz apart and that cover the range from 150 Hz to 3750 Hz. The tones are sent at two distinct signal levels (amplitudes). The probing allows the modems to determine the bandwidth and distortion of the telephone line. The modems then select their carrier frequency and signal level. At the end of this phase, the modems exchange information about their carrier frequency, transmit power, symbol rate, and maximum data rate.

Phase 3: The receiver equalizer starts its adaptation to the channel. The echo canceler is started. The function of the echo canceler is to suppress its modem's transmission signal so that the modem can hear the arriving signal.

Phase 4: The modems exchange information about the specific modem parameters that are to be used, for example, signal constellation, trellis encoding, and other encoding parameters.

The modems are now ready to work for you!

---

[10]Indeed, when we discuss transmission medium in section 3.7, we show that much higher bit rates are attainable over twisted-wire pairs.

Modem standards have also been developed for providing error control as well as data compression capabilities. The V.42bis standard specifies the Link Access Procedure for Modems (LAPM) for providing error control. LAPM is based on the HDLC data link control, which is discussed in Chapter 5. V.42bis also specifies the use of the Lempel-Ziv data compression scheme for the compression of information prior to transmission. This scheme can usually provide compression ratios of two or more, thus providing an apparent modem speedup of two or more. The data compression scheme is explained in Chapter 12.

In this section we have considered only telephone modem applications. Digital modulation techniques are also used extensively in other digital transmission systems such as digital cellular telephony and terrestrial and satellite communications. These systems are discussed further in the next section.

3.7 PROPERTIES OF MEDIA AND DIGITAL TRANSMISSION SYSTEMS

For transmission to occur, we must have a transmission medium that conveys the energy of a signal from a sender to a receiver. A communication system places transmitter and receiver equipment on either end of a transmission medium to form a communications channel. In previous sections we discussed how communications channels are characterized in general. In this section we discuss the properties of the transmission media that are used in modern communication networks.

We found that the capability of a channel to carry information reliably is determined by several properties. First, the manner in which the medium transfers signals at various frequencies (that is, the amplitude-response function $A(f)$ and the phase-shift function $\varphi(f)$) determines the extent to which the input pulses are distorted. The transmitter and receiver equipment must be designed to remove enough distortion to make reliable detection of the pulses possible. Second, the transmitted signal is attenuated as it propagates through the medium and noise is also introduced in the medium and in the receiver. These phenomena determine the SNR at the receiver and hence the probability of bit errors. In discussing specific transmission media, we are therefore interested in the following characteristics:

- The amplitude-response function $A(f)$ and the phase-shift function $\varphi(f)$ of the medium and the associated bandwidth as a function of distance.
- The susceptibility of the medium to noise and interference from other sources.

A typical communications system transmits information by modulating a sinusoidal signal of frequency f_0 that is inserted into a guided medium or radiated through an antenna. The sinusoidal variations of the modulated signal propagate in a medium at a speed of v meters/second, where

$$v = \frac{c}{\sqrt{\varepsilon f_0}}$$

and where $c = 3 \times 10^8$ meters/second is the speed of light in a vacuum and ε is the dielectric constant of the medium. In free space, $\varepsilon = 1$, and $\varepsilon \geq 1$ otherwise. The **wavelength** λ of the signal is given by the length in space spanned by one period of the sinusoid:

$$\lambda = v/f_0 \text{ meters}$$

As shown in Figure 3.35, a 100 MHz carrier signal, which corresponds to FM broadcast radio, has a wavelength of 3 meters, whereas a 3 GHz carrier signal has a wavelength of 10 cm. Infrared light covers the range from 10^{12} to 10^{14} Hz, and the visible light used in optical fiber occupies the range 10^{14} to 10^{15} Hz.

The speed of light c is a maximum limit for propagation speed in free space and cannot be exceeded. Thus if a pulse of energy enters a communications channel of distance d at time $t = 0$, then none of the energy can appear at the output before time $t = d/c$ as shown in Figure 3.36. Note that in copper wire the speed of light is 2.3×10^8 meters/second, and in optical fiber systems the speed of light v is approximately 2×10^8 meters/second.

The two basic types of media are wired media, in which the signal energy is contained and guided within a solid medium, and wireless media, in which the signal energy propagates in the form of unguided electromagnetic signals. Copper pair wires, coaxial cable, and optical fiber are examples of wired media. Radio and infrared light are examples of wireless media.

Wired and wireless media differ in a fundamental way. In its most basic form, wired media provide communications from point to point. By interconnecting wires at various repeater or switching points, wired media lead to well-defined *discrete* network topologies. Since the energy is confined within the medium, additional transmission capacity can be procured by adding more wires. Unguided media, on the other hand, can achieve only limited directionality and can be transmitted, as in the case of broadcast radio, in all directions

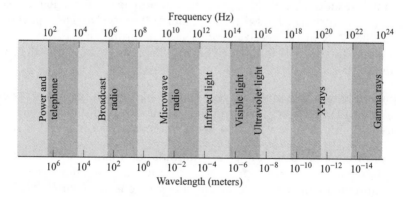

FIGURE 3.35 Electromagnetic spectrum

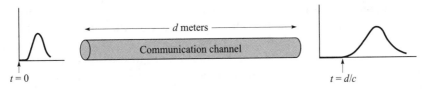

FIGURE 3.36 Propagation delay of a pulse over the communication channel

making the medium broadcast in nature. This condition leads to a network topology that is *continuous* in nature. In addition, all users within receiving range of each other must share the frequency band that is available and can thus interfere with each other. Unlike wired media, the radio spectrum is finite, and it is not possible to procure additional capacity. A given frequency band can be reused only in a sufficiently distant geographical area. To maximize its utility, the radio spectrum is closely regulated by government agencies.

Another difference between wired and wireless media is that wired media require establishing a right-of-way through the land that is traversed by the cable. This process is complicated, costly, and time-consuming. On the other hand, systems that use wireless media do not require the right-of-way and can be deployed by procuring only the sites where the antennas are located. Wireless systems can therefore be deployed more quickly and at lower cost.

Finally, we note that for wired media the attenuation has an exponential dependence on distance; that is, the attenuation at a given frequency is of the form 10^{kd} where the constant k depends on the specific frequency and d is that distance. The attenuation for wired media in dB is then

$$\text{attenuation for wired media} = kd \text{ dB}$$

That is, the attenuation in dB increases linearly with the distance. For wireless media the attenuation is proportional to d^n where n is the **path loss exponent**. For free space $n = 2$, and for environments where obstructions are present $n > 2$. The attenuation for wireless media in dB is then

$$\text{attenuation for wireless media is proportional to } n \log_{10} d \text{ dB}$$

and so the attenuation in dB only increases logarithmically with the distance. Thus in general the signal level in wireless systems can be maintained over much longer distances than in wired systems.

3.7.1 Twisted Pair

The simplest guided transmission medium consists of two parallel insulated conducting (e.g., copper) wires. The signal is transmitted through one wire while a ground reference is transmitted through the other. This two-wire system is susceptible to crosstalk and noise. *Crosstalk* refers to the picking up of electrical signals from other adjacent wires. Because the wires are unshielded, there is also a tendency to pick up noise, or *interference*, from other electromagnetic sources such as broadcast radio. The receiver detects the information signal by the voltage difference between the ground reference signal and the information signal. If either one is greatly altered by interference or crosstalk, then the chance of error is increased. For this reason parallel two-wire lines are limited to short distances.

A **twisted pair** consists of two wires that are twisted together to reduce the susceptibility to interference. The close proximity of the wires means that any interference will be picked up by both and so the difference between the pair of wires should be largely unaffected by the interference. Twisting also helps to reduce (but not eliminate) the crosstalk interference when multiple pairs are placed within one cable. Much of the wire in the telephone system is twisted-pair wire. For example, it is used between the customer and the central office, also called the *subscriber loop*, and often between central offices, called the *trunk plant*. Because multiple pairs are bundled together within one telephone cable, the amount of crosstalk is still significant, especially at higher frequencies.

A twisted pair can pass a relatively wide range of frequencies. The attenuation for twisted pair, measured in decibels/mile, can range from a 1 to 4 dB/mile at 1 kHz to 10 to 20 dB/mile at 500 kHz, depending on the gauge (diameter) of the wire as shown in Figure 3.37. Thus the bandwidth of twisted pair decreases with distance. Table 3.5 shows practical limits on data rates that can be achieved in a unidirectional link over a 24-gauge (0.016-inch-diameter wire) twisted pair for various distances.

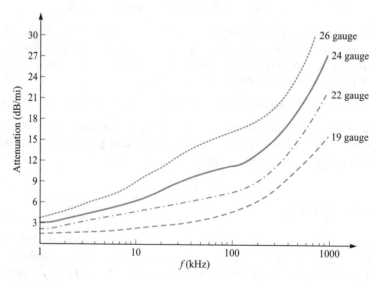

FIGURE 3.37 Attenuation versus frequency for twisted pair [after Smith 1985]

Standard	Data rate	Distance
T-1	1.544 Mbps	18,000 feet, 5.5 km
DS2	6.312 Mbps	12,000 feet, 3.7 km
1/4 STS-1	12.960 Mbps	4500 feet, 1.4 km
1/2 STS-1	25.920 Mbps	3000 feet, 0.9 km
STS-1	51.840 Mbps	1000 feet, 300 m

TABLE 3.5 Data rates of 24-gauge twisted pair

The first digital transmission system used twisted pair and was used in the trunk portion of the telephone network. This T-1 carrier system achieved a transmission rate of 1.544 Mbps and could carry 24 voice channels. The T-1 carrier system used baseband pulse transmission with bipolar encoding. The T-1 carrier system is discussed further in Chapter 4 in the context of the telephone network. Twisted pair in the trunk portion of the telephone network is being replaced by optical fibers. However, twisted pair constitutes the bulk of the access network that connects users to the telephone office, and as such, is crucial to the evolution of future digital networks. In the remainder of this section, we discuss how new systems are being introduced to provide high-speed digital communications in the access network.

Originally, in optimizing for the transmission of speech, the telephone company elected to transmit frequencies within the range of 0 to 4 kHz. Limiting the frequencies at 4 kHz reduced the crosstalk that resulted between different cable pairs at the higher frequencies and provided the desired voice quality. Within the subscriber loop portion, **loading coils** were added to further improve voice transmission within the 3 kHz band by providing a flatter transfer function. This loading occurred in lines longer than 5 kilometers. While improving the quality of the speech signal, the loading coils also increased the attenuation at the higher frequencies and hence reduced the bandwidth of the system. Thus the choice of a 4 kHz bandwidth for the voice channel and the application of loading coils, not the inherent bandwidth of twisted pair, are the factors that limit digital transmission over telephone lines to approximately 40 kbps.

Application—Digital Subscriber Loops

Several digital transmission schemes were developed in the 1970s to provide access *to Integrated Services Digital Network (ISDN)* using the twisted pair (without loading coils) in the subscriber loop network. These schemes provide for two Bearer (B) 64 kbps channels and one data (D) 16 kbps channel from the user to the telephone network. These services were never deployed on a wide scale.

To handle the recent demand from consumers for higher speed data transmission, the telephone companies plan to introduce a new technology called **asymmetric digital subscriber line (ADSL)**. The objective of this technology is to use existing twisted-pair lines to provide the higher bit rates that are possible with unloaded twisted pair. The frequency spectrum is divided into two regions.

The lower frequencies are used for conventional analog telephone signals. The region above is used for bidirectional digital transmission. The system is asymmetric in that the user can transmit upstream into the network at speeds ranging from 64 kbps to 640 kbps but can receive information from the network at speeds from 1.536 Mbps to 6.144 Mbps, depending on the distance from the telephone central office. This asymmetry in upstream/downstream transmission rates is said to match the needs of current applications such as upstream requests and downstream page transfers in the World Wide Web application.

The ITU-T G.992.1 standard for ADSL uses the Discrete Multitone (DMT) system that divides the available bandwidth into a large number of small subchannels. The binary information is distributed among the subchannels, each of which uses QAM. DMT can adapt to line conditions by avoiding subchannels with poor SNR. ITU-T has also approved standard G.992.2 as a "lite" version that provides access speeds of up to 512 kbps from the user and download speeds of up 1.5 Mbps. The latter is the simpler and less expensive standard because it does not require a "splitter" to separate telephone voice signals from the data signal and can instead be plugged directly into the PC by the user as is customary for most voiceband modems.

Application—Local Area Networks

Twisted pair is installed during the construction of most office buildings. The wires that terminate at the wall plate in each office are connected to wiring closets that are placed at various locations in the building. Consequently, twisted pair is a good candidate for use in local area computer networks where the maximum distance between a computer and a network device is in the order of 100 meters. As a transmission medium, however, high-speed tranmission over twisted pairs poses serious challenges. Several categories of twisted-pair cable have been defined for use in LANs. Category 3 **unshielded twisted pair (UTP)** corresponds to ordinary voice-grade twisted pair and can be used at speeds up to 16 Mbps. Category 5 UTP is intended for use at speeds up to 100 Mbps. Category 5 twisted pairs are twisted more tightly than are those in category 3, resulting in much better crosstalk immunity and signal quality. *Shielded twisted pair* involves providing a metallic braid or sheath to cover each twisted pair. It provides better performance than UTP but is more expensive and more difficult to use.

10BASE-T ETHERNET LAN

The most widely deployed version of Ethernet LAN uses the 10BASE-T physical layer. The designation 10BASE-T denotes *10* Mbps operation using *base*band transmission over *t*wisted-pair wire. The NIC card in each computer is connected to a hub in a star topology as shown in Figure 3.38. Two category 3 UTP cables provide the connection between computer and hub. The transmissions use Manchester line coding, and the cables are limited to a maximum distance of 100 meters.

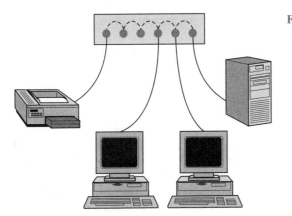

FIGURE 3.38 Ethernet hub

100BASE-T ETHERNET LAN

The 100BASE-T Ethernet LAN is also known as *Fast Ethenet*. As indicated by the designation, 100BASE-T Ethernet operates at a speed of 100 Mbps using twisted-pair wire. The computers are connected to a *hub* or a *switch* in a star topology, and the distance of the twisted pairs is limited to 100 meters. Operating 100 Mbps on UTP is challenging, and so three options for doing so were developed, one for category 3 UTP, one for shielded twisted pair, and one for category 5 UTP. One problem with extending the 10BASE-T transmission format is that Manchester line coding is inefficient in its use of bandwidth. Recall from the section on line coding that Manchester coding pulses vary at twice the information rate, so the use of Manchester coding would have required operation at 200 Mpulses/second. Another problem is that higher pulse rates result in more electromagnetic interference. For this reason, new and more efficient line codes were used in the new standards.

In the 100BASE-T4 format, four category 3 twisted-pair wires are used. At any given time three pairs are used to jointly provide 100 Mbps in a given direction; that is, each pair provides 33 1/3 Mbps. The fourth pair is used for collision detection. The transmission uses ternary signaling in which the transmitted pulses can take on three levels, $+A$, 0, or $-A$. The line code maps a group of eight bits into a corresponding group of six ternary symbols that are transmitted over the three parallel channels over two pulse intervals, or equivalently four bits into three ternary symbols/pulse interval. This mapping is possible because $2^4 = 16 < 3^3 = 27$. The transmitter on each pair sends 25 Mpulses/second, which gives a bit rate of 25 Mp/s $\times$ 4 bits/3 pulses = 33 1/3 Mbps as required. As an option, four category 5 twisted pairs can be used instead category 3 twisted pairs.

In the 100BASE-TX format, two category 5 twisted pairs are used to connect to the hub. Transmission is full duplex with each pair transmitting in one of the directions at a pulse rate of 125 Mpulses/second. The line code used takes a group of four bits and maps it into five binary pulses, giving a bit rate of 125

Mpulses/second × 4 bits/pulse = 100 Mbps. An option allows two pairs of shielded twisted wire to be used instead of the category 5 pairs.

3.7.2   Coaxial Cable

In **coaxial cable** a solid center conductor is located coaxially within a cylindrical outer conductor. The two conductors are separated by a solid dielectric material, and the outer conductor is covered with a plastic sheath as shown in Figure 3.39. The coaxial arrangement of the two conductors provides much better immunity to interference and crosstalk than twisted pair does. By comparing Figure 3.40 with Figure 3.37, we can see that coaxial cable can provide much higher bandwidths (hundreds of MHz) than twisted pair (a few MHz). For example, existing cable television systems use a bandwidth of 500 MHz.

Coaxial cable was initially deployed in the backbone of analog telephone networks where a single cable could be used to carry in excess of 10,000 simultaneous analog voice circuits. Digital transmission systems using coaxial cable were also deployed in the telephone network in the 1970s. These systems operate in the range of 8.448 Mbps to 564.992 Mbps. However, the deployment of coaxial cable transmission systems in the backbone of the telephone network was discontinued because of the much higher bandwidth and lower cost of optical fiber transmission systems.

Application—Cable Television Distribution

The widest use of coaxial cable is for distribution of television signals in cable TV systems. Existing coaxial cable systems use the frequency range from 54 MHz to 500 MHz. A National Television Standards Committee (NTSC) analog television signal occupies a 6 MHz band, and a phase alternation by line (PAL) analog television signal occupies 8 MHz, so 50 to 70 channels can be accommodated.[11] Existing cable television systems are arranged in a tree-and-branch topology as shown in Figure 3.41. The master television signal originates at a head end office,

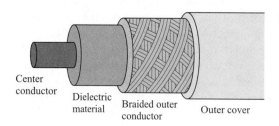

Center conductor

Dielectric material

Braided outer conductor

Outer cover

FIGURE 3.39 Coaxial cable

---

[11] The NTSC and PAL formats are two standards for analog television. The NTSC format is used in North America and Japan, and the PAL format is used in Europe.

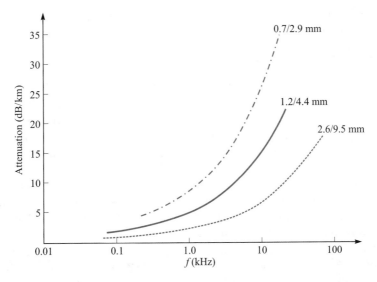

FIGURE 3.40 Attenuation versus frequency for coaxial cable [after Smith 1985]

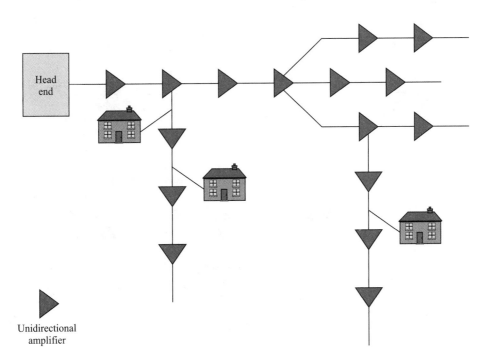

FIGURE 3.41 Tree-and-branch topology of conventional cable TV systems

and unidirectional analog amplifiers maintain the signal level. The signal is split along different branches until all subscribers are reached. Because all the information flows from the head end to the subscribers, cable television systems were designed to be unidirectional.

Application—Cable Modem

The coaxial cable network has a huge bandwidth flowing from the network to the user. For example, a single analog television channel will provide approximately 6 MHz of bandwidth. If QAM modulation is used with a 64-point constellation, then a bit rate of 6 Mpulses/second × 6 bits/pulse = 36 Mbps is possible. However, the coaxial network was not designed to provide communications from the user to the network. Figure 3.42 shows how coaxial cable networks are being modified to provide upstream communications through the introduction of bidirectional split-band amplifiers that allow information to flow in both directions.

Figure 3.43a shows the existing cable spectrum that uses the band from 54 MHz to 500 MHz for the distribution of analog television signals. Figure 3.43b shows the proposed spectrum for hybrid fiber-coaxial systems. The band from 550 MHz to 750 MHz would be used to carry new digital video and data signals as well as downstream telephone signals. In North America channels are 6 MHz wide, so these downstream channels can support bit rates in the range of 36 Mbps. The band from 5 MHz to 42 MHz, which was originally intended for

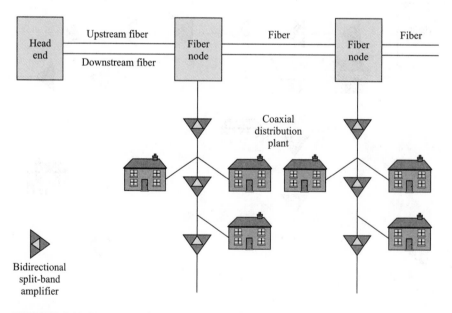

FIGURE 3.42 Topology of hybrid fiber-coaxial systems

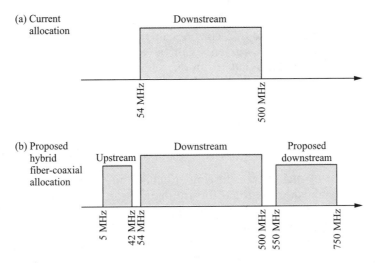

FIGURE 3.43 Frequency allocation in cable TV systems

pay-per-view signaling, would be converted for **cable modem** upstream signals as well as for *cable telephony*. This lower band is subject to much worse interference and noise than the downstream channels. Using channels of approximately 2 MHz, upstream transmission rates from 500 kbps to 4 Mbps could be provided. As in the case of ADSL, we see that the upstream/downstream transmission rates are asymmetric.

Both the upstream and downstream channels need to be *shared* among subscribers in the feeder line. The arrangement is similar to that of a local area network in that the cable modems from the various users must communicate with a *cable modem termination system (CMTS)* at the operator's end of the cable. The cable modems must listen for packets destined to them on an assigned downstream channel. They must also contend to obtain time slots to transmit their information in an assigned channel in the upstream direction.

Application—Ethernet LAN

The original design of the Ethernet LAN used coaxial cable for the shared medium (see Figure 1.17a.) Coaxial cable was selected because it provided high bandwidth, offered good noise immunity, and led to a cost-effective transceiver design. The original standard specified 10Base5, which uses thick (10 mm) coaxial cable operating at a data rate of *10* Mbps, using *base*band transmission and with a maximum segment length of *500* meters. The transmission uses Manchester coding. This cabling system required the use of a *transceiver* to attach the NIC card to the coaxial cable. The thick coaxial cable Ethernet was typically deployed along the ceilings in building hallways, and a connection from a workstation in an office would tap onto the cable. Thick coaxial cable is awkward to handle and install. The 10Base2 standard uses thin (5 mm) coaxial cable operating 10 Mbps and with a maximum segment of 185 meters. The cheaper and easier to handle thin coaxial cable makes use of T-shaped connectors. 10Base5 and 10Base2 segments can be combined through the use of a *repeater* that forwards the signals from one segment to the other.

3.7.3   Optical Fiber

The deployment of digital transmission systems using twisted pair and coaxial cable systems established the trend toward digitization of the telephone network during the 1960s and 1970s. These new digital systems provided significant economic advantages over previous analog systems. Optical fiber transmission systems, which were introduced in the 1970s, offered even greater advantages over copper-based digital transmission systems and resulted in a dramatic acceleration of the pace toward digitization of the network. Figure 1.9 of Chapter 1 showed that optical fiber systems represented an acceleration in the long-term rate of improvement in transmission capacity.

The typical T-1 or coaxial transmission system requires repeaters about every 2 km. Optical fiber systems, on the other hand, have maximum repeater spacings in the order of tens to hundreds of kilometers. The introduction of optical fiber systems has therefore resulted in great reductions in the cost of digital transmission. Optical fiber systems have also allowed dramatic reductions in the space required to house the cables. A single fiber strand is much thinner than twisted pair or coaxial cable. Because a single optical fiber can carry much higher transmission rates than copper systems, a single cable of optical fibers can replace many cables of copper wires. In addition, optical fibers do not radiate significant energy and do not pick up interference from external sources. Thus compared to electrical transmission, optical fibers are more secure from tapping and are also immune to interference and crosstalk.

Optical fiber consists of a very fine cylinder of glass (core) surrounded by a concentric layer of glass (cladding) as shown in Figure 3.44. The information itself is transmitted through the core in the form a fluctuating beam of light. The core has a slightly higher optical density (index of refraction) than the cladding. The ratio of the indices of refraction of the two glasses defines a critical angle θ_c. When a ray of light from the core approaches the cladding at an angle less than θ_c, the ray is completely reflected back into the core. In this manner the ray of light is guided within the fiber.

The attenuation in the fiber can be kept low by controlling the impurities that are present in the glass. When it was invented in 1970, optical fiber had a loss of 20 dB per kilometer. Within 10 years systems with a loss of 0.2 dB/km had become available. Figure 3.45 shows that minimum attenuation of optical fiber varies with the wavelength of the signal. It can be seen that the systems that operate at wavelengths of 850 nanometer (nm), 1300 nm, and 1550 nm occupy regions of low attenuation. The attenuation peaks are due to residual water

(a) Geometry of optical fiber

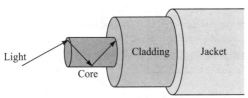

FIGURE 3.44 Transmission of light waves in optical fiber

(b) Reflection in optical fiber

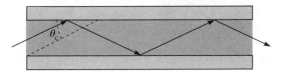

vapor in the glass fiber. Early optical fiber transmission systems operated in the 850 nm region at bit rates in the tens of megabits per second and used relatively inexpensive light emitting diodes (LEDs) as the light source. Second- and third-generation systems use laser sources and operate in the 1300 nm and 1500 nm region achieving gigabits/second bit rates.

A **multimode fiber** has an input ray of light reach the receiver over multiple paths, as shown in Figure 3.46a. Here the first ray arrives in a direct path, and the second ray arrives through a reflected path. The difference in delay between the two paths causes the rays to interfere with each other. The amount of interference depends on the duration of a pulse relative to the delays of the

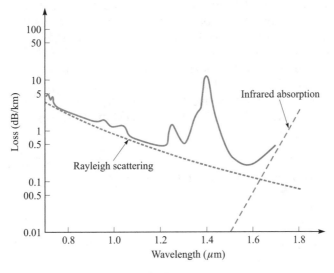

FIGURE 3.45 Attenuation versus wavelength for optical fiber

(a) Multimode fiber: multiple rays follow different paths

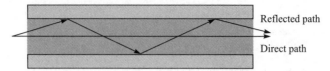

Reflected path

Direct path

(b) Single mode: only direct path propagates in fiber

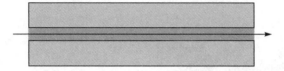

FIGURE 3.46 Single-mode and multimode optical fiber

paths. The presence of multiple paths limits the maximum bit rates that are achievable using multimode fiber. By making the core of the fiber much narrower, it is possible to restrict propagation to the single direct path. These **single-mode fibers** can achieve speeds of many gigabits/second over hundreds of kilometers.

Figure 3.47 shows an optical fiber transmission system. The transmitter consists of a light source that can be modulated according to an electrical input signal to produce a beam of light that is inserted into the fiber. Typically the binary information sequence is mapped into a sequence of on/off light pulses at some particular wavelength. An optical detector at the receiver end of the system converts the received optical signal into an electrical signal from which the original information can be detected.

The region around 1300 nm contains a band with attenuation less than 0.5 dB/km. This region has a bandwidth of 25 terahertz. One terahertz is 10^{12} Hz, that is, 1 million MHz! The region around 1550 nm has another band with attenuation as low as 0.2 dB/km [Mukherjee 1997]. This region has a bandwidth of about 25 THz. Clearly, existing optical transmission systems do not come close to utilizing this bandwidth.

Wavelength-division multiplexing (WDM) is one approach for attempting to exploit more of the available bandwidth. In WDM multiple wavelengths are used

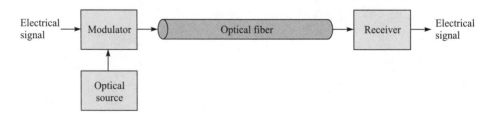

FIGURE 3.47 Optical transmission system

to simultaneously carry several information streams over the same fiber. WDM is a form of multiplexing and is covered in Chapter 4. Early WDM systems handled 16 wavelengths each transmitting 2.5 Gbps for a total of 40 Gbps. These systems could span distances in excess of 300 km. Dense WDM systems can provide 160 wavelengths, each operating at 10 Gbps for a total of 1600 Gbps.

As the light pulses propagate through the fiber, the pulses become spread out. This dispersion limits the minimum time between consecutive pulses and hence the bit rate. A special pulse shape, called a *soliton*, that retains its shape as it propagates through the fiber has been discovered. Experiments have demonstrated that solitons can achieve speeds of 80 Gbps over distances of 10,000 km. Early field trials have demonstrated soliton transmission at 10 Gbps over 200 km. Soliton-based systems promise extremely high-speed repeaterless digital transmission systems.

Application—Network Backbone

Optical fiber transmission systems are widely deployed in the backbone of networks. In Chapter 4 we present the digital multiplexing hierarchy that has been developed for electrical and optical digital transmission systems. Current optical fiber transmissions sytems provide transmission rates from 45 Mbps to 9.6 Gbps using single wavelength transmission and 40 Gbps to 1600 Gbps using WDM. Optical fiber systems are very cost-effective in the backbone of networks because the cost is spread over a large number of users. Optical fiber transmissions systems provide the facilities for long-distance telephone communications. Repeaterless optical fiber transmission systems are also used to interconnect telephone offices in metropolitan areas.

The cost of installing optical fiber in the subscriber portion of the network remains higher than the cost of using the existing installed base of twisted pair and coaxial cable. Fiber-to-the-home proposals that would provide huge bandwidths to the user remain too expensive. Fiber-to-curb proposals attempt to reduce this cost by installing fiber to a point that is sufficiently close to the subscriber that twisted pair or coaxial cable can provide high data rates to the user.

Application—Local Area Networks

Optical fiber is used as the physical layer of several LAN standards. The 10BASE-FP Ethernet physical layer standard uses optical fiber operating with an 850 nm source. The transmission system uses Manchester coding and intensity-light modulation and allows distances up to 2 km. The Fiber Distributed Data Interface (FDDI) ring-topology LAN uses optical fiber transmission at a speed of 100 Mbps, using LED light sources at 1300 nm with repeater spacings of up to 2 km. The binary information is encoded using a 4B5B code followed by NRZ inverted line coding. The 100BASE-FX Fast Ethernet physical layer standard uses two fibers, one for send and one for receive. The maximum distance is limited to 100 meters. The transmission format is the same as that of FDDI with slight modifications.

Optical fiber is the preferred medium for Gigabit Ethernet. As has become the practice in the development of physical layer standards, the 1000BASE-X standards are based on the preexisting fiber channel standard. The pulse transmission rate is 1.25 gigapulses/second, and an 8B10B code is used to provide the 1 Gbps transmission rate. There are two variations of the 1000BASE-X standard. The 1000BASE-SX uses a "shortwave" light source, nominally 850 nm, and multimode fiber. The distance limit is 550 meters. The 1000BASE-LX uses a "longwave" light source, nominally at 1300 nm, single mode or multimode fiber. For multimode fiber the distance limit is 550 meters, and for single mode fiber the distance is 5 km.

3.7.4   Radio Transmission

Radio encompasses the electromagnetic spectrum in the range of 3 kHz to 300 GHz. In radio communications the signal is transmitted into the air or space, using an antenna that radiates energy at some carrier frequency. For example, in QAM modulation the information sequence determines a point in the signal constellation that specifies the amplitude and phase of the cosine wave that is transmitted. Depending on the frequency and the antenna, this energy can propagate in either a unidirectional or omnidirectional fashion. In the unidirectional case a properly aligned antenna receives the modulated signal, and an associated receiver in the direction of the transmission recovers the original information. In the omnidirectional case any receiver with an antenna in the area of coverage can pick up the signal.

Radio communication systems are subject to a variety of transmission impairments. We indicated earlier that the attenuation in radio links varies logarithmically with the distance. Attenuation for radio systems also increases with rainfall. Radio systems are subject to multipath fading and interference. Multipath fading refers to the interference that results at a receiver when two or more versions of the same signal arrive at slightly different times. If the arriving signals differ in polarity, then they will cancel each other. Multipath fading can result in wide fluctuations in the amplitude and phase of the received signal. Interference refers to energy that appears at the receiver from sources other than the transmitter. Interference can be generated by other users of the same frequency band or by equipment that inadvertently transmits energy outside its band and into the bands of adjacent channels. Interference can seriously affect the performance of radio systems, and for this reason regulatory bodies apply strict requirements on the emission properties of electronic equipment.

Figure 3.48 gives the range of various frequency bands and their applications. The frequency bands are classified according to wavelengths. Thus the low frequency (LF) band spans the range 30 kHz to 300 kHz, which corresponds to a

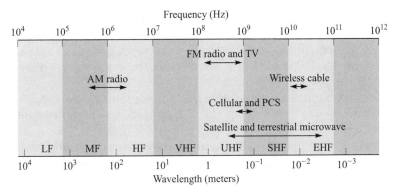

FIGURE 3.48 Radio spectra

wavelength of 1 km to 10 km, whereas the extremely high frequency (EHF) band occupies the range from 30 to 300 GHz corresponding to wavelengths of 1 millimeter to 1 centimeter. Note that the progression of frequency bands in the logarithmic frequency scale have increasingly larger bandwidths, for example, the "band" from 10^{11} to 10^{12} Hz has a bandwidth of 0.9×10^{12} Hz, whereas the band from 10^5 to 10^6 Hz has a bandwidth of 0.9×10^6 Hz.

The propagation properties of radio waves vary with the frequency. Radio waves at the VLF, LF, and MF bands follow the surface of the earth in the form of ground waves. VLF waves can be detected at distances up to about 1000 km, and MF waves, for example, AM radio, at much shorter distances. Radio waves in the HF band are reflected by the ionosphere and can be used for long-distance communications. These waves are detectable only within certain specific distances from the transmitter. Finally, radio waves in the VHF band and higher are not reflected back by the ionosphere and are detectable only within line-of-sight.

In general, radio frequencies below 1 GHz are more suitable for omnidirectional applications, such as those shown in Table 3.6. For example, paging systems ("beepers") are an omnidirectional application that provides one-way communications. A high-power transmission system is used to reach simple, low-power pocket-size receivers in some geographic area. The purpose of the system is to alert the owner that someone wishes to communicate with him or her. The

System	Description	Distance
Paging	Short message	10s of kilometers
Cordless telephone	Analog/digital voice	10s of meters
Cellular telephone	Analog/digital voice and data	kilometers
Personal Communication Services	Digital voice and data	100s of meters
Wireless LAN	High-speed data	100 meters

TABLE 3.6 Examples of omnidirectional systems

system may consist of a single high-powered antenna, or a network of interconnected antennas, or be a nationwide satellite-based transmission system. These systems deliver the calling party's telephone number and short text messages. Paging systems have operated in a number of frequency bands. Most systems currently use the 930 to 932 MHz band.

Cordless telephones are an example of an omnidirectional application that provides two-way communications. Here a simple base station connects to a telephone outlet and relays signaling and voice information to a cordless phone. This technology allows the user to move around in an area of a few tens of meters while talking on the phone. The first generation of cordless phones used analog radio technology and subsequent generations have used digital technology.

Application—Cellular Communications

Analog cellular telephone systems were introduced in 1979 in Japan. This system provided for 600 two-way channels in the 800 MHz band. In Europe the Nordic Mobile Telephone system was developed in 1981 in the 450 MHz band. The U.S. Advanced Mobile Phone System (AMPS) was deployed in 1983 in a frequency band of 50 MHz in the 800 MHz region. This band is divided into 30 kHz channels that can each carry a single FM-modulated analog voice signal.

Analog cellular phones quickly reached their capacity in large metropolitan areas because of the popularity of cellular phone service. Several digital cellular telephone systems based on digital transmission have been introduced. In 1991 Interim Standard IS-54 in the United States allowed for the replacement of a 30 kHz channel with a digital channel that can support three users. This digital channel uses differential QAM modulation in place of the analog GM modulation. A cellular standard based on code division multiple access (CDMA) was also standardized as IS-95. This system, based on direct sequence spread spectrum transmission, can handle more users than earlier systems could. In Europe the Global System for Mobile (GSM) standard was developed to provide for a pan-European digital cellular system in the 900 MHz band. These cellular systems are discussed further in Chapter 6.

In 1995 personal communication services (PCS) licenses were auctioned in the U.S. for spectrum in the 1800/1900 MHz region. PCS is intended to extend digital cellular technology to a broader community of users by using low-power transmitters that cover small areas, "microcells." PCS thus combines aspects of conventional cellular telephone service with aspects of cordless telephones. The first large deployment of PCS is in the Japanese Personal Handiphone system that operates in the 1800/1900 band. This system is now very popular. In Europe the GSM standard has been adapted to the 1800/1900 band.

Application—Wireless LANs

Wireless LANs are another application of omnidirectional wireless communications. The objective here is to provide high-speed communications among a number of computers located in relatively close proximity. Most standardization efforts in the United States have focused in the Industrial/Scientific/Medical (ISM) bands, which span 902 to 928 MHz, 2400 to 2483.5 MHz, and 5725 to 5850 MHz, respectively. Unlike other frequency bands, the ISM band is designated for unlicensed operation so each user must cope with the interference from other users. In Europe, the high-performance radio LAN (HIPERPLAN) standard was developed to provide high-speed (20 Mbps) operation in the 5.15 to 5.30 GHz band. In 1996 the Federal Communications Commission (FCC) in the United States announced its intention to make 350 MHz of spectrum in the 5.15 to 5.35 GHz and 5.725 to 5.825 GHz bands available for unlicensed use in LAN applications. These developments are significant because these systems will provide high-speed communications to the increasing base of portable computers. This new spectrum allocation will also enable the development of ad hoc digital radio networks in residential and other environments.

Application—Point-to-Point and Point-to-Multipoint Radio Systems

Highly directional antennas can be built for microwave frequencies that cover the range from 2 to 40 GHz. For this reason point-to-point wireless systems use microwave frequencies and were a major component of the telecommunication infrastructure introduced several years ago. Digital microwave transmission systems have been deployed to provide long-distance communications. These systems typically use QAM modulation with fairly large signal constellations and can provide transmission rates in excess of 100 Mbps. The logarithmic, rather than linear, attenuation gave microwave radio systems an advantage over coaxial cable systems by requiring repeater spacings in the tens of kilometers. In addition, microwave systems did not have to deal with right-of-way issues. Microwave transmission systems can also be used to provide inexpensive digital links between buildings.

Microwave frequencies in the 28 GHz band have also been licensed for point-to-multipoint "wireless cable" systems. In these systems microwave radio beams from a telephone central office would send 50 Mbps directional signals to subscribers within a 5 km range. Reflectors would be used to direct these beams so that all subscribers can be reached. These signals could contain digital video and telephone as well as high-speed data. Subscribers would also be provided with transmitters that would allow them to send information upstream into the network. The providers of this service have about 1 GHz in total bandwidth available.

Application—Satellite Communications

Early satellite communications systems can be viewed as microwave systems with a single repeater in the sky. A (geostationary) satellite is placed at an altitude of about 36,000 km above the equator where its orbit is stationary relative to the rotation of the earth. A modulated microwave radio signal is beamed to the satellite on an uplink carrier frequency. A transponder in the satellite receives the uplink signal, regenerates it, and beams it down back to earth on a downlink carrier frequency. A satellite typically contains 12 to 20 transponders so it can handle a number of simultaneous transmissions. Each transponder typically handles about 50 Mbps. Satellites operate in the 4/6, 11/14, and 20/30 GHz bands, where the first number indicates the downlink frequency and the second number the uplink frequency.

Geostationary satellite systems have been used to provide point-to-point digital communications to carry telephone traffic between two points. Satellite systems have an advantage over fiber systems in situations where communications needs to be established quickly or where deploying the infrastructure is too costly. Satellite systems are inherently broadcast in nature, so they are also used to simultaneously beam television, and other signals, to a large number of users. Satellite systems are also used to reach mobile users who roam wide geographical areas.

Constellations of low-earth orbit satellites (LEOS) are being planned for deployment. These include the Iridium and Teledesic systems. The satellites are not stationary with respect to the earth, but they rotate in such a way that there is continuous coverage of the earth. The component satellites are interconnected by high-speed links forming a network in the sky.

3.7.5   Infrared Light

Infrared light is a communication medium whose properties differ significantly from radio frequencies. Infrared light does not penetrate walls, so an inherent property is that it is easily contained within a room. This factor can be desirable from the point of view of reducing interference and enabling reuse of the frequency band in different rooms. Infrared communications systems operate in the region from 850 nm to 900 nm where receivers with good sensitivity are available. Infrared light systems have a very large potential bandwidth that is not yet exploited by existing systems. A serious problem is that the sun generates radiation in the infrared band, which can be a cause of severe interference. The infrared band is being investigated for use in the development of very high speed wireless LANs.

Application—IrDA Links

The Infrared Data Association (IrDA) was formed to promote the development of infrared light communication systems. A number of standards have been developed under its auspices. The IrDA-C standard provides bidirectional communications for cordless devices such as keyboards, mice, joysticks, and handheld computers. This standard operates at a bit rate of 75 kbps at distances of up to 8 meters. The IrDA-D standard provides for data rates from 115 kb/s to

4 Mb/s over a distance of 1 meter. It was designed as a wireless alternative to connecting devices, such as a laptop to a printer.

3.8 ERROR DETECTION AND CORRECTION

In most communication channels a certain level of noise and interference is unavoidable. Even after the design of the digital transmission system has been optimized, bit errors in transmission will occur with some small but nonzero probability. For example, typical bit error rates for systems that use copper wires are in the order of 10^{-6}, that is, one in a million. Modern optical fiber systems have bit error rates of 10^{-9} or less. In contrast, wireless transmission systems can experience error rates as high as 10^{-3} or worse. The acceptability of a given level of bit error rate depends on the particular application. For example, certain types of digital speech transmission are tolerant to fairly high bit error rates. Other types of applications such as electronic funds transfer require essentially error-free transmission. In this section we introduce **error-control** techniques for improving the error-rate performance that is delivered to an application in situations where the inherent error rate of a digital transmission system is unacceptable.

There are two basic approaches to error control. The first approach involves the detection of errors and an **automatic retransmission request (ARQ)** when errors are detected. This approach presupposes the availability of a return channel over which the retransmission request can be made. For example, ARQ is widely used in computer communication systems that use telephone lines. The second approach, **forward error correction (FEC)**, involves the detection of errors followed by processing that attempts to correct the errors. FEC is appropriate when a return channel is not available, retransmission requests are not easily accommodated, or a large amount of data is sent and retransmission to correct a few errors is very inefficient. For example, FEC is used in satellite and deep-space communications. A recent application is in audio CD recordings where FEC is used to provide tremendous robustness to errors so that clear sound reproduction is possible even in the presence of smudges and scratches on the disk surface. Error detection is the first step in both ARQ and FEC. The difference between ARQ and FEC is that ARQ "wastes" bandwidth by using retransmissions, whereas FEC requires additional redundancy in the transmitted information and incurs significant processing complexity in performing the error correction.

In this section we discuss parity check codes, the Internet checksum, and polynomial codes that are used in error detection. We also present methods for assessing the effectiveness of these codes in several error environments. These results are used in Chapter 5, in the discussion of ARQ protocols. An optional section on linear codes gives a more complete introduction to error detection and correction.

8.1 Error Detection

ın this section we discuss the idea of error detection in general terms, using the single parity check code as an example throughout the discussion. The basic idea in performing error detection is very simple. As illustrated in Figure 3.49, the information produced by an application is encoded so that the stream that is input into the communication channel satisfies a specific *pattern* or condition. The receiver checks the stream coming out of the communication channel to see whether the pattern is satisfied. If it is not, the receiver can be certain that an error has occurred and therefore sets an alarm to alert the user. This certainty stems from the fact that no such pattern would have been transmitted by the encoder.

The simplest code is the **single parity check code** that takes k information bits and appends a single **check bit** to form a **codeword**. The parity check ensures that the total number of 1s in the codeword is even; that is, the codeword has even parity.[12] The check bit in this case is called a *parity bit*. This error-detection code is used in ASCII where characters are represented by seven bits and the eighth bit consists of a parity bit. This code is an example of the so-called linear codes because the parity bit is calculated as the modulo 2 sum of the information bits:

$$b_{k+1} = b_1 + b_2 + \ldots + b_k \quad \text{modulo 2} \tag{2}$$

where $b_1, b_2, \ldots, b_k$ are the information bits.

Recall that in modulo 2 arithmetic $0 + 0 = 0, 0 + 1 = 1, 1 + 0 = 1$, and $1 + 1 = 0$. Thus, if the information bits contain an even number of 1s, then the parity bit will be 0; and if they contain an odd number, then the parity bit will be 1. Consequently, the above rule will assign the parity bit a value that will produce a *codeword* that *always contains an even number of 1s*. This pattern defines the single parity check code.

If a codeword undergoes a single error during transmission, then the corresponding binary block at the output of the channel will contain an odd number of 1s and the error will be detected. More generally, if the codeword undergoes an odd number of errors, the corresponding output block will also contain an odd number of 1s. Therefore, the single parity bit allows us to detect all error patterns that introduce an odd number of errors. On the other hand, the single parity bit will fail to detect any error patterns that introduce an even number of errors, since the resulting binary vector will have even parity. Nonetheless, the

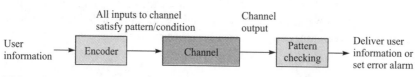

FIGURE 3.49 General error-detection system

---

[12]Some systems use odd parity by defining the check bit to be the binary complement of equation 2.

single parity bit provides a remarkable amount of error-detection capability, since the addition of a single check bit results in making half of all possible error patterns detectable, regardless of the value of k.

Figure 3.50 shows an alternative way of looking at the operation of this example. At the transmitter a checksum is calculated from the information bits and transmitted along with the information. At the receiver, the checksum is recalculated, based on the received information. The received and recalculated checksums are compared, and the error alarm is set if they disagree.

This simple example can be used to present two fundamental observations about error detection. The first observation is that error detection requires **redundancy** in that the amount of information that is transmitted is over and above the required minimum. For a single parity check code of length $k + 1$, k bits are information bits, and one bit is the parity bit. Therefore, the fraction $1/(k + 1)$ of the transmitted bits is redundant.

The second fundamental observation is that *every error-detection technique will fail to detect some errors*. In particular, an error-detection technique will always fail to detect transmission errors that convert a valid codeword into another valid codeword. For the single parity check code, an even number of transmission errors will always convert a valid codeword to another valid codeword.

The objective in selecting an error-detection code is to select the codewords that reduce the likelihood of the transmission channel converting one valid codeword into another. To visualize how this is done, suppose we depict the set of all possible binary blocks as the space shown in Figure 3.51, with codewords shown by xs in the space and noncodewords by os. To minimize the probability of error-detection failure, we want the codewords to be selected so that they are spaced as far away from each other as possible. Thus the code in Figure 3.51a is a poor code because the codewords are close to each other. On the other hand, the code in Figure 3.51b is good because the distance between codewords is maximized. The effectiveness of a code clearly depends on the types of errors that are introduced by the channel. We next consider how the effectiveness is evaluated for the example of the single parity check code.

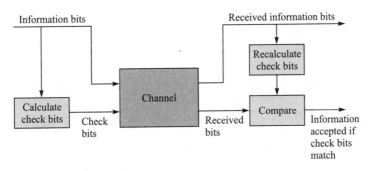

FIGURE 3.50 Error-detection system using check bits

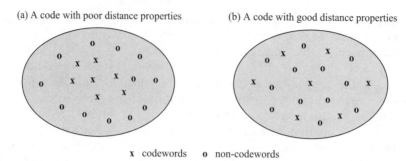

(a) A code with poor distance properties   (b) A code with good distance properties

x codewords   o non-codewords

FIGURE 3.51 Distance properties of codes

EFFECTIVENESS OF ERROR-DETECTION CODES

The effectiveness of an error-detection code is measured by the probability that the system fails to detect an error. To calculate this probability of error-detection failure, we need to know the probabilities with which various errors occur. These probabilities depend on the particular properties of the given communication channel. We will consider three models of error channels: the random error vector model, the random bit error model, and burst errors.

Suppose we transmit a codeword that has n bits. Define the error vector $\underline{e} = (e_1, e_2, \ldots, e_n)$ where $e_i = 1$ if an error occurs in the ith transmitted bit and $e_i = 0$ otherwise. In one extreme case, the **random error vector model**, all 2^n possible error vectors are equally likely to occur. In this channel model the probability of $\underline{e}$ does not depend on the number of errors it contains. Thus the error vector $(1, 0, \ldots, 0)$ has the same probability of occurrence as the error vector $(1, 1, \ldots, 1)$. The single parity check code will fail when the error vector has an even number of 1s. Thus for the random error vector channel model, the probability of error detection failure is $1/2$.

Now consider the **random bit error model** where the bit errors occur independently of each other. Satellite communications provide an example of this type of channel. Let p be the probability of an error in a single-bit transmission. The probability of an error vector that has j errors is $p^j (1 - p)^{n-j}$, since each of the j errors occurs with probability p and each of the $n - j$ correct transmissions occurs with probability $1 - p$. By rewriting this probability we obtain:

$$p[\underline{e}] = (1 - p)^{n - w(\underline{e})} p^{w(\underline{e})} = (1 - p)^n \left(\frac{p}{1 - p} \right)^{w(\underline{e})} \tag{3}$$

where the **weight** $w(\underline{e})$ is defined as the number of 1s in $\underline{e}$. For any useful communication channel, the probability of bit error is much smaller than 1, and so $p < 1/2$ and $p/(1 - p) < 1$. This implies that for the random bit error channel the probability of $\underline{e}$ decreases as the number or errors (1s) increases; that is, an error pattern with a given number of bit errors is more likely than an error pattern with a larger number of bit errors. Therefore this channel tends to map a transmitted codeword into binary blocks that are clustered around the codeword.

The single parity check code will fail if the error pattern has an even number of 1s. Therefore, in the random bit error model:

$$P[\text{error detection failure}] = P[\text{undetectable error pattern}]$$
$$= P[\text{error patterns with even number of 1s}] \quad\quad (4)$$
$$= \binom{n}{2} p^2 (1-p)^{n-2} + \binom{n}{4} p^4 (1-p)^{n-4} + \cdots$$

where the number of terms in the sum extends up to the maximum possible even number of errors. In the preceding equation we have used the fact that the number of distinct binary n-tuples with j ones and $n-j$ zeros is given by

$$\binom{n}{j} = \frac{n!}{j!(n-j!)}$$

In any useful communication system, the probability of a single-bit error p is much smaller than 1. We can then use the following approximation: $p^i (1-p)^j \approx p^i (1-pj) \approx p^i$. For example, if $p = 10^{-3}$ then $p^2(1-p)^{n-2} \approx 10^{-6}$ and $p^4 (1-p)^{n-4} \approx 10^{-12}$. Thus the probability of detection failure is determined by the first term in equation 4. For example, suppose $n = 32$ and $p = 10^{-4}$. Then the probability of error-detection failure is 5×10^{-6}, a reduction of nearly two orders of magnitude.

We see then that a wide gap exists in the performance achieved by the two preceding channel models. Many communication channels combine aspects of these two channels in that errors occur in **bursts**. Periods of low error-rate transmission are interspersed with periods in which clusters of errors occur. The periods of low error rate are similar to the random bit error model, and the periods of error bursts are similar to the random error vector model. The probability of error-detection failure for the single parity check code will be between those of the two channel models. In general, measurement studies are required to characterize the statistics of burst occurrence in specific channels.

3.8.2   Two-Dimensional Parity Checks

A simple method to improve the error-detection capability of a single parity check code is to arrange the information bits in columns of k bits, as shown in Figure 3.52. The last bit in each column is the check bit for the information bits in the column. Note that in effect the last column is a "check codeword" over the previous m columns. The right-most bit in each row is the check bit of the other bits in the row. The resulting encoded matrix of bits satisfies the pattern that all rows have even parity and all columns have even parity.

If one, two, or three errors occur anywhere in the matrix of bits during transmission, then at least one row or parity check will fail, as shown in Figure 3.53. However, some patterns with four errors are not detectable, as shown in the figure.

1	0	0	1	0	0
0	1	0	0	0	1
1	0	0	1	0	0
1	1	0	1	1	0
1	0	0	1	1	1

Last row consists of
check bit for each row

Bottom row consists of
check bit for each column

FIGURE 3.52 Two-dimensional parity check code

The two-dimensional parity check code is another example of a linear code. It has the property that error-detecting capabilities can be identified visually, but it does not have particularly good performance. Better codes are discussed in a later (optional) section on linear codes.

3.8.3 Internet Checksum

Several Internet protocols (e.g., IP, TCP, UDP) use check bits to detect errors. With IP a **checksum** is calculated for the contents of the header and included in a special field. Because the checksum must be recalculated at every router, the algorithm for the checksum was selected for its ease of implementation in software rather than for the strength of its error-detecting capabilities.

The algorithm assumes that the header consists of a certain number, say, L, of 16-bit words, $\mathbf{b}_0, \mathbf{b}_1, \mathbf{b}_2, \ldots, \mathbf{b}_{L-1}$ plus a checksum $\mathbf{b}_L$. These L words correspond to the "information" in the terminology introduced in the previous sections. The 16-bit checksum $\mathbf{b}_L$ corresponds to the parity bits and is calculated as follows:

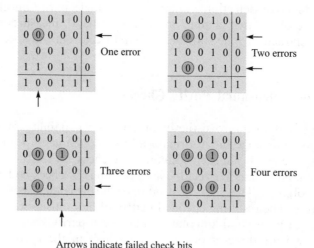

Arrows indicate failed check bits

FIGURE 3.53 Detectable and undetectable error patterns for two-dimensional code

1. Each 16-bit word is treated as an integer, and the L words are added modulo $2^{16} - 1$:

$$\mathbf{x} = \mathbf{b}_0 + \mathbf{b}_1 + \mathbf{b}_2 + \ldots + \mathbf{b}_{L-1} \text{ modulo } 2^{16} - 1$$

2. The checksum then consists of the negative value of $\mathbf{x}$:

$$\mathbf{b}_L = -\mathbf{x}$$

3. The checksum $\mathbf{b}_L$ is then inserted in a dedicated field in the header.

The contents of all headers, including the checksum field, must then satisfy the following pattern:

$$0 = \mathbf{b}_0 + \mathbf{b}_1 + \mathbf{b}_2 + \ldots + \mathbf{b}_{L-1} + \mathbf{b}_L \text{ modulo } 2^{16} - 1$$

Each router can then check for errors in the header by calculating the preceding equation for each received header.

As an example, consider two-bit words with $L = 2$. There are 16 possible combinations for $\mathbf{b}_0$, $\mathbf{b}_1$ and the corresponding checksum $\mathbf{b}_2$. For example if $\mathbf{b}_0 = 00$ and $\mathbf{b}_1 = 01$, then in modulo 3 arithmetic we have $\mathbf{b}_0 + \mathbf{b}_1 = 0 + 1 = 1$. In modulo 3 arithmetic, $-1 = 2$, so $\mathbf{b}_2 = 2$. The header then satisfies $\mathbf{b}_0 + \mathbf{b}_1 + \mathbf{b}_2 = 0 + 1 + 2 = 0$ modulo 3, as required. Note that a sequence of errors will not be detectable if it causes the sum of integers corresponding to the transmitted $\mathbf{b}_0$, $\mathbf{b}_1$, $\mathbf{b}_2$ to change by a multiple of 3. For example, if we transmit $(0, 1, 2)$ as above but receive $(1, 3, 2)$, then the receiver will calculate $1 + 3 + 2 = 0$ modulo 3, so the errors will not be detected.

The actual Internet algorithm for calculating the checksum is described in terms of 1s complement arithmetic. In this arithmetic, addition of integers corresponds to modulo $2^{16} - 1$ addition, and the negative of the integer corresponding to the 16-bit word $\mathbf{b}$ is found by taking its 1s complement; that is, every 0 is converted to a 1 and vice versa. This process leads to the peculiar situation where there are two representations for $\mathbf{b}$, $(0, 0, \ldots, 0)$ and $(1, 1, \ldots, 1)$, which in turn results in additional redundancy in the context of error detection. Given these properties of 1s complement arithmetic, step 1 above then corresponds to simply adding the 16-bit integers $\mathbf{b}_0 + \mathbf{b}_1 + \mathbf{b}_2 + \ldots + \mathbf{b}_{L-1}$ using regular 32-bit addition. The modulo $2^{16} - 1$ reduction is done by taking the 16 higher-order bits in the sum, shifting them down by 16 positions, and adding them back to the sum. Step 2 produces the negative of the resulting sum by taking the 1s complement. Figure 3.54 shows a C function for calculating the Internet checksum adapted from [RFC 1071].

Table 3.7 shows the 16 possible codewords that result when the algorithm is applied to two-bit words with $L = 2$. From the table we can determine the error-detecting capability of this code. The minimum distance of the code is 2; that is, at least two bits are required to change one codeword into another. For example the first two codewords are 000011 (003) and 000110 (012), which differ in the fourth and sixth bit. In general, error patterns that cause the words $\mathbf{b}_0 + \mathbf{b}_1 + \mathbf{b}_2 + \ldots + \mathbf{b}_{L-1} + \mathbf{b}_L$ to change by a total of a multiple of $2^{16} - 1$ will not be detected.

```
unsigned short cksum(unsigned short *addr, int count)
  {
    /* Compute Internet checksum for "count" bytes
     * beginning at location "addr".
     */
  register long sum = 0;
  while ( count > 1 ) {
    /* This is the inner loop*/
        sum += *addr++;
        count -=2;
    }

    /* Add left-over byte, if any */
  if ( count > 0 )
    sum += *addr;

    /* Fold 32-bit sum to 16 bits */
  while (sum >>16)
    sum = (sum & 0xffff) + (sum >> 16);

  return ~sum;
}
```

FIGURE 3.54 C language program for computing Internet checksum

b$_0$	b$_1$	b$_2$
0	0	3
0	1	2
0	2	1
0	3	0
1	0	2
1	1	1
1	2	0
1	3	2
2	0	1
2	1	0
2	2	2
2	3	1
3	0	0
3	1	2
3	2	1
3	3	0

TABLE 3.7 Resulting codewords when $L = 2$

3.8.4 Polynomial Codes

We now introduce the class of **polynomial codes** that are used extensively in error detection and correction. Polynomial codes are readily implemented using shift-register circuits and therefore are the most widely implemented error-control codes. Polynomial codes involve generating check bits in the form of a **cyclic redundancy check (CRC)**. For this reason they are also known as CRC codes.

In polynomial codes the information symbols, the codewords, and the error vectors are represented by polynomials with binary coefficients. The k information bits $(i_{k-1}, i_{k-2}, \ldots, i_1, i_0)$ are used to form the **information polynomial** of degree $k - 1$:

$$i(x) = i_{k-1}x^{k-1} + i_{k-2}x^{k-2} + \ldots + i_1x + i_0$$

The encoding process takes $i(x)$ and produces a codeword polynomial $b(x)$ that contains the information bits and additional check bits and that satisfies a certain *pattern*. To detect errors, the receiver checks to see whether the pattern is satisfied. Before we explain this process, we need to review polynomial arithmetic.

The polynomial code uses polynomial arithmetic to calculate the codeword corresponding to the information polynomial. Figure 3.55 gives examples of polynomial addition, multiplication, and division using binary coefficients. Note that with binary arithmetic, we have $x^j + x^j = (1 + 1)x^j = 0$. Note in particular that when the division is completed the remainder $r(x)$ will have a degree smaller than the degree of the divisor polynomial. In the example the divisor polynomial has degree 3, so the division process continues until the remainder term has degree 2 or less.

A polynomial code is specified by its **generator polynomial $g(x)$**. Here we assume that we are dealing with a code in which codewords have n bits, of which k are information bits and $n - k$ are check bits. We refer to this type of

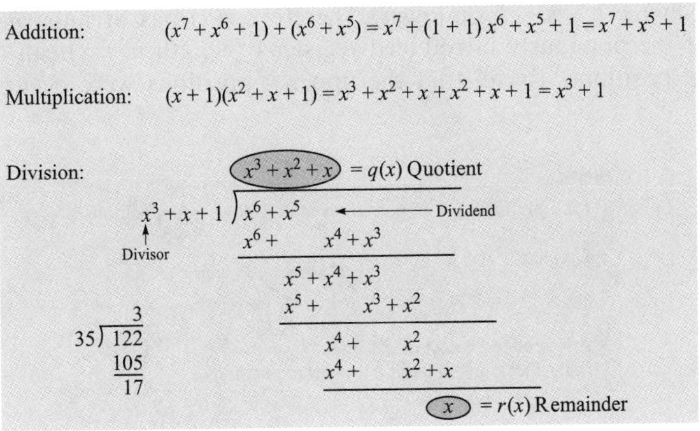

Addition: $(x^7 + x^6 + 1) + (x^6 + x^5) = x^7 + (1 + 1)x^6 + x^5 + 1 = x^7 + x^5 + 1$

Multiplication: $(x + 1)(x^2 + x + 1) = x^3 + x^2 + x + x^2 + x + 1 = x^3 + 1$

Division:

$$x^3 + x^2 + x = q(x) \text{ Quotient}$$

$$
\begin{array}{r}
x^3 + x + 1 \enclose{longdiv}{x^6 + x^5} \\
x^6 + x^4 + x^3 \\
\hline
x^5 + x^4 + x^3 \\
x^5 + x^3 + x^2 \\
\hline
x^4 + x^2 \\
x^4 + x^2 + x \\
\hline
x = r(x) \text{ Remainder}
\end{array}
$$

$$
\begin{array}{r}
3 \\
35 \overline{)122} \\
105 \\
\hline
17
\end{array}
$$

FIGURE 3.55 Polynomial arithmetic

code as an (n, k) code. The generator polynomial for such a code has degree $n - k$ and has the form

$$g(x) = x^{n-k} + g_{n-k-1}x^{n-k-1} + \ldots + g_1 x + 1$$

where $g_{n-k-1}, g_{n-k-2}, \ldots, g_1$ are binary numbers. In the following we will introduce an example that corresponds to a $(7,4)$ code with generator polynomial $g(x) = x^3 + x + 1$.

The calculation of the cyclic redundancy check bits is described in Figure 3.56. First the information polynomial is multiplied by x^{n-k}.

$$x^{n-k}i(x) = i_{k-1}x^{n-1} + i_{k-2}x^{n-2} + \ldots + i_1 x^{n-k+1} + i_0 x^{n-k}$$

If you imagine that the k information bits are in the lower k positions in a register of length n, the multiplication moves the information bits to the k highest-order positions, since the highest term of $i(x)$ can have degree $n - 1$. This situation is shown in the example in Figure 3.57. The information polynomial is $i(x) = x^3 + x^2$, so the first step yields $x^3 i(x) = x^6 + x^5$. After three shifts to the left, the contents of the shift register are $(1,1,0,0,0,0,0)$.

Step 2 involves dividing $x^{n-k}i(x)$ by $g(x)$ to obtain the remainder $r(x)$. The terms involved in division are related by the following expression:

$$x^{n-k}i(x) = g(x)q(x) + r(x)$$

The remainder polynomial $r(x)$ provides the CRCs. In the example in Figure 3.57, we have $x^6 + x^5 = g(x)(x^3 + x^2 + x) + x$; that is, $r(x) = x$. In the figure we also show a more compact way of doing the division without explicitly writing the powers of x.

The final step in the encoding procedure obtains the binary codeword $b(x)$ by adding the remainder $r(x)$ from $x^{n-k}i(x)$:

$$b(x) = x^{n-k}i(x) + r(x).$$

Because the divisor $g(x)$ had degree $n - k$, the remainder $r(x)$ can have maximum degree $n - k - 1$ or lower. Therefore $r(x)$ has at most $n - k$ terms. In terms of the previously introduced register of length n, $r(x)$ can occupy the lower $n - k$ positions. Recall that the upper k positions were occupied by the

Steps:

1. Multiply $i(x)$ by x^{n-k} (puts zeros in $(n-k)$ low-order positions).

2. Divide $x^{n-k}i(x)$ by $g(x)$.

$$\underbrace{x^{n-k}i(x) = g(x)}_{} \; \overset{\text{Quotient}}{q(x)} = \overset{\text{Remainder}}{r(x)}$$

3. Add remainder $r(x)$ to $x^{n-k}i(x)$ (puts check bits in the $n-k$ low-order positions).

$$b(x) = x^{n-k}i(x) + r(x) \longleftarrow \text{Transmitted codeword}$$

FIGURE 3.56 Encoding procedure

Generator polynomial: $g(x) = x^3 + x + 1$
Information: $(1,1,0,0) \rightarrow i(x) = x^3 + x^2$
Encoding: $x^3 i(x) = x^6 + x^5$

$$
\begin{array}{r}
x^3 + x^2 + x \\
\hline
x^3 + x + 1 \,)\, x^6 + x^5 \\
x^6 + x^4 + x^3 \\
\hline
x^5 + x^4 + x^3 \\
x^5 + x^3 + x^2 \\
\hline
x^4 + x^2 \\
x^4 + x^2 + x \\
\hline
x
\end{array}
\qquad
\begin{array}{r}
1110 \\
\hline
1011 \,)\, 1100000 \\
1011 \\
\hline
1110 \\
1011 \\
\hline
1010 \\
1011 \\
\hline
010
\end{array}
$$

Transmitted codeword:
$$b(x) = x^6 + x^5 + x$$
$$\rightarrow \underline{b} = (1,1,0,0,0,1,0)$$

FIGURE 3.57 Example of CRC encoding

information bits. We thus see that this encoding process introduces a binary polynomial in which the k higher-order terms are the information bits and in which the $n - k$ lower-order terms are the cyclic redundancy check bits. In the example in Figure 3.57, the division of $x^3 i(x)$ by $g(x)$ gives the remainder polynomial $r(x) = x$. The codeword polynomial is then $x^6 + x^5 + x$, which corresponds to the binary codeword $(1,1,0,0,0,1,0)$. Note how the first four positions contain the original four information bits and how the lower three positions contains the CRC bits.

In Figure 3.55 we showed that in normal division dividing 122 by 35 yields a quotient of 3 and a remainder of 17. This result implies that $122 = 3(35) + 17$. Note that by subtracting the remainder 17 from both sides, we obtain $122 - 17 = 3(35)$ so that $122 - 17$ is evenly divisible by 35. Similarly, the codeword polynomial $b(x)$ is divisible by $g(x)$ because

$$b(x) = x^{n-k} i(x) + r(x) = g(x)q(x) + r(x) + r(x) = g(x)q(x)$$

where we have used the fact that in modulo 2 arithmetic $r(x) + r(x) = 0$. This equation implies that *all codewords are multiples of the generator polynomial* $g(x)$. This is the pattern that must be checked by the receiver. The receiver can check to see whether the pattern is satisfied by dividing the received polynomial by $g(x)$. If the remainder is nonzero, then an error has been detected.

The familiar algorithm that is taught in elementary school for carrying out "longhand" division can be used to derive a feedback shift-register circuit that implements division. The feedback taps in this circuit are determined by the coefficients of the generator polynomial. Figure 3.58 shows the division circuit for the generated polynomial $g(x) = x^3 + x + 1$. The figure also shows the states of the registers as the algorithm implements the same division that was carried out in the previous encoding example.

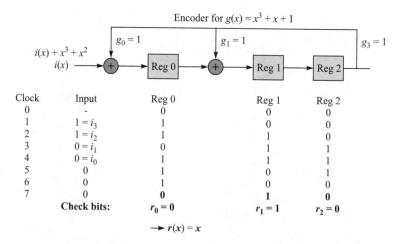

FIGURE 3.58 Shift-register circuit for generated polynomial

The same division circuit that was used by the encoder can be used by the receiver to determine whether the received polynomial is a valid codeword polynomial.

3.8.5 Standardized Polynomial Codes

Table 3.8 gives generator polynomials that have been endorsed in a number of standards. The CRC-12 and CRC-16 polynomials were introduced as part of the IBM bisync protocol for controlling errors in a communication line. The CCITT-16 polynomial is used in the HDLC standard and in XMODEM. The CCITT-32 is used in IEEE 802 LAN standards and in Department of Defense protocols, as well as in the CCITT V.42 modem standard. Finally CRC-8 and CRC-10 have recently been recommended for use in ATM networks. In the problem section we explore properties and implementations of these generator polynomials.

Name	Polynomial	Used in
CRC-8	$x^8 + x^2 + x + 1$	ATM header error check
CRC-10	$x^{10} + x^9 + x^5 + x^4 + x + 1$	ATM AAL CRC
CRC-12	$x^{12} + x^{11} + x^3 + x^2 + x + 1$	Bisync
	$= (x+1)(x^{11} + x^2 + 1)$	
CRC-16	$x^{16} + x^{15} + x^2 + x + 1$	Bisync
	$= (x+1)(x^{15} + x + 1)$	
CCITT-16	$x^{16} + x^{12} + x^5 + 1$	HDLC, XMODEM, V.41
CCITT-32	$x^{32} + x^{26} + x^{23} + x^{22} + x^{16} + x^{12} + x^{11} + x^{10}$	IEEE 802, DoD, V.42, AAL5
	$+ x^8 + x^7 + x^5 + x^4 + x^2 + x + 1$	

TABLE 3.8 Standard generator polynomials

3.8.6   Error-Detecting Capability of a Polynomial Code

We now determine the set of channel errors that a polynomial code cannot detect. In Figure 3.59 we show an additive error model for the polynomial codes. The channel can be viewed as adding, in modulo 2 arithmetic, an error polynomial, which has 1s where errors occur, to the input codeword to produce the received polynomial $R(x)$:

$$R(x) = b(x) + e(x).$$

At the receiver, $R(x)$ is divided by $g(x)$ to obtain the remainder that is defined as the **syndrome polynomial** $s(x)$. If $s(x) = 0$, then $R(x)$ is a valid codeword and is delivered to the user. If $s(x) \neq 0$, then an alarm is set, alerting the user to the detected error. Because

$$R(x) = b(x) + e(x) = g(x)q(x) + e(x)$$

we see that if an error polynomial $e(x)$ is divisible by $g(x)$, then the error pattern will be undetectable.

The design of a polynomial code for error detection involves first identifying the error polynomials we want to be able to detect and then synthesizing a generator polynomial $g(x)$ that will not divide the given error polynomials. Figures 3.60 and Figure 3.61 show the conditions required of $g(x)$ to detect various classes of error polynomials.

First consider single errors. The error polynomial is then of the form $e(x) = x^i$. Because $g(x)$ has at least two nonzero terms, it is easily shown that when multiplied by any quotient polynomial the product will also have at least two nonzero terms. Thus single errors cannot be expressed as a multiple of $g(x)$, and hence all single errors are detectable.

An error polynomial that has double errors will have the form $e(x) = x^i + x^j = x^i(1 + x^{j-i})$ where $j > i$. From the discussion for single errors, $g(x)$ cannot divide x^i. Thus $e(x)$ will be divisible by $g(x)$ only if $g(x)$ divides $(1 + x^{j-i})$. Since i can assume values from 0 to $n - 2$, we are interested in having $1 + x^m$ not be divisible by $g(x)$ for m assuming values from 1 to the maximum possible codeword length for which the polynomial will be used. The class of primitive polynomials has the property that if a polynomial has degree N, then the smallest value of m for which $1 + x^m$ is divisible by the polynomial is $2^N - 1$. [Lin 1983]. Thus if $g(x)$ is selected to be a primitive polynomial with degree $N = n - k$, then it will detect all double errors as long as the total codeword length does not exceed $2^{n-k} - 1$. Several of the generator polynomials used in practice are of the form $g(x) = (1 + x)p(x)$ where $p(x)$ is a primitive polynomial. For example, the CRC-16 polynomial is $g(x) = (1 + x)(x^{15} + x + 1)$ where $p(x) = x^{15} + x + 1$ is a

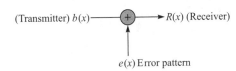

(Transmitter) $b(x)$ ——⊕——→ $R(x)$ (Receiver)

$e(x)$ Error pattern

FIGURE 3.59 Additive error model for polynomial codes

1. Single errors: $e(x) = x^i \quad 0 \leq i \leq n\text{-}1$
If $g(x)$ has more than one term, it cannot divide $e(x)$.

2. Double errors: $e(x) = x^i + x^j \quad 0 \leq i \leq j \leq n\text{-}1$
$= x^i(1 + x^{j-i})$
If $g(x)$ is primitive, it will not divide $(1 + x^{j-i})$ for $j\text{-}i \leq 2^{n-k} - 1$.

3. Odd number of errors: $e(1) = 1$ if number of errors is odd.
If $g(x)$ has $(x + 1)$ as a factor, then $g(1) = 0$ and all codewords have an even number of 1s.

FIGURE 3.60 Generator polynomials for detecting errors—part 1

primitive polynomial. Thus this $g(x)$ will detect all double errors as long as the codeword length does not exceed $2^{15} - 1 = 32{,}767$.

Now suppose that we are interested in being able to detect all odd numbers of errors. If we can ensure that all code polynomials have an even number of 1s, then we will achieve this error-detection capability. If we evaluate the codeword polynomial $b(x)$ at $x = 1$, we then obtain the sum of the binary coefficients of $b(x)$. If $b(1) = 0$ for all codeword polynomials, then $x + 1$ must be a factor of all $b(x)$ and hence $g(x)$ must contain $x + 1$ as a factor. For this reason $g(x)$ is usually chosen so that it has $x + 1$ as a factor.

Finally consider the detection of a burst of errors of length L. As shown in Figure 3.61, the error polynomial has the form $x^i d(x)$. If the error burst involves d consecutive bits, then the degree of $d(x)$ is $L - 1$. Reasoning as before, $e(x)$ will be a multiple of $g(x)$ only if $d(x)$ is divisible by $g(x)$. Now if the degree of $d(x)$ is less than that of $g(x)$, then it will not be possible to divide $d(x)$ by $g(x)$. We conclude that if $g(x)$ has degree $n - k$, then all bursts of length $n - k$ or less will be detected.

4. Error bursts of length b:

$e(x) = x^i\, d(x)$ where $\deg(d(x)) = L\text{-}1$
$g(x)$ has degree $n\text{-}k$;
$g(x)$ cannot divide $d(x)$ if $\deg(g(x)) > \deg(d(x))$

- $L = (n\text{-}k)$ or less: all will be detected
- $L = (n\text{-}k + 1)$: $\deg(d(x)) = \deg(g(x))$
  i.e. $d(x) = g(x)$ is the only undetectable error pattern,
  fraction of bursts that are undetectable $= 1/2^{L-2}$
- $L > (n\text{-}k + 1)$: fraction of bursts that are undetectable $= 1/2^{n-k}$

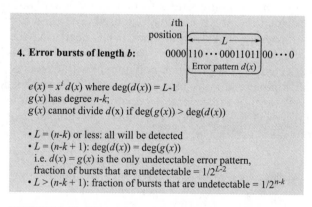

FIGURE 3.61 Generator polynomials for detecting errors—part 2

If the burst error has length $L = n - k + 1$, that is, degree of $d(x) =$ degree of $g(x)$, then $d(x)$ is divisible by $g(x)$ only if $d(x) = g(x)$. From Figure 3.61 $d(x)$ must have 1 in its lowest-order term and in its highest-order term, so it matches $g(x)$ in these two coefficients. For $d(x)$ to equal $g(x)$, it must also match $g(x)$ in the $n - k - 1$ coefficients that are between the lowest- and highest-order terms. Only one of the 2^{n-k-1} such patterns will match $g(x)$. Therefore, the proportion of bursts of length $L = n - k + 1$ that is undetectable is $1/2^{n-k-1}$. Finally, it can be shown that in the case of $L > n - k + 1$ the fraction of bursts that is undetectable is $1/2^{n-k}$.

◆3.8.7   Linear Codes[13]

We now introduce the class of linear codes that are used extensively for error detection and correction. A **binary linear code** is specified by two parameters: k and n. The linear code takes groups of k information bits, $b_1, b_2, \ldots, b_k$, and produces a binary codeword $\underline{b}$ that consists of n bits, $b_1, b_2, \ldots, b_n$. The $n - k$ check bits $b_{k+1}, \ldots, b_n$, are determined by $n - k$ linear equations:[14]

$$b_{k+1} = a_{11}b_1 + a_{12}b_2 + \ldots + a_{1k}b_k$$
$$b_{k+2} = a_{21}b_1 + a_{22}b_2 + \ldots + a_{2k}b_k$$
$$\vdots$$
$$b_n = a_{(n-k)1}b_1 + a_{(n-k)2}b_2 + \ldots + a_{(n-k)k}b_k$$

(5)

The coefficients in the preceding equations are binary numbers, and the addition is modulo 2. We say that b_{k-j} checks the information bit b_i if a_{ji} is 1. Therefore b_{k-j} is given by the modulo 2 sum of the information bits that it checks, and thus the redundancy in general linear codes is determined by parity check sums on subsets of the information bits. Note that when all of the information bits are 0, then all of the check bits will be 0. Thus the n-tuple $\underline{0}$ with all zeros is always one of the codewords of a linear code. Many linear codes can be defined by selecting different coefficients $[a_{ji}]$. Coding books contain catalogs of good codes that can be selected for various applications.

In linear codes the redundancy is provided by the $n - k$ check bits. Thus if the transmission channel has a bit rate of R bits/seconds, then k of every n transmitted bits are information bits, so the rate at which user information flows through the channel is $R_{\text{info}} = (k/n)R$ bits/second.

As an example consider the (7,4) linear **Hamming code** in which the first four bits of the codeword $\underline{b}$ consist of the four information bits b_1, b_2, b_3, and b_4 and the three check bits b_5, b_6, and b_7 are given by

---

[13]Section titles preceded by ◆ provide additional details and are not essential for subsequent sections.

[14]We require the set of linear equations to be linearly independent; that is, no equation can be written as a linear combination of the other equations.

$$b_5 = b_1 \quad\;\; + b_3 + b_4$$
$$b_6 = b_1 + b_2 \quad\;\; + b_4$$
$$b_7 = \quad\;\; + b_2 + b_3 + b_4$$

(6)

We have arranged the preceding equations so that it is clear which information bits are being checked by which check bits, that is, b_5 checks information bits b_1, b_3, and b_4. These equations allow us to determine the codeword for any block of information bits. For example, if the four information bits are $(0,1,1,0)$, then the codeword is given by $(0, 1, 1, 0, 0+1+0, 0+1+0, 0+1+0, 1+1+0) =$ $(0, 1, 1, 0, 1, 1, 0)$. Table 3.9 shows the set of 16 codewords that are assigned to the 16 possible information blocks.

Linear codes provide a very simple method for detecting errors. Before considering the general case, we illustrate the method using the Hamming code (Table 9). Suppose that in equation 6 we add b_5 to both sides of the first equation, b_6 to both sides of the second equation, and b_7 to both sides of the third equation. We then obtain

$$0 = b_5 + b_5 = b_1 \quad\;\; + b_3 + b_4 + b_5$$
$$0 = b_6 + b_6 = b_1 + b_2 \quad\;\; + b_4 \quad\;\; + b_6$$
$$0 = b_7 + b_7 = \quad\;\; + b_2 + b_3 + b_4 \quad\;\;\;\; + b_7$$

where we have used the fact that in modulo 2 arithmetic any number plus itself is always zero. The preceding equations state the conditions that must be satisfied by every codeword. Thus if $\underline{r} = (r_1, r_2, r_3, r_4, r_5, r_6, r_7)$ is the output of the transmission channel, then $\underline{r}$ is a codeword only if its components satisfy these equa-

Information				Codeword							Weight
b_1	b_2	b_3	b_4	b_1	b_2	b_3	b_4	b_5	b_6	b_7	$w(\underline{b})$
0	0	0	0	0	0	0	0	0	0	0	0
0	0	0	1	0	0	0	1	1	1	1	4
0	0	1	0	0	0	1	0	1	0	1	3
0	0	1	1	0	0	1	1	0	1	0	3
0	1	0	0	0	1	0	0	0	1	1	3
0	1	0	1	0	1	0	1	1	0	0	3
0	1	1	0	0	1	1	0	1	1	0	4
0	1	1	1	0	1	1	1	0	0	1	4
1	0	0	0	1	0	0	0	1	1	0	3
1	0	0	1	1	0	0	1	0	0	1	3
1	0	1	0	1	0	1	0	0	1	1	4
1	0	1	1	1	0	1	1	1	0	0	4
1	1	0	0	1	1	0	0	1	0	1	4
1	1	0	1	1	1	0	1	0	1	0	4
1	1	1	0	1	1	1	0	0	0	0	3
1	1	1	1	1	1	1	1	1	1	1	7

TABLE 3.9 Hamming (7,4) code

tions. If we write the equations in matrix form, we obtain a more compact representation for the conditions that all codewords must meet.

$$
\begin{bmatrix} 0 \\ 0 \\ 0 \end{bmatrix} = \begin{bmatrix} 1011100 \\ 1101010 \\ 0111001 \end{bmatrix} \begin{bmatrix} b_1 \\ b_2 \\ b_3 \\ b_4 \\ b_5 \\ b_6 \\ b_7 \end{bmatrix} = \mathbf{H}\underline{b}^t = \underline{0}
$$

In the matrix $\mathbf{H}$ is defined as the 3×7 matrix and $\underline{b}^t$ is the codeword arranged as a column vector. (Above we had previously defined $\underline{b}$ as row vector.)

By following the same procedure we have used in this example, we find that *the condition that all codewords must meet* is given by the matrix equation:

$$
\mathbf{H}\underline{b}^t = \underline{0} \tag{7}
$$

where $\underline{0}$ is a column vector with $n - k$ components all equal to zero, $\underline{b}^t$ is the codeword arranged as a column vector, and **check matrix H** has dimension $n - k$ rows by n columns and is given by

$$
\mathbf{H} = \begin{bmatrix} a_{11} & a_{12} & \cdots & a_{1k} & 10\ldots0 \\ a_{21} & a_{22} & \cdots & a_{2k} & 01\ldots0 \\ & & \vdots & & \ddots \\ a_{(n-k)1} & a_{(n-k)2} & \cdots & a_{(n-k)k} & 00\ldots1 \end{bmatrix}
$$

Errors introduced in a binary channel can be modeled by the additive process shown in Figure 3.62a. The output of the binary channel can be viewed as the modulo 2 addition of an error bit e to the binary input b. If the error bit equals 0, then the output of the channel will correspond to the input of the channel and no transmission error occurs; if the error bit equals 1, then the output will differ from the input and a transmission error occurs. Now consider the effect of the binary channel on a transmitted codeword. As shown in Figure 3.62b, the

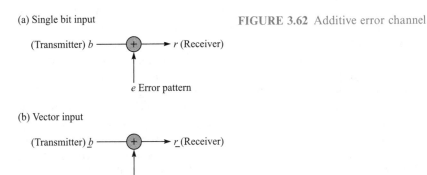

(a) Single bit input

(Transmitter) b ——→ $+$ ——→ r (Receiver)

e Error pattern

FIGURE 3.62 Additive error channel

(b) Vector input

(Transmitter) $\underline{b}$ ——→ $+$ ——→ $\underline{r}$ (Receiver)

$\underline{e}$ Error pattern

channel can be viewed as having a vector input that consists of the n bits that correspond to the codeword. The output of the channel $\underline{r}$ is given by the modulo 2 sum of the codeword $\underline{b}$ and an error vector $\underline{e}$ that has 1s in the components where an error occurs and 0s elsewhere:

$$\underline{r} = \underline{b} + \underline{e}$$

The error-detection system that uses a linear code checks the output of a binary channel $\underline{r}$ to see whether $\underline{r}$ is a valid codeword. The system does this by checking to see whether $\underline{r}$ satisfies equation 7. The result of this calculation is an $(n - k) \times 1$ column vector called the **syndrome**:

$$\underline{s} = H\underline{r} \tag{8}$$

If $\underline{s} = \underline{0}$, then $\underline{r}$ is a valid codeword; therefore, the system assumes that no errors have occurred and delivers $\underline{r}$ to the user. If $\underline{s} \neq \underline{0}$, then $\underline{r}$ is not a valid codeword and the error-detection system sets an alarm indicating that errors have occurred in transmission. In an ARQ system a retransmission is requested in response to the alarm. In an FEC system the alarm would initiate a processing based on the syndrome that would attempt to identify which bits were in error and then proceed to correct them.

The error-detection system fails when $\underline{s} = 0$ but the output of the channel is not equal to the input of the channel; that is, $\underline{e}$ is nonzero. In terms of equation 8 we have

$$\underline{0} = \underline{s} = H\underline{r} = H(\underline{b} + \underline{e}) = H\underline{b} + H\underline{e} = \underline{0} + H\underline{e} = H\underline{e} \tag{9}$$

where the fourth equality results from the linearity property of matrix multiplication and the fifth equality uses equation 7. The equality $H\underline{e} = \underline{0}$ implies that when $\underline{s} = \underline{0}$ the error pattern $\underline{e}$ satisfies equation 7 and hence must be a codeword. This implies that error detection using linear codes fails when the error vector is a codeword that transforms the input codeword $\underline{b}$ into a different codeword $\underline{r} = \underline{b} + \underline{e}$. Thus the set of all undetectable error vectors is the set of all nonzero codewords, and the probability of detection failure is the probability that the error vector equals any of the nonzero codewords.

In Figure 3.63 we show an example of the syndrome calculation using the (7,4) Hamming code for error vectors that contain single, double, and triple errors. We see from the example that if a single error occurred in the jth position, then the syndrome will be equal to the jth column of the **H** matrix. Since all the columns of **H** are nonzero, it follows that the syndrome will always be nonzero when the error vector contains a single error. Thus all single errors are detectable. The second example shows that if the error vector contains an error in location i and an error in location j, then the syndrome is equal to the sum of the ith and jth columns of **H**. We note that for the Hamming (7,4) code all columns are distinct. Thus the syndrome will be nonzero, and all error vectors with two errors are detectable. The third example shows an error vector that contains three errors. The syndrome for this particular error vector is zero. Thus we find that this Hamming code can detect all single and double errors but fails to detect some triple errors.

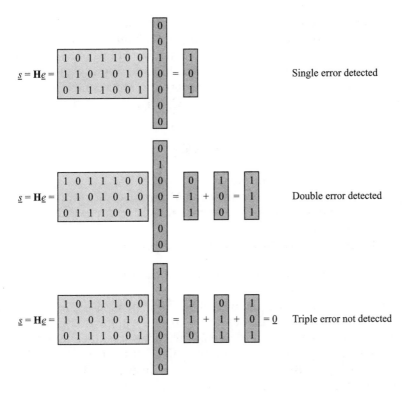

FIGURE 3.63 Syndrome calculation

A general class of Hamming codes can be defined with **H** matrices that satisfy the properties identified in the preceding example. Note that in the Hamming (7,4) code each of the $2^3 - 1$ possible nonzero binary triplets appears once and only once as a column of the H matrix. This condition enables the code to detect all single and double errors. Let m be an integer greater than or equal to 2. We can then construct an **H** matrix that has as its columns the $2^m - 1$ possible nonzero binary m-tuples. This **H** matrix corresponds to a linear code with codewords of length $n = 2^m - 1$ and with $n - k = m$ check bits. The codes that have this **H** matrix are called Hamming codes, and they are all capable of detecting all error vectors that have single and double errors. In the examples, we have been using the $m = 3$ Hamming code.

PERFORMANCE OF LINEAR CODES

In Figure 3.51 we showed qualitatively that we can minimize the probability of error-detection failure by spacing codewords apart in the sense that it is unlikely for errors to convert one codeword into another. In this section we show that the error-detection performance of a code is determined by the distances between codewords.

The **Hamming distance** $d(\underline{b}_1, \underline{b}_2)$ between the binary vectors $\underline{b}_1$ and $\underline{b}_2$ is defined as the number of components in which they differ. Thus the Hamming distance between two vectors increases as the number of bits in which they differ increases. Consider the modulo 2 sum of two binary n-tuples $\underline{b}_1 + \underline{b}_2$. The components of this sum will equal one when the corresponding components in $\underline{b}_1$ and $\underline{b}_2$ differ, and they will be zero otherwise. Clearly, this result is equal to the number of 1s in $\underline{b}_1 + \underline{b}_2$, so

$$d(\underline{b}_1, \underline{b}_2) = w(\underline{b}_1 + \underline{b}_2) \tag{10}$$

where w is the weight function introduced earlier. The extent to which error vectors with few errors are more likely than error vectors with many errors suggests that we should design linear codes that have codewords that are far apart in the sense of Hamming distance.

Define the **minimum distance** d_{min} of a code as follows:

$d_{min} =$ *distance between two closest distinct codewords*

For any given linear code, the pair of closest codewords is the most vulnerable to transmission error, so d_{min} can be used as a worst-case type of measure. From equation 9 we have that if $\underline{b}_1$ and $\underline{b}_2$ are codewords, then $\underline{b}_1 + \underline{b}_2$ is also a codeword. To find d_{min}, we need to find the pair of distinct codewords $\underline{b}_1$ and $\underline{b}_2$ that minimize $d(\underline{b}_1, \underline{b}_2)$. By equation 10, this is equivalent to finding the nonzero codeword with the smallest weight. Thus

$d_{min} =$ *weight of the nonzero codeword with the smallest number of* 1s

From Table 3.9 in section 3.8.7, we see the Hamming (7,4) code has $d_{min} = 3$.

If we start changing the bits in a codeword one at a time until another codeword is obtained, then we will need to change at least d_{min} bits before we obtain another codeword. This situation implies that all error vectors with $d_{min} - 1$ or fewer errors are detectable. We say that **a code is *t*-error detecting** if $d_{min} \geq t + 1$.

Finally, let us consider the probability of error-detection failure for a general linear code. In the case of the random error vector channel model, all 2^n possible error patterns are equally probable. A linear (n, k) code fails to detect only the $2^k - 1$ error vectors that correspond to nonzero codewords. We can state then that *the probability of error-detection failure for the random error vector channel model is $(2^k - 1)/2^n \approx 1/2^{n-k}$*. Furthermore, we can decrease the probability of detection failure by increasing the number of parity bits $n - k$.

Consider now the random bit error channel model. The probability of detection failure is given by

$$P[\text{detection failure}] = P[\underline{e} \text{ is a nonzero codeword}]$$

$$= \sum_{\text{nonzero codewords } \underline{b}} (1-p)^{n-w(\underline{b})} p^{w(\underline{b})}$$

$$= \sum_{w=d_{\min}}^{d_{\max}} N_w (1-p)^{n-w} p^w$$

$$\approx N_{d_{\min}} p^{d_{\min}} \qquad \text{for } p \ll 1$$

The second summation adds the probability of all nonzero codewords. The third summation combines all codewords of the same weight, so N_w is the total number of codewords that have weight w. The approximation results from the fact that the summation is dominated by the leading term when p is very small.

Consider the (7,4) Hamming code as an example once again. For the random error vector model, the probability of error-detection failure is $1/2^3 = 1/8$. On the other hand, for the random bit error channel the probability of error-detection failure is approximately $7p^3$, since $d_{min} = 3$ and seven codewords have this weight. If $p = 10^{-4}$, then the probability of error detection failure is 7×10^{-12}. Compared to the single parity check code, the Hamming code yields a tremendous improvement in error-detection capability.

◆3.8.8 Error Correction

In FEC the detection of transmission errors is followed by processing to determine the most likely error locations. Assume that an error has been detected so $\underline{s} \neq \underline{0}$. Equation 11 describes how an FEC system attempts to carry out the correction.

$$\underline{s} = H\underline{r} = H(\underline{b} + \underline{e}) = H\underline{b} + H\underline{e} = \underline{0} + H\underline{e} = H\underline{e}. \qquad (11)$$

The receiver uses equation 11 to find the value of the syndrome to diagnose the most likely error pattern. If $\mathbf{H}$ were an invertible matrix, then we could readily find the error vector from $\underline{e} = \mathbf{H}^{-1}\underline{s}$. Unfortunately, $\mathbf{H}$ is not invertible. Equation 11 consists of $n - k$ equations in n unknowns, $e_1, e_2, \ldots, e_n$. Because we have fewer equations than unknowns, the system is underdetermined and equation 11 has more than one solution. In fact, it can be shown that 2^k binary n-tuples satisfy equation 11. Thus for any given nonzero $\underline{s}$, equation 11 allows us to identify the 2^k possible error vectors that could have produced $\underline{s}$. The error-correction system cannot proceed unless it has information about the probabilities with which different error patterns can occur. The error-correction system uses such information to identify the most likely error pattern from the set of possible error patterns.

We provide a simple example to show how error correction is carried out. Suppose we are using the Hamming (7,4) code. Assume that the received vector is $\underline{r} = (0,0,1,0,0,0,1)$. The syndrome calculation gives $\underline{s} = (1,0,0)^k$. Because the fifth column of $\mathbf{H}$ is (1,0,0), one of the error vectors that gives this syndrome is

(0,0,0,0,1,0,0). Note from equation 11 that if we add a codeword to this error vector, we obtain another vector that gives the syndrome $(1,0,0)^t$. The $2^k = 16$ possible error vectors are obtained by adding the 16 codewords to (0,0,0,0,1,0,0) and are listed in Table 3.10. The error-correction system must now select the error vector in this set that is most likely to have been introduced by the channel. Almost all error-correction systems simply select the error vector with the smallest number of 1s. Note that this error vector also corresponds to the most likely error vector for the random bit error channel model. For this example the error correction system selects $\underline{e} = (0,0,0,0,1,0,0)$ and then outputs the codeword $\underline{r} + \underline{e} = (0,0,1,0,1,0,1)$ from which the user extracts the information bits, 0010. Algorithms have been developed that allow the calculation of the most likely error vector from the syndrome. Alternatively, the calculations can be carried out once, and then a table can be set up that contains the error vector that is to be used for correction for each possible syndrome. The error-correction system then carries out a table lookup each time a nonzero syndrome is found.

The error-correction system is forced to select only one error vector out of the 2^k possible error vectors that could have produced the given syndrome. Thus the error-correction system will successfully recover the transmitted codeword only if the error vector is the most likely error vector in the set. When the error vector is one of the other 2^k possible error vectors, the error-correction system will perform corrections in the wrong locations and actually introduce more errors! In the preceding example, assuming a random bit error channel model, the probability of the most likely error vector is $p(1 - p)^6 \approx p$ for the error vector of weight 1; the probability of the other error vectors is approximately $3p^2(1 - p)^5$ for the three error vectors of weight 2 and for where we have neglected the remainder of the error patterns. Thus when the error-correction system detects

Error vectors							Weight
e_1	e_2	e_3	e_4	e_5	e_6	e_7	$w(\underline{e})$
0	0	0	0	1	0	0	1
0	0	0	1	0	1	1	3
0	0	1	0	0	0	1	2
0	0	1	1	1	1	0	4
0	1	0	0	1	1	1	4
0	1	0	1	0	0	0	2
0	1	1	0	0	1	0	3
0	1	1	1	1	0	1	5
1	0	0	0	0	1	0	2
1	0	0	1	1	0	1	4
1	0	1	0	1	1	1	5
1	0	1	1	0	0	0	3
1	1	0	0	0	0	1	3
1	1	0	1	1	1	0	5
1	1	1	0	1	0	0	4
1	1	1	1	0	1	1	6

TABLE 3.10 Error vectors corresponding to syndrome $(1, 0, 0)^t$

an error the system's attempt to correct the error fails with probability

$$\frac{3p^2(1-p)^5}{p(1-p)^6 + 3p^2(1-p)^5} \approx 3p$$

Figure 3.64 summarizes the four outcomes that can arise from the error-correction process. We begin with the error vector that is revealed through its syndrome. If the $\underline{s} = \underline{0}$, then the received vector $\underline{r}$ is accepted as correct and delivered to the user. Two outcomes lead to $\underline{s} = \underline{0}$: the first corresponds to when no errors occur in transmission and has probability $(1-p)^7 \approx 1 - 7p$; the second corresponds to when the error vector is undetectable and has probability $7p^3$. If $\underline{s} \neq \underline{0}$, the system attempts to perform error correction. This situation occurs with probability $1 - P[\underline{s} = \underline{0}] = 1 - \{1 - 7p + 7p^3\} \approx 7p$. Two further outcomes are possible in the $\underline{s} \neq \underline{0}$ case: the third outcome is when the error vector is correctable and has conditional probability $(1 - 3p)$; the fourth is when the error vector is not correctable and has conditional probability $3p$. From the figure we see that the probability that the error-correction system fails to correct an error pattern is $21p^2$. To summarize, the first and third outcomes yield correct user information. The second and fourth outcomes result in the delivery of incorrect information to the user. Through this example we have demonstrated the analysis required to determine the effectiveness of any error-correction system.

The minimum distance of a code is useful in specifying its error-correcting capability. In Figure 3.65 we consider a code with $d_{min} = 5$, and we show two codewords that are separated by the minimum distance. If we start by changing the bits in $\underline{b}_1$, one bit at a time until we obtain $\underline{b}_2$, we find four n-tuples between the two codewords. We can imagine drawing a sphere of radius 2 around each codeword. The sphere around $\underline{b}_1$ will contain two of the n-tuples, and the sphere around $\underline{b}_2$ will contain the other n-tuples.

Note that because all pairs of codewords are separated by at least distance d_{min}, we can draw a sphere of radius 2 around every single codeword, and these spheres will all be nonoverlapping. This geometrical view gives us another way of looking at error correction. We can imagine that the error-correction system takes the vector $\underline{r}$ and looks up which sphere it belongs to; the system then

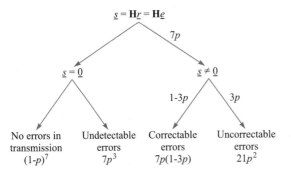

FIGURE 3.64 Summary of error-correction process outcomes

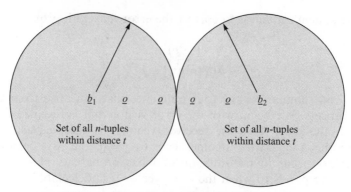

If $d_{\min} = 2t + 1$, nonoverlapping spheres of radius t can
be drawn around each codeword; $t = 2$ in the figure

FIGURE 3.65 Partitioning of n-tuples into disjoint spheres

generates the codeword that is at the center of the sphere. Note that if the error
vector introduced by the channel has two or fewer errors, then the error-correction system will always produce the correct codeword. Conversely, if the number
of errors is more than two, the error-correction system will produce an incorrect
codeword.

The discussion of Figure 3.65 can be generalized as follows. Given a linear
code with $d_{min} \geq 2t + 1$, it is possible to draw nonoverlapping spheres of radius t
around all the codewords. Hence the error-correction system is guaranteed to
operate correctly whenever the number of errors is smaller than t. For this reason
we say that **a code is t-error correcting** if $d_{min} \geq 2t + 1$.

The Hamming codes introduced above all have $d_{min} = 3$. Consequently, they
are all single-error correcting. The Hamming codes use $m = n - k$ bits of redundancy and are capable of correcting single errors. An interesting question is, If
we use $n - k = 2m$ bits in a code of length $n = 2^m - 1$, can we correct all double
errors? Similarly, if we use $n - k = 3m$, can we correct triple errors? The answer
is yes in some cases and leads to the classes of BCH and Reed-Solomon codes
[Lin 1983].

In this book we have presented only linear codes that operate on nonoverlapping blocks of information. These block codes include the classes of
Hamming codes, BCH codes, and Reed-Solomon codes that have been studied
extensively and are in wide use. These codes provide a range of choice in terms of
n, k, and d_{min} that allows a system designer to select a code for a given application. Convolutional codes are another important class of error-correcting codes.
These codes operate on overlapping blocks of information and are also in wide
use.

Finally, we consider the problem of error correction in channels that introduce bursts of errors. The codes discussed up to this point correct error vectors
that contain $(d_{min} - 1)/2$ or fewer errors. These codes can also be used in channels with burst errors if combined with **interleaving**. The user information is

L codewords
written vertically
in array; then
transmitted row
by row

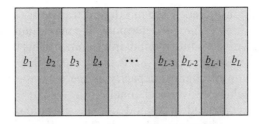

A long error
burst produces
errors in two
adjacent rows

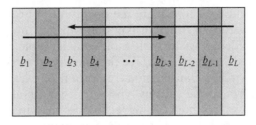

FIGURE 3.66 Interleaving

encoded using the given linear code, and the codewords are written as columns in an array as shown in Figure 3.66. The array is transmitted over the communication channels *row by row*. The interleaver depth L is selected so that the errors associated with a burst are distributed over many codewords. The error-correction system will be effective if the number of errors in each codeword is within its error-correcting capability. For example, if the linear code can correct up to two errors, then interleaving makes it possible to correct any burst of length less than $2L$.

SUMMARY

Binary information, "bits," are at the heart of modern communications. All information can be represented as blocks or streams of bits. Modern communication networks are designed to carry bits and therefore can handle *any* type of information.

We began this chapter with a discussion of the basic properties of common types of information such as text, image, voice, audio, and video. We discussed the amount of information that is required to represent them in terms of bits or bits/second. We also discussed requirements that applications impose on the network when they involve these types of information.

We described the difference between digital and analog communication and explained why digital communication has prevailed. We then considered the design of digital transmission systems. The characterization of communication channels in terms of their response to sinusoidal signals and to pulse signals was introduced. The notion of bandwidth of a channel was also introduced.

We first considered baseband digital transmission systems. We showed how the bandwidth of a channel determines the maximum rate at which pulses can be transmitted with zero intersymbol interference. This is the Nyquist signaling rate. We then showed the effect of SNR on the reliability of transmissions and developed the notion of channel capacity as the maximum reliable transmission rate that can be achieved over a channel.

Next we explained how modems use sinusoidal signals to transmit binary information over bandpass channels. The notion of a signal constellation was introduced and used to explain the operation of telephone modem standards.

The properties of different types of transmission media were discussed next. We first considered twisted-pair cable, coaxial cable, and optical fiber, which are used in "wired" transmission. We then discussed radio and infrared light, which are used in wireless transmission. Important physical layer standards were used as examples where the various types of media are used.

Finally, we presented coding techniques that are used in error control. Basic error-detection schemes that are used in many network standards were introduced first. An optional section then discussed error-correction schemes that are used when a return channel is not available.

CHECKLIST OF IMPORTANT TERMS

amplitude-response
amplitude shift keying (ASK)
asymmetric digital subscriber line
  (ADSL)
attenuation
automatic retransmission request
  (ARQ)
bandwidth
baseband transmission
◆binary linear code
bipolar encoding
bit error rate
bit rate
block-oriented information
burst error
cable modem
channel
channel capacity
check bit
◆check matrix
checksum
coaxial cable

codeword
compression ratio
cyclic redundancy check (CRC)
data compression
delay
differential encoding
digital information
digital transmission
equalizer
error control
forward error correction (FEC)
frequency shift keying (FSK)
generator polynomial
◆Hamming code
◆Hamming distance
impulse response
information polynomial
in-phase component
◆interleaving
jitter
line coding
loading coils

loss
lossless data compression
lossy data compression
Manchester encoding
◆minimum distance
modem
multilevel transmission
multimode fiber
nonreturn-to-zero (NRZ) encoding
NRZ inverted
Nyquist signaling rate
optical fiber
path loss exponent
phase shift
phase shift keying (PSK)
physical transmission medium
polynomial code
pulse code modulation (PCM)
quadrature amplitude modulation
   (QAM)

quadrature-phase component
random bit error channel model
random error vector model
redundancy
signal constellation
signal-to-noise ratio (SNR)
single parity check code
single-mode fiber
◆syndrome
syndrome polynomial
◆t-error detecting
transmission error
transmission medium
trellis modulation
twisted pair
unshielded twisted pair (UTP)
wavelength
wavelength division multiplexing
   (WDM)
weight

FURTHER READING

Ahamed, S. W. and V. B. Lawrence, *Design and Engineering of Intelligent Communication Systems*, Kluwer Academic Publishers, Boston, 1997. Detailed information about properties of transmission media.

Bell Telephone Laboratories, *Transmission Systems for Communications*, 1971. Classic on transmission aspects of telephony.

Bellamy, J., *Digital Telephony*, John Wiley & Sons, Inc., New York, 1991. Excellent coverage of digital telephony.

Dutton, H. J. R. and P. Lenhard, *High-Speed Networking Technology*, Prentice Hall PTR, Upper Saddle River, New Jersey, 1995. Good coverage of physical layer aspects of computer networks.

Gigabit Ethernet Alliance, "Accelerating the Standard for Speed," White paper, http://www.gigabit-ethernet.org, 1998.

Glover, I. A. and P. M. Grant, *Digital Communications*, Prentice-Hall, Englewood Cliffs, New Jersey, 1998. Up-to-date introduction to digital transmission systems.

Leon-Garcia, A., *Probability and Random Processes for Electrical Engineering*, Addison-Wesley, Reading, Massachusetts, 1994.

Lin, S. and D. J. Costello, *Error Control Coding: Fundamentals and Applications*, Prentice Hall, Englewood Cliffs, NJ, 1983.

Mukherjee, B., *Optical Communication Networks*, McGraw-Hill, New York, 1997.

Seifert, R., *Gigabit Ethernet*, Addison-Wesley, Reading, Massachusetts, 1998.

Smith, D. R., *Digital Transmission Systems*, Van Nostrand Reinhold Company, New York, 1985.

RFC 1071, R. Braden and D. Dorman, "Computing the Internet Checksum," September 1988.

PROBLEMS

1. Suppose the size of an uncompressed text file is 1 megabyte.
   a. How long does it take to download the file over a 32 kilobit/second modem?
   b. How long does it take to download the file over a 1 megabit/second modem?
   c. Suppose data compression is applied to the text file. How much do the transmission times in parts (a) and (b) change?

2. A scanner has a resolution of 600 × 600 pixels/square inch. How many bits are produced by an 8-inch-×-10-inch image if scanning uses 8 bits/pixel? 24 bits/pixel?

3. Suppose a computer monitor has a screen resolution of 1200 × 800 pixels. How many bits are required if each pixel uses 256 colors? 65,536 colors?

4. Explain the difference between facsimile, GIF, and JPEG coding. Give an example of an image that is appropriate to each of these three methods.

5. A digital transmission system has a bit rate of 45 Megabits/second. How many PCM voice calls can be carried by the system?

6. Suppose a storage device has a capacity of 1 gigabyte. How many 1-minute songs can the device hold using conventional CD format? using MP3 coding?

7. How many high-quality audio channels can be transmitted using an HDTV channel?

8. How many HDTV channels can be transmitted simultaneously over the optical fiber transmission systems in Table 3.3?

9. Comment on the properties of the sequence of frame images and the associated bit rates in the following examples:
   a. A children's cartoon program.
   b. A music video.
   c. A tennis game; a basketball game.
   d. A documentary on famous paintings.

10. Suppose that at a given time of the day, in a city with a population of 1 million, 1% of the people are on the phone.
    a. What is the total bit rate generated by all these people if each voice call is encoded using PCM?
    b. What is the total bit rate if all of the telephones are replaced by H.261 videoconferencing terminals?

11. Consider an analog repeater system in which the signal has power σ_x^2 and each stage adds noise with power σ_n^2. For simplicity assume that each repeater recovers the original signal without distortion but that the noise accumulates. Find the SNR after n repeater links. Write the expression in decibels: SNR dB $= 10 \log_{10}$SNR.

12. Suppose that a link between two telephone offices has 50 repeaters. Suppose that the probability that a repeater fails during a year is 0.01 and that repeaters fail independently of each other.
    a. What is the probability that the link does not fail at all during one year?
    b. Repeat (a) with 10 repeaters; with 1 repeater.

13. Suppose that a signal has twice the power as a noise signal that is added to it. Find the SNR in decibels. Repeat if the signal has 10 times the noise power? 2^n times the noise power? 10^k times the noise power?

14. A square periodic signal is represented as the following sum of sinusoids:

$$g(t) = \frac{2}{\pi} \sum_{k=0}^{\infty} \frac{(-1)^k}{2k+1} \cos(2k+1)\pi t$$

    a. Suppose that the signal is applied to an ideal low-pass filter with bandwidth 15 Hz. Plot the output from the low-pass filter and compare to the original signal. Repeat for 5 Hz; for 3 Hz. What happens as W increases?
    b. Suppose that the signal is applied to a bandpass filter that passes frequencies from 5 to 9 Hz. Plot the output from the filter and compare to the original signal.

15. Suppose that the 8 kbps periodic signal in Figure 3.15 is transmitted over a system that has an attenuation function equal to one for all frequencies and a phase function that is equal to $-90°$ for all frequencies. Plot the signal that comes out of this system. Does it differ in shape from the input signal?

16. A 10 kHz baseband channel is used by a digital transmission system. Ideal pulses are sent at the Nyquist rate, and the pulses can take 16 levels. What is the bit rate of the system?

17. Suppose a baseband transmission system is constrained to a maximum signal level of ± 1 volt and that the additive noise that appears in the receiver is uniformly distributed between $[-1/16, 1/16]$. How many levels of pulses can this transmission system use before the noise starts introducing errors?

18. What is the maximum reliable bit rate possible over a telephone channel with the following parameters:
    a. $W = 2.4$ kHz   SNR $= 20$ dB
    b. $W = 2.4$ kHz   SNR $= 40$ dB
    c. $W = 3.0$ kHz   SNR $= 20$ dB
    d. $W = 3.0$ kHz   SNR $= 40$ dB

19. Suppose we wish to transmit at a rate of 64 kbps over a 3 kHz telephone channel. What is the minimum SNR required to accomplish this?

20. Suppose that a low-pass communications system has a 1 MHz bandwidth. What bit rate is attainable using 8-level pulses? What is the Shannon capacity of this channel if the SNR is 20 dB? 40 dB?

21. Most digital transmission systems are "self-clocking" in that they derive the bit synchronization from the signal itself. To do this, the systems use the transitions between positive and negative voltage levels. These transitions help define the boundaries of the bit intervals.
    a. The nonreturn-to-zero (NRZ) signaling method transmits a 0 with a +1 voltage of duration T, and a 1 with a −1 voltage of duration T. Plot the signal for the sequence n consecutive 1s followed by n consecutive 0s. Explain why this code has a synchronization problem.
    b. In differential coding the sequence of 0s and 1s induces changes in the polarity of the signal; a binary 0 results in no change in polarity, and a binary 1 results in a change in polarity. Repeat part (a). Does this scheme have a synchronization problem?
    c. The Manchester signaling method transmits a 0 as a +1 voltage for $T/2$ seconds followed by a −1 for $T/2$ seconds; a 1 is transmitted as a −1 voltage for $T/2$ seconds followed by a +1 for $T/2$ seconds. Repeat part (a) and explain how the synchronization problem has been addressed. What is the cost in bandwidth in going from NRZ to Manchester coding?

22. Consider a baseband transmission channel with a bandwidth of 10 MHz. Which bit rates can be supported by the bipolar line code and by the Manchester line code?

23. The impulse response in a T-1 copper-wire transmission system has the idealized form where the initial pulse is of amplitude 1 and duration 1 and the afterpulse is of amplitude −0.1 and of duration 10.
    a. Let $\delta(t)$ be the narrow input pulse in Figure 3.18a. Suppose we use the following signaling method: Every second, the transmitter accepts an information bit; if the information bit is 0, then $-\delta(t)$ is transmitted, and if the information bit is 1, then $\delta(t)$ is transmitted. Plot the output of the channel for the sequence 1111000. Explain why the system is said to have "dc" or baseline wander.
    b. The T-1 transmission system uses bipolar signaling in the following fashion: If the information bit is a 0, then the input to the system is $0^*\delta(t)$; if the information bit is a 1, then the input is $\delta(t)$ for an even occurrence of a 1 and $-\delta(t)$ for an odd occurrence of a 1. Plot the output of the channel for the sequence 1111000. Explain how this signaling solves the "dc" or baseline wander problem.

24. The raised cosine transfer function, shown in Figure 3.21, has a corresponding impulse response given by

$$p(t) = \frac{\sin(\pi t/T)}{\pi t/T} \frac{\cos(\pi \alpha t/T)}{1 - (2\alpha t/t)^2}$$

    a. Plot the response of the information sequence 1010 for $\alpha = \frac{1}{2}$; $\alpha = \frac{1}{8}$.
    b. Compare this plot to the response, using the pulse in Figure 3.17.

25. Suppose a CATV system uses coaxial cable to carry 100 channels, each of 6 MHz bandwidth. Suppose that QAM modulation is used.

   a. What is the bit rate/channel if a four-point constellation is used? eight-point constellation?

   b. Suppose a digital TV signal requires 4 Mbps. How many digital TV signals can each channel handle for the two cases in part (a)?

26. Explain how ASK was used in radio telegraphy. Compare the use of ASK to transmit Morse code with the use of ASK to transmit text using binary information.

27. Suppose that a modem can transmit eight distinct tones at distinct frequencies. Every T seconds the modem transmits an arbitrary combination of tones (that is, some are present, and some are not present).

   a. What bit rate can be transmitted using this modem?

   b. Is there a relationship between T and the frequency of the signals?

28. A phase modulation system transmits the modulated signal $A\cos(2\pi f_c t + \phi)$ where the phase ϕ is determined by the two information bits that are accepted every T-second interval:

      for 00 $\phi = 0$; for 01 $\phi = \pi/2$; for 10 $\phi = \pi$; for 11 $\phi = 3\pi/2$.

   a. Plot the signal constellation for this modulation scheme.

   b. Explain how an eight-point phase modulation scheme would operate.

29. Suppose that the receiver in a QAM system is not perfectly synchronized to the carrier of the received signal; that is, the receiver multiplies the received signal by $2\cos(2\pi f_c t + \phi)$ and by $2\sin(2\pi/f_c t + \phi)$ where ϕ is a small phase error. What is the output of the demodulator?

30. In differential phase modulation the binary information determines the *change* in the phase of the carrier signal $\cos(2\pi f_c t)$. For example, if the information bits are 00, the phase change is 0; if 01, it is $\pi/2$; for 10, it is π; and for 11, it is $3\pi/2$.

   a. Plot the modulated waveform that results from the binary sequence 01100011. Compare it to the waveform that would be produced by ordinary phase modulation as described in problem 28.

   b. Explain how differential phase modulation can be demodulated.

31. A new broadcast service is to transmit digital music using the FM radio band. Stereo audio signals are to be transmitted using a digital modem over the FM band. The specifications for the system are the following: Each audio signal is sampled at a rate of 40 kilosamples/second and quantized using 16 bits; the FM band provides a transmission bandwidth of 200 kiloHertz.

   a. What is the total bit rate produced by each stereo audio signal?

   b. How many points are required in the signal constellation of the digital modem to accommodate the stereo audio signal?

32. A twisted-wire pair has an attenuation of 0.7 dB/kilometer at 1 kHz.

   a. How long can a link be if an attenuation of 20 dB can be tolerated?

   b. A twisted pair with loading coils has an attenuation of 0.2 dB/kilometer at 1 kHz. How long can the link be if an attenuation of 20 dB can be tolerated?

33. Use Figure 3.37 and Figure 3.40 to explain why the bandwidth of twisted-wire pairs and coaxial cable decreases with distance.

34. Calculate the bandwidth of the range of light covering the range from 1200 nm to 1400 nm. Repeat for 1400 nm to 1600 nm. Keep in mind that the speed of light in fiber is approximately 2×10^8 m/sec.

35. Compare the attenuation in a 100 km link for optical fibers operating at 850 nm, 1300 nm, and 1550 nm.

36. A satellite is stationed approximately 36,000 km above the equator. What is the attenuation due to distance for the microwave radio signal?

37. Suppose a transmission channel operates at 3 Mbps and has a bit error rate of 10^{-3}. Bit errors occur at random and independent of each other. Suppose that the following code is used. To transmit a 1, the codeword 111 is sent; to transmit a 0, the codeword 000 is sent. The receiver takes the three received bits and decides which bit was sent by taking the majority vote of the three bits. Find the probability that the receiver makes a decoding error.

38. An early code used in radio transmission involved codewords that consist of binary bits and contain the same number of 1s. Thus the two-out-of-five code only transmits blocks of five bits in which two bits are 1 and the others 0.
    a. List the valid codewords.
    b. Suppose that the code is used to transmit blocks of binary bits. How many bits can be transmitted per codeword?
    c. What pattern does the receiver check to detect errors?
    d. What is the minimum number of bit errors that cause a detection failure?

39. Find the probability of error-detection failure for the code in problem 38 for the following channels:
    a. The random error vector channel.
    b. The random bit error channel.

40. Suppose that two check bits are added to a group of $2n$ information bits. The first check bit is the parity check of the first n bits, and the second check bit is the parity check of the second n bits.
    a. Characterize the error patterns that can be detected by this code.
    b. Find the error-detection failure probability in terms of the error-detection probability of the single parity check code.
    c. Does it help to add a third parity check bit that is the sum of all the information bits?

41. Let $g(x) = x^3 + x + 1$. Consider the information sequence 1001.
    a. Find the codeword corresponding to the preceding information sequence.
    b. Suppose that the codeword has a transmission error in the first bit. What does the receiver obtain when it does its error checking?

42. ATM uses an eight-bit CRC on the information contained in the header. The header has six fields:

First 4 bits: GFC field
Next 8 bits: VPI field
Next 16 bits: VCI field
Next 3 bits: Type field
Next 1 bit: CLP field
Next 8 bits: CRC

a. The CRC is calculated using the following generator polynomial: $x^8 + x^2 + x + 1$. Find the CRC bits if the GFC, VPI, Type, and CLP fields are all zero and the VCI field is 00000000 00001111. Assume the GFC bits correspond to the highest-order bits in the polynomial.
b. Can this code detect single errors? Explain why.
c. Draw the shift register division circuit for this generator polynomial.

43. Suppose a header consists of four 16-bit words: (11111111 11111111, 11111111 00000000, 11110000 11110000, 11000000 11000000). Find the Internet checksum for this code.

44. Let $g_1(x) = x + 1$ and let $g_2(x) = x^3 + x^2 + 1$. Consider the information bits (1,1,0,1,1,0).
a. Find the codeword corresponding to these information bits if $g_1(x)$ is used as the generating polynomial.
b. Find the codeword corresponding to these information bits if $g_2(x)$ is used as the generating polynomial.
c. Can $g_2(x)$ detect single errors? double errors? triple errors? If not, give an example of an error pattern that cannot be detected.
d. Find the codeword corresponding to these information bits if $g(x) = g_1(x)g_2(x)$ is used as the generating polynomial. Comment on the error-detecting capabilities of $g(x)$.

45. Take any binary polynomial of degree 7 that has an even number of nonzero coefficients. Show by longhand division that the polynomial is divisible by $x + 1$.

46. A repetition code is an $(n, 1)$ code in which the $n - 1$ parity bits are repetitions of the information bit. Is the repetition code a linear code? What is the minimum distance of the code?

47. A transmitter takes K groups of k information bits and appends a single parity bit to each group. The transmitter then appends a block parity check word in which the jth bit in the check word is the modulo 2 sum of the jth components in the K codewords.
a. Explain why this code is a $((K + 1)(k + 1), Kk)$ linear code.
b. Write the codeword as a $(k + 1)$ row by $(K + 1)$ column array in which the first K columns are the codewords and the last column is the block parity check. Use this array to show how the code can detect all single, double, and triple errors. Give an example of a quadruple error that cannot be detected.
c. Find the minimum distance of the code. Can it correct all single errors? If so, show how the decoding can be done.
d. Find the probability of error-detection failure for the random bit error channel.

48. Consider the $m = 4$ Hamming code.
a. What is n, and what is k for this code?
b. Find the parity check matrix for this code.
c. Give the set of linear equations for computing the check bits in terms of the information bits.

d. Write a program to find the set of all codewords. Do you notice anything peculiar about the weights of the codewords?

49. Show that an easy way to find the minimum distance is to find the minimum number of columns of **H** whose sum gives the zero vector.

50. Suppose we take the (7,4) Hamming code and obtain an (8,4) code by adding an overall parity check bit.
   a. Find the **H** matrix for this code.
   b. What is the minimum distance?
   c. Does the extra check bit increase the error-correction capability? the error-detection capability?

51. A (7,3) linear code has check bits given by
$$b_4 = b_1 + b_2$$
$$b_3 = b_1 \qquad + b_3$$
$$b_6 = \qquad b_2 + b_3$$
$$b_7 = b_1 + b_2 + b_3$$
   a. Find the **H** matrix.
   b. Find the minimum distance.
   c. Find the set of all codewords. Do you notice anything peculiar about the set of codewords.

52. An error-detecting code takes k information bits and generates a codeword with $2k + 1$ encoded bits as follows:
   The first k bits consist of the information bits.
   The next k bits repeat the information bits.
   The next bit is the XOR of the first k bits.
   a. Find the check matrix for this code.
   b. What is the minimum distance of this code?
   c. Suppose the code is used on a channel that introduces independent random bit errors with probability 10^{-3}. Estimate the probability that the code fails to detect an erroneous transmission.

53. A (6,3) linear code has check bits given by
$$b_4 = b_1 + b_2$$
$$b_5 = b_1 \qquad + b_3$$
$$b_6 = \qquad b_2 + b_3$$
   a. Find the check matrix for this code.
   b. What is the minimum distance of this code?
   c. Find the set of all codewords.

54. (Appendix 3A). Consider an asynchronous transmission system that transfers a sequence of N bits between a start bit and a stop bit. What is the maximum value of N if the receiver clock frequency is within 1 percent of the transmitter clock frequency?

APPENDIX 3A

Asynchronous Data Transmission

The Recommended Standard (RS) 232, better known as the serial line interface, provides connections between the computer and devices such as modems and printers. RS-232 is an Electronic Industries Association (EIA) standard that specifies the interface between data terminal equipment (DTE) and data communications equipment (DCE) for the purpose of transferring serial data. Typically, DTE represents a computer or a terminal, and DCE represents a modem. CCITT recommended a similar standard called V.24.

RS-232 specifies the connectors, various electrical signals, and transmission procedures. The connectors have 9 or 25 pins, referred to as DB-9 or DB-25, respectively. The D-type connector contains two rows of pins. From the front view of a DB-25 connector, the pins at the top row are numbered from 1 to 13, and the pins at the bottom row are numbered from 14 to 25. Figure 3.67a shows a typical 25-pin connector.

The electrical specification defines the signals associated with connector pins. A voltage between $+3$ to $+25$ volts is interpreted to be a binary 0, and -3 to -25 volts a binary 1. Figure 3.67b shows the functional description of commonly used signals. DTR is used by the DTE to tell the DCE that the DTE is on. DSR is used by the DCE to tell the DTE that the DCE is also on. When the DCE detects a carrier indicating that the channel is good, the DCE asserts the CD pin. If there is an incoming call, the DCE notifies the DTE via the RI signal. The DTE asserts the RTS pin if the DTE wants to send data. The DCE asserts the CTS pin if the DCE is ready to receive data. Finally, data is transmitted in full-duplex mode, from DTE to DCE on the TXD line and from DCE to DTE on the RXD line.

In RS-232, data is transmitted asynchronously on the serial line in the sense that the receiver clock is not synchronized to the transmitter clock. For the receiver to sample the data bits correctly, the transmitter precedes the transmission of data with a start bit. When the receiver detects the leading edge of the

(a)

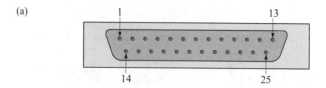

(b)

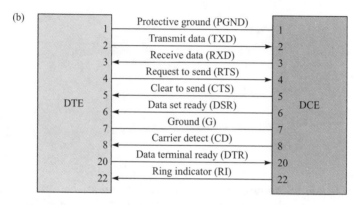

FIGURE 3.67 Commonly used pins in DB-25 connector

start bit, the receiver begins sampling the data bits after 1.5 periods of the receiver clock; thus sampling occurs near the middle of the subsequent data bits. Because the receiver clock is not synchronized to the transmitter clock, clock drift will eventually cause the receiver to sample the data bits incorrectly. This problem can be prevented by transmitting only a short sequence of data bits. Typically the sequence of bits consists of one character of seven or eight bits. A parity bit can be optionally added to enable the receiver to check the integrity of the data bits. Finally, a stop bit indicates the end of the bit sequence. Figure 3.68 illustrates the asynchronous transmission process.

Note that the receiver clock frequency should be approximately the same as the transmitter clock frequency in order to sample the data bits correctly.

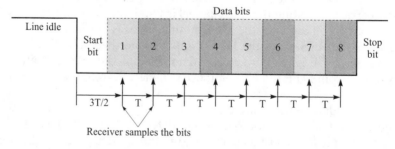

FIGURE 3.68 Framing and synchronization in asynchronous transmission

Suppose that the transmitter pulse duration is X and the receiver pulse duration is T. If the receiver clock is slower than the transmitter clock and the last sample must occur before the end of the stop bit, then we must have $9.5T < 10X$. If the receiver clock is faster than the transmitter clock and the last sample must occur after the beginning of the stop bit, then we must have $9.5T > 9X$. These two inequalities can be satisfied if $|(T - X)/X| < 5.3\%$. In other words, the receiver clock frequency must be within 5.3 percent of the transmitter clock frequency.

Fourier Series

Let $x(t)$ represent a periodic signal with period T. The Fourier series resolves this signal into an infinite sum of sine and cosine terms

$$x(t) = a_0 + 2\sum_{n=1}^{\infty}\left[a_n \cos\left(\frac{2\pi nt}{T}\right) + b_n \sin\left(\frac{2\pi nt}{T}\right)\right] \tag{A.1}$$

where the coefficients a_n and b_n represent the amplitude of the cosine and sine terms, respectively. The quantity n/T represents the nth harmonic of the fundamental frequency $f_0 = 1/T$.

The coefficient a_0 is given by the time average of the signal over one period

$$a_0 = \frac{1}{T}\int_{-T/2}^{T/2} x(t)dt \tag{A.2}$$

which is simply the time average of $x(t)$ over one period.

The coefficient a_n is obtained by multiplying both sides of equation (A.1) by the cosine function $\cos(2\pi nt/T)$ and integrating over the interval $-T/2$ to $T/2$. Using equation (A.1) and (A.2) we obtain

$$a_n = \frac{1}{T}\int_{-T/2}^{T/2} x(t)\cos\left(\frac{2\pi nt}{T}\right)dt, \qquad n = 1, 2, \ldots \tag{A.3}$$

The coefficient b_n of the sinusoid components is obtained in a similar manner:

$$b_n = \frac{1}{T}\int_{-T/2}^{T/2} x(t)\sin\left(\frac{2\pi nt}{T}\right)dt, \qquad n = 1, 2, \ldots \tag{A.4}$$

The following trigonometric identity

$$A \cos \mu + B \sin \mu = \sqrt{A^2 + B^2} \cos\left(\mu - \tan^{-1}\frac{B}{A}\right) \tag{A.5}$$

allows us to rewrite equation (A.1) as follows:

$$x(t) = a_0 + 2\sum_{n=1}^{\infty} \sqrt{a_n^2 + b_n^2} \cos\left(\frac{2\pi nt}{T} - \tan^{-1}\frac{b_n}{a_n}\right) = a_0 + 2\sum_{n=1}^{\infty} |c_n| \cos\left(\frac{2\pi nt}{T} + \theta_n\right) \tag{A.6}$$

A periodic function $x(t)$ is said to have a **discrete spectrum** with components at the frequencies, $0, f_0, 2f_0, \ldots$ The magnitude of the discrete spectrum at the frequency component nf_0 is given by

$$|c_n| = \sqrt{a_n^2 + b_n^2} \tag{A.7}$$

and the phase of the discrete spectrum at nf_0 is given by

$$\theta_n = -\tan^{-1}\frac{b_n}{a_n} \tag{A.8}$$

CHAPTER 4

Transmission Systems and the Telephone Network

In the preceding chapter we presented the basic techniques that are used in digital communication systems to transfer information from one point to another. We discussed how these techniques are used with various types of transmission media. In this chapter we consider how these individual communication systems are organized into the digital transmission systems that form the backbone of modern computer and telephone networks. We examine different approaches to multiplexing information flows, and we present the digital multiplexing hierarchy that defines the structure of modern transmission systems. In terms of the OSI reference model, these systems provide the physical layer that transfers bits.

The digital multiplexing hierarchy has developed around the telephone network, so it is natural to include the discussion of the telephone network in this chapter. We elaborate on how the telephone network provides the various network functions that were introduced in Chapter 1. We also provide an introduction to signaling in the telephone network, and we explain the basic principles of cellular telephone networks.

The chapter is organized into the following sections:

1. *Multiplexing*. We explain multiplexing techniques that are used for sharing transmission resources, in particular frequency-division multiplexing and time-division multiplexing. We introduce the digital multiplexing hierarchy.
2. *SONET*. We explain the SONET standard for optical transmission, and we discuss the application of SONET systems to provide flexible network configuration and fault tolerance.
3. *Wavelength-division multiplexing*. We discuss wavelength-division multiplexing (WDM), which can increase the transmission capacity of an optical fiber by a factor of 100 or more. We also discuss the impact of WDM on network design.

4. *Circuit switches.* We consider the design of circuit switches that can be used to set up end-to-end physical connections across a network.
5. *The telephone network.* We discuss the operation of the telephone network, and we examine how transmission and switching facilities are organized to provide end-to-end physical connections.
6. *Signaling.* We introduce the signaling system in the telephone network and explain the signaling system's layered architecture. We also explain how the signaling system is used to provide enhanced services.
7. *Traffic and overload controls.* We consider the management of traffic flows in the network and the various techniques for routing circuits in a network. We then discuss the overload control mechanisms that are required when various problem conditions arise in telephone networks.
8. *Cellular communications.* We explain the frequency-reuse concept that underlies cellular communications. We explain the operation of cellular telephone networks, and we give an overview of the various standards.
9. *Satellite cellular networks.* We show how the cellular concept is applied in constellation of satellites that are configured to provide global communications services.

4.1 MULTIPLEXING

Multiplexing involves the sharing of expensive network resources by several connections or information flows. The network resource that is of primary interest to us in this section is bandwidth, which is measured in Hertz for analog transmission systems and bits/second for digital transmission systems. In this section we consider multiplexing techniques that are used to share a set of transmission lines among a community of users. These techniques are primarily used in telephone networks and in broadcasting services.

In Figure 4.1a we show an example where three pairs of users communicate by using three separate sets of wires.[1] This arrangement, which completely dedicates network resources, that is, wires, to each pair of users, was typical in the very early days of telephony. However, this approach quickly becomes unwieldy and inefficient as the number of users increases. A better approach is to dynamically share a set of resources, that is, a set of transmission lines, among a community of users. In Figure 4.1b we show how a multiplexer allows this sharing to take place. When a customer on one end wishes to communicate with a customer at the other end, the multiplexer assigns a communication line for the duration of the call. When the call is completed, the transmission line is returned to the pool that is available to meet new connection requests.

---

[1] A telephone connection requires two sets of wires for communication in each direction. To keep the discussion simple, we deal with communication in one direction only.

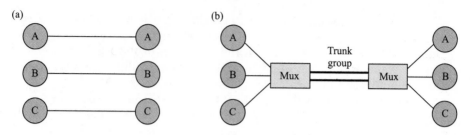

FIGURE 4.1 Multiplexing

Note that signaling between the two multiplexers is required to set up and terminate each call.

The transmission lines connecting the two multiplexers are called **trunks**. Initially each trunk consisted of a single transmission line; that is, the information signal for one connection was carried in a single transmission line. However, advances in transmission technology made it possible for a single transmission line of large bandwidth to carry multiple connections. From the point of view of setting up connections, such a line can be viewed as being equivalent to a number of trunks. In the remainder of this section, we discuss several approaches to combining the information from multiple connections into a single line.

4.1.1 Frequency-Division Multiplexing

Suppose that the transmission line has a bandwidth (measured in Hertz) that is much greater than that required by a single connection. For example, in Figure 4.2a each user has a signal of W Hz, and the channel that is available is greater than $3W$ Hz. In **frequency-division multiplexing (FDM)**, the bandwidth is divided

(a) Individual signals occupy W Hz

FIGURE 4.2 Frequency-division multiplexing

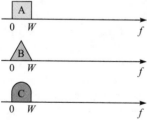

(b) Combined signal fits into channel bandwidth

into a number of frequency slots, each of which can accommodate the signal of an individual connection. The multiplexer assigns a frequency slot to each connection and uses modulation to place the signal of the connection in the appropriate slot. This process results in an overall combined signal that carries all the connections as shown in Figure 4.2b. The combined signal is transmitted, and the demultiplexer recovers the signals corresponding to each connection. Reducing the number of wires that need to be handled reduces the overall cost of the system.

FDM was introduced in the telephone network in the 1930s. The basic analog multiplexer combines 12 voice channels in one line. Each voice signal occupies 4 kHz of bandwidth. The multiplexer modulates each voice signal so that it occupies a 4 kHz slot in the band between 60 and 108 kHz. The combined signal is called a *group*. A hierarchy of analog multiplexers has been defined. For example, a *supergroup* (that carries 60 voice signals) is formed by multiplexing five groups, each of bandwidth 48 kHz, into the frequency band from 312 to 552 kHz. Note that for the purposes of multiplexing, each group is treated as an individual signal. Ten supergroups can then be multiplexed to form a *mastergroup* of 600 voice signals that occupies the band 564 to 3084 kHz. Various combinations of mastergroups have also been defined.

Familiar examples of FDM are broadcast radio and broadcast and cable television, where each station has an assigned frequency band. Stations in AM, FM, and television are assigned frequency bands of 10 kHz, 200 kHz, and 6 MHz, respectively. FDM is also used in cellular telephony where a pool of frequency slots, typically of 25 to 30 kHz each, are shared by the users within a geographic cell. Each user is assigned a frequency slot for each direction. Note that in FDM the user information can be in analog or digital form and that the information from all the users flows simultaneously.

4.1.2 Time-Division Multiplexing

In **time-division multiplexing (TDM)**, the transmission between the multiplexers is provided by a single high-speed digital transmission line. Each connection produces a digital information flow that is then inserted into the high-speed line. For example in Figure 4.3a each connection generates a signal that produces one unit of information every $3T$ seconds. This unit of information could be a bit, a byte, or a fixed-size block of bits. Typically, the transmission line is organized into frames that in turn are divided into equal-sized slots. For example, in Figure 4.3b the transmission line can send one unit of information every T seconds, and the combined signal has a frame structure that consists of three slots, one for each user. During connection setup each connection is assigned a slot that can accommodate the information produced by the connection.

TDM was introduced in the telephone network in the early 1960s. The T-1 carrier system that carries 24 digital telephone connections is shown in Figure 4.4. Recall that a digital telephone speech signal is obtained by sampling a speech waveform 8000 times/second and by representing each sample with eight bits.

(a) Each signal transmits 1 unit every $3T$ seconds

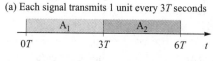

FIGURE 4.3 Time-division multiplexing

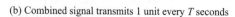

(b) Combined signal transmits 1 unit every T seconds

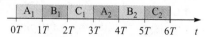

The T-1 system uses a transmission frame that consists of 24 slots of eight bits each. Each slot carries one PCM sample for a single connection. The beginning of each frame is indicated by a single bit that follows a certain perodic pattern. The resulting transmission line has a speed of

$$(1 + 24 \times 8) \text{ bits/frame} \times 8000 \text{ frames/second} = 1.544 \text{ Mbps}$$

Note how in TDM the slot size and the repetition rate determines the bit rate of the individual connections.

The T-1 carrier system was introduced in 1961 to carry the traffic between telephone central offices. The growth of telephone network traffic and the advances in digital transmission led to the development of a standard digital multiplexing hierarchy. The emergence of these digital hierarchies is analogous to the introduction of high-speed multilane expressways interconnecting major cities. These digital transmission hierarchies define the global flow of telephone traffic. Figure 4.5 shows the digital transmission hierarchies that were developed in North America and Europe. In North America and Japan, the **digital signal 1 (DS1)**, which corresponds to the output of a T-1 multiplexer, became the basic building block. The DS2 signal is obtained by combining 4 DS1 signals, and the

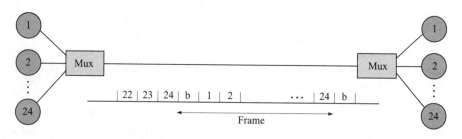

FIGURE 4.4 T-1 carrier system

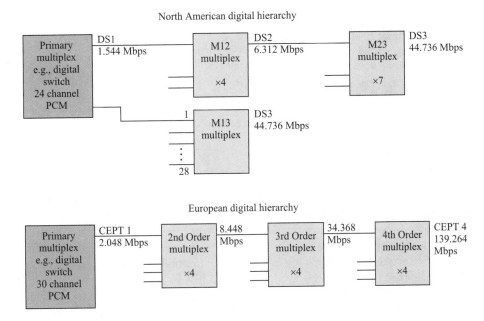

FIGURE 4.5 Basic digital hierarchies

DS3 is obtained by combining 28 DS1 signals. The DS3 signal, with a speed of 44.736 Mbps, has found extensive use in providing high-speed communications to large users such as corporations. In Europe the CCITT developed a similar digital hierarchy. The CEPT-1 (also referred to as E1) signal consisting of thirty-two 64-kilobit channels forms the basic building block.[2] Only 30 of the 32 channels are used for voice channels; one of the other channels is used for signaling, and the other channel is used for frame alignment and link maintenance. The second, third, and fourth levels of the hierarchy are obtained by grouping four of the signals in the lower level, as shown in Figure 4.5.

The operation of a time-division multiplexer involves tricky problems with the synchronization of the input streams. Figure 4.6 shows two streams, each with a nominal rate of one bit every T seconds, that are combined into a stream that sends two bits every T seconds. What happens if one of the streams is slightly slower than $1/T$ bps? Every T seconds, the multiplexer expects each input to provide a one-bit input; at some point the slow input will fail to produce its input bit. We will call this event a *bit slip*. Note that the "late" bit will be viewed as an "early" arrival in the next T-second interval. Thus the slow stream will alternate between being late, undergoing a bit slip, and then being early. Now consider what happens if one of the streams is slightly fast. Because bits are arriving faster than they can be sent out, bits will accumulate at the multiplexer and eventually be dropped.

---

[2] These standards were first developed by the Committee European de Post et Telegraph (CEPT).

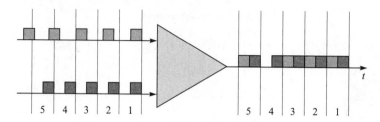

FIGURE 4.6 Relative timing of input and output streams in a TDM multiplexer

To deal with the preceding synchronization problems, time-division multiplexers have traditionally been designed to operate at a speed slightly *higher* than the combined speed of the inputs. The frame structure of the multiplexer output signal contains bits that are used to indicate to the receiving multiplexer that a slip has occurred. This approach enables the streams to be demultiplexed correctly. Note that the introduction of these extra bits to deal with slips implies that the frame structure of the output stream is not exactly synchronized to the frame structure of all the input streams. To extract an individual input stream from the combined signal, it is necessary to demultiplex the entire combined signal, make the adjustments for slips, and then remove the desired signal. This type of multiplexer is called "asynchronous" because the input frames are not synchronized to the output frame.

4.2   SONET

In 1966 Charles Kao reported the feasibility of optical fibers that could be used for communications. By 1977 a DS3 45 Mbps fiber optic system was demonstrated in Chicago, Illinois. By 1998, 40 Gbps fiber optic transmission systems had become available. The advances in optical transmission technology have occurred at a rapid rate, and the backbone of telephone networks has become dominated by fiber optic digital transmission systems. As an example Figure 4.7 shows the optical fiber network for a long-distance telephone carrier in 1998.

The first generation of equipment for optical fiber transmission was proprietary, and no standards were available for the interconnection of equipment from different vendors. The deregulation of telecommunications in the United States led to a situation in which the long-distance carriers were expected to provide the interconnection between local telephone service providers. To meet the urgent need for standards to interconnect optical transmission systems, the **Synchronous Optical Network (SONET)** standard was developed in North America. The CCITT later developed a corresponding set of standards called **Synchronous Digital Hierarchy (SDH)**. SONET and SDH form the basis for current high-speed backbone networks.

FIGURE 4.7 Optical fiber network for a long-distance telephone carrier in 1998

4.2.1 SONET Multiplexing

The SONET standard uses a 51.85 Mbps signal as a building block to extend the digital transmission hierarchy into the multigigabit range. SONET incorporates extensive capabilities for the operations, administration, and maintenance (OAM) functions that are required to operate digital transmission facilities. It also introduces a synchronous format that greatly simplifies the handling of the lower-level digital signals and that enables network topologies that are self-healing in the presence of faults.

Table 4.1 shows the SONET and SDH digital hierarchy. The **synchronous transport signal level-1 (STS-1)** is the basic building block of the SONET hierarchy. A higher-level signal in the hierarchy is obtained through the interleaving of bytes from the lower-level component signals. Each STS-n electrical signal has

SONET electrical signal	Optical signal	Bit rate (Mbps)	SDH electrical signal
STS-1	OC-1	51.84	
STS-3	OC-3	155.52	STM-1
STS-9	OC-9	466.56	STM-3
STS-12	OC-12	622.08	STM-4
STS-18	OC-18	933.12	STM-6
STS-24	OC-24	1244.16	STM-8
STS-36	OC-36	1866.24	STM-12
STS-48	OC-48	2488.32	STM-16
STS-192	OC-192	9953.28	STM-64

STS-synchronous transport signal; OC-optical channel; STM-synchronous transfer module.

TABLE 4.1 SONET digital hierarchy

a corresponding **optical carrier level-*n* (OC-*n*)** signal. The bit format of STS-*n* and OC-*n* signals is the same except for the use of scrambling in the optical signal.[3] The SDH standard refers to **synchronous transfer modules-*n* (STM-*n*)** signals and begins at a bit rate of 155.52 Mbps. The SDH STM-1 signal is equivalent to the SONET STS-3 signal. The STS-1 signal accommodates the DS3 signal from the existing digital transmission hierarchy in North America. The STM-1 signal accommodates the CEPT-4 signal in the CCITT digital hierarchy. The STS-48 signal is widely deployed in the backbone of modern communication networks.

SONET uses a frame structure that has the same 8 kHz repetition rate as traditional TDM systems. SONET was designed to be very flexible in the types of traffic that it can handle. SONET uses the term **tributary** to refer to the component streams that are multiplexed together. Figure 4.8 shows how a SONET multiplexer can handle a wide range of tributary types. A slow-speed mapping function allows DS1, DS2, and CEPT-1 signals to be combined into an STS-1 signal. As indicated above DS3 signal can be mapped into an STS-1 signal, and a CEPT-4 signal can be mapped into an STS-3 signal. A mapping has also been defined for mapping ATM streams into an STS-3 signal.[4] A SONET multiplexer can then combine STS input signals into a higher-order STS-*n* signal. Details of

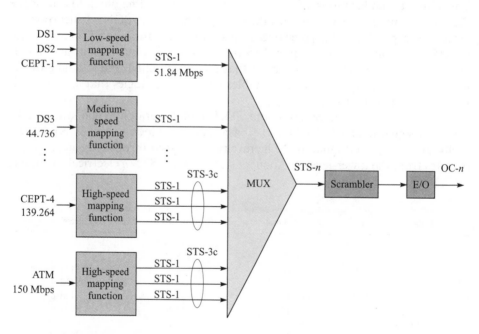

FIGURE 4.8 SONET multiplexing

---

[3]Scrambling maps long sequences of 1s or 0s into sequences that contain a more even balance of 1s and 0s to facilitate bit-timing recovery.

[4]ATM is introduced in Chapter 7 and discussed in detail in Chapter 9.

the SONET frame structure and the mappings into STS signal formats are provided in section 4.2.2.

Asynchronous multiplexing systems prior to SONET required the entire multiplexed stream to be demultiplexed to access a tributary, as shown in Figure 4.9a. Transit tributaries would then have to be remultiplexed onto the next hop. Thus every point of tributary removal or insertion required a demultiplexer-multiplexer pair. SONET produced significant reduction in cost by enabling **add-drop multiplexers (ADM)** to insert and extract tributary streams without disturbing tributary streams that are in transit as shown in Figure 4.9b. SONET accomplishes this process through the use of pointers that identify the location of a tributary within a frame. Pointers are explained in section 4.2.2.

ADMs in combination with SONET equipment allow distant switching nodes to be connected by tributaries. This arrangement allows the network operator to define networks of switching nodes with arbitrary topologies. As an example, Figure 4.10 shows three sites, a, b, and c, that are connected by three add-drop multiplexers. The ADMs are all connected in a unidirectional ring by an OC-3n optical transmission system that carries three STS-n signals. Figure 4.10 shows how, at node b, two STS-n tributaries are inserted destined for node c and for node a. The first tributary terminates at node c, and the second tributary flows across node c and terminates at node a. The ADM at each other site also removes two STS-n tributaries and inserts two STS-n tributaries, and it passes one STS-n tributary unchanged as shown in Figure 4.11a. The first inserted tributary is destined to the next node, and the other inserted tributary is destined to the remaining node. For example, the ADM at site c removes the tributaries indicated by the dotted and dashed lines that originated at nodes b and a, respectively. The ADM at site c also inserts tributaries destined from nodes a and b that are indicated by solid lines. The network in Figure 4.11a has a physical ring topology, but in fact, each pair of nodes is connected *directly* by an STS-n tributary, and so the three nodes are *logically* configured in a fully connected topology, as shown in Figure 4.11b. If switches at each of the three

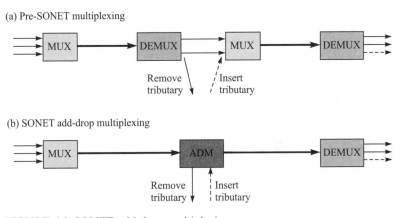

(a) Pre-SONET multiplexing

(b) SONET add-drop multiplexing

FIGURE 4.9 SONET add-drop multiplexing

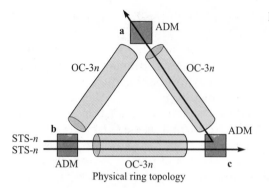

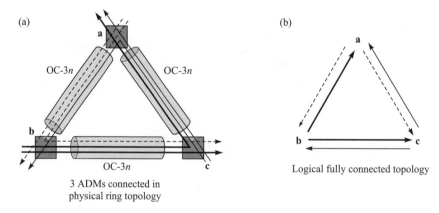

FIGURE 4.10 SONET ring network

Physical ring topology

sites are interconnected by these tributaries, then the switches would see a fully connected topology.

The preceding example shows that nodes that do not have direct physical connections can be provided with direct *logical* connections through the use of tributaries that are added at the source node and dropped at the destination node. This approach allows the configuration of arbitrary logical topologies with arbitrary link transmission rates. Furthermore, this configuration can be done using *software control*. Thus we see that the introduction of SONET equipment provides the network operator with tremendous flexibility in managing the transmission resources to meet the user requirements.

SONET systems can be deployed in the form of *self-healing* rings. Such rings provide two paths between any two nodes in the ring, thus providing for fault recovery in the case of single node or link failure. Figure 4.12a shows a two-fiber ring in which data is copied in both fibers, one traveling clockwise and the other counterclockwise. In a normal operation one fiber (clockwise) is in a working mode, while another (counterclockwise) is in a protect mode. When the fibers between two nodes are broken, the ring wraps around as shown in Figure 4.12b.

FIGURE 4.11 Configuration of logical networks using add-drop multiplexers

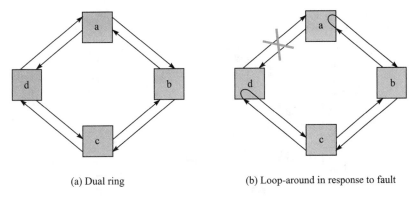

(a) Dual ring   (b) Loop-around in response to fault

FIGURE 4.12 Survivability in a SONET ring

Traffic continues to flow for all tributaries. A similar procedure is carried out in case of a node failure. In this case traffic is redirected by the two nodes adjacent to the affected node. Only traffic to the faulty node is discontinued. SONET ring networks typically recover from these types of faults in less than 50 milliseconds, depending on the length of the ring, which can span diameters of several thousand kilometers. The preceding discussion assumes a "unidirectional" ring. A SONET ring can also be bidirectional, in which case working traffic travels in both directions. Furthermore, a SONET ring can have either two fibers or four fibers per link.

The capability to manage bandwidth flexibly and to respond quickly to faults has altered the topology of long-distance and metropolitan area networks from a mesh of point-to-point links to interconnected ring networks. SONET ring networks can be deployed in a metropolitan area as shown in Figure 4.13. User traffic is collected by access networks and directed to access nodes such as a telephone office. A number of such nodes are interconnected in a first-tier ring network. Large users that cannot afford to lose service may be connected to an access node with dual paths as shown. A metropolitan area ring operating at a higher rate may in turn interconnect the first tier ring networks. To provide protection against faults, rings may be interconnected by using matched inter-ring gateways as shown between the interoffice ring and the metro ring and between the metro ring and the regional ring. The traffic flow between the rings is sent simultaneously along the primary and secondary gateway. Automated protection procedures determine whether the primary or secondary incoming traffic is directed into the ring. The metropolitan area ring, in turn, may connect to the ring of an interexchange or regional carrier as shown in the figure.

Several variations of SONET rings can be deployed to provide survivability. The merits of the approaches depend to some extent on the size of the ring and the pattern of traffic flows between nodes. In the problem section we explore some of these issues.

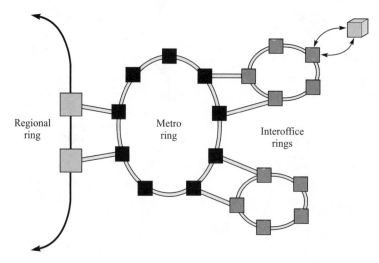

FIGURE 4.13 SONET ring structure in local, metropolitan, and regional networks

◆4.2.2   SONET Frame Structure

This section examines the SONET system and its frame structure. A SONET system is divided into three layers: sections, lines, and paths as shown in Figure 4.14a. A *section* refers to the span of fiber between two adjacent devices, such as two repeaters. The section layer deals with the transmission of an STS-*n* signal across the physical medium. A *line* refers to the span between two adjacent multiplexers and therefore in general encompasses several sections. Lines deal with the transport of an aggregate multiplexed stream of user information and the associated overhead. A *path* refers to the span between the two SONET terminals at the endpoints of the system and in general encompasses one or more lines.

In general the multiplexers associated with the path level, for example, STS-1, are lower in the hierarchy than the multiplexers in the line level, for example, STS-3 or STS-48, as shown in Figure 4.14a. The reason is that a typical information flow begins at some bit rate at the edge of the network, which is then combined into higher-level aggregate flows inside the network, and finally delivered back at the original lower bit rate at the outside edge of the network.

Figure 4.14b shows that every section has an associated optical layer. The section layer deals with the signals in their electrical form, and the optical layer deals with the transmission of optical pulses. It can be seen that every regenerator involves converting the optical signal to electrical form to carry out the regeneration function and then back to optical form. Note also in Figure 4.14b that all of the equipment implements the optical and section functions. Line functions are found in the multiplexers and end terminal equipment. The path function occurs only at the end terminal equipment.

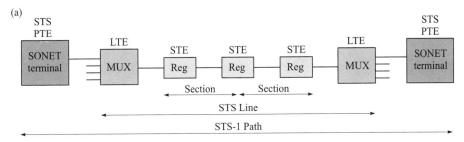

STE: Section terminating equipment, for example, a repeater
LTE: Line terminating equipment, for example, an STS-1 to STS-3 multiplexer
PTE: Path terminating equipment, for example, an STS-1 multiplexer

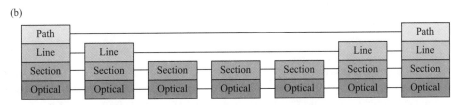

FIGURE 4.14 Section, line, and path layers of SONET

Figure 4.15 shows the structure of the SONET STS-1 frame that is defined *at the line level*. A frame consisting of a rectangular array of bytes arranged in 9 rows by 90 bytes is repeated 8000 times a second.[5] Thus each byte in the array corresponds to a bit rate of 64 kbps, and the overall bit rate of the STS-1 is

$$8 \times 9 \times 90 \times 8000 = 51.84 \text{ Mbps}$$

The first three columns of the array are allocated to *section* and *line overhead*. The section overhead is interpreted and modified at every section termination and is used to provide framing, error monitoring, and other section-related management functions. The line overhead is interpreted and modified at every line termination and is used to provide synchronization and multiplexing for the path layer, as well as protection-switching capability. We will see that the first three bytes of the line overhead play a crucial role in how multiplexing is carried out. The remaining 87 columns of the frame constitute the *information payload* that carries the path layer information. The bit rate of the information payload is

$$8 \times 9 \times 87 \times 8000 = 50.122 \text{ Mbps}$$

The information payload includes one column of *path overhead* information, but the column is not necessarily aligned to the frame for reasons that will soon become apparent.

---

[5]The bits are physically transmitted row by row and from left to right.

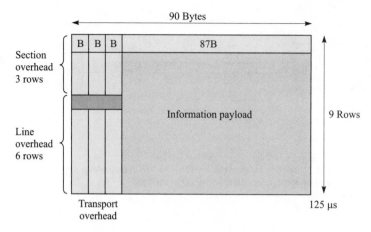

FIGURE 4.15 SONET STS-1 frame format

Consider next how the end-to-end user information is organized *at the path level*. The user data and the path overhead are included in the **synchronous payload envelope (SPE)**, which consists of a byte array of 87 columns by nine rows, as shown in Figure 4.16. The path overhead constitutes the first column of this array. This SPE is then inserted into the STS-1 frame. The SPE is not necessarily aligned to the information payload of an STS-1 frame. Instead, the first two bytes of the line overhead are used as a pointer that indicates the byte within the information payload where the SPE begins. Consequently, the SPE can be spread over two consecutive frames as shown in Figure 4.16.[6] The use of the pointer makes it possible to extract a tributary signal from the multiplexed signal. This feature gives SONET its add-drop capability.

The pointer structure shown in Figure 4.16 maintains synchronization of frames and SPEs in situations where their clock frequencies differ slightly. If the payload stream is faster than the frame rate, then a buffer is required to hold payload bits as the frame stream falls behind the payload stream. To allow the frame to catch up, an extra SPE byte is transmitted in a frame from time to time. This extra byte, which is carried within the line overhead, clears the backlog that has built up. Whenever this byte is inserted, the pointer is moved forward by one byte to indicate that the SPE starting point has been moved one byte forward. When the payload stream is slower than the frame stream, the number of SPE bytes transmitted in a frame needs to be reduced by one byte from time to time. This is done by stuffing an SPE byte with dummy information and adjusting the pointer to indicate that the SPE now starts one byte later.

---

[6]Imagine an STS-1 signal as a conveyor belt with 90-×-9-byte frames drawn on the belt. The SPEs are boxes of size 90 × 9 that are placed on the conveyor belt but are not necessarily aligned to the frame boundary. This situation occurs because boxes are transferred between conveyor belts "on the fly."

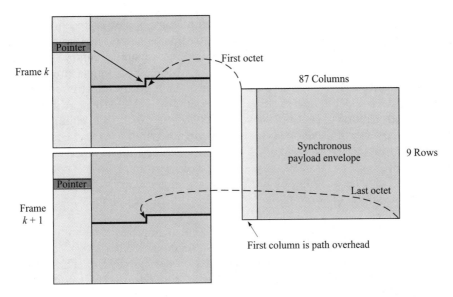

FIGURE 4.16 The synchronous payload envelope can span two consecutive frames

Now consider how n STS-1 signals are multiplexed into an STS-n signal. Each STS-1 signal is first synchronized to the local STS-1 clock of the multiplexer as follows. The section and line overhead of the incoming STS-1 signal are terminated, and its payload (SPE) is mapped into a *new* STS-1 frame that is synchronized to the local clock as shown in Figure 4.17. The pointer in the new STS-1 frame is adjusted as necessary, and the mapping is done on the fly. This procedure ensures that all the incoming STS-1 frames are mapped into STS-1 frames that are synchronized with respect to each other. The STS-n frame is produced by interleaving the bytes of the n synchronized STS-1 frames, in effect producing a frame that has nine rows, $3n$ section and line overhead columns, and $87n$ payload columns. To multiplex k STS-n signals into an STS-kn signal, the incoming signals are first de-interleaved into STS-1 signals and then the above procedure is applied.

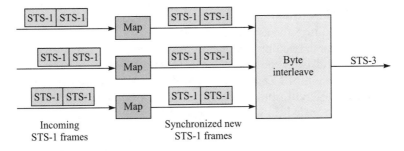

FIGURE 4.17 Synchronous multiplexing in SONET

Various mappings have also been defined to combine lower-speed tributaries of various formats into standard SONET streams as shown in Figure 4.8. For example, a SONET STS-1 signal can be divided into **virtual tributary** signals that accommodate lower-bit-rate streams. In each SPE, 84 columns are set aside and divided into seven groups of 12 columns. Each group constitutes a virtual tributary and has a bit rate of $12 \times 9 \times 8 \times 8000 = 6.912$ Mbps. Alternatively, each virtual tributary can be viewed as $12 \times 9 = 108$ voice channels. Thus mappings have been developed so that a virtual tributary can accommodate four T-1 carrier signals $(4 \times 24 = 96 < 108)$, or three CEPT-1 signals $(3 \times 32 = 96 < 108)$. The SPE can then handle any mix of T-1 and CEPT-1 signals that can be accommodated in its virtual tributaries. In particular the SPE can handle a maximum of $7 \times 4 = 28$ T-1 carrier signals or $3 \times 7 = 21$ CEPT-1 signals.

A mapping has also been developed so that a single SPE signal can handle one DS3 signal. Several STS-1 frames can be concatenated to accommodate signals with bit rates that cannot be handled by a single STS-1. The suffix c is appended to the signal designation when *concatenation* is used to accommodate a signal that has a bit rate higher than STS-1. Thus an STS-3c signal is used to accommodate a CEPT-4 139.264 Mbps signal. Concatenated STS frames carry only one column of path overhead. For example, the SPE in an STS-3 frame has $86 \times 3 = 258$ columns of user data, whereas the SPE in an STS-3c frame carries $87 \times 3 - 1 = 260$ columns of user data. A mapping has also been developed so that an STS-3c frame can carry streams of ATM cells.

4.3 WAVELENGTH-DIVISION MULTIPLEXING

Current optical fiber transmission systems can operate at bit rates in the tens of Gbps. The underlying available electronics technologies have a maximum speed limit in the tens of Gbps. Similarly, laser diodes can support bandwidths in the tens of GHz. In Figure 3.45 in Chapter 3, we can see that a range of low-attenuation wavelengths about 100 nm wide is available in the 1300 nm range. This range corresponds to a bandwidth of 18 terahertz (THz). Another band of about 100 nm in the 1550 nm range provides another 19 THz of bandwidth. Recall that 1 THz = 1000 GHz. Clearly the available technology does not come close to exploiting the available bandwidth.

The information carried by a single optical fiber can be increased through the use of **wavelength-division multiplexing (WDM)**. WDM can be viewed as an optical-domain version of FDM in which multiple information signals modulate optical signals at different optical wavelengths (colors). The resulting signals are combined and transmitted simultaneously over the same optical fiber as shown in Figure 4.18. Prisms and diffraction gratings can be used to combine and split color signals. For example, WDM systems are available that use 16 wavelengths at OC-48 to provide aggregate rates up to 16×2.5 Gbps = 40 Gbps. Figure

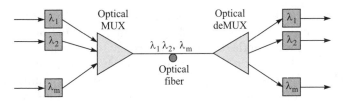

FIGURE 4.18 Wavelength-division multiplexing

4.19 shows the transmitted signal in one such system. WDM systems with 32 wavelengths at OC-192 are also available with a total bit rate of 320 Gbps. The attraction of WDM is that a huge increase in available bandwidth is obtained without the huge investment associated with deploying additional optical fiber. The additional bandwidth can be used to carry more traffic and can also provide the additional protection bandwidth required by self-healing topologies.

Early WDM systems differ in substantial ways from electronic FDM systems. In FDM the channels are separated by frequency guard bands that are small relative to the bandwidth of each channel slot. This configuration is possible because the devices for carrying out the required modulation, filtering, and demodulation are available. This narrow spacing is not the case for WDM systems. Consequently, the spacing between wavelengths in WDM systems tends to be large compared to the bandwidth of the information carried by each wavelength.

Optical add-drop multiplexers have been designed for WDM systems. The assignment of wavelengths in various multiplexer configurations can then be

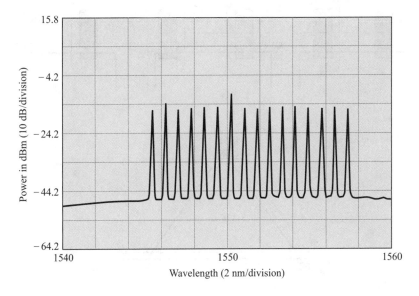

FIGURE 4.19 Optical signal in a WDM system

used to create networks with various logical topologies. In these topologies a *light path* between two nodes is created by inserting information at an assigned wavelength at the source node, bypassing intermediate nodes, and removing the information at the destination node. Figure 4.20a shows a chain of optical add-drop multiplexers in which a single fiber connects adjacent multiplexers. Each fiber contains a set of four wavelengths that are removed and inserted to provide a one-directional communication link from upstream to downstream nodes. Thus a has a link to each of b, c, and d; b has a link to each of c and d; and c has a link to d.

Figure 4.20b shows a WDM ring network in which three nodes are connected by three optical fibers that carry three wavelengths. Each node removes two wavelengths and inserts two wavelengths so that each pair of nodes is connected by an information stream flowing in one wavelength. In effect a fully connected logical network is produced. We again see that through the assignment of wavelengths, it is possible to obtain logical topologies that differ from the physical topology. This capability can be exploited to provide survivability with respect to faults and topology reconfigurability to meet changing network requirements.

The introduction of WDM and optical add-drop multiplexers into a network adds a layer of logical abstraction between the physical topology and the logical topology that is seen by the systems that send traffic flows through the network. The physical topology consists of the optical add-drop multiplexers intercon-

(a) WDM chain network

(b) WDM ring network

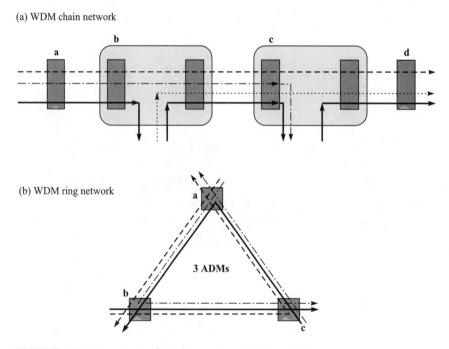

FIGURE 4.20 Network configurations using WDM multiplexers

nected with a number of optical fibers. The manner in which light paths are defined by the optical ADMs in the WDM system determines the topology that is seen by SONET ADMs that are interconnected by these light paths. The systems that input tributaries into the SONET network in turn may see a different topology that is defined by the SONET system. For example, in Figure 4.20b each node could correspond to a different metropolitan area. Each metropolitan area might have a network of interconnected SONET rings. The light paths between the areas provide a direct interconnection between these metropolitan networks.

In WDM each wavelength is modulated separately, so each wavelength need not carry information in the same transmission format. Thus some wavelengths might carry SONET formatted information streams, while others might carry Gigabit Ethernet formatted information or other transmission formats.

HISTORY REPEATS ITSELF . . . AGAIN

Optical transmission is still in the early stages of development relative to its potential, so it is interesting to examine its possible evolution given the history of networks in the 19th and 20th centuries. During this period networks went through a cycle from digital techniques in telegraphy to analog techniques in the early phase of telephony and back to digital techniques in modern networks. WDM technology is clearly in the analog phase of this development cycle, and necessarily so because of limitations in electronic and optical devices. By looking to the past, we can see clearly that optical time-division multiplexing must be on the horizon. With the development of optical implementation of simple logical operations, we can also expect some forms of optical packet switching and optical code division systems (introduced in Chapter 6). Looking further into the future, optical computing, should it become available, would affect networking as much as computer control changed signaling in telephone networks and inexpensive processing made the Internet possible. These insights are gained from looking to the past. Of course, then we have the radical ideas that (appear to) come out of nowhere to change the course of history. Stay tuned!

4.4 CIRCUIT SWITCHES

A network is frequently represented as a cloud that connects multiple users as shown in Figure 4.21a. A circuit-switched network is a generalization of a physical cable in the sense that it provides connectivity that allows information to flow between inputs and outputs to the network. Unlike a cable, however, a

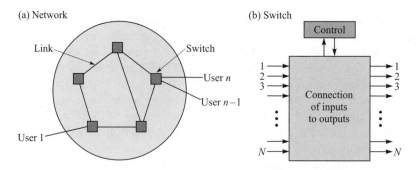

FIGURE 4.21 Network consists of links and switches

network is geographically distributed and consists of a graph of transmission lines (that is, links) interconnected by switches (nodes). As shown in Figure 4.21b, the function of a **circuit switch** is to transfer the signal that arrives at a given input to an appropriate output. The interconnection of a sequence of transmission links and circuit switches enables the flow of information between inputs and outputs in the network.

In the first part of this section we consider the design of circuit switches that transfer the information from one incoming link to one outgoing link. The first telephone switches were of this type and involved the establishment of a physical path across the switch that enabled the flow of current from an input line to an output line. The principle of circuit switches is general, however, and one could consider the design of optical circuit switches that enable the transfer of optical signals from an input line to an output line.

In many cases the input lines to a switch contain multiplexed information flows, and the purpose of the switch is to transfer each specific subflow from an input line to a specific subflow in a given output line. In principle the incoming flows must first be demultiplexed to extract the subflows that can then be transferred by the switch to the desired output links. In section 4.4.2 we consider the case where the incoming and outgoing flows are time-division multiplexed streams. The associated digital circuit switches form the basis for modern telephone switches.

4.4.1   Space-Division Switches

The first switches we consider are called **space-division switches** because they provide a separate physical connection between inputs and outputs so the different signals are separated in space. Figure 4.22 shows the **crossbar switch**, which is an example of this type of switch. The crossbar switch consists of an $N \times N$ array of *crosspoints* that can connect any input to any available output. When a request comes in from an incoming line for an outgoing line, the corresponding crosspoint is closed to enable information to flow from the input to the

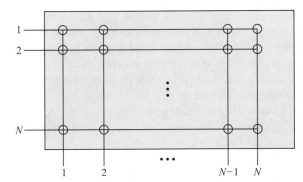

FIGURE 4.22 Crossbar switch

output. The crossbar switch is said to be **nonblocking**; in other words, connection requests are never denied because of lack of connectivity resources, that is, crosspoints. Connection requests are denied only when the requested outgoing line is already engaged in another connection.

The complexity of the crossbar switch as measured by the number of crosspoints is N^2. This number grows quickly with the number of input and output ports. Thus a 1000-input-by-1000-output switch requires 10^6 crosspoints, and a 100,000 by 100,000 switch requires 10^{10} crosspoints. In the next section we show how the number of crosspoints can be reduced by using multistage switches.

◆MULTISTAGE SWITCHES

Figure 4.23 shows a **multistage switch** that consists of three stages of smaller space-division switches. The N inputs are grouped into N/n groups of n input lines. Each group of n input lines enters a small switch in the first stage that

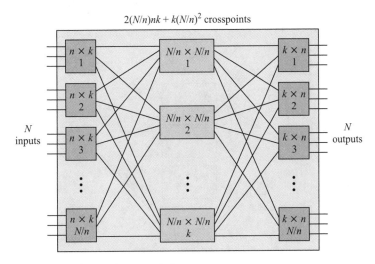

FIGURE 4.23 Multistage switch

consists of an $n \times n$ array of crosspoints. Each input switch has one line connecting it to each of k intermediate stage $N/n \times N/n$ switches. Each intermediate switch in turn has one line connecting it to each of the N/n switches in the third stage. The latter switches are $k \times n$. In effect each set of n input lines *shares* k *possible paths to any one of the switches at the last stage*; that is, the first path goes through the first intermediate switch, the second path goes through the second intermediate switch, and so on. The resulting multistage switch is not necessarily nonblocking. For example, if $k < n$, then as soon as a switch in the first stage has k connections, all other connections will be blocked.

The question of determining when a multistage switch becomes nonblocking was answered by [Clos 1953]. Consider any desired input and any desired output such as those shown in Figure 4.24. The worst case for the desired input is when all the other inputs in its group have already been connected. Similarly, the worst case for the desired output is when all the other outputs in its group have already been connected. The set of routes that maximize the number of intermediate switches already in use by the given input and output groups is shown in Figure 4.24. That is, each existing connection uses a different intermediate switch. Therefore, the maximum number of intermediate switches not available to connect the desired input to the desired output is $2(n - 1)$. Now suppose that $k = 2n - 1$; then k paths are available from any input group to any output group. Because $2(n - 1)$ of these paths are already in use, it then follows that a single path remains available to connect the desired input to the desired output. Thus the multistage switch with $k = 2n - 1$ is nonblocking.

The number of crosspoints required in a three-stage switch is the sum of the following components:

- N/n input switches $\times$ nk crosspoints/input switch.
- k intermediate switches $\times (N/n)^2$ crosspoints/intermediate switch.
- N/n output switches $\times$ nk crosspoints/output switch.

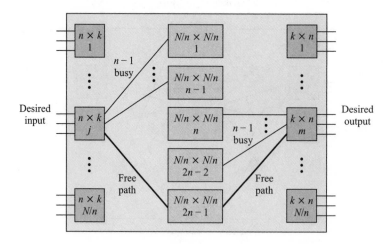

FIGURE 4.24 A multistage switch is nonblocking if $k = 2n - 1$

In this case the total number of crosspoints is $2Nk + k(N/n)^2$. The number of crosspoints required to make the switch nonblocking is $2N(2n - 1) + (2n - 1)(N/n)^2$. The number of crosspoints can be minimized through the choice of group size n. By differentiating the above expression with respect to n, we find that the number of crosspoints is minimized if $n \approx (N/2)^{1/2}$. The minimum number of crosspoints is then $4N((2N)^{1/2} - 1)$. We then see that the minimum number of crosspoints grows at a rate proportional to $N^{1.5}$, which is less than the N^2 growth rate of a crossbar switch.

When $k < 2n - 1$, there is a nonzero probability that a connection request will be blocked. The methods for calculating these probabilities can be found in [Bellamy 1991, pp. 234–242].

4.4.2   Time-Division Switches

In the first part of this chapter, we explained how TDM could replace multiple physical lines by a single high-speed line. In TDM a slot within a frame corresponds to a single connection. The **time-slot interchange (TSI)** technique replaces the crosspoints in a space switch with the reading and writing of a slot into a memory. Suppose we have a number of pairs of speakers in conversation. The speech of each speaker is digitized to produce a sequence of 8000 bytes/second. Suppose that the bytes from all the speakers are placed into a T-1 carrier, as shown in Figure 4.25. Suppose also that the first pair of speakers has been assigned slots 1 and 23. For the speakers to hear each other, we need to route slots 1 and 23 in the incoming frames to slots 23 and 1 in the outgoing frames. Similarly, if the second pair of speakers is assigned slots 2 and 24, we need to interchange incoming slots 2 and 4 with outgoing slots 24 and 2, respectively.

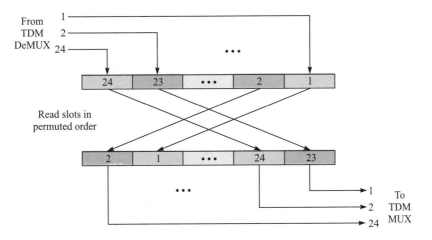

FIGURE 4.25 Time-slot interchange technique

Figure 4.25 shows the interchange technique: The octets in each incoming frame are written into a register. The call setup procedure has set a permutation table that controls the order in which the contents of the register are read out. Thus the outgoing frame begins by reading the contents of slot 23, followed by slot 24, and so on until slots 1 and 2 are read, as shown in the figure. This procedure can connect any input to any available output. Because frames come in at a rate of 8000 times a second and the time-slot interchange requires one memory write and one memory read operation per slot, the maximum number of slots per frame that can be handled is

$$\text{Maximum number of slots} = \frac{125\mu\text{sec}}{2 \times \text{memory cycle time}}$$

For example, if the memory cycle time is 125 nanoseconds, then the maximum number of slots is 500, which can accommodate 250 connections.

The development of the TSI technique was crucial in completing the digitization of the telephone network. Starting in 1961 digital transmission techniques were introduced in the trunks that interconnected telephone central offices. Initially, at each office the digital streams would be converted back to analog form and switched by using space switches of the type discussed above. The introduction of TSI in digital time-division switches led to significant reductions in cost and to improvements in performance by obviating the need to convert back to analog form. Most modern telephone backbone networks are now entirely digital in terms of transmission and switching.

◆TIME-SPACE-TIME SWITCHES

We now consider a hybrid switch design in which TSI switches are used at the input and output stages and a crossbar space switch is used at the intermediate stage. These switches are called **time-space-time switches**. *The design approach is to establish an exact correspondence between the input lines in a space-division switch in the first stage and time slots in a TSI switch.* Suppose we replace the $n \times k$ switch in the first stage of a multistage space switch by an $n \times k$ TSI switch, as shown in Figure 4.26. Each input line to the switch corresponds to a slot, so the TSI switch has input frames of size n slots. Similarly, the output frame from the TSI switch has k slots. Thus the operation of the TSI switch involves taking the n slots from the incoming frame and reading them out in a frame of size k, according to some preset permutation table. Note that for the system to operate in synchronous fashion, the transmission time of an input frame must be equal to the transmission time of an output frame. Thus, for example, if $k = 2n - 1$, then the internal speed is nearly double the speed of the incoming line.

Consider now the flow of slots between the switches in the first stage and the switches in the intermediate stage. We assume that frames coming out of the TSI switches in the first stage are synchronized. Consider what happens as the first slot in a frame comes out of the first stage. This first slot corresponds to the first output line out of each of the first stage switches. Recall from Figure 4.23 that the first line out of each first stage switch is connected to the first intermediate switch. Thus the first slot in each intermediate frame will be directed to inter-

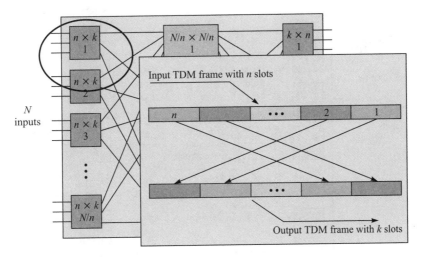

FIGURE 4.26 Hybrid switches

mediate switch 1, as shown in Figure 4.27. Recall that this switch is a crossbar switch, and so it will transfer the N/n input slots into N/n output slots according to the crosspoint settings. Note that all the other intermediate switches are idle during the first time slot.

Now consider what happens with the second slot in a frame. These slots are now directed to crossbar switch 2, and all other intermediate switches are idle. It thus becomes apparent that only one of the crossbar switches is active during any given time slot. This situation makes it possible to replace the k intermediate crossbar switches with a single crossbar switch that is time-shared among the k slots in a frame, as shown in Figure 4.28. To replace the k intermediate original crossbar switches, the time-shared crossbar switch must be reconfigured to the

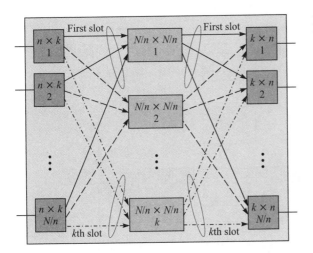

FIGURE 4.27 Flow of slots between switches

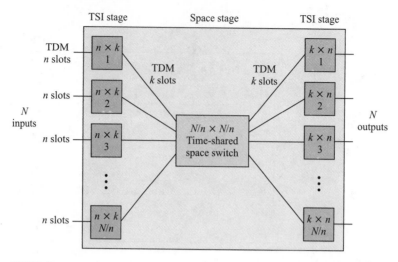

FIGURE 4.28 Time-space-time switches

interconnection pattern of the corresponding original switch at every time slot. This approach to sharing a space switch is called **time-division switching**.

Example—A 4 × 4 Time-Space-Time Switch

Figure 4.29 shows a simple 4×4 switch example that is configured for the connection pattern (A, B, C, D) to (C, A, D, B). The inputs arrive in frames of size 2 that are mapped into frames of size 3 by the input TSI switches. The interconnection pattern for each of the three slots is shown for the intermediate switches. Finally, the mapping from the three-slot frame to the output two-slot

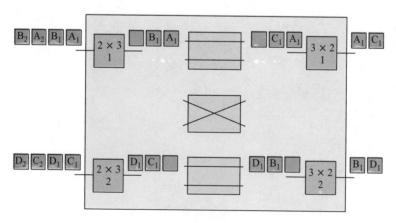

FIGURE 4.29 Example of a time-space-time switch

frame is shown. By tracing the handling of each slot, we can see how the overall interconnection pattern (A,B,C,D) to (C,A,D,B) is accomplished.

The introduction of TSI switches at the input and output stages and the introduction of a single time-shared crossbar switch result in a much more compact design than space switches. The replacement of the N original input lines by the N/n TDM lines also leads to a compact design. For example, consider the design of a nonblocking 4096×4096 time-space-time switch that has input frames with 128 slots. Because $N = 4096$ and $n = 128$, we see that $N/n = 32$ input TSI switches are required in the first stage. The nonblocking requirement implies that the frames between the input stage and the intermediate stage must be of size $k = 2n - 1 = 255$ slots. The internal speed of the system is then approximately double that of the input lines. The time-shared crossbar switch is of size 32×32.

4.5 THE TELEPHONE NETWORK

The modern telephone network was developed to provide basic telephone service, which involves the two-way, real-time transmission of voice signals. In its most basic form, this service involves the transfer of an analog signal of a nominal bandwidth of 4 kHz across a sequence of transmission and switching facilities. We saw in Chapter 3 that the digital transmission capacity of this 4 kHz channel is about 45 kbps, which is miniscule in relation to the speed of modern computers. Nevertheless, the ubiquity and low cost of the telephone network make it an essential component of computer communications.

Telephone networks operate on the basis of *circuit switching*. Initially, circuit switching involved the setting up of a physical path from one telephone all the way across the network to the other telephone, as shown in Figure 4.30. At the telephone offices operators would make physical connections that would allow electric current to flow from one telephone to the other. The physical resources, such as wires and switch connections, were dedicated to the call for its entire duration. Modern digital telephone networks combine this circuit-switching

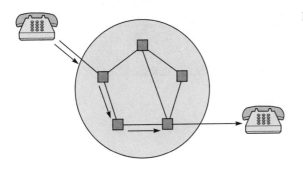

FIGURE 4.30 Circuit switching

approach to operating a network with digital transmission and digital switching. In the following discussion we explain how a telephone call is set up in a modern digital network.

Figure 4.31 shows the *three phases of connection-oriented communications.* When a user picks up the phone, a current flow is initiated that alerts equipment in a switch at the local telephone office of a call request. The switch prepares to accept the dialed digits and then provides the user with a dial tone. The user then enters the telephone number either by turning a dial that generates a sequence of pulses or by pushing a series of buttons that generates a sequence of tones. The switch equipment converts these pulses or tones into a telephone number.

In North America each telephone is given a 10-digit number according to the North American numbering plan. The first three digits are the area code, the next three digits are the central office code, and the last four digits are the station number. The first six digits of the telephone number uniquely identify the central office of the destination telephone. To set up a call, the source switch uses the telephone signaling network to find a route and allocate resources across the network to the destination office.[7] This step is indicated by the propagation of signaling messages in Figure 4.31. The destination office next alerts the destination user of an incoming call by ringing the phone. This ring is sent back to the source telephone, and conversation can begin when the destination phone is picked up. The call is now in the message transfer phase. The call is terminated and the resources released when the users hang up their telephones.

The call setup procedure involves finding a path from the source to the destination. Figure 4.32a shows a typical arrangement in a metropolitan area. Central offices are connected by high-speed digital transmission lines that correspond to a group of trunks. If the two telephones are connected to the same central office, that is, the two phones are attached to switch A, then they are connected directly by the local switch. If the two telephones are connected to different central offices (that is, A and B), then a route needs to be selected.[8]

In the United States the telephone network is divided into **local access and transport areas (LATAs)** that are served by the **local exchange carriers (LECs)**. The LECs consist of **regional bell operating companies (RBOCs)** such as BellSouth, Southwestern Bell, and Bell Atlantic, and independent carriers such

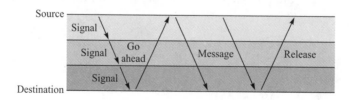

FIGURE 4.31 Telephone call setup

---

[7]The telephone signaling network is discussed in section 4.6.
[8]The routing control procedures are discussed in section 4.7.

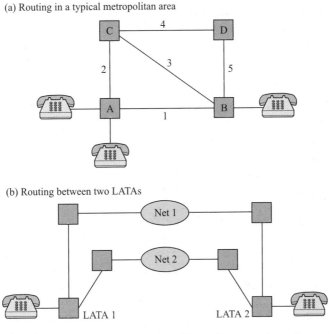

(a) Routing in a typical metropolitan area

(b) Routing between two LATAs

FIGURE 4.32 Routing in local and long-distance connections

as GTE. Communication between LATAs is provided by separate independent **interexchange carriers (IXCs)**, that is, long-distance service providers such as AT&T and MCI. Figure 4.32b shows how long distance calls are routed from a switch in the LATA to a facility that connects the call to the network of one of a number of interexchange carriers. The call is then routed through the network of the interexchange carrier to the destination LATA and then to the destination telephone.

Now let us consider the end-to-end path that is set up between the source and the destination telephones. In the majority of cases, the telephones are connected to their local telephone office with a twisted pair of copper wires. The voice signal flows in analog form from the telephone to the telephone office. The voice signal is converted into digital form using PCM at the line card interface where the copper wires connect to the local telephone switch. The digitized voice signal flows from that point onward as a sequence of PCM samples over a path that has been set up across the network. This path consists of reserved time slots in transmission links that use TDM. The transmission links connect digital switches in which TSI arrangements have been made during call set up. Finally, at the destination switch the received PCM signal is converted back to analog form and transmitted to the destination telephone over the pair of copper wires.

The pair of copper wires that connect the user telephones to their local offices is called the "last mile," which constitutes the remaining obstacle to providing end-to-end digital connectivity. In the next section we discuss the structure of the last mile in more detail.

4.5.1   Transmission Facilities

The user's telephone is connected to its local telephone office by twisted-pair copper wires that are called the *local loop*. The wire pairs are stranded into groups, and the groups are combined in cables that can hold up to 2700 pairs. Wire pairs connect to the telephone office at the *distribution frame* as shown in Figure 4.33. From the local office the feeder cables extend in a star topology to the various geographic serving areas. Each feeder cable connects to a *serving area interface*. Distribution cables in turn extend in a star topology from the serving area interface to the *pedestals* that connect to the user's telephone.

In the local loop a single pair of wires carries the information in both directions, from the user to the local switch and from the local switch to the user. Inside the network a separate pair of wires is used for each direction of signal flow. As shown in Figure 4.34, at the interface to the switch a device called a *hybrid transformer* converts the signals from the incoming wire pair to the four-wire connection that is used inside the network. These hybrid transformers tend to reflect some of the arriving signal resulting in **echoes** that under certain con-

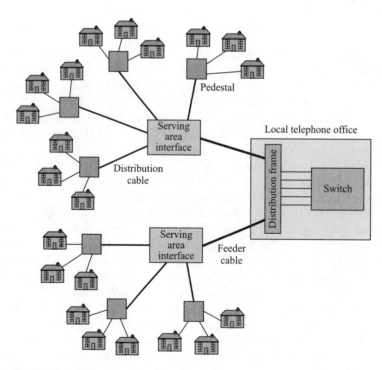

FIGURE 4.33 Access transmission facilities

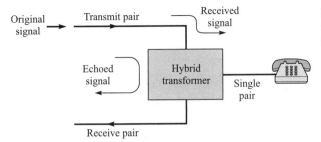

FIGURE 4.34 Two- and four-wire connections in the telephone network

ditions can be disturbing to speakers and that can impair digital transmission. For these reasons, echo-cancellation devices have been developed that estimate the echo delay and use a properly delayed and scaled version of the transmitted signal to cancel the arriving echoes.

The twisted-wire pairs that connect the user to the local switch can be used to carry high-speed digital signals using ADSL and other digital transmission technologies. In many instances the feeder cable that connects the serving area interface to the local switch is being replaced with an optical fiber that carries time-division-multiplexed traffic. In this case the service area interface carries out the conversion to and from PCM. As optical fiber extends closer to the user premises, the distance that needs to be covered by the twisted-wire pairs is reduced. Table 3.5 in Chapter 3 shows that it is then possible to deliver bit rates in the tens of megabits/second to the user premises. In the future this deployment of optical fiber is likely to be one of the approaches that will be used to provide much higher bit rates to the user premises.

In addition to handling the analog telephone traffic from its serving areas, a local telephone switching office must handle a multiplicity of other types of transmission lines. These include digital lines of various speeds from customers premises, private lines, foreign exchange lines, lines from celllar telephone networks, and high-speed digital transmission line to the backbone of the network. A digital cross-connect system is used to organize the various types of lines that arrive and depart from a telephone office. A **digital cross-connect (DCC)** system is simply a digital time-division switch that is used to manage the longer-term flows in a network. Unlike the time-division switches that deal with individual telephone calls, DCC systems are not controlled by the signaling process associated with call setup. Instead they are controlled by the network operator to meet network configuration requirements.

Figure 4.35 shows how a DCC system can be used to interconnect various types of traffic flows in a telephone central office. The local analog telephone signals on the left side are digitized and then input into the DCC system. The other traffic on the left side are local digital traffic and digital trunks connected to other offices. Some of these inputs are permanently connected as "tie lines" that interconnect customer sites; some of the other inputs may originate in other central offices and be permanently connected to foreign exchange lines that access an international gateway. Voice traffic that needs to be switched is routed

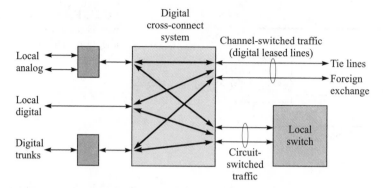

FIGURE 4.35 Digital cross-connect system

from the input lines to a digital switch that is attached to the DCC system. The DCC system provides a flexible means for managing these connections on a semipermanent basis.

The transmission facilities between switches in the telephone network are becoming dominated by SONET-based optical transmission equipment using ponit-to-point and ring topologies as discussed earlier in section 4.2. DCC systems can be combined with SONET transmission systems to configure the topology "seen" by the telephone switches during call setup. As an example, in Figure 4.36 we show a SONET network that has a "stringy" physical topology. The SONET tributaries can be configured using add-drop multiplexers and a DCC system so tributaries interconnect the switches to produce the topology shown in part (b). In particular note that the maximum number of hops between any pair of switches is two hops instead of the four hops in the physical topology. One

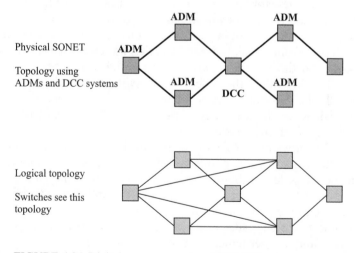

FIGURE 4.36 Digital cross-connect and SONET

benefit of decreasing the number of hops between switches is to simplify the routing procedure that is implemented by the switches.

DCC and SONET equipment can also be combined to provide network recovery from faults. For example, standby tributaries may be set up to protect against failures. Should a fault occur, the standby tributaries are activated to restore the original logical topology seen by the switches. This approach reduces the actions that the network switches must take in response to the failure.

4.5.2 End-to-End Digital Services

Since the inception of telephony, the telephone user has been connected to the telephone office by a pair of copper wires in the local loop. This last mile of the network has remained analog even as the backbone of the network was converted to all-digital transmission and switching. In the mid-1970s it became apparent that the demand for data services would necessitate making the telephone network entirely digital as well as capable of providing access to a wide range of services including voice and data. In the early 1980s the CCITT developed the **Integrated Services Digital Network (ISDN)** set of standards for providing end-to-end digital connectivity.

The ISDN standards define two *interfaces* between the user and the network as shown in Figure 4.37. The purpose of the interfaces is to provide *access* to various services that are possibly supported by different networks. The **basic rate interface (BRI)** provides the user with two 64 kbps bearer (B) channels and one 16 kbps data (D) channel. The **primary rate interface (PRI)** provides users with

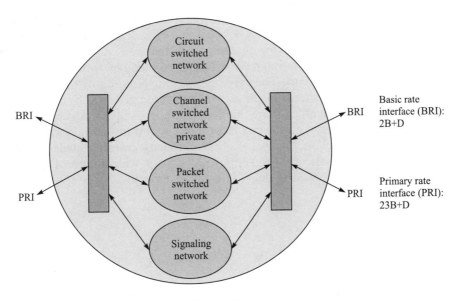

FIGURE 4.37 Integrated Services Digital Network

23B + 1 D channels in North America and Japan and 30B + 1 D channels in Europe. The primary rate formats were clearly chosen to fit the T-1 and CEPT-1 frame formats in the digital multiplexing hierarchy. Each B channel is bidirectional and provides a 64 kbps end-to-end digital connection that can carry PCM voice or data. The primary function of the D channel is to carry signaling information for the B channels at the interface between the user and the network. The D channel can also be used to access a packet network. The D channel is 64 kbps in the primary rate interface. Additional H channels have been defined by combining B channels to provide rates of 384 kbps (H0), 1.536 Mbps (H11), and 1.920 Mbps (H12).

The basic rate interface was intended as a replacement for the basic telephone service. In North America the associated transmission format over twisted-wire pairs was selected to operate at 160 kbps and use the band that is occupied by traditional analog voice. The two B channels could provide two digital voice connections, and the D channel would provide signaling and access to a packet network. In practice, the 2B channels in the basic rate interface found use in digital videoconferencing applications and in providing access to Internet service providers at speeds higher than those available over conventional modems. The primary rate interface was intended for providing access from user premises equipment such as private branch exchanges (PBXs) and multiplexers to the network.[9]

ISDN was very slow in being adopted primarily because few services could use it. The growth of the Internet as a result of the World Wide Web has stimulated the deployment of ISDN as an alternative to telephone modems. However, as discussed in Chapter 3, alternative ADSL transmission techniques for providing much higher speeds over the local loop are currently being introduced to meet the demand for high-speed Internet access. These techniques coexist with the regular analog telephone signal and do not interact with the telephone network.

As the ISDN standard was being completed, the interest in high-definition television and high-speed data interconnection prompted new work on a **broadband ISDN** (BISDN) standard. The very high bit rate required by these applications necessitated access speeds much higher than those provided by ISDN. The BISDN effort resulted in much more than an interface standard. An entirely new network architecture based on the connection-oriented transfer and switching of small fixed-length packets, known as **asynchronous transfer mode (ATM)**, emerged as a target network for supporting a very wide range of services. ATM networks are discussed in detail later in Chapter 9.

---

[9]A PBX is a switch in the user premises that provides connection services for intrapremise calls and that controls access to public and private network services.

HOW DOES THE OSI REFERENCE MODEL APPLY TO THE TELEPHONE NETWORK?

The telephone network was established well before the OSI reference model, and so there is not a perfect fit between the model and the telephone network. However, the ISDN standards do help to clarify this relationship. The telephone network (and circuit switching in general) require two basic functions: (1) signaling to establish and release the call and (2) end-to-end transmission to transfer the information between the users. ISDN views these two functions as separate and indeed as being provided by different networks (see Figure 4.37).

The signaling function is in fact a distributed computer application that involves the exchange of signaling messages between users and the network and between switches in the network to control the setup and release of calls. This process involves the use of all the layers in the OSI reference model. The set of protocols that implement signaling in the telephone network is said to constitute the *control plane*. We discuss these protocols in the next section.

The set of protocols that implement the transfer of information between users is called the *user plane*. In the case of an end-to-end 64 kbps connection, the control plane sets up the connection across the network. The user plane consists of an end-to-end 64 kbps flow and so involves only the physical layer. End users are free to use any higher-layer protocols on this digital stream.

This ISDN framework has been generalized so that the control plane is used to establish virtual connections in the user plane that involve more than just the physical layer. For example, the original ISDN standards defined a "packet mode" service in which an OSI layer-3 connection is set up across a packet-switching network. Subsequently a "frame relay mode" was defined to establish an end-to-end data-link layer connection between two terminals attached to a public network.

4.6 SIGNALING

To establish a telephone call, a series of signaling messages must be exchanged. There are two basic types of signal exchanges: (1) between the user and the network and (2) within the network. Both types of signaling must work together to establish the call. In this section we consider the signaling that takes place inside the network.

In general, the signaling messages generate control signals that determine the configuration of switches; that is, the messages direct a switch to a state in which a given input is connected to the desired output. As shown in Figure 4.38, in traditional networks signaling information would arrive in telephone lines and be routed to the control system. Initially, hard-wired electromechanical or electronic logic was used to process these signaling messages. The class of **stored-**

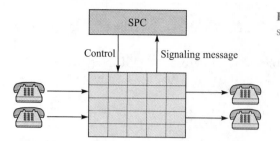

FIGURE 4.38 Stored-program control switches

program control (SPC) switches emerged when computers were introduced to control the switch, as shown in Figure 4.38. Through the intervention of the stored-program control computer, a request for a call would come in, a check would be made to see whether the destination was available, and if so, the appropriate connection would be made. The use of a program to control the switch provided great flexibility in modifying the control and in introducing new features.

As shown in Figure 4.39, setting up a call also required that the computers controlling the switches communicate with each other to exchange the signaling information. A modem and separate communication lines were introduced to interconnect these computers. This situation eventually led to the introduction of a separate computer communications network to carry the signaling information.

Consider the operation of the **signaling network**. Its purpose is to implement connectivity between the computers that control the switches in the telephone network by providing for the exchange of messages. Figure 4.40 shows the telephone network as consisting of two parts: a signaling network that carries the information to control connections and a transport network that carries the user information. Communications from the user are split into two streams at the

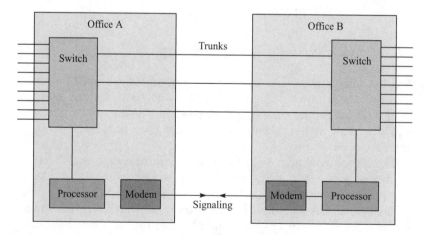

FIGURE 4.39 Common channel signaling

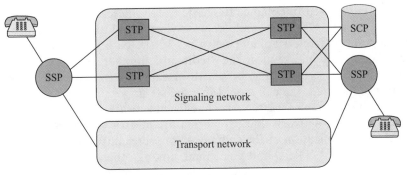

SSP = Service switching point (signal to message)
STP = Signal transfer point (message transfer)
SCP = Service control point (processing)

FIGURE 4.40 Signaling network

service switching point (SSP). The signaling information is directed toward the signaling network where it is routed and processed as required. The signaling system then issues commands to the switches to establish the desired connection. The signaling network functions much like the "nervous system" of the telephone network, directing the switches and communication lines in the network to be configured to handle the various connection requests. The second stream in the SSP consists of the user information that is directed to the transport network where it flows from one user to the other. Note that the signaling network does not extend to the user because of security concerns. Separate user-to-network signaling procedures are in place.

The function of the signaling network is to provide communications between the computers that control the switches. The computers communicate through the exchange of discrete messages. The best way of implementing such a network is through a *packet-switching network* that transfers information in the form of packets between network elements. The introduction of a packet-switching signaling network in digital telephony is important, since it is at this point that the evolution of the signaling network for telephony converges with the evolution of computer networks. In the next section we discuss the layered architecture of the telephone signaling system.

Because uninterrupted availability of telephone service is extremely important, *reliability* was built into the packet-switching network for signaling. The packet-switching nodes (signal transfer points or STPs) are interconnected as shown in Figure 4.40. Any given region has two STPs that can reach any given office, so if one STP goes down the other is still available.

The processing of a connection request may involve going to *databases* and *special purpose processors* at the service control points (SCPs) in Figure 4.40. As signaling messages enter the network, they are routed to where a decision can be made or where information that is required can be retrieved. In the early 1970s telephone companies realized that the signaling network (and its computer control) could be used to enhance the basic telephone service. Credit-card calls, long-distance calls, 800 calls, and other services could all be implemented by using this more capable signaling network. In the case of credit-card calls, a recorded message could request the credit-card number. The digits would be collected, and a message would be sent to a database to check the credit-card number; if authorized, the call would then be set up.

Telephone companies use the term **intelligent network** to denote the use of an enhanced signaling network that provides a broad array of services. These services include identification of the calling person, screening out of certain callers, callback of previous callers, and voice mail, among others. As shown in Figure 4.41, the addition of new devices, "intelligent peripherals," to the intelligent network enables other new services. For example, one such device can provide voice recognition. When making a call, your voice message might be routed to this intelligent peripheral, which then decodes what you are saying and translates it into a set of actions that the signaling network has to perform in order to carry out your transaction.

Another service that intelligent networks provide is personal mobility. *Personal mobility* allows the user who subscribes to the service to have a personal ID. Calls to the user are not directed to a specific location in the network. Instead the network dynamically keeps track of where the user is at any given time and routes calls accordingly.

4.6.1   Signaling System #7 Architecture

The Signaling System #7 (SS7) network is a packet network that controls the setting up, managing, and releasing of telephone calls. The network also provides support for intelligent networks, mobile cellular networks, and ISDN. The SS7 network architecture is shown in Figure 4.42.

This architecture uses "parts" instead of "layers." The message transfer part (MTP) corresponds to the lower three layers of the OSI reference model. Level 1 of the MTP corresponds to the physical layer of the signaling links in the SS7 networks. Physical links have been defined for the following transmission speeds:

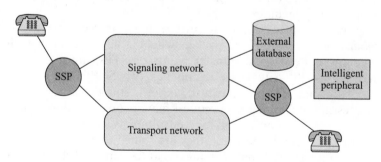

FIGURE 4.41 Intelligent network

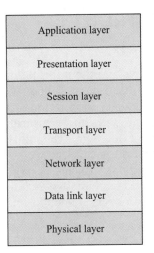 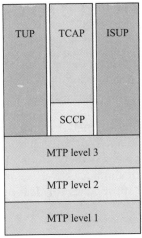

FIGURE 4.42 OSI reference model and SS7 network architecture

Application layer	
Presentation layer	
Session layer	
Transport layer	
Network layer	
Data link layer	
Physical layer	

TUP = telephone user part
TCAP = transaction capabilities part
ISUP = ISDN user part
SCCP = signaling connection control part
MTP = message transfer part

E-1 (2.048 Mbps = 32 channels at 64 kbps each), T-1 (1.544 Mbps = 24 channels at 64 kbps each), V-35 (64 kbps), DS-0 (64 kbps), and DS-0A (56 kbps). These telephone transmission standards were discussed earlier in this chapter.

MTP level 2 ensures that messages are delivered reliably across a signaling link. This level corresponds to the data link layer in the OSI reference model. The MTP level 3 ensures that messages are delivered between signaling points across the SS7 network. Level 3 provides routing and congestion control that reroutes traffic away from failed links and signaling points.

The ISDN user part (ISUP) protocols perform the basic setup, management, and release of telephone calls. The telephone user part (TUP) is used instead in some countries.

The MTP addresses the signaling points, but is not capable of addressing the various applications that may reside within signaling point. These applications include 800-call processing, calling-card processing, call management services, and other intelligent network services. The signaling connection control part (SCCP) allows these applications to be addressed by building on the MTP to provide connectionless and connection-oriented service. The SCCP can also translate "global titles" (e.g., a dialed 800 number or a mobile subscriber number) into an application identifier at a destination signaling point. This feature of SCCP is similar to the Internet DNS and assists in the routing of messages to the appropriate processing point.

The transaction capabilities part (TCAP) defines the messages and protocols that are used to communicate between applications that use the SS7 network. The TCAP uses the connectionless service provided by the SCCP to support database queries that are used in intelligent networks.

4.7 TRAFFIC AND OVERLOAD CONTROL IN TELEPHONE NETWORKS

In this section we consider the *dynamic* aspects of multiplexing the information flows from various users into a single high-speed digital transmission line. We begin by looking at the problem of **concentration** that involves the sharing of a number of trunks by a set of users. Here we examine the problem of ensuring that there are sufficient resources, namely, trunks, to provide high availability, that is, low probability of blocking. We find that concentration can lead to very efficient usage of network resources if the volume of traffic is sufficiently large. We next discuss how this result influences routing methods in circuit-switching networks. Finally we consider approaches for dealing with overload conditions.

4.7.1 Concentration

In Figure 4.43 numerous users at a given site, each with a communication line, need to use expensive trunks provided by a high-speed digital transmission line to connect to another location, for example, a telephone central office or another user site. The number of trunks in use varies randomly over time but is typically much smaller than the total number of lines. For this reason, a multiplexer is introduced to concentrate the requests for connections over a smaller number of trunks. While the objective is to maximize the use of the trunks, typically a maximum acceptable probability of blocking is specified. We say that a connection request is **blocked** when no trunks are available. Thus the system design problem involves selecting the number of trunks so that the blocking probability is kept below the specified level.

Figure 4.44 shows the occupancy of a set of seven trunks over time. The shaded rectangles indicate periods when a given trunk is in use. The upper part of the figure shows the corresponding $N(t)$, the number of trunks in use at time t. In this example the system is in a blocking state when $N(t) = 7$.

The users require trunk connections in a sporadic and unscheduled manner. Nevertheless, the statistical behavior of the users can be characterized. In particular it has been found that user requests for connections take place according to a Poisson process with connection request rate λ calls/second. A Poisson process is characterized by the following two properties:

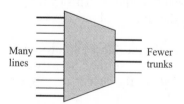

Many lines   Fewer trunks

FIGURE 4.43 Concentration (bold lines indicate lines/ trunks that are in use)

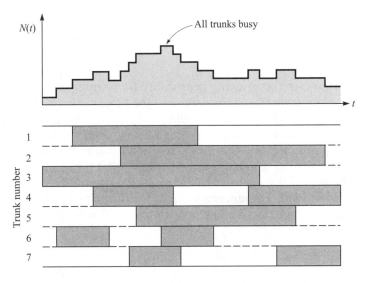

FIGURE 4.44 Number of trunks in use as a function of time

1. In a very small time interval Δ, only two things can happen: There is a request for one call, with probability $\lambda\Delta$, or there are no requests for calls, with probability $1 - \lambda\Delta$.
2. The arrivals of connection requests in different intervals are statistically independent.

The analysis of the trunk concentration problem is carried out in Appendix A. We present only the results of the analysis here.

The time that a user maintains a connection is called the **holding time**. In general, the holding time X is a random variable. The average holding time $E[X]$ can be viewed as the amount of "work" that the transmission system has to do for a typical user. In telephone systems typical conversations have a mean holding time of several minutes. The **offered load a** is defined as the total rate at which work is offered by the community of users to the multiplexing system:

$$a = \lambda \text{ calls/second } {}^*E[X] \text{ seconds/call (Erlang)}$$

One Erlang corresponds to an offered load that would occupy a single trunk 100 percent of the time, for example, an **arrival rate** of $\lambda = 1$ calls/second and a call holding time of $E[X] = 1$ would occupy a single trunk all of the time. Typically telephone systems are designed to provide a certain grade of service during the **busy hour** of the day. Measurements of call attempts reveal clear patterns of activity and relatively stable patterns of call attempts. In the subsequent discussion the offered load should be interpreted as the load during the busy hour.

The blocking probability P_b for a system with c trunks and offered load a is given by the **Erlang B formula**:

$$P_b = \frac{a^c/c!}{\displaystyle\sum_{k=0}^{c} a^k/k!}, \text{ where } k! = 1 \cdot 2 \cdot 3 \ldots \cdot (k-1) \cdot k$$

Figure 4.45 shows the blocking probability for various offered loads as the number of trunks c is increased. As expected, the blocking probability decreases with the number of trunks. A 1% blocking probability is typical in the design of trunk systems. Thus from the figure we can see that four trunks are required to achieve this P_b requirement when the offered load is one Erlang. On the other hand, only 16 trunks are required for an offered load of nine Erlangs. This result shows that the system becomes more efficient as the size of the system increases, in terms of offered load. The efficiency can be measured by trunk **utilization** that is defined as the average number of trunks in use divided by the total number of trunks. The utilization is given by

$$\text{Utilization} = \lambda(1 - P_b)E[X]/c = (1 - P_b)a/c$$

Table 4.2 shows the trunk utilization for the various offered loads and $P_b = 0.01$. Note that for small loads the utilization is relatively low. In this case extra trunks are required to deal with surges in connection requests. However, the utilization increases as the size of the systems increases in terms of offered load. For a load of two Erlangs, a total of 7 trunks is required; however, if the load is tripled to six Erlangs, the number of trunks required, 13, is less than double. The entry in Table 4.2 for offered loads of 50 and 100 Erlangs shows that high utilization is possible when the offered loads are large. These examples demonstrate how the sharing of network resources becomes more efficient as the scale or size of the system increases. The improvement in system performance that results from aggregating traffic flow is called *multiplexing gain*.

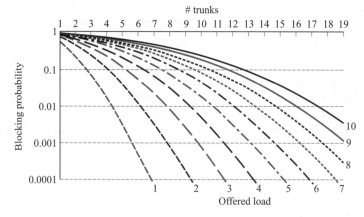

FIGURE 4.45 Blocking probability versus number of trunks

Load	Trunks@1%	Utilization
1	5	0.20
2	7	0.29
3	8	0.38
4	10	0.40
5	11	0.45
6	13	0.46
7	14	0.50
8	15	0.53
9	17	0.53
10	18	0.56
30	42	0.71
50	64	0.78
60	75	0.80
90	106	0.85
100	117	0.85

TABLE 4.2 Trunk utilization

4.7.2 Routing Control

Routing control refers to the procedures for assigning paths in a network to connections. Clearly, connections should follow the most direct route, since this approach uses the fewest network resources. However, we saw in section 4.7.1 that when traffic flows are not large the required set of resources to provide high availability, that is, a blocking probability of 1%, will be used inefficiently. Economic considerations lead to an approach that provides direct trunks between switches that have large traffic flows between them and that provide indirect paths through tandem switches for smaller flows.

A hierarchical approach to routing is desirable when the volume of traffic between switches is small. This approach entails aggregating traffic flows onto paths that are shared by multiple switches. Consider the situation in Figure 4.46 where switches A, B, and C have 10 Erlangs of traffic to D, E, and F. Suppose that switches A, B, and C are close to each other and that they have access to tandem switch 1. Similarly, suppose that switches D, E, and F are close to each

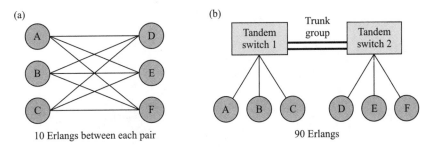

FIGURE 4.46 Hierarchical routing control

other and have access to tandem switch 2. Furthermore, suppose that the distances between switches A, B, and C and D, E, and F are large. From Table 4.2, each pair of switches requires 18 long distance trunks to handle the 10 Erlangs of traffic at 1% blocking probability. Thus the approach in Figure 4.46a requires $9 \times 18 = 162$ trunks. Concentrating the traffic flows through the tandems reduces to 106 the number of trunks required to handle the combined 90 Erlangs of traffic. The second approach does require the use of local trunks to the tandem switch, and so the choice depends on the relative costs of local and long-distance trunks.

The increase in efficiency of trunk utilization that results from the increased offered load introduces a sensitivity problem. The higher efficiency implies that a smaller number of spare circuits is required to meet the 1% blocking probability. However, the smaller number of spare circuits makes the system more sensitive to traffic overload conditions. For example, if each trunk group in part (a) is subjected to an overload of 10% the resulting offered load of 11 Erlangs on the 17 trunks results in an increased blocking probability of 2.45%. On the other hand, the 10% overload on the trunk group in part (b) results in an offered load of 99 Erlangs to 106 trunks. The blocking probability of the system increases dramatically to 9.5%. In other words, the blocking probability for large systems is quite sensitive to traffic overloads, and therefore the selection of the trunk groups must provide a margin for some percentage of overload.

Figure 4.47 shows a typical approach for routing connections between two switches that have a significant volume of traffic between them. A set of trunks is provided to directly connect these two switches. A request for a connection between the two switches first attempts to engage in a trunk in the direct path. If no trunk is available in the direct path, then an attempt is made to secure an alternative path through the tandem switch. The number of trunks in the direct route is selected to have a high usage and hence a blocking probability higher than 1% say, 10%. The number of trunks available in the alternative route needs to be selected so that the overall blocking probability is 1%. Note that because only 10% of the traffic between the switches attempts the alternative route, a 10% blocking probability on the alternative path is sufficient to bring the overall blocking probability to 1%. It should be noted also that the Erlang formula cannot be applied directly in the calculation of blocking probability on the alternative route. The reason is that the requests for routes to the tandem switch

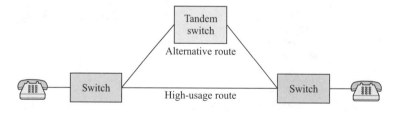

FIGURE 4.47 Alternative routing

arrive only during periods when the high-usage route is unavailable. The method for handling this case is discussed in [Cooper 1981].

Figure 4.48 shows a more realistic scenario where the tandem switch handles overflow traffic from some high-usage trunk groups and direct traffic between swtiches that have small volumes of traffic between them. Note that in this case the traffic between switches A and D must be provided with a 1% blocking probability, while a 10% blocking probability is sufficient for the other pairs of switches. To achieve this blocking probability the traffic from A to D must receive a certain degree of preferential access to the trunks between the tandem switches.

Traffic flows vary according to the time of day, the day of the week, and even the time of year. The ability to determine the state of network links and switches provides an opportunity to assign routes in more dynamic fashion. For example, time/day differences between the East Coast and the West Coast in North America allow the network resources at one coast to provide alternative routes for traffic on the other coast during certain times of the day. *Dynamic non-hierarchical routing (DNHR)* is an example of this type of dynamic approach to routing calls. As shown in Figure 4.49, the first route attempt between two switches consists of a direct route. A certain number of tandem switches is capable of providing a two-hop alternative route. The order in which tandem switches are attempted as alternative routes is determined dynamically according to the state of the network. The AT&T long-distance network consists of approximately 100 switches almost interconnected entirely with direct links. This topology allows the use of DNHR [Carne 1995].

4.7.3 Overload Controls

Traffic and routing control are concerned with the handling of traffic flows during normal predictable network conditions. Overload control addresses the handling of traffic flows during unexpected or unusual conditions, such as occur

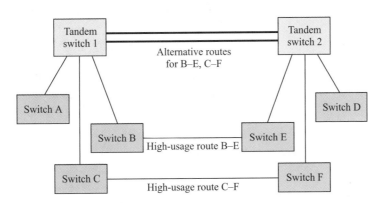

FIGURE 4.48 Typical routing scenario

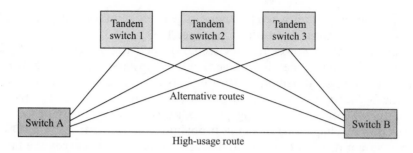

FIGURE 4.49 Dynamic nonhierarchical routing

during holidays (Christmas, New Year's Day, and Mother's Day), catastrophes (e.g., earthquakes), or equipment failures (e.g., a fire in a key switch or a cut in a key large-capacity optical fiber).

Overload conditions result in traffic levels that the network equipment has not been provisioned for and if not handled properly can result in a degradation in the level of service offered to all network customers. The situation can be visualized as shown in Figure 4.50. Under normal conditions the traffic carried by the network increases or decreases with the traffic that is offered to it. As the offered traffic approaches network capacity, the carried traffic may begin to fall. The reason for this situation is that as network resources become scarce, many call attempts manage to seize only some of the resources they need and ultimately end up uncompleted. One purpose of overload control is to ensure that a maximum number of calls are completed so that the carried load can approach the network capacity under overload conditions.

Network monitoring is required to identify overload conditions. Clearly the traffic loads at various links and switches need to be measured and tracked. In addition, the success ratio of call attempts to a given destination also needs to be monitored. The answer/bid ratio measures this parameter. The traffic load measurement in combination with the answer/bid ratio is useful in diagnosing fault conditions. For example, the failure of switch A will result in an increased traffic level at other switches due to reattempts from callers to switch A. The increased

FIGURE 4.50 Traffic overload

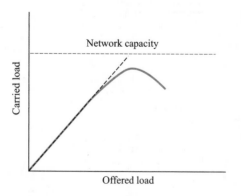

traffic load indicates a problem condition but is not sufficient to identify the problem. The answer/bid ratio provides the information that identifies switch A as the location of the problem. Network monitoring software is used to process alarms that are set by the monitoring system to diagnose problems in the network.

Once an overload condition has been identified, several types of actions can be taken, depending on the nature of the problem. One type of overload control addresses problems by allocating additional resources. Many transmission systems include backup redundant capacity that can be activated in response to failures. For example, SONET transmission systems use a ring topology of add-drop multiplexers to provide two paths between any two stations on the ring. Additional redundancy can be provided by interconnecting SONET rings using DCCs. Dynamic alternative routing provides another approach for allocating resources between areas experiencing high levels of traffic.

Certain overload conditions cannot be addressed by allocation of additional resources. The overload controls in this case act to maximize the efficiency with which the available resources are utilized. For example, in the case of network-wide congestion the routing procedures could be modified so that all call attempts, if accepted, are met using direct routes. Alternative routes are disallowed because they require more resources to complete calls. As a result, the traffic carried by the network is maximized.

Another overload condition occurs when a certain area experiences extreme levels of inbound and outbound traffic as may result from the occurrence of a natural disaster. A number of overload controls have been devised to deal with this situation. One approach involves allowing only outbound traffic to seize the available trunks. This approach relieves the switches in the affected area from having to process incoming requests for calls while allowing a maximum of outbound calls to be completed. A complementary control involves code blocking, where distant switches are instructed to block calls destined for the affected area. A less extreme measure is to pace the rate at which call requests from distant switches to the affected area are allowed to proceed.

It should be noted that all of the above overload controls make extensive use of the signaling system. This dependence on the signaling system is another potential source of serious problem conditions. For example, faulty signaling software can result in abnormal levels of signaling traffic that in turn can incapacitate the network. Clearly, overload control mechanisms are also essential for the signaling system.

4.8   CELLULAR TELEPHONE NETWORKS

Cellular telephone networks extend the basic telephone service to *mobile* users with *portable* telephones. Unlike conventional telephone service where the call to a telephone number is directed to a specific line that is connected to a specific switch, in cellular telephony the telephone number specifies a specific subscriber's mobile station (telephone). Much of the complexity in cellular telephony results from the need to track the location of the mobile station. In this section we discuss how radio transmission systems and the telephone network infrastructure are organized to make this service possible.

Radio telephony was first demonstrated in 1915 when an analog voice signal was modulated onto a radio wave. Because electromagnetic waves propagate over a wide geographical area, they are ideally suited for a *radio broadcasting service* where information from a source or station is transmitted to a community of receivers that is within range of the signal. The economics of this type of communication dictate that the cost can be high for the station equipment but that the cost of the receivers must be low so that the service can become available to a large number of users. Commercial broadcast radio was introduced in the early 1920s, and within a few years the service was used by millions of homes.

The introduction of commercial radio resulted in intense competition for frequency bands. The signals from different stations that use the same frequency band will interfere with each other, and neither signal will be received clearly. A limited number of frequencies are available, so in the 1930s it became clear the *regulation* was needed to control the use of frequency bands. Governments established agencies responsible for determining the use and the allocation of frequency bands to various users.

Radio transmission makes communications possible to *mobile* users. Early mobile radio telephone systems used a radio antenna installed on a hill and equipped with a high-power multichannel transmitter. Transmission from the mobile users to the antenna made use of the power supplied by the car battery. These systems provided communications for police, taxis, and ambulance services. The limited amount of available bandwidth restricted the number of calls that could be supported and hence the number of subscribers that could use such systems was limited. For example, in the late 1940s, the mobile telephone service for the entire New York City could only support 543 users [CSTB 1997].

The scarcity of the radio frequency bands that are available and the high demand for their use make frequency spectrum a precious resource. Transmission of a radio signal at a certain power level results in a coverage area composed of the region where the signal power remains significant. By reducing the power level, the coverage area can be reduced and the frequency band can then be reused in adjacent areas. The **frequency-reuse** principle forms the basis for **cellular radio communications**, shown in Figure 4.51.

In cellular telephony, a region, for example, a city, is divided into a number of geographical areas called **cells**.[10] Figure 4.51 shows how a region can be partitioned in a honeycomb pattern using hexagonal cells. Cell areas are established based on the density of subscribers. Large cells are used in rural areas, and

---

[10]Unfortunately, the term *cell* appears in two separate and unrelated areas in networks. The geographic "cells" in a cellular network have nothing to do with the fixed-packet "cells" that are found in ATM networks. The context is usually sufficient to determine what kind of cell is being discussed.

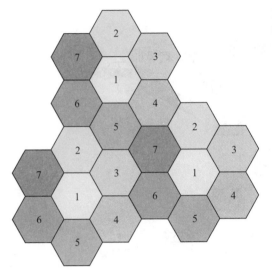

FIGURE 4.51 Cellular network structure

small cells are used in urban areas. As shown in Figure 4.52, a **base station** is placed near the center of each cell. The base station has an antenna that is used to communicate with mobile users in its vicinity. Each base station has a number of *forward channels* available to transmit to its mobile users and an equal number of *reverse channels* to receive from its mobile users.[11]

Base stations are connected by a wireline transmission link or by point-to-point microwave radio to a telephone switch, called the **mobile switching center (MSC)**, which is sometimes also called the *mobile telephone switching office (MTSO)*. The MSC handles connections between cells as well as to the public switched telephone network. As a mobile user moves from one cell to another, a **handoff** procedure is carried out that transfers the connection from one base station to the other, allowing the call to continue without interruption.

In general, immediately adjacent cells cannot use the same set of frequency channels because doing so may result in interference in transmissions to users near their boundary.[12] The set of available radio channels are reused following the *frequency-reuse pattern*. For example, Figure 4.51 shows a seven-cell reuse pattern in which seven disjoint sets of frequency channels are reused as shown. This pattern introduces a minimum distance of one cell between cells using the same frequency channels. Other reuse patterns have reuse factors of 4 and 12. As traffic demand grows, additional capacity can be provided by splitting a cell into several smaller cells.

As an example consider the **Advanced Mobile Phone Service (AMPS)**, which is an analog cellular system in use in North America. In this system the frequency

---

[11]The channels are created by using frequency-division multiple access (FDMA), time-division multiple access (TDMA), or code-division multiple access (CDMA). These techniques are explained in Chapter 6.
[12]An exception is CDMA, which allows the same "code division" channel to be reused in adjacent cells.

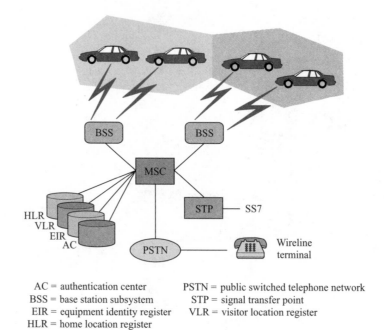

AC = authentication center   PSTN = public switched telephone network
BSS = base station subsystem   STP = signal transfer point
EIR = equipment identity register   VLR = visitor location register
HLR = home location register
MSC = mobile switching center

FIGURE 4.52 Components of a cellular network

band 824 to 849 MHz is allocated to transmissions from the mobile to the base station, and the band 869 to 894 MHz is allocated to transmissions from the base station to the mobile. AMPS uses a 30 kHz channels to carry one voice signal, so the total number of channels available in each direction is 25 MHz/30 kHz = 832 channels. The bands are divided equally between two independent service providers, so each cellular network has 416 bidirectional channels. Each forward and reverse channel pair has frequency assignments that are separated by 45 MHz. This separation between transmit and receive channels reduces the interference between the transmitted signal and the received signal.

A small number of channels within each cell have been designated to function as **setup channels**. For example, the AMPS system allocates 21 channels for this purpose. These channels are used in the setting up and handing off of calls as follows. When a mobile user turns on his or her unit, the unit scans the setup channels and selects the one with the strongest signal. It then monitors this setup channel as long as the signal remains above a certain threshold. To establish a call from the public telephone network or from another mobile user to a mobile user, the MSC sends the call request to all of its base stations, which in turn broadcast the request in all the forward setup channels, specifying the mobile user's telephone number. When the desired mobile station receives the request message, it replies by identifying itself on a reverse setup channel. The corresponding base station forwards the reply to the MSC and assigns a forward and reverse voice channel. The base station instructs the mobile station to begin using these channels, and the mobile telephone is rung.

To initiate a call, the mobile station sends a request in the reverse setup channel. In addition to its phone number and the destination phone number, the mobile station also transmits a serial number and possible password information that is used by the MSC to validate the request. This call setup involves consulting the *home location register*, which is a database that contains information about subscribers for which this is the home area. The validation involves the *authentication center*, which contains authentication information about subscribers. The MSC then establishes the call to the public telephone network by using conventional telephone signaling, and the base station and mobile station are moved to the assigned forward and reverse voice channels.

As the call proceeds, the signal level is monitored by the base station. If the signal level falls below a specified threshold, the MSC is notified and the mobile station is instructed to transmit on the setup channel. All base stations in the vicinity are instructed to monitor the strength of the signal level in the prescribed setup channel. The MSC uses this information to determine the best cell to which the call should be handed off. The current base station and the mobile station are instructed to prepare for a handoff. The MSC then releases its connection to the first base station and establishes a connection to the new base station. The mobile station changes its channels to those selected in the new cell. The connection is interrupted for the brief period that is required to execute the handoff.[13]

When **roaming users** enter an area outside their home region, special procedures are required to provide the cellular phone service. First, business arrangements must be in place between the home and visited cellular service providers. To automatically provide roaming service, a series of interactions is required between the home network and the visited network, using the telephone signaling system. When the roamer enters a new area, the roamer registers in the area by using the setup channels. The MSC in the new area uses the information provided by the roamer to request authorization from the roamer's home location register. The *visitor location register* contains information about visiting subscribers. After registering, the roamer can receive and place calls inside the new area.

Two sets of standards have been developed for the signaling required to support cellular telephone service. The **Global System for Mobile Communications (GSM)** signaling was developed as part of a pan-European public land mobile system. The **Interim Standard 41 (IS-41)** was developed later in North America, using much of the GSM framework. In the following section we describe the protocol layered architecture of GSM.

In the GSM system the *base station subsystem (BSS)* consists of the *base transceiver station (BTS)* and the *base station controller (BSC)*. The BTS consists of the antenna and transceiver to communicate with the mobile telephone.

---

[13]Again CDMA systems differ in that they carry out a "soft" handoff that uses a "make before break" connection approach.

The BTS is also concerned with the measurement of signal strength. The BSC manages the radio resources of one or more BTSs. The BSC is concerned with the setup of frequency channels as well as with the handling of handoffs. Each BTS communicates with the mobile switching center through the BSC, which provides an interface between the radio segment and the switching segment as shown in Figure 4.53.

The GSM signaling protocol stack has three layers as shown in Figure 4.53. Layer 1 corresponds to the physical layer, and layer 2 to the data link layer. GSM layer 3 corresponds to the application layer and is divided into three sublayers: radio resources management (RRM), mobility management (MM), and call management (CM). Different subsets of these layers/sublayers are present in different elements in the GSM network. We discuss these proceeding from the mobile station to the MSC in Figure 4.53.

The radio air interface between the mobile station and the BTS is denoted as U_m. The physical layer across the U_m interface is provided by the radio transmission system. the LAPD protocol is a data link protocol that is part of the ISDN protocol stack and is similar to asynchronous balanced mode in HDLC discussed in Chapter 5. $LAPD_m$ denotes a "mobile" version of LAPD. The radio resources management sublayer between the mobile station and the BTS deals with setting up the radio channels and with handover (the GSM term for handoff).

The interface between the BTS and its BSC is denoted as the A_{bis} interface. The physical layer consists of a 64 kbps link with LAPD providing the data link layer. A BSC can handle a handover if the handover involves two cells under its control. This approach relieves the MSC of some processing load. The interface

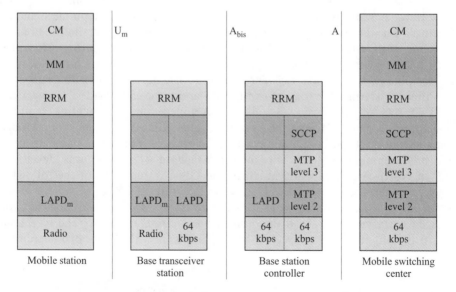

FIGURE 4.53 Protocol stacks in the cellular network

WHAT IS PERSONAL COMMUNICATIONS?

You have probably heard of PCS and of PDAs, two types of "personal" devices. PCS stands for Personal Communication Services and it indicates the class of digital cellular telephone service that has become available in the 1850–1990 MHz band in North America. The *personal* in PCS attempts to contrast the *portability* of the service from that provided by the early analog cellular telephone service that was initially designed for car phones. The PCS user can carry the phone and hence enjoys greater *mobility*. PDA stands for Personal Digital Assistant, and it denotes a small, portable, handheld computer that can be used for personal and business information, such as electronic schedules and calendars, address and phone books, and with the incorporation of modems, e-mail transmission and retrieval. The "personal" in PDA denotes the *private* nature of the use of the device. Therefore, we find that personal communications involves *terminal portability, personal mobility, communications in a variety of modes (voice, data, fax, and so on) and personal profile/customization/privacy.*

What's involved in providing personal communications in the broad sense of the term? An essential component of personal communications is terminal portability, that is, the ability to carry the phone/device with you, and this feature is clearly provided by radio communications and cellular technology. However, this is not all. Personal mobility, the ability to access the network at any point possible by using different terminal devices, is provided by intelligence built into a network. In the case of telephone networks, the intelligence is provided by the signaling system. In the case of other networks, for example, the Internet, this intelligence may be provided in other ways. In any case, personal mobility implies that it is the *individual* who accesses the communication service, not the terminal. Thus there is a need to provide the individual with the *universal personal ID*. Such an ID could be a form of telephone number or address or it could be provided by a smart card such as in GSM. The mode of communications would depend on the capabilities of the terminal device and access network that is in use at a particular time. It would also depend on the personal profile of the individual that would describe a personalized set of communications and other services to which the individual has subscribed. The personal profile could be stored in a database akin to the home location register in cellular networks and/or in a personal smart card.

PCS phones and PDAs have already become consumer market products. In the near future they will be incorporated into a broad array of *wearable* devices, much as radios and tape and CD players in the 1980s and 1990s. We may find them in earrings, in tie clasps, or for those who hate ties, in cowboy-style turquoise-clasp string ties!

between the BSC and the MSC is denoted as the A interface that uses the protocol stack of SS7. The RRM sublayer in the MSC is involved in handovers between cells that belong to different BSCs but that are under the control of the MSC.

Only the mobile station and the MSC are involved in the mobility management and call management sublayers. Mobility management deals with the procedures to locate mobile stations so that calls can be completed. In GSM, cells are grouped into location areas. A mobile station is required to send update messages to notify the system when it is moving between location areas. When a call arrives for a mobile station, it is paged in all cells in the current location area. The MSC, the HLR, and the VLR are involved in the procedures for updating location and for routing incoming calls. The mobility sublayer also deals with the authentication of users. In GSM, unlike other standards, the mobile station includes a smart card, called the Subscriber Identity Module (SIM), which identifies the subscriber, independently of the specific terminal device and provides a secret authorization key. The call management sublayer deals with the establishment and release of calls. It is based on the signaling procedures of ISDN with modifications to deal with the routing of calls to mobile users.

In this section we have focused on the network aspects of cellular telephony. A very important aspect of cellular networks is the access technique that is used to provide the channels that are used for individual connections. Here we have discussed an example of the AMPS standard that is based on frequency division multiplexing of analog signals. In Chapter 6 we consider the various types of access techniques that are used in digital cellular telephone networks.

4.9   SATELLITE CELLULAR NETWORKS

Satellite-based networks using the cellular concept are being deployed to provide global communications. In these systems the entire planet is covered by a constellation of satellites that allow communications from one point in the world to another. These systems can be designed to provide true global personal communications where users with a mobile terminal communicate with the nearest satellite in the constellation and from there to other users anywhere in the world. These systems can also be designed to provide a broadband network in the sky.

The period of time T that it takes a satellite to rotate around the earth is given by the equation

$$T = 2\pi[A^3/g]^{1/2} \text{ seconds}$$

where A is the earth's radius (6378 km) plus the altitude of the satellite, and the gravitational constant is $g = 3.99 \times 10^5 \text{km}^3/\text{s}^2$. Traditional geostationary earth orbit satellites (GEOS) are placed in an orbit approximately 35,786 km above

the earth's surface at the equator. The rotation of the satellite is then synchronized to that of the earth, so the satellite appears stationary with respect to the
earth. At this altitude, the spatial beam of a GEO satellite can cover a large part
of the word, and earth station antennas do not need to track the satellite because
it is stationary. However, the high altitude implies a round-trip transmission time
to the satellite of approximately 270 ms.

In satellites two separate frequency bands are used for communications in
the uplink and downlink direction to minimize interference between the transmitter and receiver. In early satellites the frequency band available was divided
into multiple channels using frequency-division multiplexing. At the satellite
each channel in the uplink would be translated to a different frequency prior
to being broadcast in the downlink. Later time-division multiplexing was used to
share the access to the transmission medium.

Advances in antenna directionality allowed the introduction of **spot beam**
tramsission where the uplink and downlink signals can be focused in smaller
geographical areas. The use of spot beams allows the frequency band to be
reused in geographically separate areas. The introduction of an **onboard switch**
in the satellite allows information to be switched between different spot beams.
However, the number of spot beams that can be implemented in a single satellite
is limited, so the area that can be covered is also limited.

Low-earth orbit satellites (LEOS) use a cellular architecture to provide
global coverage. The altitude of these satellites is typically in the range of 750
to 2000 km, corresponding to rotation periods of approximately two hours and a
satellite speed of about 25,000 km/hour. A group of satellites rotate inside a
plane around the earth in a polar orbit as shown in Figure 4.54a. Consider a
slice of the area covered by these satellites as shown in Figure 4.54b. There are
two basic approaches to how the cells in a slice are defined. In the *satellite-fixed*
approach, the cells are defined with respect to the satellites. Here the satellite
beam directions are fixed, and the earth stations must adjust to the passing

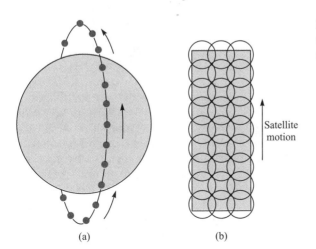

FIGURE 4.54 Cellular
structure in low-earth orbit
satellite network

Satellite
motion

(a)   (b)

satellite. In the *earth-fixed* approach, the cells are fixed with respect to the earth. The satellite closest to the center of a cell steers its beam to a fixed location in the cell. As a result, the satellite appears stationary with respect to the cell for the duration that the satellite covers it. As a satellite completes its pass over a cell, a handoff is carried out to the next cell in the plane. In practice, each LEO satellite has multiple spot beams so its area of coverage at any given time consists of multiple cells. Frequencies can then be reused in nonadjacent cells. Complete coverage of the world is provided by deploying a sufficient number of satellite groups in different orbital planes.

The constellation of satellites in a LEO system is designed to function as a network. Each satellite in a LEO constellation acts as a switching node, and each satellite is connected to nearby satellites by *intersatellite links (ISLs)*. These links route information that is received from a user terminal through the satellite network and eventually toward a gateway to an earth-based network or to a mobile user. In general, each satellite in an orbital plane maintains its position in relation to the other satellites in the plane. However, satellites in different planes are not coordinated. This situation implies that the routing algorithm must adapt to the relative positions of the different satellites and earth stations, as well as to changes in traffic flows.

Each satellite node must also implement signaling and control functions to set up and release connections. The signaling protocols in LEO sytems are similar to ISDN and SS7 protocols used in conventional telephone cellular systems. As in conventional cellular systems, databases must also be maintained to provide the service profiles of users and terminals, as well as authentication and various types of administrative data.

Example—Iridium System

The Iridium system was the first LEO system to be proposed and was deployed in 1998. The primary purpose of the system is to provide global mobile telephone and paging service. The initial proposal involved a constellation of 77 satellites, motivating the name Iridium, for the chemical element 77. The current system consists of 66 satellites in six orbital planes. Each satellite has 48 spot beams, so the global total number of beams is 3168. As the satellite approaches the poles, some of the beams are turned off so that a total of 2150 beams are active at any given time.

The terminal sets are portable telephones that transmit in the 1.616 to 1.626 GHz frequency band. Iridium provides 2.4 or 4.8 kbps voice transmission as well as 2.4 kbps data transmission. Each satellite is connected to four neighboring satellites, and the intersatellite links operate in the 23.18 to 23.38 GHz band. Communications between the satellite system and the terrestrial network take place through ground station gateways. The downlink and uplink frequency bands to these gateways are in the 19.4 to 19.6 GHz band and 29.1 to 29.3 GHz band, respectively.

In mid-1999, the Iridium consortium faced serious financial problems due to the lack of demand for the high cost Iridium service.

Example—Teledesic Network

The Teledesic network is intended to provide high-speed data access to a satellite-based backbone network. Access to the network will be provided through earth-based gateways. Like ISDN, the Teledesic Network defines a basic voice channel and a basic data channel for signaling and control, consisting of 16 kbps and 2 kbps, respectively. Other channel rates are available up to 2.048 Mbps (E-1). Special applications can be provided with up to 1.24416 Gbps (STS-12).

Plans call for 288 satellites to be deployed in a number of polar orbital planes. Each satellite is interconnected to eight adjacent satellites to provide tolerance to faults and adaptability to congestion. All communication inside the satellite network is in the form of 512-bit fixed-length packets. The intersatellite links operate at 155.52 Mbps (STS-3), with multiples of this speed up to 1.24416 Gbps possible. These speeds are also available for communication with the fixed earth stations.

In the Teledesic network the earth is divided into approximately 20,000 square "supercells," each consisting of nine square cells. Each supercell is 160 km long. A satellite uses its beams to cover up to 64 supercells, but the number it is responsible for depends on its orbital position and distance to other satellites. Each cell in a supercell is assigned a time slot, and the satellite beam focuses on each cell in the supercell according to the time slot. When the beam is directed at a cell, each terminal in the cell transmits on the uplink, using one or more frequency channels that have been allocated to it. During this time, the satellite transmits a sequence of packets to terminals in the cell. Terminals receive all packets and select those destined for them by using the address in the header.

SUMMARY

The purpose of this chapter was to show how individual communication systems can be configured into transmission systems that can be controlled to provide end-to-end physical layer connections.

A long-term trend in communications is availability of communication systems with higher bit rates and greater reliability. We began the chapter with a discussion of multiplexing techniques that allow the bandwidth of such systems to be shared among multiple users. We focused on time-division multiplexing and introduced the digital multiplexing hierarchy that forms the backbone of current transmission systems.

Optical fiber technology figures prominently in current and future backbone networks. We introduced the SONET standard for optical transmission, and we discussed its role in the design of transmission systems that are flexible in terms of configuration and robust with respect to faults. We also discussed wavelength-division multiplexing that will increase the usable bandwidth of many existing optical fiber systems by factors of a hundred and more. The availability of these huge bandwidths will change both the access and the backbone architecture of future networks.

In the next few sections of the chapter, we discussed the basic concepts of circuit switching and the modern telephone network. We discussed the design of circuit switches and explained their role in providing end-to-end physical connections on demand. We discussed the structure of the telephone network and paid particular attention to the local loop that represents a major challenge to providing very high bandwidth to the user. The role of signaling in setting up telephone connections was discussed, and the layered architecture of the signaling system was examined. The management of traffic flows in telephone networks through routing plans as well as overload controls was also discussed.

The student may ask why one should bother studying "old" networks and "old" ways of doing things. The reason is that certain fundamental concepts underlie all networks, and that many of these concepts are built into the telephone network. Nevertheless, it is also true that the current way of operating the telephone network is not the only way and not always the right way. The trick, of course, is to know which concepts apply in which context, and only a good grasp of the fundamentals and a careful examination of any given situation can provide the best solution.

As an example of the generality of telephony concepts consider the cellular networks that were discussed in the last sections of the chapter. The value of terminal portability and user mobility and the relatively low infrastructure cost are all factors in the explosive growth of cellular communications. We saw how the key control functions that enable mobility are built on the signaling infrastructure that was developed for telephone networks. Furthermore, we saw how these same concepts can be applied in satellite cellular networks. This technology is a clear demonstration of the application of the same fundamental concepts in different network contexts.

CHECKLIST OF IMPORTANT TERMS

access network
add-drop multiplexer (ADM)
addressing
Advanced Mobile Phone Service
  (AMPS)
allocation of resources
arrival rate
basic rate interface
blocked

busy hour
cellular radio communications
circuit switching
concentration
congestion control
connection oriented
connectionless
connectivity
crossbar switch

digital cross-connect (DCC)
DS1 (digital signal 1)
Erlang B formula
frequency-division multiplexing
  (FDM)
handoff
holding time
intelligent network
Integrated Services Digital Network
  (ISDN)
interexchange carrier (IXC)
local access transport areas (LATA)
local exchange carrier (LEC)
local loop
metropolitan network
mobile telephone switching office
  (MTSO)
multiplexer
multiplexing
◆multistage switch
national network
network management
nonblocking
offered load *a*
onboard switch
optical add-drop multiplexer
optical carrier level (OC)
overload control
primary rate interface
regional Bell operating company
  (RBOC)

regional network
roaming user
routing
routing control
setup channels
signaling
space switch
space-division switch
spot beam
stored program control
switch
synchronous digital hierarchy (SDH)
synchronous optical network
  (SONET)
◆synchronous payload envelope
  (SPE)
synchronous transfer module (STM)
synchronous transport signal level
  (STS)
time-division multiplexing (TDM)
◆time-division switching
time-slot interchange (TSI)
◆time-space-time (TST) switch
  tributary
trunk
utilization
◆virtual tributary
wavelength-division multiplexing
  (WDM)

FURTHER READING

Bellamy, J., *Digital Telephony*, John Wiley & Sons, Inc., New York, 1991.

Carne, E. B., *Telecommunications Primer: Signals, Building, Blocks, and Networks*, Prentice-Hall PTR, Upper Saddle River, New Jersey, 1995.

Clark, M. P., *Network and Telecommunications: Design and Operation*, John Wiley & Sons, New York, 1997.

Clos, C., "A Study of Non-Blocking Switching Networks," *Bell System Technical Journal*, Vol. 32: 406–424, March 1953.

ComLinks.com, "Teledesic Satellite Network," http://www.comlinks.com/satcom/teled.htm, 1999.

Cooper, R. B., *Introduction to Queueing Theory*, North Holland, Amsterdam, 1981.

Computer Science and Telecommunications Board (CSTB), *The Evolution of Untethered Communications*, National Academy of Sciences, Washington D.C., 1997.

Gallagher, M. D. and R. A. Snyder, *Mobile Telecommunications Networking with IS-41*, McGraw-Hill, New York, 1997.

Goodman, D. J., *Wireless Personal Communications Systems*, Addison-Wesley, Reading, Massachusetts, 1997.

Iridium LCC, "Iridium System Facts," http://www.apspg.com/whatsnew/iridium/facts.html, 1997.

Liu, M., *Principles and Applications of Optical Communications*, Irwin, Burr Ridge, 1996.

Mukherjee, B., *Optical Communication Networks*, McGraw-Hill, New York, 1997.

Nortel Networks, "SONET 101," PDF file available through the Nortel Networks site http://www.nortelnetworks.com (Customer Solutions > Optical Solutions > SONET Solutions). Also available through http://www.itprc.com/physical.htm.

Nortel Networks, "S/DMS TransportNode Overview," Appendixes A–G, PDF file available through the Nortel Networks site http://www.nortelnetworks.com (Customer Solutions > Optical Solutions > SONET Solutions > OC-192), 1996.

Pecar, J. A., R. J. O'Connor, and D. A. Garbin, *The McGraw-Hill Telecommunications Factbook*, McGraw-Hill, New York, 1993.

Ramaswami, R. and K. N. Sivaraian, *Optical Networks: A Practical Perspective*, Morgan Kaufmann, San Francisco, 1998.

Russell, T., *Signaling System #7*, McGraw-Hill, New York, 1995.

Scourias, J., "Overview of the Global System for Mobile Communications," Department of Computer Science, University of Waterloo, Waterloo, Ontario, Canada. Technical Report CS-96-15, March 1996. Available from http://www.shoshin.uwaterloo.ca/papers.html.

Sexton, M. and A. Reid, *Transmission Networking: SONET and the Synchronous Digital Hierarchy*, Artech House, Boston, 1992.

SprintBiz News, "Focus on SONET Network Survivability," a PDF file available from http://sprintbiz.com/news/techctr.html.

Teledesic, "Technical Overview of the Teledesic Network," http://www.teledesic.com/tech/details.html, 1998.

PROBLEMS

1. A television transmission channel occupies a bandwidth of 6 MHz.
    a. How many two-way 30 kHz analog voice channels can be frequency-division multiplexed in a single television channel?
    b. How many two-way 200 kHz GSM channels can be frequency-division multiplexed in a single television channel?
    c. Discuss the trade-offs involved in converting existing television channels to cellular telephony channels?

2. A cable sheath has an inner diameter of 2.5 cm.
    a. Estimate the number of wires that can be contained in the cable if the wire has a diameter of 5 mm.
    b. Estimate the diameter of a cable that holds 2700 wire pairs.

3. Suppose that a frequency band W Hz wide is divided into M channels of equal bandwidth.
   a. What bit rate is achievable in each channel? Assume all channels have the same SNR.
   b. What bit rate is available to each of M users if the entire frequency band is used as a single channel and TDM is applied?
   c. How does the comparison of (a) and (b) change if we suppose that FDM requires a guard band between adjacent channels? Assume the guard band is 10% of the channel bandwidth.

4. In a cable television system (see section 3.7.2), the frequency band from 5 MHz to 42 MHz is allocated to upstream signals from the user to the network, and the band from 550 MHz to 750 MHz is allocated for downstream signals from the network to the users.
   a. How many 2 MHz upstream channels can the system provide? What bit rate can each channel support if a 16-point QAM constellation modem is used?
   b. How many 6 MHz downstream channels can the system provide? What bit rates can each channel support if there is an option of 64-point or 256-point QAM modems?

5. Suppose a system has a large band of available bandwidth, say, 1 GHz, that is to be used by a central office to transmit and receive from a large number of users. Compare the following two approaches to organizing the system:
   a. A single TDM system.
   b. A hybrid TDM/FDM system in which the frequency band is divided into multiple channels and TDM is used within each channel.

6. Suppose an organization leases a T-1 line between two sites. Suppose that 32 kbps speech coding is used instead of PCM. Explain how the T-1 line can be used to carry twice the number of calls.

7. A basic rate ISDN transmission system uses TDM. Frames are transmitted at a rate of 4000 frames/second. Sketch a possible frame structure. Recall that basic rate ISDN provides two 64 kbps channels and one 16 kbps channel. Assume that one-fourth of the frame consists of overhead bits.

8. The T-1 carrier system uses a framing bit to identify the beginning of each frame. This is done by alternating the value of the framing bit at each frame, assuming that no other bits can sustain an alternating pattern indefinitely. Framing is done by examining each of 193 possible bit positions successively until an alternating pattern of sufficient duration is detected. Assume that each information bit takes a value of 0 or 1 independently and with equal probability.
   a. Consider an information bit position in the frame. Calculate the average number of times this bit position needs to be observed before the alternating pattern is found to be violated.
   b. Now suppose that the frame synchronizer begins at a random bit position in the frame. Suppose the synchronizer observes the given bit position until it observes a violation of the alternating pattern. Calculate the average number of bits that elapse until the frame synchronizer locks onto the framing bit.

9. The CEPT-1 carrier system uses a framing *byte* at the beginning of a frame.
   a. Suppose that all frames begin with the same byte pattern. What is the probability that this pattern occurs elsewhere in the frame? Assume that each information bit takes a value of 0 or 1 independently and with equal probability.
   b. Consider an arbitrary information bit position in the frame. Calculate the average number of times that the byte beginning in this bit position needs to be observed before it is found to not be the framing byte.
   c. Now suppose that the frame synchronizer begins at a random bit position in the frame. Suppose the synchronizer observes the byte beginning in the given bit position until it observes a violation of the alternating pattern. Calculate the average number of bits that elapse until the frame synchronizer locks onto the framing byte.

10. Suppose a multiplexer has two input streams, each at a nominal rate of 1 Mbps. To accommodate deviations from the nominal rate, the multiplexer transmits at a rate of 2.2 Mbps as follows. Each group of 22 bits in the output of the multiplexer contains 18 positions that always carry information bits, nine from each input. The remaining four positions consist of two flag bits and two data bits. Each flag bit indicates whether the corresponding data bit carries user information or a stuff bit because user information was not available at the input.
    a. Suppose that the two input lines operate at exactly 1 Mbps. How frequently are the stuff bits used?
    b. How much does this multiplexer allow the input lines to deviate from their nominal rate?

11. Calculate the number of voice channels that can be carried by an STS-1, STS-3, STS-12, STS-48, and STS-192. Calculate the number of MPEG2 video channels that can be carried by these systems.

12. Consider a SONET ring with four stations. Suppose that tributaries are established between each pair of stations to produce a fully connected logical topology. Find the capacity required in each hop of the SONET ring in the following three cases, assuming first that the ring is unidirectional and then that the ring is bidirectional.
    a. The traffic between each pair of stations is one STS-1.
    b. Each station produces three STS-1's worth of traffic to the next station in the ring and no traffic to other stations.
    c. Each station produces three STS-1's worth of traffic to the farthest station along the ring and no traffic to other stations.

13. Consider a set of 16 sites organized into a two-tier hierarchy of rings. At the lower tier a bidirectional SONET ring connects four sites. At the higher tier, a bidirectional SONET ring connects the four lower-level SONET rings. Assume that each site generates traffic that requires an STS-3.
    a. Discuss the bandwidth requirements that are possible if 80% of the traffic generated by each site is destined to other sites in the same tier ring.
    b. Discuss the bandwidth requirements that are possible if 80% of the traffic generated by each site is destined to sites in other rings.

14. Consider a four-station SONET ring operated using the four-fiber, bidirectional, line-switched approach. Adjacent stations are connected by two pairs of optical fibers. One

fiber pair carries working traffic, and one fiber pair provides protection against faults. Traffic can be routed between any two stations in either direction, and no limit is placed on the occupancy in each working fiber.

    a. Sketch a break in the ring and explain how the system can recover from this fault.

    b. Suppose a failure affects only one of the fiber pairs connecting two adjacent stations. Explain how the system can recover from this fault.

    c. How does this ring operation differ from the ring operation in Figure 4.12 in terms of available bandwidth.

15. Consider the operation of the dual gateways for interconnecting two bidirectional SONET rings shown in Figure 4.13. The primary gateway transmits the desired signal to the other ring and simultaneously transmits the signal to the secondary gateway that also routes the signal across the ring and then to the primary gateway. A service selector switch at the primary gateway selects between the primary and secondary signals. Explain how this setup recovers from failures in the link between the primary gateways.

16. Consider the synchronous multiplexing in Figure 4.17. Explain how the pointers in the outgoing STS-1 signals are determined.

17. Draw a sketch to explain the relationship between a virtual tributary and the synchronous payload envelope. Show how 28 T-1 signals can be carried in an STS-1.

18. Show that the 100 nm window in the 1300 nm optical band corresponds to 18 terahertz of bandwidth. Calculate the bandwidth for the 100 nm window in the 1550 band.

19. Consider WDM systems with 100, 200, and 400 wavelengths operating at the 1550 nm region and each carrying an STS-48 signal.

    a. How close do these systems come to using the available bandwidth in the 1550 nm range?

    b. How many telephone calls can be carried by each of these systems? How many MPEG2 television signals?

20. Consider the SONET fault protection schemes described in problem 14 and also earlier in the chapter. Explain whether these schemes can be used with WDM rings.

21. Calculate the spacing between the WDM component signals in Figure 4.19. What is the spacing in hertz, and how does it compare to the bandwidth of each component signal?

22. How does WDM technology affect the hierarchical SONET ring topology in Figure 4.13? In other words, what are the consequences of a single fiber providing a large number of high-bandwidth channels?

23. WDM and SONET can be used to create various logical topologies over a given physical topology. Discuss how WDM and SONET differ and explain what impact these differences have in the way logical topologies can be defined.

24. Compare the operation of a multiplexer, an add-drop multiplexer, a switch, and a digital cross-connect.

25. Consider a crossbar switch with n inputs and k outputs.
   a. Explain why the switch is called a concentrator when $n > k$? Under what traffic conditions is this switch appropriate?
   b. Explain why the switch is called an expander when $n < k$? Under what traffic conditions is this switch appropriate?
   c. Suppose an $N \times N$ switch consists of three stages: an $N \times k$ concentration stage; a $k \times k$ crossbar stage; and a $k \times N$ expansion stage. Under what conditions is this arrangement appropriate?
   d. When does the three-stage switch in part (c) fail to provide a connection between an idle input and an idle output line?

26. Consider the multistage switch in Figure 4.23 with $N = 16$, $n = 4$, $k = 2$.
   a. What is the maximum number of connections that can be supported at any given time? Repeat for $k = 4$ and $k = 10$.
   b. For a given set of input-output pairs, is there more than one way to arrange the connections over the multistage switch?

27. In the multistage switch in Figure 4.23, an input line is busy 10% of the time.
   a. Estimate the percent of time p that a line between the first and second stage is busy.
   b. How is p affected by n and k?
   c. How does this p affect the blocking performance of the intermediate crossbar switch?
   d. Supposing that the blocking probability of the intermediate crossbar is small, what is the proportion of time p' that a line between the second and third stage is busy?
   e. For a given input and output line, what is the probability that none of the N/n paths between the input and output lines are available?

28. Consider the multistage switch in Figure 4.23 with $N = 32$. Compare the number of crosspoints required by a nonblocking switch with $n = 16$, $n = 8$, $n = 4$, and $n = 2$.

29. A multicast connection involves transmitting information from one source user to several destination users.
   a. Explain how a multicast connection may be implemented in a crossbar switch.
   b. How does the presence of multicast connections affect blocking performance of a crossbar switch? Are unicast calls adversely affected?
   c. Explain how multicast connections should be implemented in a multistage switch. Should the multicast branching be done as soon as possible or as late as possible?

30. Show that the minimum number of crosspoints in a three-stage nonblocking Clos switch is given by n equal to approximately $(N/2)^{1/2}$.

31. What is the delay incurred in traversing a TSI switch?

32. Explain how the TSI method can be used to build a time-division multiplexer that takes four T-1 lines and combines them into a single time-division multiplexed signal. Be specific in terms of the number of registers required and the speeds of the lines involved.

33. Suppose that the frame structure is changed so that each frame carries two PCM samples. Does this change affect the maximum number of channels that can be supported using TSI switching?

34. Examine the time-space-time circuit-switch architecture and explain the elements that lead to greater compactness, that is, smaller physical size in the resulting switching system.

35. Consider the three-stage switch in problem 25c. Explain why a space-time-space implementation of this switch makes sense. Identify the factors that limit the size of the switches that can be built using this approach.

36. Consider n digital telephones interconnected by a unidirectional ring. Suppose that transmissions in the ring are organized into frames with slots that can hold one PCM sample.
   a. Suppose each telephone has a designated slot number into which it inserts its PCM sample in the outgoing direction and from which it extracts its received PCM sample from the incoming direction. Explain how a TSI system can provide the required connections in this system.
   b. Explain how the TSI system can be eliminated if the pairs of users are allowed to share a time slot?
   c. How would connections be established in parts (a) and (b)?

37. Consider the application of a crossbar structure for switching optical signals.
   a. Which functions are the crosspoints required to implement?
   b. Consider a 2×2 crossbar switch and suppose that the switch connection pattern is ($1 \rightarrow 1, 2 \rightarrow 2$) for T seconds and ($1 \rightarrow 2, 2 \rightarrow 1$) for T seconds. Suppose it takes τ seconds to change between connection patterns, so the incoming optical signals must have guard bands to allow for this gap. Calculate the relationship between the bit rate R of the information in the optical signals, the number of bits in each "frame," and the values T and τ. For R in the range from 1 gigabit per second to 1 terabit per second and τ in the range of 1 microsecond to 1 millisecond, find values of T that yield 50% efficiency in the use of the transmission capacity.

38. Suppose an optical signal contains n wavelength-division multiplexed signals at wavelengths $\lambda_1, \lambda_2, \ldots, \lambda_n$. Consider the multistage switch structures in Figure 4.23 and suppose that the first-stage switches consist of an element that splits the incoming optical signal into n separate optical signals each at one of the incoming wavelengths. The jth such signal is routed to the jth crossbar switch in the middle stage. Explain how the resulting switch structure can be used to provide a rearrangeable optical switch.

39. Consider the equipment involved in providing a call between two telephone sets.
   a. Sketch a diagram showing the various equipment and facilities between the originating telephone through a single telephone switch and on to the destination telephone. Suppose first that the local loop carries analog voice; then suppose it carries digital voice.
   b. Repeat part (a) in the case where the two telephone calls are in different LATAs.
   c. In parts (a) and (b) identify the points at which a call request during setup can be blocked because resources are unavailable.

40. Suppose that an Internet service provider has a pool of modems located in a telephone office and that a T-1 digital leased line is used to connect to the ISP's office. Explain how the 56 K modem (that was discussed in Chapter 3) can be used to provide a 56 kbps transfer rate from the ISP to the user. Sketch a diagram showing the various equipment and facilities involved.

41. Why does the telephone still work when the electrical power is out?

42. In Figure 4.32b, how does the network know which interexchange carrier is to be used to route a long-distance call?

43. ADSL was designed to provide high-speed digital access using existing telephone facilities.
   a. Explain how ADSL is deployed in the local loop.
   b. What happens after the twisted pairs enter the telephone office?
   c. Can ADSL and ISDN services be provided together? Explain why or why not.

44. In this problem we compare the local loop topology of the telephone network with the coaxial cable topology of cable television networks (discussed in Chapter 3).
   a. Explain how telephone service may be provided by using the cable television network.
   b. Explain how cable television service may be provided by using the local loop.
   c. Compare both topologies in terms of providing Internet access service.

45. The local loop was described as having a star topology in the feeder plant and a star topology in the distribution plant.
   a. Compare the star-star topology with a star-ring topology and a ring-ring topology. Explain how information flows in these topologies and consider issues such as efficiency in use of bandwidth and robustness with respect to faults.
   b. What role would SONET transmission systems play in the above topologies?

46. Suppose that the local loop is upgraded so that optical fiber connects the central office to the pedestal and twisted pair of length at most 1000 feet connects the user to the pedestal.
   a. What bandwidth can be provided to the user?
   b. What bit rate does the optical fiber have to carry if each pedestal handles 500 users?
   c. How can SONET equipment be used in the setting?

47. Let's consider an approach for providing fiber-to-the-home connectivity from the central office to the user. The telephone conversations of users are time-division multiplexed at the telephone office and *broadcast* over a "passive optical network" that operates as follows. The TDM signal is broadcast on a fiber up to a "passive optical splitter" that transfers the optical signal to N optical fibers that are connected to N users. Each user receives the entire TDM signal and retrieves its own signal.
   a. Trace the flow of PCM samples to and from the user. What bit rates are required if $N = 10$? 100? Compare these to the bit rate that is available?
   b. Discuss how Internet access service might be provided by using this approach.
   c. Discuss how cable television service might be provided by using this approach.
   d. What role could WDM transmission play in the connection between the central office and the optical splitter? in the connection all the way to the user?

48. Explain where the following fit in the OSI reference model:
   a. A 4 kHz analog connection across the telephone network.
   b. A 33.6 kbps modem connection across the telephone network.
   c. A 64 kbps digital connection across the telephone network.

49. Refer to Figure 1.1 in Chapter 1 and explain the signaling events that take place inside the network as a call is set up and released. Identify the equipment and facilities involved in each phase of the call.

50. Sketch the sequence of events that take place in the setting up of a credit-card call over the intelligent network. Identify the equipment involved in each phase.

51. Explain how the intelligent network can provide the following services:
   a. Caller identification—A display on your phone gives the telephone number or name of the incoming caller.
   b. Call forwarding—Allows you to have calls transferred from one phone to another where you can be reached.
   c. Call answer—Takes voice mail message if you are on the phone or do not answer the phone.
   d. Call waiting—If you are on the phone, a distinctive tone indicates that you have an incoming local or long-distance call. You can place your current call on hold, speak briefly to the incoming party, and return to the original call.
   e. Called-party identification—Each member of a family has a different phone number. Incoming calls have a different ring for each phone number.

52. Consider a 64 kbps connection in ISDN.
   a. Sketch the layers of the protocol stack in the user plane that are involved in the connection. Assume that two switches are involved in the connection.
   b. Sketch the layers of the protocol stack in the control plane that are involved in setting up the connection. Assume that the two switches in part (a) are involved in the call setup.

53. Explain how the control plane from ISDN can be used to set up a virtual connection in a packet-switching network. Is a separate signaling network required in this case?

54. Identify the components in the delay that transpires from when a user makes a request for a telephone connection to when the connection is set up. Which of these components increase as the volume of connection requests increases?

55. Discuss the fault tolerance properties of the STP interconnection structure in the signaling system in Figure 4.40.

56. A set of trunks has an offered load of 10 Erlangs. How many trunks are required to obtain a blocking probability of 2%? Use the following recursive expression for the Erlang B blocking probability:

$$B(c, a) = \frac{aB(c-1, a)}{c + aB(c-1, a)} \qquad B(0, a) = 1$$

57. Compare the blocking probabilities of a set of trunks with offered load $a = 9$ and $c = 10$ trunks to a system that is obtained by scaling up by a factor of 10, that is, $a = 90$ and $c = 100$. Hint: Use the recursion in problem 56.

58. Calls arrive to a pool of 50 modems according to a Poisson process. Calls have an average duration of 25 minutes.
   a. What is the probability an arriving call finds all modems busy if the arrival rate is two calls per minute?
   b. What is the maximum arrival rate that can be handled if the maximum acceptable blocking probability if 1%? 10%?

59. Consider dynamic nonhierarchical routing (DNHR).
   a. Explain how DNHR can be used to exploit the time differences between different time zones in a continent.
   b. Explain how DNHR can be used to exploit different business and residential activity patterns during the day.

60. Suppose that setting up a call requires reserving N switch and link segments.
   a. Suppose that each segment is available with probability p. What is the probability that a call request can be completed?
   b. In allocating switch and transmission resources, explain why it makes sense to give priority to call requests that are almost completed rather than to locally originating call requests.

61. Consider a cellular telephone system with the following parameters: B is the total bandwidth available for the system for communications in both directions; b is the bandwidth required by each channel, including guard bands; R is the reuse factor; and a is the fraction of channels used for set up.
   a. Find an expression for the number of channels available in each cell.
   b. Evaluate the number of channels in each cell for the AMPS system.

62. Consider the AMPS system in problem 61.
   a. How many Erlangs of traffic can be supported by the channels in a cell with a 1% blocking probability? 5%?
   b. Explain why requests for channels from handoffs should receive priority over requests for channels from new calls. How does this change the Erlang load calculations?

63. Suppose that an analog cellular telephone system is converted to digital format by taking each channel and converting it into n digital telephone channels.
   a. Find an expression for the number of channels that can be provided in the digital system using the parameters introduced in problem 61.
   b. Consider the AMPS system and assume that it is converted to digital format using $n = 3$. How many Erlangs of traffic can the new system support at 1% blocking probability? 5%?

64. Suppose that a CDMA system has the same number of channels as the digital system in problem 63, but with a reuse factor of 1.
   a. How many Erlangs of traffic can be supported in each cell by this system at 1% blocking probability? 5%?
   b. Suppose the per capita traffic generated in a city is 0.10 Erlangs during the busiest hour of the day. The city has a population of 1 million residents, and the traffic is generated uniformly throughout the city. Estimate the number of cells required to meet the city's traffic demand using the system in part (a).

65. Consider the equipment involved in providing a call between mobile and wireline telephones.
   a. Sketch a diagram showing the various equipment and facilities between an originating mobile telephone to a wireline destination telephone. Suppose first that the mobile station carries analog voice; then suppose it carries digital voice.
   b. Repeat part (a) in the case where the two telephone calls are mobile.
   c. In parts (a) and (b) identify the points at which a call request can be blocked during call setup because resources are unavailable.

66. Explain the signaling events that take place when a call is set up and released in a cellular telephone system. Identify the equipment and facilities involved in each phase of the call. Consider the following cases.
   a. The source and destination mobile phones are in the same cell.
   b. The source and destination mobile phones are in different cells but in the same MSC.
   c. The source and destination mobile phones are in different cellular networks.

67. Explain the signaling events that take place when a call is handed off from one cell to another cell. Suppose first that the two cells are under the same MSC and then that they are under different MSCs.

68. Bob Smiley, star salesman, has just arrived in Los Angeles from his home base in Chicago. He turns on his cell phone to contact his Los Angeles client.
   a. Sketch the sequence of events that take place to set up his call. Assume he subscribes to roaming service.
   b. Next he calls home in Chicago to inform his wife that he forgot to take out his son's hockey gear from the trunk of his car and to give her the parking spot where he left the car in the airport ("somewhere on the third level of parking lot A"). Sketch the sequence of events that take place to set up his call. (Don't concern yourself with the specifics of the conversation.)
   c. In the meantime, Bob's college roommate, Kelly, who now works in Hollywood, calls Bob's cell phone. Note that Bob's cell phone begins with the Chicago area code. Sketch the sequence of events that take place to set up this call. Should this call be billed as local or long distance?

69. Compare cellular wireless networks to the local loop and to the coaxial cable television system in terms of their suitability to serve as an integrated access network. In particular, comment on the ability to support telephone service, high-speed Internet access, and digital television service. Consider the following two cases:
   a. The band of frequencies available spans 300 MHz.
   b. The band of frequencies available spans 2 GHz.

70. Compare the capabilities of the Iridium system with conventional cellular telephony networks.

71. Compare the capabilities of the Teledesic network with terrestrial fiber networks.

CHAPTER 5

Peer-to-Peer Protocols

In Chapter 2 we presented a top-down examination of the protocol stack and introduced application layer protocols that make use of underlying network services. As an example, we saw how the HTTP protocol makes use of the services provided by the TCP protocol to transfer messages in an arrangement such as that shown in Figure 5.1. We also saw that such an arrangement is central to the notion of layering. As shown in Figure 5.1, two entities or **peer processes** carry out a conversation at each layer. The communications between the layer n + 1 entities is virtual and in fact is carried out by a service provided by layer n. In this chapter we examine the peer-to-peer protocols that are carried out by the layer n peer processes to provide the desired services.

We examine the processing that takes place from when layer n + 1 requests a transfer of a **service data unit (SDU)** until the SDU is delivered to the destination layer n + 1. In particular, we examine how the layer n peer processes construct **protocol data units (PDUs)** and convey control information through the **headers**. We show how each peer process maintains a state that dictates what actions are to be performed when certain events occur.

We are particularly interested in examining the peer-to-peer protocols in two cases. In the first case peer-to-peer data link protocols are used to provide reliable communication across noisy digital transmission lines. We examine these protocols in enough detail to illustrate the general functioning of peer-to-peer protocols. In the second case we are interested in peer-to-peer protocols that operate end to end across a network. In this case the protocols are used to provide reliable stream services and other adaptation functions to deal with transmission impairments in the network.

The chapter is organized as follows:

1. *Peer-to-peer protocols and service models.* We examine where peer-to-peer protocols occur: across a single hop in a network or end to end across an

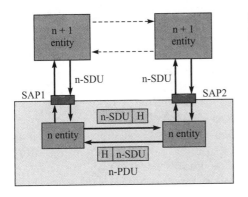

FIGURE 5.1 Layer n entities carry out a protocol to provide service to layer n + 1

entire network. We discuss the various types of services that can be provided by peer-to-peer protocols, especially in terms of adaptation between the network and applications.

2. *ARQ.* We consider Automatic Request (ARQ) protocols that provide reliable transfer of information over a connection-oriented network or over a data link. These protocols are essential when transmitting over channels that are prone to errors. This detailed section is very important because we use it to highlight the essential challenge in network protocols, namely, *the coordination of the actions of two or more geographically separate machines with different state information.* We also examine the efficiency of these protocols in various scenarios and identify the delay-bandwidth product as a key performance parameter.

3. *Other adaptation functions.* We consider other adaptation functions that are provided by peer-to-peer protocols. We discuss pacing and flow control techniques that can be used across a data link or a network so that fast end systems do not overrun the buffers of slow end systems. We also discuss peer-to-peer protocols that provide synchronization and timing recovery. These capabilities are very important for applications that involve voice, audio, and video information. We also preview how the Transmission Control Protocol (TCP) uses ARQ techniques to provide reliable stream service and flow control end to end across connectionless packet networks.[1]

4. *Data link layer.* We examine the data link layer and its essential functions. We discuss two data link control standards that are in widespread use: High-level Data Link Control (HDLC) and Point-to-Point Protocol (PPP).

5. *Statistical multiplexing.* We examine the performance of statistical multiplexers that are used when packets from multiple flows share a common data link. We show how these multiplexers allow links to be used efficiently despite the fact that the individual packet flows are bursty.

---

[1]TCP is discussed in detail in Chapter 8.

5.1 PEER-TO-PEER PROTOCOLS AND SERVICE MODELS

In this section we consider peer-to-peer protocols and the services they provide. A **peer-to-peer protocol** involves the interaction of two processes or entities through the exchange of messages, called protocol data units (PDUs). The purpose of a protocol is to provide a service to a higher layer. Typically the service involves the sending and receiving of information, possibly with features such as confirmation of delivery and guarantees regarding the order in which the information is delivered. Other features may include guarantees regarding the delay or jitter incurred by the PDUs. The service provided by a protocol is described by a **service model**.

Peer-to-peer protocols occur in two basic settings: across a single hop in the network or end to end across an entire network. These two settings can lead to different characteristics about whether PDUs arrive in order, about how long it takes for the PDUs to arrive, or about whether they arrive at all. The design of the corresponding protocols must take these characteristics into account.

Figure 5.2 shows a peer-to-peer protocol that operates across a single hop in a network. Part (a) of the figure uses the lower two layers of the OSI reference

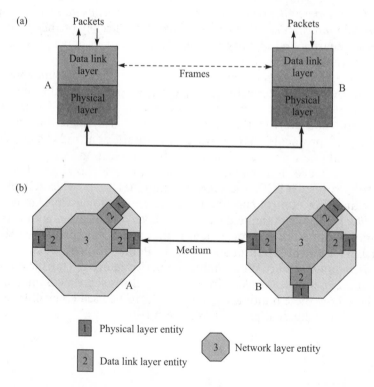

FIGURE 5.2 Peer-to-peer protocol across a single hop

model to show how the data link layer protocol provides for the transfer of packets across a single link in the network. The data link layer takes packets from the network layer, encapsulates them in frames that it transfers across the link, and delivers them to the network layer at the other end. Figure 5.2b shows a more detailed description of the situation in part (a). Each octagon represents a switch in a packet-switching network. Each switch has a single network layer entity and several pairs of data link and physical link entities. Packets arrive from neighboring packet switches and are passed by the data link layers to the network layer entity. The network layer entity decides how each packet is to be routed and then passes the packet to the data link layer entity that connects the switch to the next packet switch in the route. In Figure 5.2b we focus on the data link that connects packet switch A and packet switch B. For example, the peer processes in this data link could be providing a reliable and sequenced transfer service between the two switches.

Figure 5.3 shows a peer-to-peer protocol that operates end to end across a network. In the figure the transport layer peer processes at the end systems accept messages from their higher layer and transfer these messages by exchanging segments end to end across the network. The exchange of segments is accomplished by using network layer services. We use Figure 5.4 to show that the task of the peer-to-peer protocols in this case can be quite complicated. The figure shows the two end systems α and β operating across a three-node network. The segments that are exchanged by the end systems are encapsulated in packets that traverse the three-node network. Suppose that the network operates in datagram node where packets from the same end system are routed independently. It is then possible that packets will follow different paths across the network, and so their corresponding segments may arrive out of order at their destination. Some packets and their segments may also be delayed for long periods or even lost if

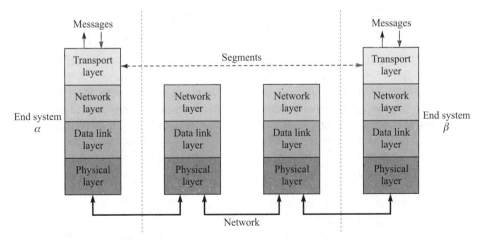

FIGURE 5.3 Peer-to-peer protocols operating end to end across a network—protocol stack view

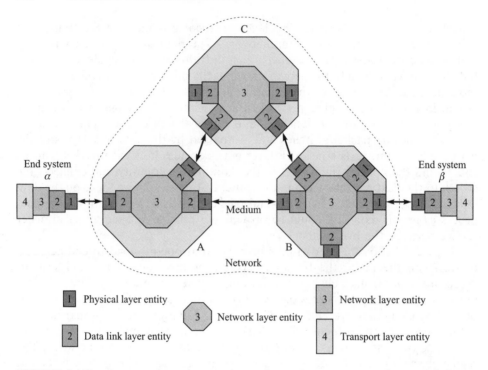

FIGURE 5.4 Peer-to-peer protocols operating end to end across a network—spatial view

they are routed to a congested packet switch. The end-to-end peer transport layer processes must take into account all the characteristics of the segment transfer to be able to provide the desired service to its higher layer. Note that end-to-end peer protocols can also be implemented at layers higher than the transport layer, for example, HTTP at the application layer.

Before discussing service models, let us consider a number of typical applications that involve communication networks and identify their end-to-end requirements. The **end-to-end requirements** of an application are the set of qualitative and quantitative measures that define an acceptable level of performance. The most familiar application is the telephone call. The objective here is to provide for the bidirectional flow of speech information. From the end user's point of view, the communications should be clear and nearly instantaneous to enable natural human interaction. Another familiar application is electronic mail (e-mail), from which in its simplest form involves the unidirectional transmission of a text message. From the end user's point of view, error-free transmission is an essential requirement, but relatively long delays are acceptable.

Video on demand is an emerging application that is based on the familiar television and VCR technologies. Here the user requires the delivery of high-quality video from a distant server. Because video on demand involves the playback of recorded material, the transmission need not be instantaneous. Interactive television, on the other hand, is more likely to be based on electronic video games, and so a greater degree of responsiveness is required.

Finally, consider the World Wide Web (WWW) application that potentially includes the requirements of many of the previous applications. This application builds on a page of text and image information to include audio, video, and varying degrees of interactivity. The point-and-click nature of the application implies responsiveness requirements that have yet to be met by existing networks. In addition, the use of the Web for electronic commerce brings security concerns to the forefront. Together these requirements pose a major challenge in the development of modern communication networks. This situation also underscores a major point of this text—that network design must take into account both the current and future requirements of end-user applications.

5.1.1 Service Models

The **service model** of a given layer specifies the manner in which information is transmitted. There are two broad categories of service models: connection oriented and connectionless. In **connection-oriented services** a connection *setup* procedure precedes the transfer of information. This connection setup initializes state information in the two peer processes. During the *data transfer phase*, this state information provides a context that the peer processes use to track the exchange of PDUs between themselves as well as the exchange of SDUs with the higher layer. Connection-oriented service also involves a *connection release* procedure that removes the state and releases the resources allocated to the connection.

Connectionless services do not involve a connection setup procedure. Instead, individual self-contained blocks of information are transmitted and delivered using address information in each PDU. Information blocks transmitted from the same user to the same destination are transmitted independently. In the simplest case the service does not provide an acknowledgment for transmitted PDUs. Thus if a PDU is lost during transmission, no effort is made to retransmit it. This type of service is appropriate when the transfer of each PDU is reliable or when the higher layer is more sensitive to delay than to occasional losses. Other applications require reliable transfer, so each PDU needs to be acknowledged and retransmitted if necessary. Note that the service is still connectionless so the order in which SDUs are delivered may differ from the order in which they were transmitted.

A service model may also specify a type of transfer capability. For example, connectionless services necessarily involve the transfer of clearly defined *blocks* of information. On the other hand, connection-oriented services can transfer both a *stream* of information in which the individual bits or bytes are not grouped into blocks as well as sequences of blocks of information. Furthermore, some service models may be intended to transfer information at a *constant bit rate*, while others are designed to transfer information at a *variable bit rate*.

The service model can also include a **quality-of-service (QoS)** requirement that specifies a level of performance that can be expected in the transfer of information. For example, QoS may specify levels of reliability in terms of probability of errors, probability of loss, or probability of incorrect delivery. QoS may also address transfer delay. For example, a service could guarantee a fixed (nearly) constant transfer delay, or it could guarantee that the delay will not exceed some given maximum value. The variation in the transfer delay is also an important QoS measure for services such as real-time voice and video. The term **best-effort service** describes a service in which every effort is made to deliver information but without any guarantees.

Up to this point we have discussed the notion of a service model as it may apply to any layer. A particularly important case is the **network service model** that specifies the service provided by a network layer to the transport layer. Recall from Chapter 2 that the transport layer plays the critical role of providing applications with communication services in a manner that is independent of the underlying network(s). To do so, the transport layer must mediate between the service provided by the network layer and the requirements of a given application. This type of mediation is the topic of the next section.

5.1.2   End-to-End Requirements and Adaptation Functions

When attempting to provide a certain application across a network, the network service frequently does not meet the end-to-end requirements of the application. Figure 5.5 shows how **adaptation functions** can be introduced between the network and the applications to provide the required level of service. In this section we consider the following end-to-end requirements and the possible adaptation functions that can be used to meet the level of service required:

- Arbitrary message size.
- Reliability and sequencing.
- Pacing and flow control.
- Timing.
- Addressing.
- Privacy, integrity, and authentication.

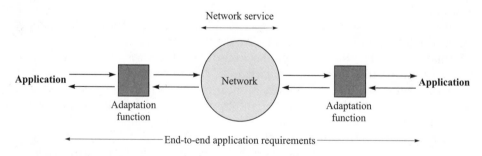

FIGURE 5.5 Adaptation functions

A common element of all user-to-user communication systems and indeed all communication networks is the transfer of a message from one point to another. To accommodate a wide range of services, *networks must be able to handle user messages of arbitrary size*. In the case of e-mail, the message is a discrete, well-defined entity. In the case of computer data files, the size of the files can be very large, suggesting that the files be broken into smaller units that are sent as data blocks and reassembled at the far end to recreate the complete file.

In the case of telephony or video, the notion of a message is less clear. As shown in Figure 5.6a, in telephony one could view the telephone call as generating a single message that consists of the entire sequence of speech samples, but this approach fails to take into account the real-time nature of the application. A more natural view of telephony information is that of a stream of digital speech samples, as shown in Figure 5.6b, which motivates the view of the telephone call as consisting of a sequence of one-byte messages corresponding to each speech sample. (Note that a very large file can also be viewed as a stream of bits or bytes that, unlike telephony, does not have a real-time requirement.)

Video can be seen as somewhat intermediate between telephony and discrete messages, since on the one hand video can be viewed as a stream of bits representing a digitized video signal, while on the other hand it can also be seen as a sequence of messages where each corresponds to a frame or image in the television signal. We can accommodate both of these views by supposing that the network transfers information in blocks. As shown in Figure 5.6c, long messages need to be segmented into a sequence of blocks. In general, sequences of small messages can be handled either by converting each small message into a block or by combining one or more small messages into a block. We will encounter various examples of this **segmentation and reassembly** procedure when we discuss specific network architectures.

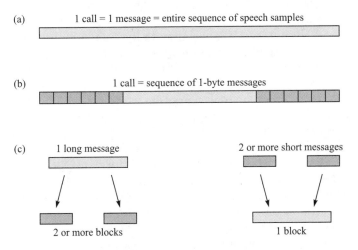

FIGURE 5.6 Message, stream, and sequence of blocks

Many applications involve the transfer of one or more messages or the transfer of streams of information that must be delivered error free, in order, and without duplication. On the other hand, many networks are unreliable in the sense of introducing errors, delivering messages out of order, or losing and even duplicating messages. By combining error-detection coding, automatic retransmission, and sequence numbering, it is possible to obtain protocols by which a transmitter and receiver can *achieve reliable and sequenced communication over unreliable networks*. In the next section we present the details of such protocols.

Reliable end-to-end communication involves not only the arrival of messages to the destination but also the actual delivery of the messages (e.g., to the listener for voice or to the computer program for a data file). A problem that can arise here is that the receiving system does not have sufficient buffering available to store the arriving message. In this case the message is lost. This problem tends to arise when the transmitter can send information at a rate higher than the receiver can accept it. The sliding-window protocols developed in the next section can also be used to provide the receiver with the means of pacing or controlling the rate at which the transmitter sends new messages.

Applications such as speech, audio, and video involve the transfer of a stream of information in which a temporal relationship exists between the information elements. In particular, each of these applications requires a procedure for playing back the information at the receiver end. The system that carries out this playback needs to have appropriate timing information in order to reconstruct the original information signal. For example, in the case of digital speech the receiver must know the appropriate rate at which samples should be entered into the digital-to-analog converter. The situation is similar but quite a bit more complex in the case of digital video. In section 5.3.2, we show how **sequence numbering** and **timestamps** can be used to reconstruct the necessary timing information.

In many applications a network connection is shared by several users. For example, in computer communications different processes in a host may simultaneously share a given network connection. In these situations the user messages need to include addressing information to allow the host to separate the messages and forward them to the appropriate process. The sharing of connections is referred to as *multiplexing*.

Note that adaptation functions cannot be used to meet all application end-to-end requirements. For example, applications such as telephony and videoconferencing are **real time** in the sense that the information must be delivered within a certain delay to be useful. The capability to deliver information within the given delay requirement is determined by the performance of the network and cannot be reduced by adaptation functions.

Public networks increasingly have a new type of "impairment" in the form of security threats. Imposters attempt to impersonate legitimate clients or servers. Attempts to deny service to others are made by flooding a server with requests. In this context the adaptation functions may be used to act as packet-level guards at the gateways to the public network. We deal with these issues in Chapter 11.

5.1.3   End to End versus Hop by Hop

In many situations an option exists for introducing the above adaptation functions on an end-to-end basis or on a hop-by-hop basis, as shown in Figure 5.7. For example, to provide reliable communication, error-control procedures can be introduced at every hop, that is, between every pair of adjacent nodes in a path across the network. Every node is then required to implement a protocol that checks for errors and requests retransmission using ACK and NAK messages until a block of information is received correctly. Only then is the block forwarded along the next hop to the next node. An end-to-end approach, on the other hand, removes the error-recovery responsibility from the intermediate nodes. Instead blocks of information are forwarded across the path, and only the end systems are responsible for initiating error recovery.

There is a basic trade-off in choosing between these two approaches. The hop-by-hop approach initiates error recovery more quickly and gives more reliable service. On the other hand, the processing in each node is more complex. In addition, for the hop-by-hop approach to be effective on an end-to-end basis *every* element in the end-to-end chain must operate correctly. For example, the hop-by-hop approach in Figure 5.7 is vulnerable to the introduction of errors within the switches. The possibility of such errors would necessitate introducing end-to-end error-recovery procedures.[2]

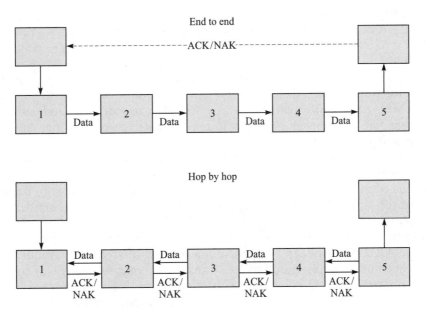

FIGURE 5.7 End-to-end versus hop-by-hop approaches

---

[2]We return to a discussion of the end-to-end argument for system design in Chapter 7.

In the case of reliable service, both approaches are implemented. In situations where errors are infrequent, end-to-end mechanisms are preferred. The TCP reliable stream service introduced later in this chapter provides an example of such an end-to-end mechanism. In situations where errors are likely, hop-by-hop error recovery becomes necessary. The HDLC data link control is an example of a hop-by-hop mechanism.

The end-to-end versus hop-by-hop option appears in other situations as well. For example, flow control and congestion control can be exercised on a hop-by-hop or an end-to-end basis. Security mechanisms provide another example in which this choice needs to be addressed. The approach selected typically determines which layer of the protocol stack provides the desired function. Thus, for example, mechanisms for providing congestion control and mechanisms for providing security are available at the data link layer, the network layer, and the transport layer.

5.2 ARQ PROTOCOLS

Automatic Repeat Request (ARQ) is a technique used to ensure that a data stream is delivered accurately to the user despite errors that occur during transmission. ARQ forms the basis for peer-to-peer protocols that provide for the reliable transfer of information. In this section we develop the three basic types of ARQ protocols, starting with the simplest and building up to the most complex. We also discuss the settings in which the three ARQ protocol types are applied.

This discussion assumes that the user generates a sequence of information blocks for transmission. The ARQ mechanism requires the block to contain a header with control information that is essential to proper operation, as shown in Figure 5.8. The transmitter will also append CRC check bits that cover the header and the information bits to enable the receiver to determine whether errors have occurred during transmission. We assume that the design of the CRC ensures that transmission errors can be detected with very high probability, as discussed in Chapter 3. Recall from Chapter 2 that we use the term **frame** to refer to the binary block that results from the combination of the header, user information, and CRC at the data link layer. We refer to the set of rules that govern the operation of the transmitter and receiver as the **ARQ protocol**.

The ARQ protocol can be applied in a number of scenarios. Traditionally, the most common application involves transmission over a single noisy communication channel. ARQ is introduced here to ensure a high level of reliability across a single transmission hop. As communication lines have become less noisy, the ARQ protocol has been implemented more often at the edges of the network to provide end-to-end reliability in the transmission of packets over multiple hops in a network, that is, over multiple communication channels and other network equipment. In this section we assume that the channel or

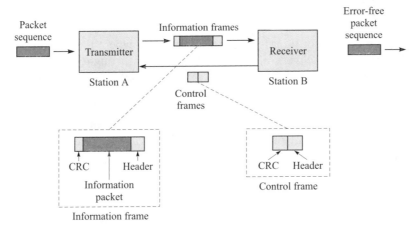

FIGURE 5.8 Basic elements of ARQ

sequence of channels is "wirelike" in the sense that the frames arrive at the receiver, if they arrive at all, in the same order in which they were sent. In the case of multiple hops over a network, this assumption would hold when a connection is set up and where all frames follow the same path, as in frame relay or ATM networks. In particular, we assume that frames, while in transit, cannot pass previously transmitted frames.[3] In situations where these assumptions hold, the objective of the ARQ protocol is to ensure that packets are delivered error free to the destination, exactly once without duplicates, in the same order in which they were transmitted.

In addition to the error-detection code, the other basic elements of ARQ protocols consist of **information frames (I-frames)** that transfer the user packets, control frames, and time-out mechanisms, as shown in Figure 5.8. **Control frames** are short binary blocks that consist of a header that provides the control information followed by the CRC. The control frames include **ACKs**, which acknowledge the correct receipt of a given frame or group of frames; **NAKs**, which indicate that a frame has been received in error and that the receiver is taking certain action; and an **enquiry frame ENQ**, which commands the receiver to report its status. The time-out mechanisms are required to prompt certain actions to maintain the flow of frames. We can visualize the transmitter and receiver as working jointly on ensuring the correct and orderly delivery of the sequence of packets provided by the sender.

We begin with the simple case where information flows only in one direction, from the transmitter to the receiver. The reverse communication channel is used only for the transmission of control information. Later in the section we consider the case of bidirectional transfer of information.

---

[3]We consider the case where frames can arrive out of order later in this chapter.

5.2.1   Stop-and-Wait ARQ

The first protocol we consider is **Stop-and-Wait ARQ** where the transmitter and receiver work on the delivery of one frame at a time through an alternation of actions. In Figure 5.9a we show how ACKs and time-outs can be used to provide recovery from transmission errors, in this case a lost frame. At the initial point in the figure, stations A and B are working on the transmission of frame 0. Note that each time station A sends an I-frame, it starts an **I-frame timer** that will expire after some time-out period. The time-out period is selected so that it is greater than the time required to receiver the corresponding ACK frame. Figure 5.9a shows the following sequence of events:

1. Station A transmits frame 0 and then waits for an ACK frame from the receiver.
2. Frame 0 is transmitted without error, so station B transmits an ACK frame.
3. The ACK from station B is also received without error, so station A knows the frame 0 has been received correctly.
4. Station A now proceeds to transmit frame 1 and then resets the timer.
5. Frame 1 undergoes errors in transmission. It is possible that station B receives frame 1 and detects the errors through the CRC check; it is also possible that frame 1 was so badly garbled that station B is unaware of the transmission.[4] In either case station B does not take any action.
6. The time-out period expires, and frame 1 is retransmitted.

(a) Frame 1 lost

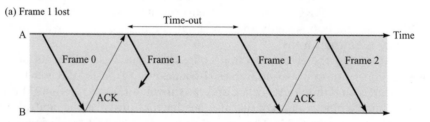

(b) ACK lost

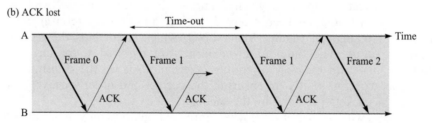

In parts (a) and (b) transmitting station A acts the same way, but part (b) receiving station B accepts frame 1 twice.

FIGURE 5.9 Possible ambiguities when frames are unnumbered

---

[4]In general, when errors are detected in a frame, the frame is ignored. The receiver cannot trust any of the data in the frame and, in particular, cannot take any actions based on the contents of the frame header.

The protocol continues in this manner until frame 1 is received and acknowledged. The protocol then proceeds to frame 2, and so on.

Transmission errors in the reverse channel lead to ambiguities in the Stop-and-Wait protocol that need to be corrected. Figure 5.9b shows the situation that begins as in Figure 5.9a, but where frame 1 is received correctly, and its acknowledgment undergoes errors. After receiving frame 1 station B delivers its contents to the destination. Station A does not receive the acknowledgment for frame 1, so the time-out period expires. Note that at this point station A cannot distinguish between the sequence of events in parts (a) and (b) of Figure 5.9. Station A proceeds to retransmit the frame. If the frame is received correctly by station B, as shown in the figure, then station B will accept frame 1 as a new frame and redeliver it to the user. Thus we see that the loss of an ACK can result in the delivery of a duplicate packet. The ambiguity can be eliminated by including a sequence number in the header of each I-frame. Station B would then recognize that the second transmission of frame 1 was a duplicate, discard the frame, and resend the ACK for frame 1.

A second type of ambiguity arises if the ACKs do not contain a sequence number. In Figure 5.10 frame 0 is transmitted, but the time-out expires prematurely. Frame 0 is received correctly, and the (unnumbered) ACK is returned. In the meantime station A has resent frame 0. Shortly thereafter, station A receives an ACK and assumes it is for the last frame. Station A then proceeds to send frame 1, which incurs transmission errors. In the meantime the second transmission of frame 0 has been received and acknowledged by station B. When station A receives the second ACK, the station assumes the ACK is for frame 1 and proceeds to transmit frame 2. The mechanism fails because frame 1 is not delivered. This example shows that *premature time-outs (or delayed ACKs) combined with loss of I-frames can result in gaps in the delivered packet sequence.* This ambiguity is resolved by providing a sequence number in the acknowledgment frames that enables the transmitter to determine which frames have been received.

The sequence numbers cannot be allowed to become arbitrarily large because only a finite number of bits are available in the frame headers. We now show that a one-bit sequence number suffices to remove the above ambiguities in the Stop-and-Wait protocol. Figure 5.11 shows the information or "state" that is main-

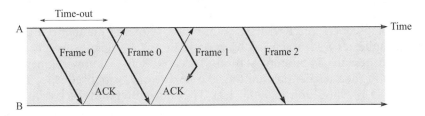

Transmitting station A misinterprets duplicate ACKs

FIGURE 5.10 Possible ambiguities when ACKs are unnumbered

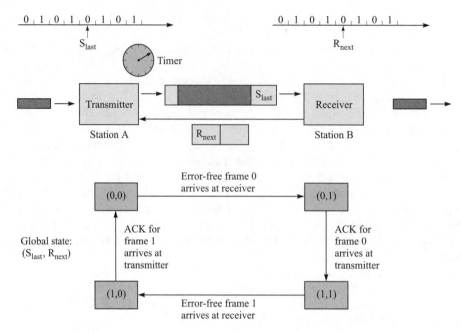

FIGURE 5.11 System state information in Stop-and-Wait ARQ

tained by the transmitter and receiver. The transmitter must keep track of the sequence number S_{last} of the frame being sent, its associated timer, and the frame itself in case retransmission is required. The receiver keeps track only of the sequence number R_{next} of the next frame it is expecting to receive.

Suppose that initially the transmitter and receiver are synchronized in the sense that station A is about to send a frame with $S_{last} = 0$ and station B is expecting $R_{next} = 0$. In Figure 5.11 the global state of the system is defined by the pair (S_{last}, R_{next}), so initially the system is in state (0,0).

The system state will not change until station B receives an error-free version of frame 0. That is, station A will continue resending frame 0 as dictated by the time-out mechanism. Eventually station B receives frame 0, station B changes R_{next} to 1 and sends an acknowledgment to station A with $R_{next} = 1$ implicitly acknowledging the receipt of frame 0. At this point the state of the system is (0,1). Any subsequent received frames that have sequence number 0 are recognized as duplicates and discarded by station B, and an acknowledgment with $R_{next} = 1$ is resent. Eventually station A receives an acknowledgment with $R_{next} = 1$ and then begins transmitting the next frame, using sequence number $S_{last} = 1$. The system is now in state (1,1). The transmitter and receiver are again synchronized, and they now proceed to work together on the transfer of frame 1. Therefore, a protocol that implements this mechanism that follows the well-defined sequence of states shown in Figure 5.11 can ensure the correct and orderly delivery of frames to the destination.

A number of modifications can be made to the above ARQ mechanism by the use of additional control frames. For example, in **checkpointing** the error recovery can be expedited through the use of a short control frame called the enquiry frame (ENQ). When a time-out expires, if the frame that needs to be retransmitted is very long, the transmitter can send an ENQ. When a station receives such a frame that station is compelled to retransmit its previous frame. As shown in Figure 5.12, station B then proceeds to retransmit its last ACK frame, which when received by station A resolves the ambiguity. Station A can then proceed with the transmission of the appropriate frame, as shown in the figure. By convention ACK and NAK frames contain R_{next}, indicating the next frame that is expected by the receiver and implicitly acknowledging delivery of all prior frames.

Stop-and-Wait ARQ becomes inefficient when the propagation delay is much greater than the time to transmit a frame. For example, suppose that we are transmitting frames that are 1000 bits long over a channel that has a speed of 1.5 megabits/second and suppose that the time that elapses from the beginning of the frame transmission to the receipt of its acknowledgment is 40 ms. The number of bits that can be transmitted over this channel in 40 ms is $40 \times 10^{-3} \times 1.5 \times 10^6 = 60,000$ bits. However, Stop-and-Wait ARQ can transmit only 1000 bits in this period time. This severe inefficiency is due to the requirement that the transmitter wait for the acknowledgment of a frame before proceeding with other transmissions. Later in the section we show that the situation becomes much worse in the presence of transmission errors that trigger retransmissions.

The **delay-bandwidth product** is the product of the bit rate and the delay that elapses before an action can take place. In the preceding example the delay-

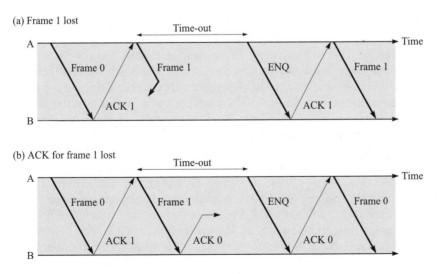

FIGURE 5.12 Stop-and-Wait ARQ enquiry frame

bandwidth product is 60,000 bits. In Stop-and-Wait ARQ the delay-bandwidth product can be viewed as a measure of lost opportunity in terms of transmitted bits. This factor arises as a fundamental limitation in many network problems.

Example—Bisync

Stop-and-wait ARQ was used in IBM's Binary Synchronous Communications (Bisync) protocol. Bisync is a character-oriented data link control that uses the ASCII character set, including several control characters. In Bisync, error detection was provided by two-dimensional block coding where each seven-bit character has a parity bit attached and where an overall block-check character is appended to a frame. Bisync has been replaced by data link controls based on HDLC, which is discussed later in this chapter.

Example—Xmodem

Xmodem, a popular file transfer protocol for modem, incorporates a form of Stop-and-Wait ARQ. Information is transmitted in fixed-length blocks consisting of a 3-byte header, 128 bytes of data, and a one-byte checksum. The header consists of a special start-of-header character, a one-byte sequence number, and a 2s complement of the sequence number. The check character is computed by taking the modulo 2 sum of the 128 data bytes. The receiver transmits an ACK or NAK character after receiving each block.

5.2.2 Go-Back-N ARQ

In this section we show that the inefficiency of Stop-and-Wait ARQ can be overcome by allowing the transmitter to continue sending enough frames so that the channel is kept busy while the transmitter waits for acknowledgments. We develop the Go-Back-N protocol that forms the basis for the HDLC data link protocol, which is discussed at the end of this chapter. Suppose for now that frames are numbered 0, 1, 2, 3, ... The transmitter has a limit on the number of frames W_S that can be outstanding. W_S is chosen larger than the delay-bandwidth product to ensure that the channel can be kept busy.

The idea of the **Basic Go-Back-N ARQ** is as follows: Consider the transfer of a reference frame, say, frame 0. After frame 0 is sent, the transmitter sends $W_S - 1$ additional frames into the channel, optimistic that frame 0 will be received correctly and not require retransmission. If things turn out as expected, an ACK for frame 0 will arrive in due course while the transmitter is still busy sending frames into the channel, as shown in Figure 5.13. The system is now done with frame 0. Note, however, that the handling of frame 1 and subsequent frames is already well underway. A procedure where the processing of a new task is begun before the completion of the previous task is said to be **pipelined**. In

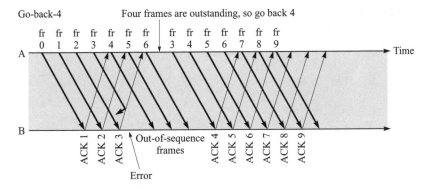

FIGURE 5.13 Basic Go-Back-N ARQ

effect Go-Back-N ARQ pipelines the processing of frames to keep the channel busy.

Go-Back-N ARQ gets its name from the action that is taken when an error occurs. As shown in Figure 5.13, after frame 3 undergoes transmission errors, the receiver ignores frame 3 and all subsequent frames. Eventually the transmitter reaches the maximum number of outstanding frames. It is then forced to "go back N" frames, where $N = W_S$, and begin retransmitting all packets from 3 onwards.

In the previous discussion of Stop-and-Wait, we used the notion of a global state (S_{lasts}, R_{next}) to demonstrate the correct operation of the protocol in the sense of delivering the packets error free and in order. Figure 5.14 shows the similarities between Go-Back-N ARQ and Stop-and-Wait ARQ in terms of error recovery. In Stop-and-Wait ARQ, the occurrence of a frame-transmission error results in the loss of transmission time equal to the duration of the time-out period. In Go-Back-N ARQ, the occurrence of a frame-transmission error results in the loss of transmission time corresponding to W_S frames. In Stop-and-Wait the receiver is looking for the frame with sequence number R_{next}; in Go-Back-N the receiver is looking for a frame with a specific sequence number, say, R_{next}. If we identify the oldest outstanding (transmitted but unacknowledged) frame in Go-Back-N with the S_{last} frame in Stop-and-Wait, we see that the correct operation of the Go-Back-N protocol depends on ensuring that the oldest frame is eventually delivered successfully. The protocol will trigger a retransmission of S_{last} and the subsequent $W_S - 1$ frames each time the send window is exhausted. Therefore, as long as there is a nonzero probability of error-free frame transmission, the eventual error-free transmission of S_{last} is assured and the protocol will operate correctly. We next modify the basic protocol we have developed to this point.

The Go-Back-N ARQ as stated above depends on the transmitter exhausting its maximum number of outstanding frames to trigger the retransmission of a frame. Thus this protocol works correctly as long as the transmitter has an unlimited supply of packets that need to be transmitted. In situations where

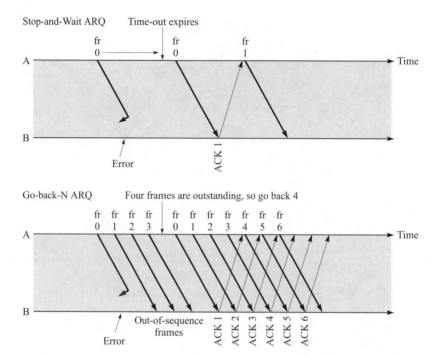

FIGURE 5.14 Relationship of Stop-and-Wait ARQ and Go-Back-N ARQ

packets arrive sporadically, there may not be $W_S - 1$ subsequent transmissions. In this case retransmissions are not triggered, since the window is not exhausted. This problem is easily resolved by modifying Go-Back-N ARQ such that a timer is associated with each transmitted frame.

Figure 5.15 shows how the resulting Go-Back-N ARQ protocol operates. The transmitter must now maintain a list of the frames it is processing, where S_{last} is the number of the last transmitted frame that remains unacknowledged and S_{recent} is the number of the most recently transmitted frame. The transmitter must also maintain a timer for each transmitted frame and must also buffer all frames that have been transmitted but have not yet been acknowledged. At any point in time the transmitter has a **send window** of available sequence numbers. The lower end of the window is given by S_{last}, and the upper limit of the transmitter window is $S_{last} + W_S - 1$. If S_{recent} reaches the upper limit of the window, the transmitter is not allowed to transmit further new frames until the send window slides forward with the receipt of a new acknowledgment.

The Go-Back-N protocol is an example of a **sliding-window protocol**. The receiver maintains a **receive window** of size 1 that consists of the next frame R_{next} it expects to receive. If an arriving frame passes the CRC check and has the correct sequence number, that is, R_{next}, then it is accepted and R_{next} is incremented. We say that the receive window slides forward. The receiver then sends an acknowledgment containing the incremented sequence number R_{next}, which

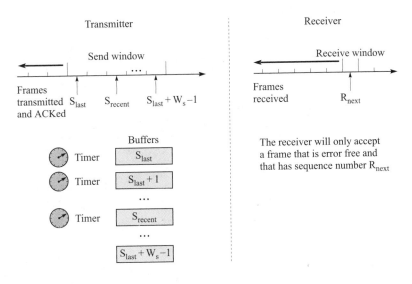

FIGURE 5.15 Go-Back-N ARQ

implicitly acknowledges receipt of all frames prior to R_{next}. Note how we are making use of the assumption that the channel is wirelike: When the transmitter receives an ACK with a given value R_{next}, it can assume that all prior frames have been received correctly, even if it has not received ACKs for those frames, either because they were lost or because the receiver chose not to send them. Upon receiving an ACK with a given value R_{next}, the transmitter updates its value of S_{last} to R_{next} and in so doing the send window slides forward. (Note that this action implies that $S_{last} \leq R_{next}$. Note as well that $R_{next} \leq S_{recent}$, since S_{recent} is the last in the transmission frames.)

The number of bits that can be allotted within a header per sequence number is limited to some number, say, m, which then allows us to represent at most 2^m possible sequence numbers, so the sequence numbers must be counted using modulo 2^m. Thus if $m = 3$, then the sequence of frames would carry the sequence numbers 0, 1, 2, 3, 4, 5, 6, 7, 0, 1, 2, 3, ... When an error-free frame arrives, the receiver must be able to unambiguously determine which frame has been received, taking into account that the sequence numbers wrap around when the count reaches 2^m. We will next show that the receiver can determine the correct frame if the window size is smaller than 2^m.

The example in Figure 5.16 shows the ambiguities that arise when the above inequality is not satisfied. The example uses $2^m = 2^2 = 4$ sequence numbers. The transmitter initially sends four frames in a row. The receiver sends four corresponding acknowledgments, which are all obliterated in the return channel. When the transmitter exhausts its available frame numbers, it goes back four and begins retransmitting from frame 0. When frame 0 arrives at the receiver, the receiver has $R_{next} = 0$, so it accepts the frame. However, the receiver does not know whether the ACK for the previous frame 0 was received. Consequently, the

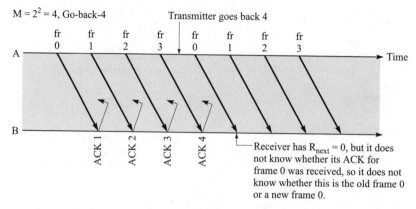

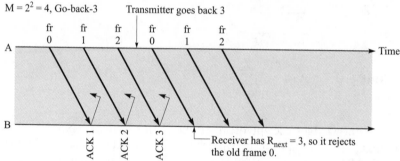

FIGURE 5.16 The window size should be less than 2^m

receiver cannot determine whether this is a new frame 0 or an old frame 0. The second example in Figure 5.16 uses $2^m = 4$ sequence numbering, but a window size of 3. In this case we again suppose that the transmitter sends three consecutive frames and that acknowledgments are lost in the return channel. When the retransmitted frame 0 arrives at the receiver, $R_{next} = 3$, so the frame is recognized to be an old one.

We can generalize Figure 5.16 as follows. In general, suppose the window size W_S is $2^m - 1$ or less and assume that the current send window is 0 up to $W_S - 1$. Suppose that frame 0 is received, but the acknowledgment for frame 0 is lost. The transmitter can only transmit new frames up to frame $W_S - 1$. Depending on which transmissions arrive without error, R_{next} will be in the range of 1 to W_S. Crucially, the receiver will not receive frame W_S until the acknowledgment for frame 0 has been received at the transmitter. Therefore, any receipt of frame 0 prior to frame W_S indicates a duplicate transmission of frame 0.

Figure 5.17 shows that the performance of Go-Back-N ARQ can be improved by having the receiver send a NAK message immediately after the first out-of-sequence frame is received. The NAK with sequence number R_{next} acknowledges all frames up to $R_{next} - 1$ and informs the transmitter that an error

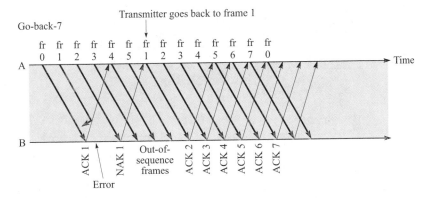

FIGURE 5.17 NAK error recovery

condition has been detected in frame R_{next}. The NAK informs the transmitter that the receiver is now discarding frames that followed R_{next} and instructs the transmitter to go back and retransmit R_{next} and all subsequent packets. By sending the NAK, the receiver sets a condition that cannot be cleared until frame R_{next} is received. For this reason, the receiver is allowed to send only one NAK for any given frame. If the NAK request fails, that is, the NAK is lost or the response transmission fails, then the receiver will depend on the time-out mechanism to trigger additional retransmissions. Note that in general the NAK procedure results in having the transmitter go back less than W_S frames.

Finally we consider the case where the information flow is bidirectional. The transmitter and receiver functions of the modified Go-Back-N protocol are now implemented by both stations A and B. In the direct implementation the flow in each direction consists of information frames and control frames. Many of the control frames can be eliminated by **piggybacking** the acknowledgments in the headers of the information frames as shown in Figure 5.18.

This use of piggybacking can result in significant improvements in the use of bandwidth. When a receiver accepts an error-free frame, the receiver can insert the acknowledgment into the next departing information frame. If no information frames are scheduled for transmission, the receiver can set an **ACK timer** that defines the maximum time it will wait for the availability of an information frame. When the time expires a control frame will be used to convey the acknowledgment.

In the bidirectional case the receiver handles out-of-sequence packets a little differently. A frame that arrives in error is ignored. Subsequent frames that are out of sequence but error free are discarded *after* the ACK (i.e., R_{next}) has been examined. Thus R_{next} is used to update the local S_{last}.

The I-frame time-out value should be selected so that it exceeds the time normally required to receive a frame acknowledgment. As shown in Figure 5.19, this time period includes the round-trip propagation delay $2T_{prop}$, two maximum-length frame transmission times on the reverse channel $2T_f^{max}$, and the frame

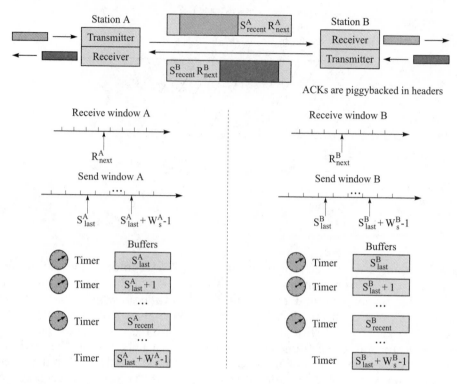

FIGURE 5.18 System parameters in bidirectional Go-Back-N ARQ

processing time T_{proc}.[5] The transmission time of the frame in the forward direction and the ACK time-out are absorbed by the aforementioned frame transmission times.

$$T_{out} = 2T_{prop} + 2T_f^{max} + T_{proc}$$

As in the case of Stop-and-Wait ARQ, the performance of the modified Go-Back-N ARQ protocol is affected by errors in the information frames and errors in the acknowledgments. The transmission of long information frames at the receiver end can delay the transmission of acknowledgments, which can result in the expiry of time-outs at the transmitter end and in turn trigger unnecessary retransmissions. These long delays can be avoided by not using piggybacking and instead sending the short control frames ahead of the long information frames.

---

[5]Note that Figure 5.19 is more detailed than previous figures in that both the beginning and end of each frame transmission need to be shown explicitly.

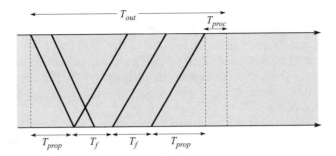

FIGURE 5.19 Calculation of time-out values

Example—HDLC

The High-level Data Link Control (HDLC) is a data link control standard developed by the International Standards Organization. HDLC is a bit-oriented protocol that uses the CRC-16 and CRC-32 error-detection codes discussed in Chapter 3. HDLC can be operated so that it uses Go-Back-N ARQ, as discussed in this section, or Selective Repeat ARQ, which is discussed in section 5.2.3. HDLC is discussed in detail later in this chapter.

The most common use of HDLC is the Link Access Procedure—Balanced (LAPB) discussed later in this chapter. Several variations of the LAPB are found in different applications. For example, Link Access Procedure—Data (LAPD) is the data link standard for the data channel in ISDN. Another example is Mobile Data Link Protocol (MDLP), which is a variation of LAPD that was developed for Cellular Digital Packet Data systems that transmit digital information over AMPS and digital cellular telephone networks.

Example—V.42 Modem Standard

Error control is essential in telephone modem communications to provide protection against disturbances that corrupt the transmitted signals. The ITU-T V.42 standard was developed to provide error control for transmission over modem links. V.42 specifies the Link Access Procedure for Modems (LAPM) to provide the error control. LAPM is derived from HDLC and contains extensions for modem use. V.42 also supports the Microcom Network Protocol (MNP) error-correction protocols as an option. Prior to the development of V.42, these early protocols were de facto standards for error control in modems.

5.2.3 Selective Repeat ARQ

In channels that have high error rates, the Go-Back-N ARQ protocol is inefficient because of the need to retransmit the frame in error and all the subsequent frames. A more efficient ARQ protocol can be obtained by adding two new features: first, the receive window is made larger than one frame so that the receiver can accept frames that are out of order but error free; second, the retransmission mechanism is modified so that only individual frames are retransmitted. We refer to this protocol as **Selective Repeat ARQ**. We continue to work under the constraint that the ARQ protocol must deliver an error-free and ordered sequence of packets to the destination.

Figure 5.20 shows that the send window at the transmitter is unchanged but that the receive window now consists of a range of frame numbers spanning from R_{next} to $R_{next} + W_R - 1$, where W_R is the maximum number of frames that the receiver is willing to accept at a given time. As before, the basic objective of the protocol is to advance the values of R_{next} and S_{last} through the delivery of the oldest outstanding frame. Thus ACK frames carry R_{next}, the oldest frame that has not yet been received. The receive window is advanced with the receipt of an error-free frame with sequence number R_{next}. Unlike the case of Go-Back-N ARQ, the receive window may be advanced by several frames. This step occurs when one or more frames that follow R_{next} have already been received correctly and are buffered in the receiver. R_{next} and the following consecutive packets are delivered to the destination at this point.

Now consider the retransmission mechanism in Selective Repeat ARQ. The handling of timers at the transmitter is done as follows. When the timer expires, only the corresponding frame is retransmitted. There is no longer a clear correspondence between the age of the timers and the sequence numbers. This situation results in a considerable increase in complexity. Whenever an out-of-sequence frame is observed at the receiver, a NAK frame is sent with sequence number R_{next}. When the transmitter receives such a NAK frame, it retransmits

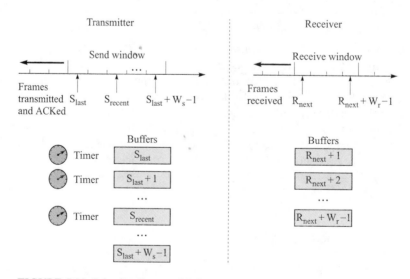

FIGURE 5.20 Selective Repeat ARQ

the specific frame, namely, R_{next}. Finally, we note that the piggybacked acknowledgment in the information frames continues to carry R_{next}.

For example, in Figure 5.21 when the frame with sequence number 2 finally arrives correctly at the receiver, frames 3, 4, 5, and 6 have already been received correctly. Consequently, the receipt of frame 2 results in a sliding of the window forward by five frames.

Consider now the question of the maximum send window size that is allowed for a given sequence numbering 2^m. In Figure 5.22 we show an example in which the sequence numbering is $2^2 = 4$ and in which the send windows and receive windows are of size 3. Initially station A transmits frames 0, 1, and 2.

All three frames arrive correctly at station B, and so the receive window is advanced to {3, 0, 1}. Unfortunately all three acknowledgments are lost, so when the timer for frame 0 expires, frame 0 is retransmitted. The inadequacy of the window size now becomes evident. Upon receiving frame 0, station B cannot determine whether it is the old frame 0 or the new frame 0. So clearly, send and receive windows of size $2^m - 1$ are too large.

It turns out that the maximum allowable window size is $W_S = W_R = 2^{m-1}$, that is, half the sequence number space. To see this we retrace the arguments used in the case of Go-Back-N ARQ. Suppose the window size W_S is 2^{m-1} or less and assume that the current send window is 0 to $W_S - 1$. Suppose also that the initial receive window is 0 to $W_S - 1$. Now suppose that frame 0 is received correctly but that the acknowledgment for frame 0 is lost. The transmitter can transmit new frames only up to frame $W_S - 1$. Depending on which transmissions arrive without error, R_{next} will be in the range between 1 and W_S while $R_{next} + W_R - 1$ will be in the range of 1 to $2W_S - 1$. The maximum value of R_{next} occurs when frames 0 through $W_S - 1$ are received correctly, so the value of R_{next} is W_S and the value of $R_{next} + W_R - 1$ increases to $2W_S - 1$. Crucially, the receiver will not receive frame $2W_S$ until the acknowledgment for frame 0 has been received at the transmitter. Any receipt of frame 0 prior to frame $2W_S$ indicates a duplicate transmission of frame 0. Therefore, the maximum size windows when $W_S = W_R$ is 2^{m-1}.

In the next section we develop performance models for the ARQ protocols. We show that Selective Repeat ARQ outperforms Go-Back-N and Stop-and-

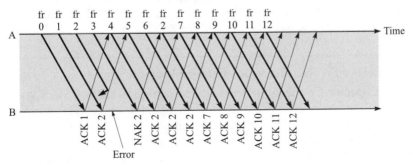

FIGURE 5.21 Error recovery in Selective Repeat ARQ

$M = 2^2 = 4$, selective repeat: send window = receive window = 3

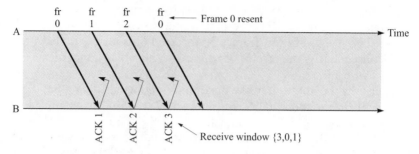

Send window = receive window = 2

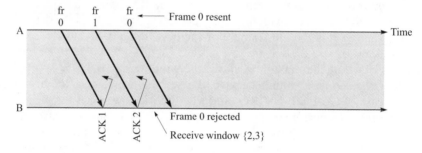

FIGURE 5.22 Maximum window size in Selective Repeat ARQ

Wait ARQ. Of course, this performance level is in exchange for significantly greater complexity.

Example—Transmission Control Protocol

The Transmission Control Protocol (TCP) uses a form of Selective Repeat ARQ to provide end-to-end error control across a network. TCP is normally used over internets that use IP to transfer packets in connectionless mode. TCP must therefore contend with the transfer of PDUs that may arrive out of order, may be lost, or may experience a wide range of transfer delays. We explain how TCP uses ARQ to provide the user of the TCP service with a reliable, connection-oriented stream service in section 5.3.3.

Example—Service Specific Connection Oriented Protocol

The Service Specific Connection Oriented Protocol (SSCOP) provides error control for signaling messages in ATM networks. SSCOP uses a variation of Selective Repeat ARQ that detects the loss of information by monitoring the sequence number of blocks of information that have been received and requesting selective retransmission. The ARQ protocol in SSCOP was originally designed for use in high-speed satellite links. These links have a large delay-bandwidth product, so the protocol was also found useful in ATM networks, which also exhibit large delay-bandwidth product. SSCOP is discussed in Chapter 9.

ARQ: ROBUSTNESS AND ADAPTIVITY

Paul Green, in providing his list of remarkable milestones in computer communications in 1984, made the following remark about ARQ: "(ARQ) is one of those disarmingly simple ideas that seems so trivial (particularly in hindsight) that it really shouldn't be on anyone's list of exciting ideas." ARQ was invented by H.C.A. van Duuren during World War II to provide reliable transmission of characters over radio. In his system each seven-bit character consisted of a combination of four marks (1s) and three spaces (0s). The reception of any other combination of bits led to a request for retransmission. This simple system led to the development of the many modern protocols that have ARQ as their basis.

Simplicity is not the only reason for the success of ARQ. Effectiveness and robustness are equally important attributes of ARQ. The retransmission mechanism gives ARQ the ability to adapt to variable and time-varying channel conditions. If a channel is clean, then ARQ operates in an efficient manner. If a channel becomes noisy, then ARQ adapts the transmission rate to the capability of the channel. This adaptivity makes ARQ relatively robust with respect to channel conditions and hence safe to deploy without detailed knowledge of the channel characteristics.

5.2.4 Transmission Efficiency of ARQ Protocols

In the previous sections we already introduced the performance differences between Stop-and-Wait, Go-Back-N, and Selective Repeat ARQ protocols. In this section we discuss performance results for these protocols and present a quantitative comparison of their transmission efficiency. We show that the delay-bandwidth product, the frame error rate, and the frame length are key parameters in determining system performance.

For simplicity we focus here on the case where the information is unidirectional and where the reverse channel is used only for acknowledgments. We also assume that all information frames have the same length and that the transmitter always has frames to transmit to the receiver. Figure 5.23 shows the components in the basic delay t_0 that transpires in Stop-and-Wait ARQ from the instant a frame is transmitted into the channel to the instant when the acknowledgment is confirmed. The first bit that is input into the channel appears at the output of the channel after a propagation time t_{prop}; the end of the frame is received at station B after t_f additional seconds. Station B requires t_{proc} seconds to prepare an

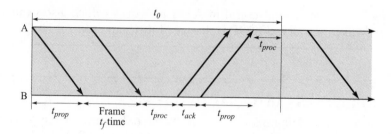

FIGURE 5.23 Delay components in Stop-and-Wait ARQ

acknowledgment frame that will require t_{ack} seconds of transmission time. After an additional propagation delay, the acknowledgment frame is received at station A. Finally, t_{proc} additional seconds are required to carry out the CRC check. The basic time to send a frame and receive an ACK, in the absence of errors, is then given by

$$t_0 = 2t_{prop} + 2t_{proc} + t_f + t_{ack} = 2t_{prop} + 2t_{proc} + \frac{n_f}{R} + \frac{n_a}{R}$$

where n_f is the number of bits in the information frame, n_a is the number of bits in the acknowledgment frame, and R is the bit rate of the transmission channel.

The effective information transmission rate of the protocol in the absence of errors is then given by

$$R_{eff}^0 = \frac{\text{number of information bits delivered to destination}}{\text{total time required to deliver the information bits}} = \frac{n_f - n_o}{t_0}$$

where n_o is the number of overhead bits in a frame and is given by the total number of bits in the header and the number of CRC check bits. The transmission efficiency of Stop-and-Wait ARQ is given by the ratio R_{eff}^0 to R:

$$\eta_0 = \frac{\frac{n_f - n_o}{t_0}}{R} = \frac{1 - \frac{n_o}{n_f}}{1 + \frac{n_a}{n_f} + \frac{2(t_{prop} + t_{proc})R}{n_f}}$$

The preceding equation identifies clearly the sources of inefficiency in transmission. In the numerator the ratio n_o/n_f represents the loss in transmission efficiency due to the need to provide headers and CRC checks. In the denominator the term n_a/n_f is the loss in efficiency due to the time required for the acknowledgment message, where n_a is the number of bits in an ACK/NAK frame. Finally, the term $2(t_{prop} + t_{proc})R$ is the delay-bandwidth product, which was introduced earlier in the section.

Table 5.1 shows the efficiencies of Stop-and-Wait ARQ in a number of scenarios. We first assume a frame size of 1024 bytes (8192 bits) and $n_o = n_a = 8$ bytes. We consider three values of reaction time $t_{prop} + t_{proc}$: 5 ms, 50 ms, and 500 ms. Because the speed of light is 3×10^8 meters/second, these

(a) $n_{frame} = 8192$
$n_{overhead} = 64$, $n_{ack} = 64$

		R (bps)			
		30 kbps	1.5 Mbps	45 Mbps	2.4 Gbps
	0.005	0.95	0.35	1.77E−0.2	3.39E−04
$t_{prop} + t_{proc}$:	**0.050**	0.72	5.14E−02	1.80E−03	3.39E−05
	0.500	0.21	5.39E−03	1.81E−04	3.39E−06

(b) $n_{frame} = 524,288$

	0.005	0.99	0.97	0.53	2.14E−02
$t_{prop} + t_{proc}$:	**0.050**	0.99	0.77	1.04E−01	2.18E−03
	0.500	0.21	0.26	1.15E−02	2.18E−04

TABLE 5.1 Efficiency of Stop-and-Wait ARQ in the absence of errors

delays correspond to distances of about 1500 km, 15,000 km, and 150,000 km links that could correspond roughly to moderate distance, intercontinental, and satellite links, respectively. We consider four values of bit rate: 30,000 bps for an ordinary telephone modem, 1.5 Mbps for a high-speed telephone line, 45 Mbps for high-speed access lines, and 2.4 Gbps for a high-speed backbone line. For the case of an ordinary telephone modem, we see that reasonable efficiencies are attained for reaction times up to 50 ms. This is why Stop-and-Wait ARQ is used extensively in data communication links. However, for the higher-speed lines the efficiencies quickly drop to the point where they are essentially zero. The efficiency can be improved by going to larger frame sizes. The second case considered in Table 5.1 is for $n_f = 64,000$ bytes = 524,288 bits. The efficiency is improved for the 1.5 Mbps line. Nevertheless, the weakness of the Stop-and-Wait protocol with respect to the delay-bandwidth product is again clearly evident in these examples.

We now consider the effect of transmission errors on the efficiency of Stop-and-Wait ARQ. If a frame incurs errors during transmission, the time-out mechanism will cause retransmission of the frame. Let P_f be the probability that a frame transmission has errors. We will assume that a frame transmission has errors independent of other frame transmissions.

Table 5.2 shows the transmission efficiencies of Stop-and-Wait, Go-Back-N, and Selective Repeat ARQ. Recall that Stop-and-Wait ARQ wastes $n_a/n_f + 2(t_{proc} + t_{prop})/n_f$ for every frame waiting for the acknowledgment. On the other hand, Go-Back-N ARQ wastes $W_S - 1$ only when an error occurs (with probability P_f). The superior efficiency of Go-Back-N can be seen by comparing the denominators in the expressions for Stop-and-Wait and Go-Back-N ARQ: For Stop-and-Wait we have $n_a/n_f + 2(t_{proc} + t_{prop})/n_f$, and for Go-Back-N we have $(W_S - 1)P_f$, which is much smaller in general. Note also that when $P_f = 0$, the efficiency of Go-Back-N ARQ is $1 - n_o/n_f$, which is the maximum possible given that the headers and CRC checks are essential.

The expressions in Table 5.2 for Go-Back-N and Selective Repeat ARQ are not affected explicitly by the delay-bandwidth product of the channel. Thus both schemes can achieve efficiency $1 - n_o/n_f$ in the absence of errors. Note, however, that this situation assumes that the window sizes have been selected to keep the

ARQ technique	Efficiency	Comments
Stop-and-Wait	$$\eta = \dfrac{\dfrac{n_f - n_0}{E[t_{total}]}}{R} = (1 - P_f)\dfrac{1 - \dfrac{n_0}{n_f}}{1 + \dfrac{n_a}{n_f} + \dfrac{2(t_{prop} + t_{proc})R}{n_f}} = (1 - P_f)\eta_0$$	Delay-bandwidth product is main factor
Go-Back-N	$$\eta = \dfrac{\dfrac{n_f - n_0}{E[t_{total}]}}{R} = (1 - P_f)\dfrac{1 - \dfrac{n_0}{n_f}}{1 + (W_S - 1)P_f}$$	Average wasted time: $(W_S - 1)P_f$
Selective-Reject	$$\eta = (1 - P_f)\left(1 - \dfrac{n_0}{n_f}\right)$$	Note impact of delay-bandwidth product through W_S

TABLE 5.2 Summary of performance results

channel busy while acknowledgments are returned. For Selective Repeat ARQ this assumption requires the buffering of up to W_S frames at the transmitter and at the receiver, and for Go-Back-N ARQ this assumption requires the buffering of up to W_S frames at the transmitter. The size W_S of the window required for each scenario considered in Table 5.2 is shown in Table 5.3. For the 30 kbps and the 1.5 Mbps channels, we obtain reasonable window sizes. For the 45 Mbps channel reasonable window sizes are obtained by going to longer frames, that is, 64 Kbytes frames instead of 1 Kbytes frames. However for the 2.4 Gbps channel, large window sizes are required even for 64 Kbytes frames.

The selection of optimum frame size must take the channel error rate into account. For a random bit error channel that introduces a bit error with probability p, we have

$$P_f = 1 - P[\text{no bit errors in a frame}]$$
$$= 1 - (1 - p)^{n_f}$$
$$\approx 1 - (1 - n_f p) \approx n_f p \text{ for } p \ll 1$$

(a) $n_{frame} = 8192$

			R (bps)		
		30 kbps	1.5 Mbps	45 Mbps	2.4 Gbps
	0.005	2	4	57	2932
$t_{prop} + t_{proc}$:	**0.050**	2	20	551	29299
	0.500	6	185	5495	292,971

(b) $n_{frame} = 524,288$

	0.005	2	2	3	48
$t_{prop} + t_{proc}$:	**0.050**	2	2	11	460
	0.500	2	5	88	4580

TABLE 5.3 Required window sizes for different combinations of delay and bit rate

The preceding equation shows that the probability of frame error increases with frame size. For burst error channels the following approximation is sometimes used:

$$P_f \approx cn_f p \text{ where } 1/10 < c < 1/3.$$

As before, p is the bit error rate. We see that given the same p, the frame error rate is lower by a factor c for the burst error channel than for the random bit error channel. The reason is that the errors in the burst error channel tend to cluster in a few frames rather than be spread out randomly over many frames.

Figure 5.24 shows the relative performance of the three ARQ protocols in the presence of random bit errors that occur with probability p. The example considers the case of a 1024 byte frame, a 1.5 Mbps channel over a moderate-distance link, that is, 5 ms, where Go-Back-N and Selective Repeat ARQ use windows of size 4. It can be seen that Stop-and-Wait ARQ cannot achieve efficiencies higher than .35 regardless of the value of p. Go-Back-N ARQ achieves efficiencies comparable to Selective Repeat ARQ for p less than 10^{-6} and deteriorates to the performance of Stop-and-Wait ARQ when p increases to about 5×10^{-5}. Selective Repeat ARQ achieves high efficiencies over a greater range of p but also deteriorates as p becomes larger than 10^{-4}. Note that at this rate the probability of one or more errors in a frame is $1 - (1 - 10^{-4})^{8192} \approx 0.56$, so about half of the frames are in error.

Finally we return to the question of selecting the frame length. As the frame length is increased, the impact of the delay-bandwidth product is reduced. However, increasing the frame length also increases the probability of frame transmission error. These factors suggest that for any given bit error rate p, an optimum value of frame length n_f maximizes the transmission efficiency. In Figure 5.25 we show the throughput efficiency for Stop-and-Wait, Go-Back-N, and Selective Repeat ARQ as a function of n_f. We consider the case where $p = 10^{-4}$ and where the other parameters are the same as in the previous example. As n_f is varied, the window sizes are adjusted so that the channel can be kept busy. These window sizes are also shown in Figure 5.25. For Selective Repeat ARQ decreasing n_f reduces P_f and hence increases the efficiency. However if n_f

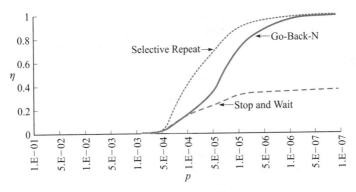

FIGURE 5.24 Transmission efficiency of ARQ protocols

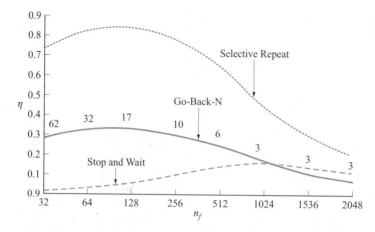

FIGURE 5.25 Optimum frame size

becomes too small, then the overhead term $(1 - n_0/n_f)$ starts decreasing the efficiency. Consequently, we see that the optimum n_f is in the range of 64 to 128 bytes. For Go-Back-N ARQ the optimum value of n_f balances the increases in efficiency due to decreasing P_f against the decrease in efficiency from the overhead term. Again the optimum value of n_f is in the range 64 to 128 bytes. Stop-and-Wait ARQ is more severely penalized for small values of n_f because of the effect of the delay-bandwidth product. Consequently, the optimum value of n_f is in the vicinity of 1024 bytes.

Example—Adapting Frame Size to Error Rate

Certain channels, such as cellular wireless telephone channels, can exhibit a wide variation in signal quality. These variations in quality are manifest in the rate at which transmitted frames incur errors. The results in Figure 5.25 suggest that significant improvements can be obtained by adjusting the frame length to the channel error rate. The MNP 10 error-control protocol was one of the first to implement this type of strategy of dynamically adjusting the frame size to increase the observed throughput in modem transmission over cellular telephone channels. More recent protocols incorporate the frame size adaptation feature as well as other physical layer adaptations to the channel conditions.

◆DERIVATION OF EFFICIENCY OF ARQ PROTOCOLS

In this section we derive the results that were presented in Table 5.2. To calculate the transmission efficiency for Stop-and-Wait ARQ, we need to calculate the average total time required to deliver a correct frame. Let n_t be the number of transmissions required to deliver a frame successfully; then $n_t = i$ transmissions are required if the first $i - 1$ transmissions are in error and the ith transmission is error free. Therefore,

$$P[n_t = i] = (1 - P_f)P_f^{i-1} \text{ for } i = 1, 2, 3 \dots$$

Each unsuccessful frame transmission will contribute a time-out period t_{out} to the total transmission time. The final successful transmission will contribute the basic delay t_0. The average total time to transmit a frame is therefore

$$E[t_{total}] = t_0 + \sum_{i=1}^{\infty} (i - 1)t_{out}P[n_t = i]$$

$$= t_0 + \sum_{i=1}^{\infty} (i - 1)t_{out}(1 - P_f)P_f^{i-1}$$

$$= t_0 + \frac{t_{out}P_f}{1 - P_f}$$

The preceding equation takes into account the fact that in general the time-out period and the basic transmission time are different. To simplify the equation we will assume that $t_{out} = t_0$, then

$$E[t_{out}] = \frac{t_0}{1 - P_f}$$

Thus the effective transmission time for Stop-and-Wait ARQ is

$$R_{eff} = \frac{n_f - n_o}{E[t_{total}]} = (1 - P_f)\frac{n_f - n_o}{t_0} = (1 - P_f)R_{eff}^0$$

and the associated transmission efficiency is

$$\eta = \frac{\frac{n_f - n_o}{E[t_{total}]}}{R} = (1 - P_f)\frac{1 - \frac{n_o}{n_f}}{1 + \frac{n_a}{n_f} + \frac{2(t_{prop} + t_{proc})R}{n_f}} = (1 - P_f)\eta_0$$

In other words, the effect of errors in the transmission channel is to reduce the effective transmission rate and the efficiency by the factor $(1 - P_f)$.

Consider the basic Go-Back-N ARQ where the window size W_S has been selected so that the channel can be kept busy all the time. As before, let n_t be the number of transmissions required to deliver a frame successfully. If $n_t = i$, then as shown in Figure 5.14, $i - 1$ retransmissions of groups of W_S frames are followed by the single successful frame transmission. Therefore, the average total time required to deliver the frame is

$$E[t_{total}] = t_f\{1 + W_S \sum_{i=1}^{\infty}(i-1)P[n_t = i]\}$$

$$= t_f\{1 + W_S \sum_{i=1}^{\infty}(i-1)(1-P_f)P_f^{i-1}\}$$

$$= t_f\{1 + W_S \frac{P_f}{1-P_f}\} = t_f\{\frac{1+(W_S-1)P_f}{1-P_f}\}$$

Thus the effective transmission time for basic Go-Back-N ARQ is

$$R_{eff} = \frac{n_f - n_o}{E[t_{total}]} = (1-P_f)\frac{n_f - n_o}{t_f\{1+(W_S-1)P_f\}} = (1-P_f)\frac{1-\frac{n_0}{n_f}}{1+(W_S-1)P_f}R$$

and the associated transmission efficiency is

$$\eta = \frac{\frac{n_f - n_o}{E[t_{total}]}}{R} = (1-P_f)\frac{1-\frac{n_o}{n_f}}{1+(W_s-1)P_f}$$

Finally, for Selective Repeat ARQ each transmission error involves only the retransmission of the specific frame. Therefore, the average time required to transmit a frame is

$$E[t_{total}] = t_f\{1 + \sum_{i=1}^{\infty}(i-1)(1-P_f)P_f^{i-1}\}$$

$$= t_f\{1 + \frac{P_f}{1-P_f}\} = t_f\frac{1}{1-P_f}$$

Thus the effective transmission time for Selective Repeat ARQ is

$$R_{eff} = (1-P_f)\left(1-\frac{n_0}{n_f}\right)R$$

and the associated transmission efficiency is

$$\eta = (1-P_f)\left(1-\frac{n_o}{n_f}\right)$$

5.3   OTHER ADAPTATION FUNCTIONS

In section 5.2 we did a detailed study of ARQ sliding-window protocols. Our objective there was to provide reliable transfer of a sequence of messages over an unreliable communication channel. In this section we show how the various elements of the ARQ protocols can be used to provide other adaptation functions. In particular, we show how adaptation functions can provide flow control, provide reliable stream service, and provide synchronization and timing information. We also introduce the TCP protocol for providing reliable stream service end-to-end across a network.

5.3.1 Sliding-Window Flow Control

In situations where a transmitter can send information faster than the receiver can process it, messages can arrive at the receiver and be lost because buffers are unavailable. **Flow control** refers to the procedures that prevent a transmitter from overrunning a receiver's buffer.

The simplest procedure for exercising flow control is to use signals that direct the sender to stop transmitting information. Suppose station A is transmitting to station B at a rate of R bps. If station B detects that its buffers are filling up, it issues a stop signal to station A. After approximately one propagation delay T_{prop}, station A stops transmitting as shown in Figure 5.26. From the instant that B sent its signal, it receives an additional $2T_{prop}R$ bits, which is equal to the delay-bandwidth product of the link. Thus station B must send the off signal when its buffer contents exceed a threshold value. This type of flow control is used in the X-ON/X-OFF protocol that is used between a terminal and a computer. This type of control is also used in various data link controls.

The sliding-window protocols that were introduced in section 5.2 on ARQ can also be used to provide flow control. In the simplest case the size of the send window W_S is made equal to the number of buffers that are available at the receiver for messages from the given transmitter. Because W_S is the maximum number of outstanding frames from the transmitter, buffer overflow cannot occur at the receiver. When the sliding window protocol is used for flow control, each acknowledgment of a frame can be viewed as an issuing of a credit by the receiver that authorizes the transmitter to send another frame.

Figure 5.27 shows an example where the receiver sends an acknowledgment after the last frame in a window has been received. In this figure t_{cycle} is the basic delay that elapses from the time the first frame is transmitted to the receipt of its acknowledgment. For this example we have $W_S t_f < t_{cycle}$ where $W_S = 2$ and t_f is the time to transmit a frame. The delay in sending acknowledgments has the effect of pacing or controlling the rate at which the transmitter sends messages to

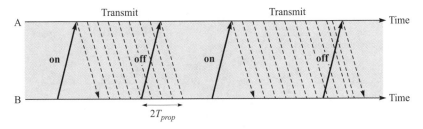

FIGURE 5.26 ON-OFF flow control

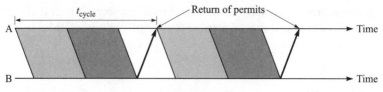

FIGURE 5.27 Sliding-window flow control

the receiver. The average rate at which the sender sends frames is then W_S/t_c frames/second.

In Figure 5.27 we do not take into account the fact that a retransmission time-out mechanism may also be in use. In this case expiry of the send timer may cause unwanted retransmissions. This situation shows how limitations can arise when the same mechanism (sliding-window control) is used for more than one purpose, namely, error control and flow control. For this reason, special control messages are also used to direct a sender to stop sending frames and later to resume sending frames.

It is also possible to decouple acknowledgments of received frames from the issuing of transmitter credits. In this approach separate fields in the header are used to acknowledge received information and to issue transmission credits. The transmission credits authorize the sender to send a certain amount of information in addition to the information that has already been acknowledged. In effect, the receiver is advertising a window of information that it is ready to accept. This approach is used by TCP.

5.3.2   Timing Recovery for Synchronous Services

Many applications involve information that is generated in a synchronous and periodic fashion. For example, digital speech and audio signals generate samples at the corresponding Nyquist sampling rate. Video signals generate blocks of compressed information corresponding to an image 30 times a second. To recover the original signal at the receiver end, the information must be input into the decoder at the same rate at which it was produced at the encoder. Figure 5.28 shows the effect of a transmission over a network on the temporal relationship of the original information blocks. In general networks will introduce a

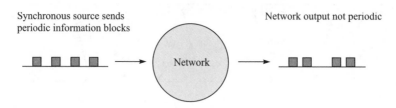

FIGURE 5.28 Timing recovery

variable amount of delay, and so certain blocks will arrive relatively earlier than others, resulting in break up of the periodicity in the sequence, as indicated in the figure. This effect is referred to as **timing jitter**. In this section we consider adaptation functions that can be used to restore the original timing relationship of the sequence of information to allow the decoder to operate correctly.

The first thing that can be done is to provide relative timing information in the sequence of transmitted blocks by inserting a timestamp in each block that enters the network. These timestamps can be used at the receiver to determine both the sequencing of the blocks as well as their relative timing. The typical approach in the "playout" procedure is to select some fixed delay target $T_{playout}$ that with high probability exceeds the total delay experienced by blocks traversing the network. The objective is to play out all the blocks of information so that the total delay, including the network delay, is constant and equal to $T_{playout}$. Thus each time a block of information arrives, an additional delay is calculated, using the difference in the timestamps of the current block and the preceding block.

Most applications have a nominal clock frequency that is known to the transmitter and the receiver, for example, 8 kHz for telephone speech, 44 kHz for audio, or 27 MHz for television. This clock frequency dictates the rate at which samples are to be played out. However, it is extremely unlikely that the clock at the receiver will be exactly synchronized to the clock at the transmitter. Figure 5.29 shows the effect of differences in clock rate. The figure shows the

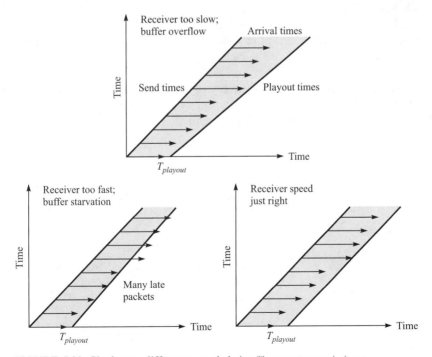

FIGURE 5.29 Clock rate differences and their effect on transmission

times at which the information blocks were produced, and the arrows indicate the total delay that was encountered in the network. When the receiver clock is slow relative to the transmitter clock, the receiver will play its samples at a rate slower than they were produced. Consequently, over a period of time the receiver buffer will grow, eventually leading to loss of information due to buffer overflow. On the other hand, when the receiver is too fast, the receiver buffer will gradually empty and the playout procedure will be interrupted due to lack of available samples. These examples show why the adaptation function must also include a clock recovery procedure that attempts to also synchronize the receiver clock to the transmitter clock.

Many techniques have been proposed for carrying out clock recovery. We will discuss two representative approaches. Figure 5.30 shows a system in which the sequence of timestamp values is used to perform clock recovery. The timestamps in the arriving blocks of information were generated by sampling a counter that is driven by the transmitter clock. The receiver system has a counter that attempts to synchronize to the transmitter clock. The sequence of timestamp values is compared to the local counter values to generate an error signal that is indicative of the difference between the transmitter and receiver clocks. This error signal is filtered and used to adjust the frequency of the local clock. If the difference signal indicates that the local clock is slow, then the frequency of the local clock is increased. If the difference indicates that the local clock is slow, then the frequency of the local clock is reduced. The recovered clock is then used to control the playout from the buffer.

When the local clock is synchronized to the transmitter clock the buffer contents will vary about some constant value. Variations in the buffer contents occur according to the delay jitter experienced by the blocks in traversing the network. Therefore, the size of the buffer must be designed to accommodate the jitter that is experienced in the given network.

Another very effective clock recovery procedure can be used when the transmitter and receiver are connected to a network in which all the elements are synchronized to a common clock, as shown in Figure 5.31. For example, net-

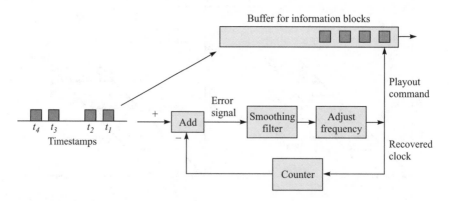

FIGURE 5.30 Adaptive clock recovery

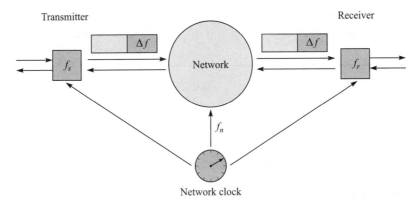

FIGURE 5.31 Clock recovery with synchronous network

works that are built using the SONET standard provide this capability. The transmitter can then compare its frequency f_s to the network frequency f_n to obtain difference Δf. The difference Δf is calculated as follows. Every time the transmitter clock completes N cycles, the number of cycles M completed by the network clock is found. The measured value M is transmitted to the receiver. The receiver deduces Δf by using the fact that the frequencies are related by the following equation:

$$\frac{f_n}{f_s} = \frac{1/M}{1/N}$$

The receiver then controls the playout using the frequency f_r, which is given by

$$f_r = f_n - \Delta f$$

The advantage of this method is that the jitter that takes place within the network does not at all affect the timing recovery procedure, as was the case in Figure 5.30. The basic approach presented here has been made more efficient through a technique called the *synchronous residual timestamp (SRTS)* method. This method has been incorporated in standards for timing recovery in ATM networks that operate over SONET.

Example—Real-Time Transport Protocol

The Real-Time Transport Protocol (RTP) is an application layer protocol designed to support real-time applications such as videoconferencing, audio broadcasting, and Internet telephony. RTP usually operates over UDP, but it can work over connection-oriented or connectionless lower-layer protocols. RTP is concerned only with providing a mechanism for the transfer of information regarding source type, sequence numbering, and timestamps. RTP itself does not implement the procedures for performing timing recovery. This function must be done by applications that operate on top of RTP. RTP is discussed in Chapter 12.

5.3.3   Reliable Stream Service

We now preview the TCP protocol that provides connection-oriented reliable stream service. As in the case of ARQ protocols discussed in previous sections, we are interested in delivering the user information so that it is error free, without duplication, and in the order produced by the sender. However, the user information does not necessarily consist of a sequence of information blocks, but instead consists of a stream of bytes,[6] that is, groups of eight bits, as shown in Figure 5.32. For example, in the transfer of a long file the sender is viewed as inserting a byte stream into the transmitter's send buffer. The task of the TCP protocol is to ensure the transfer of the byte stream to the receiver and the orderly delivery of the stream to the destination application.

TCP was designed to deliver a connection-oriented service in an Internet environment that itself offers connectionless packet transfer service. In this environment each message that is transmitted between a sender and receiver can traverse a different path and can therefore arrive out of order. Unlike "wirelike" links considered in previous sections, in the Internet it is possible for old messages from previous connections to arrive at a receiver, thus potentially complicating the task of eliminating duplicate messages. TCP deals with this problem by using long (32 bit) sequence numbers and by establishing randomly selected initial sequence numbers during connection setup. At any given time the receiver is accepting sequence numbers from a much smaller window, so the likelihood of accepting a very old message is very low. In addition, TCP enforces a time-out

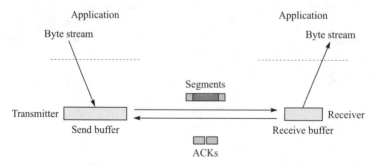

FIGURE 5.32 TCP preview

---

[6]The term *octet* is used for an eight-bit byte in RFC 793 in which TCP is defined.

period at the end of each connection to allow the network to clear old segments from the network.

TCP uses a sliding-window mechanism that implements a form of Selective Repeat ARQ. TCP uses a mechanism that advertises a receiver window size to provide flow control. Finally, TCP can also be used to perform network congestion control through mechanisms that modify a congestion window in response to network conditions. The TCP protocol is discussed in detail in Chapter 8.

5.4 DATA LINK CONTROLS

In Chapter 3 we discussed the various methods for the transmission of a binary information stream over a communications channel. A number of additional functions are required to enable the transfer of frames at the data link layer:

- *Framing* information needs to be introduced to indicate the boundaries that define the transmitted frames.
- *Error control* is required to ensure reliable transmission.
- *Flow control* may be required to prevent the transmitter from overrunning the receiver buffers.
- *Address information* may be required when the channel carries information for multiple users.

The **data link control** provides these functions as well as additional maintenance and security functions that are required to operate the data link. The main assumption in this section is that the data link is wirelike in the sense that messages arrive, if they arrive at all, in the order in which they were transmitted. Thus the data link controls are applicable to point-to-point communication lines as well as to connections over a network where messages follow the same path.

In this chapter we have already discussed the methods that can be used to provide error control and flow control. Our focus in this section is to show how these mechanisms are incorporated into a standard data link control. Here we focus on the **High-level Data Link Control (HDLC)** that was set by the International Standards Organization (ISO).

5.4.1 HDLC Data Link Control

DATA LINK SERVICES

In Figure 5.33 we show the data link control as a set of functions whose role is to provide a communication service to the network layer. The network layer entity is involved in an exchange of packets with a peer network layer entity located at a neighbor packet-switching node. To exchange these packets the network layer must rely on the service that is provided by the data link layer. The **data link**

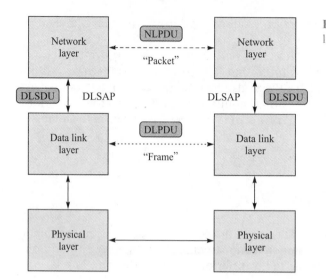

FIGURE 5.33 The data link layer

layer itself transmits frames and makes use of the bit transport service that is provided by the physical layer, that is, the actual digital transmission system.

The network layer passes its packets (network layer PDU or NLPDU) to the data link layer in the form of a data link SDU (DLSDU). The data link layer adds a header and CRC check bits to the SDU (packet) to form a data link PDU (DLPDU) frame. The frame is transmitted using the physical layer. The data link layer at the other end recovers the frame from the physical layer, performs the error checking, and when appropriate delivers the SDU (packet) to its network layer.

Data link layers can be designed to provide several types of services to the network layer. For example, a *connection-oriented service* can provide error-free, ordered delivery of packets. Connection-oriented services involve three phases. The first phase involves setting up the connection. Each network layer has a **service access point (SAP)** through which it accesses its data link layer. These SAPs are identified by addresses, and so the connection setup involves setting up variables and allocating buffers in the data link layers so that packets can flow from one SAP to the other. The second phase of a connection-oriented service involves the actual transfer of packets encapsulated in data link frames. The third phase releases the connection and frees up the variables and buffers that have been allocated to the connection.

Data link layers can also be designed to provide *connectionless service*. In this case there is no connection setup, and the network layer is allowed to pass a packet across its local SAP together with the address of the destination SAP to which the packet is being sent. The data link layer transmits a frame that results in the delivery of a packet to the destination network layer. The connectionless service can be an *acknowledged* service in which case the destination network layer must return an acknowledgment that is delivered to the sending network layer. The connectionless service can also be *unacknowledged* in which case no acknowledgment is issued to the sending network layer. We show later that the local area networks generally provide unacknowledged connectionless service.

HDLC CONFIGURATIONS AND TRANSFER MODES

HDLC provides for a variety of data transfer modes that can be used in a number of different configurations. The **normal response mode (NRM)** of HDLC defines the set of procedures that are to be used with the *unbalanced* configurations shown in Figure 5.34. This mode uses a command/response interaction whereby the primary station sends command frames to the secondary stations and interrogates or polls the secondaries to provide them with transmission opportunities. The secondaries reply using response frames. In the *balanced* point-to-point link configuration, two stations implement the data link control, acting as peers. This configuration is currently in wide use. HDLC has defined the **asynchronous balanced mode (ABM)** for data transfer for this configuration. In this mode information frames can be transmitted in full-duplex manner, that is, simultaneously in both directions.

HDLC FRAME FORMAT

We saw in the discussion of ARQ that the functionality of a protocol depends on the control fields that are defined in the header. The format of the HDLC frame is defined so that it can accommodate the various data transfer modes. Figure 5.35 shows the format of an HDLC frame. Each frame is delineated by two 8-bit

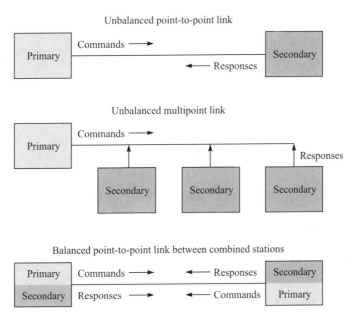

FIGURE 5.34 HDLC configurations

Flag	Address	Control	Information	FCS	Flag

FIGURE 5.35 HDLC frame format

flags. The frame has a field for only one address. Recall that in the unbalanced configuration, there is always only one primary, but there can be more than one secondary. For this reason, the address field always contains the address of the secondary. The frame also contains an 8- or 16-bit control field. We discuss the various types of controls fields below. The information field contains the user information, that is, the SDU. Finally, a 16- or 32-bit CRC calculated over the control, address, and information fields is used to provide error-detection capability. The ITU-CRC polynomials discussed in Chapter 3 are used with HDLC.

The flag in HDLC consists of the byte 01111110. **Bit stuffing** is used to prevent the occurrence of the flag inside the HDLC frame. The transmitter examines the contents inside the frame and inserts an extra 0 after each instance of five consecutive 1s; it then attaches the flag at the beginning and end of the resulting bit-stuffed frame. The receiver looks for five consecutive 1s in the received sequence.

Five 1s followed by a 0 indicate that the 0 is a stuffing bit, and so the bit is removed. Five consecutive 1s followed by 10 indicate the presence of a flag. As an example, suppose that the sequence

0110111111111100

is to be transmitted. The corresponding sequence after bit stuffing is

011011111011111000

The two underlined 0s are stuffing bits. Now suppose that the receiver encounters the sequence

0001110111110111110110

The resulting sequence after destuffing is

000111011111-11111-110

The "-" in the preceding sequence indicates a location where a stuffing 0 has been removed.

There are three types of control fields. Figure 5.36 shows the general format of the control field. A 0 in the first bit of the control field identifies an **information frame (I-frame)**. A 10 in the first two bits of the control field identifies a **supervisory frame**. Finally, a 11 in the first two bits of the control field identifies an **unnumbered frame**. The information frame and supervisory frames implement the main functions of the data link control, which is to provide error and flow control.

Each control field contains a Poll/Final bit indicated by P/F in Figure 5.36. In unbalanced mode this bit indicates a poll when being sent from a primary to a

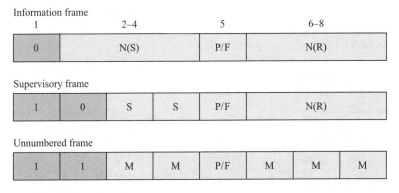

FIGURE 5.36 Control field format

secondary. The bit indicates a final frame when being sent from a secondary to a primary. Thus to poll a given secondary, a host sends a frame to the secondary, indicated by the address field with the P/F bit set to 1. The secondary responds to such a frame by transmitting the frames it has available for transmission. Only the last frame transmitted from the secondary has the P/F bit set to 1 to indicate that it is the final frame.

In balanced mode the P/F bit implements the checkpointing procedure that was introduced in the discussion on ARQ protocols.

The N(S) field in the I-frame provides the send sequence number of the I-frame. The N(R) field is used to piggyback acknowledgments and to indicate the next frame that is expected at the given station. N(R) acknowledges the correct receipt of all frames up to and including N(R) − 1.

There are four types of supervisory frames, corresponding to the four possible values of the S bits in the control field. A value of SS = 00 indicates a **receive ready (RR)** frame. RR frames are used to acknowledge frames when no I-frames are available to piggyback the acknowledgment. A value of SS = 01 corresponds to a **reject (REJ)** frame. REJ frames are used by the receiver to send a negative acknowledgment. As discussed in the section on ARQ, a REJ frame indicates that an error has been detected and that the transmitter should go back and retransmit frames from N(R) onwards (for Go-Back-N). A value of SS = 10 indicates a **receive not ready (RNR)** frame. The RNR frame acknowledges all frames up to N(R) − 1 and informs the transmitter that the receiver has temporary problems, that is, no buffers, and will not accept any more frames. Thus RNR can be used for flow control. Finally, SS = 11 indicates a **selective reject (SREJ)** frame. SREJ indicates to the transmitter that it should retransmit the frame indicated in the N(R) subfield (for Selective Repeat ARQ). Note that this frame is defined only in HDLC, but not necessarily in other variations of HDLC.

The combination of the I-frames and supervisory frames allows HDLC to implement Stop-and-Wait, Go-Back-N, and Selective Repeat ARQ. The discussions earlier in the chapter regarding the operation of these ARQ protocols apply here. In particular, we note that HDLC has two options for sequence numbering.

In the default case HDLC uses a three-bit sequence numbering. This scheme implies that the maximum send window size is $2^3 - 1 = 7$ for Stop-and-Wait and Go-Back-N ARQ. In the extended sequence numbering option, the control field is increased to 16-bits, and the sequence numbers are increased to seven bits. The maximum send window size is $2^7 - 1 = 127$ for Stop-and-Wait and Go-Back-N ARQ. For Selective Repeat ARQ the maximum send and receive window sizes are 4 and 64, respectively.

The unnumbered frames implement a number of control functions. Each type of unnumbered frame is identified by a specific set of M bits. During call setup or release, specific unnumbered frames are used to set the data transfer mode. For example, the **set asynchronous balanced mode (SABM)** frame indicates that the sender wishes to set up an asynchronous balanced mode connection; similarly, a **set normal response mode (SNRM)** frame indicates a desire to set up a normal response mode connection. Unnumbered frames are also defined to set up connections with extended, that is, seven-bit sequence numbering. For example, a **set asynchronous balanced mode extended (SABME)** frame indicates a request to set up an asynchronous balanced mode connection with seven-bit sequence numbering. The **disconnect (DISC)** frame indicates that a station wishes to terminate a connection. An **unnumbered acknowledgment (UA)** frame acknowledges frames during call setup and call release. The **frame reject (FRMR)** unnumbered frame reports receipt of an unacceptable frame. Such a frame passes a CRC check but is not acceptable. For instance, it could have invalid values of the S bits in the case of supervisory frames, for example, 11 when SREJ is not defined, or invalid values of the M bits for an unnumbered frame. The FRMR frame contains a field where additional information about the error condition can be provided. Finally, we note that additional unnumbered frame types are defined for such tasks as initialization, status reporting, and resetting.

TYPICAL FRAME EXCHANGES

We now consider a number of simple examples to show how the frames that are defined for HDLC can be used to carry out various data link control procedures. We use the following convention to specify the frame types and contents in the following figures. The first entry indicates the contents of the address field; the second entry specifies the type of frame, that is, I for information, RR for receive ready, and so on; the third entry is N(S) in the case of I-frames only, the send sequence number; and the following entry is N(R), the receive sequence number. A P or F at the end of an entry indicates that the Poll or Final bit is set.

Figure 5.37 shows the exchange of frames that occurs for connection establishment and release. Station A sends an SABM frame to indicate that it wishes to set up an ABM mode connection. Station B sends an unnumbered acknowledgment to indicate its readiness to proceed with the connection. A bidirectional flow of information and supervisory frames then takes place. When a station wishes to disconnect, it sends a DISC frame and the other station sends UA.

Figure 5.38 shows a typical exchange of frames, using normal response mode. In this example the primary station A is communicating with the secondary stations B and C. Note that all frames contain the address of the secondary

FIGURE 5.37 Exchange of frames for connection establishment and release

station. Station A begins by sending a frame to station B with the poll bit set and the sequence numbering $N(R) = 0$, indicating that the next frame it is expecting from B is frame 0. Station B receives the polling frame and proceeds to transmit three information frames with $N(S) = 0, 1, 2$. The last frame has the final bit set. Frame 1 from station B incurs transmission errors, and subsequently station A receives an out-of-sequence frame with $N(S) = 2$. Station A now sends an SREJ frame with $N(R) = 1$ but without setting the poll bit. This frame indicates to station B that it should be prepared to retransmit frame 1. Station A proceeds to poll station C, which replies that it has no I-frames to transmit. Station A now sends an SREJ with $N(R) = 1$, and the poll bit set. Station B responds by resending frame 1 and then skipping to frames 3 and 4. In the last frame in the figure, station A sends to station B an information frame with $N(S) = 0$ and with $N(R) = 5$, acknowledging receipt of all frames from B up to 4.

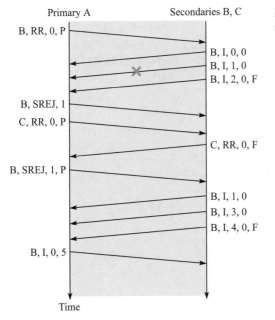

FIGURE 5.38 Exchange of frames using normal response mode

Finally, in Figure 5.39 we consider the case of bidirectional flow of informa-tion using ABM. The following convention is used for the addressing: The address field always contains the address of the secondary station. Thus if a frame is a command, then the address field contains the address of the receiving station. If a frame is a response, then the address field contains the address of the sending station. Information frames are always commands. RR and RNR frames can be either command or response frame; REJ frames are always response frames.

In the example, station A begins by sending frames 0 and 1 in succession. Station B begins slightly later and transmits frame 0. Shortly thereafter station B receives frame 0 from A. When station B transmits its I-frame with $N_B(S) = 1$, an acknowledgment is piggybacked by setting $N_B(R) = 1$. In the meantime, station A has received frame 0 from station B so when it transmits its frame 2, it piggy-backs an acknowledgment by using $N_A(R) = 1$. Now frame 1 from station A has undergone transmission errors, and so when station B receives a frame with $N_A(S) = 2$, it finds that the frame is out of sequence. Station B then sends a negative acknowledgment by transmitting an REJ frame with $N_B(R) = 1$. Meanwhile station A is happily proceeding with the transmission of frames 2, 3, and 4, which carry piggybacked acknowledgments. After receiving the REJ frame, station A goes back and begins retransmitting from frame 1 onward. Note that the value of $N(R)$ is unaffected by retransmission. In the figure we use solid lines to indicate response frames that are sent from B to A. Thus we see that when station B does not have additional I-frames for transmission, it sends acknowledgments by using RR frames in response form.

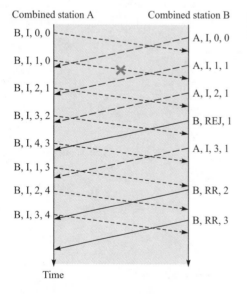

Combined station A   Combined station B

B, I, 0, 0

B, I, 1, 0

B, I, 2, 1

B, I, 3, 2

B, I, 4, 3

B, I, 1, 3

B, I, 2, 4

B, I, 3, 4

A, I, 0, 0

A, I, 1, 1

A, I, 2, 1

B, REJ, 1

A, I, 3, 1

B, RR, 2

B, RR, 3

Time

FIGURE 5.39 Exchange of frames using asynchronous balanced mode

5.4.2   Point-to-Point Protocol

The **Point-to-Point protocol (PPP)** provides a method for encapsulating IP packets over point-to-point links. PPP can be used as a data link control to connect two routers or can be used to connect a personal computer to an Internet service provider (ISP) using a telephone line and a modem. The PPP protocol can operate over almost any type of full-duplex point-to-point transmission link. It can also operate over traditional asynchronous links,[7] bit synchronous links, and new transmission systems such as ADSL and SONET.

The PPP protocol uses an HDLC-like frame format to encapsulate datagrams over point-to-point links, as shown in Figure 5.40. The PPP frame always begins and ends with the standard HDLC flag. Unlike HDLC, PPP frames consist of an integer number of bytes. For this reason, the bit stuffing technique of HDLC is not used. Instead a byte insertion method that makes use of an escape character is applied.

The second field in the PPP frame normally contains the "all 1s" address field that indicates that all stations are to accept the frame. The control field is usually set to 00000011 because PPP is normally run in connectionless mode. The 00000011 indicates an unnumbered HDLC frame, and so sequence numbers are not used. For noisy links a numbered mode option for using the ABM mode of HDLC to provide sequenced and reliable communication is available. In this case the initial flag in the HDLC frame is followed by a one- or two-byte address field and a one- or two-byte control field.

PPP was designed to support multiple network protocols simultaneously; that is, PPP can transfer packets that are produced by different network layer protocols. This situation arises in *multiprotocol* routers that can simultaneously support several network layer protocols. The protocol field is 1 or 2 bytes long and is used to identify the network layer protocol of the packet contained in the information field. Note that the protocol and PPP information field together correspond to the information field in the normal HDLC frame (see Figure 5.35).[8] Finally, the CRC field can use the CCITT 16 or CCITT 32 generator polynomials presented in Chapter 3.

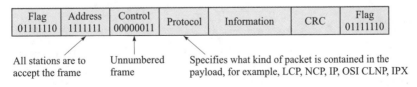

FIGURE 5.40 PPP frame format

---

[7] In asynchronous links the transmission of each character is preceded by a "start" bit and followed by a "stop" bit. Synchronous links provide long-term bit synchronization, making start and stop bits unnecessary.

[8] Note also how the bits in the PPP control field appear in reverse order relative to the HDLC control field.

The PPP protocol provides many useful capabilities through a link control protocol and a family of network control protocols. The **link control protocol (LCP)** is used to set up, configure, test, maintain, and terminate a link connection. As shown in Figure 5.41, the LCP begins by establishing the physical connection. The LCP involves an exchange of messages between peers to negotiate the link configuration. During the negotiation, a PPP endpoint may also indicate to its peer that it wants to multilink (i.e., combine multiple physical links into one logical link). Multilink PPP allows a high-speed data link to be built from multiple low-speed physical links. Once the peers have agreed on a configuration, an authentication process, if selected in the configuration, is initiated. We discuss the authentication option below.

After authentication has been completed, a **network control protocol (NCP)** is used to configure each network layer protocol that is to operate over the link. PPP can subsequently transfer packets from these different network layer protocols (such as IP, IPX, Decnet, AppleTalk) over the same data link. The destination peer can then direct the encapsulated packet to the appropriate network layer protocol by reading the protocol field in the frame. The reason this capability is important is that routers have evolved to simultaneously support packets from different network protocols.

When a PC is connecting to an IP network, as in Figure 5.41, the NCP for IP negotiates a dynamically assigned IP address for the PC. In low-speed lines it may also negotiate TCP and IP header compression schemes that reduce the number of bits that need to be transmitted.

A particular strength of PPP is that it includes authentication protocols, which is a major issue when the computer connects to a remote network. After the LCP has set up the link, these protocols can be used to authenticate the user. The **Password Authentication Protocol (PAP)** requires the initiator to

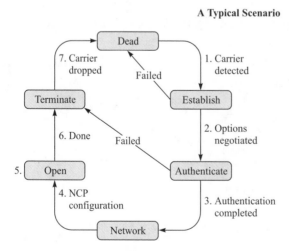

A Typical Scenario

Home PC to Internet Service Provider

1. PC calls router via modem.

2. PC and router exchange LCP packets to negotiate PPP parameters.

3. Check on identities.

4. NCP packets exchanged to configure the network layer, for example, TCP/IP (requires IP address assignment).

5. Data transport, for example, send/receive IP packets.

6. NCP used to tear down the network layer connection (free up IP address); LCP used to shut down data link connection.

7. Modem hangs up.

FIGURE 5.41 PPP phase diagram

send an ID and a password. The peer process then responds with a message indicating that the authentication has been successful or has failed. Depending on the type of situation, a system may allow a few retries. When PAP decides that a request has failed, it instructs the LCP to terminate the link. PAP is susceptible to eavesdropping because the ID and password are sent in plain text; PAP is therefore vulnerable to many different types of security attacks.

The **Challenge-Handshake Authentication Protocol (CHAP)** provides greater security by having the initiator and the responder go through a challenge-response sequence. CHAP assumes that the peer processes have somehow established a shared secret key. After LCP has established the link, the authenticator sends a challenge to its peer. The challenge consists of a random number and an ID. The peer process responds with a cryptographic checksum of the challenge value that makes use of the shared secret. The authenticator verifies the cryptographic checksum by using the shared secret key. If the two checksums agree, then the authenticator sends an authentication message. The CHAP protocol allows an authenticator to reissue periodically the challenge to reauthenticate the process. Security protocols are discussed in detail in Chapter 11.

◆5.5   LINK SHARING USING PACKET MULTIPLEXERS

In Chapter 1 we discussed how applications in early terminal-oriented networks were found to generate data in a bursty fashion: Message transmissions would be separated by long idle times. Assigning a dedicated line to transmission from a single terminal resulted in highly inefficient use of the line. Such inefficiency became a serious issue when the transmission lines involved were expensive, as in the case of long-distance lines. Statistical multiplexers were developed to concentrate data traffic from multiple terminals onto a shared communication line, leading to improved efficiency. Framing, addressing, and link control procedures were developed to structure the communications in the shared transmission link, eventually leading to the development of data link control protocols such as HDLC and PPP.

Current networks support a very broad array of applications many of which generate traffic in highly bursty fashion, making statistical multiplexing an essential component in the operation of packet networks. Indeed in Figure 5.4 we see that in modern networks each packet switch takes packets that it receives from users and from other packet switches and multiplexes them onto shared data links. In this section we present an introduction to classical modeling techniques for analyzing the performance of packet multiplexers.[9] We begin by looking at

---

[9]The characterization of the traffic generated by current applications is now well understood and is currently an active area of research. The reader is referred to recent issues of the *IEEE Journal on Selected Areas in Communications*.

the problem of multiplexing the packet flows from various data sources. We find that the approach leads to significant improvements in the utilization of the transmission line. We also find that the aggregation of traffic flows results in improved performance. We then consider the multiplexing of packetized voice traffic and again find benefits in the aggregation of multiple flows.

5.5.1   Statistical Multiplexing

Computer applications tend to generate data for transmission in a bursty manner. Bursts of information are separated by long idle periods, and so dedicating a transmission line to each computer is inefficient. This behavior led to the development of statistical multiplexing techniques for sharing a digital transmission line. The information generated by a computer is formatted into packets that contain headers that identify the source and destination, as shown in Figure 5.42. The packets are then transmitted in a shared communications line. In general the packets are variable in length.

To see the benefit of multiplexing, consider Figure 5.43 where the packets from three terminals are to be multiplexed. Part (a) shows the times when the packets would have been transmitted if each terminal had its own line at speed R bps. Part (b) shows the time when the packets are transmitted if the packets from the three flows are combined by a multiplexer. Because the terminals generate packets in bursty fashion, it is possible to combine the packet streams into a single line of speed R bps. Note that because the packet generation times overlap, the multiplexer must buffer and delay some of the packets. Nevertheless, all the packets get transmitted in the order in which they were generated. Thus by aggregating the packet flows into a single transmission line, the multiplexer reduces the system cost by reducing the number of lines.

In general, packet multiplexers are used in two situations. In the first situation packets arrive from multiple lines to a statistical multiplexer for transmis-

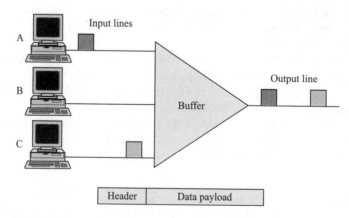

FIGURE 5.42 Statistical multiplexing of data

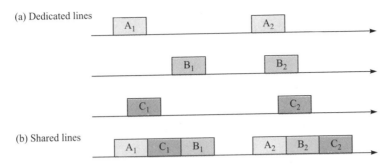

(a) Dedicated lines

(b) Shared lines

FIGURE 5.43 Lines with and without multiplexing

sion to a remote site. In the second situation packets arrive to a packet switch that then routes some of the packets for transmission over a specific data link. In effect, these packets are statistically multiplexed onto the given link. In both situations the multiplexer buffers the packets and arranges them in a queue. As the transmission line becomes available, packets are then transmitted according to their position in the queue. Typically packets are transmitted in first-in, first-out (FIFO) fashion, but increasingly multiplexers use priorities and scheduling of various types to determine the order of packet transmission. In general, most computer data applications do not tolerate loss, but they can tolerate some delay. Consequently, the multiplexers are operated so that they trade off packet delay versus utilization of the transmission line. However, the amount of buffering available is limited, so packet losses can occur from time to time when a packet arrives to a full system. End-to-end protocols are responsible for recovering from such losses.

The behavior of a statistical multiplexer is characterized by $N(t)$, the number of packets in the statistical multiplexer at time t. Figure 5.44 shows $N(t)$ for the example in Figure 5.43. The variations in $N(t)$ are determined by the arrival times and the departure times of the packets. Suppose that the average length of a packet is $E[L]$ bits and that the transmission line has a speed of R bits per second. The average packet transmission time is then $E[L]/R$ seconds. Over the long run the transmission line can handle at most $\mu = R/E[L]$ packets/second, so μ is the **maximum departure rate** at which packets can be transmitted out of the system. For example, suppose that the average packet size is 1000 bytes and that the transmission line speed is 64,000 bps. The maximum packet transmission rate is then $\mu = 64,000$ bps/(1000 bytes * 8 bits per byte) $= 8$ packets/second.

Let λ be the average packet **arrival rate** to a multiplexer in packets/second. If λ is higher than μ, then the buffer will build up on the average and many packet losses will occur. However, when λ is less than μ, the number of packets in the multiplexer will fluctuate because packet arrivals can bunch up or build up during the transmission of particular long packets. Consequently, packets can still overflow from time to time. We can reduce the incidence of this type of

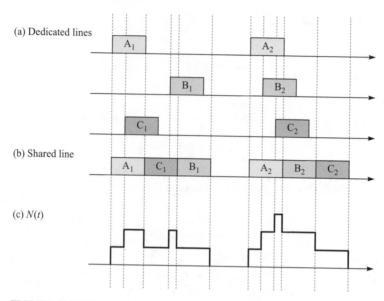

FIGURE 5.44 The number in the statistical multiplexer $N(t)$

packet loss by increasing the number of the buffers. We define the **load** ρ to be given by $\rho = \lambda/\mu$. Clearly, we want the arrival rate λ to be less than the departure rate μ, and hence $\rho < 1$.

As an example we present the results for a statistical multiplexing system that is modeled by the so-called M/M/1/K queueing model.[10] In this model packets arrive at a rate of λ packets/second, and the times between packet arivals are random, are statistically independent of each other, and have an exponential density function with mean $1/\lambda$ seconds as shown in Figure 5.45. This arrival process is usually called the **Poisson arrival process**.

Note from the figure that short interarrival times are more likely than are long interarrival times. The model is also assumes that the packet transmission times are random and have an exponential density with mean $1/\mu = E[L]/R$. The model assumes that there is enough buffering to hold up to K packets in the statistical multiplexer. The packet delay T is defined as the total time a packet spends in the multiplexer and is given by the sum of the time spent waiting in queue and the packet transmission time. We show in Appendix A that the **packet loss probability** is given by

---

[10]The M/M/1/K queueing system is analyzed in Appendix A. The first M refers to the exponentially distributed interarrival times, the second M refers to the exponentially distributed transmission times, the l refers to the fact that there is a single server, that is, transmission line, and the K refers to the maximum number of packets allowed in the system. For this discussion we only need the formulas that result from the analysis.

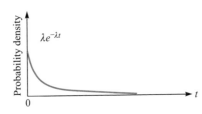

Probability density

$\lambda e^{-\lambda t}$

0

t

FIGURE 5.45 Exponential density function for interarrival times; the likelihood of an interarrival time near *t* is given by the value of the density function

$$P_{loss} = \frac{(1 - \rho)\rho^K}{1 - \rho^{K+1}} \tag{1}$$

and the average packet delay is given by

$$E[T] = \frac{E[N]}{\lambda(1 - P_{loss})} \tag{2}$$

where $E[N]$ is the average number of packets in the multiplexer

$$E[N] = \frac{\rho}{1 - \rho} - \frac{(K + 1)\rho^{K+1}}{1 - \rho^{K+1}} \tag{3}$$

Figure 5.46 and Figure 5.47 show the average delay and the packet loss probability as a function of the load ρ. Note that in Figure 5.46 the average delay $E[T]$ is normalized to multiples of the average packet transmission time $E[L]/R$. Figure 5.46 assumes a value of $K = 10$ and shows both the desired case where $\rho < 1$ and overload case where $\rho > 1$. We discuss these two cases below.

When the arrival rate is very small, that is, ρ is approximately zero, an arriving packet is unlikely to find any packets ahead of it in queue, and so the delay in the statistical multiplexer is simply one packet transmission time. Note that the packet loss probability is very low at small loads. As the load approaches about 0.7, the average delay and the packet loss probability begin increasing.

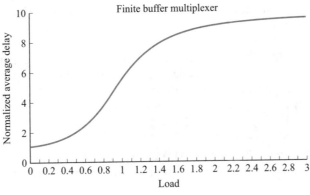

FIGURE 5.46 Average packet delay as a function of load in M/M/1/10 system; delay is normalized to multiples of a packet transmission time

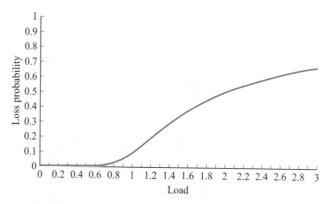

FIGURE 5.47 Packet loss probability as a function of load for M/M/1/10

Note that if $\rho < 1$, then ρ^K decreases with K, so the loss probability P_{loss} can be made arbitrarily small by increasing K.

Now suppose that we have the overload condition $\rho > 1$. The packet loss probability increases steadily as the load increases beyond 1 because the system is usually full, and so the excess packet arrivals are lost. However, note from Figure 5.46 that the average delay approaches an asymptotic value of 10. Because the system is almost always full, most of the packets that actually enter the system experience close to the maximum delay of 10 transmission times.

The behavior of the average delay and the loss probability as a function of load shown in the two figures is typical of a statistical multiplexing system. The exact behavior will depend on the distribution of the packet interarrivals, on the distribution of the packet transmission times, and on the size of the buffer. For example, consider systems that are "more random" that the M/M/1/K system, that is, because arrivals are more bursty or because the packet length distribution is such that long packets are more probable. For such systems the average delay and the loss probabilities will increase more quickly as a function of load ρ than in the figures. On the other hand, systems that are "less random" will increase less quickly as a function of load.

Now suppose that the buffer size K is made arbitrarily large; that is, $K \to \infty$ when $\rho < 1$. The result is the so-called **M/M/1** model. The loss probability P_{loss} goes to zero, and the average delay becomes

$$E[T_M] = \frac{1}{\lambda}\left[\frac{\rho}{1-\rho}\right] = \left[\frac{1}{1-\rho}\right]\frac{1}{\mu} = \left[\frac{\rho}{1-\rho}\right]\frac{1}{\mu} + \frac{1}{\mu} \text{ for M/M/1 model} \quad (4)$$

Figure 5.48 shows $E[T]$ for the M/M/1 system. The average packet transmission time is $1/\mu$, so the average time spent waiting in queue prior to transmission is

$$E[W_M] = E[T] - \frac{1}{\mu} = \left[\frac{\rho}{1-\rho}\right]\frac{1}{\mu} \text{ for M/M/1 model} \quad (5)$$

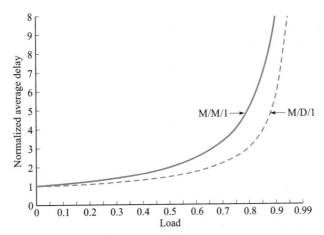

FIGURE 5.48 Average delay for systems with infinite buffers

Figure 5.48 also shows $E[T]$ for the M/D/1 system that has exponential interarrivals, *constant* service times L/R corresponding to fixed-length packets, and infinite buffer size. For the M/D/1 system, we have

$$E[T_D] = \left[1 + \frac{\rho}{2(1-\rho)}\right]\frac{1}{\mu} = \left[\frac{\rho}{2(1-\rho)}\right]\frac{1}{\mu} + \frac{1}{\mu} \text{ for M/D/1 system} \qquad (6)$$

so the average time spent waiting in queue is

$$E[W_D] = \left[\frac{\rho}{2(1-\rho)}\right]\frac{1}{\mu} \text{ for M/D/1 system} \qquad (7)$$

Note that both systems have delays that become arbitrarily large as the load approaches 1, that is, as the arrival rate approaches the maximum packet transmission rate. However, the system with constant service times has half the average waiting time of the system with exponential service times. In general, the packet delay increases as systems become more random, that is, more variable in terms of packet arrivals and packet transmission times.

Example—M/M/1 versus M/D/1

Consider a statistical multiplexer that has a transmission line with a speed of $R = 64$ kbps. Suppose that the average packet length is $E[L] = 1000$ bytes = 8000 bits and that the average arrival rate is four packets/second. Compare the average packet delay for constant-length packets to exponentially distributed packets.

The packet service rate is $\mu = 64{,}000$ bps/8000 bits/packet = 8 packets/second. Since $\lambda = 4$ packets/second, the load is $\rho = \lambda/\mu = 4/8 = 1/2$. If packets have an exponential density function, then $E[T_M] = 2/8 = 250$ ms. If packets are constant, we then have $E[T_D] = 1.5/8 = 187$ ms.

Example—Effect of Header Overhead on Goodput

Consider a statistical multiplexer that has a transmission line with a speed of $R = 64$ kbps. The *goodput* is the amount of actual user information that is transmitted. Suppose that each packet has 40 bytes of IP and TCP header and that packets are constant in length. Find the useful throughput if the total packet length is 200 bytes, 400 bytes, 800 bytes, and 1200 bytes.

Let L be the packet length in bytes. The packet service rate is then $\mu = 64,000/8L$ packets/second. Let λ be the packet arrival rate; then the load is $\rho = \lambda/\mu = 8\lambda L/64,000$, and the goodput is $\gamma = 8\lambda(L - 40)$ bps. Figure 5.49 shows the average packet delay versus the goodput. Note that the delay is given in seconds. Thus when the multiplexer has longer packets, the delay at low goodput is higher, since it takes longer to transmit each packet. On the other hand, the longer packets also incur less header overhead in terms of percentage and hence have a higher maximum achievable goodput. Indeed, it is easy to show that the maximum goodput is given by $\gamma = (1 - 40/L)\,64,000$ bps.

PERFORMANCE IMPROVEMENTS FROM FLOW AGGREGATION

Suppose that we initially have 24 individual statistical multiplexers, each with a 64 kbps line, and that we aggregate the packet arrivals into one stream *t* and apply it to a statistical multiplexer with a 24*64 kbps line. Suppose that each multiplexer has an arrival rate $\lambda = 4$ packets/second, a service rate $\mu = 8$ packets/second, and hence a load $\rho = 1/2$. If we use an M/M/1 model the average packet delay in each individual multiplexer is $2/8 = 250$ ms. On the other hand, the combined system has arrival rate $\lambda' = 24*4$ packets/second, a service rate $\mu' = 24*8$ packets/second, and hence a load $\rho' = \frac{1}{2} = \rho$. A key property of exponential interarrivals is that the merged arrival streams also have exponential interarrivals. Thus the same expression for the average delay holds, and the average packet delay in the combined system is $2/(8*24) = 250/24 \approx 10$ ms.

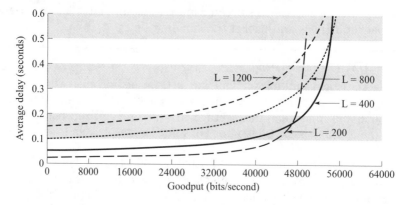

FIGURE 5.49 Effect of header overhead on packet delay and goodput

MEASUREMENT, MODELS, AND REAL PERFORMANCE

The flow of traffic in modern networks is extremely complex. The various interacting layers of protocols, for example, HTTP, TCP/IP, Ethernet and PPP, provide a framework for the interaction of the information flows generated by a multiplicity of applications. The times when applications make requests are not scheduled, and the pattern of packets transmitted is never quite the same. The challenge of the network operator is to have enough resources (bandwidth, buffering, processing) in the right configuration to handle the traffic at any given time. Measurement and performance modeling are essential tools in carrying out this task.

Measurement helps identify patterns in the apparent chaos that is network traffic. Time-of-day and day-of-week cycles in traffic levels and traffic patterns tend to persist, and their changes can be tracked by measurement. The traffic flows generated by specific types of applications can also be characterized and tracked over time. Sudden surges and changes in traffic patterns can also be recognized, given the template provided by what is "normal." Longer-term trends in traffic levels and patterns are also used to plan the deployment of network equipment.

Traffic models like the ones introduced in this section are useful in understanding the dynamics and interplay between the basic parameters that determine performance. Models are intended to simplify and capture only the essential features of a situation. In doing so, models can be used to predict the (approximate) performance in a given situation, and so they can form the basis for making decisions regarding traffic management. However models are merely our attempt to characterize what is going on "out there." In fact, models that do not capture all the relevant features of a situation can lead to incorrect conclusions. Measurement can be used to close the loop between models and real performance; by comparing predictions of the models with actual observations, we can modify and fine-tune the models themselves.

The average packet delay has been reduced by a factor of 24! We leave as an exercise the task of showing that a factor of 24 improvement also results when packets are constant.

The improved performance that results when the arrival rate and the transmission rate are increased by the same factor k is simple to explain. In effect the time scale of the system is reduced by a factor k. The interarrivals and service times of packets with respect to each other remain unchanged. Packets in the system find the same number of packets ahead of them in queue as in the old system. The only difference is that the packets are moving k times faster.

Suppose instead that each individual system can only hold 10 packets, and so we model it with an M/M/1/K system with $K = 10$. The packet loss probability for each individual system is given by $P_{loss} = (1/2)(1/2)^{10}/(1 - (1/2)^{11}\lambda(1/2)^{11}$. Suppose that the combined multiplexer has the same total buffers as the individual multiplexers; that is, $K' = 24*10 = 240$. The combined multiplexer then has loss probability $P'_{loss} = (1/2)(1/2)^{240}/(1 - (1/2)^{241}\lambda(1/2)^{241}$. This is a huge improvement in packet loss performance!

Thus we find that for Poisson arrivals (that is, exponential packet inter-arrivals) increasing the size or scale of the system by aggregating packet flows leads to improved performance in terms of delay and loss.

5.5.2 Speech Interpolation and the Multiplexing of Packetized Speech

In telephone networks a connection is set up so that speech samples traverse the network in a single uninterrupted stream. Normal conversational speech, however, is moderately bursty as it contains silence periods. In this section we consider the multiplexing of packetized speech.

First, consider n one-way established telephone connections. In a typical conversation a person is actively speaking less than half of the time. The rest of the time is taken up by the pauses inherent in speech and by listening to the other person. This characteristic of speech can be exploited to enable m telephone lines to carry the n conversations, where m is less than n. Because a connection produces active speech only about half of the time, we expect that the minimum number of trunks required is approximately $n/2$. This problem was first addressed in speech interpolation systems in transatlantic undersea cable telephone communications.

As shown in Figure 5.50, the n speakers generate bursts of active speech that are separated by periods of silence. The early systems monitored the speech activity in each line and multiplexed active bursts onto available lines. Associated signaling allowed the receiving end to reconstitute the original speech signal by reinserting the appropriate silence periods. Clearly, the technique involves juggling the n calls, using the m available trunks. It is inevitable that from time to time there will be more active bursts than the number of lines. The early systems simply discarded the excess bursts, which then resulted in "clipping" in the recovered speech signal. The resulting speech quality was acceptable as long as the amount of lost speech was kept below a certain level.

This concentration technique can also be implemented and enhanced by using digital signal processing. The digitized speech signal corresponding to

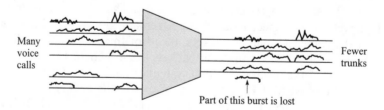

Part of this burst is lost

FIGURE 5.50 Multiplexing of bursty speech

some duration, say, 10 ms, is used to form a packet to which a header is attached. Packets are classified as active or silent, depending on the speech activity. Silence packets are discarded. Suppose that one line can transmit all the packets, active and silent, for a single conversation. The digital system now tries to use m lines to transmit the active packets from $n > m$ calls; that is, the digital transmission line can send m packets per 10 ms period. Whenever the number of active packets exceeds m, the excess number is discarded. Given n speakers, we need to find the number of trunks required so that the **speech loss**, that is, the fraction of active speech that is discarded, is kept below a certain level.

Let p be the probability that a packet is active. Then it can be shown that the speech loss is given by

$$\text{speech loss} = \frac{\sum_{k=m+1}^{n} (k-m)\binom{n}{k} p^k (1-p)^{n-k}}{np} \quad \text{where} \quad \binom{n}{k} = \frac{n!}{k!(n-k)!} \quad (8)$$

The denominator in equation 8 is the average number of active packets in a 10 ms period. The numerator is the average number of active packets in excess of m. Figure 5.51 shows the speech loss for various numbers of speakers, that is, $n = 24, 32, 40,$ and 48. We assume that $p = 0.4$; that is, 40 percent of packets are active. As expected, the speech loss decreases as the number of trunks increases. The acceptable level of speech loss is approximately 1%.

Table 5.4 shows the number of trunks needed to meet a 1% speech loss requirement. The **multiplexing gain** is defined as the ratio of the number of connections, that is, speakers, to the number of trunks actually provided. We note that because $p = 0.4$, the maximum multiplexing gain is approximately 2.5. Table 5.4 also shows the utilization which is the percentage of time that the

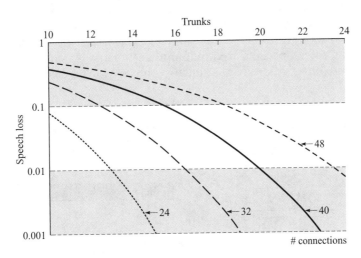

FIGURE 5.51 Speech loss—number of speakers versus number of trunks

Speakers	Trunks	Multiplexing gain	Utilization
24	13	1.85	0.74
32	16	2.00	0.80
40	20	2.00	0.80
48	23	2.09	0.83

TABLE 5.4 Trunks required for 1% speech loss

trunks are in use transmitting active packets. If the number of connections is n, then each 10 ms period produces np active packets on the average. Because the speech loss is 1%, the number of active packets transmitted per 10 ms period is $.99\ np$. Thus the utilization is given by $.99\ np/m$.

Let us now consider the statistical multiplexing of a packetized speech using delay and loss. The speech signal is digitized, say, at a rate of 8000 samples/second and eight bits/sample. As each sample is obtained, it is inserted into a fixed-length packet that has appropriate header information and that holds the samples produced in some time interval, say, 10 ms. Note that the first sample inserted into the packet must wait 10 ms until the packet is filled. This initial delay is called the **packetization delay**.

Once the packet is full, and if it contains active speech, it is passed to a statistical multiplexer that operates in the same way as the data packet multiplexer discussed Figure 5.42. As shown in Figure 5.52, the multiplexer accepts speech packets from various conversations and holds each packet in the queue until the transmission line becomes available. Note that the addition of buffering reduces the number of packets that need to be discarded in comparison to the digital speech interpolation system discussed earlier. However, this reduction is at the expense of additional delay while waiting in queue for transmission. Note also that in packet speech, packet arrivals are periodic when the conversation is generating active speech. This situation differs from the exponential interarrivals discussed in section 5.5.1 for data traffic. Nevertheless the queueing delays experienced by speech packets are random in nature and depend on the arrivals from the other conversations. Consequently, the delay experienced in traversing

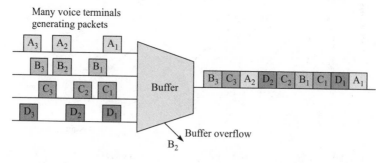

FIGURE 5.52 Statistical multiplexing of packetized speech

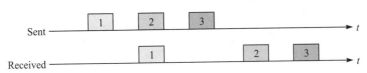

FIGURE 5.53 Packets experience delay jitter during transmission through multiplexer

the multiplexing link is not constant, and so the packets experience delay jitter, as shown in Figure 5.53. A playout procedure along the lines of the timing recovery methods discussed earlier in this chapter is required to compensate for the delay jitter.

The real-time nature of speech requires that the packets not experience an end-to-end delay greater than some value, usually around 250 ms. Hence excessive buffering should be avoided for packet speech multiplexers. If the buffers become too large, packets will experience greater delays. Packets that arrive after

ON-OFF MODELS, LOSS AND DELAY, AND MEMORY IN THE SYSTEM

The ON-OFF nature of the voice model is typical of many applications that alternate between periods of activity when data is generated and idle periods when no traffic is produced. Suppose that a system is operated on a *loss* basis—that is, any information in excess of what can be handled immediately is discarded—and suppose that the source *does not* retransmit or in any other way regenerate the information. For these systems the result given by equation 8 applies for n identical users as long as they generate information independently of each other. The number of packets discarded at a particular time slot is characterized completely by the probabilities in the equation. In particular, the result *does not* depend on the length of the ON periods or the length of the OFF periods. Once the packets are discarded, the system can completely forget about the lost packets because they are irrelevant to future behavior.

If the system is operated on a *delay* basis, then any packets that cannot be transmitted at a given time are buffered for later transmission. In other words, delay systems must "remember" all excess packets. This memory causes past events (surges in arrivals) to influence the future (congested queues and longer delays). In this case the durations of ON and OFF periods *do* matter. The co-occurrence of many long ON periods implies a prolonged period of higher than average arrivals and correspondingly large queue buildups and delays.

The buffering in the multiplexer is not the only source of memory in the packet arrival process. Protocols that involve retransmission or adjustment of transmission rate to network conditions are another source of memory. Finally, the users themselves have memory and will reattempt a transmission at a later time when their requests are not met. This behavior can be another source of long-term dependencies in the traffic arrival process.

the required end-to-end delay are useless and will be discarded. Therefore, no advantage accrues from increasing the buffer size beyond a certain point.

SUMMARY

This chapter had two primary objectives. The first was to introduce a discussion of the *protocols* that operate within a layer and to the *services* that they provide to the layer above them. We considered the simplest case in which only two peer processes are involved in executing a protocol to provide a service, namely, the transfer of information in a manner that satisfies certain conditions. The second primary objective was to introduce the two important examples of such peer-to-peer protocols: *TCP*, which provides reliable stream service end to end across a network, and *HDLC*, which can provide for the reliable transfer of blocks of information across an unreliable data link.

The operation of a protocol involves the exchange of *PDUs* that consist of *headers* with protocol control information and of user SDUs. To show the operation of a protocol, we considered the use of ARQ to provide a reliable, sequenced transfer of blocks of information. We showed how a protocol is specified in terms of a *state* machine that dictates what actions each protocol entity is to take when an event occurs. We saw how the protocol entities *exchange control information* through specific control PDUs (to set up a connection, to release a connection, to inquire about status) as well as through control information that is piggybacked onto data PDUs (sequence numbers, CRC check sums, ACKs). We also saw how ARQ protocols depend on *timers* to keep the protocol alive.

We introduced the *sliding-window* mechanism as a means of providing sequence numbering within a finite sequence space. For the ARQ protocols to operate correctly, only a portion of the sequence space can be in use at any given time. The sliding-window mechanism can also provide *flow control* to regulate the rate at which a sender transmits information to a receiver. The use of timestamps by peer-to-peer protocols to assist in the transfer of information with timing requirements was also discussed.

We introduced the Transmission Control Protocol (TCP) that uses a form of Selective Repeat ARQ to provide *connection-oriented, reliable stream service* and flow control end to end across connectionless packet networks.

In the next section we considered the relatively simple task of data link control protocols, which is to provide for the *connection-oriented or connectionless transfer of blocks of information across a data link*. We introduced the basic, but essential, function of *bit stuffing* to provide *framing* that demarcates the boundary of data link PDUs. We also examined the structure of data link frames, focusing on the control information in the header that directs the operation of the protocol. We found that *HDLC* does not provide only *one* data link control, but instead provides a toolkit of control frames and protocol mechanisms that can be used to provide a wide range of data link protocols, from simple, unacknowledged, connectionless transfer to full-fledged, connection-oriented, reliable, sequenced transfer. Finally, we considered *PPP* that builds on the HDLC structure to provide a versatile data link protocol with enhanced link monitoring, with authentication capability, and with the ability to simultaneously support several network layer protocols.

An essential feature of networks is the use of resource sharing to achieve economies of scale. In the final section we introduced simple models that offer insight into the performance of statistical multiplexers that allow packets from different flows to share a common data link. Bandwidth sharing at this level is a key feature of packet networks. We identified the packet arrival pattern and the packet length distribution as key parameters that determine delay and loss performance. We also indicated the challenges posed to network designers by rapidly evolving network-based applications that can result in sudden changes in network traffic that can have dramatic impact on network performance.

CHECKLIST OF IMPORTANT TERMS

ACK timer
acknowledgment frame (ACK)
adaptation function
ARQ protocol
◆arrival rate λ
asynchronous balanced mode (ABM)
Automatic Repeat Request (ARQ)
best-effort service
bit stuffing
Challenge-Handshake
  Authentication Protocol (CHAP)
checkpointing
connectionless service
connection-oriented service
control frame
data link control
data link layer
delay-bandwidth product
Disconnect (DISC)
end-to-end requirement
enquiry frame (ENQ)
FIN segment
flow control
frame
Frame Reject (FRMR)

Go-Back-N ARQ
header
High-level Data Link Control
  (HDLC)
I-frame timer
information frame (I-frame)
link control protocol (LCP)
◆load ρ
◆maximum departure rate μ
◆multiplexing gain
negative acknowledgment frame
  (NAK)
network control protocol (NCP)
network service model
normal response mode (NRM)
◆packet loss probability
◆packetization delay
Password Authentication Protocol
  (PAP)
peer process
peer-to-peer protocol
piggybacking
pipeline
Point-to-Point protocol (PPP)
◆Poisson arrival process

quality of service (QoS)
real time
Receive Not Ready (RNR)
Receive Ready (RR)
receive window
Reject (REJ)
segmentation and reassembly
Selective Reject (SREJ)
Selective Repeat (ARQ)
send window
sequence number
service model
Set Asynchronous Balanced Mode
   (SABM)

Set Normal Response Mode
   (SNRM)
sliding-window protocol
◆speech loss Stop-and-Wait ARQ
supervisory frame
SYN segment
timestamp
timing jitter
transmission control protocol (TCP)
unnumbered acknowledgment (UA)
unnumbered frame

FURTHER READING

Bertsekas, D. and R. Gallager, *Data Networks*, Prentice-Hall, Englewood Cliffs, 1992.

Carlson, J., *PPP Design and Debugging*, Addison-Wesley, Reading, Massachusetts, 1998.

Davies, D. W., D. L. A. Barber, W. L. Price, and C. M. Solomonides, *Computer Networks and Their Protocols*, John Wiley & Sons, New York, 1979.

Halsall, F. *Data Communications, Computer Networks, and Open Systems*, Addison-Wesley, Reading, Massachusetts, 1992.

Jain, B. N. and A. K. Agrawala, *Open Systems Interconnection: Its Architecture and Protocol*, McGraw-Hill, New York, 1993.

Schwartz, M., *Telecommunication Networks: Protocols, Modeling, and Analysis*, Addison-Wesley, Reading, Massachusetts, 1987.

RFC 1661, W. Simpson, "The Point-to-Point Protocol (PPP)", July 1994.

PROBLEMS

1. Explain the difference between connectionless unacknowledged service and connectionless acknowledged service. How do the protocols that provide these services differ?

2. Explain the difference between connection-oriented acknowledged service and connectionless acknowledged service. How do the protocols that provide these services differ?

3. Suppose that the two end systems α and β in Figure 5.3 communicate over a connection-oriented packet network. Suppose that station α sends a 10-kilobytes message to station β and that all packets are restricted to 1000 bytes (neglect headers); assume that each packet can be accommodated in a data link frame. For each of the links, let p be the probability that a frame incurs errors during transmission.

a. Suppose that the data link control just transfers frames and does not implement error control. Find the probability that the *message* arrives without errors at station β.

b. Suppose that error recovery is carried out end to end and that if there are any errors, the entire message is retransmitted. How many times does the message have to be retransmitted on average?

c. Suppose that the error recovery is carried out end to end on a packet by packet basis. What is the total number of packet transmissions required to transfer the entire message?

4. Suppose that two peer-to-peer processes provide a service that involves the transfer of discrete messages. Suppose that the peer processes are allowed to exchange PDUs that have a maximum size of M bytes, including H bytes of header. Suppose that a PDU is not allowed to carry information from more than one message.

a. Develop an approach that allows the peer processes to exchange messages of arbitrary size.

b. What essential control information needs to be exchanged between the peer processes?

c. Now suppose that the message transfer service provided by the peer processes is shared by several message source-destination pairs. Is additional control information required, and if so, where should it be placed?

5. Suppose that two peer-to-peer processes provide a service that involves the transfer of a stream of bytes. Suppose that the peer processes are allowed to exchange PDUs that have a maximum size of M bytes, including H bytes of header.

a. Develop an approach that allows the peer processes to transfer the stream of bytes in a manner that uses the transmission line efficiently. What control information is required in each PDU?

b. Suppose that the bytes in the stream arrive sporadically. What is a reasonable way to balance efficiency and delay at the transmitter? What control information is required in each PDU?

c. Suppose that the bytes arrive at a constant rate and that no byte is to be delayed by more than T seconds. Does this requirement have an impact on the efficiency?

d. Suppose that the bytes arrive at a variable rate and that no byte is to be delayed by more than T seconds. Is there a way to meet this requirement?

6. Suppose that two peer-to-peer processes provide a service that involves the transfer of a stream of bytes. Develop an approach that allows the stream transfer service to be shared by several pairs of users in the following cases:

a. The bytes from each user pair arrive at the same constant rate.

b. The bytes from the user pairs arrive sporadically and at different rates.

7. Consider the transfer of a single real-time telephone voice signal across a packet network. Suppose that each voice sample should not be delayed by more than 20 ms.

a. Discuss which of the following adaptation functions are relevant to meeting the requirements of this transfer: handling of arbitrary message size; reliability and sequencing; pacing and flow control; timing; addressing; and privacy, integrity, and authentication.

b. Compare a hop-by-hop approach to an end-to-end approach to meeting the requirements of the voice signal.

8. Suppose that a packet network is used to transfer *all* the voice signals that arrive at the base station of a cellular telephone network to a telephone office. Suppose that each voice sample should not be delayed by more than 20 ms.
   a. Discuss which of the following adaptation functions are relevant to meeting the requirements of this transfer: handling of arbitrary message size; reliability and sequencing; pacing and flow control; timing; addressing; and privacy, integrity, and authentication.
   b. Are the requirements the same in the opposite direction from the telephone office to the base station?
   c. Do the answers to parts (a) and (b) change if the signals arriving at the base station include e-mail and other short messages?

9. Suppose that streaming video information is transferred from a server to a user over a packet network.
   a. Discuss which of the following adaptation functions are relevant to meeting the requirements of ths transfer: handling of arbitrary message size; reliability and sequencing; pacing and flow control; timing; addressing; and privacy, integrity, and authentication.
   b. Suppose that the user has basic VCR features through control messages that are transferred from the user to the server. What are the adaptation requirements for the control messages?

10. Discuss the merits of the end-to-end versus hop-by-hop approaches to providing a constant transfer delay for information transferred from a sending end system to a receiving end system.

11. Consider the Stop-and-Wait protocol as described in the chapter. Suppose that the protocol is modified so that each time a frame is found in error at either the sender or receiver, the last transmitted frame is immediately resent.
    a. Show that the protocol still operates correctly.
    b. Does the state transition diagram need to be modified to describe the new operation?
    c. What is the main effect of introducing the immediate-retransmission feature?

12. In Stop-and-Wait ARQ why should the receiver always send an acknowledgment message each time it receives a frame with the wrong sequence number?

13. Discuss the factors that should be considered in deciding whether an ARQ protocol should act on a frame in which errors are detected.

14. Suppose that a network layer entity requests its data link layer to set up a connection to another network layer entity. In order to set up a connection in a data link, the initiating data link entity sends a SETUP frame, such as SABM in Figure 5.37. Upon receiving such a frame, the receiving data link entity sends an acknowledgment frame confirming receipt of the SETUP frame. Upon receiving this acknowledgment, the initiating entity can inform its network layer that the connection has been set up and is ready to transfer information. This situation provides an example of how *unnumbered acknowledgments* can arise for confirmed services.

a. Reexamine Figure 5.9 and Figure 5.10 with respect to error events that can take place and explain how these events are handled so that connection setup can take place reliably.

b. To terminate the connection, either data link layer can send a DISC frame that is then acknowledged by an unnumbered acknowledgment. Discuss the effect of the above error events and how they can be dealt with.

c. Suppose that an initiating station sends a SETUP frame twice but that the corresponding ACK times are delayed a long time. Just as the ACK frames from the original transmissions are about to arrive, the initiating station gives up and sends a DISC frame followed by another SETUP frame. What goes wrong if the SETUP frame is lost?

15. A 1 Mbyte file is to be transmitted over a 1 Mbps communication line that has a bit error rate of $p = 10^{-6}$.

a. What is the probability that the entire file is transmitted without errors? Note for n large and p very small, $(1 - p)^n \approx e^{-np}$.

b. The file is broken up into N equal-sized blocks that are transmitted separately. What is the probability that all the blocks arrive without error? Is dividing the file into blocks useful?

c. Suppose the propagation delay is negligible, explain how Stop-and-Wait ARQ can help deliver the file in error-free form. On the average how long does it take to deliver the file if the ARQ transmits the entire file each time?

d. Now consider breaking up the file into N blocks. (Neglect the overhead for the header and CRC bits.) On the average how long does it take to deliver the file if the ARQ transmits the blocks one at a time? Evaluate your answer for $N = 80$, 800, and 8000.

e. Explain qualitatively what happens to the answer in part (d) when the overhead is taken into account.

16. Consider the state transition diagram for Stop-and-Wait ARQ in Figure 5.11. Let P_f be the probability of frame error in going from station A to station B and let P_a be the probability of ACK error in going from B to A. Suppose that information frames are two units long, ACK frames are one unit long, and propagation and processing delays are negligible. What is the average time that it takes to go from state $(0,0)$ to state $(0,1)$? What is the average time that it then takes to go from state $(0,1)$ to state $(1,1)$? What is the throughput of the system in information frames/second?

17. Write a program for the transmitter and the receiver implementing Stop-and-Wait ARQ over a data link that can introduce errors in transmission. Assume station A has an unlimited supply of frames to send to station B. Only ACK frames are sent from station B to station A. Hint: Identify each event that can take place at the transmitter and receiver and specify the required action.

18. A 64-kilobyte message is to be transmitted from the source to the destination, as shown on the next page. The network limits packets to a maximum size of two kilobytes, and each packet has a 32-byte header. The transmission lines in the network have a bit error rate of 10^{-6}, and Stop-and-Wait ARQ is used in each transmission line. How long does it take on the average to get the message from the source to the destination? Assume that the signal propagates at a speed of 2×10^5 km/second.

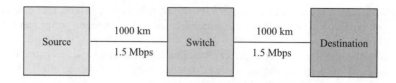

19. Suppose that a Stop-and-Wait ARQ system has a time-out value that is less than the time required to receive an acknowledgment. Sketch the sequence of frame exchanges that transpire between two stations when station A sends five frames to station B and no errors occur during transmission.

20. The Trivial File Transfer Protocol (RFC 1350) is an application layer protocol that uses the Stop-and-Wait protocol. To transfer a file from a server to a client, the server breaks the file into blocks of 512 bytes and sends these blocks to the client using Stop-and-Wait ARQ. Find the efficiency in transmitting a 1 MB file over a 10 Mbps Ethernet LAN that has a diameter of 300 meters. Assume that the transmissions are error free and that each packet has 60 bytes of header attached.

21. Compare the operation of Stop-and-Wait ARQ with bidirectional Go-Back-N ARQ with a window size of 1. Sketch out a sequence of frame exchanges using each of these protocols and observe how the protocols react to the loss of an information frame and to the loss of an acknowledgment frame.

22. Consider the various combinations of communication channels with bit rates of 1 Mbps, 10 Mbps, 100 Mbps, and 1 Gbps over links that have round-trip times of 10 msec, 1 msec, and 100 msec.
    a. Find the delay-bandwidth product for each of the 12 combinations of speed and distance.
    b. Suppose that 32-bit sequence numbers are used to transmit blocks of 1000 bytes over the above channels. How long does it take for the sequence numbers to wrap around, that is, to go from 0 up to 2^m?
    c. Now suppose the 32-bit sequence numbers are used to count individual transmitted bytes. How long does it take for the sequence numbers to wrap around?

23. Consider a bidirectional link that uses Go-Back-N with $N = 7$. Suppose that all frames are one unit long and that they use a time-out value of 2. Assume the propagation is 0.5 unit and the processing time is negligible. Assume the ACK timer is one unit long. Assuming stations A and B begin with their sequence numbers set to zero, show the pattern of transmissions and associated state transitions for the following sequences of events:
    a. Station A sends six frames in a row, starting at $t = 0$. All frames are received correctly.
    b. Station A sends six frame in a row, starting at $t = 0$. All frames are received correctly, but frame 3 is lost.
    c. Station A sends six frames in a row, starting at $t = 0$. Station B sends six frames in a row starting at $t = 0.25$. All frames are received correctly.

24. Consider a bidirectional link that uses Go-Back-N with $N = 3$. Suppose that frames from station A to station B are one unit long and use a time-out value of 2. Frames in the

opposite directions are 2.5 units long and use a time-out value of 4. Assume that propagation and processing times are negligible, that the stations have an unlimited number of frames ready for transmission, and that all ACKs are piggybacked onto information frames. Assuming stations A and B begin with their sequence numbers set to zero, show the transmissions and associated state transitions that result when there are no transmission errors.

25. Consider the Go-Back-N ARQ protocol.
    a. What can go wrong if the ACK timer is not used?
    b. Show how the frame timers can be maintained as an ordered list where the time-out instant of each frame is stated relative to the time-out value of the previous frame.
    c. What changes if each frame is acknowledged individually instead of by using a cumulative acknowledgment (R_{next} acknowledges all frames up to $R_{next} - 1$)?

26. Suppose that instead of Go-Back-N ARQ, N simultaneous Stop-and-Wait ARQ processes are run in parallel over the same transmission channel. Each SDU is assigned to one of the N processes that is currently idle. The processes that have frames to send take turns transmitting in round-robin fashion. The frames carry the binary send sequence number as well as an ID identifying which ARQ process the frame belongs to. Acknowledgments for *all* ARQ processes are piggybacked onto *every* frame.
    a. Qualitatively, compare the relative performance of this protocol with Go-Back-N ARQ and with Stop-and-Wait ARQ.
    b. How does the service offered by this protocol differ from that of Go-Back-N ARQ?

27. Write a program for the transmitter and the receiver implementing Go-Back-N ARQ over a data link that can introduce errors in transmission.
    a. Identify which variables need to be maintained.
    b. The program loops continuously waiting for an event to occur that requires some action to take place. Identify the main events that can occur in the transmitter. Identify the main events that can occur in the receiver.

28. Modify the program in problem 27 to implement Selective Repeat ARQ.

29. Three possible strategies for sending ACK frames in a Go-Back-N setting are as follows: send an ACK frame immediately after each frame is received, send an ACK frame after every other frame is received, and send an ACK frame when the next piggyback opportunity arises. Which of these strategies are appropriate for the following situations?
    a. An interactive application produces a packet to send each keystroke from the client; the server echoes each keystroke that it receives from the client.
    b. A bulk data transfer application where a server sends a large file that is segmented in a number of full-size packets that are to be transferred to the client.

30. Consider a bidirectional link that uses Selective Repeat ARQ with a window size of $N = 4$. Suppose that all frames are one unit long and use a time-out value of 2. Assume that the one-way propagation delay is 0.5 time unit, the processing times are negligible, and the ACK timer is one unit long. Assuming stations A and B begin with their sequence numbers set to zero, show the pattern of transmissions and associated state transitions for the following sequences of events:
    a. Station A sends six frames in a row, starting at $t = 0$. All frames are received correctly.

b. Station A sends six frames in a row, starting at $t = 0$. All frames are received correctly, but frame 3 is lost.

c. Station A sends six frames in a row, starting at $t = 0$. Station B sends six frames in a row, starting at $t = 0.25$. All frames are received correctly.

31. In the chapter we showed that if the transmit and receive maximum window sizes are both equal to the available sequence number space, then Selective Repeat ARQ will work correctly. Rework the arguments presented in the chapter to show that if the sum of the transmit and receive maximum window sizes equals the available sequence number space, then Selective Repeat ARQ will work correctly.

32. Suppose that Selective Repeat ARQ is modified so that ACK messages contain a list of the next m frames that the transmitter expects to receive.
a. How does the protocol need to be modified to accommodate this change?
b. What is the effect of the change on protocol performance?

33. A telephone modem is used to connect a personal computer to a host computer. The speed of the modem is 56 kbps and the one-way propagation delay is 100 ms.
a. Find the efficiency for Stop-and-Wait ARQ if the frame size is 256 bytes; 512 bytes. Assume a bit error rate of 10^{-4}.
b. Find the efficiency of Go-Back-N if three-bit sequence numbering is used with frame sizes of 256 bytes; 512 bytes. Assume a bit error rate of 10^{-4}.

34. A communication link provides 1 Mbps for communications between the earth and the moon. The link sends color images from the moon. Each image consists of $10,000 \times 10,000$ pixels, and 16 bits are used for each of the three color components of each pixel.
a. How many images/second can be transmitted over the link?
b. If each image is transmitted as a single block, how long does it take to get an acknowledgment back from earth? The distance between earth and the moon is approximately 375,000 km.
c. Suppose that the bit error rate is 10^{-5}, compare Go-Back-N and Selective Repeat ARQ in terms of their ability to provide reliable transfer of these images from the moon to earth. Optimize your frame size for each ARQ protocol.

35. Two computers are connected by an intercontinental link with a one-way propagation delay of 100 ms. The computers exchange 1-Megabyte files that they need delivered in 250 ms or less. The transmission lines have a speed of R Mbps, and the bit error rate is 10^{-8}. Design a transmission system by selecting the bit rate R, the ARQ protocol, and the frame size.

36. Find the optimum frame length that maximizes transmission efficiency by taking the derivative and setting it to zero for the following protocols:
a. Stop-and-Wait ARQ.
b. Go-Back-N ARQ.
c. Selective Repeat ARQ.

37. Suppose station A sends information to station B on a data link that operates at a speed of 10 Mbps and that station B has a 1-Megabit buffer to receive information from A. Suppose that the application at station B reads information from the receive buffer at a

rate of 1 Mbps. Assuming that station A has an unlimited amount of information to send, sketch the sequence of transfers on the data link if Stop-and-Wait ARQ is used to prevent buffer overflow at station B. Consider the following cases:

a. One-way propagation delay is 1 microsecond.

b. One-way propagation delay is 1 ms.

c. One-way propagation delay is 100 ms.

38. Redo problem 37 using Xon/Xoff flow control.

39. Suppose station A sends information to station B over a two-hop path. The data link in the first hop operates at a speed of 10 Mbps, and the data link in the second hop operates at a speed of 100 kbps. Station B has a 1-Megabit buffer to receive information from A, and the application at station B reads information from the receive buffer at a rate of 1 Mbps. Assuming that station A has an unlimited amount of information to send, sketch the sequence of transfers on the data link if Stop-and-Wait ARQ is used on an end-to-end basis to prevent buffer overflow at station B.

a. One-way propagation delay in data link 1 and in data link 2 is 1 ms.

b. One-way propagation delay in data link 1 and in data link 2 is 100 ms.

40. A sequence of fixed-length packets carrying digital audio signal is transmitted over a packet network. A packet is produced every 10 ms. The transfer delays incurred by the first 10 packets are 45 ms, 50 ms, 53 ms, 46 ms, 30 ms, 40 ms, 46 ms, 49 ms, 55 ms and 51 ms.

a. Sketch the sequence of packet transmission times and packet arrival times.

b. Find the delay that is inserted at the receiver to produce a fixed end-to-end delay of 75 ms.

c. Sketch the contents of the buffer at the receiver as a function of time.

41. Consider an application in which information that is generated at a constant rate is transferred over a packet network so timing recovery is required at the receiver.

a. What is the relationship between the maximum acceptable delay and the playout buffer?

b. What is the impact of the bit rate of the application information stream on the buffer requirements?

c. What is the effect of jitter on the buffer size design?

42. A speech signal is sampled at a rate of 8000 samples/second. Packets of speech are formed by packing 10 ms worth of samples into each payload. Timestamps are attached to the packets prior to transmission and used to perform error recovery at the receiver. Suppose that the timestamp is obtained by sampling a clock that advances every Δ seconds. Is there a minimum value that is required for Δ? If so, what is it?

43. Suppose that PDUs contain both timestamps and sequence numbers. Can the timestamps and sequence numbers be combined to provide a larger sequence number space? If so, should the timestamps or the sequence numbers occupy the most significant bit locations?

44. Consider the timestamp method for timing recovery discussed in the chapter.

a. Find an expression that relates the difference frequency Δf to the number of cycles M and N.

b. Explain why only M needs to be sent.

c. Explain how the receiver uses this value of M to control the playout procedure.

45. A 1.5 Mbps communications link is to use HDLC to transmit information to the moon. What is the smallest possible frame size that allows continuous transmission? The distance between earth and the moon is approximately 375,000 km, and the speed of light is 3×10^8 meters/second.

46. Perform the bit stuffing procedure for the following binary sequence: 110111111101111110101.

47. Perform bit destuffing for the following sequence: 11101111101111101111110.

48. Suppose HDLC is used over a 1.5 Mbps geostationary satellite link. Suppose that 250-byte frames are used in the data link control. What is the maximum rate at which information can be transmitted over the link?

49. In HDLC how does a station know whether a received frame with the fifth bit set to 1 is a P or an F bit?

50. Which of the following statements are incorrect?

a. A transmitting station puts its own address in command frames.

b. A receiving station sees its own address in a response frame.

c. A response frame contains the address of the sending station.

51. In HDLC suppose that a frame with $P = 1$ has been sent. Explain why the sender should not send another frame with $P = 1$. What should be done if the frame is lost during transmission.

52. HDLC specifies that the N(R) in a SREJ frame requests the retransmission of frame N(R) and also acknowledges all frames up to N(R)−1. Explain why only one SREJ frame can be outstanding at a given time.

53. The following corresponds to an HDLC ABM frame exchange with no errors.

a. Complete the diagram by completing the labeling of the frame exchanges.

b. Write the sequence of state variables at the two stations as each event takes place.

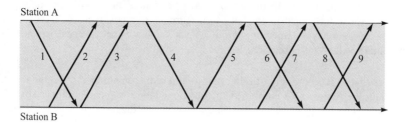

1. BI00   2. AI00   3. xIxx   4. xIxx   5. xRRxx
6. xIxx   7. xIxx   8. xRRxx   9. xRRxx

54. Assume station B is awaiting frame 2 from station A.
   a. Complete the diagram in HDLC ABM by completing the labeling of the frame exchanges.
   b. Write the sequence of state variables at the two stations as each event takes place.

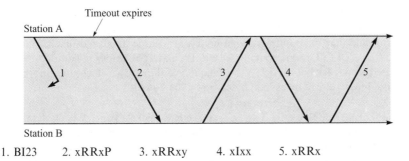

1. BI23   2. xRRxP   3. xRRxy   4. xIxx   5. xRRx

55. The following corresponds to an HDLC ABM frame exchange.
   a. Complete the diagram by completing the labeling of the frame exchanges.
   b. Write the sequence of state variables at the two stations as each event takes place.

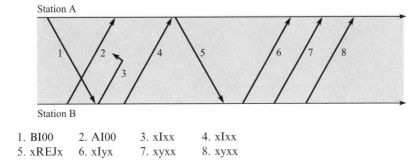

1. BI00   2. AI00   3. xIxx   4. xIxx
5. xREJx   6. xIyx   7. xyxx   8. xyxx

56. The PPP byte stuffing method uses an escape character defined by $0 \times 7D$ (01111101). When the flag is observed inside the frame, the escape character is placed in front of the flag and the flag is exclusive-ORed with 0×20. That is, $0 \times 7E$ is encoded as $0 \times 7D$ $0 \times 5E$. An escape character itself ($0 \times 7D$) is encoded as $0 \times 7D$ $0 \times 5D$. What are the contents of the following received sequence of bytes after byte destuffing: $0 \times 7D$ $0 \times 5E$ $0 \times FE$ 0×24 $0 \times 7D$ $0 \times 5D$ $0 \times 7D$ $0 \times 5D$ 0×62 $0 \times 7D$ $0 \times 5E$

57. Suppose that a 1-Megabyte message is sent over a serial link using TCP over IP over PPP. If the speed of the line is 56 kbps and the maximum PPP payload is 500 bytes, how long does it take to send the message?

58. Suppose that packets arrive from various sources to a statistical multiplexer that transmits the packets over 64 kbps PPP link. Suppose that the PPP frames have lengths that follow an exponential distribution with mean 1000 bytes and that the multiplexer can hold up to 100 packets at a time. Plot the average packet delay as a function of the packet arrival rate.

59. Suppose that the traffic is directed to a statistical multiplexer is controlled so that ρ is always less than 80%. Suppose that packet arrivals are modeled by a Poisson process and that packet lengths are modeled by an exponential distribution. Find the minimum number of packet buffers required to attain a packet loss probability of 10^{-3} or less.

60. Suppose that packets arrive from various sources to a statistical multiplexer that transmits the packets over a 1 Mbps PPP link. Suppose that the PPP frames have a constant length of L bytes and that the multiplexer can hold a very large number of packets at a time. Assume that each PPP frame contains a PPP, IP, and TCP header in addition to the user data. Plot the average packet delay as a function of the rate at which user information is transmitted for $L = 250$ bytes, 500 bytes, and 1000 bytes.

61. Suppose that a multiplexer receives constant-length packets from $N = 60$ data sources. Each data source has a probability $p = 0.1$ of having a packet in a given T-second period. Suppose that the multiplexer has one line in which it can transmit eight packets every T seconds. It also has a second line where it directs any packets that cannot be transmitted in the first line in a T-second period. Find the average number of packets that are transmitted in the first line and the average number of packets that are transmitted in the second line.

62. Discuss the importance of queueing delays in multiplexers that operate at bit rates of 1 Gbps or higher.

CHAPTER 6

Local Area Networks and Medium Access Control Protocols

There are two basic types of networks: switched networks and broadcast networks. **Switched networks** interconnect users by means of transmission lines, multiplexers, and switches. The transfer of packets across such networks requires routing tables to direct the packets from source to destination. To scale to a very large size, the addressing scheme in switched networks must be hierarchical to help provide location information that assists the routing algorithm in carrying out its task.[1] **Broadcast networks**, in contrast, are much simpler. Because all information is received by all users, routing is not necessary. A flat addressing scheme is sufficient to indicate which user a given packet is destined to. However, broadcast networks do require a medium access control protocol to orchestrate the transmissions from the various users. Local area networks (LANs), with their emphasis on low-cost and simplicity have been based on the broadcast approach. In this chapter we consider local area networks and medium access control protocols.

In broadcast networks a single transmission medium is shared by a community of users. For this reason, we also refer to these networks as **multiple access networks**. Typically, the information from a user is broadcast into the medium, and all the stations attached to the medium listen to all the transmissions. There is potential for user transmissions interfering or "colliding" with each other, and so a protocol has to be in place to prevent or minimize such interference. The role of the **medium access control (MAC)** protocols is to coordinate the access to the channel so that information gets through from a source to a destination in the same broadcast network.

It is instructive to compare MAC protocols with the peer-to-peer protocols discussed in Chapter 5. The basic role of both protocol classes is the same: to

---

[1]Switched networks are discussed in Chapter 7.

transfer blocks of user information despite transmission impairments. In the case of peer-to-peer protocols, the main concern is loss, delay, and resequencing of PDUs during transmission. In the case of MAC protocols, the main concern is interference from other users. The peer-to-peer protocols in Chapter 5 use sequence number information to detect and react to impairments that occur during transmission. We show that MAC protocols use a variety of mechanisms to adapt to collisions in the medium. The peer-to-peer protocol entities exchange control information in the form of ACK PDUs to coordinate their actions. Some MAC entities also make use of mechanisms for explicitly exchanging information that can be used to coordinate access to the channel. For peer-to-peer protocols we found that the delay-bandwidth product of the channel was a key parameter that determines system performance. The delay-bandwidth product plays the same fundamental role in medium access control. Finally, we note one basic difference between peer-to-peer protocols and MAC protocols: The former are involved with the interaction of only two peer processes, whereas the latter require the coordinated action from *all* of the MAC protocol entities within the same broadcast network. The lack of cooperation from a single MAC entity can prevent any communication from taking place over the shared medium.

The chapter is organized as follows:

1. *A general introduction to broadcast networks* and where they occur. In particular we discuss wireless networks and LANs, and we indicate the general approaches that are used to address the sharing of the medium.
2. *Overview of LANs*. We discuss the structure of the frames that are used in LANs, and we discuss the placement of LAN protocols in the OSI reference model.
3. *Random access*. We take a broader look at MAC algorithms, and we examine performance issues that help determine under what conditions a given MAC algorithm is appropriate. First, we consider MAC protocols that involve the "random" transmission of user packets into a shared medium. We begin with the seminal ALOHA protocol and proceed to the Carrier-Sense Multiple Access with Collision Detection (CSMA-CD) protocol, which forms the basis for the Ethernet LAN standard. We show that the delay-bandwidth product has a dramatic impact on protocol performance.
4. *Scheduling*. We consider MAC protocols that use scheduling to coordinate the access to the shared medium. We present a detailed discussion of ring-topology networks that make use of these protocols. We again show the impact of delay-bandwidth product on performance.
5. *Channelization*. Many shared medium networks operate through the assignment of channels to users. In this optional section we discuss three approaches to the creation of such channels: frequency-division multiple access, time-division multiple access, and code-division multiple access. We use various cellular telephone network standards to illustrate the application of these techniques. We also present results that show why these channelized approaches are not suitable for the transfer of bursty data.

6. *LAN standards.* We discuss several important LAN standards, including the IEEE 802.3 Ethernet LAN, the IEEE 802.5 token-ring LAN, the FDDI LAN, and the IEEE 802.11 wireless LAN. The MAC protocols associated with each LAN standard are also described.

7. *LAN bridges.* A LAN is limited in the number of stations that it can handle, so multiple LANs are typically deployed to handle a large number of stations with each LAN serving a reasonable number of stations. Devices to interconnect LANs become necessary to enable communications between users in different LANs. Bridges provide an approach for the interconnection of LANs.

6.1 MULTIPLE ACCESS COMMUNICATIONS

Figure 6.1 shows a generic multiple access communications situation in which a number of user stations share a transmission medium. *M* denotes the number of stations. The transmission medium is broadcast in nature, and so all the other stations that are attached to the medium can hear the transmission from any given station. When two or more stations transmit simultaneously, their signals will collide and interfere with each other.

There are two broad categories of schemes for sharing a transmission medium. The first category involves a static and collision-free sharing of the medium. We refer to these as **channelization schemes** because they involve the partitioning of the medium into separate channels that are then dedicated to particular users. Channelization techniques are suitable when stations generate a steady stream of information that makes efficient use of the dedicated channel. The second category involves a dynamic sharing of the medium on a per packet basis that is better matched to situations in which the user traffic is bursty. We refer to this category as **MAC schemes**. The primary function of the MAC is to minimize or eliminate the incidence of collisions to achieve a reasonable utilization of the medium. The two basic approaches to medium access control are random access and scheduling. Figure 6.2 summarizes the various approaches to sharing a transmission medium.

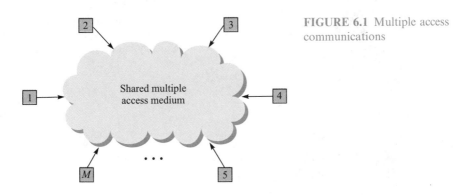

FIGURE 6.1 Multiple access communications

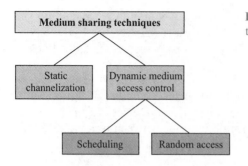

FIGURE 6.2 Approaches to sharing a transmission medium

Networks based on radio communication provide examples where a medium is shared. Typically, several stations share two frequency bands, one for transmitting and one for receiving. For example, in satellite communications each station is assigned a channel in an uplink frequency band that it uses to transmit to the satellite. As shown in Figure 6.3 the satellite is simply a repeater that takes the uplink signals that it receives from many earth stations and broadcasts them back on channels that occupy a different downlink frequency band. Cellular telephony is another example involving radio communications. Again we have two frequency bands shared by a set of mobile users. In Chapter 4 we showed that a cellular telephone connection involves the assignment of an inbound channel from the first band and an outbound channel from the second band to each mobile user. Later in this chapter we present different approaches to the creation of such channels.

Channel sharing techniques are also used in wired communications. For example, multidrop telephone lines were used in early data networks to connect a number of terminals to a central host computer, as shown in Figure 6.4. The set of M stations shares an inbound and outbound transmission line. The host computer broadcasts information to the users on the outbound line. The stations transmit information to the host using the inbound line. Here there is a potential for interference on the inbound line. In the MAC protocol developed for this

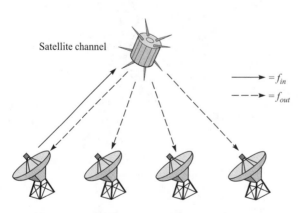

FIGURE 6.3 Satellite communication involves sharing of uplink and downlink frequency bands

Multidrop telephone lines

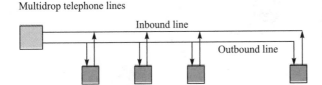

FIGURE 6.4 Multidrop telephone line requires access control

system, the host computer issues polling messages to each terminal, providing it with permission to transmit on the inbound line.

Ring networks also involve the sharing of a medium. Here a number of ring adapter interfaces are interconnected in a ring topology by point-to-point transmission lines, as shown in Figure 6.5. Typically, information packets are transmitted in the ring in cut-through fashion, with only a few bits delay per ring adapter. All stations can monitor the passing signal and extract packets intended for them. A MAC procedure is required here to orchestrate the insertion and removal of packets from the shared ring, since obviously only one signal can occupy a particular part of the ring.

Shared buses are another example of shared broadcast media. For example, in coaxial cable transmission systems users can inject a signal that propagates in both directions along the medium, eventually reaching all stations connected to the cable. All stations can listen to the medium and extract transmissions intended for them. If two or more stations transmit simultaneously, the signal in the medium will be garbled, so again some means are needed to coordinate access to the medium. In section 6.6.1 we show that hub topology networks lead to a situation identical to that of shared buses.

Finally, we return to a more current example involving wireless communications, shown in Figure 6.6. Here a set of devices such as workstations, laptop computers, cordless telephones, and other communicating appliances share a wireless medium, for example, 5 GHz radio or infrared light. These devices could be transmitting short messages, real-time voice, or video monitoring

Ring networks

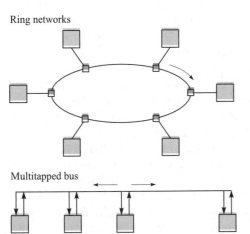

FIGURE 6.5 Ring networks and multitapped buses require MAC

Multitapped bus

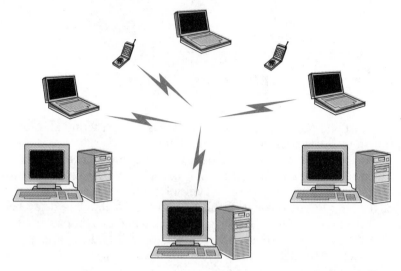

FIGURE 6.6 Wireless LAN

information, or they could be accessing Web pages. Again a MAC procedure is required to share the medium. In this example the diversity of the applications requires the medium access control to also provide some degree of quality-of-service (QoS) guarantees.

In the examples discussed so far, we can discern two basic approaches. In the first approach communications between the M stations makes use of *two* channels. A base station or controller communicates to all the other stations by broadcasting over the "outbound," "forward," or "downstream" channel. The stations in turn share an "inbound," "reverse," or "upstream" channel in the direction of the base station. In the preceding examples the satellite system, cellular telephone systems, and the multidrop telephone line take this approach. It is also the approach in cable modem access systems and certain types of wireless data access systems. In this approach the base station or controller can send coordination information to the other stations to orchestrate the access to the shared reverse channel. The second approach to providing communications uses direct communications between all M stations. This is the case for ad hoc wireless LANs as well as for the multitapped bus. Note that the ring network has features of both approaches. On the one hand, ring networks involve all stations communicating directly with other stations; on the other hand, each time a station has the token that entitles it to transmit it, can be viewed as acting as a temporary controller that directs how traffic enters the ring at that moment.

The preceding examples have the common feature that the shared medium is the *only* means available for the stations to communicate with each other. Thus any explicit or implicit coordination in accessing the channel must be done through the channel itself. This situation implies that some of the transmission resource will be utilized implicitly or explicitly to transfer coordination information. We saw in Chapter 5 that the *delay-bandwidth product* plays a key role in the performance of ARQ protocols. The following example indicates how the delay-bandwidth product affects performance in medium access control.

Example—Delay-Bandwidth Product and MAC Performance

Consider the situation shown in Figure 6.7 in which two stations are trying to share a common medium. Let's develop an access protocol for this system. Suppose that when a station has a packet to send, the station first listens to the channel to see whether it is busy with a transmission from the other station. If it is not busy, then the station begins to transmit, but it continues observing the signal in the channel to make sure that its signal is not corrupted by a signal from the other station. The signal from station A does not reach station B until time t_{prop}. If station B has not begun a transmission by that time, then station A is assured that station B will refrain from transmitting thereafter and so station A has captured the channel and its entire message will get through.

Figure 6.7 shows what happens when a collision of packet transmissions takes place. In this case station B must have begun its transmission sometime between times $t = 0$ and $t = t_{prop}$. By time $t = 2t_{prop}$, at the latest, station A will find out about the collision. At this point both stations are aware that they are competing for the channel. Some mechanism for resolving this contention is required. We will suppose for simplicity that both stations know the value of the propagation delay t_{prop} and that they measure the time from when they began transmitting to when a collision occurs. The station that began transmitting earlier (which measures the aforementioned time to be greater than $t_{prop}/2$) is declared to be the "winner" and proceeds to retransmit its packet as soon as the channel goes quiet. The "losing" station defers and remains quiet until the packet transmission from the other station is complete. For the sake of fairness, we suppose that the winning station is compelled to remain quiet for time $2t_{prop}$ after it has completed its packet transmission. This interval gives the losing station the opportunity to capture the channel and transmit its packet.

Thus we see for this example that a time approximately equal to $2t_{prop}$ is required to coordinate the access for each packet transmitted. If the bit rate of the channel is R, then the number of transmission bits "wasted" in access

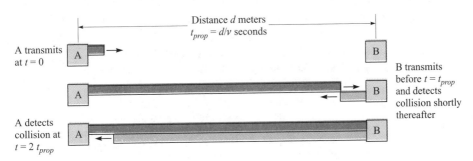

FIGURE 6.7 Channel capture and delay-bandwidth product where v is the speed of light in the medium, typically 2 to 3×10^8 meters/second

coordination is approximately $2t_{prop}R$. If packets are an average of L bits long, the **efficiency** in the use of the channel is

$$\text{Efficiency} = \frac{L}{L + 2t_{prop}R} = \frac{1}{1 + \dfrac{2t_{prop}R}{L}} = \frac{1}{1 + 2a}$$

where $a = t_{prop}R/L$ is the ratio of the (one way) delay-bandwidth product to the average packet length. When a is much smaller than 1, the channel can be used very efficiently by using the above protocol. For example, if $a = 0.01$, then the efficiency is $1/1.02 = 0.98$. As a becomes larger, the channel becomes more inefficient. For example, if $a = 0.5$, then the efficiency is $1/2 = 0.50$.

The efficiency result in the preceding example is typical of MAC protocols. Later in the chapter we explain that the CSMA-CD protocol that originally formed the basis for the Ethernet LAN standard has an efficiency or maximum normalized **throughput** of approximately $1/(1 + 6.44a)$, where $a = t_{prop}R/L$. We also show that for token-ring networks, such as in the IEEE 802.5 standard, the maximum efficiency is approximately given by $1/(1 + a')$ where a' is the ratio of the latency of the ring in bits to the average packet length. The ring latency has two components: The first component is the sum of the bit delays introduced at every ring adapter; the second is the delay-bandwidth product where the delay is the time required for a bit to circulate around the ring. In both of these important LANs, we again see that the relative value of the delay-bandwidth product to the average packet length is a key parameter in system performance.

Table 6.1 shows the number of bits in transit in a one-way propagation delay for various distances and transmission speeds. Possible network types range from a "desk area network" with a diameter of 1 meter to a global network with a diameter of 100,000 km (about two times around the earth). It can be seen that the reaction time as measured by the number of bits in transit can become quite large for high transmission speeds and for long distances. To estimate some values of the key parameter a, we note that Ethernet packets have a *maximum* length of approximately 1500 bytes = 12,000 bits. TCP segments have a maximum length of approximately 65,000 bytes = 520,000 bits. For desk area and local area networks, we see that these packet lengths can provide acceptable

Distance	10 Mbps	100 Mbps	1 Gbps	Network type
1 m	3.33×10^{-02}	3.33×10^{-01}	3.33×10^{0}	Desk area network
100 m	3.33×10^{01}	3.33×10^{02}	3.33×10^{03}	Local area network
10 km	3.33×10^{02}	3.33×10^{03}	3.33×10^{04}	Metropolitan area network
1000 km	3.33×10^{04}	3.33×10^{05}	3.33×10^{06}	Wide area network
100000 km	3.33×10^{06}	3.33×10^{07}	3.33×10^{08}	Global area network

TABLE 6.1 Number of bits in transit in one-way propagation delay assuming propagation speed of 3×10^{8} meters/second

values of *a*. For higher speeds or longer distances, the packet sizes are not long enough to overcome the large delay-bandwidth products. Consequently, we find that networks based on broadcast techniques are used primarily in LANs and other networks with small delay-bandwidth products.

The selection of a MAC protocol for a given situation depends on delay-bandwidth product and efficiency, as well as other factors. The *transfer delay* experienced by packets is an important performance measure. *Fairness* in the sense of giving stations equitable treatment is also an issue. The concern here is that stations receive an appropriate proportion of bandwidth and that their packets experience an appropriate transfer delay. Note that fairness does not necessarily mean equal treatment; it means providing stations with treatment that is consistent with policies set by the network adminstrator. *Reliability* in terms of robustness with respect to failure and faults in equipment is important as it affects the availability of service. An issue that is becoming increasingly important is the capability to carry *different types of traffic*, that is, stream versus bursty traffic, a well as the capability to provide *quality-of-service* guarantees. *Scalability* with respect to number of stations and the user bandwidth requirements are also important in certain settings. And of course *cost* is always an issue.

The performance of a MAC algorithm is described by its transfer delay and its throughput. The **frame transfer delay** is defined as the time that elapses from when the first bit of the frame arrives at the source MAC to when the last bit of the frame is delivered to the destination MAC. The **throughput** is defined as the actual rate at which information is transmitted through the shared medium. The throughput is measured in frames/second or bits/second. Suppose that the transmission rate in the medium is R bits/second and suppose that the average frame length is L bits/frame. The maximum possible throughput is then R/L frames/second. This result assumes a perfectly coordinated medium access control in which the different stations can transmit their frames back to back without colliding. In practice each MAC protocol has a maximum throughput less than R/L, since some channel time is wasted in collisions or in sending coordination information. In some situations the throughput includes only actual user information and excludes overhead such as MAC layer headers and trailers.

The throughput depends on the rate at which the stations attempt to transmit into the medium as well as on the capability of the medium (and the access control) to sustain such a rate. Figure 6.8 shows how the average transfer delay increases with throughput. First we need to introduce the notation in the curve. The abscissa (x-axis) ρ represents that **normalized throughput** or **load**, which is defined as the actual throughput in frames/second divided by the maximum possible throughput R/L. The ordinate (y-axis) gives the *average frame transfer delay* **E[T]** in multiples of a single *frame transmission time* $E[X] = L/R$ seconds. When the load ρ is small, the stations transmit infrequently. Because there is likely to be little or no contention for the channel, the transfer delay is close to one frame transmission time, as shown in the figure. As the load increases, there is more contention for the channel, and so frames have to wait increasingly longer times before they are transferred. Finally, the rate of frame arrivals

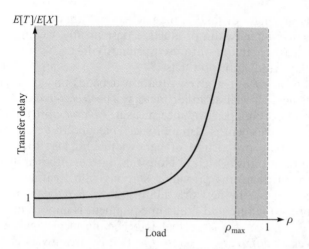

from all the stations reaches a point that exceeds the rate that the channel can sustain. At this point the buffers in the stations build up with a backlog of frames, and if the arrivals remain unchecked, the transfer delays grow without bound as shown in the figure.

Note from Figure 6.8 that the *maximum achievable load* ρ_{max} is usually less than 1. The reason is that most MAC algorithms consume some transmission time in collisions and coordination. Consequently, the maximum actual throughput is less than R/L, and the achievable load is less than 1. Figure 6.9 shows that the transfer delay versus load curve varies with parameter a, which is defined as the ratio of the one-way delay-bandwidth product to the average frame size. When a is small, the relative cost of coordination is low and ρ_{max} can be close to 1. However, when a increases, the relative cost of coordination becomes high and ρ_{max} can be significantly less than 1. Figure 6.9 is typical of the delay-throughput performance of medium access controls. The exact shape of the

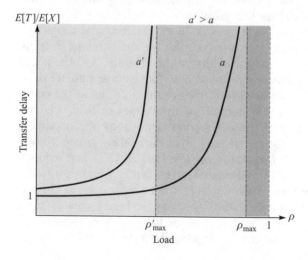

FIGURE 6.9 Dependence of delay-throughput performance on $a = Rt_{prop}/L$

curve depends on the particular medium and MAC protocol, the number of stations, the arrival pattern of frames at the stations, and the distribution of lengths of the frames.

6.2 LOCAL AREA NETWORKS

In Chapter 1 we noted that the development of LANs was motivated by the need to share resources and information among workstations in a department or workgroup. We also noted that the requirements for LANs are different than those in a wide area or in a public network. The short distances between computers imply that low-cost, high-speed, reliable communications is possible. The emphasis on low-cost implies a broadcast network approach that does not use equipment that switches information between stations. Instead, the stations cooperate by executing a MAC protocol that minimizes the incidence of collisions in a shared medium (e.g., a wire).

In this section we discuss general aspects of LAN standards. Most LAN standards have been developed by the IEEE 802 committee of the Institute of Electrical and Electronic Engineers (IEEE), which has been accredited in the area of LANs by the American National Standards Institute (ANSI). The set of standards includes the CSMA-CD (Ethernet) LAN and the token-passing ring LAN. It also includes the definition of the logical link control, which places LANs within the data link layer of the OSI reference model.

6.2.1 LAN Structure

The structure of a typical LAN is shown in Figure 6.10a. A number of computers and network devices such as printers are interconnected by a shared transmission medium, typically a cabling system, which is arranged in a bus, ring, or star topology. The cabling system may use twisted-pair cable, coaxial cable, or optical fiber transmission media. In some cases the cabling system is replaced by wireless transmission based on radio or infrared signals. The Ethernet bus topology is shown in Figure 6.10a. LAN standards define physical layer protocols that specify the physical properties of the cabling or wireless system, for example, connectors and maximum cable lengths, as well as the digital transmission system, for example, modulation, line code, and transmission speed.

The computers and network devices are connected to the cabling system through a **network interface card (NIC)** or **LAN adapter card** (Figure 6.10b). For desktop computers the NIC is inserted into an expansion slot or built into the system. Laptop computers typically use the smaller PCMCIA card, which is inserted into a slot that can also be used by a modem or other device.

The NIC card coordinates the transfer of information between the computer and the network. The NIC card transfers information in parallel format to and

(a)

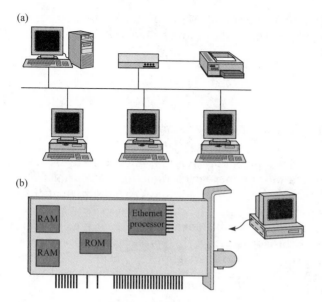

FIGURE 6.10 Typical LAN structure and network interface card

(b)

from main memory (RAM) in the computer. On the other hand, the NIC card transfers information in serial format to and from the network, so parallel-to-serial conversion is one of the NIC's functions. The speed of the network and the computer are not matched, so the NIC card must also buffer data.

The NIC card has a port that meets the connector and transmission specifications of physical layer standards. The NIC card includes read-only memory (ROM) containing firmware that allows the NIC to implement the MAC protocol of a LAN standard. This process involves taking network layer PDUs, encapsulating them inside MAC frames, and transferring the frames by using the MAC algorithm, as well as receiving MAC frames and delivering the network layer PDUs to the computer.

Each NIC card is assigned a *unique physical address* that is burned into the ROM. Typically, the first three bytes of the address specify the card vendor, and the remaining bytes specify a unique number for that vendor. The NIC card contains hardware that allows it to recognize its physical address, as well as the broadcast address. The hardware can also be set to recognize multicast addresses that direct frames to groups of stations. The card can also be set to run in "promiscuous" mode where it listens to all transmissions. This mode is used by system administrators to troubleshoot the network. It is also used by hackers to intercept unencrypted passwords and other information that can facilitate unauthorized access to computers in the LAN.[2]

---

[2]LANs were developed to operate in a private environment where an element of trust among users could be assumed. This assumption is no longer valid, so network security protocols such as those presented in Chapter 11 are now required.

6.2.2   The Medium Access Control Sublayer

The layered model in Figure 6.11 shows how the LAN functions are placed within the two lower layers of the OSI reference model. The data link layer is divided into two sublayers: the logical link control (LLC) sublayer and the medium access control (MAC) sublayer. The MAC sublayer deals with the problem of coordinating the access to the shared physical medium. Figure 6.11 shows that the IEEE has defined several MAC standards, including IEEE 802.3 (Ethernet) and IEEE 802.5 (token ring). Each MAC standard has an associated set of physical layers over which it can operate.

The MAC layer provides for the connectionless transfer of datagrams. Because transmissions in LANs are relatively error free, the MAC protocols usually do not include procedures for error control. The MAC entity accepts a block of data from the LLC sublayer or directly from the network layer. This entity constructs a PDU that includes source and destination MAC addresses as well as a frame check sequence (FCS), which is simply a CRC checksum. The MAC addresses specify the physical connections of the workstations to the LAN. The main task of the MAC entities is to execute the MAC protocol that directs when they should transmit the frames into the shared medium.

In Figure 6.12 we show the protocol stacks of three workstations interconnected through a LAN. Note how all three MAC entities must cooperate to provide the datagram transfer service to the LLC sublayer. In other words, the interaction between MAC entities is not between pairs of peers, but rather

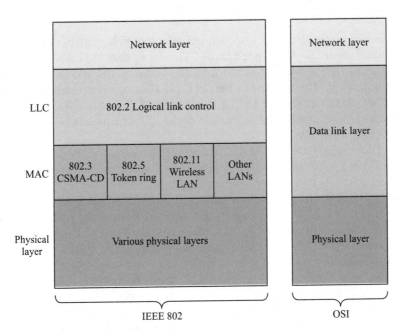

FIGURE 6.11 IEEE 802 LAN standards

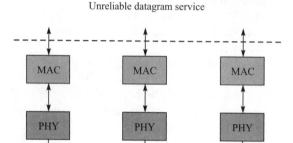

FIGURE 6.12 The MAC sublayer provides unreliable datagram service

all entities must monitor all frames that are transmitted onto the shared medium. We defer the discussion of the specific MAC algorithms to the sections on individual LAN standards.

6.2.3   The Logical Link Control Layer

The IEEE 802 committee has also defined a **logical link control (LLC)** sublayer that operates over all MAC standards. The LLC can enhance the datagram service offered by the MAC layer to provide some of the services of HDLC at the data link layer. This approach makes it possible to offer the network layer a standard set of services while hiding the details of the underlying MAC protocols. The LLC also provides a means for exchanging frames between LANs that use different MAC protocols.

The LLC builds on the MAC datagram service to provide three HDLC services. Type 1 LLC service is *unacknowledged connectionless service* that uses unnumbered frames to transfer unsequenced information. Recall from Chapter 5 that the HDLC protocols use unnumbered frames in some of their message exchanges. Type 1 LLC service is by far the most common in LANs. Type 2 LLC service uses information frames and provides *reliable connection-oriented service* in the form of the asynchronous balanced mode of HDLC. A connection setup and release is required, and the connection provides for error control, sequencing, and flow control. Figure 6.13 shows two type 2 LLC entities at stations A and C providing reliable packet transfer service. Type 2 operation is useful when the endsystems do not use a transport layer protocol to provide reliable service. For example, type 2 is used in several proprietary PC LAN software products. Type 3 LLC service provides *acknowledged connectionless*

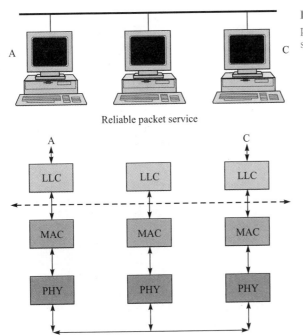

FIGURE 6.13 The LLC can provide reliable packet transfer service

service, that is, connectionless transfer of individual frames with acknowledgments. To provide type 3 LLC service, two additional unnumbered frames were added to the set defined by HDLC.

Additional addressing is provided by the LLC to supplement the addressing provided by the MAC. A workstation in the LAN has a single MAC (physical) address. However, at any given time such a workstation might simultaneously handle several data exchanges originating from different upper-layer protocols but operating over this same physical connection. These logical connections are distinguished by their *service access point (SAP)* in the LLC, as shown in Figure 6.14. For example, frames that contain IP packets are identified by hexadecimal 06 in the SAP, frames that contain Novell IPX are identified by E0, frames with OSI packets by FE, and frames with SNAP PDUs (discussed below) by AA. In practice, the LLC SAP specifies in which memory buffer the NIC places the frame contents, thus allowing the appropriate higher-layer protocol to retrieve the data.

Figure 6.14 first shows the LLC PDU structure that consists of one byte each for source and destination SAP addresses, one or two bytes for control information, and the information itself, that is, the network layer packet. The second part of Figure 6.14 shows the details of the address bytes. In general a seven-bit address is used. The first bit of the destination SAP address byte indicates whether it is an individual or group address. The first bit of the source SAP address byte is not used for addressing and instead is used to indicate whether a frame is a command or response frame. The control field is one byte long if

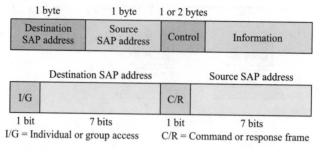

FIGURE 6.14 LLC PDU structure

three-bit sequence numbering is used. The control field is two bytes long when extended sequence numbering is used.

The LLC PDU is encapsulated in IEEE MAC frames as shown in Figure 6.15. The MAC adds both a header and a trailer. Note the accumulation of header overhead: After TCP and IP have added their minimum of 20 bytes of headers, the LLC adds 3 or 4 bytes, and then the MAC adds its header and trailer. In the next several sections we consider MAC protocols. We then consider specific frame formats as we introduce several of the IEEE 802 LAN standards.

6.3 RANDOM ACCESS

In this and the next section we consider the two main classes of MAC algorithms. In particular we investigate the system parameters that affect MAC performance. An understanding of these issues is required in the design and selection of MAC algorithms, especially in situations that involve large delay-bandwidth products.

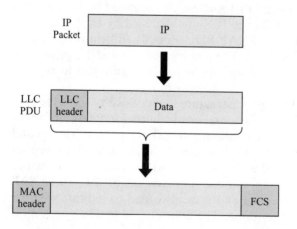

FIGURE 6.15 LLC PDU and MAC frame

Random access methods constitute the first major class of MAC procedures. In the beginning of this chapter, we indicated that the reaction time in the form of propagation delay and network latency is a key factor in the effectiveness of various medium access techniques. This factor should not be surprising, as we have already observed the influence of reaction time on the performance of ARQ retransmission schemes. In the case of Stop-and-Wait ARQ, we found that the performance was good as long as the reaction time was small. However, Stop-and-Wait ARQ became ineffective when the propagation delay, and hence the reaction time, became very large. The weakness of Stop-and-Wait is that the transmission of packets is closely tied to the return of acknowledgments, which cannot occur sooner than the reaction time. The solution to this problem involved proceeding with transmission without waiting for acknowledgments and dealing with transmission errors after the fact.

A similar situation arises in the case of scheduling approaches to MAC that are considered in section 6.4. When the reaction time is small, the scheduling can be done quickly, based on information that is current. However, when the reaction time becomes very large, by the time the scheduling takes effect the state of the system might have changed considerably. The experience with ARQ systems suggests an approach that involves proceeding with transmissions without scheduling. It also suggests dealing with collisions after the fact. The class of random access MAC procedures discussed in this section takes this approach.

6.3.1 ALOHA

As the name suggest, the ALOHA random access scheme had its origins in the Hawaiian Islands. The University of Hawaii needed a means to interconnect terminals at campuses located on different islands to the host computer on the main campus. The solution is brilliant in its simplicity. A radio transmitter is attached to the terminals, and messages are transmitted as soon as they become available, thus producing the smallest possible delay. From time to time packet transmissions will collide, but these can be treated as transmission errors, and recovery can take place by retransmission. When traffic is very light, the probability of collision is very small, and so retransmissions need to be carried out infrequently.

There is a significant difference between normal transmission errors and those that are due to packet collisions. Transmission errors that are due to noise affect only a single station. On the other hand, in the packet collisions more than one station is involved, and hence more than one retransmission is necessary. This interaction by several stations produces a positive feedback that can trigger additional collisions. For example, if the stations use the same time-out values and schedule their retransmissions in the same way, then their future retransmissions will also collide. For this reason, the ALOHA scheme requires stations to use a random retransmission time. This randomization is intended to spread out the retransmissions and reduce the likelihood of additional collisions between the stations. Nevertheless, the likelihood of collisions is increased after each packet collision.

Figure 6.16 shows the basic operation of the ALOHA scheme. It can be seen that the first transmission is done without any scheduling. Information about the outcome of the transmission is obtained after the reaction time $2t_{prop}$, where t_{prop} is the maximum one-way propagation time between two stations. If no acknowledgment is received after the time-out period, a backoff algorithm is used to select a random retransmission time.

It is clear that in the ALOHA scheme the network can swing between two modes. In the first mode packet transmissions from the station traverse the network successfully on the first try and collide only from time to time. The second mode is entered through a snowball effect that occurs when there is a surge of collisions. The increased number of backlogged stations, that is, stations waiting to retransmit a message, increases the likelihood of additional collisions. This situation leads to a further increase in the number of backlogged stations, and so forth.

Abramson provided an approximate analysis of the ALOHA system that gives an insight into its behavior. Assume that packets are a constant length L and constant transmission time $X = L/R$. Consider a reference packet that is transmitted starting at time t_0 and completed at time $t_0 + X$, as shown in Figure 6.16. This packet will be successfully transmitted if no other packet collides with it. Any packet that begins its transmission in the interval t_0 to $t_0 + X$ will collide with the reference packet. Furthermore, any packet that begins its transmission in the prior X seconds will also collide with the reference packet. Thus the probability of a successful transmission is the probability that there are no additional packet transmissions in the **vulnerable period** $t_0 - X$ to $t_0 + X$.

Abramson used the following approach to find the probability that there is no collision with the reference packet. Let S be the *arrival rate of new packets* to the system in units of packets/X seconds. We assume that all packets eventually make it through the system, so S also represents the throughput of the system. Now the actual arrival rate to the system consists of new arrivals and retransmissions. Let G be the *total arrival rate* in units of packets/X seconds. G is also called the total load. The probability of transmissions from new packets and retransmitted packets is not straightforward to calculate. Abramson made the key simplifying assumption that the aggregate arrival process that results from the new and retransmitted packets has a Poisson distribution with an average number of arrivals of $2G$ arrivals/$2X$ seconds, that is:

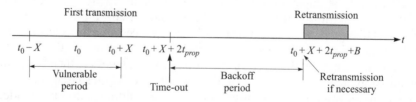

FIGURE 6.16 ALOHA random access scheme

$$P[k \text{ transmissions in } 2X \text{ seconds}] = \frac{(2G)^k}{k!} e^{-2G}, k = 0, 1, 2, \ldots$$

The throughput S is equal to the total arrival rate G times the probability of a successful transmission, that is:

$$S = GP[\text{no collision}] = GP[0 \text{ transmissions in } 2X \text{ seconds}]$$

$$= G \frac{(2G)^0}{0!} e^{-2G}$$

$$= Ge^{-2G}$$

Figure 6.17 shows a graph of S versus G. Starting with small G, we show that S increases, reaches a peak value of $1/2e$ at $G = 0.5$, and then declines back toward 0. For a given value of S, say, $S = 0.05$, there are two associated values of G. This is in agreement with our intuition that the system has two modes: one associating a small value of G with S, that is, $S \approx G$, and another associating a large value of G with S, that is, $G \gg S$ when many stations are backlogged. The graph also shows that values of S beyond $1/2e$ are not attainable. This condition implies that the ALOHA system cannot achieve throughputs higher than $1/2e = 18.4$ percent.

The average delay in an ALOHA system can be approximated as follows. The average number of transmission attempts/packet is

$$G/S = e^{2G} \text{ attempts per packet.}$$

Therefore, the average number of *unsuccessful attempts* per packet is

$$\varepsilon = G/S - 1 = e^{2G} - 1$$

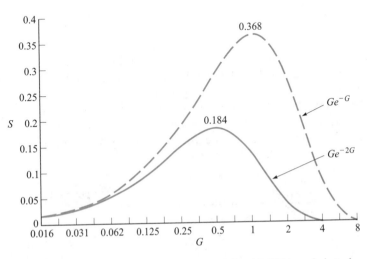

FIGURE 6.17 Throughput S versus load G for ALOHA and slotted ALOHA

The first transmission requires $X + t_{prop}$ seconds, and each subsequent retransmission requires $2t_{prop} + X + B$, where B is the average backoff time and t_{prop} is the one-way propagation delay. Thus the average packet transmission time is approximately given by:

$$E[T_{aloha}] = X + t_{prop} + (e^{2G} - 1)(X + 2t_{prop} + B)$$

Following our practice of expressing delay in multiples of X, we have

$$E[T_{aloha}]/X = 1 + a + (e^{2G} - 1)(1 + 2a + B/X)$$

where $a = t_{prop}/X$ is the one-way normalized propagation delay. For example, if the backoff time is uniformly distributed between 1 and K packet transmission times, then $B = (K + 1)X/2$.

6.3.2 Slotted ALOHA

The performance of the ALOHA scheme can be improved by reducing the probability of collisions. The slotted ALOHA scheme shown in Figure 6.18 reduces collisions by constraining the stations to transmit in synchronized fashion. All the stations keep track of transmission time slots and are allowed to initiate transmissions only at the beginning of a time slot. Packets are assumed to be constant and to occupy one time slot. The packets that can collide with our reference packet must now arrive in the period $t_0 - X$ to t_0. The vulnerable period in the system is now X seconds long.

Proceeding as before, we obtain

$$S = GP[\text{no collision}] = GP[0 \text{ transmissions in } X \text{ seconds}]$$

$$= G\frac{(G)^0}{0!}e^{-G}$$

$$= Ge^{-G}$$

Figure 6.17 shows the behavior of the throughput S versus load G for the slotted ALOHA system. The system still exhibits its bimodal behavior and has a maximum throughput of $1/e = 36.8$ percent. The ALOHA and the slotted ALOHA systems show how low-delay packet transmission is possible using essentially uncoordinated access to a medium. However, this result is at the expense of significant wastage due to collisions, which limits the maximum

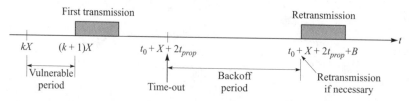

FIGURE 6.18 Slotted ALOHA random access scheme, $(t_0 = (k + 1)X)$

achievable throughput to low values. Furthermore, unlike the other MAC schemes discussed in this chapter, the maximum throughput of the ALOHA schemes is not sensitive to the reaction time because stations act independently.

Following the same procedure as in the ALOHA case, we can develop the following estimate for the average packet delay in slotted ALOHA:

$$E[T_{slotaloha}]/X = 1 + a + (e^G - 1)(1 + 2a + B/X)$$

If the backoff time is uniformly distributed between 1 and K packet times, then $B = (K + 1)X/2$.

Example—ALOHA and Slotted ALOHA

Suppose that a radio system uses a 9600 bps channels for sending call setup request messages to a base station. Suppose that packets are 120 bits long, that the timeout is 20 ms, and that the backoff is uniformly distributed between 1 and 7. What is the maximum throughput possible with ALOHA and with slotted ALOHA? Compare the average delay in ALOHA and slotted ALOHA when the load is 40 percent of the maximum possible throughput of the ALOHA system.

The system transmits packets at a rate of 9600 bits/second × 1 packet/120 bits = 80 packets/second. The maximum throughput for ALOHA is then 80(0.184) ≈ 15 packets/second. The maximum throughput for slotted ALOHA is then ≈ 3 packets/second.

The load at 40 percent of the maximum of the ALOHA system is 15 × 0.40 = 6 packets/second. G is expressed in packet arrivals/packet transmission time, therefore, $G = 6/80$. Assuming that the propagation delay is negligible, the average packet delay for ALOHA in multiples of X is then

$$1 + (e^{12/80} - 1)(1 + (1 + 7)/2) = 1.81 \text{ packet transmission times.}$$

The slotted ALOHA system with a load of 6 packets/second

$$1 + (e^{6/80} - 1)(1 + (1 + 7)/2) = 1.39 \text{ packet transmission times.}$$

6.3.3 CSMA

The low maximum achievable throughput of the ALOHA schemes are due to the wastage of transmission bandwidth because of packet collisions. This wastage can be reduced by avoiding transmissions that are certain to cause collisions. By sensing the medium for the presence of a carrier signal, a station can determine whether there is an ongoing transmission. The class of **carrier sensing multiple access (CSMA)** MAC schemes uses this strategy.

In Figure 6.19 we show how the vulnerable period is determined in a CSMA system. At time $t = 0$, station A begins at one extreme end of a broadcast

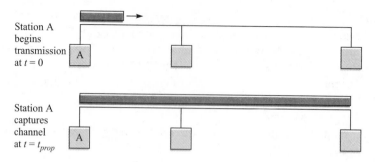

FIGURE 6.19 CSMA random access scheme

medium. As the signal propagates through the medium, stations become aware of the transmission from station A. At time $t = t_{prop}$, the transmission from station A reaches the other end of the medium. By this time all stations are aware of the transmission from station A. Thus the vulnerable period consists of one propagation delay, and if no other station initiates a transmission during this period, station A will in effect capture the channel because no other stations will transmit thereafter.

CSMA schemes differ according to the behavior of stations that have a packet to transmit when the channel is busy. In **1-Persistent CSMA**, stations with a packet to transmit sense the channel. If the channel is busy, they sense the channel continuously, waiting until the channel becomes idle. As soon as the channel is sensed idle, they transmit their packets. If more than one station is waiting, a collision will occur. In addition, stations that have a packet arrive within t_{prop} of the end of the preceding transmission will also transmit and possibly be involved in a collision. Stations that are involved in a collision perform the backoff algorithm to schedule a future time for resensing the channel. In a sense, in 1-Persistent CSMA stations act in a "greedy" fashion, attempting to access the medium as soon as possible. As a result, 1-Persistent CSMA has a relatively high collision rate.

Non-Persistent CSMA attempts to reduce the incidence of collisions. Stations with a packet to transmit sense the channel. If the channel is busy, the stations immediately run the backoff algorithm and reschedule a future resensing time. If the channel is idle, the stations transmit. By immediately rescheduling a resensing time and not persisting, the incidence of collisions is reduced relative to 1-Persistent CSMA. This immediate rescheduling also results in longer delays than are found in 1-Persistent CSMA.

The class of **p-Persistent CSMA** schemes combines elements of the above two schemes. Stations with a packet to transmit sense the channel, and if the channel is busy, they persist with sensing until the channel becomes idle, as shown in Figure 6.20. If the channel is idle, the following occurs: with probability p, the station transmits its packet; with probability $1 - p$ the station decides to wait an additional propagation delay t_{prop} before again sensing the channel. This behavior is intended to spread out the transmission attempts by the stations that have

FIGURE 6.20 p-Persistent
CSMA random access scheme

been waiting for a transmission to be completed and hence to increase the like-lihood that a waiting station successfully seizes the medium.

All of the variations of CSMA are sensitive to the end-to-end propagation delay of the medium that constitutes the vulnerable period. An analysis of the throughput S versus load G for CSMA is beyond the scope of this text. Figure 6.21 shows the throughput S versus load G for 1-Persistent and Non-Persistent CSMA for three values of a. It can be seen that the throughput of 1-Persistent CSMA drops off much more sharply with increased G. It can also be seen that the normalized propagation delay $a = t_{\text{prop}}/X$ has a significant impact on the maximum achievable throughput. Non-Persistent CSMA achieves a higher throughput than 1-Persistent CSMA does over a broader range of load values G. For very small values of a, Non-Persistent CSMA has a relatively high max-imum achievable throughput. However, as a approaches 1, both 1-Persistent and Non-Persistent CSMA have maximum achievable throughputs that are even lower than the ALOHA schemes.

6.3.4   CSMA-CD

The CSMA schemes improve over the ALOHA schemes by reducing the vulner-able period from one- or two-packet transmission times to a single propagation delay t_{prop}. In both ALOHA and CSMA schemes, collisions involve entire packet transmissions. If a station can determine whether a collision is taking place, then the amount of wasted bandwidth can be reduced by aborting the transmission when a collision is detected. The **carrier sensing multiple access with collision detection (CSMA-CD)** schemes use this approach.

Figure 6.22 shows the basic operation of CSMA-CD. At time $t = 0$, station A at one extreme end of the network begins transmitting a packet. This packet transmission reaches station B at another extreme end of the network, at time t_{prop}. If no other station initiates a transmission in this time period, then station A will have captured the channel. However, suppose that station B initiates a transmission just before the transmission arrival from station A. Station A will not become aware of the collision until time $= 2t_{prop}$. Therefore, station A requires $2t_{prop}$ seconds to find out whether it has successfully captured the channel.

In CSMA-CD a station with a packet first senses the channel and transmits if the channel is idle. If the channel is busy, the station uses one of the possible strategies from CSMA; that is, the station can persist, backoff immediately, or persist and attempt transmission with probability p. If a collision is detected during transmission, then a short jamming signal is transmitted to ensure that

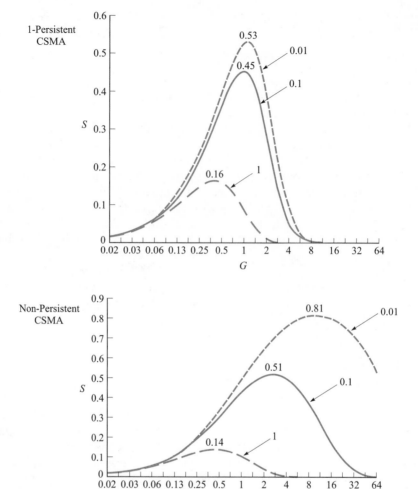

FIGURE 6.21 Throughput S versus load for G for 1-Persistent and Non-Persistent CSMA. The curves are for different values of a'

other stations know that a collision has occurred before aborting the transmission, and the backoff algorithm is used to schedule a future resensing time.

The channel can be in three states: busy transmitting a frame, idle, or in a contention period where stations attempt to capture the channel. The throughput performance of 1-Persistent CSMA-CD can be analyzed by assuming that time is divided into minislots of length $2t_{prop}$ seconds to ensure that stations can always detect a collision. Each time the channel becomes idle, stations contend for the channel by transmitting and listening to the channel to see whether they have successfully captured the channel. Each such contention interval takes $2t_{prop}$ seconds. Next we calculate the mean time required for a station to successfully capture the channel.

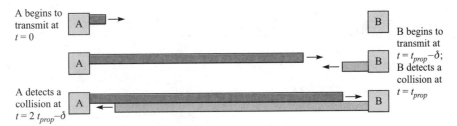

It takes $2 t_{prop}$ to find out whether the channel has been captured

FIGURE 6.22 The reaction time in CSMA-CD is $2t_{prop}$

Suppose that n stations are contending for the channel and suppose that each station transmits during a contention minislot with probability p. The probability of a successful transmission is given by the probability that only one station transmits:

$$P_{success} = np(1-p)^{n-1}$$

since there are n possible ways in which one station transmits and the other do not. To find the maximum achievable throughput, assume that stations use the value of p that maximizes $P_{success}$. By taking a derivative of $P_{success}$ with respect to p and setting it to zero, we find that the probability is maximized when $p = 1/n$. The maximum probability of success is then

$$P_{success}^{max} = n\frac{1}{n}\left(1 - \frac{1}{n}\right)^{n-1} = \left(1 - \frac{1}{n}\right)^{n-1}$$

Figure 6.23 shows that the maximum probability of success approaches $1/e$ as n increases.

The average number of minislots that elapse until a station successfully captures the channel is calculated as follows. The probability that j minislots are required is given by

$$P[j \text{ minislots in contention interval}] = (1 - P_{success}^{max})^{j-1} P_{success}^{max} \text{ for } j = 1, 2, \ldots$$

The average number of minislots in a contention period is then

$$E[J] = \sum_{j=1}^{\infty} j(1 - P_{success}^{max})^{j-1} P_{success}^{max} = \frac{1}{P_{success}^{max}}$$

If we assume that the value of n is large, then we have $P_{success}^{max} = 1/e$, and

$$E[J] = \frac{1}{P_{success}^{max}} = e = 2.718 \text{ minislots}$$

The maximum throughput in the CSMA-CD system occurs when all of the channel time is spent in packet transmissions followed by contention intervals. Each packet transmission time $E[X]$ is followed by a period t_{prop} during which

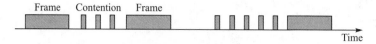

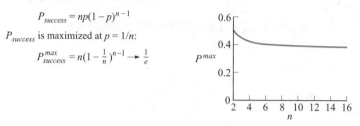

Probability of one successful transmission:

$$P_{success} = np(1-p)^{n-1}$$

$P_{success}$ is maximized at $p = 1/n$:

$$P_{success}^{max} = n(1 - \tfrac{1}{n})^{n-1} \longrightarrow \frac{1}{e}$$

FIGURE 6.23 Packet transmission times and contention periods

stations find out that the packet transmission is completed and then a contention interval of duration $2e\ t_{prop}$; so the maximum throughput is then

$$\rho_{max} = \frac{E[X]}{E[X] + t_{prop} + 2et_{prop}} = \frac{1}{1 + (2e+1)a} = \frac{1}{1 + (2e+1)Rd/vE[L]}$$

where $a = t_{prop}/E[X]$ is the propagation delay normalized to the packet transmission time. The right-most term in the preceding expression is in terms of the bit rate of the medium R, the diameter of the medium d, the propagation speed over the medium v, and the average packet length $E[L]$. The expression shows that CSMA-CD can achieve throughputs that are close to 1 when a is much smaller than 1. For example, if $a = 0.01$, then CSMA-CD has a maximum throughput of 94 percent. The CSMA-CD scheme provides the basis for the Ethernet LAN protocol that was discussed earlier in the chapter.

The calculation of the average delay in CSMA-CD is quite involved. The following complicated expression from [Schwartz 1987] is adapted to the case of constant packet lengths:

$$\frac{E[T]}{X} = \rho \frac{1 + (4e+2)a + 5a^2 + 4e(2e-1)a^2}{2\{1 - \rho(1 + (2e+1)a)\}}$$

$$+ 1 + 2ea - \frac{(1 - e^{-2a\rho})\left(\frac{2}{\rho} + 2ae^{-1} - 6a\right)}{2(e^{-\rho}e^{-\rho a-1} - 1 + e^{-2\rho a})} + \frac{a}{2}$$

This equation was used to prepare the CSMA-CD delay-throughput curves shown in Figure 6.51. The equation assumes that the time between packet arrivals has an exponential distribution. It should be emphasized that CSMA-CD does not provide the orderly packet access of a statistical multiplexer. The random backoff mechanism and the random occurrence of collisions imply that packets need not be transmitted in the order that they arrived. Indeed once a packet is involved in a collision, it is quite possible for other later arrivals to be transferred ahead of it. This behavior implies that in CSMA-CD packet delays

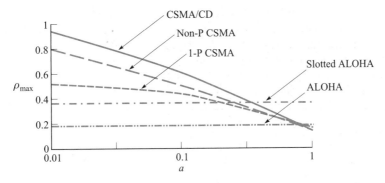

FIGURE 6.24 Maximum achievable throughputs of random access schemes

exhibit higher variability than in statistical multiplexers. In the next section we consider scheduling approaches to medium access control that provide a more orderly access to the shared medium.

Figure 6.24 shows a comparison of the maximum throughput of the main random access MAC techniques discussed in this section. For small values of a, CSMA-CD has the highest maximum throughput followed by CSMA. As a increases, the maximum throughput of CSMA-CD and CSMA decrease, since the reaction times are approaching the packet transmission times. When a approaches 1, the throughput of these schemes becomes less than that of ALOHA and slotted ALOHA. As indicated before, ALOHA and slotted ALOHA are not sensitive to a because their operation does not depend on the reaction time.

Example—Cellular Digital Packet Data (CDPD) MAC protocol

CDPD uses a 30 kHz channel of the AMPS analog cellular telephone system to provide packet data service to multiple mobile stations at speeds of up to 19,200 bps. The base stations are connected to interface nodes that connect to wired packet networks. The MAC protocol in CDPD is called *Digital Sense Multiple Access* and is a variation of CSMA-CD. The base station broadcasts frames in the forward channel, and mobile stations listen for packets addressed to them. The frames are in HDLC format and are segmented into blocks of 274 bits. Reed-Solomon error-correction coding increases the block to 378 bits. This block is transmitted in seven groups of 54 bits. A 6-bit synchronization and flag word is inserted in front of each 54-bit group to form a microblock. The flag word is used to provide coordination information in the reverse link.

To transmit on the reverse link, a station prepares a frame in HDLC format and prepares 378-bit blocks as in the forward channel. The station observes the flag words in the forward channel to determine whether the reverse link is idle or busy. This step is the "digital sensing." If the channel is busy, the station schedules a backoff time after which it will attempt again. If the station again senses the channel busy, it selects a random backoff time again but over an interval that is twice as long as before. Once it senses the channel idle, the station begins transmission. The base station provides feedback information, with a two-micro-block delay, in another flag bit to indicate that a given 54-bit block has been received correctly. If a station finds out that its transmission was involved in a collision, the station aborts the transmission as in CSMA-CD.

6.4   SCHEDULING APPROACHES TO MEDIUM ACCESS CONTROL

In section 6.3 we considered random access approaches to sharing a transmission medium. These approaches are relatively simple to implement, and we found that under light traffic they can provide low-delay packet transfer in broadcast networks. However, the randomness in the access can limit the maximum achievable throughput and can result in large variability in packet delays under heavier traffic loads. In this section we look at scheduling approaches to medium access control. These approaches attempt to produce an orderly access to the transmission medium. We first consider reservation systems and then discuss polling systems as a special form of reservation systems. Finally, we consider token-passing ring networks.

6.4.1   Reservation Systems

Figure 6.25 shows a basic reservation system. The stations take turns transmitting a single packet at the full rate R bps, and the transmissions from the stations are organized into frames that can be variable in length. Each frame is preceded by a corresponding reservation interval. In the simplest case the reservation

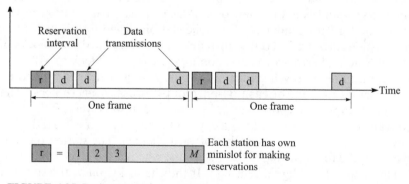

FIGURE 6.25 Basic reservation system

interval consists of M minislots. Stations use their corresponding minislot to indicate that they have a packet to transmit in a corresponding frame. The stations announce their intention to transmit a packet by broadcasting their reservation bit during the appropriate minislot. By listening to the reservation interval, the stations can determine the order of packet transmissions in the corresponding frame. The length of the frame will then correspond to the number of stations that have a packet to transmit. Note that variable-length packets can be handled if the reservation message includes packet-length information.

The basic reservation system described above generalizes and improves on a time-division multiplexing scheme by taking slots that would have gone idle and making them available to other stations. Figure 6.26a shows an example of the operation of the basic reservation system. In the initial portion only stations 3 and 5 have packets to transmit. In the middle portion of the example, station 8 becomes active, and the frame is expanded from two slots to three slots.

Let us consider the maximum attainable throughput for this system. Assume that the propagation delay is negligible, that packet transmission times are $X = 1$ time unit, and that a reservation minislot requires v time units where $v < 1$. Assume also that one minislot is required per packet reservation. Each packet transmission then requires $1 + v$ time units. The maximum throughput occurs when all stations are busy, and hence the maximum throughput is

$$\rho_{\text{max}} = \frac{1}{1 + v} \text{ for one packet reservation/minislot}$$

It can be seen that very high throughputs are achievable when v is very small in comparison to 1. Thus, for example, if $v = 5\%$, then $\rho_{\text{max}} = 95\%$.

Suppose that the propagation delay is not negligible. As shown in Figure 6.26b, the stations transmit their reservations in the same way as before, but the reservations do not take effect until some fixed number of frames later. For example, if the frame length is constrained to have some minimum duration that is greater than the round-trip propagation delay, then the reservations would take effect in the second following frame.

The basic reservation system can be modified so that stations can reserve more than one slot per packet transmission per minislot. Suppose that a minislot

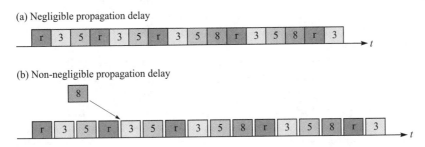

FIGURE 6.26 Operation of reservation system with negligible and non-negligible delays

can reserve up to k packets. The maximum frame size occurs when all stations are busy and is given by $Mv + Mk$ time units. One such frame transmits Mk packets, and so we see that the maximum achievable throughput is now

$$\rho_{\max} = \frac{Mk}{Mv + Mk} = \frac{1}{1 + v/k}, \text{ for } k \text{ packet reservations/minislot}$$

Now let us consider the impact of the number of stations on the performance of the system. The effect of the reservation intervals is to introduce overhead that is proportional to M, that is, the reservation interval is Mv. If M becomes very large, this overhead can become significant. This situation becomes a serious problem when a very large number of stations transmit packets infrequently. The reservations slots are incurred every frame, even though most stations do not transmit. The problem can be addressed by *not* allocating a minislot to each station and instead making stations contend for a reservation minislot by using a random access technique such as ALOHA or slotted ALOHA. If slotted ALOHA is used, then each successful reservation will require $1/0.368 = 2.71$ minislots. Therefore, the maximum achievable throughput for a reservation ALOHA system is

$$\rho_{\max} = 1/(1 + 2.71v)$$

If again we assume that $v = 5\%$, then we have $\rho_{\max} = 88\%$.

If the propagation delay is not negligible, then it is possible for slots to go unused because reservations cannot take effect quickly enough. This situation results in a reduction in the maximum achievable throughput. For this reason reservation systems are sometimes modified so that packets that arrive during a frame can attempt to "cut ahead of the line" by being transmitted during periods that all stations know have not been reserved. If a packet is successfully transmitted this way, its reservation in a following frame is canceled.

6.4.2   Polling

The reservation systems in the previous section required that stations make explicit reservations to gain access to the transmission medium. We now consider **polling systems** in which stations *take turns* accessing the medium. At any given time only one of the stations has the right to transmit into the medium. When a station is done transmitting, some mechanism is used to pass the right to transmit to another station.

There are different ways for passing the right to transmit from station to station. Figure 6.27a shows the situation in which M stations communicate with a host computer. The system consists of an outbound line in which information is transmitted from the host computer to the stations and an inbound line that must be shared with the M stations. The inbound line is a shared medium that requires a medium access control to coordinate the transmissions from the stations to the host computer. The technique developed for this system involves the host computer acting as a central controller that issues control messages to

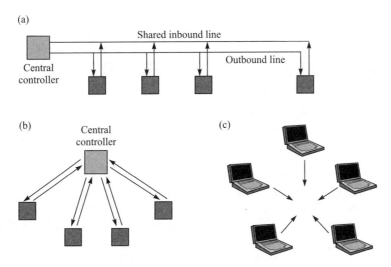

FIGURE 6.27 Examples of polling systems

coordinate the transmissions from the stations. The central controller sends a *polling message* to a particular station. When polled, the station sends its inbound messages and indicates the completion of its transmission through a *go-ahead message*. The central controller might poll the stations in round-robin fashion, or according to some other pre-determined order. The normal response mode of HDLC, which was discussed in Chapter 5, was developed for this type of system.

Figure 6.27b shows another situation where polling can be used. Here the central controller may use radio transmissions in a certain frequency band to transmit outbound messages, and stations may share a different frequency band to transmit inbound messages. This technique is called the frequency-division duplex (FDD) approach. Again the central controller can coordinate transmissions on the inbound channel by issuing polling messages. Another variation of Figure 6.27b involves having inbound and outbound transmissions share one frequency band. This is the time-division duplex (TDD) approach. In this case we would have an alternation between transmissions from the central controller and transmissions from polled stations.

Figure 6.27c shows a situation where polling is used without a central controller. In this particular example we assume that the stations have developed a polling order list, using some protocol. We also assume that all stations can receive the transmissions from all other stations. After a station is done transmitting, it is responsible for transferring a polling message to the next station in the polling list.

Figure 6.28 shows the sequence of polling messages and transmissions in the inbound line that are typical for the preceding systems. In the example, station 1 is polled first. A certain time, called the **walk time**, elapses while the polling message propagates and is received and until station 1 begins transmission.

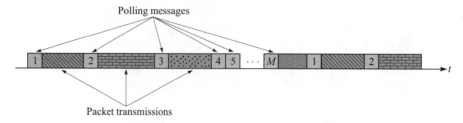

FIGURE 6.28 Interaction of polling messages and transmissions in a polling sytem

The next period is occupied by the transmission from station 1. This period is followed by the walk time that elapses while station 2 is polled and then by the transmissions from station 2. This process continues until station M is polled and has completed its transmissions. At this point the polling cycle is begun again by polling station 1. In some systems a station is allowed to transmit as long as it has information in its buffers. In other systems the transmission time for each station is limited to some maximum duration.

The walk times required to pass control of the access right to the medium can be viewed as a form of overhead. The **total walk time** τ' is the sum of the walk times in one cycle and represents the minimum time for one round of polling of all the stations. The **cycle time** T_c is the total time that elapses from when a station relinquishes access control to when it is polled next. The cycle time is the sum of the M walk times and the M station transmission times. The normalized overhead per cycle is then given by the ratio of the total walk time to the cycle time.

The average cycle time $E[T_c]$ can be found as follows. Let $E[N_c]$ be the average number of message arrivals to a station in one cycle time. If we assume that all messages that arrive in a cycle time are transmitted the next time the station is polled, then $E[N_c] = (\lambda/M)E[T_c]$. We assume that all stations have the same arrival rate λ/M packets/second and the same average packet transmission time $E[X]$. Therefore, the time spent at each station is $E[N_c]E[X] + t'$, where t' is the walk time. The average cycle time is then M times the average time spent at each station:

$$E[T_c] = M\{E[N_c]E[X] + t'\} = M\left\{\frac{\lambda}{M}E[T_c]E[X] + t'\right\}$$

The preceding equation can be solved for $E[T_c]$:

$$E[T_c] = \frac{Mt'}{1 - \lambda E[X]} = \frac{\tau'}{1 - \rho}$$

Note the behavior of the mean cycle time as a function of load ρ which is defined as the product of λ and $E[X]$. Under light load the cycle time is simply required to poll the full set of stations, and the mean cycle time is approximately τ', since most stations do not have messages to transmit. However, as the load approaches 1, the cycle time can increase without bound. Note also that the average cycle time is proportional to the number of stations M.

The total delay incurred by a packet from the instant when it arrives at a station to when its transmission is completed has the following components: (1) The packet must first wait for the transmission of all packets that it finds ahead of it in queue. (2) The packet must also wait for the time that must elapse from when it arrives at the station to when the station is polled. (3) The packet must be transmitted, requiring $E[X]$. (4) Finally, the packet must propagate from its station to the receiving station. If we assume packet arrivals have exponential interarrival times and constant length, then the total packet delay is

$$E[T] = \frac{\rho}{2(1 - \rho)} E[X] + \frac{\tau'(1 - \rho/M)}{2(1 - \rho)} + E[X] + \tau_{average}$$

where each term in the equation corresponds to the delay component in the preceding list and where $\tau_{average}$ is the average time required for a packet to propagate from the source station to the destination station [Bertsekas 1992, p. 201]. The total packet delay normalized to $E[X]$ is then

$$\frac{E[T]}{E[X]} = \frac{\rho}{2(1 - \rho)} + \frac{a'(1 - \rho/M)}{2(1 - \rho)} + 1 + \frac{\tau_{average}}{E[X]}$$

In the preceding expression, a' is the ratio of the total walk time to the service time. Neither the average waiting time (the first term) nor the transmission time (the third term) is proportional to M. The only dependence on M is through $a' = Mt'/E[X]$ and ρ/M.

Figure 6.29 shows the average packet delay for a polling system with $M = 32$ stations and for $a' = 0$, 0.5, 1, 5, and 10. In the figure we assume that $\tau_{average}$ is negligible. The effect of the normalized walk time a' is clearly evident in the higher average packet delays. As long as a' is less than 1, the average packet

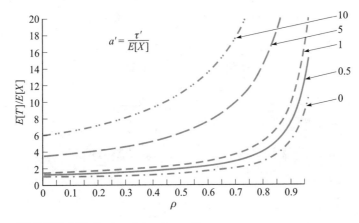

FIGURE 6.29 Packet delay for polling

delay does not differ significantly from that of the ideal system which corresponds to the $a' = 0$ case.

The walk time t' is determined by several factors. The first factor is the propagation time required for a signal to propagate from one station to another. This time is clearly a function of distance. Another factor is the time required for a station to begin transmitting after it has been polled. This time is an implementation issue. A third factor is the time required to transmit the polling message. These three factors combine to determine the total walk time of the system.

6.4.3 Token-Passing Rings

Polling can be implemented in a distributed fashion on networks with a ring topology. As shown in Figure 6.30, such ring networks consist of station interfaces that are connected by point-to-point digital transmission lines. Each interface acts like a repeater in a digital transmission line but has some additional functions. An interface in the listen mode reproduces each bit that is received at its input at its output after some constant delay, ideally in the order of one bit time. This delay allows the interface to monitor the passing bit stream for certain patterns. For example, the interface will be looking for the address of the attached station. When such an address is observed, the associated packet of information is copied bit by bit to the attached station. The interface also monitors the passing bit stream for the pattern corresponding to a "free token."

When a free token is received and the attached station has information to send, the interface changes the passing token to busy by changing a bit in the

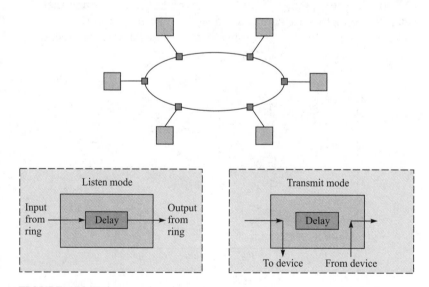

FIGURE 6.30 Token-passing rings

passing stream. In effect, receiving a free token corresponds to receiving a polling message. The station interface then changes to the transmit mode where it proceeds to transmit packets of information from the attached station. These packets circulate around the ring and are copied at the destination station interfaces.

While the station is transmitting its information, it is also receiving information at the input of the interface. If the time to circulate around the ring is less than the time to transmit a packet, then this arriving information corresponds to bits of the same packet that the station is transmitting. When the ring circulation time is greater than a packet transmission time, more than one packet may be present in the ring at any given time. In such cases the arriving information could correspond to bits of a packet from a different station, so the station must buffer these bits for later transmission.

A packet that is inserted into the ring must be removed. One approach to packet removal is to have the destination station remove the packet from the ring. Another approach is to allow the packet to travel back to the transmitting station. This approach is usually preferred because the transmitting station interface can then forward the arriving packet to its attached station, thus providing a form of acknowledgment.

Token rings can also differ according to the method used to reinsert the token after transmission has been completed. There are three approaches to token reinsertion, as shown in Figure 6.31. The main differences between the methods arise when the ring latency is larger than the packet length. The **ring latency** is defined as the number of bits that can be simultaneously in transit around the ring. In the *multitoken operation*, the free token is transmitted immediately after the last bit of the data packet. This approach minimizes the time required to pass a free token to the next station. It also allows several packets to be in transit in different parts of the ring.

The second approach, the *single-token operation*, involves inserting the token after the last bit of the busy token is received back. If the packet is longer than the ring latency, then the free token will be inserted immediately after the last bit of the packet is transmitted, so the operation is equivalent to multitoken

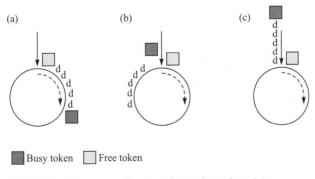

■ Busy token □ Free token

FIGURE 6.31 Approaches to token reinsertion: (a) multitoken, (b) single token, and (c) single packet

operation. However, if the ring latency is greater than the packet length, then a gap will occur between the time of the last bit transmission and the reinsertion of the free token as shown in Figure 6.31. The recovery from errors in the token is simplified by allowing only one token to be present in the ring at any given time.

In the third approach, a *single-packet operation*, the free token is inserted after the transmitting station has received the last bit of its packet. This approach allows the transmitting station to check the return packet for errors before relinquishing control of the token. Note that this approach corresponds to multi-token operation if the packet length is augmented by the ring latency.

The token-ring operation usually also specifies a limit on the time that a station can transmit. One approach is to allow a station to transmit an unlimited number of packets each time a token is received. This approach minimizes the delay experienced by packets but allows the time that can elapse between consecutive arrivals of a free token to a station to be unbounded. For this reason, a limit is usually placed either on the number of packets that can be transmitted each time a token is received or on the total time that a station may transmit information into the ring. These limits have the effect of placing a bound on the time that elapses between consecutive arrivals of a free token at a given station.

The introduction of limits on the number of packets that can be transmitted per token affects the maximum achievable throughput. Suppose that a maximum of one packet can be transmitted per token. Let τ' be the ring latency (in seconds) and a' be the ring latency normalized to the packet transmission time. We then have

$$\tau' = \tau + \frac{Mb}{R} \qquad a' = \frac{\tau'}{E[X]}$$

where τ is the total propagation delay around the ring, b is the number of bit delays in an interface, Mb is the total delay introduced by the M station inter-faces, and R is the speed of the transmission lines. The maximum throughput occurs when all stations transmit a packet. If the system uses multitoken operation, the total time taken to transmit the packets from the M stations is $ME[X] + \tau'$. Because $ME[X]$ of this time is spent transmitting information, the maximum throughput is then

$$\rho_{max} = \frac{ME[X]}{ME[X] + \tau'} = \frac{1}{1 + \tau'/ME[X]} = \frac{1}{1 + a'/M} \text{ for multitoken.}$$

Now suppose that the ring uses single-token operation. Assume that packets are of constant length L and that their transmission time is $X = L/R$. From Figure 6.31 we can see that the effective packet duration is the maximum of X and τ'. Therefore, the maximum throughput is then

$$\rho_{max} = \frac{MX}{M \max\{X, \tau'\} + \tau'} = \frac{1}{\max\{1, a'\} + \tau'/MX}$$

$$= \frac{1}{\max\{1, a'\} + a'/M} \text{ for single token.}$$

When the packet transmission time is greater than the ring latency, we see that the single-token operation has the same maximum throughput as multitoken operation. However, when the ring latency is larger than the packet transmission time, that is, $a' > 1$, then the maximum throughput is less than that of multi-token operation.

Finally, in the case of single-packet operation the effective packet transmission time is always $E[X] + \tau'$. Therefore, the maximum throughput is given by

$$\rho_{\max} = \frac{ME[X]}{M(E[X] + \tau') + \tau'} = \frac{1}{1 + a'\left(1 + \dfrac{1}{M}\right)} \quad \text{for single-packet.}$$

We see that the maximum throughput for single-packet operation is the lowest of the three approaches. Note that when the ring latency is much bigger than the packet transmission time, the maximum throughput of both the single-token and single-packet approaches is approximately $1/a'$. Recall from Figure 6.31 that this situation occurs when the distance of the ring becomes large or the transmission speed becomes very high. Figure 6.32 shows the maximum throughput for the three approaches for different values of a'. It is clear that single-packet operation has the lowest maximum throughput for all values of a'. Multitoken operation, on the other hand, has the highest maximum throughput for all values of a'. In fact, multitoken operation is sensitive to the per hop latency a'/M, not the overall ring latency a'. The figure also shows how single-token operation approaches single-packet operation as a' becomes large.

We conclude this section with a comparison of the average waiting time incurred by packets in several variations of token ring. These waiting time results are taken from [Bertsekas 1992]. The following expressions assume that the packet transmission time X includes the total time required to transmit the token. The normalized mean waiting time for a token ring in which there is no limit on the number of packet transmissions/token is given by

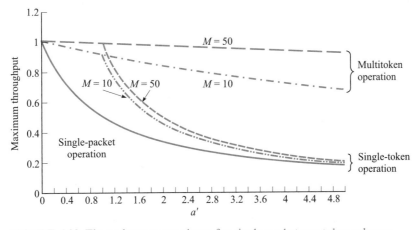

FIGURE 6.32 Throughput comparisons for single packet per token schemes

$$\frac{E[W]}{E[X]} = \frac{\rho}{2(1-\rho)} + \frac{a'(1-\rho/M)}{2(1-\rho)}.$$

Note that this expression corresponds to the average packet delay expression presented earlier for polling systems.

Now consider a token ring in which there is a limit of one packet/token and in which token reinsertion is done according to the multitoken operation. The normalized mean waiting time for the token ring is given by [Bertsekas 1992, p. 201]:

$$\frac{E[W]}{E[X]} = \frac{\rho + a'(1 + \rho/M)}{2\left(1 - \left(1 + \dfrac{a'}{M}\right)\rho\right)}.$$

Note that the mean waiting time approaches infinity as ρ approaches $1/(1 + a'/M)$, which agrees with our previous result for $\rho_{\max}$.

Finally, consider a token ring in which there is a limit of one packet/token and in which the token reinsertion is done according to the single-packet operation. The mean waiting time is [Bertsekas 1992, p. 202]:

$$\frac{E[W]}{E[X]} = \frac{\rho(1 + 2a' + a'^2) + a'\left(1 + \dfrac{\rho}{M}(1 + a')\right)}{2\left(1 - \left(1 + a'\left(1 + \dfrac{1}{M}\right)\right)\rho\right)}$$

Note again that the mean waiting time approaches infinity as ρ approaches $1/(1 + a'(1 + 1/M)$ in agreement with our previous result for $\rho_{\max}$.

Figure 6.33 shows the mean waiting time for the system with $M = 32$ stations and unlimited service per token. The maximum throughput for this system is always 1. It can be seen that when the normalized ring latency a' is less than 1 the system has performance that does not differ significantly from the ideal system which corresponds to the $a' = 0$ case. As a' becomes much larger than 1, the average waiting time can be seen to increase significantly.

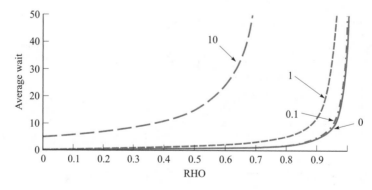

FIGURE 6.33 Mean waiting time token ring, $M = 32$ stations, unlimited service/token

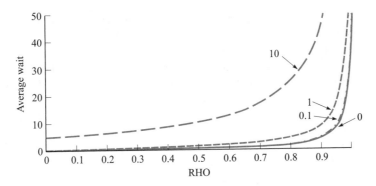

FIGURE 6.34 Mean waiting time for multitoken ring, $M = 32$, one packet/token

Figure 6.34 shows the mean waiting time for a system that places a limit of one packet transmission/token and uses multitoken operation. The number of stations is assumed to be $M = 32$. Recall that the multitoken operation reinserts the free token in the minimum time possible. When the normalized ring latency is less than 0.1, the mean waiting time does not differ significantly from that of an ideal system. However, as a' increases, the maximum throughput decreases, and so when $a' = 10$, the maximum throughput is 0.76. The figure shows that this behavior results in increased waiting times at lighter loads.

Figure 6.35 shows the waiting time for a token ring with a limit of one packet transmission/token and single-packet operation. In this case the free token is not reinserted until after the entire packet is received back at a station. Again the number of stations is $M = 32$. It can be seen that the mean waiting time for this system is much more sensitive to the normalized ring latency. Performance comparable to an ideal system is possible only when $a' < .01$. The maximum throughput decreases rapidly with increasing a' so that by the time $a' = 1$, the maximum throughput is only 0.47. A comparison of Figure 6.34 and Figure 6.35

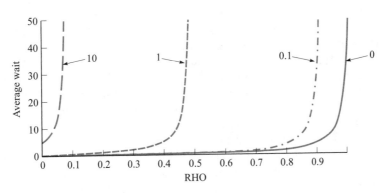

FIGURE 6.35 Mean waiting time for single-packet token ring, $M = 32$

shows how multitoken operation is essential when the normalized ring latency becomes much larger than 1.

The preceding results for mean waiting time do not include the case of single-token operation. From Figure 6.32 we know that for $a' < 1$, the mean waiting times are the same as for a system with multitoken operation. On the other hand, as a' becomes much larger than 1, the system behaves like a system with single-token operation.

6.4.4   Comparison of Scheduling Approaches to Medium Access Control

We have discussed two basic scheduling approaches to medium access control: reservations and polling. Token-passing rings are essentially an extension of the polling concepts to ring-topology networks. The principal strength of these approaches is that they provide a relatively fine degree of control in accessing the medium. These approaches can be viewed as an attempt to make time-division multiplexing more efficient by making idle slots available to other users.

Reservation systems are the most direct in obtaining the coordination in medium access that is inherent in a multiplexer. Reservation systems can be modified to implement the various scheduling techniques that have been developed to provide quality-of-service guarantees in conventional multiplexers. However, unlike centralized multiplexers, reservation systems must deal with the overheads inherent in multiple access communications, for example, time gaps between transmissions and reaction-time limitations. In addition, the decentralized nature of the system requires the reservation protocols to be robust with respect to errors and to be conducive to simple error-recovery procedures.

Polling systems and token-ring systems in their most basic form can be viewed as dynamic forms of time-division multiplexing where users transmit in round-robin fashion, but only when they have information to send. In polling systems the overhead is spread out in time in the form of walk times. The limitations on transmission time/token can lead to different variations. At one extreme, allowing unlimited transmission time/token minimizes delay but also makes it difficult to accommodate packets with stringent delay requirements. At the other extreme, a limit of one packet/token leads to a more efficient form of time-division multiplexing. Polling systems however can be modified so that the polling order changes dynamically. When we reach the extreme where the polling order is determined by the instantaneous states of the different stations, we obtain what amounts to a reservation system. From this viewpoint polling systems can be viewed as an important special case of reservation systems.

All the scheduling approaches were seen to be sensitive to the reaction time as measured by the propagation delay and the network latency normalized by the average packet transmission time. The reaction time of a scheme is an unavoidable limitation of scheduling approaches. Thus in reservation systems with long propagation delays, there is no way for the reservations to take effect until after a full propagation delay time. However, in some cases there is some flexibility in what constitutes the minimum reaction time. For example, in token-passing rings with single-packet operation the reaction time is a full ring latency, whereas in multitoken operation the reaction time is the latency of a single hop.

In keeping with the title of this book, this section has focused on fundamental aspects of medium access control. The current key standards for local and wide area networks are discussed in a later section. As an aside we would like to point out to the student and designer of future networks that as technologies advance, the key standards will change, but the fundamental concepts will remain. In other words, there will be many new opportunities to apply the fundamental principles to develop the key standards of the future, especially in the area of wireless networks and optical networks.

6.4.5 Comparison of Random Access and Scheduling Medium Access Controls

The two classes of medium access control schemes, random access and scheduling, differ in major ways, but they also share many common features. Their differences stem primarily from their very different points of departure. Scheduling techniques have their origins in reservation systems that attempt to emulate the performance of a centrally scheduled system such as a multiplexer. Random access techniques, on the other hand, have their origins in the ALOHA scheme that involves transmitting immediately, and subsequently at random times in response to collisions. The scheduling approach provides methodical orderly access to the medium, whereas random access provides a somewhat chaotic, uncoordinated, and unordered access. The scheduling approach has less variability in the delays encountered by packets and therefore has an edge in supporting applications with stringent delay requirements. On the other hand, when bandwidth is plentiful, random access systems can provide very small delays as long as the systems are operated with light loads.

Both random access and scheduling schemes have the common feature that channel bandwidth is used to provide information that controls the access to the channel. In the case of rescheduling systems, the channel bandwidth carries explicit information that allows stations to schedule their transmissions. In the case of random access systems, channel bandwidth is used in collisions to alert stations of the presence of other transmissions and of the need to spread out their transmissions in time. Indeed, the contention process in CSMA-CD amounts to a distributed form of scheduling to determine which station should transmit next.

Any attempt to achieve throughputs approaching 100 percent involves using some form of coordination, either through polling, token-passing, or some form of contention resolution mechanism. All such systems can be very sensitive to the reaction time in the form of propagation delay and network latency. The comparison of the single-packet and multitoken approaches to operating a token ring shows that a judicious choice of algorithm can result in less demanding reaction times. Truly random access schemes such as ALOHA and slotted ALOHA do not attempt coordination and are not sensitive to the reaction time, but they also do not achieve high throughputs.

◆6.5   CHANNELIZATION

Consider a network in which the M stations that are sharing a medium produce the same steady rate of information, for example, digital voice or audio streams. It then makes sense to divide the transmission medium into M channels that can be allocated for the transmission of information from each station. In this section we first present two channelization schemes that are generalizations of frequency-division multiplexing and time-division multiplexing: frequency-division multiple access and time-division multiple access. We will take care to indicate how the distributed nature of multiple access leads to a much more difficult situation than in the conventional multiplexers that were discussed in Chapter 4. The third channelization technique, code-division multiple access, involves coding the transmitted signal to produce a number of separate channels. We explain the basic features of this technique and compare it to the other two. Next we discuss how all three of these techniques have been applied in telephone cellular networks. Finally, we show how channelization approaches are not well matched to bursty packet traffic.

6.5.1   FDMA

In **frequency-division multiple access (FDMA)** the transmission medium is divided into M separate frequency heads. Each station transmits its information continuously on an assigned band. Because practical transmitters cannot completely eliminate the out-of-band energy, guard bands are introduced between the assigned bands to reduce the co-channel interference. We will suppose that the total bandwidth available to the station is W, which can support a total bit rate of R bits/second. For simplicity we neglect the effect of the guard bands and assume that each station can transmit at a rate of R/M bits/second on its assigned band. As shown in Figure 6.36, in FDMA a station uses a fixed portion of the frequency band *all the time*. For this reason, FDMA is suitable for stream traffic and finds use in connection-oriented systems such as cellular telephony where each call uses a forward and reverse channel to communicate to and from a base station.

If the traffic produced by a station is bursty, then FDMA will be inefficient in its use of the transmission resource. The efficiency can be improved by allocating a frequency band to a group of stations in which each station generates bursty traffic. However, because the stations are uncoordinated, they will also need to use a dynamic sharing technique, that is, a medium access control, to access the given frequency band.

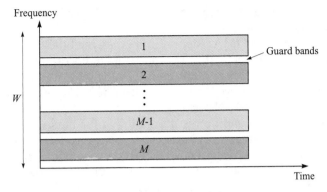

FIGURE 6.36 Frequency-division multiple access

6.5.2 TDMA

In **time-division multiple access (TDMA)** stations take turns making use of the *entire* transmission channel. The stations transmit according to a frame that consists of M time slots, as shown in Figure 6.37. Each station transmits during its assigned time slot and uses the entire frequency band during its transmission. Thus each station transmits at R bits/second $1/M$ of the time for an average rate of R/M bits/second. Each station spends most of the time accumulating packets and preparing them for transmission in a burst during the assigned time slot. *Guard times* are required to ensure that the transmissions from different stations do not overlap. In the case where the stations transmit to a common base station, the stations need to be synchronized with the base station to within a fraction of the guard time. Another source of overhead in TDMA is a *preamble* signal that is

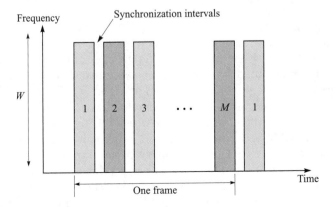

FIGURE 6.37 Time-division multiple access

required at the beginning of each time slot to allow the receiver to synchronize to the transmitted bit stream.

In the basic form of TDMA, each station is assigned the same size time slot, so each station has a channel with the same average bit rate. However, TDMA can accommodate a wider range of bit rates by allowing a station to be allocated several slots or by allowing slots to be variable in duration. In this sense TDMA is more flexible than FDMA. Nevertheless, in TDMA the bit rate allocated to a station is static, and this condition is not desirable for bursty traffic.

TDMA is used to connect to a base station or controller in two ways. In the first approach, **frequency-division duplex (FDD)**, two separate frequency bands are used. One band is used for communication in the "forward" direction from the base station to the other stations. The second band is used for communications in the "reverse" direction from the stations to the base station. Each station has an assigned time slot in the forward channel to receive from the base station, as well as an assigned time slot to transmit to the base station. Typically the slot assignments are staggered to allow a station to divide its time between transmit and receive processing functions. A second approach to communication with a base station is **time-division duplex (TDD)** in which the base station and the other stations take turns transmitting over the same shared channel. This approach requires the coordination of transmissions in the forward and reverse directions. FDD requires that the frequency bands be allocated ahead of time, and so it cannot be adapted to changes in the ratio between forward and reverse traffic. TDD can more readily adapt to these types of changes. We will see in Section 6.5.4 that TDMA is used with FDD in several digital cellular telephone systems. TDMA is also used with TDD in cordless telephone systems for in-building communications.

6.5.3   CDMA

Code-division multiple access (CDMA) provides another type of channelization technique. In TDMA and FDMA the transmissions from different stations are clearly separated in either time or in frequency. In CDMA the transmission from different stations occupy the entire frequency band at the same time. The transmissions are separated by the fact that different codes are used to produce the signals that are transmitted by the different stations. The receivers use these codes to recover the signal from the desired station.

Suppose that the user information is generated at R_1 bits/second. As shown in Figure 6.38, each user bit is transformed into G bits by multiplying the user bit value (as represented by a $+1$ or a -1) by G "chip" values (again represented by $+1$s and -1s) according to a unique binary pseudorandom sequence that has been assigned to the station. This sequence is produced by a special code and appears to be random except that it repeats after a very long period. The resulting sequence of $+1$s and -1s is then digitally modulated and transmitted over the medium. The **spreading factor G** is selected so that the transmitted signal

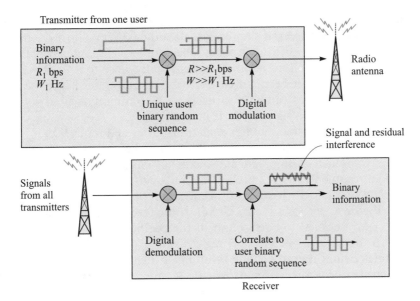

FIGURE 6.38 Code-division multiple access

occupies the entire frequency band of the medium.[3] Thus we have a situation in which a user transmits over *all the frequency band all the time*. Other stations transmit in the same manner at the same time but use different binary random sequences to spread their binary information.

Let us see why this type of modulation works. Suppose that the spreading sequence is indeed random and selected ahead of time by the transmitter and receiver by flipping a fair coin G times. If the transmitter wants to send a $+1$ symbol, the modulator multiplies this $+1$ by a sequence of G chip values, say, $c_1, c_2, \ldots, c_G$, each of which takes on a $+1$ or -1 according to a random coin flip. When the signal arrives at the receiver, the received signal is correlated with the known chip sequence; that is, the arriving chips are multiplied by the known sequence $c_1, c_2, \ldots, c_G$, and the resulting products are added. The resulting output of the correlator is $c_1^2 + c_2^2 + \ldots + c_G^2 = G$, since $(-1)^2 = (+1)^2 = 1$. Now suppose that another receiver attempts to detect the sequence $c_1, c_2, \ldots, c_G$ using another random chip sequence, say, $d_1, d_2, \ldots, d_G$. The output of this correlator is $c_1 d_1 + c_2 d_2 + \ldots + c_G d_G$. Because each c_j is equally likely to have value $+1$ or -1 and each d_j is also equally likely to have value $+1$ or -1, then each product term is equally likely to be $+1$ and -1. Consequently, the average value of the correlator output is 0. Indeed for large G, the vast majority of combinations of c_j and d_j will result in approximately half the terms in the

---

[3]We warn the reader that the throughput term G in the discussion of ALOHA is different from the spreading factor G in the discussion of spread spectrum systems. In situations where both terms arise, we suggest using the term G_{proc} for the spread spectrum term.

sum being $+1$ and the other half -1, so the correlator output is almost always close to zero.[4] In conclusion, if the receiver uses the correct chip sequence then the correlator output is G; if it uses some other random chip sequence, then the correlator output is a random number that is usually close to zero. Thus as the value of G is increased, it becomes easier for the receiver to detect the signal. Consequently, it becomes possible to decrease the amplitude (and lower the power) of the transmitted signal as G is increased.

The preceding discussion requires each transmitter and receiver pair to select very long random chip sequences ahead of time before they communicate. Clearly this approach is impractical, so an automated means of generating pseudorandom sequences is required. The shift-register circuits that we introduced when we discussed error-detection codes can be modified to provide these sequences. Figure 6.39 shows such a circuit. If the feedback taps in the feedback shift register are selected to correspond to the coefficients of a primitive polynomial, then the contents of the shift register will cycle over all possible $2^n - 1$ nonzero states before repeating. The resulting sequence is the maximum length possible and can be shown to approximate a random binary sequence in the sense that shifted versions of itself are approximately uncorrelated. Thus these sequences are suitable for spread spectrum communications.

In Chapter 3 we showed that transmitted signals can occupy the same frequency band and can still be separated by a receiver. In particular, in QAM modulation a sine and cosine can be used to produce signals that occupy the same frequency band but that can be separated at the receiver. In CDMA the spreading sequences that are used by the stations are approximately uncorrelated so that the correlators at the receiver can separate the signals from different transmitters. Thus to receive the information from a particular station, a receiver uses the same binary spreading sequence synchronized to the chip level to recover the original information, as shown in Figure 6.38. In the case of QAM modulation, the original signal could be recovered exactly. In the case of CDMA, the

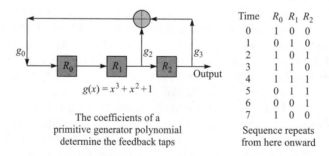

FIGURE 6.39 A maximum-length sequence generator

[4] The probability that there are k appearances of $+1$s in a sequence is the probability that there are k heads in G tosses of a fair coin. This probability is given by the binomial distribution.

signals from the other stations appear as residual noise at the receiver. From time to time the correlator output will yield an incorrect value, so error-correction coding needs to be used. Nevertheless, low error rates can be attained as long as the residual noise is kept below a certain threshold. This situation in turn implies that the number of active transmitters needs to be kept below some value.

In our analysis of the correlator output, we assumed that the signal level for each received signal was the same at a given receiver. This is a crucial assumption, and in order to work properly, CDMA requires all the signals at the receiver to have approximately the same power. Otherwise, a powerful transmission from a nearby station could overwhelm the desired signal from a distant station, which is called the *near-far* problem. For this reason CDMA systems implement a power control mechanism that dynamically controls the power that is transmitted from each station.

Figure 6.40 shows how CDMA can be viewed conceptually as dividing the transmission medium into M channels in "codespace." Each channel is distinguished by its spreading sequence. Unlike FDMA and TDMA, CDMA provides a graceful trade-off between number of users and residual interference. As the number of users is increased, the residual interference that occurs at the receiver increases. Unlike FDMA and TDMA, CDMA degrades only gradually when the channels begin to overlap; that is, the interference between channels becomes evident through a gradual increase in the bit error rate. This behavior provides the system with flexibility in terms of how to service different types of traffic, for example, allow higher error rate for voice traffic but provide more error correction for data traffic.

The set of spreading codes that are to be used for different channels must be selected so that any pair of spreading sequences will have low cross-correlation. Otherwise, the correlator in the receiver in Figure 6.38 will not always be able to separate transmissions from stations using these codes.[5] In the discussion so far we have assumed that the transmissions from the stations are not synchronized.

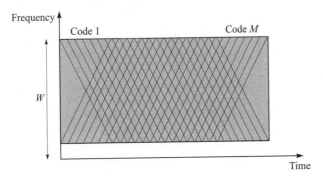

FIGURE 6.40 Conceptual view of CDMA

---

[5]See [Stüber 1996, Chapter 8] or [Gibson 1999, Chapter 8] for a discussion on the selection of spreading sequences.

We will now show that when it is possible to synchronize transmissions for the different channels, then it is possible to eliminate the interference between channels by using **orthogonal sequences** in the spreading. This situation is possible, for example, in the transmission from a base station to the different mobile stations. It is also possible when a mobile station sends several subchannels in its transmissions back to the base. We will use a simple example to show how this is done.

Example—Orthogonal Spreading by Using Walsh Functions

Suppose that a four-station system uses the following four orthogonal spreading sequences to produce four channels: $\{(-1, -1, -1, -1)$, $(-1, 1, -1, 1)$, $(-1, -1, 1, 1)$, and $(-1, 1, 1, -1)\}$. These sequences are examples of the Walsh orthogonal functions. To transmit an information bit, a station converts a binary 0 to a -1 symbol and a binary 1 to a $+1$ symbol. The station then transmits the symbol times its spreading sequence. Thus station 2 with code $(-1, 1, -1, 1)$ transmits a binary 0 at $(1, -1, 1, -1)$ and a 1 as $(-1, 1, -1, 1)$.

Now suppose that stations 1, 2, and 3 transmit the following respective binary information sequences: 110, 010, and 001. Figure 6.41 shows how these binary information bits are converted into symbols, how the orthogonal spreading is applied, and the resulting individual channel signals as well as the aggregate signal.

Figure 6.42 shows how the signal from channel 2 is recovered from the aggregate signal. The receiver must be synchronized to the time slot that corresponds to each symbol. The receiver then multiplies the aggregate signal by the four chip values of the spreading code for channel 2. The resulting signal is then integrated. Each integrated signal gives either $+4$ or -4, depending on whether the original symbol was $+1$ or -1. The same procedure can be used to recover the signals for channels 1 and 3. It should also be noted that if we try to recover the channel 2 signal with the spreading sequence for channel 4 $(-1, 1, -1, 1)$, we will obtain zero for each integrated period, indicating that no signal from channel 4 was present.

We also emphasize that the bit transmission times for the different channels must be aligned precisely. An error of a fraction of a single chip period can cause the receiver to fail.

The preceding example shows that when orthogonal sequences are used for spreading then the signal for each channel can be recovered from the aggregate signal without any interference from the other channel signals. The reason for this behavior is in the orthogonality property itself. Let $a = (a_1, a_2, \ldots, a_n)$ and $b = (b_1, b_2, \ldots, b_n)$ be two spreading sequences. We say that the two sequences are orthogonal if their inner product (also called the dot product) is zero:

$$a * b = \Sigma a_j b_j = a_1 b_1 + a_2 b_2 + \ldots + a_n b_n = 0$$

Because the spreading sequences consist of $+1$s and -1s, we have

Channel 1: 110 ⟶ +1+1+1 ⟶ (-1,-1,-1,-1), (-1,-1,-1,-1), (+1,+1,+1,+1)
Channel 2: 010 ⟶ -1+1-1 ⟶ (+1,-1,+1,-1), (-1,+1,-1,+1), (+1,-1,+1,-1)
Channel 3: 001 ⟶ -1-1+1 ⟶ (+1,+1,-1,-1), (+1,+1,-1,-1), (-1,-1,+1,+1)
Sum signal: (+1,-1,-1,-3), (-1,+1,-3,-1), (+1,-1,+3,+1)

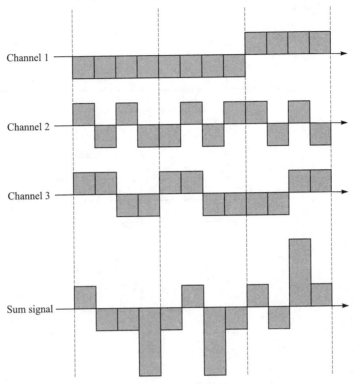

FIGURE 6.41 Example of orthogonal coding for channelization

$$\boldsymbol{a} * \boldsymbol{a} = \Sigma a_j^2 = a_1^2 + a_2^2 + \ldots + a_n^2 = n$$

and

$$\boldsymbol{b} * \boldsymbol{b} = \Sigma b_j^2 = b_1^2 + b_2^2 + \ldots + b_n^2 = n$$

Now suppose that we transmit a binary 0 in the a channel and a binary 1 in the b channel, then the signal in the a channel is $-\boldsymbol{a} = -(a_1, a_2, \ldots, a_n)$, and the signal in the b channel is $\boldsymbol{b} = (b_1, b_2, \ldots, b_n)$, so the aggregate channel signal is $\boldsymbol{r} = (r_1, r_2, \ldots, r_n) = (-a_1 + b_1, -a_2 + b_2, \ldots, -a_n + b_n) = -\boldsymbol{a} + \boldsymbol{b}$. When a receiver attempts to recover the a channel, the receiver multiplies the jth received chip r_j by a_j and then integrates, or adds, the terms over all j:

$$\Sigma a_j r_j = a_1 r_1 + a_2 r_2 + \ldots + a_n r_n = \boldsymbol{a} * \boldsymbol{r} = \boldsymbol{a} * (-\boldsymbol{a} + \boldsymbol{b})$$
$$= -\boldsymbol{a} * \boldsymbol{a} + \boldsymbol{a} * \boldsymbol{b} = -n + 0 = -n$$

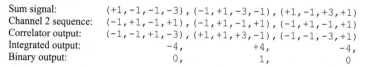

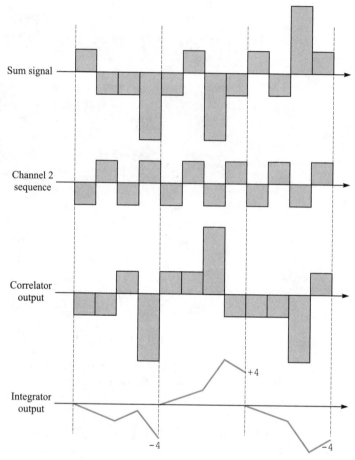

FIGURE 6.42 Example of channel signal recovery using orthogonal coding

Therefore, the receiver will conclude that the symbol in channel a was -1 and the information bit was 0. Similarly, if a receiver tries to detect channel b, the receiver will obtain

$$\Sigma b_j r_j = b_1 r_1 + b_2 r_2 + \ldots + b_n r_n = \boldsymbol{b} * \boldsymbol{r} = \boldsymbol{b} * (-\boldsymbol{a} + \boldsymbol{b})$$
$$= -\boldsymbol{b} * \boldsymbol{a} + \boldsymbol{b} * \boldsymbol{b} = 0 + n = n$$

And so the receiver will conclude that the symbol and information bit in channel b were $+1$ and 1, respectively.

$$W_1 = \big(0\big) \qquad W_2 = \begin{pmatrix} 0 & 0 \\ 0 & 1 \end{pmatrix} \qquad W_4 = \begin{pmatrix} 0 & 0 & 0 & 0 \\ 0 & 1 & 0 & 1 \\ 0 & 0 & 1 & 1 \\ 0 & 1 & 1 & 0 \end{pmatrix}$$

FIGURE 6.43 Construction of Walsh-Hadamard matrices

$$W_8 = \begin{pmatrix} 0 & 0 & 0 & 0 & 0 & 0 & 0 & 0 \\ 0 & 1 & 0 & 1 & 0 & 1 & 0 & 1 \\ 0 & 0 & 1 & 1 & 0 & 0 & 1 & 1 \\ 0 & 1 & 1 & 0 & 0 & 1 & 1 & 0 \\ 0 & 0 & 0 & 0 & 1 & 1 & 1 & 1 \\ 0 & 1 & 0 & 1 & 1 & 0 & 1 & 0 \\ 0 & 0 & 1 & 1 & 1 & 1 & 0 & 0 \\ 0 & 1 & 1 & 0 & 1 & 0 & 0 & 1 \end{pmatrix}$$

The **Walsh-Hadamard matrix** provides orthogonal spreading sequences of length $n = 2^m$. These matrices have binary coefficients and are defined recursively

$$W_1 = [0]$$

$$W_{2n} = \begin{bmatrix} W_n & W_n \\ W_n & W_n^c \end{bmatrix}$$

where W_n^c is obtained by taking the complement of the elements of W_n. Figure 6.43 shows the construction of the Walsh-Hadamard matrices with $n = 2, 4, 8$. The n rows of a Walsh-Hadamard matrix provide a set of n orthogonal spreading sequences by replacing each 0 by a -1 and each 1 by a $+1$. Note from Figure 6.43 that not all spreading sequences alternate quickly between $+1$ and -1. These sequences will not produce a transmitted signal that is spread over the available spectrum. For this reason the purpose of using Walsh sequences is primarily to provide channelization. In practice additional spreading using other sequences is combined with Walsh sequences. The resulting spread signal is robust with respect to multipath fading and interference in general.

6.5.4   Channelization in Telephone Cellular Networks

In this section we given an overview of how FDMA, TDMA, and CDMA channelization techniques have been implemented in various telephone cellular networks.

The **Advanced Mobile Phone System (AMPS)**, which was developed in the United States, is an example of a first-generation telephone cellular system. The system had an initial allocation of 40 MHz that was divided between two service providers (A and B) as shown in Figure 6.44. The allocation was later increased to 50 MHz as shown in the figure. AMPS uses FDMA operation for transmission between a base station and mobile stations. One band is used for forward channels from the base station to the mobile stations and the other band is used

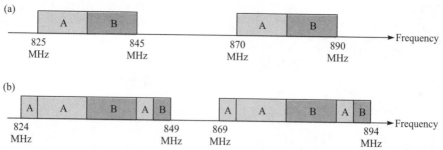

FIGURE 6.44 AMPS frequency allocation and channel structure: (a) initial allocation; (b) extended allocation

for reverse channels from the mobile stations to the base station. Each channel pair is separated by 45 MHz. AMPS uses analog frequency modulation to send a single voice signal over a 30 kHz transmission channel. Thus the 50 MHz allocation provides for $(50 \times 10^6)/(2 \times 30 \times 10^3) = 832$ two-way channels. Of these channels 42 are set aside for control purposes, such as call setup, and the remainder are used to carry voice traffic. AMPS uses a seven-cell frequency reuse pattern so only one-seventh of the channels are available in a given cell. A measure of the **spectrum efficiency** in a cellular system is the number of calls/MHz/cell that can be supported. For AMPS each service provider has $416 - 21$ traffic channels that are divided over seven cells and 25 MHz, thus

$$\text{Spectrum efficiency for AMPS} = 395/(7 \times 25) = 2.26 \text{ calls/cell/MHz}$$

The success of the cellular telephone service led to an urgent need to increase the capacity of the systems. This need was met through the introduction of digital transmission technologies. The **Interim Standard 54 (IS-54)** was developed in North America to meet the demand for increased capacity. IS-54 uses a hybrid channelization technique that retains the 30 kHz structure of AMPS but divides each 30 kHz channel into several digital TDMA channels. This approach allows cellular systems to be operated in dual mode, AMPS and TDMA. Each 30 kHz channel carries a 48.6 kbps digital signal organized into six-slot frames as shown in Figure 6.45. Each frame has a duration of 40 ms, and each slot contains 324 bits. Thus each slot corresponds to a bit rate of 324 bits/40 ms = 8.1 kbps. Typically a *full-rate* channel which consists of two slots per frame, and hence 16.2 kbps, is used to carry a voice call.[6] Thus IS-54 supports three digital voice channels in one analog AMPS channel. Half-rate channels (8.1 kbps), double full-rate channels (32.4 kbps), and triple full-rate channels (48.6 kbps) are also defined. Note that the time slots in the forward and reverse directions are offset with respect to each other to allow a mobile terminal to operate without having to transmit and receive at the same time. The 30 kHz spacing and hybrid TDMA/FDMA structure is also used in the 1.9 GHz PCS band. Interim

---

[6]The actual bit rate for a voice signal is about 13 kbps, since only 260 of the 324 bits carry data.

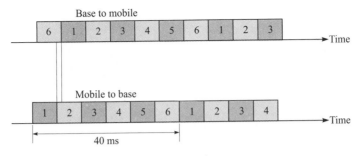

FIGURE 6.45 IS-54 frame structure

Standard 136 is a revision of IS-54 that takes into account the availability of fully digital control channels.

To calculate the spectrum efficiency of IS-54, we note that the 416 analog channels available provide $3 \times 416 = 1248$ digital channels. If we suppose that 21 of these channels are used for control purposes and that the frequency reuse factor is 7, then we have

$$\text{Spectrum efficiency of IS-54} = 1227/(7 \times 25) = 7 \text{ calls/cell/MHz}$$

The **Global System for Mobile Communications (GSM)** is a European standard for cellular telephony that has gained wide acceptance. GSM was designed to operate in the band 890 to 915 MHz for the reverse channels and the band 935 to 960 MHz for the forward channel (See Figure 6.46a). Initially the upper 10 MHz of this band is used for GSM, and the other portion of the band is used for existing analog services. The GSM technology can also be used in the PCS bands, 1800 MHz in Europe and 1900 MHz in North America.

GSM uses a hybrid TDMA/FDMA system. The available frequency band is divided into carrier signals that are spaced 200 kHz apart. Thus the 25 MHz bandwidth can support 124 one-way carriers. Each base station is assigned one or more carriers to use in its cell. Each carrier signal carries a digital signal that provides traffic and control channels. The carrier signal is divided into 120 ms multiframes, where each multiframe consists of 26 frames, and each frame has eight slots as shown in Figure 6.46b. Two frames in a multiframe are used for control purposes, and the remaining 24 frames carry traffic. In GSM the slots in the frames in the reverse direction lag the corresponding slots in the forward direction to allow the mobile station to alternate between receive and transmit processing.

A *full-rate traffic channel* uses one slot in every traffic frame in a multiframe. Therefore, the bit rate of a full-rate channel is

$$\text{traffic channel bit rate} = 24 \text{ slots/multiframe} \times 114 \text{ bits/slot}$$
$$\times (1 \text{ multiframe}/120 \text{ ms}) = 22{,}800 \text{ bps}$$

A substantial number of bits are used to provide error correction for voice calls. The full-rate traffic channel actually carries a digital voice signal that is

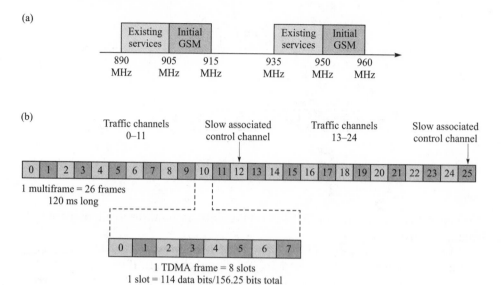

FIGURE 6.46 (a) GSM channel structure; (b) GSM TDMA structure

13 kbps, prior to the addition of error-correction bits. The hefty error-correcting capability allows GSM to operate with a frequency reuse factor of 3 or 4. If we assume 124 carriers in the 50 MHz band, then we obtain a total of $12 \times 8 = 992$ traffic channels. Assuming a frequency reuse factor of 3, we then have

$$\text{Spectrum efficiency of GSM} = 992/(3 \times 50) = 6.61 \text{ calls/cell/MHz}$$

The higher frequency reuse factor allows GSM to achieve a spectrum efficiency close to that IS-54.

The **Interim Standard 95 (IS-95)** is a second standard developed in North America to meet the demand for increased capacity. IS-95 is based on spread spectrum communication, which represents a very different approach to partitioning the available bandwidth. Spread spectrum communication introduces into cellular communication new features that provide higher voice quality and capacity than IS-54 TDMA systems. IS-95 systems can operate in the original AMPS frequency bands as well as in the 1900 MHz PCS bands.

IS-95 supports dual-mode operation with AMPS. Each channel signal is spread into a 1.23 MHz band, so the conversion of spectrum of AMPS to IS-95 must be done in chunks of 41 AMPS channel $\times$ 30 kHz/channel = 1.23 MHz. Recall that each service provider has 12.5 MHz of bandwidth, so these chunks represent about 10 percent of the total band.

For reasons that will become apparent IS-95 requires that all base stations be synchronized to a common clock. This synchronization is achieved by using the Global Positioning System (GPS), which is a network of satellites that can provide accurate timing information to a precision of 1 microsecond. In addition, all base stations use a common **short code** pseudorandom sequence in the spreading of signals and in the production of pilot signals that are used in the forward channel as well as in the handoff between cells.[7] All the base stations transmit the same sequence that is produced by this short code, but each base station has a unique phase or timing offset of the signal. The use of the same sequence reduces the task of synchronizing to this pilot signal.

In IS-95 the forward and reverse direction use *different* transmission techniques. First we consider the transmission in the forward direction, from the base station to the mobile stations. The IS-95 supports a basic user information bit rate of 9600 bps. After error-correction coding and interleaving, a 19,200 bps binary sequence is produced that is converted into a symbol sequence of $+1$s and -1s. This 19,200 symbol/second stream is then multiplied by a 19,200 symbol/second stream that is derived by taking every 64th symbol from a **long code** pseudorandom sequence that depends on the user electronic serial number (ESN) and operates at 1.2288 Msymbol/second as shown in Figure 6.47a.[8] Each symbol in the resulting 19,200 symbol/second sequence is then multiplied by a 64-chip Walsh orthogonal sequence that corresponds to the given channel.

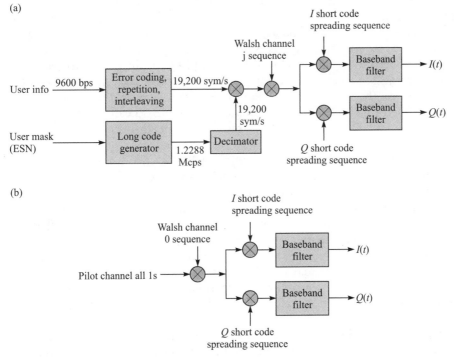

FIGURE 6.47 IS-95 modulator for forward channel

---

[7]The short code produces a pseudorandom sequence that repeats every $2^{15} - 1$ chip times, which translates into $80/3 = 26.667$ ms at the 1.2288 MHz spreading rate.

[8]The long code repeats every $2^{42} - 1$ chips, which translates into every 41.4 days.

Because the base station simultaneously handles the transmissions to all the base stations, it can synchronize its transmissions so that the signals to the different channels are orthogonal. This feature allows the receivers at the mobile stations to eliminate interference from the transmissions that are intended for other stations in the cell as discussed in section 6.5.3. The chip rate of the resulting signal is 19,200 × = 1.2288 Mchips/second. Note that in the forward channel the spreading is divided into two parts: the first part is provided by the channel-specific Walsh sequences; the second part is provided by the spreading sequence provided by the short code. The symbol sequence that results from the Walsh spreading is spread by using the short code and is then modulated using QPSK, a form of QAM with four constellation points.

The all-zeros Walsh sequence is not used as a traffic channel but is reserved to produce a pilot channel as shown in Figure 6.47b. The pilot signal is received by all mobile stations, and it is used to recover the timing information required to demodulate the received signal. Note in particular that the pilot signal enables the receiver to synchronize to the short code sequence of the signal that arrives from a base station. A mobile station can also detect the pilot signal from more than one base station and then can compare these signals and decide to initiate a **soft handoff** procedure during which the mobile station can receive and transmit to two base stations simultaneously while moving from one cell to another.

Once a mobile station has synchronized to the short code spreading sequence, the station can synchronize to the phase of the carrier to recover the Walsh spread sequence. The correlator detector introduced in the section 6.5.3 then produces the scrambled information sequence, which is then descrambled to produce the original 9600 bps information signal. IS-95 can accommodate user bit rates of 4800, 2400, and 1200 bps by simply repeating a user information bit several times. For example, a 4800 bps rate is handled by repeating each information bit twice and feeding the resulting 9600 bps stream into the system. An option for bit rates in the set {14400, 7200, 3600, and 1800} bps is available by changing the type of error-correction coding.

In the forward channel the pilot signal is "affordable" because the synchronization required for orthogonal spreading (and channelization) is possible at the base station and because it greatly simplifies the job of the mobile receivers. The situation is different in the reverse channel. Here it is not feasible to synchronize the transmissions of the many mobile stations, so orthogonal spreading is not possible. Consequently, the more conventional spread spectrum transmission technique based on nonorthogonal spreading sequences is implemented in the reverse channel.

As shown in Figure 6.48 the transmitter in the mobile station takes a basic 9600 bps user information sequence and applies error-correction coding, interleaving, and modulation to produce a 307,200 symbol/second sequence that consists of + 1s and −1s.[9] This sequence is spread by a factor of 4 by multiplying

---

[9]A confusing point here is that the coding/modulation uses a code that maps blocks of six binary symbols into strings of 64-symbol Walsh sequences. The role of the Walsh sequences here is to facilitate "noncoherent" detection, *not* to produce orthogonal channels. See [Viterbi 1995, Section 4.5].

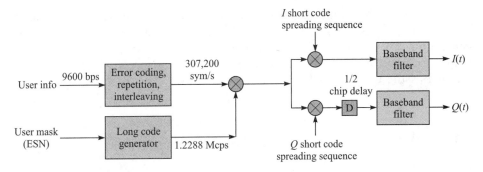

FIGURE 6.48 IS-95 modulator for reverse channel

it by the 1.2288 Msymbol/second sequence produced by the station's long code spreading sequence. This sequence is subsequently multiplied by the short code sequence that is common to all stations in the cell and then modulated using QPSK. The base station detects the spread spectrum signals from its various mobile stations in the usual manner. Note that, in principle, the reverse channel can produce up to $2^{42} - 1$ code channels, one for each of its possible phases. In fact at most 63 such channels are active in a cell at any given time.

The CDMA approach to cellular communications is clearly very different from either TDMA or FDMA. It did not fit the conventional mode of doing things, and so not surprisingly the calculation of spectrum efficiency proved to be quite controversial. The first major difference is that CDMA can operate with a frequency reuse factor of 1. Recall that TDMA and FDMA operate with a reuse factor of 7; that is, only one-seventh of the channels can be used in a given cell. This reuse factor is required to control the amount of interference between stations in different cells. The use of spread spectrum transmission greatly reduces the severity of intercell interference. The signals that arrive at any base station, whether from mobile stations from its cell or elsewhere, are uncorrelated because the associated transmitters use different long code spreading sequences. The signals that arrive at a mobile station from different base stations are also uncorrelated, since they all use the same short code sequence but with different phase. These features lead to the frequency reuse factor of 1.

An additional factor that contributes to the efficiency of CDMA is its ability to exploit variations in the activity of the users. For example, silence intervals in speech are exploited by reducing the bit rate, say, from 9600 bps to 1200 bps. In effect this step increases the spreading factor G by a factor of 8, so the transmitted power can be reduced. But this reduces the interference that is caused to other receivers and thus makes it possible to handle more calls. [Goodman 1997] develops a simplified analysis of spectrum efficiency and arrives at the following bounds:

12.1 calls/cell/MHz < spectrum capacity of IS-95 < 45.1 calls/cell/MHz

Even the lower bound is substantially larger than the spectrum efficiencies of IS-54 or GSM. IS-95, IS-54, and GSM are considered examples of second-generation cellular systems. The third-generation cellular system will provide higher bit rates and support a broader range of services. CDMA has been selected as the technology for third-generation systems.

6.5.5   Performance of Channelization Techniques with Bursty Traffic

In this section we compare the delay performance of FDMA, TDMA, and CDMA in the direction from remote stations to a central site. We show that channelization techniques are not effective in dealing with bursty traffic.

We suppose that each station that is connected to the transmission medium has its own buffer and is modeled as a separate multiplexer. We assume that packets arrive at each station with exponential interarrival times with mean λ/M packets/second. We also assume that the packet are always L bits long and that the time to transmit a packet at the full rate of the medium R bps is $X = L/R$ seconds. We assume that for FDMA the transmission rate available to one station is R/M, and so the transmission time of one packet is MX seconds. For TDMA a station gets access to the entire bandwidth $1/M$ of the time, so its average transmission rate is R/M time units/second. We assume that CDMA transmits at a constant rate and does not exploit variations in the activity of the information. In this case the multiplexer in the CDMA behaves in the same way as the FDMA station. Therefore, we focus on FDMA and TDMA after this point.

Figure 6.49 shows the sequence of transmissions as observed by a single station in a system with $M = 3$ stations and assuming $X = 1$. In the FDMA system we assume that the transmissions in each channel are slotted. Therefore, the transmissions consist of a sequence of slots that are three time units long, as shown in the figure. When a packet arrives at a given station, the packet must wait for the transmission of all packets in the queue. If a packet arrives to an empty system, the packet must still wait until the beginning of the next slot. Thus our packet arrival in Figure 6.29a must wait till the beginning of the next time slot at $t = 0$ and then for the transmission of the two packets in the queue. Finally, at time $t = 6$ our packet begins transmission that is completed at time $t = 9$.

Figure 6.49 also shows the transmissions as viewed by a station in the TDMA case. Here the station transmits at the full rate for one slot out of the M slots in the frame. A packet that arrives at a station must wait for the transmission of all packets in the queue. Note that each such packet found in queue implies M time units of waiting time to the arriving packet. Thus a packet that arrives at the same time as in part (a) would have to wait until the beginning of the next frame at $t = 0$ and then for the transmission of the two packets it found in queue. At time $t = 6$ our packet enters service. In TDMA the packet is transmitted at *full speed*, so the packet finishes transmission at $t = 7$. The last

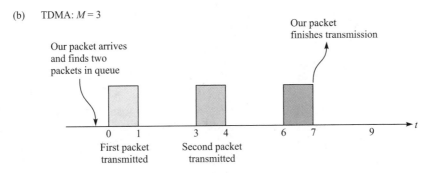

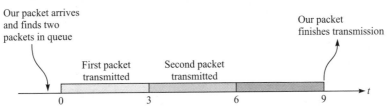

FIGURE 6.49 Comparison of FDMA, TDMA, and CDMA

packet transmission time is the main difference between the TDMA and FDMA systems.

This example shows that the TDMA and FDMA systems have the same time T_{access} from when a packet arrvies at a station to when the packet begins transmission. The access time T_{access} has two components in delay: (1) the time τ_0 until the beginning of the next frame and (2) the time W waiting for the packets found in queue upon arrival. In Appendix A we show that the average access time is given by

$$\frac{E[T_{access}]}{X} = \frac{M}{2} + \frac{\rho M}{2(1 - \rho)}$$

where the first term after the first equality is $E[\tau_0]$ and the next term is $E[W]$. The term ρ is called the *load* of a station and is defined by $\rho =$ arrival rate at a station $\times$ transmission time $= (\lambda/M)(MX) = \lambda X$.

The total packet delay in each station is obtained by adding the packet transmission time to the average access time. For FDMA the packet transmission time is MX, and so the total normalized packet delay is

$$\frac{E[T_{FDMA}]}{E[X]} = \frac{\rho M}{2(1 - \rho)} + \frac{M}{2} + M$$

For TDMA the packet transmission time is X, and to the average total packet delay is

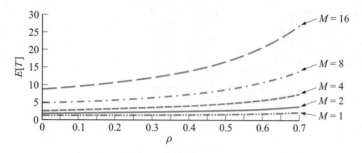

FIGURE 6.50 Average delay for TDMA

$$\frac{E[T_{TDMA}]}{E[X]} = \frac{\rho M}{2(1-\rho)} + \frac{M}{2} + 1$$

Thus we see that TDMA outperforms FDMA because of the faster packet transmission time. However, both TDMA and FDMA have the undesirable feature that the average total packet delay grows with the number of stations M. In this respect both TDMA and FDMA compare poorly to an ideal system that would combine all the traffic from the stations into one multiplexer and transmit always at the full rate R. Such a system corresponds to the $M = 1$ case. The average delay in such an ideal system would be

$$\frac{E[T_{ideal}]}{E[X]} = \frac{\rho}{2(1-\rho)} + \frac{1}{2} + 1$$

Figure 6.50 shows the average delay for TDMA as the number of stations is varied. The $M = 1$ case gives the performance of an ideal statistically multiplexed system. The undesirable effect of large M is evident here. We note that the root cause of this inferior packet delay performance is due to the fact that when the medium is divided among several stations, transmission slots can go unused when a station has no packets to transmit, whereas in a combined system these slots would be used by other stations. The problem, of course, is that we are dealing with a multiaccess system in which stations are geographically distributed. One way to view MAC algorithms discussed in previous sections is as an attempt to achieve packet delay performance that approaches that of an ideal $M = 1$ system.

6.6   LAN STANDARDS

In this section we discuss the structure of the important LAN standards. We introduce the MAC algorithms used by each protocol and the associated frame structures. We also discuss the various physical layer options that are available with each LAN standard.

6.6.1   Ethernet and IEEE 802.3 LAN Standard

The Ethernet LAN protocol was developed in the early 1970s by Xerox as a means of connecting workstations. In the early 1980s DEC, Intel, and Xerox completed the "DIX" Ethernet standard for a 10 Mbps LAN based on coaxial cable transmission. This standard formed the basis for the IEEE 802.3 LAN standard that was first issued in 1985 for "thick" coaxial cable. The Ethernet and IEEE 802.3 standards differ primarily in the definition of one header field, which we discuss below. The IEEE 802.3 standard has been revised and expanded every few years. Specifications have been issued for operation using "thin" coaxial cable, twisted-pair wires, and single-mode and multimode optical fiber. Higher-speed versions were approved in 1995 (100 Mbps Fast Ethernet) and in 1998 (1000 Mbps Gigabit Ethernet).

ETHERNET PROTOCOL

The original 802.3 standard was defined for a bus-based coaxial cable LAN in which terminal transmissions are broadcast over the bus medium using Carrier Sensing Multiple Access with Collision Detection (CSMA-CD) for the MAC protocol. A station with a frame to transmit waits until the channel is silent. When the channel goes silent, the station transmits but continues to listen for collisions that can occur if other stations also begin to transmit. If a collision occurs, the station aborts the transmission and schedules a later random time when it will reattempt to transmit its frame. If a collision does not occur within two propagation delay times, then the station knows that it has captured the channel, as the station's transmission will have reached all stations and so they will refrain from transmitting until the first station is done. A **minislot time** defines a time duration that is at least as big as two propagation delays.

The critical parameter in the CSMA-CD system is the minislot time that forms the basis for the contention resolution that is required for a station to seize control of the channel. The original 802.3 was designed to operate at 10 Mbps over a maximum distance of 2500 meters. When allowances are made for four repeaters, the delay translates into a maximum end-to-end propagation delay of 51.2 microseconds. At 10 Mbps this propagation delay equals 512 bits, or 64 bytes, which was selected as the minimum frame length or minislot.

The IEEE 802.3 standard specifies that the rescheduling of retransmission attempts after a collision uses a truncated binary exponential backoff algorithm. If a frame is about to undergo its nth retransmission attempt, then its retransmission time is determined by selecting an integer in the range between 0 and $2^k - 1$, where $k = \min(n, 10)$. That is, the first retransmission time involves zero or one minislot times; the second retransmission time involves 0, 1, 2, or 3 minislot times; and each additional slot retransmission extends the range by a factor of 2 until the maximum range of 12^{10}. The increased retransmission range after each collision is intended to increase the likelihood that retransmissions will succeed. Up to 16 retransmissions will be attempted, after which the system gives up.

The typical activity in the Ethernet channel consists of idle periods, contention periods during which stations attempt to capture the channel, and successful frame transmission times. When the channel approaches saturation, there are few idle periods and mostly frame transmissions alternate with contention periods. Therefore, at or near saturation the time axis consists of the following three subintervals: a period L/R seconds long during which frames are transmitted, a period t_{prop} seconds long during which all the other stations find out about the end of the transmission, and a contention period consisting of an integer number of minislot times, each of duration $2t_{prop}$ seconds. In section 6.3.4 we showed that the average number of minislots in a contention period is approximately $e = 2.71$. Therefore, the fraction of time that the channel is busy transmitting frames is

$$\frac{L/R}{L/R + t_{prop} + 2et_{prop}} = \frac{1}{1 + (1 + 2e)t_{prop}R/L} = \frac{1}{1 + (1 + 2e)a} = \frac{1}{1 + 6.44a},$$

where $a = t_{prop}R/L$

Example—Effect of a on Ethernet Performance

Let us assess the impact of the parameter a on the performance of an Ethernet LAN. Suppose that $a = 0.01, 0.1$, and 0.2. The corresponding maximum possible normalized throughputs are then 0.94, 0.61, and 0.44. Thus we see that a has a dramatic impact on the throughput that can be achieved.

Earlier in the chapter we presented an expression for the average frame transfer delay in Ethernet under the following assumptions: Frame arrival times are independent of each other; frame interarrival times have an exponential distribution; all frames are of the same length. Figure 6.51 shows the average transfer delays for $a = 0.01, 0.1$, and 0.2. It can be seen that the transfer delays grow very large as the load approaches the maximum possible value for the given value of a.

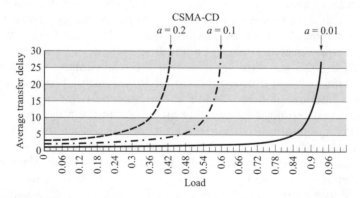

FIGURE 6.51 Frame transfer delay for Ethernet example

FRAME STRUCTURE

Figure 6.52 shows the MAC frame structure for the IEEE 802.3. The frame begins with a seven-octet preamble that repeats the octet 10101010. This pattern produces a square wave that allows the receivers to synchronize to the beginning of the frame. The preamble is followed by the start frame delimiter that consists of the pattern 10101011. The two consecutive 1s in the delimiter indicate the start of the frame.

The destination and source address fields follow. The address fields are six bytes long. (Two-byte address fields have been defined but are not used). The first bit of the destination address distinguishes between single addresses and group addresses that are used to multicast a frame to a group of users. The next bit indicates whether the address is a local address or a global address. Thus in the case of six-byte addresses, the standard provides for 2^{46} global addresses. The first three bytes specify the NIC vendor, so this scheme allows up to $2^{24} = 16,777,215$ addresses per vendor. For example, Cisco has addresses in which the first three bytes are 00-00-0C and 3Com has addresses that begin with 02-60-8C, where the numbers are in hexadecimal notation.

There are three types of physical addresses. **Unicast addresses** are the unique address permanently assigned to a NIC card. The card normally matches transmissions against this address to identify frames destined to it. **Multicast addresses** identify a group of stations that are to receive a given frame. NIC cards are set by their host computer to accept specific multicast addresses. Multicasting is an efficient way of distributing information in situations where multiple entities or processes require a piece of information as, for example, in the spanning tree algorithm (discussed in section 6.7.1 on bridges). The **broadcast address**, indicated by the all 1s physical address, indicates that all stations are to receive a given packet.

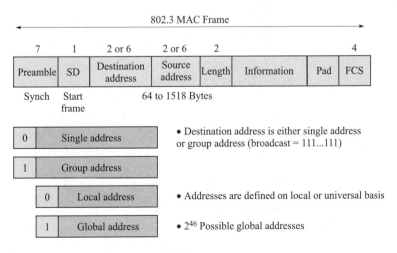

FIGURE 6.52 IEEE 802.3 MAC frame

The length field indicates the number of bytes in the information field. The longest allowable 802.3 frame is 1518 bytes, including the 18-byte overhead but excluding the preamble and SD. The pad field ensures that the frame size is always at least 64 bytes long. The maximum data field size of 1500 bytes translates into the hexadecimal code 05DC, so the length field always has a number smaller than 0600.

The last field is the CCITT 32-bit CRC check discussed in Chapter 3. The CRC field covers the address, length information, and pad fields. Upon receiving a frame, the NIC card checks to see that the frame is of an acceptable length and then checks the received CRC for errors. If errors are detected, the frame is discarded and not passed to the network layer.

Figure 6.53 shows the frame structure for the Ethernet (DIX) standard. The Ethernet frame has a type field that identifies the upper-layer protocol in the same location as the 802.3 field has its length field. For example, type field values are defined for IP, Address Resolution Protocol, and Reverse ARP (which are discussed in Chapter 8). The Ethernet standard assigns type field values starting at 0600. Recall that the length field in IEEE 802.3 never takes on values larger than 0600. This field tells an Ethernet controller whether it is handling an Ethernet frame or an IEEE 802.3 frame.

Nevertheless, the IEEE standard assumes that the LLC is always used, which provides the upper-layer protocol indication through the SAP field in the LLC header as shown in Figure 6.54 and discussed next.

Upper-layer software programs developed to work with DIX Ethernet expect a "Type" field. To allow Ethernet-standard software to work with IEEE 802.3 frames, the **Subnetwork Access Protocol (SNAP)** provides a way of encapsulating Ethernet-standard frames inside a Type 1 LLC PDU, as shown in Figure 6.54. Recall from section 6.2.3 that Type 1 corresponds to unacknowledged connectionless service. The DSAP and SSAP fields in the LLC header (see Figure 6.14) are set to AA to notify the LLC layer that an Ethernet frame is enclosed and should be processed accordingly. The value 03 in the control field indicates Type 1 service. The SNAP header consists of a three-byte vendor code (usually set to 0) and the two-byte type field required for compatibility.

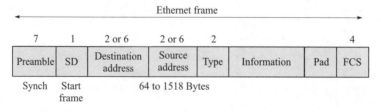

FIGURE 6.53 Ethernet frame (DIX standard)

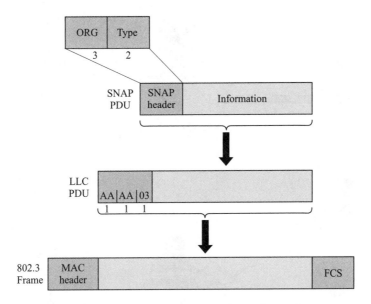

FIGURE 6.54 SNAP frame for encapsulating Ethernet frames

PHYSICAL LAYERS

Table 6.2 shows the various physical layers that have been defined for use with IEEE 802.3. Each of these medium alternatives are designated by three parameters that specify the bit rate, the signaling technique, and the maximum segment length. For example, the original standard specified 10Base5, which made use of thick (10 mm) coaxial cable operating at a data rate of *10* Mbps, using *base*band transmission and with a maximum segment length of *500* meters. The transmission uses Manchester coding, which is discussed in Chapter 3. This cabling system required the use of a *transceiver* to attach the NIC card to the coaxial cable. The thick coaxial cable Ethernet was typically deployed along the ceilings in building hallways, and a connection from a workstation in an office would tap onto the cable as shown in Figure 6.55a. Thick coaxial cable is awkward to handle and install. The 10Base2 standard uses thin (5 mm) coaxial cable operating at 10 Mbps and with a maximum segment of 185 meters. The cheaper and easier-to-handle thin coaxial cable makes use of T-shaped BNC junctions as shown in Figure 6.55b. 10Base5 and 10Base2 segments can be combined through the use of a *repeater* that forwards the signals from one segment to the other.

	10Base5	10Base2	10BaseT	10BaseF
Medium	Thick coax	Thin coax	Twisted pair	Optical fiber
Maximum segment length	500 m	200 m	100 m	2 km
Topology	Bus	Bus	Star	Point-to-point link

TABLE 6.2 IEEE 802.3 10 Mbps medium alternatives

(a)

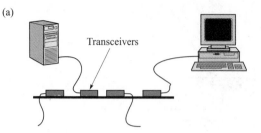

Transceivers

FIGURE 6.55 Ethernet cabling using thick and thin coaxial cable (Note: The T junction typically attaches to the NIC.)

(b)

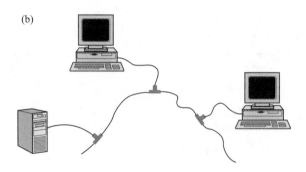

The 10BaseT standard involves the use of two unshielded twisted pairs of copper wires operating at 10 Mbps and connected to a **hub** as shown in Figure 6.56a. The T designates the use of twisted pair. The advantage of twisted pair is low cost and its prevalence in existing office wiring where it is used for telephones. Existing wiring arrangements allow the hubs to be placed in telephone wiring closets. However, twisted pair cannot support high bit rates over long distances, so the maximum cable length is 100 meters. The use of the 10BaseT standard also involves a move toward a star topology in which the stations connect the twisted pair to a hub where the collisions take place.

The star topology of 10BaseT provides three approaches to operating the LAN. In all three approaches the stations implement the CDMA-CD protocol. The difference is in the operation of the hub at the center of the star. In the first approach, the hub monitors all transmissions from the stations. When there is only one transmission, the hub repeats the transmission on the other lines. If there is a collision, that is, more than one transmission, then the hub sends a jamming signal to all the stations. This action causes the stations to implement the backoff algorithm. In this approach the stations are said to be in the same **collision domain**.

A second approach involves operating the hub as an **Ethernet switch**, as shown in Figure 6.56b. Each input port buffers incoming transmissions. The incoming frames are examined and transferred to the appropriate outgoing ports. Each incoming line in this case is in its own collision domain, so collisions will not occur if only a single station is attached to the line. It is possible, however, to have several stations share an input line using another hub, for example. In this case the group of stations will constitute a collision domain.

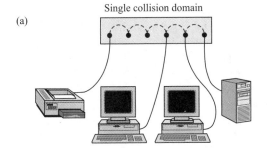

(a)

Single collision domain

FIGURE 6.56 Ethernet hub-and-switch topologies using twisted-pair cabling

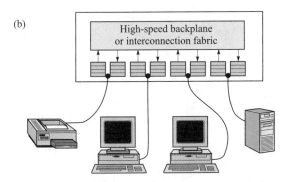

(b)

High-speed backplane or interconnection fabric

The number of stations in a LAN cannot be increased indefinitely. Eventually the traffic generated by stations will approach the limit of the shared transmission medium. The introduction of switching LANs provides a means of interconnecting larger numbers of stations without reaching this limit.

A third approach involves having stations transmit in **full-duplex** mode. Consider the case where each port in the switch has only a single station attached to it. Introducing a dedicated transmission line for each direction enables transmissions to take place in both directions simultaneously without collisions. Note that the stations can continue to operate the CSMA-CD algorithm, but they will never encounter collisions. All three of these approaches have been implemented in LAN products.

FAST ETHERNET

The IEEE 802.3u standard was approved in 1995 to provide Ethernet LANs operating at 100 Mbps. We refer to systems that operate under this standard as Fast Ethernet. To maintain compatibility with existing standards, the frame format, interfaces, and procedures have been kept the same. Recall that the performance of the CSMA-CD medium access control is sensitive to the ratio of the round-trip propagation delay and the frame transmission time. To obtain good performance, this ratio must be small. In addition, the correct operation of the protocol itself requires the minimum frame size transmission time to be larger than the round-trip propagation delay. When the transmission speed is increased from 10 Mbps to 100 Mbps, the packet transmission time is reduced by a factor of 10. For the MAC protocol to operate correctly, either the size of the minimum frame must be increased by a factor of 10 to 640 bytes or the maximum length between stations is reduced by a factor of 10 to, say, 250 meters.

The decision in developing the 100 Mbps IEEE 802.3u standard was to keep frame sizes and procedures unchanged and to define a set of physical layers that were entirely based on a hub topology involving twisted pair and optical fiber, as shown in Table 6.3. Coaxial cable was not included in the standard. (Note that the 100BaseF option can extend up to 2000 m because it operates in full-duplex mode and operates with buffered switches only.)

The standard involves stations that use unshielded twisted-pair (UTP) wiring to connect to hubs in a star topology. To obtain a bit rate of 100 Mbps, the 100BaseT4 standard uses four UTP 3 wires (UTP, category 3, that is, ordinary telephone-grade twisted pair). The 100 Mbps transmission is divided among three of the twisted pairs and flows in one direction at a time.

The 100BaseTX uses two UTP 5 wires. The category 5 twisted pair involves more twists per meter than UTP 3, which provides greater robustness with respect to interference thus enabling higher bit rates. One pair of wires is for transmission and one for reception, so 100BaseTX can operate in full-duplex mode.

A 100BaseFX standard has also been provided that uses two strands of multimode optical to provide full-duplex transmission at 100 Mbps in each direction. The 100BaseFX system can reach over longer distances than the twisted pair options, and so it is used in interconnecting wiring closets and buildings in a campus network.

The 100 Mbps IEEE 802.3 standards provide for two modes of operations at the hubs. In the first mode all incoming lines are logically connected into a single collision domain, and the CSMA-CD MAC procedure is applied. In the second mode the incoming frames are buffered and then switched internally within the hub. In the latter approach the CSMA-CD procedure is not used, and instead the IEEE 802.3 standard simply provides a means of accessing the first stage in a LAN that is based on multiplexing and switching.

Fast Ethernet LANs can be used to provide higher bandwidth in campus backbones and in the access portions of the network where packet flows are aggregated. Figure 6.57 shows a scenario in which Fast Ethernet hubs are used to (1) aggregate traffic from shared 10 Mbps LANs, (2) provide greater bandwidth to a server, and (3) provide greater bandwidth to individual users.

	100BaseT4	100BaseT	100BaseF
Medium	Twisted pair category 3 UTP four pairs	Twisted pair category 5 UTP two pairs	Optical fiber multimode two strands
Maximum segment length	100 m	100 m	2 km
Topology	Star	Star	Star

TABLE 6.3 IEEE 802.3u Fast Ethernet medium alternatives

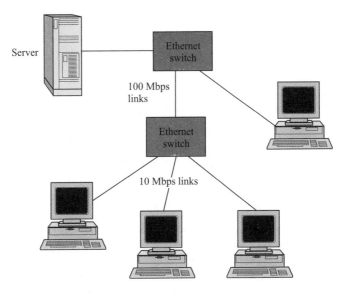

FIGURE 6.57 Application of Fast Ethernet

GIGABIT ETHERNET

The IEEE 802.3z **Gigabit Ethernet** standard was completed in 1998 and established an Ethernet LAN that increased the transmission speed over that of Fast Ethernet by a factor of 10. The goal was to define new physical layers but to again retain the frame structure and procedures of the 10 Mbps IEEE 802.3 standard.

The increase in speed by another factor of 10 put a focus on the limitations of the CSMA-CD MAC algorithm. For example, at a 1 Gbps speed, the transmission of a minimum size frame of 64 bytes can result in the transmission being completed before the sending station senses a collision. For this reason, the slot time was extended to 512 bytes. Frames smaller than 512 bytes must be extended with an additional carrier signal, in effect resulting in the same overhead as in padding the frame. In addition, an approach called *packet bursting* was introduced to address this scaling problem. Stations are allowed to transmit a burst of small packets, in effect to improve the key ratio *a*. Nevertheless, it is clear that with Gigabit Ethernet the CSMA-CD access control reached the limits of efficient operation. In fact, the standard preserves the Ethernet frame structure but operates primarily in a switched mode.

Gigabit Ethernet physical layer standards have been defined for multimode fiber with maximum length of 550 m, single-mode fiber with maximum length of 5 km, and four-pair category 5 UTP at a maximum length of up to 100 m. Table 6.4 lists the different medium alternatives.

	1000BaseSX	1000BaseLX	1000BaseCX	1000BaseT
Medium	Optical fiber multimode two strands	Optical fiber single mode two strands	Shielded copper cable	Twisted pair category 5 UTP
Maximum segment length	550 m	5 km	25 m	100 m
Topology	Star	Star	Star	Star

TABLE 6.4 IEEE 802.3z Gigabit Ethernet medium alternatives

6.6.2   Token-Ring and IEEE 802.5 LAN Standard

Several versions of token-ring networks were developed in the 1970s and 1980s; in token rings a number of stations are connected by point-to-point transmission links in a ring topology. Information flows in one direction along the ring from the source to the destination and back to the source. The key notion is that medium access control is provided via a small frame called a **token** that circulates around a ring-topology network. Only the station that has possession of the token is allowed to transmit at any given time.

The ring topology brings certain advantages to medium access control. The flow of the token along the ring automatically provides each station with a turn to transmit. Thus the ring topology provides for fairness in access and for a fully distributed implementation. The token mechanism also allows for the introduction of access priorities as well as the control of the token circulation time.

The ring topology, however, is seriously flawed when it comes to faults. The entire network will fail if there is a break in any transmission link or a failure in the mechanism that relays a signal from one point-to-point link to the next. This problem is overcome by using a star topology to connect stations to a wiring closet where the wires from the stations can be connected to form a ring as shown in Figure 6.58. Reliability is provided by relays that can bypass the wires of stations that are deemed to have failed. Thus, for example, station E in Figure 6.58 has been bypassed by the relay circuit in the wiring center. The star topology also has the advantage that it can use existing telephone wiring arrangements that are found in office buildings.

The IEEE 802.5 LAN standard defines token-ring networks operating at 4 Mbps and 16 Mbps transmission. The rings are formed by twisted-pair cables using differential Manchester line coding. The maximum number of stations is set to 250.

TOKEN-RING PROTOCOL

To transmit a frame, a station must wait for a "free" token to arrive at the interface card. When such a token arrives, the station claims the token by removing it from the ring.[10] The station then proceeds to transmit its frame into its

---

[10]In fact, the situation "claims" the token by flipping a specific bit from 0 to 1; this process converts the token frame into a data frame.

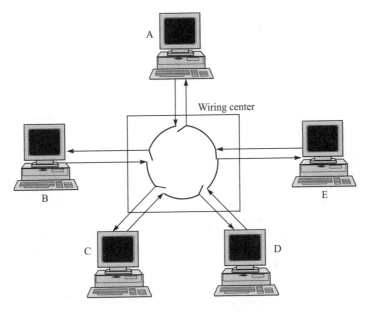

FIGURE 6.58 Token-ring network implemented using a star topology

outgoing line. The frame travels along the ring over every point-to-point link and across every interface card. Each station examines the destination address in each passing frame to see whether it matches the station's own address. If not, the frame is forwarded to the next link after a few bits delay. If the frame is intended for the station, the frame is copied to a local buffer, several status bits in the frame are set, and the frame is forwarded along the ring. The sending station has the responsibility of removing the frame from the ring and of reinserting a free token into the ring.

When the traffic on the ring is light, the token spends most of the time circulating around the ring until a station has a frame to transmit. As the traffic becomes heavy, many stations have frames to transmit, and the token mechanism provides stations with a fair round-robin access to the ring.

The approach that is used to reinsert the free token into the ring can have a dramatic effect on the performance when the delay-bandwidth product of the ring is large. To show why this happens, we first have to examine how a frame propagates around the ring. Suppose that the ring has M stations. Each station interface introduces b bits of delay between when the interface receives a frame and forwards it along the outgoing line, so the interfaces introduce Mb bits of delay. A typical value of b is 2.5.[11] If the total length of the links around the ring is d meters, then an additional delay of d/v seconds or dR/v bits is incurred because of propagation delay, where v is the propagation speed in the medium.

---

[11]The half-bit delay is possible because token ring uses Manchester line coding.

For example, $v = 2 \times 10^8$ meters/second in twisted-pair wires, or equivalently it takes 5 microseconds to travel 1 kilometer. The **ring latency** is defined as the time that it takes for a bit to travel around the ring and is given by

$$\tau' = d/v + Mb/R \text{ seconds} \qquad \text{and} \qquad \tau'R = dR/v + Mb \text{ bits}$$

Example—Ring Latency and Token Reinsertion

Let us investigate the interplay between ring latency and the token reinsertion method. First suppose that we have a ring that operates at a speed of $R = 4$ Mbps with $M = 20$ stations separated by 100 meters and $b = 2.5$ bits. The ring latency (in bits) is then $20 \times 100 \times 4 \times 10^6/(2 \times 10^8) + 20(2.5) = 90$ bits. Thus the first bit in a frame returns to the sending station 90 bit times after being inserted. On the other hand, if the speed of the ring is 16 Mbps and the number of stations is 80, then the ring latency is $80 \times 100 \times 16 \times 10^6/ (2 \times 10^8) + 80(2.5) = 840$ bits.

Now suppose that we are transmitting a frame that is $L = 400$ bits long. Suppose that the token reinsertion strategy is to reinsert the token after the frame transmission is completed but not until after the last bit of the frame returns to the sending station. Figure 6.59a shows that the last bit in the frame returns after 490 bits in the first ring. Thus the sending station must insert an "idle" signal for 90 additional bit times before that station can reinsert the token into the ring. In the second ring the token returns after 1240 bit times, as shown in Figure 6.59b. In this case the sending station has to insert an idle signal for 840 bits times before reinserting the token. Thus we see that this token

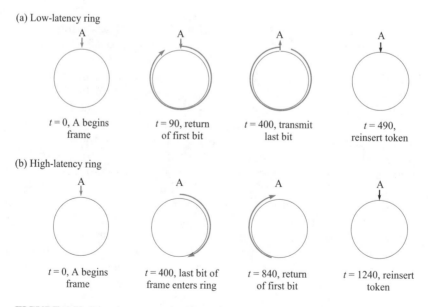

(a) Low-latency ring

$t = 0$, A begins frame $t = 90$, return of first bit $t = 400$, transmit last bit $t = 490$, reinsert token

(b) High-latency ring

$t = 0$, A begins frame $t = 400$, last bit of frame enters ring $t = 840$, return of first bit $t = 1240$, reinsert token

FIGURE 6.59 Ring latency and token reinsertion strategies

reinsertion method extends the effective length of each frame by the ring latency. For the first ring the efficiency is 400/490 = 82 percent; for the second ring the efficiency drops to 400/1240 = 32 percent.

Now suppose that the token reinsertion strategy is to reinsert the token after the frame transmission is completed but not until after the header of the frame returns to the sending station. Suppose that the header is 15 bytes = 120 bits long. The header returns after 90 + 120 = 210 bits in the first ring, as shown in Figure 6.60a. The sending station can therefore reinsert the token immediately after transmitting bit 400 of the frame. Figure 6.60b shows that in the second ring the header returns after 840 + 120 = 960 bits. Consequently, the sending station must send an idle signal for 560 bit times before that station can reinsert the token into the ring. The first ring now operates efficiently, but the second ring has an efficiency of 400/960 = 42 percent.

Finally suppose that the token reinsertion strategy had been to reinsert the token immediately after the frame transmission is completed. The need for the idle signal is completely eliminated and so is the associated inefficiency.

All three of the token reinsertion strategies introduced in the preceding example have been incorporated into token-ring LAN standards. The first strategy is part of the MAC protocol of the IEEE 802.5 standard for a 4 Mbps token-ring LAN. The reason for waiting until the last bit in the frame is that the last byte in the frame contains response information from the destination station. The IBM token-ring LAN for 4 Mbps uses the second strategy, where the token is reinserted after the header is returned. Both the IEEE 802.5 standard and the

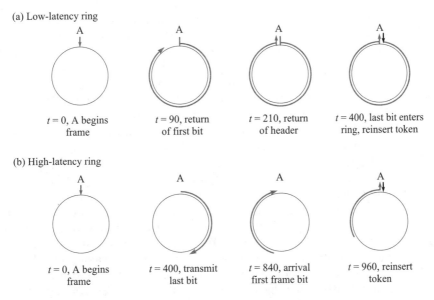

(a) Low-latency ring

$t = 0$, A begins frame

$t = 90$, return of first bit

$t = 210$, return of header

$t = 400$, last bit enters ring, reinsert token

(b) High-latency ring

$t = 0$, A begins frame

$t = 400$, transmit last bit

$t = 840$, arrival first frame bit

$t = 960$, reinsert token

FIGURE 6.60 Reinsert token after header of frame returns

IBM token-ring LAN for 26 Mbps use the third strategy because of its higher efficiency. Each of the token reinsertion strategies has a different maximum achievable throughput leading to dramatic differences in frame transfer delay performance. These differences are discussed in section 6.4.3.

Once the token has been reinserted into the ring, the token must travel to the next station that has a frame to transmit. The "walk" time that elapses from when the token is inserted to when it is captured by the next active station is also a form of overhead that can affect the maximum achievable throughput.

Finally, we note that different variations of MAC protocols are obtained according to how long a station is allowed to transmit once it captures a free token. One possibility is to allow a station to transmit only a single frame per token. This rather strict rule implies that each frame transmission is extended by a walk time. On the other hand, the rule also guarantees that a token will return to a station after at most M frame transmissions. At the other extreme, a station could be allowed to transmit until it empties its buffers of all frames. This approach is more efficient in that it amortizes the walk-time overhead over several frame transmissions. However, this approach also allows the token return time to grow without bound. An intermediate approach limits the time that a station can hold a token. For example, the IEEE 802.5 standard imposes a maximum token-holding-time limit of 10 ms.

FRAME STRUCTURE

The structure of the token and data frames for the IEEE 802.5 standard is shown in Figure 6.61. The token frame consists of three bytes. The first and last bytes are the *starting delimiter* (SD) and *ending delimiter* (ED) fields. The standard uses differential Manchester line coding. Recall from Chapter 3 (Figure 3.25) that this line coding has transitions in the middle of each bit time. The SD and ED bytes are characterized by the fact that they contain symbols that violate this pattern: the J symbol begins as a 0 but has no transition in the middle; the K symbol begins as a 1 and has no transition in the middle. The second byte in the token frame is the *access control* (AC) field. The T bit in the access control field is the **token bit**: T = 0 indicates a token frame, and T = 1 indicates a data frame. A station can convert an available token frame (T = 0) into a data frame (T = 1) by simply flipping the T bit. This feature explains why token ring interfaces can pass from an incoming link onto an outgoing link with only a one-bit delay (although a delay of 2.5 bits is usually implemented).

The data frame begins with SD and AC fields. The PPP and RRR bits in the AC field implement eight levels of priority in access to the ring. The monitor M bit is used by a designated monitor station to identify and remove "orphan" frames that are not removed from the ring by their sending station, for example, as a result of a station crash. The *frame control* (FC) field indicates whether a frame contains data or MAC information. Data frames are identified by FF = 01, and the Z bits are then ignored. MAC control frames are identified by FF = 00, and the Z bits then indicate the type of MAC control frame. The IEEE 802.5 standard specifies both 16-bit and 48-bit addressing, using the same format as the IEEE 802.3 Ethernet standard uses. The address fields are followed

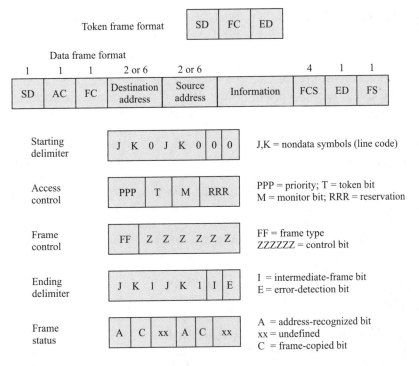

FIGURE 6.61 IEEE 802.5 Token and data frame structure

by the information field that is limited in length only by the maximum token holding time. The *frame check sequence* (FCS) field contains a CRC checksum as in IEEE 802.3. The ED field contains an E bit that indicates that a station interface has detected an error such as a line code violation or a frame check sequence error. The I bit indicates the last frame in a sequence of frames exchanged between two stations.

The *frame status* field in the data frame allows the receiving station to convey transfer status information to the sending station through A and C bits that are repeated within the field. An A = 1 bit indicates that the destination address was recognized by the receiving station. A C = 1 bit indicates that the frame was copied onto the receiving station's buffer. Therefore, an A = 1, C = 1 frame status field indicates that the frame was received by the intended destination station.

The IEEE 802.5 standard allows the token ring to be operated with a priority access mechanism. To transmit a frame of a given priority, a station must wait to capture a token of equal or lower priority. The station can reserve a token of the desired level by setting the RRR field in passing frames to the level of priority of its frame *if* the RRR level is *lower* than the priority the station is seeking. In effect, the RRR field allows stations to bid up the priority of the next token. When the token arrives at a station that has a frame of higher or equal priority, the token is removed and a data frame is inserted into the ring. The RRR field in the data frame is set to 0, and the priority field is kept at the same value as the token frame. When the station is done transmitting its frames, it issues a token at the reserved priority level.

Ring maintenance procedures are necessary to ensure the continued operation of the ring. Problem conditions can lead to the circulation of orphan data frames in the ring, the disappearance of tokens from the ring, the corruption of the frame structure within the ring, or the incidence of breaks in the link between stations. The IEEE 802.5 standard provides a procedure for the selection of a station to become the *active monitor* that is assigned the task of detecting and removing orphan frames, as well as identifying and replacing lost tokens. Additional procedures are defined to deal with other problem conditions. For example a "beacon" MAC control frame can be used by any station to determine whether its incoming link has become broken. Other MAC control frames indicate the presence of an active monitor, elect a new monitor, identify duplicate addresses, clear the ring of all frames, and identify neighbor stations in the ring.

6.6.3   FDDI

The **Fiber Distributed Data Interface (FDDI)** is a token-based LAN standard developed by the American National Standards Institute. FDDI uses a ring-topology network in which station interfaces are interconnected by optical fiber transmission links operating at 100 Mbps in a ring that spans up to 200 kilometers and accommodates up to 500 stations. FDDI has found application as a campus backbone network to interconnect various Ethernet LAN subnetworks.

FDDI can operate over multimode or single-mode optical fiber systems. FDDI can also operate over twisted-pair cable at lengths of less than 100 meters. The high bit rate in FDDI precludes the use of the bandwidth-inefficient Manchester line code. Instead FDDI uses a 4B5B binary line code and NRZ-inverted signaling that requires a symbol rate of 125 Msymbols/second. The 4B5B code lacks the self-clocking property of the Manchester code. For this reason FDDI frames begin with a longer preamble that serves to synchronize the receiver to the transmitter clock. Each FDDI station transmits into its outgoing link according to its own local clock. To accommodate differences between the local and incoming clock, each station uses a 10-bit elastic buffer to absorb timing differences. FDDI specifies that all clocks must meet a tolerance of 0.005 percent $= 5 \times 10^{-5}$ of 125 MHz. Assuming the maximum clock difference between a fast transmitter and a slow receiver, the worst-case clock difference is 0.01 percent. Symbols accumulate in the buffer at a rate of 125 Msymbols/second $\times 1 \times 10^{-4} = 12.5 \times 10^3$ bits/second. The five-bit buffer will then fill up in 5 bits/$(12.5 \times 10^3$ bps$) = 0.4$ ms. For this reason the maximum allowable FDDI frame length is 125 Msymbols/second $\times .4$ ms $= 50,000$ symbols. The 4B5B code implies this is equivalent to 40,000 bits. Because of this timing consideration, the FDDI frame has a maximum size of 4500 bytes $= 36,000$ bits.

To provide reliability with respect to link breakages and interface failure, a dual-ring arrangement is used, as shown in Figure 6.62. A break in the ring is handled by redirecting the flow in the opposite direction at the last station before the break. This action has the effect of converting the dual ring into a single ring.

From Figure 6.63 we can see that the FDDI frame structure is very similar to that of IEEE 802.5. The frame begins with 16 or more idle control signals that generate a square wave signal that serves to synchronize the receiver. The SD and ED fields contain distinct signal violations that help identify them. The FDDI frame does not contain an AC field or a token bit. Instead the FC field is used to indicate the presence of a token and to provide information about the type of frame. A token frame is indicated by either 10000000 or 11000000 in the FC field. The capture of the token is done, not by flipping a bit, but by removing the token transmission from the ring and replacing it with a data frame. The other remaining fields function as in IEEE 802.5.

TIMED-TOKEN RING PROTOCOL

The FDDI medium access control was designed to operate in a high-ring-latency environment. If we assume 500 stations each introducing a latency of 10 bits and a maximum length 200 km ring, then the ring latency is $500 \times 10 + 100$ Mbps $\times (200 \times 10^3$ m$)/(2 \times 10^8$ m/sec$) = 5000 + 100{,}000 = 105{,}000$ bits. Thus FDDI LANs can have a very high ring latency, making it essential that the token reinsertion strategy be immediate insertion after completion of each frame transmission. Even with this strategy, however, a maximum length ring can hold more

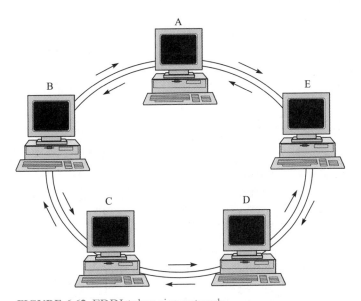

FIGURE 6.62 FDDI token-ring network

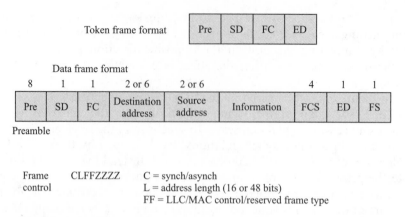

FIGURE 6.63 FDDI frame structure

than two maximum length frames! For this reason FDDI also provides an option of having more than one token circulating the ring at a given time.

The FDDI MAC protocol can handle two types of traffic: synchronous traffic that has a tight transfer delay requirement, such as voice or video, and asynchronous traffic that has a greater delay tolerance as in many types of data traffic. To meet the timing requirements of synchronous traffic, the MAC uses a timed-token mechanism that ensures that the token rotation value is less than some value. In particular, all the stations in an FDDI ring must agree to operate according to a given **target token rotation time (TTRT)**. Each station i is allocated a certain amount of time, S_i seconds, that specifies the maximum duration the station is allowed to send synchronous traffic each time it has captured the token. If the sum of S_i times is smaller than the TTRT, then the operation of the FDDI MAC guarantees that the token will return to every node in less than 2 TTRT seconds. This property allows FDDI to meet the delay requirements of synchronous traffic.

Each station maintains a **token rotation timer (TRT)** that measures the time that has elapsed since the station last received a token. When a station receives a token, the station first calculates the **token holding time (THT)**, defined by THT = TTRT − TRT, which is a measure of the degree of activity in the ring.

When traffic is light, the token will rotate quickly around the ring and the TRT will be much smaller than the TTRT. In this case the MAC need not be restrictive in providing access to the ring. As the traffic becomes heavy, the TRT will approach the TTRT, indicating that the MAC must begin restricting access to the ring. The FDDI MAC implements these notions as follows:

If THT > 0, then the station can transmit all its synchronous traffic S. In addition, if the THT timer has not expired after S_i seconds, the station is allowed to transmit asynchronous traffic for the balance of its THT time, that is, up to THT − S_i seconds. The station must then release the token.

If THT < 0, then the station is allowed to transmit only its synchronous traffic S_i and must then release the token.

This timed-token mechanism throttles the asynchronous traffic when the ring becomes congested. In combination with the globally agreed upon TTRT, this mechanism assures the timely delivery of synchronous traffic. The timed-token mechanism also ensures fairness in access to the ring. Because the TRT is reset upon *arrival* of a token at a station, a station that has a lot of traffic to send will find it has a large TRT the next time it receives the token. Consequently, the THT will be small, and that station will be prevented from transmitting for an excessive period of time. Thus the station will be forced to relinquish the token sooner so other stations can have an opportunity to transmit.

6.6.4   Wireless LANs and IEEE 802.11 Standard[12]

The case for wireless LANs is quite compelling. All you have to do is look under the desks in a typical small business or home office. You will find a rat's nest of wires: in addition to a variety of power cords and adapters, you have cables for a telephone modem, a printer, a scanner, a mouse, and a keyboard. In addition, there is the need for communications to synchronize files with laptop computers and personal organizers. And, of course, there is still the need to connect to other computers in the office. Wireless technology, in the form of digital radio and infrared transmission, can eliminate many of these wires and, in the process, simplify the installation and movement of equipment, as well as provide connectivity between computers. Wireless technology, however, must overcome significant challenges:

- Radio and infrared transmission is susceptible to noise and interference, so such transmission is not very reliable.
- The strength of a radio transmission varies in time and in space because of fading effects that result from multipath propagation and from uneven propagation due to physical barriers and geographic topology, and so coverage is inconsistent and unpredictable.
- In the case of radio, the transmitted signal cannot easily be contained to a specific area, so signals can be intercepted by eavesdroppers.
- The spectrum is finite and must be shared with other users (your neighbor's wireless LAN), and devices (e.g., microwave ovens and florescent lamps!).
- In the case of radio, the limited spectrum also makes it difficult to provide the high transmission speeds that are easily attained using wired media.
- Radio spectrum has traditionally been regulated differently by different government administrations, so it can be difficult to design products for a global market.

---

[12]The IEEE 802.11 is a complex standard, and so its explanation requires more space than the previous LAN standards. The discussion on physical layers for 802.11 can be skipped if necessary.

The most compelling reason for wireless networks, however, is that they enable *user mobility*. Many of us have already been conditioned to the convenience of television remote controls and cordless telephones in the "local area" of the home, as well as to the convenience of cellular phones over a "wider" area. In the context of wireless LANs, user mobility is particularly significant in situations where users carry portable computers or devices that need to communicate to a server or with each other. One example is a doctor or nurse in a hospital accessing up-to-date information on a patient, even as the patient is wheeled down a corridor. In this case the hospital may have an infrastructure of *wireless access points* that the portable devices can communicate with to access a backbone (wired) network. Another example is a meeting where the participants can create a temporary *ad hoc LAN* simply by turning on their laptop computers. To provide mobility, further challenges must be overcome:

- Mobile devices operate on batteries, so the MAC protocols must incorporate power management procedures.
- Protocols need to be developed that enable a station to discover neighbors in the local network and to provide seamless connections even as users roam from one coverage area to another.

The development of wireless LAN products was initially stimulated mildly by the allocation of spectrum in the industrial, scientific, and medical (ISM)

CSMA-CA? WHY NOT WIRELESS ETHERNET?

Given the dominance of the Ethernet standards in wired LANs, an obvious question is, Why not use wireless Ethernet? After all, Ethernet was designed for broadcast networks, and wireless networks are certainly broadcast in nature. There are several reasons why CSMA-CD cannot be used. The first reason is that it is difficult to detect collisions in a radio environment, so it is not possible to abort transmissions that collide. A second reason is that the radio environment is not as well controlled as a wired broadcast medium, and transmissions from users in other LANs can interfere with the operation of CSMA-CD. A third reason is that radio LANs are subject to the *hidden-station problem* that occurs when two stations, say, A and C, attempt to transmit to a station that is located between them, say, B, as shown in Figure 6.64. The two stations may be sufficiently distant from each other that they cannot hear each other's transmission. Consequently, when they sense the channel, they may detect it as idle even as the other station is transmitting. This condition will result in the transmissions from the two stations proceeding and colliding at the intermediate station. The Carrier-Sense Multiple Access with Collision Avoidance (CSMA-CA) medium access control was developed to prevent this type of collision. CSMA-CA is incorporated in IEEE 802.11 and is described in the subsection on medium access control later in this section.

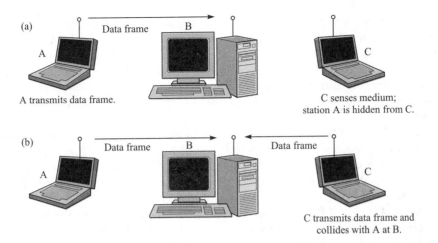

FIGURE 6.64 The hidden-station problem

bands of 902 to 928 MHz, 2400 to 2483.5 MHz, and 5725 to 5850 MHz in the United States under part 15 of the FCC rules. These rules, however, require that users accept interference from other users already using these frequencies, such as microwave ovens. Furthermore, users of the ISM band were required to use spread spectrum transmission techniques that limit the bit rates that can be attained. A stronger stimulus for the development of wireless LANs was the development of a HIPERLAN standard in Europe for a 20 Mbps wireless LAN operating in the 5 GHz band. This development was reinforced in the United States by the FCC's designation of 300 MHz of spectrum in the 5 GHz band for the development of unlicensed LAN applications operating at speeds of 20 Mbps or higher. In addition, the Infrared Data Association (IrDA) has been promoting the development of MAC and physical layer standards for high-speed infrared systems for interconnecting computers to other devices at short range.

In this section we focus on the IEEE 802.11 LAN standard that specifies a MAC layer that is designed to operate over a number of physical layers. We show that the standard is quite complex. The standard's complexity is rooted in the challenges indicated above. In particular, we show that in addition to the basic issue of coordinating access, the standard must incorporate error control to overcome the inherent unreliability of the channel, modified addressing and association procedures to deal with station portability and mobility, and inter-connection procedures to extend the reach of a wireless stations as well as to accommodate users who move while communicating.

AD HOC AND INFRASTRUCTURE NETWORKS

The **basic service set (BSS)** is the basic building block of the IEEE 802.11 architecture. A BSS is defined as a group of stations that coordinate their access to the medium under a given instance of the medium access control. The geographical area covered by the BSS is known as the *basic service area* (BSA), which is analogous to a cell in a cellular communications network. A BSA may extend over an area with a diameter of tens of meters. Conceptually, all stations in a BSS can communicate directly with all other stations in a BSS. Note that two unrelated BSSs may be colocated. IEEE 802.11 provides a means for these BSSs to coexist.

A single BSS can be used to form an **ad hoc network**. An ad hoc network consists of a group of stations within range of each other. Ad hoc networks are typically temporary in nature. They can be formed spontaneously anywhere and be disbanded after a limited period of time. Figure 6.65 is an illustration of an ad hoc network. Two stations can make an ad hoc network.

In 802.11 a set of BSSs can be interconnected by a **distribution system** to form an **extended service set (ESS)** as shown in Figure 6.66. The BSSs are like cells in a cellular network. Each BSS has an **access point (AP)** that has station functionality and provides access to the distribution system. The AP is analogous to the base station in a cellular communications network. An ESS can also provide gateway access for wireless users into a wired network such as the Internet. This access is accomplished via a device known as a **portal**. The term **infrastructure network** is used informally to refer to the combination of BSSs, a distribution system, and portals.

The distribution system provides the *distribution service*, which is

1. The transfer of MAC SDUs (MSDUs) between APs of BSSs within the ESS.
2. The transfer of MSDUs between portals and BSSs within ESS.

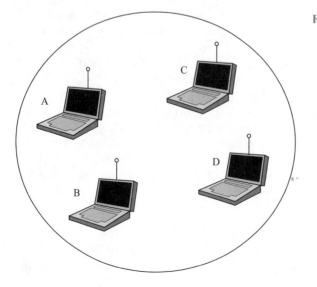

FIGURE 6.65 Ad hoc network

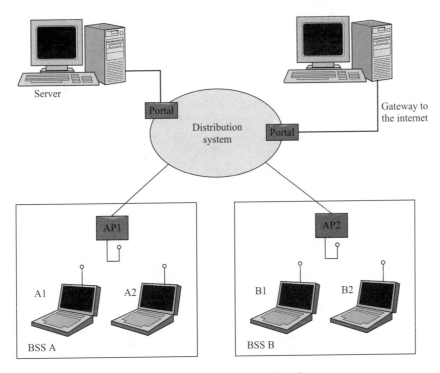

FIGURE 6.66 Infrastructure network and extended service set

3. The transport of MSDUs between stations in the same BSS when either the MSDU has a multicast or broadcast address or the sending station chooses to use the distribution service.

The role of the distribution service is to make the ESS appear as a single BSS to the logical link control (LLC) that operates above the MAC in any of the stations in the ESS. IEEE 802.11 defines the distribution service but not the distribution system. The distribution system can be implemented by using wired or wireless networks.

To join an infrastructure BSS, a station must select an AP and establish an **association** with it. This procedure establishes a mapping between the station and the AP that can be provided to the distribution system. The station can then send and receive data messages via the AP. A **reassociation** service allows a station with an established association to *move* its association from one AP to another AP. The **dissociation** service is used to terminate an existing association. Stations have the option of using an **authentication** service to establish the identity of other stations. Stations also have the option of using a **privacy** service that prevents the contents of messages from being read by anyone other than the intended recipient(s).

The dynamic nature of the LAN topologies under the scope of the IEEE 802.11 implies several fundamental differences between wireless and wired

LANs. In wired LANs the MAC address specifies the physical location of a station, since users are stationary. In wireless LANs, the MAC address identifies the station but *not* the location, since the standard assumes that stations can be portable or mobile. A station is *portable* if it can move from one location to another but remains fixed while in use. A *mobile* station moves while in use. The 802.11 MAC sublayer is required to present the same set of standard services that other IEEE 802 LANs present to the LLC. This requirement implies that mobility has to be handled within the MAC sublayer.

FRAME STRUCTURE AND ADDRESSING

IEEE 802.11 supports three types of frames: management frames, control frames, and data frames. The management frames are used for station association and disassociation with the AP, timing and synchronization, and authentication and deauthentication. Control frames are used for handshaking and for positive acknowledgments during the data exchange. Data frames are used for the transmission of data. The MAC header provides information on frame control, duration, addressing, and sequence control. Figure 6.67 shows that the format of the MAC frame consists of a MAC header, a frame body, and a CRC checksum.

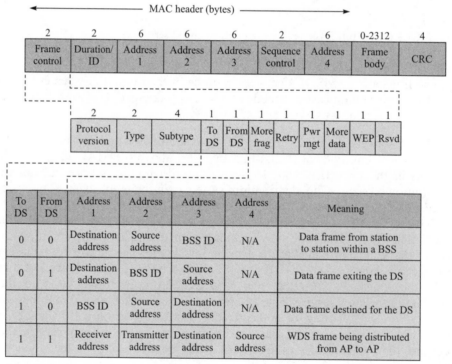

To DS	From DS	Address 1	Address 2	Address 3	Address 4	Meaning
0	0	Destination address	Source address	BSS ID	N/A	Data frame from station to station within a BSS
0	1	Destination address	BSS ID	Source address	N/A	Data frame exiting the DS
1	0	BSS ID	Source address	Destination address	N/A	Data frame destined for the DS
1	1	Receiver address	Transmitter address	Destination address	Source address	WDS frame being distributed from AP to AP

DS = distribution system   AP = access point

FIGURE 6.67 IEEE 802.11 frame structure

The *frame control field* in the MAC header is 16 bits long, and it specifies the following items:

- The 802.11 protocol version (the current version is 0).
- The type of frame, that is, management (00), control (01), or data (10).
- The subtype within a frame type, for example, type = "management," subtype = "association request" or type = "control," subtype = "ACK."
- The To DS field is set to 1 in Data type frames destined for the distribution system, including Data type frames from a station associated with the AP that have broadcast or multicast addresses.
- The From DS field is set to 1 in Data type frames exiting the distribution system.
- The more fragments field is set to 1 in frames that have another fragment of the current MSDU to follow.
- The retry field is set to 1 in Data or Management type frames that are retransmissions of an earlier frame; this helps the receiver deal with duplicate frames.
- The power management bit is set to indicate the power management mode of a station.
- The more data field is set to 1 to indicate to a station in power save mode that more MSDUs are buffered for it at the AP.
- The Wired Equivalent Privacy (WEP) field is set to 1 if the frame body field contains information that has been processed by the cryptographic algorithm.

The *duration/ID field* in the MAC header is 16 bits long and is used in two ways. It usually contains a duration value (net allocation vector) that is used in the MAC algorithm. The only exception is in Control type frames of subtype PS-Poll, where this field carries the ID of the station that transmitted the frame.

The use of the four *address fields* is specified by the To DS and From DS fields in the frame control field as shown in Figure 6.67. Addresses are 48-bit long IEEE 802 MAC addresses and can be individual or group (multicast/broadcast). The Address 1 field in this case contains the destination address. The *BSS identifier* (BSS ID) is a 48-bit field of the same format as IEEE 802 MAC addresses, uniquely identifies a BSS, and is given by the MAC address of the station in the AP of the BSS. The *destination address* is an IEEE MAC individual or group address that specifies the MAC entity that is the final recipient of the MSDU that is contained in the frame body field. The *source address* is a MAC individual address that identifies the MAC entity from which the MSDU originated. The *receiver address* is a MAC address that identifies the intended immediate recipient station for the MPDU in the frame body field. The *transmitter address* is a MAC individual address that identifies the station that transmitted the MPDU contained in the frame body field. This description is confusing so let's consider the four cases shown in the figure.

- To DS = 0, From DS = 0. This case corresponds to the transfer of a frame from one station in the BSS to another station in the same BSS. The station in the BSS look at the address 1 field to see whether the frame is intended for them. The address 2 field contains the address that the ACK frame is to be sent to. The address 3 field specifies the BSS ID.

- To DS = 0, From DS = 1. This case corresponds to the transfer of a frame from the distribution system to a station in the BSS. The stations in the BSS look at the address 1 field to see whether the frame is intended for them. The address 2 field contains the address that the ACK frame is to be addressed to, in this case the AP. The address 3 field specifies the source address.
- To DS = 1, From DS = 0. This case corresponds to the transfer of a frame from a station in the BSS to the distribution system. The stations in the BSS, including the AP, look at the address 1 field to see whether the frame is intended for them. The address 2 field contains the address that the ACK frame is to be addressed to, in this case the source address. The address 3 field specifies the destination address that the distribution system is to deliver the frame to.
- To DS = 1, From DS = 1. This special case applies when we have a *wireless distribution system* (WDS) transferring frames between BSSs. The address 1 field contains the receiver address of the station in the AP in the WDS that is the next immediate intended recipient of the frame. The address 2 field contains the destination address of the station in the AP in the WDS that is transmitting the frame and should receive the ACK. The address 3 field specifies the destination address of the station in the ESS that is to receive the frame, and the address 4 field specifies the source address of the station in the ESS that originated the frame.

The *sequence control field* is 16 bits long, and it provides 4 bits to indicate the number of each fragment of an MSDU and 12 bits of sequence numbering for a sequence number space of 4096. The *frame body field* contains information of the type and subtype specified in the frame control field. For Data type frames, the frame body field contains an MSDU or a fragment of an MSDU. Finally, the *CRC field* contains the 32-bit cyclic redundancy check calculated over the MAC header and frame body field.

MEDIUM ACCESS CONTROL

The MAC sublayer is responsible for the channel access procedures, protocol data unit (PDU) addressing, frame formatting, error checking, and fragmentation and reassembly of MSDUs. The MAC layer also provides options to support security services through authentication and privacy mechanisms. MAC management services are also defined to support roaming within an ESS and to assist stations in power management.

The IEEE 802.11 MAC protocol is specified in terms of *coordination functions* that determine when a station in a BSS is allowed to transmit and when it may be able to receive PDUs over the wireless medium. The **distributed coordination function (DCF)** provides support for asynchronous data transfer of MAC SDUs on a best-effort basis. Under the DCF, the transmission medium operates in the *contention mode* exclusively, requiring all stations to contend for the channel for each packet transmitted. IEEE 802.11 also defines an optional **point**

coordination function (PCF), which may be implemented by an AP, to support connection-oriented time-bounded transfer of MAC SDUs. Under PCF the medium can alternate between the *contention period* (CP), during which the medium uses contention mode, and a *contention-free period* (CFP). During the CFP, the medium usage is controlled by the AP, thereby eliminating the need for stations to contend for channel access.

Distributed coordination function

The DCF is the basic access method used to support asynchronous data transfer on a best-effort basis. All stations are required to support the DCF. The access control in ad hoc networks uses only the DCF. Infrastructure networks can operate using just the DCF or a coexistence of the DCF and PCF. The 802.11 MAC architecture is depicted in Figure 6.68, which shows that the DCF sits directly on top of the physical layer and supports contention services. Contention services imply that each station with an MSDU queued for transmission must contend for the channel and, once the given MSDU is transmitted, must recontend for the channel for all subsequent frames. Contention services are designed to promote fair access to the channel for all stations.

The DCF is based on the **carrier sensing multiple access with collision avoidance (CSMA-CA)** protocol. Carrier sensing involves monitoring the channel to determine whether the medium is idle or busy. If the medium is busy, it makes no sense for a station to transmit its frame and cause a collision and waste bandwidth. Instead the station should wait until the channel becomes idle. When this happens, there is another problem: Other stations may have also been waiting for the channel to become idle. If the algorithm is to transmit immediately after the channel becomes idle, then collisions are likely to occur; and because collision detection is not possible, the channel will be wasted for an entire frame duration. A solution to this problem is to randomize the times at which the contending stations attempt to seize the channel. This approach reduces the likelihood of simultaneous attempts and hence the likelihood that a station can seize the channel.

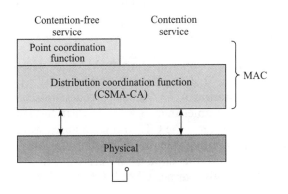

FIGURE 6.68 IEEE 802.11 MAC architecture

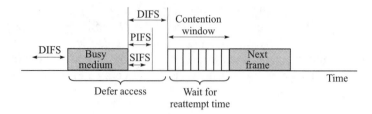

FIGURE 6.69 Basic CSMA-CA operation

Figure 6.69 shows the basic CSMA-CA operation. *All* stations are obliged to remain quiet for a certain minimum period after a transmission has been completed, called the **interframe space (IFS)**. The length of the IFS depends on the type of frame that the station is about to transmit. High-priority frames must only wait the **short IFS (SIFS)** period before they contend for the channel. Frame types that use SIFS include ACK frames, CTS frames, data frames of a segmented MSDU, frames from stations that are responding to a poll from an AP, and any frame from an AP during the CFP. All of these frame types complete frame exchanges that are already in progress. The **PCF interframe space (PIFS)** is intermediate in duration and is used by the PCF to gain priority access to the medium at the start of a CFP. The **DCF interframe space (DIFS)** is used by the DCF to transmit data and management MDPUs.

A station is allowed to transmit an *initial* MAC PDU under the DCF method if the station detects the medium idle for a period DIFS or greater. However, if the station detects the medium busy, then it must calculate a random backoff time to schedule a reattempt. A station that has scheduled a reattempt monitors the medium and decrements a counter each time an idle contention slot transpires. The station is allowed to transmit when its backoff timer expires during the contention period. If another station transmits during the contention period before the given station, then the backoff procedure is suspended and resumed the next time a contention period takes place. When a station has successfully completed a frame transmission and has another frame to transmit, the station must first execute the backoff procedure. Stations that had already been contending for the channel tend to have smaller remaining backoff times when their timers are resumed, so they tend to access the medium sooner than stations with new frames to transmit. This behavior introduces a degree of fairness in accessing the channel.

A handshake procedure was developed to operate with CSMA-CA when there is a hidden-station problem. Figure 6.70 shows that if a station, say, A, wants to send a data frame to station B, station A first sends a **request-to-send (RTS)** frame. If station B receives the RTS frame, then B issues a **clear-to-send (CTS)** frame. *All* stations within range of B receive the CTS frame and are aware that A has been given permission to send, so they remain quiet while station A proceeds with its data frame transmission. If the data frame arrives without error, station B responds with an ACK. In this manner CSMA-CA coordinates

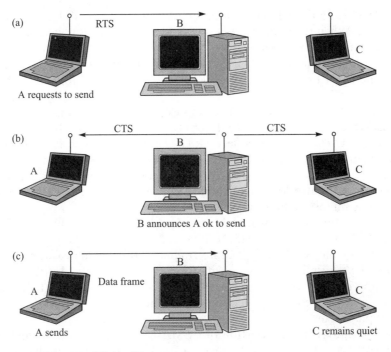

FIGURE 6.70 CSMA-CA

stations, even in the presence of hidden stations, so that collisions are avoided. It is still possible for two stations to send RTS frames at the same time so that they collide at B. In this case the stations must execute a backoff to schedule a later attempt. Note that having RTS frames collide is preferable to having data frames collide, since RTS frames are much shorter than data frames. For example, RTS is 20 bytes and CTS is 14 bytes, whereas an MPDU can be 2300 bytes long.

In IEEE 802.11, carrier sensing is performed at both the air interface, referred to as *physical carrier sensing*, and at the MAC sublayer, referred to as *virtual carrier sensing*. Physical carrier sensing detects the presence of other IEEE 802.11 stations by analyzing all detected packets and also detects activity in the channel via relative signal strength from other sources. Virtual carrier sensing is used by a source station to inform all other stations in the BSS of how long the channel will be utilized for the successful transmission of a MAC protocol data unit (MPDU). The source stations set the *duration field* in the MAC header of data frames or in RTS and CTS control frames. The duration field indicates the amount of time (in microseconds) after the end of the present frame that the channel will be utilized to complete the successful transmission of the data or management frame. Stations detecting a duration field in a transmitted MSDU adjust their **network allocation vector (NAV)**, which indicates the amount of time that must elapse until the current transmission is complete and the channel can be sampled again for idle status. The channel is marked busy if either the physical or virtual carrier-sensing mechanism indicates that the channel is busy.

Figure 6.71 is a timing diagram that illustrates the successful transmission of a data frame. When the data frame is transmitted, the duration field of the frame lets all stations in the BSS know how long the medium will be busy. All stations hearing the data frame adjust their NAV based on the duration field value, which includes the SIFS interval and the acknowledgment frame following the data frame.

Figure 6.72 illustrates the transmission of an MPDU using the RTS/CTS mechanism. Stations can choose to never use RTS/CTS, to use RTS/CTS whenever the MSDU exceeds the value of RTS_Threshold (which is a manageable parameter), or to always use RTS/CTS. If a collision occurs with an RTS or CTS MPDU, far less bandwidth is wasted in comparison to a large data MPDU. However, for a lightly loaded medium the overhead of the RTS/CTS frame transmissions imposes additional delay.

Wireless channels cannot handle very long transmissions due to their relatively large error rates. Large MSDUs handed down from the logical link sublayer (LLC) to the MAC may require fragmentation to increase transmission reliability. To determine whether to perform fragmentation, MDPUs are compared to the manageable parameter, Fragmentation_Threshold. If the MPDU size exceeds the value of Fragmentation_Threshold, then the MSDU is broken into multiple fragments.

The collision avoidance portion of CSMA-CA is performed through a random backoff procedure. If a station with a frame to transmit initially senses the channel to be busy, then the station waits until the channel becomes idle for a DIFS period and then computes a random backoff time. For IEEE 802.11 time is slotted in time periods that correspond to a Slot_Time. The Slot_Time used in IEEE 802.11 is much smaller than an MPDU and is used to define the IFS intervals and to determine the backoff time for stations in the CP. The random backoff time is an integer value that corresponds to a number of time slots. Initially, the station computes a backoff time uniformly in the range 0 to 7.

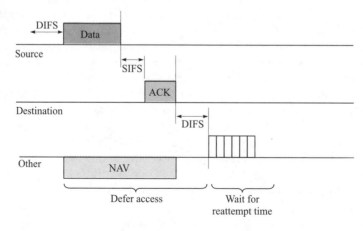

FIGURE 6.71 Transmission of MPDU without RTS/CTS

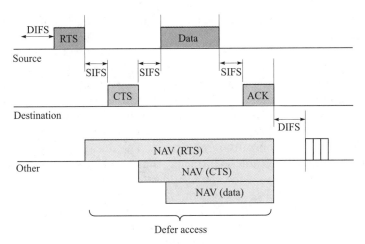

FIGURE 6.72 Transmission of MPDU with RTS/CTS

When the medium becomes idle after a DIFS period, stations decrement their backoff timer until either the medium becomes busy again or the timer reaches zero. If the timer has not reached zero and the medium becomes busy, the station freezes its timer. When the timer is finally decremented to zero, the station transmits its frame. If two or more stations decrement to zero at the same time, then a collision will occur and each station will have to generate a new backoff time in the range 0 to 15. For each retransmission attempt, the number of available backoff slots grows exponentially as 2^{2+i}. The idle period after a DIFS period is referred to as the contention window (CW).

The operation of the DCF includes mechanisms for dealing with lost or errored frames. Receiving stations are required to transmit an ACK frame if the CRC of the frame they receive is correct. The sending station expects an ACK frame and interprets the failure to receive such a frame as an indication of loss of the frame. Note, however, that the lack of an ACK frame may also be due to loss of the ACK frame itself, not the original data frame. The sending station maintains an ACK_Timeout, equal to an ACK frame time plus a SIFS, for each data frame. If the ACK is not received, the station executes the backoff procedure to schedule a reattempt time. Receiver stations use the sequence numbers in the frame to detect duplicate frames. 802.11 does not provide MAC-level recovery for the broadcast of multicast frames except when these frames are sent with the To_DS bit set.

Point coordination function

The PCF is an optional capability that can be used to provide connection-oriented, contention-free services by enabling polled stations to transmit without contending for the channel. The PCF function is performed by the *point coordinator* (PC) in the AP within a BSS. Stations within the BSS that are capable of operating in the CFP are known as *CF-aware stations*. The method by which polling tables are maintained and the polling sequence is determined by the PC is left to the implementor.

The PCF is required to coexist with the DCF and logically sits on top of the DCF (see Figure 6.68). The *CFP repetition interval* (CFP_Rate) determines the frequency with which the PCF occurs. Within a repetition interval, a portion of the time is allotted to contention-free traffic, and the remainder is provided for contention-based traffic. The CFP repetition interval is initiated by a *beacon frame*, where the beacon frame is transmitted by the AP. One of the AP's primary functions is synchronization and timing. The duration of the CFP repetition interval is a manageable parameter that is always an integer-multiple number of beacon frames. Once the CFP_Rate is established, the duration of the CFP is determined. The maximum size of the CFP is determined by the manageable parameter, CFP_Max_Duration. At a minimum, time must be allotted for at least one MPDU to be transmitted during the CP. It is up to the AP to determine how long to operate the CFP during any given repetition interval. If traffic is very light, the AP may shorten the CFP and provide the remainder of the repetition interval for the DCF. The CFP may also be shortened if DCF traffic from the previous repetition interval carries over into the current interval. The maximum amount of delay that can be incurred is the time it takes to transmit an RTS/CTS handshake, maximum MPDU, and an acknowledgment. Figure 6.73 is a sketch of the CFP repetition interval, illustrating the coexistence of the PCF and DCF.

At the nominal beginning of each CFP repetition interval, the so-called target beacon transmission time (TBTT), all stations in the BSS update their NAV to the maximum length of the CFP (i.e., CFP_Max_Duration). During the CFP, stations may transmit only to respond to a poll from the PC or to transmit

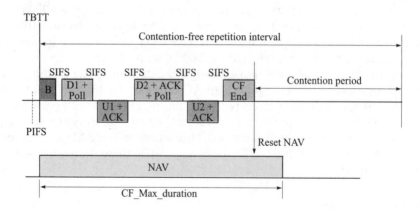

D1, D2 = frames sent by point coordinator
U1, U2 = frames sent by polled station
TBTT = target beacon transmission time
B = beacon frame

FIGURE 6.73 Point coordination frame transfer

an acknowledgment one SIFS interval after receipt of an MPDU. At the nominal start of the CFP, the PC senses the medium. If the medium remains idle for a PIFS interval, the PC transmits a beacon frame to initiate the CFP. In case the CFP is lightly loaded, the PC can foreshorten the CFP and provide the remaining bandwidth to contention-based traffic by issuing a CF-End or CF-End + Ack control frame. This action causes all stations that receive the frame in the BSS to reset their NAV values.

RTS/CTS frames are not used by the point coordinator or by CF-aware stations during the CFP. After the PC issues a poll, the intended CF-aware station may transmit one frame to *any* station as well as piggyback an ACK of a frame received from the PC by using the appropriate subtypes of a Data type frame. When a frame is transmitted to a non-CF-aware station, the station sends its ACK using DCF rules. The PC keeps control of the medium by only waiting the PIFS duration before proceeding with its contention-free transmissions.

◆ PHYSICAL LAYERS

The IEEE 802.11 LAN has several physical layers defined to operate with its MAC layer. Each physical layer is divided into two sublayers that correspond to two protocol functions as shown in Figure 6.74. The **physical layer convergence procedure (PLCP)** is the upper sublayer, and it provides a convergence function that maps the MAC PDU into a format suitable for transmission and reception over a given physical medium. The **physical medium dependent (PMD)** sublayer is concerned with the characteristics and methods for transmitting over the wireless medium.

Figure 6.74 shows that the MAC PDU is mapped into a PLCP frame that consists of three parts. The first part is a preamble that provides synchronization and start-of-frame information. The second part is a PLCP header that provides transmission bit rate and other initialization information as well as frame-length information and a CRC. The third part consists of the MAC PDU possibly

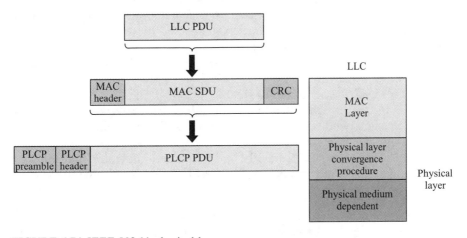

FIGURE 6.74 IEEE 802.11 physical layer

modified (scrambled) to meet requirements of the transmission system. The specific structure of each PLCP depends on the particular physical layer definition. We next discuss three physical layers that have been defined for IEEE 802.11.

Frequency-hopping spread spectrum for the 2.4 GHz ISM band

Spread spectrum transmission is a form of digital modulation technique that takes a data signal of certain bit rate and modulates it onto a transmitted signal of much larger bandwidth. In essence, spread spectrum systematically spreads the energy of the data signal over a wide frequency band.[13] The spread spectrum receiver uses its knowledge of how the spreading was done to compress the received signal and recover the original data signal. Spread spectrum provides great robustness with respect to interference as well as other transmission impairments such as fading that results from multipath propagation. Frequency hopping is one type of spread spectrum technique.

Frequency hopping involves taking the data signal and modulating it so that the modulated signal occupies different frequency bands as the transmission progresses. It is analogous to transmitting a song over a large number of FM radio channels. To recover the signal, the receiver must know the sequence of channels that it should tune to as well as the "dwell" time in each channel. The obvious question is, Why not give each user its own dedicated channel? One reason is that multipath fading affects narrow frequency bands so some of the channels can provide very poor transmission. Frequency hopping minimizes the time spent on each channel.

The 802.11 frequency-hopping physical layer standard uses 79 nonoverlapping 1 MHz channels to transmit a 1 Mbps data signal over the 2.4 GHz ISM band. An option provides for transmission at a rate of 2 Mbps. This band occupies the range 2400 to 2483.5 MHz, providing 83.5 MHz of bandwidth. A channel hop occurs every 224 microseconds. The standard defines 78 hopping patterns that are divided into three sets of 26 patterns each. Each hopping pattern jumps a minimum of six channels in each hop, and the hopping sequences are derived via a simple modulo 79 calculation. The hopping patterns from each set collide three times on the average and five times in the worst case over a hopping cycle. Each 802.11 network must use a particular hopping pattern. The hopping patterns allow up to 26 networks to be colocated and still operate simultaneously.

Figure 6.75 shows the format of the PLCP frame. The PLCP preamble starts with 80 bits of 0101 synchronization pattern that the receiver uses to detect the presence of a signal and to acquire symbol timing. The preamble ends with a 16-bit start frame delimiter that consists of the pattern 0000 1100 1011 1101. The PLCP header consists of a 12-bit PLCP_PDU length indicator that allows for PLCP total lengths of up to 4095 bytes. The PLCP header also contains a four-

---

[13]The origins of spread spectrum transmission are in military communications. The fixed energy in a spread spectrum signal is spread over such a big band that an enemy eavesdropper cannot even detect the presence of the signal.

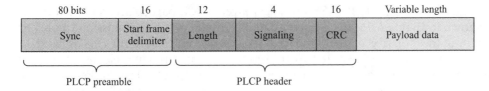

80 bits	16	12	4	16	Variable length
Sync	Start frame delimiter	Length	Signaling	CRC	Payload data

PLCP preamble   PLCP header

FIGURE 6.75 Frequency-hopping spread spectrum PLCP frame format

bit field in which the first three bits are reserved and the last bit indicates operation at 1 Mbps or 2 Mbps. The last 16 bits of the PLCP header are a 16-bit CRC using the CCITT-16 generator polynomial that covers the preceding 16 bits in the header. The PLCP header is always transmitted at the base rate of 1 Mbps. The PLCP_PDU is formed by scrambling the binary sequence of the MAC PDU and converting it into the sequence of symbols that are suitable for the frequency shift keying modulation scheme that is used in the frequency hopping.

When we discussed the 802.11 MAC, we found that the operation depended on the values of certain key time parameters. In Table 6.5 we show the default values of some of the key parameters for the frequency-hopping physical layer.

Direct sequence spread spectrum for the 2.4 GHz ISM band

Direct sequence spread spectrum (DSSS) is another method for taking a data signal of a given bit rate and modulating it into a signal that occupies a much larger bandwidth. DSSS represents each data 0 and 1 by the symbols -1 and $+1$ and then multiplies each symbol by a binary pattern of $+1$s and -1s to obtain a digital signal that varies more rapidly and hence occupies a larger frequency

Parameter	Value μsec	Definition
Air propagation time	1	Time for transmitted signal to go from transmitter to receiver.
RxTx turnaround time	20	Time for a station to transmit a symbol after request from MAC.
CCA assessment time	29	Time for the receiver to determine the state of the channel.
Slot time	50	Time used by MAC to determine PIFS and DIFS periods = CCA assessment + RxTx turnaround + air propagation.
SIFS time	28 $+2/-3$	Time required by MAC and physical sublayers to receive the last symbol of a frame at the air interface, process the frame, and respond with the first symbol of a preamble on the air interface.
Preamble length	96	Time to transmit the PLCP preamble.
PLCP header	32	Time required to transmit the PLCP header.

TABLE 6.5 Default time parameters in IEEE 802.11 frequency-hopping spread spectrum physical layer

band. The IEEE 802.11 DSSS physical layer uses a particularly simple form as shown in Figure 6.76: Each binary data bit results in the transmission of plus or minus the polarity of the 11-chip Barker sequence. The term *chip* is used to distinguish the time required to transmit a $+1$ or -1 signal element from the time required to transmit a data bit ($=11$ chip times). The Barker sequence provides good immunity against interference and noise as well as some protection against multipath propagation.

The DSSS transmission system in 802.11 takes the 1 Mbps data signal and converts it into an 11 Mbps signal using differential binary phase shift keying (DBPSK) modulation. Eleven channels have been defined to operate in the 2.4 GHz ISM band in the United States. Nine channels have been defined to operate in the 2.4 GHz band in Europe. Channels can operate without interfering with each other if their center frequencies are separated by at least 30 MHz. The 802.11 DSSS physical layer also defines an option for 2 Mbps operation using differential quaternary PSK (DQPSK).

Figure 6.77 shows the format of the PLCP frame. The PLCP preamble starts with 128 scrambled bits of synchronization that the receiver uses to detect the presence of a signal. The preamble ends with a 16-bit start frame delimiter (hF3A0) that is used for bit synchronization. The PLCP header consists of an 8-bit signal field that indicates to the physical layer the modulation that is to be used for transmission and reception of the MPDU (h0A for 1 Mbps DBPSK, h14 for 2 Mbps DQPSK); an 8-bit service field that is reserved for future use; a 16-bit field that indicates the number of bytes in the MPDU, from 4 to 2^{16}; and a 16-bit CRC using the CCITT-16 generator polynomial. The PLCP header is always transmitted at the base rate of 1 Mbps.

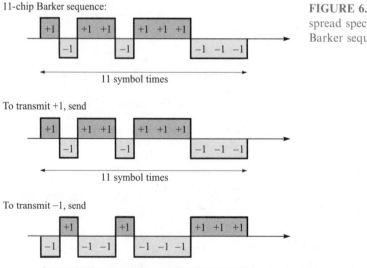

FIGURE 6.76 Direct sequence spread spectrum using 11-chip Barker sequence

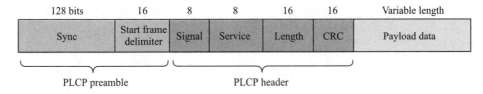

128 bits	16	8	8	16	16	Variable length
Sync	Start frame delimiter	Signal	Service	Length	CRC	Payload data

PLCP preamble   PLCP header

FIGURE 6.77 Direct sequence spread spectrum PLCP frame format

The 802.11 MAC operation depends on the values of certain key time parameters. In Table 6.6 we show the default values of some of the key parameters for the DSSS physical layer.

Infrared physical layer

The IEEE 802.11 infrared physical layer operates in the near-visible light range of 850 to 950 nanometers. Diffuse transmission is used so the transmitter and receivers do not have to point to each other. The transmission distance is limited to the range 10 to 20 meters, and the signal is contained by walls and windows. This feature has the advantage of isolating the transmission systems in different rooms. The system cannot operate outdoors.

The transmission system uses pulse-position modulation (PPM) in which the binary data is mapped into symbols that consist of a group of slots. The 1 Mbps data system uses a 16 L-PPM slot that uses "symbols" that consist of 16 time slots and in which only 1 slot can contain a pulse. A slot is 250 nanoseconds in duration. The modulation takes four data bits to determine an integer in the range 1 to 16, which determines the corresponding symbol. The 2 Mbps data system uses a 4 L-PPM in which groups of two data bits are mapped into symbols that consist of four slots.

Parameter	Value μsec	Definition
RxTx turnaround time	< 5	Time it takes a station to transmit a symbol after request from MAC.
CCA assessment time	< 15	Time it takes the receiver to determine the state of the channel.
Slot time	20	Time used by MAC to determine PIFS and DIFS periods = CCA assessment + RxTx turnaround + air propagation.
SIFS time	10	Time required by MAC and physical sublayers to receive the last symbol of a frame at the air interface, process the frame, and respond with the first symbol of a preamble on the air interface.
Preamble length	144 bits	Time to transmit the PLCP preamble.
PLCP header	48 bits	Time required to transmit the PLCP header.

TABLE 6.6 Default time parameters in IEEE 802.11 direct sequence spread spectrum physical layer

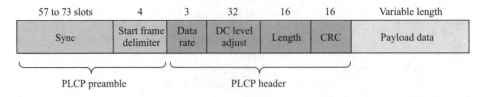

FIGURE 6.78 Infrared PLCP frame format

Figure 6.78 shows the format of the PLCP frame. The PLCP preamble starts with a minimum of 57 and a maximum of 73 L-PPM slots of alternating presence and absence of pulse in consecutive slots, which must terminate in the absence of pulse. The receiver uses this sequence to perform slot synchronization and other optional initialization procedures. The preamble ends with a 4 L-PPM slot start frame delimiter (1001) to indicate the start of frame and to perform bit and symbol synchronization. The PLCP header consists of a 3 L-PPM slot data rate field to indicate the data rate (000 for 1 Mbps; 001 for 2 Mbps); a 32 L-PPM slot sequence of pulses that stabilizes the DC level of the received signal; a 16-bit integer (modulated using L-PPM) that indicates the length of the PSDU; and a 16-bit CRC calculated over the length field using the CCITT-16 generator polynomial and modulated using L-PPM. The PSDU field consists of 0 to 2500 octets modulated using the L-PPM format. The PLCP length, CRC, and PSDU fields can be transmitted at either 1 Mbps or 2 Mbps. The fields prior to these are defined in terms of slots, not symbols.

The 802.11 MAC operation depends on the values of certain key time parameters. In Table 6.7 we show the default values of some of the key parameters for the infrared physical layer.

Parameter	Value μsec	Definition
RxTx turnaround time	0	Time it takes a station to transmit a symbol after request from MAC.
CCA assessment time	5	Time it takes the receiver to determine the state of the channel.
Slot time	6	Time used by MAC to determine PIFS and DIFS periods = CCA assessment + RxTx turnaround + air propagation
SIFS time	7	Time required by MAC and physical sublayers to receive the last symbol of a frame at the air interface, process the frame, and respond with the first symbol of a preamble on the air interface.

TABLE 6.7 Default time parameters in IEEE 802.11 infrared physical layer

6.7   LAN BRIDGES

There are several ways of interconnecting networks. When two or more networks are interconnected at the physical layer, the type of device is called a **repeater**. When two or more networks are interconnected at the MAC or data link layer, the type of device is called a **bridge**. When two or more networks are interconnected at the network layer, the type of device is called a **router**. Interconnection at higher layers is done less frequently. The device that interconnects networks at a higher level is usually called a **gateway**. Gateways usually perform some protocol conversion and security functions. In this section we will focus on bridges.

When range extension is the only problem, repeaters may solve the problem as long as the maximum distance between two stations is not exceeded. Local area networks (LANs) that involve sharing of media, such as Ethernet and token ring, can only handle up to some maximum level of traffic. As the number of stations in the LAN increases, or as the traffic generated per station increases, the amount of activity in the medium increases until it reaches a saturation point. As we saw earlier in this chapter, the point at which saturation occurs depends on the particular MAC algorithm as well as the ratio a of delay-bandwidth product to frame size. Figure 6.51 shows a typical performance curve for a LAN system. As a LAN reaches saturation, it becomes necessary to somehow extend the LAN. A typical approach is to segment the user group into two or more LANs and to use bridges to interconnect the LANs to form a **bridged LAN** or an **extended LAN**.

Another typical scenario involves large organizations in which LANs are initially introduced by different departments to meet their particular needs. Eventually, the need arises to interconnect these departmental LANs to enable the exchange of information and the sharing of certain resources. This scenario is frequently complicated by the following factors:

1. The departmental LANs use different network layer protocols that are packaged with the applications that they require.
2. The LANs may be located in different buildings.
3. The LANs differ in type.

These three requirements can be met by bridges. Because bridges exchange frames at the data link layer, the frames can contain any type of network layer PDUs. If necessary, bridges can be connected by point-to-point links. However, we have seen that the MAC PDUs do differ in structure in operation and in the size of the frames they allow, and so at the very least, the third requirement involves some form of frame conversion process. By extending the LAN, bridges provide the plug-and-lay convenience of operating a single LAN. However, bridges need to deal with security and broadcast storm concerns.

LANs originally assume an element of trust between the users in the LAN. As a LAN grows, this assumption breaks down, and security concerns become prominent. The fact that most LANs are broadcast in nature implies that eavesdropping can be done easily by operating the NIC in *promiscuous mode* where every frame in the LAN is captured and examined. This behavior opens the door for the various security threats that are discussed in Chapter 11. Bridges can contribute to this problem by extending the reach of users. However, bridges can also help deal with security problems because of their ability to *filter* frames. By examining the contents of frame and packet headers, bridges can control the flow of traffic allowed in and out of a given LAN segment.

LANs inherently involve the broadcasting of frames in the shared medium. A problem in all LANs is that a faulty station may become stuck sending broadcast traffic. This process can bring down the entire LANs. The problem becomes more severe with the introduction of bridges that can potentially distribute these broadcast frames over all segments of the LAN.

Consider two LANs connected by a bridge, as shown in Figure 6.79. When station 1 transmits a frame to station 3 (local traffic), the frame is broadcast only on LAN1. The bridge can prevent the signal from propagating to LAN2 because it knows that station 3 is connected to the same LAN as station 1. If station 1 transmits a frame to a remote station, say, station 5, then both LAN1 and LAN2 will be busy during the frame transmission. Thus if most traffic is local, the load on each LAN will be reduced. In contrast, if the bridge is replaced with a repeater, both LANs will be busy when a station is transmitting a frame, independent of where the destination station is.

To have a frame filtering capability, a bridge has to monitor the MAC address of each frame. For this reason, a bridge cannot work with physical layers. On the other hand, a bridge does not perform a routing function, which is why a bridge is a layer 2 relay. Because bridges are mostly used for extending LANs of the same type, they usually operate at the MAC layer as shown in Figure 6.80.

Two types of bridges are widely used: **transparent bridges** and **source routing bridges**. Transparent bridges are typically used in Ethernet LANs, whereas source routing bridges are typically used in token-ring and FDDI networks.

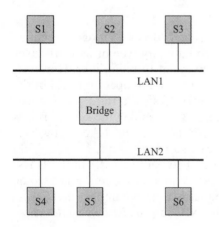

FIGURE 6.79 A bridged LAN

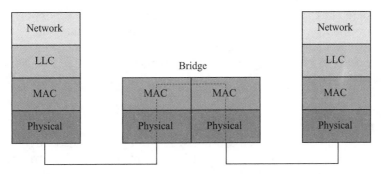

FIGURE 6.80 Interconnection by a bridge

6.7.1   Transparent Bridges

Transparent bridges were defined by the IEEE 802.1d committee. The term *transparent* refers to the fact that stations are completely unaware of the presence of bridges in the network. Thus introducing a bridge does not require the stations to be configured. A transparent bridge performs the following three basic functions:

1. Forwards frames from one LAN to another.
2. Learns where stations are attached to the LAN.
3. Prevents loops in the topology.

A transparent bridge is configured in a promiscuous mode so that each frame, independent of its destination address, can be received by its MAC layer for further processing.

BRIDGE LEARNING

When a frame arrives on one of its ports, the bridge has to decide whether or not to forward the incoming frame to another port based on the destination address of the frame. To do so, the bridge needs a table to indicate which side of the port the destination station is attached to, whether indirectly or directly. The table is called a *forwarding table*, or *forwarding database*, and associates each station address with a port number, as shown in Table 6.8. In practice, the table can have a few thousand entries to handle an interconnection of multiple LANs.

The question then is how the entries in the forwarding table are filled. One way is to have the network administrator record these entries and load them up

MAC address	Port

TABLE 6.8 Forwarding table (with no data)

during system startup. Although this approach is theoretically possible, it is not desirable in practice, as it requires the system administrator to change the entry manually when a station is moved from one LAN to another and when a new station is added or removed. It turns out that there is a simple and elegant way for a bridge to "learn" the location of the stations as it operates and to build the forwarding table automatically.

We first look at the basic learning process that is used by the bridge. Further improvements are needed to handle the dynamics of the network. The basic process works as follows. When a bridge receives a frame, the bridge first compares the source address of the frame with each entry in the forwarding table. If the bridge does not find a match, it adds to the forwarding table the *source address* together with the port number on which the frame was received. The bridge then compares the *destination address* of the frame with each entry in the forwarding table. If the bridge finds a match, it forwards the frame to the port indicated in the entry; however, if the port is the one on which the frame was received, no forwarding is required and the frame is discarded. If the bridge does not find a match, the bridge "floods" the frame on all ports except the one on which the frame was received.

To see how bridges use this procedure to learn station locations, consider an example of a bridged LAN comprising three LANs, as shown in Figure 6.81. Assume that the forwarding tables are initially empty.

Suppose now S1 (station) sends a frame to S5. The frame carries the MAC address of S5 as the destination address and the MAC address of S1 as the source address. When B1 (bridge 1) receives the frame, it finds the table empty and adds S1's source address and the port number on which the frame arrived (which is port 1). The destination address is also not found in the table, and so the frame is forwarded to port 2 and transmitted on LAN2. When B2 receives the frame, it performs the same process, adding the source address and forwarding the frame to LAN3. S5 eventually receives the frame destined to it. Figure 6.82 shows the current state of both forwarding tables. Both bridges have learned the location of S1.

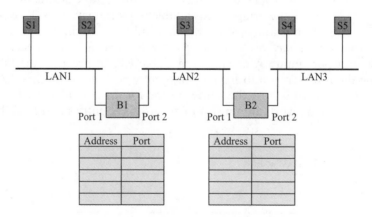

FIGURE 6.81 Initial configuration

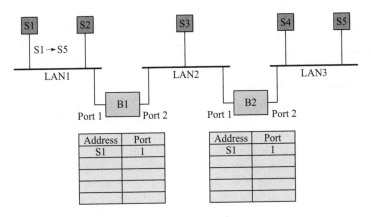

FIGURE 6.82 S1 sends a frame to S5

Next S3 sends a frame to S2. Both B1 and B2 receive the frame, since they are connected to the same LAN as S3. B1 cannot find the address of S3 in its table, so it adds (S3, 2) to its table. It then forwards the frame through port 1, which S2 finally receives. B2 also does not find the source address, adding the new information in its table and forwarding the frame on LAN 3. The traffic on LAN 3 is wasted, since the destination is located on the opposite side. But at this point the bridges are still in the learning process and need to accumulate more information to make intelligent decisions. At the end of this process, the forwarding tables gain one more entry (see Figure 6.83).

Now assume that S4 sends a frame to S3. First B2 records the address of S4 and the port number on which the frame arrived, since the address of S4 is not found. Then B2 checks the destination address of the frame in the forwarding table. The destination address of the frame matches one of the entries, so the bridge forwards the frame to the port indicated in the entry (which is port 1). When B1 receives the frame, it adds the source address and the port number on

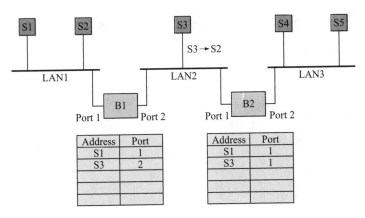

FIGURE 6.83 S3 sends a frame to S2

which the frame arrived into the forwarding table. The bridge, however, finds the destination address. Because the port number in the entry is the same as that on which the frame arrived, the frame is discarded and not transmitted to LAN1. Thus the traffic is confined to LAN2 and LAN3 only. Figure 6.84 shows the forwarding tables after this point.

Now assume that S2 sends a frame to S1. B1 first adds the address of S2 in its forwarding table. Since the bridge has learned the address of S1, it discards the frame after finding out that S1 is also connected to the same port. We see that the traffic now is completely isolated in LAN1. At the same time, transmission can occur on different LANs without interfering with each other, thus increasing the *aggregate* throughput. Note also that because the frame is not transmitted to LAN2, B2 cannot learn the address of S2, as indicated by its forwarding table in Figure 6.85.

It is now obvious that if the learning process continues indefinitely, both tables eventually store the address of each station in the bridged LAN. At this point the bridges stop learning. Unfortunately, nothing stays static in real life. For example, stations may be added to a LAN or moved to another LAN. To have a bridge that can adapt to the dynamics of the network, we need two additional minor changes. First, the bridge adds a timer associated with each entry. When the bridge adds an address to its table, the timer is set to some value (typically on the order of a few minutes). The timer is decremented periodically. When the value reaches zero, the entry is erased so that a station that has been removed from the LAN will eventually have its address removed from the table as well. When the bridge receives a frame and finds that the source address of the frame matches with the one in the table, the corresponding entry is "refreshed" so that the address of an active station will be retained in the table. Second, the bridge could update address changes quickly by performing the following simple task. When the bridge receives a frame and finds a match in the source address but the port number in the entry is different from the port number on which the

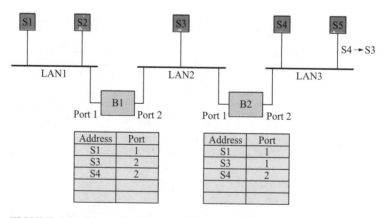

FIGURE 6.84 S4 sends a frame to S3

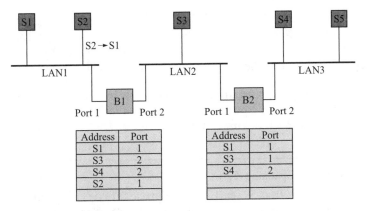

FIGURE 6.85 S2 sends a frame to S1

frame arrived, the bridge updates the entry with the new port number. Thus a station that has moved to another LAN will be updated as soon as it transmits.

SPANNING TREE ALGORITHM

The learning process just described works as long as the network does not contain any loops, meaning that there is only one path between any two LANs. In practice, however, loops may be created accidentally or intentionally to increase redundancy. Unfortunately, loops can be disastrous during the learning process, as each frame from the flooding triggers the next flood of frames, eventually causing a *broadcast storm* and bringing down the entire network.

To remove loops in a network, the IEEE 802.1 committee specified an algorithm called the *spanning tree algorithm*. If we represent a network with a graph, a spanning tree maintains the connectivity of the graph by including each node in the graph but removing all possible loops. This is done by automatically disabling certain bridges. It is important to understand that these bridges are not physically removed, since a topology change may require a different set of bridges to be disabled, thus reconfiguring the spanning tree dynamically.

The spanning tree algorithm requires that each bridge have a unique bridge ID, each port within a bridge have a unique port ID, and all bridges on a LAN recognize a unique MAC group address. Together, bridges participating in the spanning tree algorithm carry out the following procedure:

1. Select a *root bridge* among all the bridges in the bridged LAN. The root bridge is the bridge with the lowest bridge ID.
2. Determine the *root port* for each bridge except the root bridge in the bridged LAN. The root port is the port with the least-cost path to the root bridge. In case of ties the root port is the one with lowest port ID. Cost is assigned to each LAN according to some criteria. One criterion could be to assign higher costs to lower speed LANs. A path cost is the sum of the costs along the path from one bridge to another.

3. Select a *designated bridge* for each LAN. The designated bridge is the bridge that offers the least-cost path from the LAN to the root bridge. In case of ties the designated bridge is the one with the lowest bridge ID. The port that connects the LAN and the designated bridge is called a *designated port*.

Finally, all root ports and all designated ports are placed into a "forwarding" state. These are the only ports that are allowed to forward frames. The other ports are placed into a "blocking" state.

Example—Creating a Spanning Tree

To see how a spanning tree is created by the procedure described above, consider the topology of a bridged LAN shown in Figure 6.86. For simplicity the bridges are identified by B1, B2, ..., B5, and each port ID is indicated by the number in parentheses. Costs assigned to each LAN are assumed to be equal.

The resulting spanning tree configuration is shown in Figure 6.87. First bridge 1 is selected as the root bridge, since it has the lowest bridge ID. Next the root port is selected for each bridge except for B1, as indicated by the letter *R*. Then the designated bridge is selected for each LAN, as indicated by the letter *D* on the corresponding designated port. Finally, the root ports and the designated ports are put into a forwarding state, as indicated by the solid lines. The broken lines represent the ports that are in a blocking state. You should verify that the final topology contains no loops.

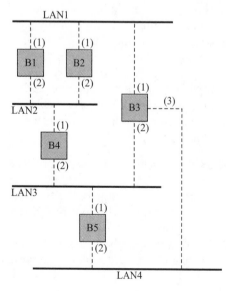

FIGURE 6.86 Sample topology

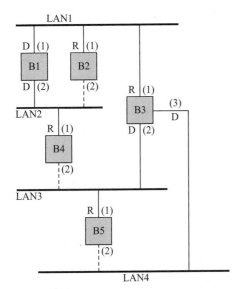

FIGURE 6.87 The corresponding spanning topology

The procedure to discover a spanning tree can be implemented by using a distributed algorithm. To do so, each bridge exchanges special messages called *configuration bridge protocol data units* (configuration BPDUs). A configuration BPDU contains the bridge ID of the transmitting bridge, the root bridge ID, and the cost of the least-cost path from the transmitting bridge to the root bridge. Each bridge records the best configuration BPDU it has so far. A configuration BPDU is "best" if it has the lowest root bridge ID. If there is a tie, the configuration BPDU is best if it has the least-cost path to the root bridge. If there is still a tie, the configuration BPDU is best if it has the lowest bridge ID of the transmitting bridge.

Initially, each bridge assumes that it is the root bridge and transmits configuration BPDUs periodically on each of its ports. When a bridge receives a configuration BPDU from a port, the bridge adds the path cost to the cost of the LAN that this BPDU was received from. The bridge then compares the configuration BPDU with the one recorded. If the bridge receives a better configuration BPDU, it stops transmitting on that port and saves the new configuration BPDU. Eventually, only one bridge on each LAN (the designated bridge) will be transmitting configuration BPDUs on that LAN, and the algorithm stabilizes.

To detect a bridge failure after the spanning tree is discovered, each bridge maintains an aging timer for the saved configuration BPDU, which is incremented periodically. The timer is reset when the bridge receives a configuration BPDU. If a designated bridge fails, one or more bridges will not receive the configuration BPDU. When the timer expires, the bridge starts the algorithm again by assuming that it is the root bridge, and the distributed algorithm should eventually configure a new spanning tree.

6.7.2 Source Routing Bridges

Source routing bridges were developed by the IEEE 802.5 committee. They are primarily used to interconnect token-ring networks. Unlike transparent bridges that place the implementation complexity in bridges, source routing bridges put the burden more on the end stations. The main idea of source routing is that each station should determine the route to the destination when it wants to send a frame and therefore include the route information in the header of the frame. Thus the problem boils down to finding good routes efficiently.

A source routing frame introduces additional routing information in the frame, as shown in Figure 6.88. The routing information field is inserted only if the two communicating stations are on different LANs. The presence of the routing information field is indicated by the individual/group address (I/G) bit in the source address field. If the routing information field is present, the I/G bit is set to 1[14]. If a frame is sent to a station on the same LAN, that bit is 0. The routing control field defines the type of frame, the length of the routing informa-tion field, the direction of the route given by the route designator fields (from left to right or right to left), and the largest frame supported over the path. The route designator field contains a 12-bit LAN number and a 4-bit bridge number.

As an example, if S1 (station 1) wants to send a frame to S2 (Figure 6.89), then a possible route is LAN1→B1→LAN2→B4→LAN4. The bridge number in the final route designator field is not used. Many more routes are available for this source-destination pair, making it possible to share the load among several routes or to choose an alternative route if one should fail. In general, when a station wants to transmit a frame to another station on a different LAN, the station consults its routing table. If the route to the destination is found, the station simply inserts the routing information into the frame. Otherwise, the station performs a route discovery procedure. Once the route is found, the station adds the route information to its routing table for future use.

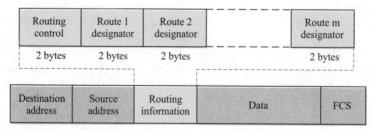

FIGURE 6.88 Frame format for source routing

---

[14]The I/G bit was originally defined to indicate a multicast source address. But because the bit has never been used for that purpose, the IEEE 802.5 committee decided to use it for indicating the presence of a routing information field.

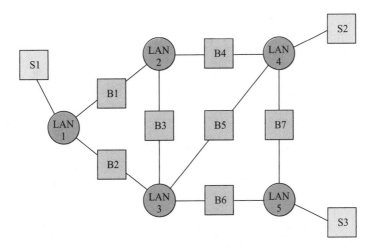

FIGURE 6.89 LAN interconnection with source routing bridges

The basic idea for a station to discover a route is as follows. First the station broadcasts a special frame, called the *single-route broadcast* frame. The frame visits every LAN in the bridged LAN exactly once, eventually reaching the destination station. Upon receipt of this frame, the destination station responds with another special frame, called the *all-routes broadcast* frame, which generates all possible routes back to the source station. After collecting all routes, the source station chooses the best route and saves it.

The detailed route discovery procedure is as follows. First the source station transmits the single-route broadcast frame on its LAN without any route designator fields. To ensure that this frame appears on each LAN exactly once, selected bridges are configured to form a spanning tree, which can be done manually or automatically. When a selected bridge at the first hop receives a single-route broadcast frame, that bridge inserts an incoming LAN number, its bridge number, and the outgoing LAN number to the routing information field and forwards the frame to the outgoing LAN. When a selected bridge at other hops receives a single-route broadcast frame, the bridge inserts its bridge number and the outgoing LAN number to the routing information field and then forwards the frame to the outgoing LAN. Nonselected bridges simply ignore the single-route broadcast frame. Because the spanning tree maintains the full connectivity of the original topology, one frame should eventually reach the destination station.

Upon receipt of a single-route broadcast frame, the destination station responds with an all-routes broadcast frame containing no route designator fields. When a bridge at the first hop receives a single-route broadcast frame, the bridge inserts an incoming LAN number, its bridge number, and the outgoing LAN number to the routing information field and forwards the frame to the outgoing LAN. Other bridges insert their bridge number and the outgoing LAN number to the routing information field and forward the frame to the outgoing LAN. To prevent all-routes broadcast frames from circulating in the network, a bridge first checks whether the outgoing LAN number is already recorded in the route designator field. The bridge will not forward the frame if the outgoing LAN number is already recorded. The all-routes broadcast received by the source station eventually should list all possible routes to the destination station.

Example—Determining Routes for Broadcast Frames

Consider a bridged LAN as shown in Figure 6.89. Assume that B1, B3, B4, and B6 are part of the spanning tree. Suppose S1 wants to send a frame to S3 but has not learned the route yet. Sketch the routes followed by the single-route and all-routes broadcast frames during route discovery.

Figure 6.90 and Figure 6.91 show the routes followed by single-route broadcast frames and all-routes broadcast frames, respectively.

Each possible route from S1 to S3 starts from LAN1 and goes back to LAN5. There are a total of seven possible routes.

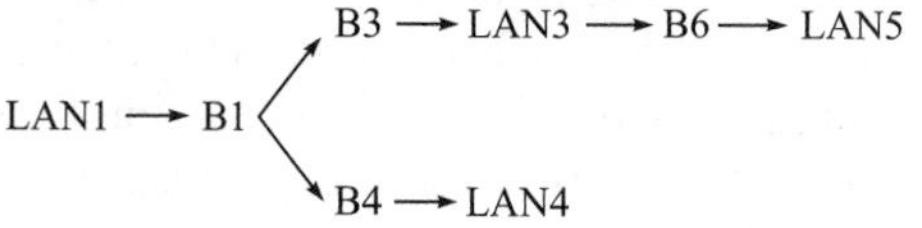

B3 → LAN3 → B6 → LAN5
LAN1 → B1
B4 → LAN4

FIGURE 6.90 Routes followed by single-route broadcast frames

B2 → LAN1 → B1 → LAN2
B5
B4 → LAN4
B5
B7
B6 → LAN3
B3 → LAN2
B1 → LAN1 → B2
B4 → LAN4
B5
B7
B5 → LAN4
B4 → LAN2
B1 → LAN1 → B2
B3
B7
LAN5
B4 → LAN2
B1 → LAN1 → B2 → LAN3
B3
B5
B6
B3 → LAN3
B2 → LAN1 → B1
B5
B6
B7 → LAN4
B5 → LAN3
B2 → LAN1 → B1 → LAN2
B3
B4
B3 → LAN2
B1 → LAN1 → B2
B4
B6

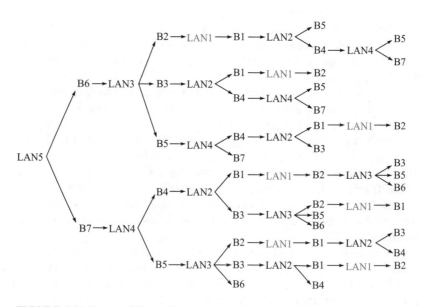

FIGURE 6.91 Routes followed by all-routes broadcast frames

6.7.3   Mixed-Media Bridges

Bridges that interconnect LANs of different type are referred to as **mixed-media bridges**. This type of interconnection is not simple. We discuss mixed-media bridges in terms of the interconnection of Ethernet and token-ring LANs. These two LANs differ in their frame structure, their operation, and their speeds, and the bridge needs to take these differences into account.

Both Ethernet and token-ring LANs use six-byte MAC addresses but they differ in the hardware representation of these addresses. Token rings consider the first bit in a stream to be the high-order bit in a byte. Ethernet considers such a bit to be the low-order bit. A bridge between these two must convert between the two representations.

Ethernet and token-ring LANs differ in terms of the maximum size frames that are allowed. Ethernet has an upper limit of approximately 1500 bytes, whereas token ring has no explicit limit. Bridges do not typically include the ability to do frame fragmentation and reassembly, so frames that exceed the maximum limit are just dropped.

Token ring has the three status bits, A and C in the frame status field and E in the end delimiter field. Recall from Figure 6.61 that the A and C bits indicate whether the destination address is recognized and whether the frame is copied. The E bit indicates errors. Ethernet frames do not have corresponding fields to carry such bits. There is no clear solution on how to handle these bits in either direction, from Ethernet to token ring and from token ring to Ethernet. Similar questions arise regarding what to do with the monitor bit, the reservation, and the priority bits that are present in the token-ring header.

Another problem in interconnecting LANs is that they use different transmission rates. The bridge that is interposed between two LANs must have sufficient buffering to be able to absorb frames that may accumulate when a burst of frames arriving from a fast LAN is destined to a slow LAN.

Two approaches to bridging between transparent (Ethernet) bridging domains and source routing (token ring) domains have been proposed. **Translational bridging** carries out the appropriate reordering of address bits between Ethernet and token-ring formats. It also provides approaches for dealing with differences in maximum transfer unit size and for handling status bits. **Source-route transparent bridging** combines source route and transparent operations in a bridge that can forward frames produced from transparent and source route nodes.

SUMMARY

In this chapter we considered the class of broadcast networks. These networks are characterized by the sharing of a transmission medium by many users or stations. They typically arise in situations where a low-cost communications system is required to interconnect a relatively small number of users. They also occur in the access portion of backbone networks.

We introduced static and dynamic approaches to channel sharing. The static approaches lead to channelization techniques, and the dynamic approaches lead to medium access control. We also discussed the effect of delay-bandwidth product on protocol performance. We then introduced the IEEE 802 layered architecture that places LANs in the data link layer and defines a common set of services that the MAC sublayer can provide to the logical link layer, which in turn supports the network layer.

We introduced the random access class of MAC protocols and explained how ALOHA and slotted ALOHA can provide low-delay packet transfer but with a limited maximum throughput. We also introduced carrier sensing and collision detection as techniques for improving delay throughput performance. We explored the dependence of delay and throughput performance on delay-bandwidth product.

We then considered reservation systems that can provide significant improvements in performance. We also introduced polling approaches to medium access control, and we investigated how latency affects system performance. We also discussed ring networks, and we compared the performance of different approaches to handling token reinsertion.

We introduced the FDMA, TDMA, and CDMA approaches to creating channels in shared medium networks. We explained why channelization approaches are suitable for situations in which user traffic is constant, but not appropriate when it is bursty. We also explained how CDMA differs from conventional channelization approaches. We also introduced the AMPS, IS-54, GSM, and IS-95 cellular telephone standards as examples of how the channelization techniques are applied in practice.

We discussed the IEEE 802.3 and the Ethernet LAN standards and their variations, noting in particular the trend toward switched implementations. The IEEE 802.5 token-ring and the FDDI ring standards were introduced, and the various possible ways of handling the token were discussed. We also explained the IEEE 802.11 wireless LAN standard. We introduced system design constraints that are particular to wireless networks. We saw how the requirement that movable and mobile users be accommodated within an extended LAN and the requirements to provide time-bounded and asynchronous services led to a standard that is broader than previous wired LAN standards.

LANs are limited in the number of stations that they can handle, so bridges are required to build extended LANs. We introduced the transparent bridge and source routing bridge approaches to building these extended LANs.

CHECKLIST OF IMPORTANT TERMS

1-Persistent CSMA
access point (AP)
ad hoc network
◆Advanced Mobile Phone System
   (AMPS)
association
authentication
basic service set (BSS)
bridge
bridged LAN
broadcast address
broadcast network
carrier sensing multiple access
   (CSMA)
carrier sensing multiple access with
   collision avoidance (CSMA-CA)
carrier sensing multiple access with
   collision detection (CSMA-CD)
channelization scheme
clear-to-send frame (CTS)
◆code division multiple access
   (CDMA)
collision domain
contention-free period (CFP)
cycle time
DCF interframe space (DIFS)
disassociation
distributed coordination function
   (DCF)
distribution system
efficiency
Ethernet
Ethernet switch
extended LAN
extended service set (ESS)
Fiber Distributed Data Interface
   (FDDI)
frame transfer delay
◆frequency-division duplex (FDD)
◆frequency-division multiple access
   (FDMA)
full duplex
gateway
Gigabit Ethernet

◆Global System for Mobile
   Communications (GSM)
hub
interframe space
◆Interim Standard 55 (IS-55)
◆Interim Standard 95 (IS-95)
infrastructure network
LAN adapter card
load
logical link control (LLC)
long code
medium access control (MAC)
medium access control scheme
minislot time
mixed-media bridge
multicast address
multiple access network
network allocation vector (NAV)
network interface card (NIC)
Non-Persistent CSMA
◆orthogonal sequences
PCF interframe space (PIFS)
◆physical layer convergence
   procedure (PCLP)
◆physical medium dependent (PMD)
point coordination function (PCF)
polling systems
p-Persistent CSMA
portal
privacy
reassociation
repeater
request-to-send (RTS)
ring latency
router
short code
short IFS (SIFS)
◆soft handoff
source routing bridge
source-route transparent bridging
spectrum efficiency
◆spreading factor G
Subnetwork Access Protocol (SNAP)
switched network

target token rotation time (TTRT)
◆time-division duplex (TDD)
◆time-division multiple access
   (TDMA)
throughput
token
token bit
token holding time (THT)

token rotation timer (TRT)
total walk time
translational bridging
transparent bridge
unicast address
vulnerable period
walk time
Walsh-Hadamard matrix

FURTHER READING

Backes, F., "Transparent Bridges for Interconnection of IEEE 802 LANs," *IEEE Network*, Vol. 2, No. 1, January 1988, pp. 5–9.

Bertsekas, D. and R. Gallager, *Data Networks*, Prentice-Hall, Englewood Cliffs, New Jersey, 1992.

Dixon, R. C. and Pitt, D. A., "Addressing, Routing, and Source Routing," *IEEE Network*, Vol. 2, No. 1, January 1988, pp. 25–32.

Garg, V. K. and J. E. Wilkes, *Wireless and Personal Communications Systems*, Prentice-Hall PTR, Upper Saddle River, New Jersey, 1996.

Gibson, J. D., ed., *The Mobile Communications Handbook*, CRC Press, Boca Raton, Florida, 1999.

Goodman, D. J., *Wireless Personal Communications Systems*, Addison-Wesley, Reading, Massachusetts, 1997.

IEEE, *IEEE 802.11 Draft Standard—Wireless LAN*, January 1996.

Perlman, R., *Interconnections: Bridges and Routers*, Addison-Wesley, Reading, Massachusetts, 1992 (the most comprehensive book on bridges and routers).

Rappaport, T. S., *Wireless Communications: Principles and Practice*, Prentice-Hall PTR, Upper Saddle River, New Jersey, 1996.

Schwartz, M., *Telecommunication Networks: Protocols, Modeling, and Analysis*, Addison-Wesley, Reading, Massachusetts, 1987.

Stüber, G. L., *Principles of Mobile Communication*, Kluwer Academic Publishers, Boston, 1996.

Viterbi, A. J., *CDMA: Principles of Spread Spectrum Communication*, Addison-Wesley, Reading, Massachusetts, 1995.

PROBLEMS

1. Why do LANs tend to use broadcast networks? Why not use networks consisting of multiplexers and switches?

2. Explain the typical characteristics of a LAN in terms of network type, bit rate, geographic extent, delay-bandwidth product, addressing, and cost. For each characteristic, can you find a LAN that deviates from the typical? Which of the above characteristics is most basic to a LAN?

3. Compare the two-channel approach (Figure 6.4) with the single-channel approach (Figure 6.5) in terms of the types of MAC protocols they can support.

4. Consider an "open concept" office where 64 carrels are organized in an 8 × 8 square array of 3 m × 3 m space per carrel with a 2 m alley between office rows. Suppose that a conduit runs in the floor below each alley and provides the wiring for a LAN to each carrel.
   a. Estimate the distance from each carrel to a wiring closet at the side of the square office.
   b. Does it matter whether the LAN is token ring or Ethernet? Explain.

5. Suppose that a LAN is to provide each worker in problem 4 with the following capabilities: digital telephone service; H.261 video conferencing; 250 ms retrieval time for a 1 Mbyte file from servers in the wiring closet; 10 e-mails/hour sent and received by each worker (90 percent of e-mails are short, and 10 percent contain a 100-kilobyte attachment).
   a. Estimate the bit rate requirements of the LAN.
   b. Is it worthwhile to assign the users to several LANs and to interconnect these LANs with a bridge?

6. Consider a LAN that connects 64 homes arranged in rows of 8 homes on 20 m × 30 m lots on either side of a 10-meter-wide street. Suppose that an underground conduit on either side of the street connects the homes to a pedestal at the side of this rectangular array.
   a. Estimate the distance from each house to the pedestal.
   b. Estimate the bit rate requirements of the LAN. Assume two telephone lines, three MPEG2 televisions, and intense peak-hour Web browsing, say two Web page retrievals/minute at an average of 20 kilobytes/page.
   c. Can a single LAN meet the service requirements of the 64 homes? Explain.

7. Use HDLC and Ethernet to identify three similarities and three differences between medium access control and data link control protocols. Is HDLC operating as a LAN when it is used in normal response mode and multipoint configuration?

8. An application requires the transfer of network layer packets between clients and servers in the same LAN. Explain how reliable connection-oriented service can be provided over an Ethernet LAN. Sketch a diagram that shows the relationship between the PDUs at the various layers that are involved in the transfer.

9. Suppose that the ALOHA protocol is used to share a 56 kbps satellite channel. Suppose that packets are 1000 bits long. Find the maximum throughput of the system in packets/second.

10. Let G be the total rate at which packets are transmitted in a slotted ALOHA system. What proportion of slots go empty in this system? What proportion of slots go empty when the system is operating at its maximum throughput? Can observations about channel activity be used to determine when stations should transmit?

11. Modify the state transition diagram of Stop-and-Wait ARQ to handle the behavior of a station that implements the ALOHA protocol.

12. Suppose that each station in an ALOHA system transmits its packets using spread spectrum transmission. Assume that the spreading sequences for the different stations have been selected so that they have low cross-correlations. What happens when transmissions occur at the same time? What limits the capacity of this sytem?

13. Consider four stations that are attached to two different bus cables. The stations exchange fixed-size packets of length 1 sec. Time is divided into slots of 1 sec. When a station has a packet to transmit, the station chooses either bus with equal probability and transmits at the beginning of the next slot with probability p. Find the value of p that maximizes the rate at which packets are successfully transmitted.

14. In a LAN, which MAC protocol has a higher efficiency: ALOHA or CSMA-CD?

15. A channel using random access protocols has three stations on a bus with end-to-end propagation delay τ. Station A is located at one end of the bus, and stations B and C are together located at the other end of the bus. Frames arrive at the three stations and are ready to be transmitted at stations A, B, and C at the respective times $t_A = 0$, $t_B = \tau/2$, and $t_C = 3\tau/2$. Frames require transmission times of 4τ. In appropriate figures, with time as the horizontal axis, show the transmission activity of each of the three stations for
    a. ALOHA
    b. Non-Persistent CSMA
    c. Non-Persistent CSMA-CD

16. Estimate the maximum throughput of the CDPD system assuming a packet length of 1096 bytes. Hint: What is a for this system?

17. Can the Digital Sense Multiple Access protocol, which is used by CDPD, also be used on the digitial carrier of GSM? If yes, explain how.

18. M terminals are attached by a dedicated pair of lines to a hub in a star topology. The distance from each terminal to the hub is d meters, the speed of the transmission lines is R bits/second, all packets are of length 12,500 bytes, and the signal propagates on the line at a speed of $2.5\ (10^8)$ meters/second. For the four combinations of the following parameters ($d = 25$ meters or $d = 2500$ meters; $R = 10$ Mbps or $R = 10$ Gbps), compare the maximum network throughput achievable when the hub is implementing slotted ALOHA and CSMA-CD.

19. Consider the star-topology network problem 18 when the token-ring protocol is used for medium access control. Assume single-packet operation, eight-bit latency at each station, $M = 125$ stations. Assume a free token is three bytes long.
    a. Find the effective packet transmission time for the four combinations of d and R.
    b. Assume that each station can transmit up to a maximum of k packets per token. Find the maximum network throughput for the four cases of d and R.

20. A wireless LAN uses polling to provide communications between M workstations and a central base station. The system uses a channel operating at 25 Mbps. Assume that all stations are 100 meters from the base station and that polling messages are 64 bytes long. Assuming that all packets are of constant length of 1250 bytes. Assume that stations indicate that they have no packets to transmit with a 64-byte message.

a. What is the maximum possible arrival rate that can be supported if stations are allowed to transmit an unlimited number of packets/poll?

b. What is the maximum possible arrival rate that can be supported if stations are allowed to transmit N packets/poll?

c. Repeat parts (a) and (b): if the transmission speed is 2.5 Gbps.

21. A token-ring LAN interconnects M stations using a star topology in the following way. All the input and output lines of the token-ring station interfaces are connected to a cabinet where the actual ring is placed. Suppose that the distance from each station to the cabinet is 100 meters and that the ring latency per station is eight bits. Assume that all packets are 1250 bytes and that the ring speed is 25 Mbps.

a. What is the maximum possible arrival rate that can be supported if stations are allowed to transmit an unlimited number of packets/token?

b. What is the maximum possible arrival rate that can be supported if stations are allowed to transmit 1 packet/token using single-packet operation? using multitoken operation?

c. Repeat parts (a) and (b) if the transmission speed is 2.5 Gbps.

22. Suppose that a LAN is to carry voice and packet data traffic. Discuss what provisions if any are required to handle the voice traffic in the reservation, polling, token ring, ALOHA, and CSMA-CD environments. What changes if any are required for the packet data traffic?

23. Calculate the difference in header overhead between a DIX Ethernet frame and an IEEE 802.3 frame with SNAP encapsulation.

24. Suppose that a group of 10 stations is serviced by an Ethernet LAN. How much bandwidth is available to each station if (a) the 10 stations are connected to a 10 Mbps Ethernet hub; (b) the 10 stations are connected to a 100 Mbps Ethernet hub; (c) the 10 stations are connected to a 10 Mbps Ethernet switch.

25. Suppose that an Ethernet LAN is used to meet the requirements of the office in problem 5.

a. Can the requirements of one row of carrels be met by a 10 Mbps Ethernet hub? by a 10 Mbps Ethernet switch?

b. Can the requirements of the office be met by a hierarchical arrangement of Ethernet switches as shown in Figure 6.57?

26. Suppose that 80 percent of the traffic generated in the LAN is for stations in the LAN and 20 percent is for stations outside the LAN. Is an Ethernet hub preferable to an Ethernet switch? Does the answer change if the percentages are reversed?

27. Calculate the parameter a and the maximum throughput for a Gigabit Ethernet switch with stations at a 100-meter distance and average packet size of 512 bytes; 1500 bytes; and 64,000 bytes.

28. Provide a brief explanation for each of the following equations:
    a. Under a light load, which LAN has a smaller delay: Ethernet or token ring?
    b. Under a high load, which LAN has a smaller delay: Ethernet or token ring?

29. Suppose that a token-ring LAN is used to meet the requirements of the office in problem 5.
    a. Calculate the ring latency if all carrels are to be connected in a single ring as shown in Figure 6.58. Repeat for a ring for a single row or carrels.
    b. Can the requirements of one row of carrels be met by a 16 Mbps token ring?
    c. Can the requirements of the office be met by a FDDI ring?

30. Suppose that a group of 32 stations is serviced by a token-ring LAN. For the following cases calculate the time it takes to transfer a packet using the three token reinsertion strategies: after completion of transmission, after return of token and after return of packet.
    a. 1000-bit packet; 10 Mbps speed; 2.5-bit latency/adapter; 50 meters between stations.
    b. Same as (a) except 100 Mbps speed and eight-bit latency/adapter.
    c. Same as (a) except 1 km distance between stations.

31. Suppose that an FDDI LAN is used to meet the packet voice requirements of a set of users. Assume voice information uses 64 kbps coding and that each voice packet contains 20 ms worth of speech.
    a. Assume that each station handles a single voice call and that stations are 100 meters apart. Suppose that the FDDI ring is required to transfer each voice packet within 10 ms. How many stations can the FDDI accommodate while meeting the transfer requirement?
    b. How many simultaneous calls can be handled if each station is allowed to handle up to eight calls?

32. Use IEEE 802.3 and IEEE 802.11 to discuss three differences between wired and wireless LANs.

33. Consider the distributed coordination function in IEEE 802.11. Suppose that all packet transmissions are preceded by an RTS-CTS handshake. Find the capacity of this protocol following the analysis used for CSMA-CD.

34. Suppose one station sends a frame to another station in an IEEE 802.11 ad hoc network. Sketch the data frame and the return ACK frame that are exchanged, showing the contents in the relevant fields in the headers.

35. Suppose one station sends a frame to another station in a different BSS in an IEEE 802.11 infrastructure network. Sketch the various data frames and ACK frames that are exchanged, showing the contents in the relevant fields in the headers.

36. Why is error control (ARQ and retransmission) included in the MAC layer in IEEE 802.11 and not in IEEE 802.3?

37. Consider the exchange of CSMA-CA frames shown in Figure 6.70. Assume the IEEE 802.11 LAN operates at 2 Mbps using a frequency-hopping physical layer. Sketch a time diagram showing the frames transmitted including the final ACK frame. Show the appropriate interframe spacings and NAV values. Use Table 6.5 to obtain the appropriate time parameters. Assume that the data frame is 2000 bytes long.

38. Suppose that four stations in an IEEE 802.11 infrastructure network are in a polling list. The stations transmit 20 ms voice packets produced by 64 kbps speech encoders. Suppose that the contention-free period is set to 20 ms. Sketch a point-coordination frame transfer with the appropriate values for interframe spacings, NAV, and data and ACK frames.

39. Can a LAN bridge be used to provide the distribution service in an IEEE 802.11 extended service set? If so, explain how the service is provided and given an example of how the frames are transferred between BSSs.

40. Can a router be used to provide the distribution service in an IEEE 802.11 extended service set? If so, explain how addressing is handled and give an example of how the frames are transferred between BSSs.

41. A wireless LAN has mobile stations communicating with a base station. Suppose that the channel available has W Hz of bandwidth and suppose that the inbound traffic from the mobile stations to the base is K times smaller than the outbound traffic from the base to the workstations. Two methods are considered for dealing with the inbound/outbound communications. In frequency-division duplexing the channel is divided into two frequency bands, one for inbound and one for outbound communications. In time-division duplexing all transmissions use the full channel but the transmissions are time-division multiplexed for inbound and outbound traffic.
    a. Compare the advantages and disadvantages of the two methods in terms of flexibility, efficiency, complexity, and performance.
    b. How is the ratio K taken into account in the two methods?

42. Consider the following variation of FDMA. Each station is allotted two frequency bands: a band on which to transmit and a band in which to receive reservations from other stations directing it to listen to a transmission from a certain station (frequency band) at a certain time. To receive a packet, a station tunes in the appropriate channel at the appropriate time. To make a reservation, a station transmits at the receiving station's reservation channel. Explain how transmitting and receiving stations can use the reservation channels to schedule packet transmissions.

43. Suppose that a 1 MHz channel can support a 1 Mbps transmission rate. The channel is to be shared by 10 stations. Each station receives packets with exponential interarrivals and rate $\lambda = 50$ packets/second, and packets are constant length $L = 1000$ bits. Compare the total packet delay of a system that uses FDMA to a system that uses TDMA.

44. Compare FDMA, TDMA, and CDMA in terms of their ability to handle groups of stations that produce information flows that are produced at constant but different bit rates.

45. Calculate the autocorrelation function of the pseudorandom sequence in Figure 6.39 as follows. Replace each 0 by −1 and each 1 by +1. Take the output sequence of the generator and shift it with respect to itself; take the product of seven (one period) symbol

pairs and add. Repeat this calculation for shift values of 0, 1, ..., 7. In what sense does the result approximate the autocorrelation of a random sequence?

46. Construct the Walsh orthogonal spreading sequences of length 16.

47. Decode the sum signal in Figure 6.42 using the Walsh sequence for channel 4. What do you get? Explain why.

48. Compare IS-54 and GSM in terms of their handling of speech and the effect on spectrum efficiency.

49. Suppose that the A provider in the 800 MHz cellular band uses GSM and the B provider uses IS-95. Explain how a call from a mobile user in system B to a user in system A is accomplished.

50. Six stations (S1-S6) are connected to an extended LAN through transparent bridges (B1 and B2), as shown in the figure below. Initially, the forwarding tables are empty. Suppose the following stations transmit frames: S2 transmits to S1, S5 and S4, S3 transmits to S5, S1 transmits to S2, and S6 transmits to S5. Fill in the forwarding tables with appropriate entries after the frames have been completely transmitted.

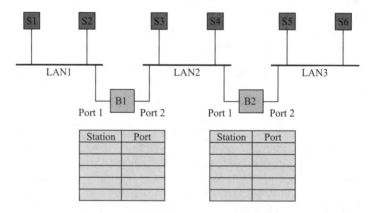

51. Suppose N stations are connected to an extended Ethernet LAN, as shown below, operating at the rate of 10 Mbps. Assume that the efficiency of each Ethernet is 80 percent. Also assume that each station transmits frames at the average rate of R bps, and each frame is equally likely to be destined to any station (including itself). What is the max-

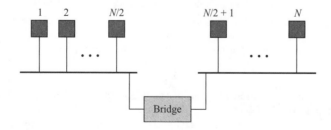

imum number of stations N that can be supported if R is equal to 100 Kbps? If the bridge is replaced with a repeater, what is the maximum number of stations that can be supported? (Assume that the efficiency of the entire Ethernet is still 80 percent.)

52. The LANs in the figure below are interconnected by using source routing bridges. Assume that bridges 3 and 4 are not part of the initial spanning tree.
    a. Show the paths of the single route broadcast frames when S1 wants to learn the route to S2.
    b. Show the paths of all routes broadcast frames returned by S2.
    c. List all possible routes from S1 to S2 from part (b)
    d. How many LAN frames are required to learn the possible routes?

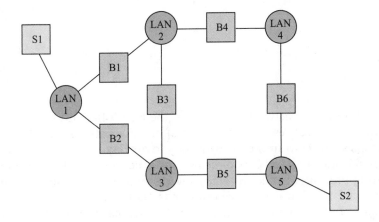

CHAPTER 7

Packet-Switching Networks

Traditional telephone networks operate on the basis of circuit switching. A call setup process reserves resources (time slots) along a path so that the stream of voice samples can be transmitted with very low delay across the network. The resources allocated to a call cannot be used by other users for the duration of the call. This approach is inefficient when the amount of informtion transferred is small or if information is produced in bursts, as is the case in many computer applications. In this chapter we examine networks that transfer blocks of information called **packets**. **Packet-switching networks** are better matched to computer applications and can also be designed to support real-time applications such as telephony.

We can view packet networks from two perspectives. One perspective involves an *external view* of the network and is concerned with the services that the network provides to the transport layer that operates above it at the end systems. Here we are concerned with whether the network service requires the setting up of a connection and whether the transfer of user data is provided with quality-of-service guarantees. Ideally the definition of the network services is independent of the underlying network and transmission technologies. This approach allows the transport layer and the applications that operate above it to be designed so that they can function over any network that provides the given services.

A second perspective on packet networks is concerned with the *internal* operation of a network. Here we look at the physical topology of a network, the interconnection of links, switches, and routers. We are concerned with the approach that is used to direct information across the network: datagrams, or virtual circuits. We are also concerned with addressing and routing procedures, as well as with dealing with congestion inside the network. We must also manage traffic flows so that the network can deliver information with the quality of service it has committed to.

It is useful to compare these two perspectives in the case of broadcast networks and LANs from the previous chapter and the switched packet networks considered here. The first perspective, involving the services provided to the layer above, does not differ in a fundamental way between broadcast and switched packet networks. The second perspective, however, is substantially different. In the case of LANs, the network is small, addressing is simple, and the frame is transferred in one hop so no routing is required. In the case of packet-switching networks, addressing must accommodate extremely large-scale networks and must work in concert with appropriate routing algorithms. These two challenges, addressing and routing, are the essence of the network layer.

In this chapter we deal with the general issues regarding packet-switching networks. Later chapters deal with specific architectures, namely, Internet Protocol (IP) packet networks and asynchronous transfer mode (ATM) packet networks. The chapter is organized as follows:

1. *Network services and internal network operation.* We elaborate on the two perspectives on networks, and we discuss the functions of the network layer, including internetworking.
2. *Physical view of networks.* We examine typical configurations of packet-switching networks. This section defines the role of multiplexers, LANs, switches, and routers in network and internetwork operation.
3. *Datagrams and virtual circuits.* We introduce the two basic approaches to operating a packet network, and we use IP and ATM as examples of these approaches.
4. *Routing.* We introduce the basic approaches for selecting routes across the network.
5. *Shortest path algorithms.* We continue our discussion of routing, focusing on two shortest-path routing algorithms: the Bellman-Ford algorithm and Dijskstra's algorithm.
6. *ATM networks.* We introduce ATM networks as an example of an advanced virtual-circuit packet-switching network that can support many services.
7. *Traffic management.* We introduce traffic shaping, scheduling and call admission control as methods for providing Quality-of-Service.
8. *Congestion control.* We introduce techniques to deal with congestion due to surges in traffic or equipment failures.

The material on ATM, traffic management, and congestion control is relatively advanced. The corresponding sections (7.6, 7.7, and 7.8 respectively) can be skipped and the reader may proceed to Chapter 8, depending on their background or interest.

7.1 NETWORK SERVICES AND INTERNAL NETWORK OPERATION

The essential function of a network is to transfer information among the users that are attached to the network or internetwork. In Figure 7.1 we show that this transfer may involve a single block of information or a sequence of blocks that are temporally related. In the case of a single block of information, we are interested in having the block delivered correctly to the destination, and we may also be interested in the delay experienced in traversing the network. In the case of a sequence of blocks, we may be interested not only in receiving the blocks correctly and in the right sequence but also in delivering a relatively unimpaired temporal relation.

Figure 7.2 shows a transport protocol that operates end to end across a network. The transport layer peer processes at the end systems accept messages from their higher layer and transfer these messages by exchanging segments end to end across the network. The figure shows the interface at which the network service is visible to the transport layer. The network service is all that matters to the transport layer, and the manner in which the network operates to provide the service is irrelevant.

The network service can be connection-oriented or connectionless. A connectionless service is very simple, with only two basic interactions between the transport layer and the network layer: a request to the network that it send a packet and an indication from the network that a packet has arrived. The user can request transmission of a packet at any time, and *does not need to inform the network layer* that the user intends to transmit information ahead of time. A connectionless service puts total responsibility for error control, sequencing, and flow control on the end-system transport layer.

The network service can be connection-oriented. In this case the transport layer cannot request transmission of information until a connection has been set up. The essential points here are that *the network layer must be informed* about the new flow that is about to be applied to the network and that the network layer maintains state information about the flows it is handling. During call setup, parameters related to usage and quality of service may be negotiated and network resources may be allocated to ensure that the user flow can be handled as required. A connection-release procedure may also be required to

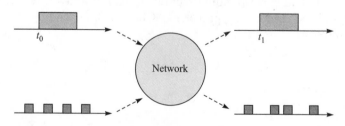

FIGURE 7.1 A network transfers information among users

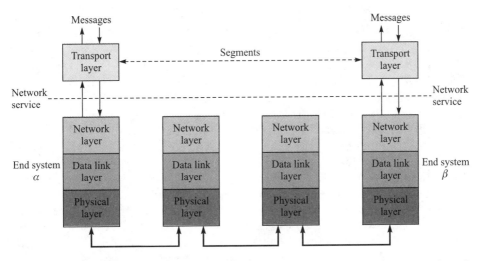

FIGURE 7.2 Peer-to-peer protocols operating end to end across a network—protocol stack view

terminate the connection. It is clear that providing connection-oriented service entails greater complexity than connectionless service in the network layer.

It is also possible for a network layer to provide a choice of services to the user of the network. For example, the network layer could offer: (1) best-effort connectionless service; (2) low-delay connectionless service; (3) connection-oriented reliable stream service; and (4) connection-oriented transfer of packets with delay and bandwidth guarantees. It is easy to come up with examples of applications that can make use of each of these services. However, it does not follow that all the services should be offered by the network layer. Two inter-related reasons can be given for keeping the set of network services to a minimum: the end-to-end argument and the need for network scalability.

When applied to the issue of choice of network services, the end-to-end argument suggests that functions should be placed as close to the application as possible, since it is the application that is in the best position to determine whether a function is being carried out completely and correctly. This argument suggests that as much functionality as possible should be located in the transport layer or higher and that the network services should provide the minimum functionality required to meet application performance.

Up to this point we have considered only the services offered by the network layer. Let us now consider the internal operation of the network. Figure 7.3 shows the relation between the service offered by the network and the internal operation. We say that the internal operation of a network is *connectionless* if packets are transferred within the network as datagrams. Thus in the figure each packet is routed independently. Consequently packets may follow different paths from α to β and so may arrive out of order. We say that the internal operation of a network is *connection-oriented* if packets follow virtual circuits that have been

THE END-TO-END ARGUMENT FOR SYSTEM DESIGN

The *end-to-end argument* in system design articulated in [Saltzer 1984] states that an end-to-end function is best implemented at a higher level than at a lower level. The reason is that the correct end-to-end implementation requires *all* intermediate low-level components to operate correctly. This feature is difficult and sometimes impossible to ensure and is frequently too costly. The higher-level components at the ends are in a better position to determine that a function has been carried out correctly and in better position to take corrective action if they have not. Low-level actions to support the end-to-end function are justified only as performance enhancements.

We already encountered the end-to-end argument in the comparison of end-to-end error control and hop-by-hop error control in Chapter 5. The argument here is that the end system will have to implement error control on an end-to-end basis regardless of lower-level error-control mechanisms that may be in place because the individual low-level mechanisms cannot cover all sources of errors, for example, errors introduced within a node. Consequently, lower-level mechanisms are not essential and should be introduced only to enhance performance. Thus the transmission of a long file over a sequence of nearly error-free links does not require per link error control. On the other hand, the transmission of such files over a sequence of error-prone links does argue for per link error control.

established from a source to a destination. Thus to provide communications between α and β, routing to set up a virtual circuit is done once, and thereafter packets are simply forwarded along the established path. If resources are reserved during connection setup, then bandwidth, delay, and loss guarantees can be provided.

The fact that a network offers connection-oriented service, connectionless service, or both does not dictate how the network must operate internally. In discussing TCP and IP, we have already seen that a connectionless packet network (e.g., IP) can support connectionless service (UDP) as well as connection-oriented service (TCP). We will also see that a connection-oriented network (e.g., ATM) can provide connectionless service as well as connection-oriented service. We discuss virtual-circuit and datagram network operation in more detail in a later section. However, it is worthwhile to compare the two at this point at a high level.

The approach suggested by the end-to-end argument keeps the network service (and the network layer that provides the service) as simple as possible while adding complexity at the edge only as required. This strategy fits very well with the need to grow networks to very large scale. We have seen that the value of a network grows with the community of users that can be reached and with the range of applications that can be supported. Keeping the core of the network simple and adding the necessary complexity at the edge enhances the scalability of the network to larger size and scope.

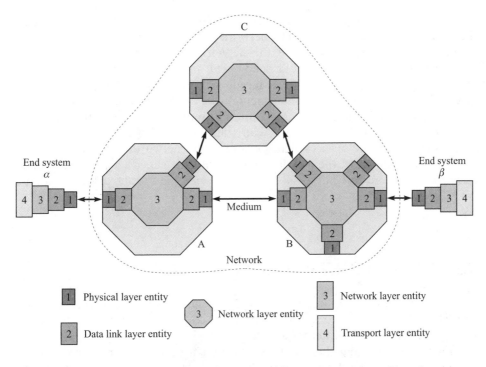

FIGURE 7.3 Layer 3 entities work together to provide network service to layer 4 entities

This reasoning suggests a preference for a connectionless network, which has much lower complexity than a connection-oriented network. The reasoning does allow the possibility for some degree of "connection orientation" as a means to ensure that applications can receive the proper level of performance. Indeed current research and standardization efforts (discussed in Chapter 10) can be viewed as an attempt in this direction to determine an appropriate set of network services and an appropriate mode of internal network operation.

We have concentrated on high-level arguments up to this point. What do these arguments imply about the functions that should be in the network layer? Clearly, functions that need to be carried out at every node in the network must be in the network layer. Thus functions that route and forward packets need to be done in the network layer. Priority and scheduling functions that direct how packets are forwarded so that quality of service is provided also need to be in the network layer. Functions that belong in the edge should, if possible, be implemented in the transport layer or higher. A third category of functions can be implemented either at the edge or inside the network. For example, while congestion takes place inside the network, the remedy involves reducing input flows at the edge of the network. We will see that congestion control has been implemented in the transport layer and in the network layer.

Another set of functions is concerned with making the network service independent of the underlying transmission systems. For example, different transmissions sytems (e.g., optical versus wireless) may have different limits on the frame size they can handle. The network layer may therefore be called upon to carry out segmentation inside the network and reasssembly at the edge. Alternatively, the network could send error messages to the sending edge, requesting that the packet size be reduced. A more challenging set of functions arises when the "network" itself may actually be an internetwork. In this case the network layer must also be concerned not only about differences in the size of the units that the component networks can transfer but also about differences in addressing and in the services that the component networks provide.

In the remainder of the chapter we deal with the general aspects of internal network operation. In Chapters 8 and 9 we discuss the specific details of IP and ATM networks.

7.2   PACKET NETWORK TOPOLOGY

This section considers existing packet-switching networks. We present an end-to-end view of existing networks from a personal computer, workstation, or server through LANs and the Internet and back.

First let us consider the way in which users access packet networks. Figure 7.4 shows an *access multiplexer* where the packets from a number of users share a transmission line. This system arises for example, in X.25, frame relay, and ATM networks, where a single transmission line is shared in the access to a wide area packet-switching network. The multiplexer combines the typically bursty flows of the individual computers into aggregated flows that make efficient use of the transmission line. Note that different applications within a single computer can generate multiple simultaneous flows to different destinations. From a logical point of view, the link can be viewed as carrying either a single aggregated flow or a number of separate packet flows. The network access node forwards packets into a backbone packet network.

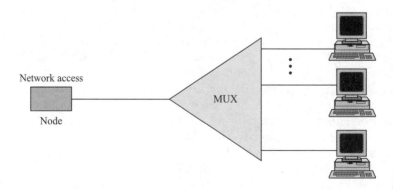

FIGURE 7.4 Access multiplexer

Local area networks (LANs) provide the access to packet-switching networks in many environments. As shown in Figure 7.5a, computers are connected to a shared transmission medium. Transmissions are broadcast to all computers in the network. Each computer is identified by a unique physical address, and so each station listens for its address to receive transmissions. Broadcast and multicast transmissions are easily provided in this environment.

LANs allow the sharing of resources such as printers, databases, and software among a small community of users. LANs can be extended through the use of *bridges* or *LAN switches*, as shown in Figure 7.5b. Here the LAN switch forwards inter-LAN traffic based on the physical address of the frames. Traffic local to each LAN stays local, and broadcast transmissions are forwarded to the other attached LANs. Switches can interconnect more than two LANs.

Multiple LANs in an organization, in turn, are interconnected into *campus networks* with a structure such as that shown in Figure 7.6. LANs for a large group of users such as a department are interconnected in an extended LAN through the use of LAN switches, identified by lowercase *s* in the figure. Resources such as servers and databases that are primarily of use to this department are kept within the subnetwork. This approach reduces delays in accessing the resources and contains the level of traffic that leaves the subnetwork. Each subnetwork has access to the rest of the organization through a router *R* that accesses the campus backbone network. A subnetwork also uses the campus backbone to reach the "outside world" such as the Internet or other sites belonging to the organization through a gateway router. Depending on the type of organization, the gateway may implement firewall functions to control the traffic that is allowed into and out of the campus network.

Servers containing critical resources that are required by the entire organization are usually located in a data center where they can be easily maintained and

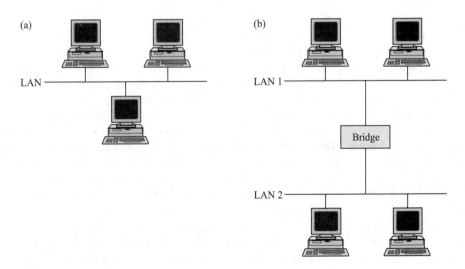

FIGURE 7.5 Local area networks

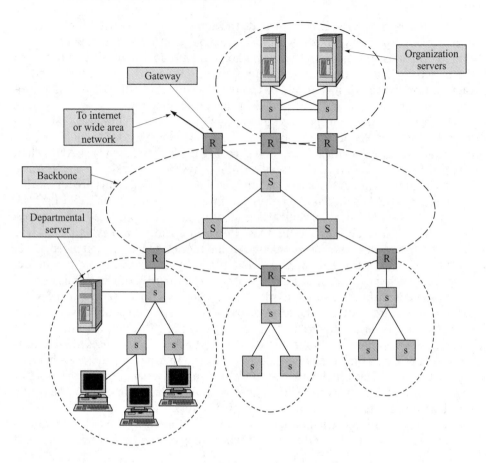

FIGURE 7.6 Campus network

where security can be enforced. As shown in Figure 7.6, the critical servers may be provided with redundant paths to the campus backbone network. These servers are usually placed near the backbone network to minimize the number of hops required to access them from the rest of the organization.

The traffic within an extended LAN is delivered based on the *physical* LAN addresses. However, applications in host computers operate on the basis of *logical* IP addresses. Therefore, the physical address corresponding to an IP address needs to be determined every time an IP packet is to be transmitted over a LAN. This *address resolution* problem can be solved by using IP address to physical address translation tables. In the next chapter we discuss the *Address Resolution Protocol* that IP uses to solve this problem.

The routers in the campus network are interconnected to form the campus backbone network, depicted by the mesh of switches, designated *S*, in Figure 7.6. Typically, for large organizations such as universities these routers are interconnected by using very high speed LANs, for example, Gigabit Ethernet or an

ATM network. The routers use the Internet Protocol (IP), which enables them to operate over various data link and network technologies. The routers exchange information about the state of their links to dynamically calculate routing tables that direct packets across the campus network. This approach allows the network to adapt to changes in traffic pattern as well as changes in topology due to faults in equipment.

The routers in the campus network form a *domain* or *autonomous system.* The term *domain* indicates that the routers run the same routing protocol. The term *autonomous* system is used for one or more domains under a single administration. All routing decisions inside the autonomous system are independent of any other network.

Organizations with multiple sites may have their various campus networks interconnected through routers interconnected by leased digital transmission lines or frame relay connections. In this case access to the *wide area network* may use an access multiplexer such as the one shown in Figure 7.4. In addition the campus network may be connected to an *Internet service provider* through one or more border routers as shown in Figure 7.7. To communicate with other networks, the autonomous system must provide information about its network routes in the border routers. The border router communicates on an interdomain level, whereas other routers in a campus network operate at the intradomain level.

A national ISP provides points of presence (POPs) in various cities where customers can connect to their network. The ISP has its own national network for interconnecting its POPs. This network could be based on ATM; it might use IP over SONET; or it might use some other network technology. The ISPs in turn exchange traffic as *network access points (NAPs),* as shown in Figure 7.8a. A NAP is a high-speed LAN or switch at which the routers from different ISPs

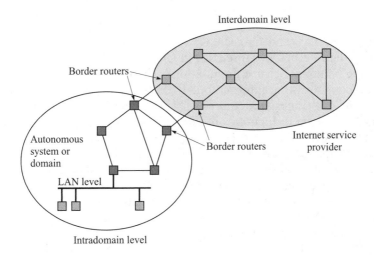

FIGURE 7.7 Intradomain and interdomain levels

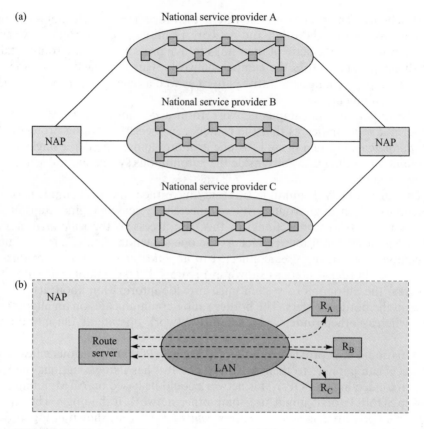

FIGURE 7.8 National ISPs exchange traffic at NAPs; routing information is exchanged through route servers

can exchange traffic, and as such NAPs are crucial to the interconnectivity provided by the Internet. (As discussed in Chapter 1, four NAPs were originally set up by the National Science Foundation). The ISPs interconnected to a NAP need to exchange routing information. If there are n such ISPs, then $n(n-1)$ pairwise route exchanges are required. This problem is solved by introducing a route server as shown in Figure 7.8b. Each ISP sends routing information to the route server, which knows the policies of every ISP. The route server in turn delivers the processed routing information to the ISPs.

Note that a national service provider also has the capability of interconnecting a customer's various sites by using its own IP network, so the customer's sites appear as a single private network. This configuration is an example of a virtual private network (VPN).

Small office and home (SOHO) users obtain packet access through ISPs. The access is typically through modem dial-up, but it could be through ADSL, ISDN, or cable modem. When a customer connects to an ISP, the customer is assigned an IP address for the duration of the connection.[1] Addresses are shared in this way because the ISP has only a limited number of addresses. If the ISP is only a local provider, then it must connect to a regional or national provider and eventually to a NAP.

Thus we see that a multilevel hierarchical network topology arises for the Internet which is much more decentralized than traditional telephone networks. This topology comprises multiple domains consisting of routers interconnected by point-to-point data links, LANs, and wide area networks such as ATM.

The principal task of a packet-switching network is to provide connectivity among users. The preceding description of the existing packet-switching network infrastructure reveals the magnitude of this task. Routers exchange information among themselves and use routing protocols to build a consistent set of routing tables that can be used in the routes to direct the traffic flows in these networks. The routing protocols must adapt to changes in network topology due to the introduction of new nodes and links or to failures in equipment. Different routing algorithms are used within a domain and between domains. A key concern here is that the routing tables result in stable traffic flows that make efficient use of network resources. Another concern is to keep the size of routing tables manageable even as the size of the network continues to grow at a rapid pace. In this chapter we show how hierarchical addressing structures can help address this problem. A third concern is to deal with congestion that inevitably occurs in the network. It makes no sense to accept packets into the network when they are likely to be discarded. Thus when congestion occurs inside the network, that is, buffers begin filling up as a result of a surge in traffic or a fault in equipment, the network should react by applying congestion control to limit access to the network only to traffic that is likely to be delivered. A final concern involves providing the capability to offer Quality-of-Service guarantees to some packet flows. We deal with these topics also in the remainder of the chapter.

7.3 DATAGRAMS AND VIRTUAL CIRCUITS

A network is usually represented as a cloud with multiple input sources and output destinations as shown in Figure 7.9. A network can be viewed as a generalization of a physical cable in the sense of providing connectivity between multiple users. Unlike a cable, a switched network is geographically distributed and consists of a graph of transmission lines (i.e., links) interconnected by switches (nodes). These transmission and switching resources are configured to enable the flow of information among users.

---

[1]The dynamic host configuration protocol (DHCP) provides users with temporary IP addresses and is discussed in Chapter 8.

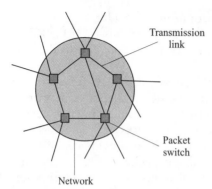

FIGURE 7.9 Switched network

Transmission
link

Packet
switch

Network

Networks provide for the interconnection of sources to destinations on a dynamic basis. Resources are typically allocated to an information flow only when needed. In this manner the resources are shared among the community of users resulting in efficiency and lower costs. There are two fundamental approaches to transferring information over a packet-switched network. The first approach, called **connection-oriented**, involves setting up a connection across the network before information can be transferred. The setup procedure typically involves the exchange of signaling messages and the allocation of resources along the path from the input to the output for the duration of the connection. The second approach is **connectionless** and does not involve a prior allocation of resources. Instead a packet of information is routed independently from switch to switch until the packet arrives at its destination. Both approaches involve the use of switches or routers to direct packets across the network.

7.3.1   Structure of Switch/Router

Figure 7.10 shows a generic switch consisting of input ports, output ports, an interconnection fabric, and a switch controller/processor. Input ports and output ports are usually paired. A line card typically handles several input/output ports. The line card implements physical and data link layer functions. Thus the card is concerned with symbol timing and line coding. It is also concerned with framing, physical layer addressing, and error checking. For widely deployed standards, the line card also implements medium access control and data link protocols in hardware with a special-purpose chip set. The line card also contains some buffering to handle the speed mismatch between the transmission line and the interconnection fabric. The controller/processor can carry out a number of functions depending on the type of packet switching. The function of the interconnection fabric is to transfer packets between the line cards. Note that Figure 7.10 shows an "unfolded" version of the switch in which the line cards appear twice, once with input ports and again with output ports. In the actual implementation the transmit and receive functions take place in a single line card. However, the function of various types of switch architectures is easier to visualize this way.

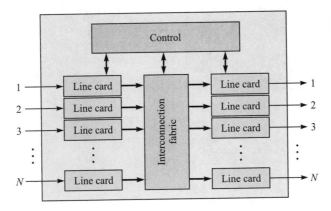

FIGURE 7.10 Components of generic switch/router

We elaborate on the operation of the switches as we develop the two approaches to packet switching.

A simple switch can be built by using a personal computer or a workstation and inserting several network interface cards (NICs) in the expansion slots as shown in Figure 7.11. The frames that arrive at the NICs are de-encapsulated, and the packets are transferred by using the I/O bus from the NIC to main memory. The processor performs the required routing and protocol processing, formats the packet header, and then forwards the packet by transferring it from main memory to the appropriate NIC.

The simple setup in Figure 7.11 reveals the three basic resources and potential bottlenecks in switches: processing, memory, and bus (interconnection) bandwidth. Processing is required to implement the protocols, and hence the processing capacity places a limit on the maximum rate at which the switch can operate. Memory is required to store packets, and hence the amount of memory available determines the rate at which packets are lost, thus placing another limit on the load at which the switch can be operated. In this approach

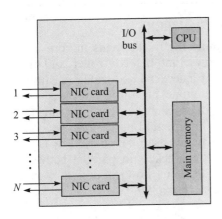

FIGURE 7.11 Building a switch from a general purpose computer

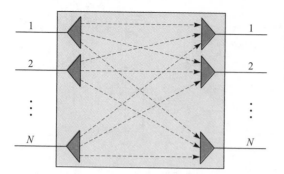

FIGURE 7.12 Input port demultiplexes incoming packet stream; packets are routed to output port; output port multiplexes outgoing packet stream

the memory bandwidth, which is the rate at which information can be read in and out of RAM, also places a limit on the aggregate rate of the switch. Finally, the I/O bus bandwidth places a limit on the total rate at which information can be transferred between ports. Different switch architectures configure these basic resources so that target aggregate switch capacities are met in a cost-effective manner.

Each input and output port in a switch/router typically contains multiplexed streams of packets. Figure 7.12 shows that the flows that enter the switch in effect are demultiplexed at the input port. The switch or router then directs the packets to output ports. Each output port can be viewed as a multiplexer that precedes the outgoing transmission line. Thus we see that switches and routers play a key role in controlling where the packet flows are placed in a network. By controlling packet flows, the network bandwidth can be used efficiently and the performacne can be optimized. We return to this discussion when we discuss Quality-of-Service mechanisms later in the chapter.

HOW TO MAKE BIG, FAST SWITCHES/ROUTERS

Big switches and routers are needed to handle the traffic loads in core networks. An examination of Figure 7.10 shows that the controller and the interconnection fabric are likely to be the bottlenecks. Two strategies can be used to increase switch size. First, as the volume of traffic increases, the placement of a dedicated controller/processor in each line card is justified. This step removes a centralized controller as a potential bottleneck. Second, bus and broadcast type of interconnection structures can be replaced by large bandwidth interconnection fabrics that transfer packets in parallel between input and output ports. A large literature explains how to design switch interconnection fabrics; for example, see [Robertazzi 1994].

7.3.2   Connectionless Packet Switching

Packet switching has its origin in **message switching**, where a message is relayed from one station to another until the message arrives at its destination. At the source each message has a header attached to it to provide source and destination addresses. CRC checkbits are attached to detect errors. As shown in Figure 7.13, the message is transmitted in a store-and-forward fashion. The message is transmitted in its entirety from one switch to the next switch. Each switch performs an error check, and if no errors are found, the switch examines the header to determine the next hop in the path to the destination. If errors are detected, a retransmission may be requested. After the next hop is determined, the message waits for transmission over the corresponding transmission link. Because the transmission links are shared, the message may have to wait until previously queued messages are transmitted. Message switching does not involve a call setup. Message switching can achieve a high utilization of the transmission line. This increased utilization is achieved at the expense of queueing delays. Loss of messages may occur when a switch has insufficient buffering to store the arriving message.[2] End-to-end mechanisms are required to recover from these losses.

Figure 7.14 shows the total delay that is incurred when a message is transmitted over a path that involves two intermediate switches. The message must first traverse the link that connects the source to the first switch. We assume that this link has a propagation delay of p seconds.[3] We also assume that the message has a transmission time of T seconds. The message must next traverse the link connecting the two switches, and then it must traverse the link connecting the second switch and the destination. For simplicity we assume that the

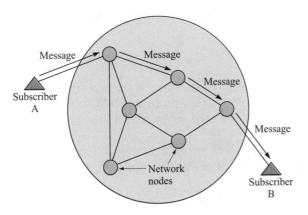

FIGURE 7.13 Message switching

---

[2]The trade-offs between delay and loss are explored in Chapter 5, section 5.6.1.

[3]The propagation delay is the time that elapses from when a bit enters a transmission line to when it exits the line at the other end.

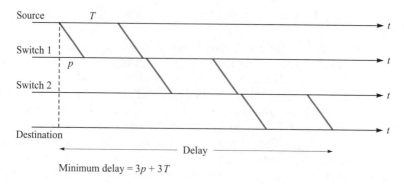

FIGURE 7.14 Delays in message switching

propagation delay and the bit rate of the transmission lines are the same. It then follows that the minimum end-to-end messge delay is $3p + 3T$. Note that this delay does not take into account any queueing delays that may be incurred in the various links waiting for prior messages to be transmitted. It also does not take into account the times required to perform the error checks or any associated retransmissions.

Example—Long Messages versus Packets

Suppose that we wish to transmit a large message ($L = 10^6$ bits) over two hops. Suppose that the transmission line in each hop has an error rate of $p = 10^{-6}$ and that each hop does error checking and retransmission. How many bits need to be transmitted using message switching?

If we transmit the message in its entirety, the probability that the message arrives correctly after the first hop is

$$P_c = (1 - p)^L = (1 - 10^{-6})^{1000000} \approx e^{-Lp} = e^{-1} \approx 1/3$$

Therefore, on the average it will take three tries to get the message over the first hop. Similarly, the second hop will require another three full message transmissions on the average. Thus 6 Mbits will need to be transmitted to get the 1 Mbit message across.

Now suppose that the message is broken up into ten 10^5-bit packets. The probability that a packet arrives correctly after the first hop is

$$P'_c = (1 - 10^{-6})^{100000} \approx e^{-1/10} \approx 0.90$$

Thus each packet needs to be transmitted $1/0.90 = 1.1$ times on the average. The message gets transmitted over each hop by transmitting an average of 1.1 Mbit. The total number of bits transmitted over the two hops is then 2.2 Mbits.

The preceding example reiterates our observation on ARQ protocols that the probability of error in a transmitted block increases with the length of the block. Thus very long messages are not desirable if the transmission lines are noisy because they lead to a larger rate of message retransmissions. This situation is one reason that it is desirable to place a limit on the maximum size of the blocks that can be transmitted by the network. Thus long messages should be broken into smaller blocks of information, or *packets*.

Message switching is also not suitable for interactive applications because it allows the transmission of very long messages that can impose very long waiting delays on other messages. By placing a maximum length on the size of the blocks that are transmitted, packet switching limits the maximum delay that can be imposed by a single packet on other packets. Thus packet switching is more suitable than message switching for interactive applications.

In the **datagram**, or **connectionless packet-switching** approach, each packet is routed independently through the network. Each packet has an attached header that provides all of the information required to route the packet to its destination. When a packet arrives at a packet switch, the destination address (and possibly other fields) in the header are examined to determine the next hop in the path to the destination. The packet is then placed in a queue to wait until the given transmission line becomes available. By sharing the transmission line among multiple packets, packet switching can achieve high utilization at the expense of packet queueing delays. We note that **routers** in the Internet are packet switches that operate in datagram mode.

Because each packet is routed independently, packets from the same source to the same destination may traverse different paths through the network as shown in Figure 7.15. For example, the routes may change in response to a network fault. Thus packets may arrive out of order, and resequencing may be required at the destination.

Figure 7.16 shows the delay that is incurred by transmitting a message that is broken into three separate packets. Here we assume that the three packets follow

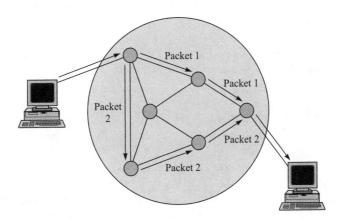

FIGURE 7.15 Datagram packet switching

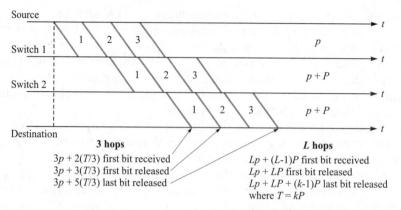

FIGURE 7.16 Delays in packet switching

the same path and are transmitted in succession. We neglect the overhead due to headers and suppose that each packet requires $P = T/3$ seconds to transmit. The three packets are transmitted successively from the source to the first packet switch.

The first packet in Figure 7.16 arrives at the first switch after $p + P$ seconds. Assuming that the packet arrives correctly, it can begin transmission over the next hop after a brief processing time. The first packet is received at the second packet switch at time $2p + 2P$. Again we assume that the packet begins transmission over the final hop after a brief processing time. The first packet then arrives at the final link at time $3p + 3P$. As the first packet traverses the network, the subsequent packets follow immediately, as shown in the figure. In the absence of transmission errors, the final packet will arrive at the destination at time $3p + 3P + 2P = 3p + 5P = 3p + T + 2P$, which is less than the delay incurred in the message switching example in Figure 7.14. In general, if the path followed by a sequence of packets consists of L hops with identical propagation delays and transmission speeds, then the total delay incurred by a message that consists of k packets is given by

$$Lp + LP + (k - 1)P$$

In contrast, the delay incurred using message switching is

$$Lp + LT = Lp + L(kP)$$

Thus message switching involves an additional delay of $(L - 1)(k - 1)P$. We note that the above delays neglect the queueing and processing times at the various hops in the network.

Figure 7.17 shows a routing table that contains an entry for each possible destination for a small network. This entry specifies the next hop that is to be taken by packets with the given destination. When a packet arrives, the destination address in the header is used to perform a table lookup. The result of the lookup is the number of the output port to which the packet must be forwarded.

Destination address	Output port
0785	7
1345	12
1566	6
2458	12

FIGURE 7.17 Routing table in connectionless packet switching

When the size of the network becomes very large, this simple table lookup is not feasible, and the switch/router processor needs to execute a route lookup algorithm for each arriving packet.

In datagram packet switching, the packet switches have no knowledge of a "connection" even when a source and destination exchange a sequence of packets. This feature makes datagram packet switching robust with respect to faults in the network. If a link or packet switch fails, the neighboring packet switches react by routing packets along different links and by sharing the fault information with other switches. This process results in the setting up of a new set of routing tables. Because no connections are set up, the sources and destinations need not be aware of the occurrence of a failure in the network. The processors in the switch/routers execute a distributed algorithm for sharing network state information and for synthesizing routing tables.

The design of the routing table is a key issue in the proper operation of a packet-switching network. This design requires knowledge about the topology of the network as well as of the levels of traffic in various parts of the network. Another issue is that the size of the tables can become very large as the size of the network increases. We discuss these issues further later in the chapter.

Example—IP Internetworks

The Internet Protocol provides for the connectionless transfer of packets across an interconnected set of networks called an internet. In general the component networks may use different protocols so the objective of IP is to provide communications across these dissimilar networks. Each device that is attached to an IP internet has a two-part address: a network part and a host part. To transmit an IP packet, a device sends an IP packet encapsulated using its local network protocol to the nearest router. The routers are packet switches that act as gateways between the component networks. The router performs a route lookup algorithm on the network part of the destination address of the packet to determine whether the destination is in an immediately accessible network or, if not, to determine the next router in the path to the destination. The router then forwards the IP packet across the given network by encapsulating the IP packet using the format and protocol of the given network. In other words, IP treats the component networks as data link layers whose role is to transfer the packet to the next router or to the destination. IP packets are routed in connectionless fashion from router to router until the destination is reached.

7.3.3 Virtual-Circuit Packet Switching

Virtual-circuit packet switching involves the establishment of a fixed path between a source and a destination prior to the transfer of packets, as shown in Figure 7.18. As in circuit switching, the call setup procedure usually takes place before any packets can flow through the network as shown in Figure 7.19.[4]

The connection setup procedure establishes a path through the network and then sets parameters in the switches along the path as shown in Figure 7.20. The controller/processor in every switch is involved in the exchange of signaling messages to set up the path. As in the datagram approach, the transmission links are shared by packets from many flows. In general, in virtual-circuit packet switching, buffer and transmission resources need not be dedicated explicitly for the use of the connection, but the number of flows admitted may be limited to control the load on certain links. All packets for the connection then follow the same path.

In datagram packet switching each packet must contain the full address of the source and destination. In large networks these addresses can require a large number of bits and result in significant packet overhead and hence wasted transmission bandwidth. One advantage of virtual-circuit packet switching is that

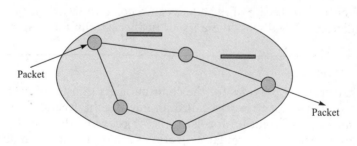

FIGURE 7.18 Virtual-circuit packet switching

---

[4]In some cases *permanent* virtual circuits are established a priori.

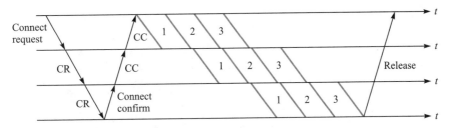

FIGURE 7.19 Delays in virtual-circuit packet switching

abbreviated headers can be used. The connection setup procedure establishes a number of entries in *forwarding tables* located in the various switches along the path. At the input to every switch, the connection is identified by a **virtual-circuit identifier (VCI)**. When a packet arrives at an input port, the VCI in the header is used to access the table, as shown in the example in Figure 7.21. The table lookup provides the output port to which the packet is to be forwarded and the VCI that is to be used at the input port of the next switch. Thus the call setup procedure sets up a chain of pointers across the network that direct the flow of packets in a connection. The table entry for a VCI can also specify the type of priority that is to be given to the packet by the scheduler that controls the transmissions in the next output port.

The number of bits required in the header in virtual-circuit switching is reduced to the number required to represent the maximum number of simultaneous connections over an input port. This number is much smaller than the number required to provide full destination network addresses. This factor is one of the advantages of virtual-circuit switching relative to datagram switching. In addition, the use of abbreviated headers and hardware-based table lookup allows fast processing and forwarding of packets. Virtual-circuit switching does a table lookup and immediately forwards the packet to the output port; connectionless packet switching traditionally was much slower because it required software processing of the header before the next hop in the route could be determined. (This situation has changed with the development of hardware-based routing techniques.)

Another advantage of virtual-circuit packet switching is that resources can be allocated during connection setup. For example, a certain number of buffers may be reserved for a connection at every switch along the path, and a certain amount of bandwidth can be allocated at each link in the path. In addition, the

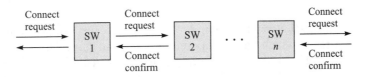

FIGURE 7.20 Signaling message exchanges in virtual-circuit setup

	Identifier	Output port	Next identifier
	12	13	44
Entry for packets → with identifier 15	15	15	23
	27	13	16
	58	7	34

FIGURE 7.21 Example of virtual-circuit routing table for an input port

connection setup process ensures that a switch is able to handle the volume of traffic that is allowed over every transmission link. In particular, a switch may refuse a connection over a certain link when the delays or link utilization exceed certain thresholds.

However, virtual-circuit packet switching does have disadvantages relative to the datagram approach. The switches in the network need to maintain information about the flows they are handling. The amount of required "state" information grows very quickly with the number of flows. Another disadvantage is evident when failures occur. In the case of virtual-circuit packet switching, when a fault occurs in the network all affected connections must be set up again.

If virtual-switching packet switching is used then the minimum delay for transmitting a message that consists of k packets is the same as in Figure 7.16, in addition to the time required to set up the connection. A modified form of virtual-circuit packet switching, called **cut-through packet switching**, can be used when retransmissions are not used in the underlying data link control. It is then possible for a packet to be forwarded as soon as the header is received and the table lookup is carried out. As shown in Figure 7.22, the minimum delay in transmitting the message is then reduced to approximately the sum of the propagation delays in the various hops plus the one-message transmission time. (This scenario assumes that all lines are available to transmit the packet immediately.)

Cut-through packet switching may be desirable for applications such as speech transmission, which has a delay requirement but can tolerate some errors. Cut-through packet switching is also appropriate when the transmission is virtually error free, as in the case of optical fiber transmission, so that hop-by-hop error checking is unnecessary.

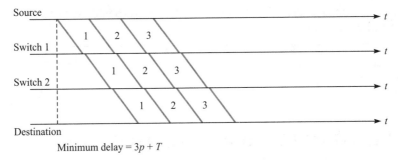

FIGURE 7.22 Cut-through packet switching

Example—ATM Networks

ATM networks provide for the connection-oriented transfer of information across a network. ATM requires all user information to be converted into fixed-length packets called cells. A connection setup phase precedes the transfer of information. During this setup a negotiation takes place in which the user specifies the type of flow that is to be offered to the network, and the network commits to some quality of service that is to be provided to the flow. The connection setup involves setting up a path across the network and allocating appropriate resources along the path.

An ATM connection is defined in terms of a chain of local identifiers called VCIs that identify the connection in each link along the path. Cells are forwarded by ATM switches that perform a table lookup on the VCI to determine the next output port and the VCI in the next link. ATM assumes low-error rate

---

FLOWS, RESERVATIONS, AND SHORTCUTS

Here we note the emergence of packet-switching approaches that combine features of datagrams and virtual circuits. These hybrid approaches are intended for packet-switching networks that handle a mix of one-time packet transfers (for which datagram mode is appropriate) and sustained packet flows such as long file transfers, Web page downloads, or even steady flows as in audio or video streaming (for which virtual-circuit forwarding is appropriate). In essence these systems attempt to identify longer-term packet flows and to set up shortcuts by using forwarding tables so that packets in a flow are forwarded immediately without the need for route lookup processing. This approach reduces the delay experienced in the packet switch and is discussed further in Chapter 10. Resource reservation procedures for allocating resources to long-term flows have also been developed for datagram networks. We also discuss this in Chapter 10.

optical connections so error control is done only end to end. We discuss ATM in more detail in section 7.6.

7.4   ROUTING IN PACKET NETWORKS

A packet-switched network consists of nodes (routers or switches) interconnected by communication links in an arbitrary meshlike fashion as shown in Figure 7.23. As suggested by the figure, a packet could take several possible paths from host A to host B. For example, three possible paths are 1-3-6, 1-4-5-6, and 1-2-5-6. However, which path is the "best"one? Here the meaning of the term *best* depends on the objective function that the network operator tries to optimize. If the objective is to minimize the number of hops, then path 1-3-6 is the best. If each link incurs a certain delay and the objective function is to minimize the end-to-end delay, then the best path is the one that gives the end-to-end minimum delay. Yet a third objective function involves selecting the path with the greatest available bandwidth. The purpose of the routing algorithm is to identify the set of paths that are best in a sense defined by the network operator. Note that a routing algorithm must have global knowledge about the network state in order to perform its task.

The main ingredients of a good routing algorithm depend on the objective function that one is trying to optimize. However, in general a routing algorithm should seek one or more of the following goals:

1. *Rapid and accurate delivery of packets.* A routing algorithm must operate correctly; that is, it must be able to find a route to the destination if it exists. In addition, the algorithm should not take an unreasonably long time to find the route to the destination.
2. *Adaptability to changes in network topology resulting from node or link failures.* In a real network equipment and transmission lines are subject to failures.

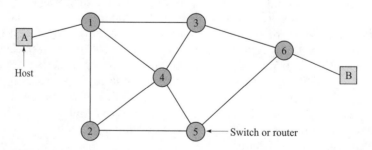

FIGURE 7.23 An example of a packet-switch network

Thus a routing algorithm must be able to adapt to this situation and reconfigure the routes automatically when equipment fails.

3. *Adaptability to varying source-destination traffic loads.* Traffic loads are quantities that are changing dynamically. In a period of 24 hours, traffic loads may go into cycles of heavy and light periods. An adaptive routing algorithm would be able to adjust the routes based on the current traffic loads.

4. *Ability to route packets away from temporarily congested links.* A routing algorithm should try to avoid heavily congested links. Often it is desirable to balance the load on each link.

5. *Ability to determine the connectivity of the network.* To find optimal routes, the routing system needs to know the connectivity or reachability information.

6. *Low overhead.* A routing system typically obtains the connectivity information by exchanging control messages with other routing systems. These messages represent an overhead that should be minimized.

7.4.1   Routing Algorithm Classification

One can classify routing algorithms in several ways. Based on their responsiveness, routing can be static or dynamic (or adaptive). In **static routing** the network topology determines the initial paths. The precomputed paths are then manually loaded to the routing table and remain fixed for a relatively long period of time. Static routing may suffice if the network topology is relatively fixed and the network size is small. Static routing becomes cumbersome as the network size increases. The biggest disadvantage of static routing is its inability to react rapidly to network failures. In **dynamic (adaptive) routing** each router continuously learns the state of the network by communicating with its neighbors. Thus a change in a network topology is eventually propagated to all the routers. Based on the information collected, each router can compute the best paths to desired destinations. One disadvantage of dynamic routing is the added complexity in the router.

Routing algorithms can be centralized or distributed. In **centralized routing** a network control center computes all routing paths and then uploads this information to the routers in the network. In **distributed routing** routers cooperate by means of message exchanges and perform their own routing computations. Distributed routing algorithms generally scale better than centralized algorithms but are more likely to produce inconsistent results. If the routes calculated by different routers are inconsistent, loops can develop. That is, if A thinks that the best route to Z is through B and B thinks that the best route to Z is through A, then packets destined for Z that have the misfortune of arriving at A or B will be stuck in a loop between A and B.

A routing decision can be made on a per packet basis or during the connection setup time. With virtual-circuit packet switching, the path (virtual circuit) is determined during the connection setup phase. Once the virtual circuit is established, all packets belonging to the virtual circuit follow the same route.

Datagram packet switching does not require a connection setup. The route followed by each packet is determined independently.

7.4.2   Routing Tables

Once a routing decision is made, the information has to be stored in a routing table so that the switch (or router) knows how to forward a packet. The specific routing information stored depends on the network type. With virtual-circuit packet switching, the routing table translates each incoming virtual circuit number to an outgoing virtual circuit number and identifies the output port to which to forward the packet. With datagram networks, the routing table identifies the next hop to which to forward the packet based on the destination address of the packet.

Consider a virtual-circuit packet-switching network as shown in Figure 7.24. There are two virtual circuits between host A and switch 1. A packet from host A with VCI 1 in the header will eventually reach host B, while a packet with VCI 5 will eventually reach host D. For each source-destination pair, the VCI has local significance only. At each link the identifier may be translated to a different identifier, depending on the availability of the virtual-circuit numbers at the given switch. In our example VCI 1 from host A gets translated to 2, and then to 7, and finally to 8 at host B. When switch 1 receives a packet with VCI 1, that switch should replace the identifier with 2 and then forward the packet to switch 3. Other switches perform similarly.

Using a local VCI rather than a global one has two advantages. First, more virtual circuits can be assigned, since the virtual-circuit numbers have to be unique only on a link basis rather than on a global basis. If the virtual circuit field in the packet header is two bytes long, then up to 64K virtual circuits can be accommodated on a single link. Second, searching for an available VCI is simple, since a switch has to guarantee uniqueness only on its local link—the information that the switch has in its own routing table. If global virtual-circuit numbers are used, the switch has to make sure that the number if not currently being used by any link along the path, a very time-consuming chore.

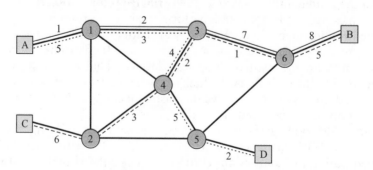

FIGURE 7.24 Virtual-circuit packet switching

The corresponding routing table at each switch is shown in Figure 7.25. We assume that the links are bidirectional and that the VCI is the same for both directions. If a packet with VCI 5 arrives at node 1 from node A, the packet is forwarded to node 3 after the VCI is replaced with 3. After arriving at node 3, the packet receives the outgoing VCI 4 and is then forwarded to node 4. Node 4 translates the VCI to 5 and forwards the packet to node 5. Finally, node 5 translates the VCI to 2 and delivers the packet to the destination, which is host D.

With datagram packet switching, no virtual circuit has to be set up, since no connection exists between a source and a destination. Figure 7.26 shows the routing tables for the network topology in Figure 7.23, assuming that a minimum-hop routing is used. If a packet destined to node 6 arrives at node 1, the packet is first forwarded to node 3 based on the corresponding entry in the routing table at node 1. Node 3 then forwards the packet to node 6. In general, the destination address may be long (32 bits for IPv4), and thus a hash table may be employed to yield a match quickly.

Now suppose that a packet arrives at node 1 from node A and is destined to host D, which is attached to node 5. The routing table in node 1 directs the packet to node 2. The routing table in node 2 directs the packet to node 5, which then delivers the packet to host D.

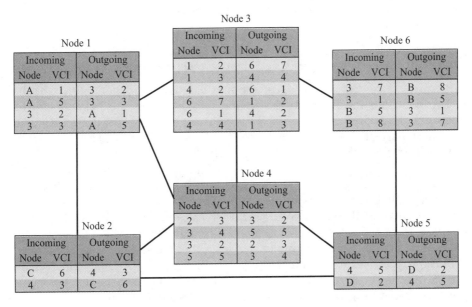

FIGURE 7.25 Routing tables for the network in Figure 7.24

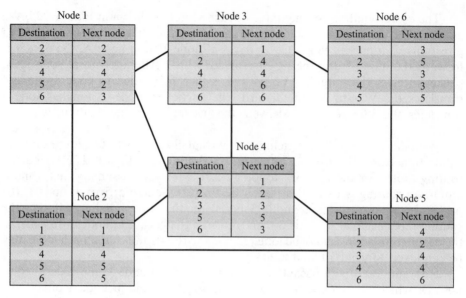

FIGURE 7.26 Routing tables for datagram network

7.4.3 Hierarchical Routing

The size of the routing tables that routers need to keep can be reduced if a hierarchical approach is used in the assignment of addresses. Essentially, hosts that are near each other should have addresses that have common prefixes. In this way routers need to examine only part of the address (i.e., the prefix) in order

---

HIERARCHICAL ADDRESSES IN THE INTERNET

IP addresses consist of two parts: the first part is a unique identifier for the network within the Internet; the second part identifies the host within the network. IP addresses are made hierarchical in two ways. Within a domain the host part of the address may be further subdivided into two parts: an identifier for a *subnetwork* within the domain and a host identifier within the subnet. Outside the domain, routers route packets according to the network part of the destination address. Once a packet arrives to the domain, further routing is done based on the subnetwork address.

The Internet also uses another hierarchy type for addressing, called *supernetting*. Here networks that connect to a common regional network are given addresses that have a common prefix. This technique allows distant routers to route packets that are destined to networks connected to the same region based on the single routing table entry for the prefix. We explain the details of this procedure when we discuss CIDR addressing in Chapter 8.

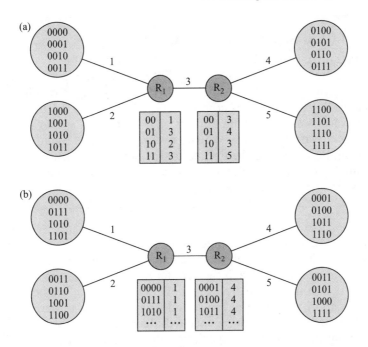

FIGURE 7.27 Hierarchical routing

to decide how a packet should be routed. Figure 7.27 gives an example of hierarchical address assignment and a flat address assignment. In part (a) the hosts at each of the four sites have the same prefix. Thus the two routers need only maintain tables with four entries as shown. On the other hand, if the addresses are not hierarchical (Figure 7.27b), then the routers need to maintain 16 entries in their routing tables.

7.4.4 Link State versus Distance Vector Routing

The set of best paths are invariably found by using a shortest-path algorithm that identifies the set of shortest paths according to some metric. The metric reflects the objective function of the network operator, for example, hops, cost, delay, available bandwidth. To perform the shortest-path calculations, the values of the metrics for different links in the networks are required. Routers must cooperate and exchange information to obtain the values of these metrics. They then use one of the two types of shortest-path algorithms to compute the set of current best routes.

   In the *distance vector routing* approach, neighboring routers exchange routing tables that state the set or vector of known distances to other destinations. After neighboring routers exchange this information, they process it to see whether they can find new better routes through the neighbor that provided

---

UNFINISHED BUSINESS: MULTICASTING

Multicasting involves the delivery of packets to a group of users in a network. Many applications can use multicasting, but the most familiar and suggestive involves receiving the "live" transmission from an audio or video studio. Multicasting has a number of components: addressing to identify multicast groups; mechanisms for joining and leaving a multicast group (i.e., how to "tune in" to a station!); and of course, routing protocols for forwarding the packets from the source to the destinations. Issues in multicasting relating to routing, quality of service, reliability, and security are not completely resolved. We discuss multicast routing in Chapter 8.

---

the information. Distance vector algorithms adapt to changes in network topology gradually as the information on the changes percolates through the network. In the *link state routing* approach each router floods information about the state of the links that connect it to its neighbors. This action allows each router to construct a map of the entire network and from this map to derive the routing table. Both approaches and the associated algorithms are discussed in section 7.5. The application of these algorithms in Internet routing is discussed in Chapter 8.

7.5   SHORTEST-PATH ALGORITHMS

Network routing is a major component at the network layer and is concerned with the problem of determining feasible paths (or routes) from each source to each destination. A router or a packet-switched node performs two main functions: *routing* and *forwarding*. In the routing function an algorithm finds an optimal path to each destination and stores the result in a routing table. In the forwarding function a router forwards each packet from an input port to the appropriate output port based on the information stored in the routing table. In this section we present two commonly implemented shortest-path routing algorithms: the Bellman-Ford algorithm and Dijkstra's algorithm. We then present several other routing approaches, including flooding, deflection routing, and source routing.

Most routing algorithms are based on variants of **shortest-path algorithms**, which try to determine the shortest path for a packet according to some cost criterion. To better understand the purpose of these algorithms, consider a communication network as a graph consisting of a set of *nodes* (or *vertices*) and a set of links (or *edges*, *arcs*, or *branches*), where each node represents a router or a packet switch and each link represents a communication channel between two routers. Figure 7.28 shows such an example. Associated with each link is a value that represents the *cost* (or *metric*) of using that link. For simplicity, it is assumed

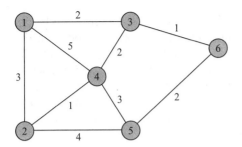

FIGURE 7.28 A sample network with associated link costs

that each link is nondirected. If a link is directed, then the cost must be assigned to each direction. If we define the path cost to be the sum of the link costs along the path, then the shortest path between a pair of nodes is the path with the least cost. For example, the shortest path from node 2 to node 6 is 2-4-3-6, and the path cost is 4.

Many metrics can be used to assign a cost to each link, depending on which function is to be optimized. Examples include

1. *Cost* ~ *1/capacity.* The cost is inversely proportional to the link capacity. Here one assigns higher costs to lower-capacity links. The objective is to send a packet through a path with the highest capacity. If each link has equal capacity, then the shortest path is the path with the minimum number of hops.
2. *Cost* ~ *packet delay.* The cost is proportional to an average packet delay, which includes queueing delay in the switch buffer and propagation delay in the link. The shortest path represents the fastest path to reach the destination.
3. *Cost* ~ *congestion.* The cost is proportional to some congestion measure, for example, traffic loading. Thus the shortest path tries to avoid congested links.

7.5.1   The Bellman-Ford Algorithm

The **Bellman-Ford algorithm** (also called the Ford-Fulkerson algorithm) is based on a principle that is intuitively easy to understand: If a node is in the shortest path between A and B, then the path from the node to A must be the shortest path and the path from the node to B must also be the shortest path. As an example, suppose that we want to find the shortest path from node 2 to node 6 (the destination) in Figure 7.28. To reach the destination, a packet from node 2 must first go through node 1, node 4, or node 5. Suppose that someone tells us that the shortest paths from nodes 1, 4, and 5 to the destination (node 6) are 3, 3, and 2, respectively. If the packet first goes through node 1, *the total distance* (also called total cost) is 3 + 3, which is equal to 6. Through node 4, the total distance is 1 + 3, equal to 4. Through node 5, the total distance is 4 + 2, equal to 6. Thus the shortest path from node 2 to the destination node is achieved if the packet first goes through node 4.

To formalize this idea, let us first fix the destination node. Define D_j to be the current estimate of the minimum cost (or minimum distance) from node j to the destination node and C_{ij} to be the link cost from node i to node j. For example, $C_{12} = C_{21} = 2$, and $C_{45} = 3$ in Figure 7.28. The link cost from node i to itself is defined to be zero (that is, $C_{ii} = 0$), and the link cost between node i and node k is infinite if node i and node k are not directly connected. For example, $C_{15} = C_{23} = \infty$ in Figure 7.28. With all these definitions, the minimum cost from node 2 to the destination node (node 6) can be calculated by

$$
\begin{aligned}
D_2 &= \min\{C_{21} + D_1, C_{24} + D_4, C_{25} + D_5\} \\
&= \min\{3 + 3, 1 + 3, 4 + 2\} \\
&= 4
\end{aligned}
\tag{1}
$$

Thus the minimum cost from node 2 to node 6 is equal to 4, and the next node to visit is node 4.

One problem in our calculation of the minimum cost from node 2 to node 6 is that we have assumed that the minimum costs from nodes 1, 4, and 5 to the destination were known. In general, these nodes would not know their minimum costs to the destination without performing similar calculations. So let us apply the same principle to obtain the minimum costs for the other nodes. For example,

$$
D_1 = \min\{C_{12} + D_2, C_{13} + D_3, C_{14} + D_4\}
\tag{2}
$$

and

$$
D_4 = \min\{C_{41} + D_1, C_{42} + D_2, C_{43} + D_3, C_{45} + D_5\}
\tag{3}
$$

A discerning reader will note immediately that these equations are circular, since D_2 depends on D_1 and D_1 depends on D_2. The magic is that if we keep iterating and updating these equations, the algorithm will eventually converge to the correct result. To see this outcome, assume that initially $D_1 = D_2 = \ldots = D_5 = \infty$. Observe that at each iteration, $D_1, D_2, \ldots, D_5$ are nonincreasing. Because the minimum distances are bounded below, eventually $D_1, D_2, \ldots, D_5$ must converge.

Now if we define the destination node, we can summarize the Bellman-Ford algorithm as follows:

1. Initialization

$$
\begin{aligned}
D_i &= \infty, \forall i \neq d \\
D_d &= 0
\end{aligned}
\tag{4}
$$

2. Updating: For each $i \neq d$,

$$
D_i = \min_j\{C_{ij} + D_j\}, \forall j \neq i
\tag{5}
$$

Repeat step 2 until no more changes occur in the iteration.

Example—Minimum Cost

Using Figure 7.28, apply the Bellman-Ford algorithm to find both the minimum cost from each node to node 6 (the destination) and the next node along the shortest path.

Let us label each node i by (n, D_i), where n is the next node along the current shortest path and D_i is the current minimum cost from node i to the destination. The next node is found from the value of j in equation 5, which gives the minimum cost. If the next node is not defined, we set n to -1. Table 7.1 shows the execution of the Bellman-Ford algorithm at the end of each iteration. The algorithm terminates after the third iteration, since no more changes are observed. The last row records the minimum cost and the next node along the shortest path from each node to node 6.

Iteration	Node 1	Node 2	Node 3	Node 4	Node 5
Initial	$(-1, \infty)$	$(-1, \infty)$	$(-1, \infty)$	$(-1, \infty)$	$(-1, \infty)$
1	$(-1, \infty)$	$(-1, \infty)$	$(6, 1)$	$(3, 3)$	$(6, 2)$
2	$(3, 3)$	$(4, 4)$	$(6, 1)$	$(3, 3)$	$(6, 2)$
3	$(3, 3)$	$(4, 4)$	$(6, 1)$	$(3, 3)$	$(6, 2)$

TABLE 7.1 Sample processing of Bellman-Ford algorithm. Each entry for node j represents the next node and cost of the current shortest path to destination 6.

Example—Shortest-Path Tree

From the preceding example, draw the shortest path from each node to the destination node. From the last row of Table 7.1, we see the next node of node 1 is node 3, the next node of node 2 is node 4, the next node of node 3 is node 6, and so forth. Figure 7.29 shows the shortest-path tree rooted at node 6.

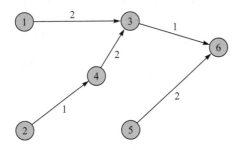

FIGURE 7.29 Shortest-path tree to node 6

One nice feature of the Bellman-Ford algorithm is that it lends itself readily to a distributed implementation. The process involves having each node independently compute its minimum cost to each destination and periodically broadcast the vector of minimum costs to its neighbors. Changes in the routing table can also trigger a node to broadcast the minimum costs to its neighbors to speed up convergence. This mechanism is called *triggered updates*. It turns out that the distributed algorithm would also converge to the correct minimum costs under mild assumptions. Upon convergence, each node would know the minimum cost to each destination and the corresponding next node along the shortest path. Because only cost vectors (or distance vectors) are exchanged among neighbors, the distributed Bellman-Ford algorithm is often referred to as a *distance vector algorithm*. Each node i participating in the distance vector algorithm computes the following equation:

$$D_{ii} = 0$$
$$D_{ij} = \min_k\{C_{ik} + D_{kj}\}, \forall\, k \neq i \tag{6}$$

where D_{ij} is the minimum cost from node i to the destination node j. Upon updating, node i broadcasts the vector $\{D_{i1} D_{i2} D_{i3} \ldots\}$ to its neighbors. The distributed version can adapt to changes in link costs or topology as the next example shows.

Example—Recomputing Minimum Cost

Suppose that after the distributed algorithm stabilizes for the network shown in Figure 7.28, the link connecting node 3 and node 6 breaks. Compute the minimum cost from each node to the destination node (node 6), assuming that each node immediately recomputes its cost after detecting changes and broadcasts its routing updates to its neighbors. The new network topology is shown in Figure 7.30.

For simplicity assume that the computation and transmission are synchronous. As soon as node 3 detects that link (3,6) breaks, node 3 recomputes the minimum cost to node 6 and finds that the new minimum cost is 5 via node 4 (as indicated in the first update in Table 7.2). It then sends the new routing update to its neighbors, which are nodes 1 and 4. These nodes then recompute their minimum costs (update 2). Node 1 transmits its routing table to nodes 2, 3, and 4, and node 4 transmits its routing table to nodes 1, 2, 3, and 5. After the messages from nodes 1 and 4 are received, nodes 2 and 3 will update their minimum costs (update 3).

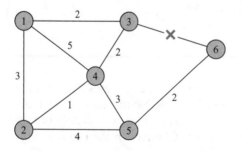

FIGURE 7.30 New network topology following break from node 3 to 6

Update	Node 1	Node 2	Node 3	Node 4	Node 5
Before break	(3, 3)	(4, 4)	(6, 1)	(3, 3)	(6, 2)
1	(3, 3)	(4, 4)	(4, 5)	(3, 3)	(6, 2)
2	(3, 7)	(4, 4)	(4, 5)	(2, 5)	(6, 2)
3	(3, 7)	(4, 6)	(4, 7)	(2, 5)	(6, 2)
4	(2, 9)	(4, 6)	(4, 7)	(5, 5)	(6, 2)
5	(2, 9)	(4, 6)	(4, 7)	(5, 5)	(6, 2)

TABLE 7.2 Next node and cost of current shortest path to node 6

Next nodes 1 and 4 update their minimum costs and send the update messages to their neighbors (update 4). In the last row no more changes are detected, and the algorithm converges. Note that during the calculation, packets already in transit may loop among nodes but eventually find the destination.

Example—Reaction to Link Failure

This example shows that the distributed Bellman–Ford algorithm may react slowly to a link failure. To see this, consider the topology shown in Figure 7.31a with node 4 as the destination. Suppose that after the algorithm stabilizes, link (3,4) breaks, as shown in Figure 7.31b. Recompute the minimum cost from each node to the destination node (node 4).

The computation of minimum costs is shown in Table 7.3. As the table shows, each node keeps updating its cost (in increments of 2 units). At each

(a)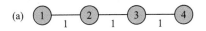

(b)

FIGURE 7.31 Topology before and after link failure

Update	Node 1	Node 2	Node 3
Before break	(2, 3)	(3, 2)	(4, 1)
After break	(2, 3)	(3, 2)	(3, 3)
1	(2, 3)	(3, 4)	(3, 3)
2	(2, 5)	(3, 4)	(3, 5)
3	(2, 5)	(3, 6)	(3, 5)
4	(2, 7)	(3, 6)	(3, 7)
5	(2, 7)	(3, 8)	(3, 7)
...	...	...	...

Note: Dots in the last row indicate that the table continues to infinity

TABLE 7.3 Routing table for Figure 7.31

iteration, node 2 thinks that the shortest path to the destination is through node 3. Likewise, node 3 thinks the best path is through node 2. As a result, a packet in either of these two nodes bounces back and forth until the algorithm stops updating. Unfortunately, in this case the algorithm keeps iterating until the minimum cost is infinite (or very large, in practice), at which point, the algorithm realizes that the destination node is unreachable. This problem is often called **counting to infinity**. It is easy to see that if link (3,4) is restored, the algorithm will converge very quickly. Therefore: Good news travels quickly, bad news travels slowly.

To avoid the counting-to-infinity problem, several changes to the algorithm have been proposed, but unfortunately, none of them work satisfactorily in all situations. One particular method that is widely implemented is called the **split horizon**, whereby the minimum cost to a given destination is not sent to a neighbor if the neighbor is the next node along the shortest path. For example, if node X thinks that the best route to node Y is via node Z, then node X should not send the corresponding minimum cost to node Z. Another variation called **split horizon with poisoned reverse** allows a node to send the minimum costs to all its neighbors; however, the minimum cost to a given destination is set to infinity if the neighbor is the next node along the shortest path. Here, if node X thinks that the best route to node Y is via node Z, then node X should set the corresponding minimum cost to infinity before sending it to node Z.

Example—Split Horizon with Poisoned Reverse

Consider again the topology shown in Figure 7.31a. Suppose that after the alogirthm stabilizes, link (3,4) breaks. Recompute the minimum cost from each node to the destination node (node 4), using the split horizon with poisoned reverse.

The computation of minimum costs is shown in Table 7.4. After the link breaks, node 3 sets the cost to the destination equal to infinity, since the minimum cost node 3 has received from node 2 is also infinity. When node 2 receives the update message, it also sets the cost to infinity. Next node 1 also learns that the destination is unreachable. Thus split horizon with poisoned reverse speeds up convergence in this case.

Update	Node 1	Node 2	Node 3
Before break	(2, 3)	(3, 2)	(4, 1)
After break	(2, 3)	(3, 2)	$(-1, \infty)$
1	(2, 3)	$(-1, \infty)$	$(-1, \infty)$
2	$(-1, \infty)$	$(-1, \infty)$	$(-1, \infty)$

TABLE 7.4 Minimum costs by using split horizon with poisoned reverse

7.5.2   Dijkstra's Algorithm

Dijkstra's algorithm is an alternative algorithm for finding the shortest paths from a source node to all other nodes in a network. It is generally more efficient than the Bellman-Ford algorithm but requires each link cost to be positive, which is fortunately the case in communication networks. The main idea of Dijkstra's algorithm is to keep identifying the closest nodes from the source node in order of increasing path cost. The algorithm is iterative. At the first iteration the algorithm finds the closest node from the source node, which must be the neighbor of the source node if link costs are positive. At the second iteration the algorithm finds the second-closest node from the source node. This node must be the neighbor of either the source node or the closest node to the source node; otherwise, there is a closer node. At the third iteration the third-closest node must be the neighbor of the first two closest nodes, and so on. Thus at the kth iteration, the algorithm will have determined the k closest nodes from the source node.

The algorithm can be implemented by maintaining a set N of *permanently labeled nodes*, which consists of those nodes whose shortest paths have been determined. At each iteration the next-closest node is added to the set until all nodes are used. To formalize the algorithm, let us define D_i to be the current minimum cost from the source node (labeled s) to node i. Dijkstra's algorithm can be described as follows:

1. Initialization:

$$N = \{s\}$$
$$D_j = C_{sj}, \forall\, j \neq s \qquad (7)$$
$$D_s = 0$$

2. Finding the next closest node: Find node $i \notin N$ such that

$$D_i = \min_{j \notin N} D_j \qquad (8)$$

   Add i to N.
   If N contains all the nodes, stop.

3. Updating minimum costs: For each node $j \notin N$

$$D_j = \min\{D_j, D_i + C_{ij}\} \qquad (9)$$

   Go to step 2.

Example—Finding the Shortest Path

Using Figure 7.28, apply Dijkstra's algorithm to find the shortest paths from the source node (assumed to be node 1) to other nodes.

Table 7.5 shows the execution of Dijkstra's algorithm at the end of the initialization and each iteration. At each iteration the value of the minimum cost of the next closest node is underlined. In case of a tie, the closest node

Iteration	N	D_2	D_3	D_4	D_5	D_6
Initial	{1}	3	2	5	∞	∞
1	{1,3}	3	2	4	∞	3
2	{1,2,3}	3	2	4	7	3
3	{1,2,3,6}	3	2	4	5	3
4	{1,2,3,4,6}	3	2	4	5	3
5	{1,2,3,4,5,6}	3	2	4	5	3

TABLE 7.5 Execution of Dijkstra's algorithm

can be chosen randomly. The minimum cost for each node not permanently labeled is then updated sequentially. The last row records the minimum cost to each node.

If we also keep track of the predecessor node of the next-closest node at each iteration, we can obtain a shortest-path tree rooted at node 1, such as shown in Figure 7.32. When the algorithm stops, is knows the minimum cost to each node and the next node along the shortest path. For a datagram network, the routing table at node 1 looks like Table 7.6.

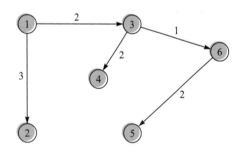

FIGURE 7.32 Shortest-path tree from node 1 to other nodes

Destination	Next node	Cost
2	2	3
3	3	2
4	3	4
5	3	5
6	3	3

TABLE 7.6 Routing table for Figure 7.32

To calculate the shortest paths, the Dijkstra algorithm requires the costs of all links to be available to the algorithm. Thus these link values must be communicated to the processor that is carrying out the computation. The class of *link state* routing algorithms uses this approach to calculating shortest paths. We discuss these algorithms further in Chapter 8 when we consider how the Internet handles routing.

7.5.3   Other Routing Approaches

Various other routing techniques may be used for other purposes. In this section we look at three common approaches: flooding, deflection routing, and source routing.

FLOODING

The principle of **flooding** calls for a packet switch to forward an incoming packet to all ports except the one the packet was received from. If each switch performs this flooding process, the packet will eventually reach the destination as long as at least one path exists between the source and the destination. Flooding is a very effective routing approach when the information in the routing tables is not available, such as during system startup, or when survivability is required, such as in military networks. Flooding is also effective when the source needs to send a packet to all hosts connected to the network (i.e., broadcast delivery). For example, we will see that the link state routing algorithm uses flooding to distribute the link state information.

However, flooding may easily swamp the network as one packet creates multiple packets that in turn create multiples of multiple packets, generating an exponential growth rate as illustrated in Figure 7.33. Initially one packet arriving at node 1 triggers three packets to nodes 2, 3, and 4. In the second phase nodes 2, 3, and 4 send two, two, and three packets, respectively. These packets arrive at nodes 2 through 6. In the third phase 15 more packets are generated, giving a total of 25 packets after three phases. Clearly, flooding needs to be controlled so that packets are not generated excessively. To limit such a behavior, one can implement a number of mechanisms.

One simple method is to use a time-to-live field in each packet. When the source sends a packet, the time-to-live field is initially set to some small number (say, 10 or smaller). Each switch decrements the field by one before flooding the packet. If the value reaches zero, the switch discards the packet. To avoid unnecessary waste of bandwidth, the time-to-live should ideally be set to the minimum hop number between two furthest nodes (called the diameter of the network). In Figure 7.33 the diameter of the network is two. To have a packet reach any destination, it is sufficient to set the time-to-live field to two.

In the second method, each switch adds its identifier to the header of the packet before it floods the packet. When a switch encounters a packet that contains the identifier of the switch, it discards the packet. This method effectively prevents a packet from going around a loop.

The third method is similar to the second method in that they both try to discard old packets. The only difference lies in the implementation. Here each packet from a given source is identified with a unique sequence number. When a switch receives a packet, the switch records the source address and the sequence

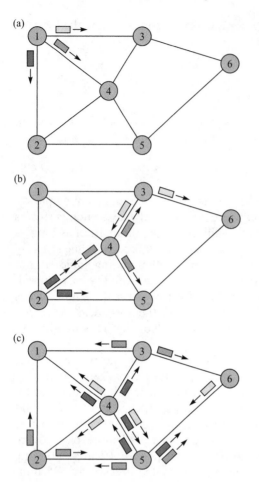

FIGURE 7.33 Flooding is initiated from node 1

number of the packet. If the switch discovers that the packet has already visited the switch, based on the stored source address and sequence number, it will discard the packet.

DEFLECTION ROUTING

Deflection routing was first proposed by Paul Baran in 1964 under the name of *hot-potato routing*. To work effectively, this approach requires the network to provide multiple paths for each source-destination pair. Each switch first tries to forward a packet to the preferred port. If the preferred port is busy or congested, the packet is deflected to another port. Deflection routing usually works well in a regular topology. One example of a regular topology is shown in Figure 7.34, which is called the **Manhattan street network**, since it resembles the streets of New York City. Each column represents an avenue, and each row represents a street. Each switch is labeled (i, j) where i denotes the row number and j denotes the column number. The links have directions that alternate for each column or row. If switch (0,2) would like to send a packet to switch (1,0), the packet could go

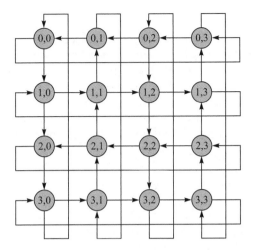

FIGURE 7.34 Manhattan street network

two left and one down. However, if the left port of switch (0,1) is busy (see Figure 7.35), the packet will be deflected to switch (3,1). Then it can go through switches (2,1), (1,1), (1,2), (1,3) and eventually reach the destination switch (1,0).

One advantage of deflection routing is that the switch can be bufferless, since packets do not have to wait for a specific port to become available. If the preferred port is unavailable, the packet can be deflected to another port, which will eventually find its own way to the destination. Since packets can take alternative paths, deflection routing cannot guarantee in-sequence delivery of packets. Deflection routing is a very strong candidate in optical networks where optical buffers are currently expensive and difficult to build. Deflection routing is also used to implement many high-speed packet switches where the topology is usually very regular and high-speed buffers are relatively expensive compared to deflection routing logic.

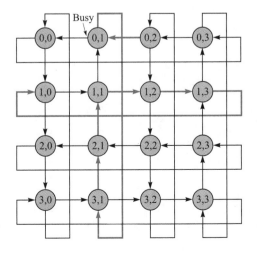

FIGURE 7.35 Deflection routing in Manhattan street network

SOURCE ROUTING

Source routing is a routing approach that does not require an intermediate node to maintain a routing table, but rather puts more burden at the source hosts. Source routing works in either datagram or virtual-circuit packet switching. Before a source host can send a packet, the host has to know the complete route to the destination host in order to include the route information in the header of the packet. The route information contains the sequence of nodes to traverse and should give the intermediate node sufficient information to forward the packet to the next node until the packet reaches the destination. Figure 7.36 shows how source routing works.

Each node examines the header, strips off the label identifying the node, and forwards the packet to the next node. The source (host A) initially includes the entire route (1,3,6,B) in the packet to be destined to host B. Switch 1 strips off its label and forwards the packet to switch 3. The route specified in the header now contains 3,6,B. Switch 3 and switch 6 perform the same function until the packet reaches host B, which finally verifies that it is the intended destination.

In certain situations it may be useful to preserve the complete route information while the packet is progressing. With complete route information host B can send a packet back to host A by simply reversing the route. Thus host B does not have to learn the route to host A a priori. Route preservation can easily be implemented by introducing another field in the header that keeps track of the next node to be visited in the route so that a node knows which specific label to read.

The current version of Internet Protocol, IPv4, and the next version, IPv6, provide an option for source routing of IP packets.

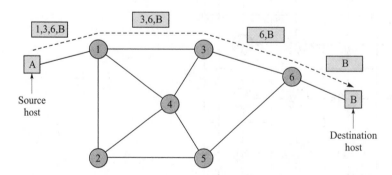

FIGURE 7.36 Example of source routing

7.6 ATM NETWORKS

Asynchronous transfer mode (ATM) is a method for multiplexing and switching that supports a broad range of services. ATM is a connection-oriented packet-switching technique that generalizes the notion of a virtual connection to one that provides Quality-of-Service guarantees.

ATM combines several desirable features of packet switching and time-division multiplexing (TDM) circuit switching. Table 7.7 compares four features of TDM and packet multiplexing. The first comparison involves the capability to support services that generate information at a variable bit rate. Packet multiplexing easily handles variable bit rates. Because the information generated by the service is simply inserted into packets, the variable-bit-rate nature of the service translates into the generation of the corresponding packets. Variable-bit-rate services can therefore be accommodated as long as packets are not generated at a rate that exceeds the speed of the transmission line. TDM systems, on the other hand, have significantly difficulty supporting variable-bit-rate services. Because TDM systems transfer information at a constant bit rate, the bit rates that TDM can support are multiples of some basic rate, for example, 64 kbps for telephone networks.

The second comparison involves the delay incurred in traversing the network. In the case of TDM, once a connection is set up the delays are small and nearly constant. Packet multiplexing, on the other hand, has inherently variable transfer delays because of the queueing that takes place in the multiplexers. Packet multiplexing also has difficulty in providing particular services with low delay. For example, because packets can be of variable length, when a long packet is undergoing transmission, all other packets including urgent ones must wait for the duration of the transmission.

The third criterion for comparison is the capability to support bursty traffic. TDM dedicates the transmission resources, namely, slots, to a connection. If the connection is generating information in a bursty fashion, then many of the dedicated slots go unused. Therefore, TDM is inefficient for services that generate bursty information. Packet multiplexing, on the other hand, was developed specifically to handle bursty traffic and can do so in an efficient way.

Processing is the fourth comparison criterion. In TDM, hardware handles the transfer of slots, so the processing is minimal and can be done at very high speeds. Packet multiplexing, on the other hand, traditionally uses software to

	Variable bit rate	Delay	Bursty traffic	Processing
TDM	Multirate only	Low, fixed	Inefficient	Minimal, very high speed
Packet	Easily handled	Variable	Efficient	Header and packet processing required

TABLE 7.7 TDM versus packet multiplexing

process the information in the packet headers. Consequently, at the time that ATM was formulated, packet multiplexing was slow and processing intensive. However, the development of hardware techniques has reduced packet-processing times and has made very high speed, variable-length packet systems viable.

ATM was developed in the mid-1980s to combine the advantages of TDM and packet multiplexing. ATM involves the conversion of all information flows into short fixed-length packets called **cells**. Cells contain abbreviated headers, or labels, which are essentially pointers to tables in the switches. In terms of the four criteria. ATM has the following features. Because it is packet based, ATM can easily handle services that generate information in bursty fashion or at variable bit rates. The abbreviated header of ATM and the fixed length facilitate hardware implementations that result in low delay and high speeds.

Figure 7.37 shows the operation of an ATM multiplexer. The information flows generated by various users are converted into cells and sent to an ATM multiplexer. The multiplexer arranges the cells into one or more queues and implements some scheduling strategy that determines the order in which cells are transmitted. The purpose of the scheduling strategy is to provide for the different qualities of service required by the different flows. ATM does not reserve transmission slots for specific information flows, and so it has the efficiencies of packet multiplexing. The reason for the term *asynchronous* is that the transmission of cells is not synchronized to any frame structure as in the case of TDM systems.

ATM networks are connection-oriented and require a connection setup prior to the transfer of cells. The connection setup is similar to that described for virtual-circuit packet-switched networks. The connection setup procedure requires the source to provide a *traffic descriptor* that describes the manner in which cells are produced, for example, peak cell rate in cells/second, sustainable (long-term average) cell rate in cells/second, and maximum length of a burst of cells. The source also specifies a set of Quality-of-Service (QoS) parameters that

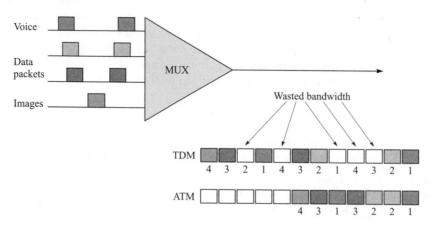

FIGURE 7.37 ATM multiplexing

the connection must provide, for example, cell delay, cell loss, and cell delay jitter. The connection setup procedure involves identifying a path through the network that can meet these requirements. A connection admission control procedure is carried out at every multiplexer along the path. This path is called a **virtual channel connection (VCC)**.

The VCC is established by a chain of *local identifiers* that are defined during the connection setup at the input port to each switch between the source and the destination. The generic packet-switch structure in Figure 7.10 can be modified to carry out ATM switching. Figure 7.38 shows the tables associated with two of the input ports to an ATM switch. In input port 5 we have cells from a voice stream arriving with identifier 32 in the header. We also have cells from a video stream arriving with identifier 25. When a cell with identifier 32 arrives at input port 5, the table lookup for entry 32 indicates that the cell is to be switched to output port 1 and that the identifier in the header is to be changed to 67. Similarly, cells arriving at port 5 with identifier 25 are switched to output port N with new identifier 75. Note that the identifier is locally defined for each input port. Thus input port 6 uses identifier 32 for a different VCC.

At this point it is clear that ATM has a strong resemblance to virtual-circuit packet switching. One major difference is ATM's use of short, fixed-length packets. This approach simplifies the implementation of switches and makes very high speed operation possible. Indeed, ATM switches have proven to be scalable to very large sizes, that is, switches of 10,000 ports with each port running at 150 Mbps are possible. The use of short fixed-length packets also gives a finer degree of control over the scheduling of packet transmissions, since the shorter packets imply a smaller minimum waiting time until the transmission line becomes available for the next transmission.

To understand how the local identifiers are defined in ATM, we first need to see how ATM incorporates some of the concepts used in SONET. SONET allows flows that have a common path through the network to be grouped together. ATM uses the concept of a **virtual path** to achieve this bundling. Figure 7.39 shows five VCCs in an ATM network. The VCCs a, b, and c enter

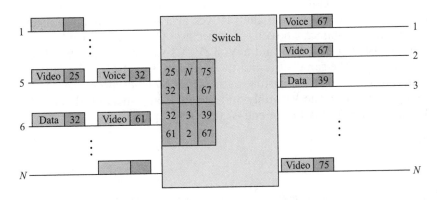

FIGURE 7.38 ATM switching

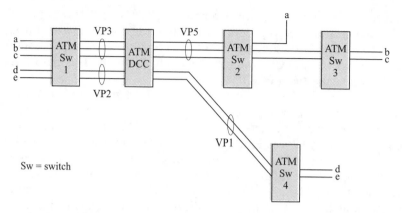

FIGURE 7.39 Role of virtual paths in an ATM network

the network at switch 1, share a common path up to switch 2, and are bundled together into a virtual path connection (VPC) that connects switch 1 to switch 2.[5] This VPC happens to pass through an ATM cross-connect switch whose role in this example is to switch only virtual paths. The VPC that contains VCCs a, b, and c has been given **virtual path identifier (VPI)** 3 between switch 1 and the cross-connect. The cross-connect switches all cells with VPI 3 to the link connecting it to switch number 2 and changes the VPI to 5, which identifies the virtual path between the cross-connect and ATM switch 2. This VPC terminates at switch 2 where the three VCCs are unbundled; cells from VCC a are switched out to a given output port, whereas cells from VCCs b and c proceed to switch 3. Figure 7.39 also shows VCCs d and e entering at switch 1 with a common path to switch 4. These two channels are bundled together in a virtual path that is identified by VPI 2 between switch 1 and the cross-connect and by VPI 1 between the cross-connect and switch 4.

The preceding discussion clearly shows that a virtual circuit in ATM requires two levels of identifiers: an identifier for the VPC, the VPI; and a local identifier for the VCC, the so-called virtual channel identifier, VCI. Figure 7.40 shows a cross-section of the cell stream that arrives at a given input port of an ATM switch or a cross-connect. The cells of a specific VCC are identified by a two-part identifier consisting of a VPI and a VCI. VCCs that have been bundled into a virtual path have the same VPI, and their cells are switched in the same manner over the entire length of the virtual path. At all switches along the virtual path, switching is based on the VPI only and the VCIs are unchanged. The VCIs are used and translated only at the end of the virtual path.

---

[5]We use letters to identify the end-to-end virtual connection. In ATM the network identifies each virtual connection by a chain of locally defined identifiers.We use numbers to indicate these identifiers.

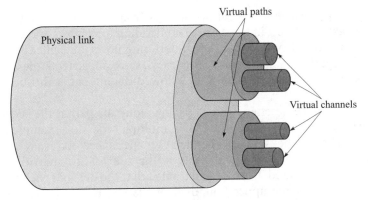

FIGURE 7.40 ATM virtual connections

The details of the VPIs and VCIs are discussed in Chapter 9. However, it is worth noting here that the VCI/VPI structure can support a very large number of connections and hence provides scalability to very large networks.

ATM provides many of the features of SONET systems that facilitate the configuration of the network topology and the management of the bandwidth. The virtual path concept combined with the use of ATM cross-connect switches allows the network operator to dynamically reconfigure the topology seen by the ATM switches using software control. This concept also allows the operator to change the bandwidth allocated to virtual paths. Furthermore, these bandwidth allocations can be done to any degree of granularity, that is, unlike SONET, ATM is not restricted to multiples of 64 kbps.

7.7 Traffic Management and QoS

Traffic management is concerned with the delivery of QoS to specific packet flows. Traffic management entails mechanisms for managing the flows in a network to control the load that is applied to various links and switches. Traffic management also involves the setting of priority and scheduling mechanisms at switches, routers, and multiplexers to provide differentiated treatment for packets and cells belonging to different classes, flows, or connections. It also may involve the policing and shaping of traffic flows as they enter the network.[6]

When we discussed the general structure of switches and routers, we noted in Figure 7.12 that a switch can be viewed as a node where multiplexed packet streams arrive and are then demultiplexed, routed, and remultiplexed onto outgoing lines. Thus the path traversed by a packet through a network can be modeled as a sequence of multiplexers and transmission links as shown in

---

[6]Traffic management also encompasses congestion control, the topic of the next section.

Figure 7.41. The dashed arrows show packets from other flows that "interfere" with the packet of interest in the sense of contending for buffers and transmission along the path. We also note that these interfering flows may enter at one multiplexer and depart at some later multiplexer, since in general they belong to different source-destination pairs and follow different paths through the network.

The performance experienced by a packet along the path is the accumulation of the performance experienced at the N multiplexers. For example, the total *end-to-end delay* is the sum of the delays experienced at each multiplexer. Therefore, the average end-to-end delay is the sum of the individual average delays. On the other hand, if we can guarantee that the delay at each multiplexer can be kept below some upper bound, then the end-to-end delay can be kept below the sum of the upper bounds at the various multiplexers. The *jitter* experienced by packets is also of interest. The jitter measures the variability in the packet delays and is typically measured in terms of the difference of the minimum delay and some maximum value of delay.

Packet loss performance is also of interest. Packet loss occurs when a packet arrives at a multiplexer that has no more buffers available. Causes of packet loss include surges in packet arrivals to a multiplexer and decreased transmission rates out of a multiplexer due to faults in equipment or congestion downstream. The end-to-end probability of packet loss is the probability of packet loss somewhere along the path and is bounded above by the sum of the packet loss probabilities at each multiplexer.

Note that the discussion here is not limited solely to connection-oriented packet transfer. In the case of connectionless transfer of packets, each packet will experience the performance along the path traversed. If each packet is likely to traverse a different path, then it is difficult to make a statement about packet performance. On the other hand, this analysis will hold in connectionless packet-switching networks for the period of time during which a single path is used between a source and a destination. If these paths can be "pinned down" for certain flows in a connectionless network, then the end-to-end analysis is valid.

Packet-switching networks are called upon to support a wide range of services with diverse QoS requirements. To meet the QoS requirements of multiple services, an ATM or packet multiplexer must implement strategies for managing how cells or packets are placed in the queue or queues, as well as control the transmission bit rates that are provided to the various information flows. We now consider a number of these strategies.

FIGURE 7.41 The end-to-end QoS of a packet along a path traversing N hops

7.7.1 FIFO and Priority Queues

The simplest approach to managing a multiplexer involves **first-in**, **first-out** **(FIFO) queueing** where all arriving packets are placed in a common queue and transmitted in order of arrival, as shown in Figure 7.42a. Packets are discarded when they arrive at a full buffer. The delay and loss experienced by packets in a FIFO system depend on the interarrival times and on the packet lengths. As interarrivals become more bursty or packet lengths more variable, performance will deteriorate. Because FIFO queueing treats all packets in the same manner, it is not possible to provide different information flows with different qualities of service. FIFO systems are also subject to *hogging*, which occurs when a user sends packets at a high rate and fills the buffers in the system, thus depriving other users of access to the multiplexer.

A FIFO queueing system can be modified to provide different packet-loss performance to different traffic types. Figure 7.42b shows an example with two classes of traffic. When the number of packets reaches a certain threshold, arrivals of lower access priority (Class 2) are not allowed into the system. Arrivals of higher access priority (Class 1) are allowed as long as the buffer is not full. As a result, packets of lower access priority will experience a higher packet-loss probability.

Head-of-line (HOL) priority queueing is a second approach that involves defining a number of priority classes. A separate queue is maintained for each priority class. As shown in Figure 7.43, each time the transmission line becomes available the next packet for transmission is selected from the head of the line of the highest priority queue that is not empty. For example, packets requiring low delay may be assigned a high priority, whereas packets that are not urgent may be assigned a lower priority. The size of the buffers for the different priority classes can be selected to meet different loss probability requirements. While priority queueing does provide different levels of service to the different classes, it still has shortcomings. For example, it does not allow for providing some

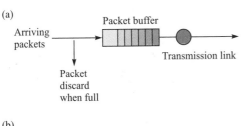

(a)

Arriving packets

Packet buffer

Transmission link

Packet discard when full

(b)

Arriving packets

Packet buffer

Transmission link

Class 1 discard when full

Class 2 discard when threshold exceeded

FIGURE 7.42 (a) FIFO queueing; (b) FIFO queueing with discard priority

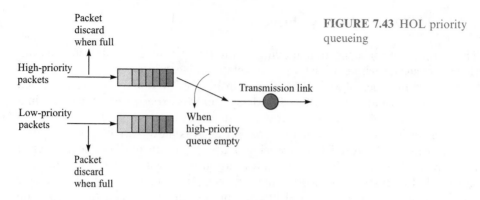

FIGURE 7.43 HOL priority queueing

degree of guaranteed access to transmission bandwidth to the lower priority classes. Another problem is that it does not discriminate between users of the same priority. Fairness problems can arise here when a certain user hogs the bandwidth by sending an excessive number of packets.

A third approach to managing a multiplexer, shown in Figure 7.44, involves sorting packets in the queue according to a priority tag that reflects the urgency with which each packet needs to be transmitted. This system is very flexible because the method for defining priority is open and can even be defined dynamically.[7] For example, the priority tag could consist of a priority class followed by the arrival time of a packet to a multiplexer. The resulting system implements the HOL priority system discussed above. In a second example the priority tag corresponds to a due date. Packets without a delay requirement get indefinite or very long due dates, and are transmitted after all time-critical packets have been transmitted. A third important example that can be implemented by the approach is fair queueing and weighted fair queueing, which are discussed next.

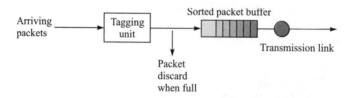

FIGURE 7.44 Sorting packets according to priority tag

---

[7]See [Hashemi 1997] for a discussion on the various types of scheduling schemes that can be implemented by this approach.

7.7.2 Fair Queueing

Fair queueing attempts to provide equitable access to transmission bandwidth. Each user flow has its own logical queue. In an ideal system the transmission bandwidth, say, C bits/second, is divided equally among the queues that have packets to transmit.[8] The contents of each queue can then be viewed as a *fluid* that is drained continuously. Fair queueing prevents the phenomenon of *hogging*, which occurs when an information flow receives an unfair share of the bit rate. The size of the buffer for each user flow can be selected to meet specific loss probability requirements so that the cells or packets of a given user will be discarded when that buffer is full.

Fair queueing is "fair" in the following sense. In the ideal fluid flow situation, the transmission bandwidth is divided equally among all nonempty queues. Thus if the total number of flows in the system is n and the transmission capacity is C, then each flow is guaranteed at least C/n bits/second. In general, the actual transmission rate experienced may be higher because queues will be empty from time to time, so a share larger than C/n bps is received at those times.

In practice, dividing the transmission capacity exactly equally is not possible. As shown in Figure 7.45 one approach could be to service each nonempty queue one bit at a time in round-robin fashion. However, decomposing the resulting bit stream into the component packets would require the introduction of framing information and extensive processing at the demultiplexer. In the case of ATM, fair queueing can be approximated in a relatively simple way. Because in ATM all packets are the same length, the multiplexer need only service the nonempty queues one packet at a time in round-robin fashion. User flows are then guaranteed equal access to the transmission bandwidth.

Figure 7.46 illustrates the differences between ideal or "fluid flow" and packet-by-packet fair queueing. The figure assumes that queue 1 and queue 2 each has a single L-bit packet to transmit at $t = 0$ and that no subsequent packets arrive. Assuming a capacity of $C = L$ bits/second $= 1$ packet/second, the fluid-flow system transmits each packet at a rate of $1/2$ and therefore

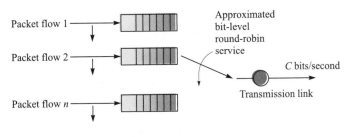

FIGURE 7.45 Fair queueing

---

[8]This technique is called *processor sharing* in the computing literature.

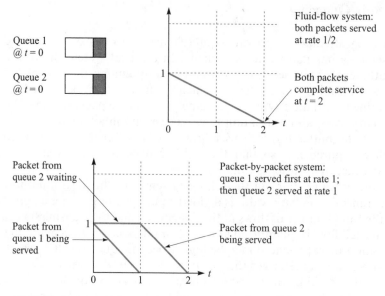

FIGURE 7.46 Fluid-flow and packet-by-packet fair queueing (two packets of equal length)

completes the transmission of both packets exactly at time $t = 2$ seconds. The bit-by-bit system (not shown in the figure) would begin by transmitting one bit from queue 1, followed by one bit from queue 2, and so on. After the first bit each subsequent bit from queue 1 would require $2/L$ seconds to transmit. Therefore, the transmission of the packet from queue 1 would be completed after $1 + 2(L - 1) = 2L - 1$ bit-transmission times, which equals $2 - 1/L$ seconds. The packet from queue 2 is completed at time 2 seconds. The transmission of both packets in the fluid-flow system would be completed at time $2L/L = 2$ seconds. On the other hand, the packet-by-packet fair-queueing system transmits the packet from queue 1 first and then transmits the packet from queue 2, so the packet completion times are 1 and 2 seconds. In this case the first packet is 1 second too early relative to the completion time in the fluid system.

Approximating fluid-flow fair queueing is not as straightforward when packets have variable lengths. If the different user queues are serviced one packet at a time in round-robin fashion, we do not necessarily obtain a fair allocation of transmission bandwidth. For example, if the packets of one flow are twice the size of packets in another flow, then in the long run the first flow will obtain twice the bandwidth of the second flow. A better approach is to transmit packets from the user queues so that the packet completion times approximate those of a fluid-flow fair queueing system. Each time a packet arrives at a user queue, the completion time of the packet is derived from a fluid-flow fair-queueing system. This number is used as a finish tag for the packet. Each time the transmission of a packet is completed, the next packet to be transmitted is the one with the smallest finish tag among all of the user queues. We refer to this system as a **packet-by-packet fair-queueing** system.

Assume that there are n flows, each with its own queue. Suppose for now that each queue is served one bit at a time. Let a *round* consist of a cycle in which all n queues are offered service as shown in Figure 7.47. The actual duration of a given round is the actual number of queues $n_{active}(t)$ that have information to transmit. When the number of active queues is large, the duration of a round is large; when the number of active queues is small, the rounds are short in duration.

Now suppose that the queues are served as in a fluid-flow system. Also suppose that the system is started at $t = 0$. Let $R(t)$ be the number of the rounds at time t, that is, the number of cycles of service to all n queues. However, we let $R(t)$ be a continuous function that increases at a rate that is inversely proportional to the number of active queues; that is:

$$dR(t)/dt = C/n_{active}(t)$$

where C is the transmission capacity. Note that $R(t)$ is a piecewise linear function that changes in slope each time the number of active queues changes. Each time $R(t)$ reaches a new integer value marks an instant at which all the queues have been given an equal number of opportunities to transmit a bit.

Let us see how we can calculate the finish tags to approximate fluid-flow fair queueing. Suppose the kth packet from flow i arrives at an empty queue at time t_k^i and suppose that the packet has length $P(i, k)$. This packet will complete its transmission when $P(i, k)$ *rounds* have elapsed, one round for each bit in the packet. Therefore, the packet completion time will be the value of time t^* when the $R(t)^*$ reaches the value:

$$F(i, k) = R(t_k^i) + P(i, k)$$

We will use $F(i, k)$ as the **finish tag** of the packet. On the other hand, if the kth packet from the ith flow arrives at a nonempty queue, then the packet will have a finish tag $F(i, k)$ equal to the finish tag of the previous packet in its queue $F(i, k - 1)$ plus its own packet length $P(i, k)$; that is:

$$F(i, k) = F(i, k - 1) + P(i, k)$$

The two preceding equations can be combined into the following compact equation:

$$F(i, k) = \max\{F(i, k - 1), R(t_k^i)\} + P(i, k) \quad \text{for fair queueing.}$$

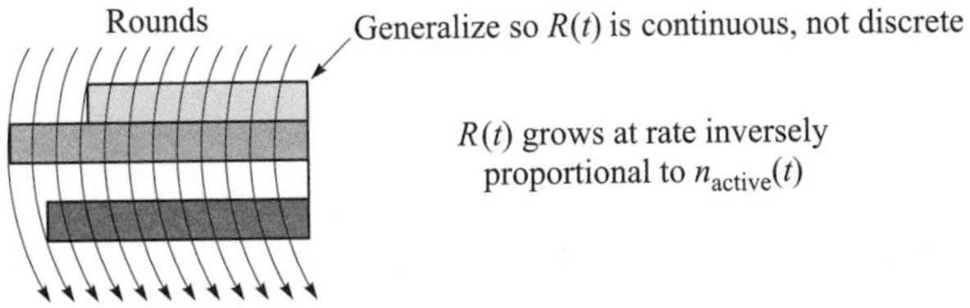

Rounds

Generalize so $R(t)$ is continuous, not discrete

$R(t)$ grows at rate inversely proportional to $n_{active}(t)$

FIGURE 7.47 Computing the finishing time in packet-by-packet fair queueing and weighted fair queueing

We reiterate: The actual packet completion time for the kth packet in flow i in a fluid-flow fair-queueing system is the time t when $R(t)$ reaches the value $F(i, k)$. The relation between the actual completion time and the finish tag is not straightforward because the time required to transmit each bit varies according to the number of active queues.

As an example, suppose that at time $t = 0$ queue 1 has one packet of length one unit and queue 2 has one packet of length two units. A fluid-flow system services each queue at rate 1/2 as long as both queues remain nonempty. As shown in Figure 7.48, queue 1 empties at time $t = 2$. Thereafter queue 2 is served at rate 1 until it empties at time $t = 3$. In the packet-by-packet fair-queueing system, the finish tag of the packet of queue 1 is $F(1, 1) = R(0) + 1 = 1$. The finish tag of the packet from queue 2 is $F(2, 1) = R(0) + 2 = 2$. Since the finish tag of the packet of queue 1 is smaller than the finish tag of queue 2, the system will service queue 1 first. Thus the packet of queue 1 completes its transmissions at time $t = 1$ and the packet of queue 2 completes its transmissions at $t = 3$.

WEIGHTED FAIR QUEUEING

Weighted fair queueing addresses the situation in which different users have different requirements. As before, each user flow has its own queue, but each user flow also has a *weight* that determines its relative share of the bandwidth. Thus if queue 1 has weight 1 and queue 2 has weight 3, then when both queues are nonempty, queue 1 will receive $1/(1 + 3) = 1/4$ of the bandwidth and queue 2 will receive 3/4 of the bandwidth. Figure 7.49 shows the completion times for the

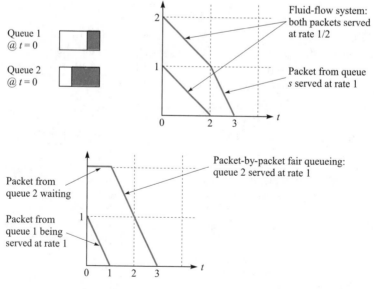

FIGURE 7.48 Fluid flow and packet-by-packet fair queueing (two packets of different lengths)

fluid-flow case where both queues have a one-unit length packet at time $t = 0$. The transmission of the packet from queue 2 is now completed at time $t = 4/3$, and the packet from queue 1 is completed at $t = 2$. The bit-by-bit approximation to weighted fair queueing would operate by allotting each queue a different number of bits/round. In the preceding example, queue 1 would receive 1 bit/round and queue 2 would receive 3 bits/round.

Weighted fair queueing is also easily approximated in ATM: in each round each nonempty queue would transmit a number of packets proportional to its weight. **Packet-by-packet weighted fair queueing** is also easily generalized from fair queueing. Suppose that there are n packet flows and that flow i has weight w_i, then the packet-by-packet system calculates its finish tag as follows:

$$F(i, k) = \max\{F(i, k - 1), R(t_k^i)\} + P(i, k)/w_i \quad \text{for weighted fair queueing.}$$

Thus from the last term in the equation, we see that if flow i has a weight that is twice that of flow j, then the finish tag for a packet from flow i will be calculated assuming a depletion rate that is twice that of a packet from flow j.

Figure 7.49 also shows the completion times for the packet-by-packet weighted fair-queueing system. The finish tag of the packet from queue 1 is $F(1, 1) = R(0) + 1/1 = 1$. The finish tag of the packet from queue 2 is $F(2, 1) + R(0) + 1/3 = 1/3$. Therefore the packet from queue 2 is served first. The packet for queue 2 is now completed at time $t = 1$, and the packet from queue 1 at time $t = 2$. Note that packet-by-packet weighted fair queueing is also applicable when packets are of different length.

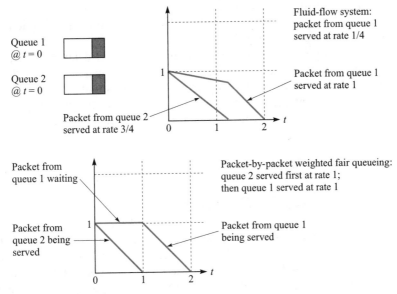

FIGURE 7.49 Fluid flow and packetized, weighted fair queueing

PROVIDING QoS IN THE INTERNET

In order to support real-time audio and video communications the Internet must provide some level of end-to-end QoS. One approach provides *differentiated service* in the sense that some classes of traffic are treated preferentially relative to other classes. This approach does not provide strict QoS guarantees. Packets are instead marked at the edge of the network to indicate the type of treatment that the packets are to receive in the routers inside the network. Modified forms of priority queueing can be used to provide the required differential treatment. A second approach provides *guaranteed service* that gives a strict bound on the end-to-end delay experienced by all packets that belong to a specific flow. This approach requires making resource reservations in the routers along the route followed by the given packet flow. Weighted fair queueing combined with traffic regulators are needed in the routers to provide this type of service. Differentiated service IP and guaranteed service IP are discussed in Chapter 10.

Weighted fair-queueing systems are a means for providing QoS guarantees. Suppose a given user flow has weight w_i and suppose that the sum of the weights of all the user flows is W. In the worst case when all the user queues are nonempty, the given user flow will receive a fraction w_i/W of the bandwidth C. When other user queues are empty, the given user flow will receive a greater share. Thus the user is guaranteed a minimum long-term bandwidth of at least $(w_i/W)C$ bps. This guaranteed share of the bandwidth to a large extent insulates the given user flow from the other user flows.

In addition, section 7.8 shows that if the user information arrival rate is *regulated* to satisfy certain conditions, then the maximum delay experienced in the multiplexer can be guaranteed to be below a certain value. In fact, it is possible to develop guaranteed bounds for the end-to-end delay across a series of multiplexers that use packet-by-packet weighted fair queueing. These bounds depend on the maximum burst that the user is allowed to submit at each multiplexer, on the weights at the various multiplexers, and on the maximum packet size that is allowed in the network. We return to the details of this scheme in section 7.8.

7.8 CONGESTION CONTROL

Congestion occurs when too many packets try to access the same buffer pool in a switch. For an example, consider the communication network shown in Figure 7.50. Suppose that nodes 1, 2, and 5 send bursts of packets to node 4 simultaneously. Assume that the aggregate incoming rate of the packets is greater than the rate at which the packets can be transmitted out. In this case the buffer in

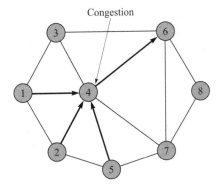

FIGURE 7.50 A congested switch

node 4 will build up. If this situation occurs sufficiently long, the buffer eventually may become full and start rejecting packets. When the destination detects the missing packets, it may ask the sources to retransmit the packets. The sources would unfortunately obey the protocol and send more packets to node 4, making the congestion even worse. In turn, node 4 discards more packets, and this effect triggers the destination to ask for more retransmissions. The net result is that the throughput at the destination will be very low, as illustrated in Figure 7.51 (uncontrolled curve). The purpose of congestion control is to eliminate or reduce congestion. If done properly, performance should improve (controlled curve).

For a novice, it is tempting to claim that congestion can be solved by just allocating a large buffer. However, this solution merely delays congestion from happening. Worse yet, when congestion kicks in, it will last much longer and will be more severe. In the worst case where the buffer size is infinite, packets can be delayed forever!

It turns out that congestion control is a very hard problem to solve. Typically, the solution depends on the application requirements (e.g., qualities of service). A variety of congestion control algorithms have been proposed in the literature. As with routing algorithms, we can classify congestion control algorithms several ways. The most logical approach is to divide them into two broad classes: *open loop* and *closed loop*. Open-loop algorithms prevent congestion from

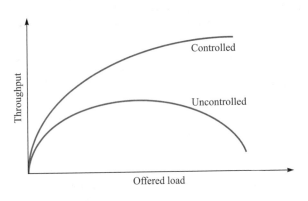

FIGURE 7.51 Throughput drops when congestion occurs

occurring by making sure that the traffic flow generated by the source will not degrade the performance of the network below the specified QoS. If the QoS cannot be guaranteed, the network has to reject the traffic flow. The function that makes the decision to accept or reject the traffic flow is usually called an *admission control*. Thus open-loop algorithms involve some type of resource reservation. Closed-loop algorithms, on the other hand, react to congestion when it is already happening or is about to happen, typically by regulating the traffic flow according to the state of the network. These algorithms are called closed loop because the state of the network has to be fed back to the point that regulates the traffic, which is usually the source. Closed-loop algorithms typically do not use any reservation.

It is important to note that congestion control algorithms are an effective way to reduce temporary overloads in the network (typically on the order of several milliseconds). If the overload lasts longer (several seconds to minutes), then adaptive routing may help by avoiding congested nodes and links. If the overload period is still longer, then the network has to be upgraded, for example, by deploying higher capacity links, faster switches, and so on.

7.8.1 Open-Loop Control

Open-loop congestion control does not rely on feedback information to regulate the traffic flow. Thus this technique assumes that once a source is accepted, its traffic flow will not overload the network. In this section we look at several promising open-loop approaches.

ADMISSION CONTROL

Admission control is an open-loop preventive congestion control scheme. It was initially proposed for virtual-circuit, packet-switched networks such as ATM but has been investigated for datagram networks as well. Admission control typically works at the connection level but can also work at the burst level. The analogy of a connection in datagram networks is a *flow*. At the connection level the function is called a connection admission control (CAC). At the burst level, it is called a burst admission control.

The main idea of CAC is very simple. When a source requests a connection setup, CAC has to decide whether to accept or reject the connection. If the QoS of all the sources (including the new one) that share the same path can be satisfied, the connection is accepted; otherwise, the connection is rejected. The QoS can be expressed in terms of maximum delay, loss probability, delay variance, and other performance parameters.

For CAC to determine whether the QoS can be satisfied, CAC has to know the traffic flow of each source. Thus each source must specify its traffic flow, described by a set of traffic parameters called the *traffic descriptor*, during the connection setup. A traffic descriptor may contain peak rate, average rate, maximum burst size, and so on, and is supposed to summarize the traffic flow compactly and accurately. Figure 7.52 shows an example of a traffic flow gen-

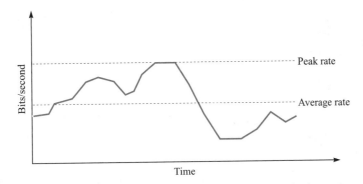

FIGURE 7.52 Example of a traffic flow

erated by a source, indicating the peak rate and the average rate. The maximum burst size usually relates to the maximum length of time the traffic is generated at the peak rate. Based on the characteristics of the traffic flow, CAC has to calculate how much bandwidth it has to reserve for the source. The amount of bandwidth typically lies between the average rate and the peak rate and is called the **effective bandwidth** of the source. The exact calculation for effective bandwidth is very complex and is beyond the scope of this book.

POLICING

Once a connection is accepted by a CAC, the QoS will be satisfied as long as the source obeys the traffic descriptor that it specified during the connection setup. However, if the traffic flow violates the initial contract, the network may not be able to maintain acceptable performance. To prevent the source from violating its contract, the network may want to monitor the traffic flow during the connection period. The process of monitoring and enforcing the traffic flow is called *traffic policing*. When the traffic violates the agreed-upon contract, the network may choose to discard or tag the nonconforming traffic. The tagged traffic will be carried by the network but given lower priority. If there is any congestion downstream, the tagged traffic is the first one to be lost.

Most implementations of traffic policing use the **leaky bucket** algorithm. To understand how a leaky bucket can be used as a policing device, imagine the traffic flows to a policing device as water being poured into a bucket that has a hole at the bottom, as illustrated in Figure 7.53. The bucket has a certain depth and leaks at a constant rate when it is not empty. A new container (that is, packet) of water is said to be *conforming* if the bucket does not overflow when the water is poured in the bucket. The bucket will spill over if the amount of water in the container is too large or if the bucket is nearly full from prior containers. The bucket depth is used to absorb the irregularities in the water flow. If we expect the traffic flow to be very smooth, then the bucket can be made very shallow. If the flow is bursty, the bucket should be deeper. The drain rate corresponds to the traffic rate that we want to police.

FIGURE 7.53 A leaky bucket

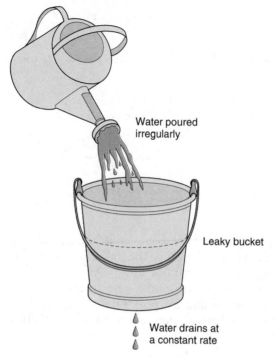

Water poured
irregularly

Leaky bucket

Water drains at
a constant rate

There are many variations of the leaky bucket algorithm. In this section we look at an algorithm that is standardized by the ATM Forum. Here packets are assumed to be of fixed length (i.e., ATM cells). A counter records the content of the leaky bucket. When a packet arrives, the value of the counter is incremented by some value I provided that the content of the bucket would not exceed a certain limit; in this case the packet is declared to be conforming. If the content would exceed the limit, the counter remains unchanged and the packet is declared to be nonconforming. The value I typically indicates the nominal inter-arrival time of the packet that is being policed (typically, in units of packet time). As long as the bucket is not empty, the bucket will drain at a continuous rate of 1 unit per packet time.

Figure 7.54 shows the leaky bucket algorithm that can be used to police the traffic flow. At the arrival of the first packet, the content of the bucket X is set to zero and the last conforming time (LCT) is set to the arrival time of the first packet. The depth of the bucket is $L + I$, where L depends on the maximum burst size. If the traffic is expected to be bursty, then the value of L should be made large. At the arrival of the kth packet, the auxiliary variable X' records the difference between the bucket content at the arrival of the last conforming packet and the interarrival time between the last conforming packet and the kth packet. The auxiliary variable is constrained to be nonnegative. If the auxiliary variable is greater than L, the packet is considered nonconforming. Otherwise, the packet is conforming. The bucket content and the arrival time of the packet are then updated.

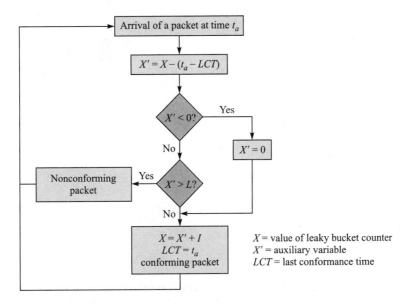

X = value of leaky bucket counter
X' = auxiliary variable
LCT = last conformance time

FIGURE 7.54 Leaky bucket algorithm used for policing

A simple example of the operation of the leaky bucket algorithm is shown in Figure 7.55. Here the value of I is four packet times, and the value of L is six packet times. The arrival of the first packet increases the bucket content by four (packet times). At the second arrival the content has decreased to three, but four more are added to the bucket resulting in a total of seven. The fifth packet is declared as nonconforming since it would increase the content to 11, which would exceed $L + I$. Packets 7, 8, 9, and 10 arrive back to back after the bucket becomes empty. Packets 7, 8, and 9 are conforming, and the last one is non-conforming. If the peak rate is one packet/packet time, then the **maximum burst size (MBS)** for this algorithm is three. Note that the algorithm does not update the content of the bucket continuously, but only at discrete points (arrival times) indicated by the asterisks. Also note that the values of I and L in general can take any real numbers.

Often the inverse of I is the *sustainable rate* which is the long-term average rate allowed for the conforming traffic. Suppose the *peak rate* of a given traffic is denoted by R and its inverse is T; that is, $T = 1/R$. Then the maximum burst size is given by

$$MBS = 1 + \left[\frac{L}{I - T}\right] \tag{10}$$

where $[x]$ gives the greatest integer less than or equal to x. To understand this formula, note that the first packet increases the bucket content to I. After the first packet the bucket content increases by the amount of $(I - T)$ for each packet arrival at the peak rate. Thus we can have approximately $L/(I - T)$

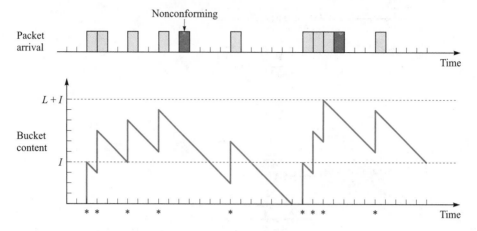

FIGURE 7.55 Behavior of leaky bucket

additional conforming packets. The relations among these quantities are pictorially depicted in Figure 7.56. MBS roughly characterizes the burstiness of the traffic. Bursty traffic may be transmitted at the peak rate for some time and then remains dormant for a relatively long period before being transmitted at the peak rate again. This type of traffic tends to stress the network.

Leaky buckets are typically used to police both the peak rate and the sustainable rate. In this situation *dual leaky buckets* such as the one shown in Figure 7.57 can be used. The traffic is first checked for the peak rate at the first leaky bucket. The cell delay variation tolerance (CDVT) is the amount of variation that is allowed in the peak cell rate. The bucket has a total capacity of T and τ and each arrival of a conforming packet increases the bucket by T. The nonconforming packets at the first bucket are dropped or tagged. The conforming (untagged) packets are then checked for the sustainable rate at the second leaky bucket. The nonconforming packets at the second leaky bucket are also dropped or tagged. The conforming packets are the ones that remain untagged after both leaky buckets.

TRAFFIC SHAPING

When a source tries to send packets, it may not know what its traffic looks like. If the source wants to ensure that the traffic conforms to the parameters specified in the leaky bucket policing device, it should first alter the traffic. The process of altering a traffic flow to another flow is called **traffic shaping**.

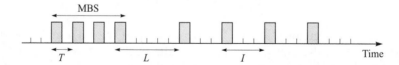

FIGURE 7.56 Relations among MBS and other parameters

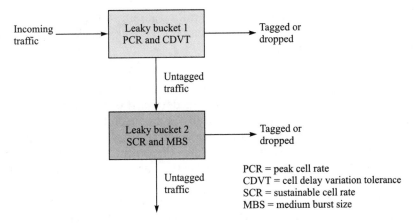

FIGURE 7.57 A dual leaky bucket configuration

Traffic shaping can also be used to make the traffic smoother. Consider an example where an application periodically generates 10 kilobits of data every second. The source can transmit the data in many ways. For example, it can transmit at the rate of 10 kbps continuously. It can transmit at the rate of 50 kbps for 0.2 seconds for each period or at the rate of 100 kbps for 0.1 second for each period, as illustrated in Figure 7.58. From the network's point of view, the traffic shown in Figure 7.58a represents the smoothest pattern, and the one least likely to stress the network. However, the destination may not want to wait for 1 second to receive the data at each period. Another use of traffic shaping is to smooth the traffic flow according to the user's specification.

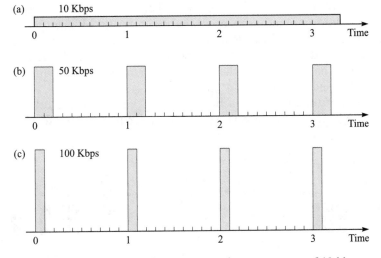

FIGURE 7.58 Possible traffic patterns at the average rate of 10 kbps

There are many implementations of traffic shaping. In this section, we will look at two possibilities. The first one is based on a leaky bucket. The second is usually called a token bucket shaper.

Leaky Bucket Traffic Shaper

A *leaky bucket traffic shaper* is a very simple device. It can be implemented by a buffer whose content is read out periodically at a constant interval, as shown in Figure 7.59. Unlike the leaky bucket policing algorithm, which only monitors the traffic, a leaky bucket traffic shaper regulates the traffic flow. The bucket in the policing algorithm is just a counter, whereas the bucket in the shaper is a buffer that stores packets.

Incoming packets are first stored in a buffer. Packets are served periodically so that the stream of packets at the output is smooth. The buffer is used to store momentary bursts of packets. The buffer size defines the maximum burst that can be accommodated. If the buffer is full, incoming packets are in violation and are thus discarded.

Token Bucket Traffic Shaper

The leaky bucket traffic shaper described above is very restricted, since the output rate is constant when the buffer is not empty. Many applications produce variable-rate traffic. If such applications have to go through the leaky bucket traffic shaper, the delay through the buffer can be unnecessarily long. Recall that the traffic that is monitored by the policing algorithm does not have to be smooth to be conforming. The policing device allows for some burstiness in the traffic as long as it is under a certain limit.

Another more flexible shaper, called the *token bucket traffic shaper*, regulates only the packets that are not conforming. Packets that are deemed conforming are passed through without further delay. In Figure 7.60 we see that the token bucket is a simple extension of the leaky bucket. Tokens are generated periodically at a constant rate and are stored in a token bucket. If the token bucket is full, arriving tokens are discarded. A packet from the buffer can be taken out only if a token in the token bucket can be drawn. If the token bucket is empty, arriving packets have to wait in the packet buffer. Thus we can think of a token as a permit to send a packet.

Imagine that the buffer has a backlog of packets when the token bucket is empty. These backlogged packets have to wait for new tokens to be generated

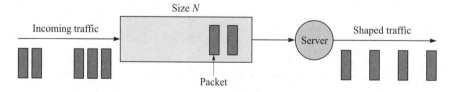

FIGURE 7.59 A leaky bucket traffic shaper

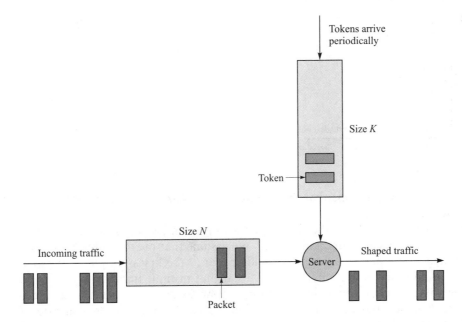

FIGURE 7.60 Token bucket traffic shaper

before they can be transmitted out. Since tokens arrive periodically, these packets will be transmitted periodically at the rate the tokens arrive. Here the behavior of the token bucket shaper is very similar to that of the leaky bucket shaper.

Now consider the case when the token bucket is not empty. Packets are transmitted out as soon as they arrive without having to wait in the buffer, since there is a token to draw for an arriving packet. Thus the burstiness of the traffic is preserved in this case. However, if packets continue to arrive, eventually the token bucket will become empty and packets will start to leave periodically. The size of the token bucket essentially limits the traffic burstiness at the output. In the limit, as the bucket size is reduced to zero, the token bucket shaper becomes a leaky bucket shaper.

QoS GUARANTEES AND SERVICE SCHEDULING

Switches and routers in packet-switched networks use buffers to absorb temporary fluctuations of traffic. Packets that are waiting in the buffer can be scheduled to be transmitted out in a variety of ways. In this section we discuss how the packet delay across a network can be guaranteed to be less than a given value. The technique makes use of a token bucket shaper and weighted fair-queueing scheduling.

Let b be the bucket size in bytes and let r be the token rate in bytes/second. Then in a time period T, the maximum traffic that can exit the traffic shaper is $b + rT$ bytes as shown in Figure 7.61. Suppose we apply this traffic to two multiplexers in tandem each served by transmission lines of speed R bytes/second

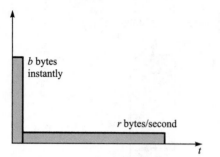

FIGURE 7.61 Maximum traffic allowed out of token bucket shaper

b bytes instantly

r bytes/second

t

with $R > r$. We assume that the two multiplexers are empty and not serving any other flows.

Figure 7.62a shows the multiplexer arrangement, and Figure 7.62b shows the buffer contents as a function of time. We assume that the token bucket allows an immediate burst of *b* bytes to exit and appear at the first multiplexer at $t = 0$, so the multiplexer buffer surges to *b* bytes at that instant. Immediately after $t = 0$, the token bucket allows information to flow to the multiplexer at a rate of *r* bytes/second, and the transmission line drains the multiplexer at a rate of *R* bytes/second. Thus the buffer occupancy falls at a rate of $R - r$ bytes/second. Note that the buffer occupancy at a given instant determines the delay that will be experienced by a byte that arrives at that instant, since the occupancy is exactly the number of bytes that need to be transmitted before the arriving byte is itself transmitted. Therefore, we conclude that the maximum delay at the first multiplexer is bounded by b/R.

Now consider the second multiplexer. At time $t = 0$, it begins receiving bytes from the first multiplexer at a rate of *R* bytes/second. The second multiplexer immediately begins transmitting the arriving bytes also at a rate of *R* bytes/second. Therefore there is no queue buildup in the second multiplexer, and the byte stream flows with zero queueing delay. We therefore conclude that the

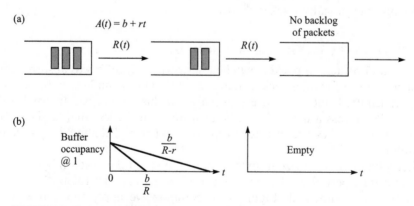

FIGURE 7.62 Delay experienced by token bucket shaped traffic

information that exits the token bucket shaper will experience a delay no greater than b/R over the chain of multiplexers.

Suppose that the output of the token bucket shaper is applied to a multiplexer that uses weighted fair queueing. Also suppose that the weight for the flow has been set so that it is guaranteed to receive at least R bytes/second. It then follows that the flow from the token bucket shaper will experience a delay of at most b/R seconds. This result, however, assumes that the byte stream is handled as a fluid flow. [Parekh 1992] showed that if packet-by-packet weighted fair queueing is used, then the maximum delay experienced by packets that are shaped by a (b, r) token bucket and that traverse H hops is bounded as follows:

$$D \le \frac{b}{R} + \frac{(H-1)m}{R} + \sum_{j=1}^{H} \frac{M}{R_j} \qquad (11)$$

where m is the maximum packet size for the given flow, M is the maximum packet size in the network, H is the number of hops, and R_j is the speed of the transmission line in link j. Also note that $r \le R$. This result provides the basis for setting up connections across a packet network that can guarantee the packet delivery time. This result forms the basis for the *guaranteed delay service* proposal for IP networks.

To establish a connection that can meet a certain delay guarantee, the call setup procedure must identify a route in which the links can provide the necessary guaranteed bandwidth so that the bound is met. This process will involve obtaining information from potential hops about their available bandwidth, selecting a path, and allocating the appropriate bandwidth in the path.

7.8.2   Closed-Loop Control

Closed-loop congestion control relies on feedback information to regulate the source rate. The feedback information can be implicit or explicit. In the implicit feedback the source may use a time-out to decide whether congestion has occurred in the network. In the explicit feedback some form of explicit message will arrive at the source to indicate the congestion state in the network. In the next two subsections, we discuss the closed-loop control used in TCP and in ATM networks. It is interesting to note that TCP exercises congestion control at the transport layer, whereas ATM operates at the network layer.

TCP Congestion Control

Recall from Chapter 5 that TCP uses a sliding-window protocol for end-to-end flow control. This protocol is implemented by having the receiver specify in its acknowledgment the amount of bytes it is willing to receive in the future, called the *advertised window*. The advertised window ensures that the receiver's buffer will never overflow, since the sender cannot transmit data that exceeds the amount that is specified in the advertised window. However, the advertised window does not prevent the buffers in the intermediate routers from overflowing—the condition is called congestion. Routers can become overloaded when they have to cope with too many packets in their buffers. Because IP does not provide any mechanism to control congestion, it is up to the higher layer to detect congestion and take proper actions. It turns out that TCP window mechanism can also be used to control congestion in the network.

The basic idea of TCP congestion control is to have each sender transmit just the right amount of data to keep the network resources utilized but not overloaded. If the senders are too aggressive by sending too many packets, the network will experience congestion. On the other hand, if TCP senders are too conservative, the network will be underutilized. The maximum amount of bytes that a TCP sender can transmit without congesting the network is specified by another window called the *congestion window*. To avoid network congestion and receiver buffer overflow, the maximum amount of data that a TCP sender can transmit at any time is the minimum of the advertised window and the congestion window.

The TCP congestion control algorithm dynamically adjusts the congestion window according to the network state. The operation of the TCP congestion control algorithm may be divided into three phases. The first phase is run when the algorithm starts or restarts, assuming that the pipe is empty. The technique is called *slow start* and is accomplished by first setting the congestion window to one maximum-size segment.[9] Each time the sender receives an acknowledgment from the receiver, the sender increases the congestion window by one segment. After sending the first segment, if the sender receives an acknowledgment before a time-out, the sender increases the congestion window to two segments. If these two segments are acknowledged, the congestion window increases to four segments, and so on. As shown in Figure 7.63, the congestion window size grows exponentially during this phase. The reason for the exponential increase is that slow start needs to fill an empty pipe as quickly as possible. The name "slow start" is perhaps a misnomer, since the algorithm ramps up very quickly.

Slow start does not increase the congestion window exponentially forever, since the pipe will be filled up eventually. Specifically, slow start stops when the congestion window reaches a value specified as the *congestion threshold* which is initially set to 65,535 bytes. At this point a *congestion avoidance* phase takes over. This phase assumes that the pipe is running close to full utilization. It is wise for the algorithm to reduce the rate of increase so that it will not overshoot excessively. Specifically, the algorithm increases the congestion window linearly rather than exponentially during congestion avoidance. This is realized by increasing the congestion window by one segment for each round-trip time.

Obviously, the congestion window cannot be increased indefinitely. The congestion window stops increasing when TCP detects that the network is congested. The algorithm now enters the third phase. At this point the congestion

---

[9] Recall from Chapters 2 and 5 that a segment is the data block or protocol data unit that is used by TCP.

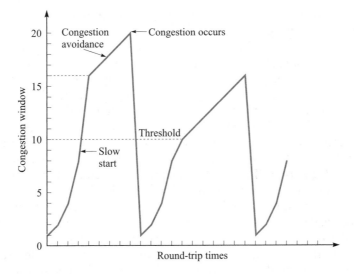

FIGURE 7.63 Dynamics of TCP congestion window

threshold is first set to one-half of the current window size (the minimum of the congestion window and the advertised window, but at least two segments). Next the congestion window is set to one maximum-sized segment. Then the algorithm restarts, using the slow start technique.

How does TCP detect that the network is congested? TCP assumes that congestion occurs in a network when an acknowledgment does not arrive before the time-out expires because of segment loss. The basic assumption the algorithm is making is that a segment loss is due to congestion rather than errors. This assumption is quite valid in a wired network where the percentage of segment losses due to transmission errors is generally low (less than 1 percent). However, it should be noted that the assumption may not be valid in a wireless network where transmission errors can be relatively high.

TCP also detects congestion when a duplicate ACK is received, which can be due to either segment reordering or segment loss. TCP reacts to duplicate ACKs by decreasing the congestion threshold to one-half the current window size, as before. However, the congestion window is *not* reset to one. If the congestion window is less than the new congestion threshold, then the congestion window is increased as in slow start. Otherwise, the congestion window is increased as in congestion avoidance.[10]

Figure 7.63 illustrates the dynamics of the congestion window as time progresses. Initially, slow start kicks in until the congestion threshold (set at 16 segments) is reached. Then congestion avoidance increases the window linearly until a time-out occurs, indicating that the network is congested. The congestion

---

[10]See RFC 2001 for details of the operation.

threshold is set to 10 segments, and the congestion window is set to 1 segment. The algorithm then starts with slow start again.

Our discussion so far assumed that the TCP entity knows when to time out. It turns out that computing the retransmission time-out value (RTO) is not a trivial problem, since network delays are highly variable. In general, we can roughly say that the queueing delay in a router is proportional to $1/(1 - \rho)$, where ρ is the traffic load. Thus the round-trip time (RTT) that is used as the basis for computing RTO varies over time, depending on the traffic load. Each time TCP receives an acknowledgment to a segment, TCP records the current measured RTT and then updates the estimate of RTT. There are two methods for estimating RTT for TCP. The older method involved estimating the time-out RTO as a fixed multiple of RTT; that is, RTO $= \beta$ RTT. Early implementations of TCP used a constant value of β equal to 2. However, the constant value was found to be inadequate as it does not take into account the delay variance. Typically, the delay variance in a queue is proportional to $1/(1 - \rho)^2$. This expression says that when the network is lightly loaded, RTT is almost constant (small variance). When the network is heavily loaded, RTT varies widely (large variance).

In the late 1980s Jacobson proposed an efficient implementation that keeps track of the delay variance [Jacobson 1988]. Specifically, the algorithm estimates a *standard deviation* using a *mean deviation*. The mean deviation gives a somewhat more conservative result than the standard deviation but is much easier to compute. The smoothed mean deviation, DEV, is computed according to

$$DEV \leftarrow \delta DEV + (1 - \delta)|RTT - M| \tag{12}$$

where δ is typically set to 3/4. Once the mean deviation of the round-trip time is found, the RTO is finally set to

$$RTO \leftarrow RTT + 4DEV \tag{13}$$

An ambiguity exists when a segment is retransmitted because the time-out has expired. When the acknowledgment eventually comes back, do we associate it with the first segment or with the retransmitted segment? Karn proposed ignoring the RTT when a segment is retransmitted because using the RTT may corrupt the estimate. This idea is generally called *Karn's algorithm*.

ABR CONGESTION CONTROL FOR ATM NETWORKS

The **available bit rate (ABR)** service in ATM is intended for non-real-time applications such as data communications. ABR service does not have a strict delay or loss constraint. However, the network would try to minimize the cell loss ratio, since a typical data payload would contain many ATM cells and a loss of an ATM cell ruins the entire payload. At connection setup an ABR source is required to specify its peak cell rate (PCR) and its minimum cell rate (MCR), which may be set to zero. During data transfer the network tries to give up as much bandwidth as possible to the PCR but never goes below the MCR. The bandwidth available to ABR sources at any time depends on the residual band-

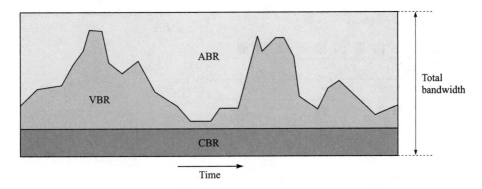

FIGURE 7.64 Bandwidth allocation to services

width in the pipe after the bandwidth for constant bit rate (CBR) and variable bit rate (VBR) sources have been allocated, as shown in Figure 7.64.

ABR congestion control in ATM network uses a rate-based control mechanism that works by continuously adjusting the source rate according to the network state. The information about the network state is carried by special control cells called **resource management (RM)** cells. An ABR source generates forward RM cells periodically at the rate of one RM cell for every $N_{RM} - 1$ of data cells, as shown in Figure 7.65. At the destination the RM cells are turned around and sent back to the source. The backward RM cells carry the feedback information used by the source to control its rate. The way the feedback information is inserted in the RM cell depends on implementations. The next two sections describe possible implementations.

Binary Feedback

The binary feedback scheme allows switches with minimal functionalities to participate in the control loop. Each source sends its data cells with the explicit forward congestion indication (EFCI) bit set to zero to indicate that no congestion is experienced. RM cells that are sent periodically also have the

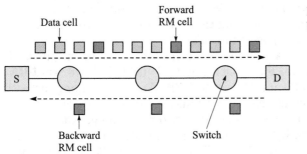

FIGURE 7.65 End-to-end rate-based control

S = source
D = destination

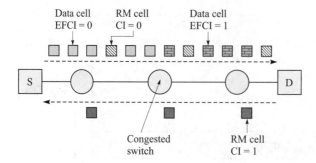

FIGURE 7.66 Binary feedback control

congestion indication (CI) bit set to zero. Each switch along the connection monitors the link congestion status continuously. A switch may decide that the link is congested if the associated queue level exceeds a certain threshold. In this case the switch would set the EFCI bit of all data cells passing through the queue to one, as shown in Figure 7.66.

When data cells received by the destination have the EFCI bit set to one, the destination should set the CI bit of the backward RM cell to one, indicating that the forward path is congested. Else, the CI bit is left unchanged. Note that the resulting technique conveys binary feedback information only, since it can only tell the source whether or not there is congestion in the path.

How does the source react to congestion? When the source receives a backward RM cell with the CI bit set to zero, that source could increase its transmission rate. If the CI bit is one, the source should instead decrease its transmission rate. Typically, the increase would be linear, whereas the decrease would be exponential. In the absence of backward RM cells, the source should decrease its transmission rate. This method, sometimes called *positive feedback*, essentially forces the source to decrease its rate in case the backward path is congested and thus provides a more robust control.

Explicit Rate Feedback

Because the binary feedback scheme can only tell the source to increase or decrease its rate, the scheme tends to converge slowly and oscillate widely around the operating point. Another method, called the *explicit rate* scheme, tries to solve the convergence problem by allowing each switch to explicitly indicate the desired rate to the RM cell that passes through.

The explicit rate scheme works as follows. Each source puts the rate at which it would like to transmit cells in the explicit rate (ER) field of the forward RM cell. The value of the ER field is initially set to PCR. Any switch along the path may reduce the ER value to the desired rate that it can support. However, a switch must not increase the ER value, since doing so would nullify the value set by the more congested switch. If the destination is congested, it may also reduce the ER value before returning the RM cell to the source. When the source receives the backward RM cell, the source adjusts its transmission rate so as not to exceed the ER value. By doing so, every switch along the path is expected

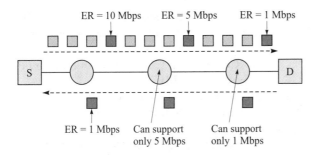

ER = 10 Mbps   ER = 5 Mbps   ER = 1 Mbps

ER = 1 Mbps   Can support   Can support
               only 5 Mbps   only 1 Mbps

FIGURE 7.67 Explicit rate control

to be able to support the rate. Figure 7.67 illustrates the operation of the ER control.

The switch can use many methods to compute its desired rate. One attractive way, which is based on the *max-min fairness* principle, assigns the available bandwidth equally among connections that are bottlenecked on the considered link. If B is the bandwidth to be shared by active connections bottlenecked on the link and n is the number of active connections bottlenecked on the link, then the fair share for connection i, $B(i)$, is

$$B(i) = \frac{B}{n} \tag{14}$$

Enhanced Proportional Rate Control Algorithm

The **enhanced proportional rate control algorithm (EPRCA)** combines both binary feedback and ER feedback schemes, allowing simple switches supporting only EFCI bit setting to interoperate with more complex switches that can compute ERs.

Switches that implement only the EFCI mechanism would ignore the content of the RM cell and would set the EFCI bit to one if the link is congested. Switches that implement the ER scheme may reduce the ER value in the RM cell accordingly if the link is congested. The destinations turn around RM cells, setting the CI bit to one if the last received data cell has the EFCI bit set to one. When the source receives the backward RM cell, the source would set its transmission rate to the minimum value calculated by the binary feedback scheme and the ER value specified in the RM cell.

In EPRCA an ABR source should adhere to the following rules:

1. The source may transmit cells at any rate up to the *allowed cell rate* (ACR). The value of the ACR should be bounded between MCR and PCR.
2. At the call setup time, the source sets ACR to the *initial cell rate* (ICR). The first cell transmitted is an RM cell. When the source has been idle for some time, ACR should also be reduced to ICR.
3. The source should send one RM cell for every $N_{RM} - 1$ data cells or when T_{RM} time (typically set to 100 msec) has elapsed.

4. If the backward RM cell does not return, the source should decrease its ACR by ACR * RDF, down to MCR. RDF stands for rate decrease factor and is typically set to 1/16.

5. When the source receives a backward RM cell with CI = 1, the source should also decrease its ACR by ACR * RDF, down to MCR.

6. When the source receives a backward RM cell with CI = 0, the source may increase the ACR by no more than RIF * PCR, up to the PCR. RIF stands for rate increase factor and is typically 1/16.

7. When the source receives any backward RM cell, the source should set the ACR to the minimum of the ER value from the RM cell and the ACR computed in 5 and 6.

An ABR destination should adhere to the following rules:

1. The destination should turn around all RM cells so that they can return to the source. The direction bit (DIR) in the RM cell should be set to one to indicate a backward RM cell.

2. If the last received data cell prior to a forward RM cell had an EFCI bit set to one, the destination should set the CI bit in the backward RM cell to one. The destination may also reduce the ER value to whatever rate it can support.

Finally, an ATM switch supporting ABR congestion control should adhere to the following rules:

1. The switch should implement either EFCI marking or ER marking. With EFCI marking the switch should set the EFCI bit of a data cell to one when the link is congested. With ER marking the switch may reduce the ER field of forward or backward RM cells.

2. The switch may set the CI bit of the backward RM cell to one to prevent the source from increasing its rate.

3. The switch may generate a backward RM cell to make the source respond faster. In such a case, the switch should set CI = 1 and BN = 1 to indicate that the RM cell is not generated by the source.

SUMMARY

In this chapter we have examined networks that transfer information in the form of blocks called packets. We began with a discussion of how packet information is transferred end-to-end across the Internet. We saw how the Internet protocol achieves this transfer over a variety of networks including LANs and ATMs. We also how the Internet involves the distributed operation of multiple autonomous domains.

Packet networks can offer either connection-oriented service or connectionless services to the transport layer. These services are supported by the internal operation of the packet network, which can involve virtual circuits and datagrams. The Internet is an example of a network that operates internally using datagrams. ATM networks provide an example of virtual-circuit operation. We discussed the advantages and disadvantages of virtual circuits and datagrams in terms of their complexity, their flexibility in dealing with failures, and their ability to provide QoS. We also discussed the structure of generic switches and routers.

We next considered the key topic of routing. We explained the use of routing tables in the selection of the paths that packets traverse in the virtual circuit and in datagram networks. We introduced approaches to synthesizing routing tables and we developed corresponding shortest path algorithms. We also discussed the interplay between hierarchical addressing and routing table size.

We then returned the topic of ATM networks and explained in more detail the use of VCIs in establishing end-to-end connections across a network. We also explained the use of VPIs in managing bandwidth and in creating logically-defined network topologies.

FIFO, priority, and weighted fair queueing mechanisms were introduced and their role in traffic management to provide QoS in packet networks was explained. The combination of traffic regulators and weighted fair queueing to provide guaranteed delay service was also discussed.

Finally, control techniques for dealing with congestion in the network were introduced. We showed how TCP provides for congestion control in the Internet using end-system mechanisms. We also showed how rate-based mechanisms provide congestion control in ATM networks.

CHECKLIST OF IMPORTANT TERMS

asynchronous transfer mode (ATM)
available bit rate (ABR)
Bellman-Ford algorithm
cell
centralized routing
connectionless
connection-oriented
counting to infinity
cut-through packet switching
datagram packet switching
deflection routing
Dijkstra's algorithm
distributed routing
dynamic (adaptive) routing
effective bandwidth
enhanced proportional rate control
    algorithm (EPRCA)

fair queueing
finish tag
first-in, first-out (FIFO) queueing
flooding
head-of-line (HOL) priority queueing
leaky bucket
load
Manhattan street network
maximum burst size (MBS)
message switching
network routing
packet switching
packet-by-packet fair queueing
packet-by-packet weighted fair
    queueing
packets
resource management (RM)

router
shortest-path algorithms
source routing
split horizon
split horizon with poisoned reverse
static routing
traffic shaping

virtual channel identifier (VCI)
virtual-circuit connection (VCC)
virtual-circuit identifier (VCI)
virtual-circuit packet switching
virtual path
virtual path identifier (VPI)
weighted fair queueing

FURTHER READING

Bertsekas, D. and R. Gallagher, *Data Networks*, Prentice-Hall, Englewood Cliffs, New Jersey, 1992.

Halabi, B., *Internet Routing Architectures*, New Riders Publishing, Indianapolis, Indiana, 1997.

Hashemi, M. R. and A. Leon-Garcia, "Implementation of Scheduling Schemes Using a Sequence Circuit," *Voice, Video, and Data Communications*, SPIE, Dallas, November 1997.

Huitema, C., *Routing in the Internet*, Prentice-Hall, Englewood Cliffs, New Jersey, 1995.

Jacobson, V., "Congestion Avoidance and Control," *Computer Communications Review*, Vol. 18, No. 4, August 1988, pp. 314–329.

Keshav, S., *An Engineering Approach to Computer Networking*, Addison-Wesley, Reading, Massachusetts, 1997.

McDysan, D. E. and D. L. Spohn, *ATM: Theory and Application*, McGraw-Hill, New York, 1995.

Parekh, A. K., "A Generalized Processor Sharing Approach to Flow Control in Integrated Services Networks," Ph.D. dissertation, Department of Electrical Engineering and Computer Science, MIT, February 1992.

Perlman, R., *Interconnections: Bridges and Routers*, Addison-Wesley, Reading, Massachusetts, 1992. (The most comprehensive book on bridges and routers.)

Robertazzi, T. G., *Performance Evaluation of High Speed Switching Fabrics*, IEEE Press, 1994.

Saltzer, J. et al., "End-to-end arguments in System Design," *ACM Transactions on Computer Systems*, Vol. 2, No. 4, November 1984, pp. 277–288.

Zhang, H., "Service Disciplines for Guaranteed Performance in Packet Switching Networks," *Proceedings of IEEE*, October 1995, pp. 1374–1396.

RFC 2001, W. Stevens, "TCP Slow Start, Congestion Avoidance, Fast Retransmit, and Fast Recovery Algorithms," January 1997.

PROBLEMS

1. Explain how a network that operates internally with virtual circuits can provide connectionless service. Comment on the delay performance of the service. Can you identify inefficiencies in this approach?

2. Is it possible for a network to offer best-effort virtual-circuit service? What features would such a service have, and how does it compare to best-effort datagram service?

3. Suppose a service provider uses connectionless operation to run its network internally. Explain how the provider can offer customers reliable connection-oriented network service.

4. Where is complexity concentrated in a connection-oriented network? Where is it concentrated in a connectionless network?

5. Comment on the following argument: Because they are so numerous, end systems should be simple and dirt cheap. Complexity should reside inside the network.

6. In this problem you compare your telephone demand behavior and your Web demand behavior.
   a. Arrival rate: Estimate the number of calls you make in the busiest hour of the day; express this quantity in calls/minute. Service time: Estimate the average duration of a call in minutes. Find the load that is given by the product of arrival rate and service time. Multiply the load by 64 kbps to estimate your demand in bits/hour.
   b. Arrival rate: Estimate the number of Web pages you request in the busiest hour of the day. Service time: Estimate the average length of a Web page. Estimate your demand in bits/hour.
   c. Compare the number of call requests/hour to the number of Web requests/hour. Comment on the connection setup capacity required if each Web page request requires a connection setup. Comment on the amount of state information required to keep track of these connections.

7. Apply the end-to-end argument to the question of how to control the delay jitter that is incurred in traversing a multihop network.

8. Compare the operation of the layer 3 entities in the end systems and in the routers inside the network.

9. In Figure 7.6 trace the transmission of IP packets from when a Web page request is made to when the Web page is received. Identify the components of the end-to-end delay.
   a. Assume that the browser is on a computer that is in the same departmental LAN as the server.
   b. Assume that the Web server is in the central organization servers.
   c. Assume that the server is located in a remote network.

10. In Figure 7.6 trace the transmission of IP packets between two personal computers running an IP telephony application. Identify the components of the end-to-end delay.
    a. Assume that the two PCs are in the same departmental LAN.
    b. Assume that the PCs are in different domains.

11. In Figure 7.6 suppose that a workstation becomes faulty and begins sending LAN frames with the broadcast address. What stations are affected by this broadcast storm? Explain why the use of broadcast packets is discouraged in IP.

12. Explain why the distance in hops from your ISP to a NAP is very important. What happens if a NAP becomes congested?

13. Consider the operation of a switch in a connectionless network. What is the source of the load on the processor? What can be done if the processor becomes the system bottleneck?

14. Consider the operation of a switch in a connection-oriented network. What is the source of the load on the processor? What can be done if the processor becomes overloaded?

15. Suppose that a computer that is used as a switch can process 20,000 packets/second. Give a range of possible bit rates that traverse the I/O bus and main memory.

16. A 64-kilobyte message is to be transmitted from over two hops in a network. The network limits packets to a maximum size of 2 kilobytes, and each packet has a 32-byte header. The transmission lines in the network are error free and have a speed of 50 Mbps. Each hop is 1000 km long. How long does it take to get the message from the source to the destination?

17. An audio-visual real-time application uses packet switching to transmit 32 kilobit/second speech and 64 kilobit/second video over the following network connection.

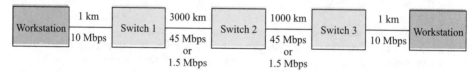

Two choices of packet length are being considered: In option 1 a packet contains 10 milliseconds of speech and audio information; in option 2 a packet contains 100 milliseconds of speech and audio information. Each packet has a 40 byte header.
a. For each option find out what percentage of each packet is header overhead.
b. Draw a time diagram and identify all the components of the end-to-end delay in the preceding connection. Keep in mind that a packet cannot be sent until it has been filled and that a packet cannot be relayed until it is completely received. Assume that bit errors are negligible.
c. Evaluate all the delay components for which you have been given sufficient information. Consider both choices of packet length. Assume that the signal propagates at a speed of 1 km/5 microseconds. Consider two cases of backbone network speed: 45 Mbps and 1.5 Mbps. Summarize your result for the four possible cases in a table with four entries.
d. Which of the preceding components would involve queueing delays?

18. Suppose that a site has two communication lines connecting it to a central site. One line has a speed of 64 kbps, and the other line has a speed of 384 kbps. Suppose each line is modeled by an M/M/1 queueing system with average packet delay given by $E[D] = E[X]/(1 - \rho)$ where $E[X]$ is the average time required to transmit a packet, λ is the arrival rate in packets/second, and $\rho = \lambda E[X]$ is the load. Assume packets have an average length of 8000 bits. Suppose that a fraction α of the packets are routed to the first line and the remaining $1 - \alpha$ are routed to the second line.

a. Find the value of α that minimizes the total average delay.

b. Compare the average delay in part (a) to the average delay in a single multiplexer that combines the two transmission lines into a single transmission line.

19. A message of size m bits is to be transmitted over an L-hop path in a store-and-forward packet network as a series of N consecutive packets, each containing k data bits and h header bits. Assume that $m \gg k + h$. The bit rate of each link is R bits/second. Propagation and queueing delays are negligible.
    a. What is the total number of bits that must be transmitted?
    b. What is the total delay experienced by the message (i.e., the time between the first transmitted bit at the sender and the last received bit at the receiver)?
    c. What value of k minimizes the total delay?

20. Suppose that a datagram network has a routing algorithm that generates routing tables so that there are two disjoint paths between every source and destination that is attached to the network. Identify the benefits of this arrangement. What problems are introduced with this approach?

21. Suppose that a datagram packet network uses headers of length H bytes and that a virtual-circuit packet network uses headers of length h bytes. Use Figure 7.19 to determine the length M of a message for which virtual-circuit switching delivers the packet in less time than datagram switching does. Assume packets in both networks are the same length.

22. Suppose a routing algorithm identifies paths that are "best" in the following sense: (1) minimum number of hops, (2) minimum delay, or (3) maximum available bandwidth. Identify the conditions under which the paths produced by the different criteria are the same? are different?

23. Suppose that the virtual circuit identifiers are unique to a switch, not to an input port. What is traded off in this scenario?

24. Consider the virtual-circuit packet network in Figure 7.24. Suppose that node 4 in the network fails. Reroute the affected calls and show the new set of routing tables.

25. Consider the datagram packet network in Figure 7.26. Reconstruct the routing tables (using minimum-hop routing) that result after node 4 fails. Repeat if node 3 fails instead.

26. Consider the following six-node network. Assume all links have the same bit rate R.

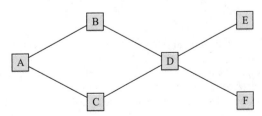

a. Suppose the network uses datagram routing. Find the routing table for each node, using minimum-hop routing.

    b. Explain why the routing tables in part (a) lead to inefficient use of network bandwidth.

    c. Can VC routing improve efficiency in the use of network bandwidth? Explain why or why not.

    d. Suggest an approach in which the routing tables in datagram routing are modified to improve efficiency. Give the modified routing tables.

27. Consider the following six-node unidirectional network. Assume all links have the same bit rate $R = 1$.

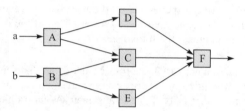

    a. If flows a and b are equal, find the maximum flow that can be handled by the network.

    b. If flow a is three times larger than flow b, find the maximum flow that can be handled by the network.

    c. Repeat (a) and (b) if the flows are constrained to use only one path.

28. Consider the network in Figure 7.30.

    a. Use the Bellman-Ford algorithm to find the set of shortest paths from all nodes to destination node 2.

    b. Now continue the algorithm after the link between node 2 and 4 goes down.

29. Consider the network in Figure 7.28.

    a. Use the Dijkstra algorithm to find the set of shortest paths from node 4 to other nodes.

    b. Find the set of associated routing table entries.

30. Suppose that a block of user information that is L bytes long is segmented into multiple cells. Assume that each data unit can hold up to P bytes of user information, that each cell has a header that is H bytes long, and that the cells are fixed in length and padded if necessary. Define the efficiency as the ratio of the L user bytes to the total number of bytes produced by the segmentation process.

    a. Find an expression for the efficiency as a function of L, H, and P. Use the ceiling function $c(x)$, which is defined as the smallest integer larger or equal to x.

    b. Plot the efficiency for the following ATM parameters: $H = 5$, $P = 48$, and $L = 24k$ for $k = 0, 1, 2, 3, 4, 5$, and 6.

31. Consider a videoconferencing application in which the encoder produces a digital stream at a bit rate of 144 kpbs. The packetization delay is defined as the delay incurred by the first byte in the packet from the instant it is produced to the instant when the packet is filled. Let P and H be defined as they are in problem 30.

    a. Find an expression for the packetization delay for this video application as a function of P.

    b. Find an expression for the efficiency as a function of P and H. Let $H = 5$ and plot the packetization delay and the efficiency versus P.

32. Suppose an ATM switch has 16 ports each operating at SONET OC-3 transmission rate, 155 Mbps. What is the maximum possible throughput of the switch?

33. Refer to the virtual circuit packet network in Figure 7.24. How many VCIs does each connection in the example consume? What is the effect of the length of routes on VCI consumption?

34. Generalize the hierarchical network in Figure 7.27 so that the 2^K nodes are interconnected in a full mesh at the top of the hierarchy and so that each node connects to two 2^L nodes in the next lower level in the hierarchy. Suppose there are four levels in the hierarchy.
    a. How many nodes are in the hierarchy?
    b. What does a routing table look like at level j in the hierarchy, $j = 1, 2, 3,$ and 4?
    c. What is the maximum number of hops between nodes in the network?

35. Assuming that the earth is a perfect sphere with radius 6400 km, how many bits of addressing are required to have a distinct address for every 1 cm × 1 cm square on the surface of the earth?

36. Suppose that 64 kbps PCM coded speech is packetized into a constant bit rate ATM cell stream. Assume that each cell holds 48 bytes of speech and has a 5 byte header.
    a. What is the interval between production of full cells?
    b. How long does it take to transmit the cell at 155 Mbps?
    c. How many cells could be transmitted in this system between consecutive voice cells?

37. Suppose that 64 kbps PCM coded speech is packetized into a constant bit rate ATM cell stream. Assume that each cell holds 48 bytes of speech and has a 5 byte header. Assume that packets with silence are discarded. Assume that the duration of a period of speech activity has an exponential distribution with mean 300 ms and that the silence periods have a duration that also has an exponential distribution but with mean 600 ms. Recall that if T has an exponential distribution with mean $1/\mu$, then $P[T > t] = e^{-\mu t}$.
    a. What is the peak cell rate of this system?
    b. What is the distribution of the burst of packets produced during an active period?
    c. What is the average rate at which cells are produced?

38. Suppose that a data source produces information according to an on/off process. When the source is on, it produces information at a constant rate of 1 Mbps; when it is off, it produces no information. Suppose that the information is packetized into an ATM cell stream. Assume that each cell holds 48 bytes of speech and has a 5 byte header. Assume that the duration of an on period has a Pareto distribution with parameter 1. Assume that the off period is also Pareto but with parameters 1 and α. If T has a Pareto distribution with parameters 1 and α, then $P[T > t] = t^{-\alpha}$ for $t > 1$. If $\alpha > 1$, then $E[T] = \alpha/(\alpha - 1)$, and if $0 < \alpha < 1$, then $E[T]$ is infinite.
    a. What is the peak cell rate of this system?
    b. What is the distribution of the burst packets produced during an on period?
    c. What is the average rate at which cells are producd?

39. An IP packet consists of 20 bytes of header and 1500 bytes of payload. Now suppose that the packet is mapped into ATM cells that have 5 bytes of header and 48 bytes of payload. How much of the resulting cell stream is header overhead?

40. Suppose that virtual paths are set up between every pair of nodes in an ATM network. Explain why connection setup can be greatly simplified in this case.

41. Suppose that the ATM network concept is generalized so that packets can be variable in length. What features of ATM networking are retained? What features are lost?

42. Explain where priority queueing and fair queueing may be carried out in the generic switch/router in Figure 7.10.

43. Consider the head-of-line priority system in Figure 7.43. Explain the impact on the delay and loss performance of the low-priority traffic under the following conditions:
a. The high-priority traffic consists of uniformly spaced, fixed-length packets.
b. The high-priority traffic consists of uniformly spaced, variable-length packets.
c. The high-priority traffic consists of highly bursty, variable-length packets.

44. Consider the head-of-line priority system in Figure 7.43. Suppose that each priority class is divided into several subclasses with different "drop" priorities. Each priority subclass has a threshold that if exceeded by the queue length results in discarding of arriving packets from the corresponding subclass. Explain the range of delay and loss behaviors that are experienced by the different subclasses.

45. Incorporate some form of weighted fair queueing in the head-of-line priority system in Figure 7.43 so that the low-priority traffic is guaranteed to receive r bps out of the total bit rate R of the transmission link. Explain why this feature may be desirable. How does it affect the performance of the high-priority traffic?

46. Consider a packet-by-packet fair-queueing system with three logical queues and with a service rate of one unit/second. Show the sequence of transmissions for this system for the following packet arrival pattern. Queue 1: arrival at time $t = 0$, length 2; arrival at $t = 4$, length 1. Queue 2: arrival at time $t = 1$, length 3; arrival at $t = 2$, length 1. Queue 3: arrival at time $t = 3$, length 5.

47. Repeat problem 46 if queues 1, 2, and 3 have weights, 2, 3, and 5, respectively.

48. Suppose that in a packet-by-packet weighted fair-queueing system, a packet with finish tag F enters service at time t. Is it possible for a packet to arrive at the system after time t and have a finish tag less than F? If yes, give an example. If no, explain why.

49. Deficit round-robin is a scheduling scheme that operates as follows. The scheduler visits the queues in round-robin fashion. A deficit counter is maintained for each queue. When the scheduler visits a queue, the scheduler adds a quantum of service to the deficit counter, and compares the resulting value to the length of the packet at the head of the line. If the counter is larger, the packet is served and the counter is reduced by the packet length. If not, the deficit is saved for the next visit. Suppose that a system has four queues and that these contain packets of length 16, 10, 12, and 8 and that the quantum is 4 units. Show the deficit counter at each queue as a function of time and indicate when the packets are transmitted.

50. Suppose that ATM cells arrive at a leaky bucket policer at times $t =$ 1, 2, 3, 5, 6, 8, 11, 12, 13, 15, and 19. Assuming the same parameters as the example in Figure 7.55, plot the bucket content and identify any nonconforming cells. Repeat if L is reduced to 4.

51. Explain the difference between the leaky bucket traffic shaper and the token bucket traffic shaper.

52. Which of the parameters in the upper bound for the end-to-end delay (equation 11) are controllable by the application? What happens as the bit rate of the transmission links becomes very large?

53. Suppose that a TCP source (with an unlimited amount of information to transmit) begins transmitting onto a link that has 1 Mbps in available bandwidth. Sketch the congestion window versus the time trajectory. Now suppose that another TCP source (also with an unlimited amount of information to transmit) begins transmitting over the same link. Sketch the congestion window versus the time for the initial source.

54. Suppose that TCP is operating in a 100 Mbps link that has no congestion.
    a. Explain the behavior of slow start if the link has RTT = 20 ms, receive window of 20 kbytes, and maximum segment size of 1 kbyte.
    b. What happens if the speed of the link is 1 Mbps? 100 kbps?

55. Random early detection (RED) is a buffer management mechanism that is intended to avoid congestion in a router by keeping average queue length small. The RED algorithm continuously compares a short-time average queue length with two thresholds: min_{th} and max_{th}. When the average queue length is below min_{th}, RED does not drop any packets. When the average queue length is between min_{th} and max_{th} RED drops an arriving packet with a certain probability that is an increasing function of the average queue length. The random packet drop is used to notify the sending TCP to reduce its rate before the queue becomes full. When the average queue length exceeds max_{th}, RED drops each arriving packet.
    a. What impact does RED have on the tendency of TCP receivers to synchronize during congestion?
    b. What is the effect of RED on network throughput?
    c. Discuss the fairness of the RED algorithm with respect to flows that respond to packet drops and nonadaptive flows, for example, UDP.
    d. Discuss the implementation complexity of the RED algorithm.

CHAPTER 8

TCP/IP

The Internet Protocol (IP) enables communications across a vast and heterogeneous collection of networks that are based on different technologies. *Any* host computer that is connected to the Internet can communicate with any other computer that is also connected to the Internet. The Internet therefore offers ubiquitous connectivity and the economies of scale that result from large deployment.

The Internet offers two basic communication services that operate on top of IP: Transmission control protocol (TCP) reliable stream service and user datagram protocol (UDP) datagram service. *Any* application layer protocol that operates on top of either TCP or UDP automatically operates across the Internet. Therefore the Internet provides a ubiquitous platform for the deployment of network-based services.

In Chapter 2 we introduced the TCP/IP protocol suite and showed how the various layers work together to provide end-to-end communications support for applications. In this chapter we examine the TCP/IP protocol suite in greater detail.

The chapter is organized as follows:

1. *TCP/IP architecture.* We have designed this book so that TCP/IP is introduced gradually throughout the text. In this section we summarize the TCP/IP concepts that are introduced in previous chapters. The remainder of the chapter then discusses various components of the TCP/IP protocol suite.
2. *Internet Protocol (IP).* We examine the structure of the network layer: the IP packet, the details of IP addressing, routing, and fragmentation and reassembly. We also discuss how IP is complemented by the Internet Control Message Protocol (ICMP).

3. *IP version 6.* We discuss the motivations for introducing a new version of IP, and we describe the features of IP version 6.
4. *Transport layer protocols.* We discuss the structure of the transport layer: TCP and UDP. We examine in detail how TCP uses ARQ techniques to provide reliable stream service and flow control end to end across connectionless packet networks. We examine the structure of a TCP and its protocol data unit (PDU), and we discuss the state transition diagram of the connection management process.
5. *DHCP and mobile IP.* We discuss two key protocols: Dynamic Host Configuration Protocol (DHCP) provides a mechanism for the temporary allocation of IP addresses to hosts; mobile IP allows a device to use the same IP address regardless of where it attaches to the network.
6. *Internet routing.* A key element of IP is the routing protocols that are used to synthesize the routing tables that direct packets in the end systems and in the routers. We introduce the Routing Information Protocol (RIP) and the Open Shortest Path First (OSPF) protocol for building routing tables within a domain. We also introduce the Border Gateway Protocol (BGP) for inter-domain routing.
7. *Multicast routing.* In the final section we introduce approaches to multicast routing. We also discuss the Internet Group Management Protocol (IGMP) that enables hosts to join a multicast group.

In Chapter 10 we return to a discussion of emerging Internet protocols that deal with Quality of Service (QoS), performance enhancement, and interworking with ATM. IP security protocols are discussed in Chapter 11. Protocols for supporting multimedia services over IP and for signaling are presented in Chapter 12.

8.1   THE TCP/IP ARCHITECTURE

The TCP/IP protocol suite usually refers not only to the two most well-known protocols called the *Transmission Control Protocol (TCP)* and the *Internet Protocol (IP)* but also to other related protocols such as the *User Datagram Protocol (UDP)*, the *Internet Control Message Protocol (ICMP)* and the basic applications such as HTTP, TELNET, and FTP. The basic structure of the TCP/IP protocol suite is shown in Figure 8.1.

We saw in Chapter 2 that application layer protocols such as FTP and HTTP send messages using TCP. Application layer protocols such as SNMP and DNS send their messages using UDP. The PDUs exchanged by the peer TCP protocols are called **TCP segments** or **segments**, while those exchanged by UDP protocols are called **UDP datagrams** or **datagrams**. IP multiplexes TCP segments and UDP datagrams and performs fragmentation, if necessary, among other tasks to be discussed below. The protocol data units exchanged by IP protocols

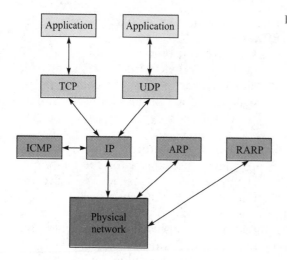

FIGURE 8.1 TCP/IP protocol suite

are called **IP packets** or **packets**.[1] IP packets are sent to the network interface for delivery across the physical network. At the receiver, packets passed up by the network interface are demultiplexed to the appropriate protocol (IP, ARP, or RARP). The receiving IP entity needs to determine whether a packet has to be sent to TCP or UDP. Finally, TCP (UDP) sends each segment (datagram) to the appropriate application based on the port number. The physical network can be implemented by a variety of technologies such as Ethernet, token ring, ATM, PPP over various transmission systems, and others.

The PDU of a given layer is encapsulated in a PDU of the layer below as shown in Figure 8.2. In this figure an HTTP GET command is passed to the TCP layer, which encapsulates the message into a TCP segment. The segment header contains an ephemeral port number for the client process and the well-known port 80 for the HTTP server process. The TCP segment in turn is passed to the IP layer where it is encapsulated in an IP packet. The IP packet header contains an IP *network* address for the sender and an IP *network* address for the destination. IP network addresses are said to be *logical* because they are defined in terms of the *logical topology* of the routers and end systems. The IP packet is then passed through the network interface and encapsulated into a PDU of the underlying network. In Figure 8.2 the IP packet is encapsulated into an Ethernet LAN frame. The frame header contains *physical* addresses that identify the physical endpoints for the sender and the receiver. The logical IP addresses need to be converted into specific physical addresses to carry out the transfer of bits from one device to the other. This conversion is done by an *address resolution protocol.*

Each host in the Internet is identified by a *globally unique IP address.* An IP address is divided into two parts: a *network ID* and a *host ID.* The network ID

---

[1]IP packets are sometimes called IP datagrams. To avoid confusion with UDP datagrams, in this text we use the term packets to refer to the PDUs at the IP layer.

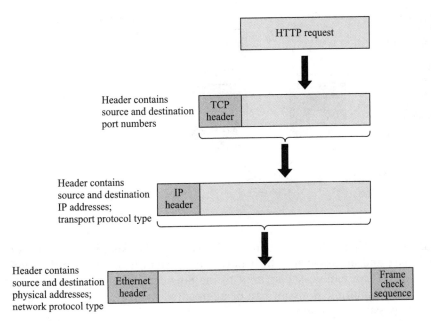

FIGURE 8.2 Encapsulation of PDUs in TCP/IP and addressing information in the headers

must be obtained from an organization authorized to issue IP addresses. The **Internet layer** provides for the transfer of information across multiple networks through the use of routers, as shown in Figure 8.3. The Internet layer provides a single service, namely, *best-effort connectionless packet transfer*. IP packets are exchanged between routers without a connection setup; they are routed independently, and may traverse different paths. The gateways that interconnect the intermediate networks may discard packets when they encounter congestion. The responsibility for recovery from these losses is passed on to the transport layer.

The network interface layer is particularly concerned with the protocols that are used to access the intermediate networks. At each gateway the network protocol is used to encapsulate the IP packet into a packet or frame of the underlying network or link. The IP packet is recovered at the exit router of the given network. This router must determine the next hop in the route to the destination and then encapsulate the IP packet or frame of the type of the next network or link. This approach provides a clear separation of the Internet layer from the technology-dependent network interface layer. This approach also allows the Internet layer to provide a data transfer service that is transparent in the sense of not depending on the details of the underlying networks. Different network technologies impose different limits on the size of the blocks that they can handle. IP must accommodate the **maximum transmission unit** of an underlying network or link by implementing segmentation and reassembly as needed.

To enhance the scalability of the routing algorithms and to control the size of the routing tables, additional levels of hierarchy are introduced in the IP

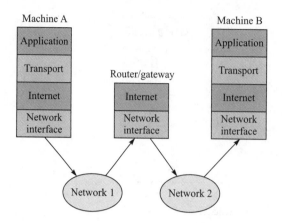

FIGURE 8.3 The Internet and network interface layers

addresses. Within a domain the host address is further subdivided into a *subnetwork* part and an associated host part. This system facilitates routing within a domain, yet can remain transparent to the outside world. At the other extreme, addresses of multiple domains can be aggregated to create *supernets*. We discuss these key issues in section 8.2

8.2   THE INTERNET PROTOCOL

The Internet Protocol (IP) is the heart of the TCP/IP protocol suite. IP corresponds to the network layer in the OSI reference model and provides a connectionless and best-effort delivery service to the transport layer. Recall that a connectionless service does not require a virtual circuit to be established before data transfer can begin. The term *best-effort* indicates that IP will try its best to forward packets to the destination, but does not guarantee that a packet will be delivered to the destination. The term is also used to indicate that IP does not make any guarantee on the QoS.[2] An application requiring high reliability must implement the reliability function within a higher-layer protocol.

8.2.1   IP Packet

To understand the service provided by the IP entity, it is useful to examine the IP packet format, which contains a header part and a data part. The format of the IP header is shown in Figure 8.4.

The header has a fixed-length component of 20 bytes plus a variable-length component consisting of options that can be up to 40 bytes. IP packets are

---

[2]In Chapter 10 we discuss work on integrated services and differentiated IP services that provide some form of QoS in IP.

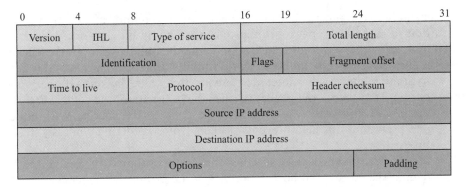

FIGURE 8.4 IP version 4 header

transmitted according to network byte order: bits 0–7 first, then bits 8–15, then bits 16–23, and finally bits 24–31 for each row. The meaning of each field in the header follows.

Version: The version field indicates the version number used by the IP packet so that revisions can be distinguished from each other. The current IP version is 4. Version 5 is used for a real-time stream protocol called ST2, and version 6 is used for the new generation IP know as IPng or IPv6 (to be discussed in the following section).

Internet header length: The Internet header length (IHL) specifies the length of the header in 32-bit words. If no options are present, IHL will have a value of 5. The length of the options field can be determined from IHL.

Type of service: The type of service (TOS) field specifies the priority of the packet based on delay, throughput, reliability, and cost requirements. Three bits are assigned for priority levels (called "precedence") and four bits for the specific requirement (i.e., delay, throughput, reliability, and cost). For example, if a packet needs to be delivered to the destination as soon as possible, the transmitting IP module can set the delay bit to one and use a high-priority level. In practice most routers ignore this field. Recent work in the Differentiated Service Working Group of IETF tries to redefine the TOS field in order to support other services that are better than the basic best effort. The differentiated services model is discussed in Chapter 10.

Total length: The total length specifies the number of bytes of the IP packet including header and data. With 16 bits assigned to this field, the maximum packet length is 65,535 bytes. In practice the maximum possible length is very rarely used, since most physical networks have their own length limitation. For example, Ethernet limits the payload length to 1500 bytes.

Identification, flags, and **fragment offset:** These fields are used for fragmentation and reassembly and are discussed below.

Time to live: The time-to-live (TTL) field is defined to indicate the amount of time in seconds the packet is allowed to remain in the network. However, most routers interpret this field to indicate the number of hops the packet is allowed to traverse in the network. Initially, the source host sets this field to some value. Each router decrements this value by one. If the value reaches zero before the packet reaches the destination, the router discards the packet and sends an error message back to the source. With either interpretation, this field prevents packets from wandering aimlessly in the Internet.

Protocol: The protocol field specifies the protocol that is to receive the IP data at the destination host. Examples of the protocols include TCP (protocol = 6), UDP (protocol = 17), and ICMP (protocol = 1).

Header checksum: The header checksum field verifies the integrity of the header of the IP packet. The data part is not verified and is left to upper-layer protocols. If the verification process fails, the packet is simply discarded. To compute the header checksum, the sender first sets the header checksum field to 0 and then applies the Internet checksum algorithm discussed in Chapter 3. Note that when a router decrements the TTL field, the router must also recompute the header checksum field.

Source IP address and **destination IP address:** These fields contain the addresses of the source and destination hosts. The format of the IP address is discussed below.

Options: The options field, which is of variable length, allows the packet to request special features such as security level, route to be taken by the packet, and timestamp at each router. The options field is rarely used. Router alert is a new option introduced to alert routers to look inside the IP packet. The option is intended for new protocols that require relatively complex processing in routers along the path [RFC 2113]. It is used by RSVP which is discussed in Chapter 10.

Padding: This field is used to make the header a multiple of 32-bit words.

When an IP packet is passed to the router by a network interface, the following processing takes place. First the header checksum is computed and the fields in the header are checked to see if they contain valid values. Next IP fields that need to be changed are updated. For example, the TTL and header checksum fields always require updating. The router then identifies the next hop for the IP packet by consulting its routing tables. The IP packet is then forwarded along the next hop.

8.2.2 IP Addressing

To identify each computer on the Internet, we have to assign a unique address to each computer. A computer (such as a router or a multihomed host) may have multiple network interfaces with each interface connected to a different network. An analogy of this situation is a house having multiple doors with one door facing a street called Main Street and another facing a different street called Broadway. In this situation an IP address is usually associated with the network interface or the network connection rather than with the computer. Thus, in our analogy, an address is assigned to each door of a house rather than to the house itself. For a computer with a single network interface (typically called a host), we can safely think of the IP address as the identity of the host.

An IP address has a fixed length of 32 bits. The address structure was originally defined to have a two-level hierarchy: **network ID** and **host ID**. The network ID identifies the network the host is connected to. Consequently, all hosts connected to the same network have the same network ID. The host ID identifies the network connection to the host rather than the actual host. An implication of this powerful aggregation concept is that a router can forward packets based on the network ID only, thereby shortening the size of the routing table significantly. The network ID is assigned by the Internet Network Information Center (InterNIC). The host ID is assigned by the network administrator at the local site. When TCP/IP is used only within an intranet (an internal and private internet), the local network administrator may wish to assign the network ID on its own. However, the address will not be recognized by a host on the global Internet. The formats of the "classful" IP address are shown in Figure 8.5. The bit position shows the number of bits from the most significant bit.

The IP address structure is divided into five address classes: Class A, Class B, Class C, Class D, and Class E, identified by the most significant bits of the address as shown in the figure. Class A addresses have seven bits for network IDs and 24 bits for host IDs, allowing up to 126 networks and about 16 million hosts per network. Class B addresses have 14 bits for network IDs and 16 bits for host IDs, allowing about 16,000 networks and about 64,000 hosts for each network. Class C addresses have 21 bits for network IDs and 8 bits for host IDs, allowing about 2 million networks and 254 hosts per network. Class D addresses are used for multicast services that allow a host to send information to a group of hosts simultaneously. Class E addresses are reserved for experiments.

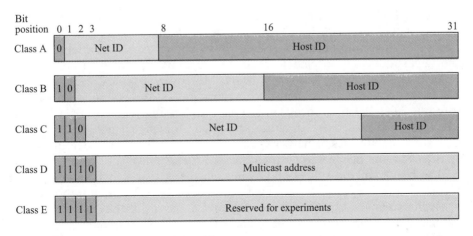

FIGURE 8.5 The five classes of IP addresses

An ID that contains all 1s or all 0s has a special purpose. A host ID that contains all 1s is meant to broadcast the packet to all hosts on the network specified by the network ID. If the network ID also contains all 1s, the packet is broadcast on the local network. A host ID that contains all 0s refers to the network specified by the network ID, rather than to a host. It is possible for a host not to know its IP address immediately after being booted up. In this case the host may transmit packets with all 0s in the source address while trying to find out the correct IP address. The machine is identified by its MAC address. Other hosts interpret the packet as originating from "this" host.

IP addresses are usually written in **dotted-decimal notation** so that they can be communicated conveniently by people. The address is broken into four bytes with each byte being represented by a decimal number and separated by a dot. For example, an IP address of

10000000 10000111 01000100 00000101

is written as

128.135.68.5

in dotted-decimal notation. The discerning student should notice immediately that this address is a Class B address. As we saw before, some of the values of the address fields (such as all 0s and all 1s) are reserved for special purposes. Another important special value is 127.X.Y.Z (X, Y, and Z can be anything), which is used for loopback. When a host sends a packet with this address, the packet is returned to the host by the IP protocol software without transmitting it to the physical network. The loopback address can be used for interprocess communication on a local host via TCP/IP protocols and for debugging purposes.

8.2.3   Subnet Addressing

The original IP addressing scheme described above has some drawbacks. Consider a typical university in the United States that has a Class B network address (which can support about 64,000 hosts connected to the Internet). With the original addressing scheme, it would be a gigantic task for the local network administrator to manage all 64,000 hosts. Moreover, a typical campus would have more than one local network, requiring the use of multiple network addresses. To solve these problems, subnet addressing was introduced in the mid-1980s when most large organizations began moving their computing platforms from mainframes to networks of workstations. The basic idea of **subnetting** is to add another hierarchical level called the "subnet" as shown in Figure 8.6. The beauty of the subnet-addressing scheme is that it is oblivious to the network outside the organization. That is, a host outside this organization would still see the original address structure with two levels. Inside the organization the local network administrator is free to choose any combination of lengths for the subnet and host ID fields.

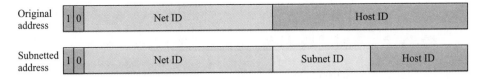

FIGURE 8.6 Introducing another hierarchical level through subnet addressing

As an illustration, consider an organization that has been assigned a Class B IP address with a network ID of 150.100. Suppose the organization has many LANs, each consisting of no more than 100 hosts. Then seven bits are sufficient to uniquely identify each host in a subnetwork. The other nine bits can be used to identify the subnetworks within the organization. If a packet with a destination IP address of 150.100.12.176 arrives at the site from the outside network, which subnet should a router forward this packet to? To find the subnet number, the router uses a **subnet mask** that consists of binary 1s for every bit position of the address except in the host ID field where binary 0s are used. For our example, the subnet mask is

11111111 11111111 11111111 10000000

which corresponds to 255.255.255.128 in dotted-decimal notation. The router can determine the subnet number by performing a binary AND between the subnet mask and the IP address. In our example the IP address is given by

10010110 01100100 00001100 10110000

Thus the subnet number becomes

10010110 01100100 00001100 10000000

corresponding to 150.100.12.128 in dotted-decimal notation. This number is used to forward the packet to the correct subnetwork inside the organization. A host connected to this subnetwork must have an IP address between 150.100.12.129 and 150.100.12.254.

Example—Subnetwork Addressing

Consider a site that has been assigned a Class B IP address of 150.100.0.1, as shown in Figure 8.7. The site has a number of subnets and many hosts (H) connected by routers (R). The figure shows only three subnets and five hosts for simplicity. Assume that the subnet ID field is nine bits long and the host ID field is seven bits long.

When a host located outside this network wants to send a packet to a host on this network, all that external routers have to know is how to get to network address 150.100.0.1. This concept is very powerful, since it hides the details of the internal network configuration. Let's see how the internal routers handle arriving packets.

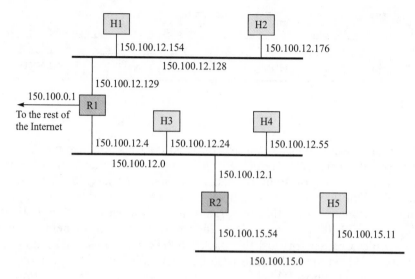

FIGURE 8.7 Example of address assignment with subnetting

Suppose a packet having a destination IP address of 150.100.15.11 arrives from the outside network. R1 has to know the next-hop router to send the packet to. The address 150.100.15.11 corresponds to the binary string 1001 0110 . 0110 0100 . 0000 1111. 0000 1011.[3] R1 knows that a nine-bit subnet field is in use, so it applies the following mask to extract the subnetwork address from the IP address: 1111 1111. 1111 1111. 1111 1111. 1000 0000. The result is then 1001 0110. 0110 0100. 0000 1111. 0000 0000, which corresponds to 150.100.15.0. This subnet number is in the routing table of R1, and the corresponding entry would specify the next-hop router address for R2, which is 150.100.12.1. When R2 receives the packet, R2 performs the same process and finds out that the destination host is connected to one of its network interfaces. It can thus send the packet directly to the destination.

8.2.4 IP Routing

The IP layer in the end-system hosts and in the routers work together to route packets from IP network sources to destinations. The IP layer in each host and router maintains a routing table that it uses to determine how to handle each IP packet. Consider the action of the originating host. If its routing table indicates that the destination host is directly connected to the originating host by a link or

---

[3] $150 = 128 + 16 + 4 + 2$, which gives 1001 0110; similarly $100 = 64 + 32 + 4$; $15 = 8 + 4 + 2 + 1$; $11 = 8 + 2 + 1$.

by a LAN, then the IP packet is sent directly to the destination host using the appropriate physical interface. Otherwise, the routing table typically specifies that the packet is to be sent to a default router that is directly connected to the originating host. Now consider the action of a router. When a router receives an IP packet from one of the network interfaces, the router examines its routing table to see whether the packet is destined to itself, and if so, delivers the packet to the appropriate higher-layer protocol. If the IP address is not the router's own address, then the router determines the next-hop router and the associated network interface.

Each row in the routing must provide the following information: destination IP address; IP address of next-hop router; several flag fields; and outgoing interface. Several types of flags may be defined. For example, the H flag indicates whether the route in the given row is to a host (H = 1), or to a network (H = 0). The G flag indicates whether the route in the given row is to a router (gateway; G = 1) or to a directly connected destination (G = 0).

Each time a packet is to be routed, the routing table is searched in the following order. First, the first column is searched to see whether the table contains an entry for the complete destination IP address. If so, then the IP packet is forwarded according to the next-hop entry and the G flag. Second, if the table does not contain the complete destination IP address, then the routing table is searched for the destination network ID. If an entry is found, the IP packet is forwarded according to the next-hop entry and the G flag. Third, if the table does not contain the destination network ID, the table is searched for a default router entry, and if one is available, the packet is forwarded there. Finally, if none of the above searches are successful, then the packet is declared undeliverable and an ICMP "host unreachable error" packet is sent back to the originating host.

Example—Routing with Subnetworks

Suppose that host H5 wishes to send an IP packet to host H2 in Figure 8.7. H2 has IP address 150.100.12.176 (1001 0110. 0110 0100. 0000 1100. 1011 0110). Let us trace the operations in carrying out this task.

The routing table in H5 may look something like this:

```
Destination      Next-Hop          Flags      Network Interface
127.0.0.1        127.0.0.1         H          1o0
default          150.100.15.54     G          emd0
150.100.15.0     150.100.15.11                emd0
```

The first entry is the loopback interface, the H indicates a host address, and 1o0 by convention is always the loopback interface. The second entry is the default entry, with next-hop router R2 (150.100.15.54), which is a router, so G = 1, and with Ethernet interface emd0. The third entry does not have H set, so it is a network address; G is also not set, so a direct route is indicated and the next-hop entry is the IP address of the outgoing network interface.

H5 first searches its routing table for the IP packet destination address 150.100.12.176. When H5 does not find the entry, it then searches for the destination network ID 150.100.12.128. Finally, H5 finds the default route to R2 and forwards the IP packet across the Ethernet.

The routing table in R2 may look something like this:

```
Destination      Next-Hop         Flag     Network Interface
127.0.0.1        127.0.0.1        H        1o0
default          150.100.12.4     G        emd0
150.100.15.0     150.100.15.11             emd1
150.100.12.0     150.100.12.1              emd1
```

R2 searches its routing table and forwards the IP packet to router R1, using the default route. R1 has the following routing table:

```
Destination      Next-Hop         Flag     Network Interface
127.0.0.1        127.0.0.1        H        1o0
150.100.12.176   150.100.12.176            emd0
150.100.12.0     150.100.12.4              emd1
150.100.15.0     150.100.12.1     G        emd1
```

R1 searches its routing table and finds a match to the host address 150.100.12.176 and sends the packet through network interface emd0, which delivers the packet to H2.

The netstat command allows you to display the routing table in your workstation. Check the manual for your system on how to use this command.

8.2.5   Classless Interdomain Routing (CIDR)

Dividing the IP address space into A, B, and C classes turned out to be inflexible. While on the one hand most organizations utilize the Class B address space inefficiently, on the other hand most organizations typically need more addresses than can be provided by a Class C address space. Giving a Class B address space to each organization would have exhausted the IP address space easily because of the rapid growth of the Internet. In 1993 the classful address space restriction was lifted. An arbitrary prefix length to indicate the network number, known as **classless interdomain routing** (CIDR),[4] was adopted in place of the classful scheme. Using a CIDR notation, a prefix 205.100.0.0 of length 22 is written as 205.100.0.0/22. The corresponding prefix range runs from 205.100.0.0 through 205.100.3.0. The /22 notation indicates that the network mask is 22 bits, or 255.255.252.0.

CIDR routes packets according to the higher-order bits of the IP address. The entries in a CIDR routing table contain a 32-bit IP address and a 32-bit

---

[4]CIDR is pronounced like "cider."

mask. CIDR uses a technique called **supernetting** so that a single routing entry covers a block of classful addresses. For example, instead of having four entries for a *contiguous* set of Class C addresses (e.g., 205.100.0.0, 205.100.1.0, 205.100.2.0, and 205.100.3.0), CIDR allows a single entry 205.100.16.0/22. To see this structure we note that

```
205.100.0.0  =   1100 1101 . 0110 0100. 0000 0000 . 0000 0000
205.100.1.0  =   1100 1101 . 0110 0100. 0000 0001 . 0000 0000
205.100.2.0  =   1100 1101 . 0110 0100. 0000 0010 . 0000 0000
205.100.3.0  =   1100 1101 . 0110 0100. 0000 0011 . 0000 0000
mask         =   1111 1111 . 1111 1111. 1111 1100 . 0000 0000
```

RFC 1518 describes address allocation policies to capitalize on CIDR's ability to aggregate routes. For example, address assignments should reflect the physical topology of the network; in this case IP address prefixes should correspond to continents or nations. This approach facilitates the aggregation of logical packet flows into the physical flows that ultimately traverse the network. Similarly, *transit routing domains* that carry traffic between domains should have unique IP addresses, and domains that are attached to them should begin with the transit routing domain's prefix. These route aggregation techniques resulted in a significant reduction in routing table growth, which was observed after the deployment of CIDR. Without the CIDR deployment, the routing table size at the core of the Internet would have easily exceeded 100,000 routes in 1996. In 1998 the routing table size was around 50,000 routes.

The use of variable-length prefixes requires that the routing tables be searched to find the *longest prefix match*. For example, a routing table may contain entries for the above supernet 205.100.0.0/22 as well as for 205.100.0.0/20. This situation may arise when a large number of destinations have been aggregated into the block 205.100.0.0/20, but packets destined to 205.100.16.0/22 are to be routed differently. A packet with destination address 205.100.1.1 will match both of these entries, so the algorithm must select the match with the longest prefix.

Routing tables can contain tens of thousands of entries, so efficient, fast, longest prefix matching algorithms are essential to implement fast routers. A number of algorithms have been developed for fast routers. For example, see [Degermark 1998] and [Waldvogel 1998].

8.2.6 Address Resolution

In section 8.2.4, we assume that a host can send a packet to the destination host by knowing the destination IP address. In reality IP packets must eventually be delivered by the underlying network technology, which uses a different address format. For a concrete example, let us assume that the underlying network technology is Ethernet, which is by far the most common situation. Recall that the Ethernet hardware can understand only its own 48-bit MAC address format. Thus the source host must also know the destination MAC address if the packet is to be delivered to the destination host successfully.

How does the host map the IP address to the MAC address? An elegant solution to find the destination MAC address is to use the **Address Resolution Protocol (ARP)**. The main idea is illustrated in Figure 8.8. Suppose H1 wants to send an IP packet to H3 but does not know the MAC address of H3. H1 first broadcasts an ARP request packet asking the destination host, which is identified by H3's IP address, to reply. All hosts in the network receive the packet, but only the intended host, which is H3, responds to H1. The ARP response packet contains H3's MAC and IP addresses. From now on H1 knows how to send packets to H3. To avoid having to send an ARP request packet each time H1 wants to send a packet to H3, H1 caches H3's IP and MAC addresses in its ARP table so that H1 can simply look up H3's MAC address in the table for future use. Each entry in the ARP table is usually "aged" so that the contents are erased if no activity occurs within a certain period, typically around 5 to 30 minutes. This procedure allows changes in the host's MAC address to be updated. The MAC address may change, for example, when an Ethernet card is broken and is replaced with a new one.

8.2.7 Reverse Address Resolution

In some situations a host may know its MAC address but not its IP address. For example, when a diskless computer such as an X terminal is being bootstrapped, it can read the MAC address from its Ethernet card. However, its IP address is usually kept separately in a disk at the server. The problem of getting an IP address from a MAC address can be handled by the **Reverse Address Resolution Protocol (RARP)**, which works in a fashion similar to ARP.

To obtain its IP address, the host first broadcasts an RARP request packet containing its MAC address on the network. All hosts on the network receive the packet, but only the server replies to the host by sending an RARP response packet containing the host's MAC and IP addresses. One limitation with RARP is that the server must be located on the same physical network as the host.

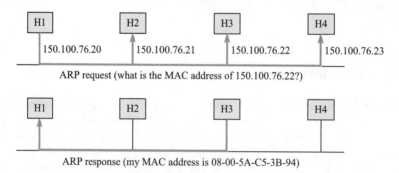

FIGURE 8.8 Address Resolution Protocol

8.2.8 Fragmentation and Reassembly

One of the strengths of IP is that it can work on a variety of physical networks. Each physical network usually imposes a certain packet-size limitation on the packets that can be carried, called the **maximum transmission unit (MTU)**. For example, Ethernet specifies an MTU of 1500 bytes, and FDDI specifies an MTU of 4464 bytes. When IP has to send a packet that is larger than the MTU of the physical network, IP must break the packet into smaller **fragments** whose size can be no larger than the MTU. Each fragment is sent independently to the destination as though it were an IP packet. If the MTU of some other network downstream is found to be smaller than the fragment size, the fragment will be broken again into smaller fragments, as shown in Figure 8.9. The destination IP is the only entity that is responsible for reassembling the fragments into the original packet. To reassemble the fragments, the destination waits until it has received all the fragments belonging to the same packet. If one or more fragments are lost in the network, the destination abandons the reassembly process and discards the rest of the fragments. To detect lost fragments, the destination host sets a timer once the first fragment of a packet arrives. If the timer expires before all fragments have been received, the host assumes the missing fragments were lost in the network and discards the other fragments.

Three fields in the IP header (identification, flags, and fragment offset) have been assigned to manage fragmentation and reassembly. At the destination IP has to collect fragments for reassembling into packets. The identification field is used to identify which packet a particular fragment belongs to so that fragments for different packets do not get mixed up. To have a safe operation, the source host must not repeat the identification value of the packet destined to the same host until a sufficiently long period of time has passed.

The flags field has three bits: one unused bit, one "don't fragment" (DF) bit, and one "more fragment" (MF) bit. If the DF bit is set to 1, it forces the router not to fragment the packet. If the packet length is greater than the MTU, the router will have to discard the packet and send an error message to the source

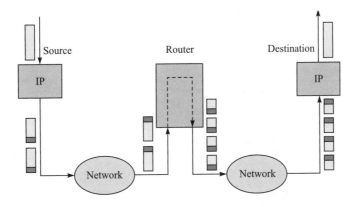

FIGURE 8.9 Packet fragmentation

host. The MF bit tells the destination host whether or not more fragments follow. If there are more, the MF bit is set to 1; otherwise, it is set to 0. The fragment offset field identifies the location of a fragment in a packet. The value measures the offset, in units of eight bytes, between the beginning of the packet to be fragmented and the beginning of the fragment, considering the data part only. Thus the first fragment of a packet has an offset value of 0. The data length of each fragment, except the last one, must be a multiple of eight bytes. The reason the offset is measured on units of eight bytes is that the fragment offset field has only 13 bits, giving a maximum count of 8192. To be able to cover all possible data lengths up to the maximum of 65,536 bytes, it is sufficient to multiply 8192 by 8. The reader should verify that these three fields give sufficient information to hosts and routers to perform fragmentation and reassembly.

Although fragmentation may seem to be a good feature to implement, it involves a subtle performance penalty. As alluded to earlier, if one of the fragments is lost, the packet cannot be reassembled at the destination and the rest of the fragments have to be discarded. This process, of course, wastes transmission bandwidth. If the upper-layer protocol requires reliability, then all the fragments for that packet would have to be retransmitted. It is possible to save some bandwidth if routers discard fragments more intelligently. Specifically, if a router has to discard a fragment, say, due to congestion, it might as well discard the subsequent fragments belonging to the same packet, since they will become useless at the destination. In fact, this same idea has been implemented in ATM networks.

Example—Fragmenting a Packet

Suppose a packet arrives at a router and is to be forwarded to an X.25 network having an MTU of 576 bytes. The packet has an IP header of 20 bytes and a data part of 1484 bytes. Perform fragmentation and include the pertinent values of the IP header of the original packet and of each fragment.

The maximum possible data length per fragment $= 576 - 20 = 556$ bytes. However, 556 is not a multiple of 8. Thus we need to set the maximum data length to 552 bytes. We can break 1484 into $552 + 552 + 380$ (other combinations are also possible).

Table 8.1 shows the pertinent values for the IP header where x denotes a unique identification value. Other values, except the header checksum, are the same as in the original packet.

	Total length	ID	MF	Fragment offset
Original packet	1504	x	0	0
Fragment 1	572	x	1	0
Fragment 2	572	x	1	69
Fragment 3	400	x	0	138

TABLE 8.1 Values of the IP header in a fragmented packet

8.2.9 ICMP: Error and Control Messages

It was noted earlier that if a router could not forward a packet for some reasons (for example, the TTL value reaches 0, or the packet length is greater than the network MTU while the DF bit is set), it would have to send an error message back to the source to report the problem. The protocol that handles error and control messages is called the **Internet Control Message Protocol (ICMP)**. Although ICMP messages are encapsulated by IP packets, ICMP is considered to be in the same layer as IP.

Each ICMP message format begins with a type field to identify the message. Some ICMP message types are echo reply, destination unreachable, source quench, redirect, echo request, time exceeded, parameter problem, timestamp request, and timestamp reply. The echo request and echo reply messages are used in the ping program that is often used to determine whether a remote host is alive. Ping is also often used to estimate the round-trip time between two hosts. The ICMP time exceeded message is exploited in the traceroute program. When a packet reaches a router with the value of the TTL equal to 0 or 1 before the packet reaches the destination, the corresponding router will send an ICMP message with type "time exceeded" back to the originating host. The time exceeded message also contains the IP address of the router that issues the message. Thus by sending messages to the destination with the TTL incremented by one per message, a source host will be able to trace the sequence of routers to the destination.

8.3 IPv6

IP version 4 has played a central role in the internetworking environment for many years. It has proved flexible enough to work on many different networking technologies. However, it has become a victim of its own success—explosive growth! In the early days of the Internet, people using it were typically researchers and scientists working in academia, high-tech companies, and research laboratories, mainly for the purpose of exchanging scientific results through e-mails. In the 1990s the World Wide Web and personal computers shifted the user of the Internet to the general public. This change has created heavy demands for new IP addresses, and the 32 bits of the current IP addresses will be exhausted sooner or later.

In the early1990s the Internet Engineering Task Force (IETF) began to work on the successor of IP version 4 that would solve the address exhaustion problem and other scalability problems. After several proposals were investigated, the new IP version was recommended in late 1994. The new version is called **IPv6** for IP version 6 (also called *IP next generation* or *IPng*). IPv6 was designed to interoperate with IPv4 since it would likely take many years to complete the transition from version 4 to version 6. Thus IPv6 should retain the most basic service provided by IPv4—a connectionless delivery service. On the other hand, IPv6 should also change the IPv4 functions that do not work well and support new emerging applications such as real-time video conferencing, etc. Some of the changes from IPv4 to IPv6 include

Longer address fields: The length of address field is extended from 32 bits to 128 bits. The address structure also provides more levels of hierarchy. Theoretically, the address space can support up to 3.4×10^{38} hosts.

Simplified header format: The header format of IPv6 is simpler than that of IPv4. Some of the header fields in IPv4 such as checksum, IHL, identification, flags, and fragment offset do not appear in the IPv6 header.

Flexible support for options: The options in IPv6 appear in optional *extension headers* that are encoded in a more efficient and flexible fashion than they were in IPv4.

Flow label capability: IPv6 adds a "flow label" to identify a certain packet "flow" that requires a certain QoS.

Security: IPv6 supports built-in authentication and confidentiality.

Large packets: IPv6 supports payloads that are longer than 64 K bytes, called *jumbo* payloads.

Fragmentation at source only: Routers are not allowed to fragment packets. If a packet needs to be fragmented, it must be done at the source.

No checksum field: The checksum field has been removed to reduce packet processing time in a router. Packets carried by the physical network such as Ethernet, token ring, X.25, or ATM are typically already checked. Furthermore, higher-layer protocols such as TCP and UDP also perform their own verification. Thus removing the checksum field probably would not introduce a serious problem in most situations.

8.3.1   Header Format

The IPv6 header consists of a required basic header and optional extension headers. The format of the basic header is shown in Figure 8.10. The packet should be transmitted in network byte order.

The description of each field in the basic header follows.

Version: The version field specifies the version number of the protocol and should be set to 6 for IPv6. The location and length of the version field stays unchanged so that the protocol software can recognize the version of the packet quickly.

Traffic class: The traffic class field specifies the traffic class or priority of the packet. The traffic class field is intended to support differentiated service.

Flow label: The flow label field can be to identify the QoS requested by the packet. In the standard, a flow is defined as "a sequence of packets sent from a particular source to a particular (unicast or multicast) destination for which the source desires special handling by the intervening routers." An example of an application that may use a flow label is a packet video

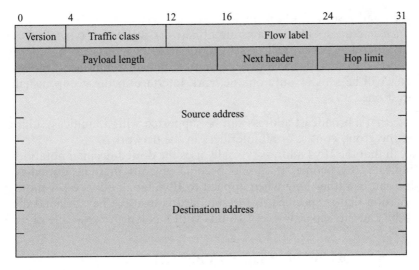

FIGURE 8.10 IPv6 basic header

system that requires its packets to be delivered to the destination within a certain time constraint. Routers that see these packets will have to process them according to their request. Hosts that do not support flows are required to set this field to 0.

Payload length: The payload length indicates the length of the data (excluding header). With 16 bits allocated to this field, the payload length is limited to 65,535 bytes. As we explain below, it is possible to send larger payloads by using the option in the extension header.

Next header: The next header field identifies the type of the extension header that follows the basic header. The extension header is similar to the options field in IPv4 but is more flexible and efficient. Extension headers are further discussed below.

Hop limit: The hop limit field replaces the TTL field in IPv4. The name now says what it means: The value specifies the number of hops the packet can travel before being dropped by a router.

Source address and **destination address:** The source address and the destination address identify the source host and the destination host, respectively. The address format is discussed below.

8.3.2 Network Addressing

The IPv6 address is 128 bits long, which increases the overhead somewhat. However, it is almost certain that the huge address space will be sufficient for many years to come. The huge address space also gives more flexibility in terms of address allocation. IPv6 addresses are divided into three categories:

1. *Unicast* addresses identify a single network interface.
2. *Multicast* addresses identify a group of network interfaces, typically at different locations. A packet will be sent to all network interfaces in the group.
3. *Anycast* addresses also identify a group of network interfaces. However, a packet will be sent to only one network interface in the group, usually the nearest one.

Note that a broadcast address can be supported with a multicast address by making the group consist of all interfaces in the network.

Recall that the IPv4 address typically uses the dotted-decimal notation when communicated by people. It should become obvious that the dotted-decimal notation can be rather long when applied to IPv6 long addresses. A more compact notation that is specified in the standard is to use a hexadecimal digit for every 4 bits and to separate every 16 bits with a colon. An example of an IPv6 address is

4BF5:AA12:0216:FEBC:BA5F:039A:BE9A:2176

Often IPv6 addresses can be shortened to a more compact form. The first shorthand notation can be exploited when the 16-bit field has some leading zeros. In this case the leading zeros can be removed, but there must be at least one numeral in the field. As an example

4BF5:0000:0000:0000:BA5F:039A:000A:2176

can be shortened to

4BF5:0:0:0:BA5F:39A:A:2176

Further shortening is possible where consecutive zero-valued fields appear. These fields can be shortened with the double-colon notation (::). Of course, the double-colon notation can appear only once in an address, since the number of zero-valued fields is not encoded and needs to be deduced from the specified total number of fields. Continuing with the preceding example, the address can be written even more compactly as

4BF5::BA5F:39A:A:2176

To recover the original address from one containing a double colon, you take the nonzero values that appear to the left of the double colons and align them to the left. You then take the number that appears to the right of the double colons and align them to the right. The field in between is set to 0s.

The dotted-decimal notation of IPv4 can be mixed with the new hexadecimal notation. This approach is useful for the transition period when IPv4 and IPv6 coexist. An example of a mixed notation is

::FFFF:128.155.12.198

Address allocations are organized by types, which are in turn classified according to prefixes (leading bits of the address). At the time of this writing, the initial allocation for prefixes is given in Table 8.2.

Binary prefix	Types	Percentage of address space
0000 0000	Reserved	0.39
0000 0001	Unassigned	0.39
0000 001	ISO network addresses	0.78
0000 010	IPX network addresses	0.78
0000 011	Unassigned	0.78
0000 1	Unassigned	3.12
0001	Unassigned	6.25
001	Unassigned	12.5
010	Provider-based unicast addresses	12.5
011	Unassigned	12.5
100	Geographic-based unicast addresses	12.5
101	Unassigned	12.5
110	Unassigned	12.5
1110	Unassigned	6.25
1111 0	Unassigned	3.12
1111 10	Unassigned	1.56
1111 110	Unassigned	0.78
1111 1110 0	Unassigned	0.2
1111 1110 10	Link local use addresses	0.098
1111 1110 11	Site local use addresses	0.098
1111 1111	Multicast addresses	0.39

TABLE 8.2 Address types based on prefixes

Less than 30 percent of the address space has been assigned. The remaining portion of the address space is for future use. Most types are assigned for unicast addresses, except the one with a leading byte of 1s which is assigned for multicast. Anycast addresses are not differentiated from unicast and share the same address space.

IPv6 assigns a few addresses for special purposes. The address 0::0 is called the *unspecified address* and is never used as a destination address. However, it may be used as a source address when the source station wants to learn its own address. The address ::1 is used for a loopback whose purpose is the same as the loopback address in IPv4. Another set of special addresses is needed during the transition period where an IPv6 packet needs to be "tunneled" across an IPv4 network. These addresses, called *IPv4-compatible* addresses, are used by routers and hosts that are directly connected to an IPv4 network. The address format consists of 96 bits of 0s followed by 32 bits of IPv4 address. Thus an IPv4 address of 135.150.10.247 can be converted to an IPv4-compatible IPv6 address of ::135.150.10.247. A similar set of special addresses is used to indicate hosts and routers that do not support IPv6. These addresses are called *IP-mapped* addresses. The format of these addresses consists of 80 bits of 0s, followed by 16 bits of 1s, and then by 32 bits of IPv4 address.

Provider-based unicast addresses are identified by the prefix 010. It appears that these addresses will be mainly used by the Internet service providers to assign addresses to their subscribers. The format of these addresses is shown in Figure 8.11.

FIGURE 8.11 Provider-based address format

Notice the hierarchical structure of this address format. The first level is identified by the registry ID, which in North America will be managed by the InterNIC. The next level identifies the Internet service provider that is responsible for assigning the subscriber IDs. Finally, each subscriber assigns the addresses according to the subnet IDs and interface IDs.

The local addresses are used for a collection of hosts that do not want to connect to the global Internet because of security and privacy concerns. There are two types of local addresses: link-local addresses and site-local addresses. The link-local addresses are used for a single link, while the site-local addresses are used for a single site. The local addresses are designed so that when an organization decides to connect the hosts to the global Internet, the move will be as painless as possible.

8.3.3   Extension Headers

To support extra functionalities that are not provided by the basic header, IPv6 allows an arbitrary number of extension headers to be placed between the basic header and the payload. Extension headers act like options in IPv4 except the former are encoded more efficiently and flexibly, as we show soon. The extension headers are daisy chained by the next header field, which appears in the basic header as well as in each extension header. Figure 8.12 illustrates the use of the next header field. A consequence of the daisy-chain formation is that the extension headers must be processed in the order in which they appear in the packet.

Six extension headers have been defined. They are listed in Table 8.3. These extension headers should appear in a packet as they are listed in the table from top to bottom.[5]

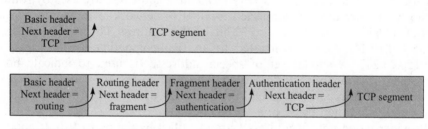

FIGURE 8.12 Daisy-chain extension headers

---

[5]The authentication header and the encapsulating security payload header are discussed in Chapter 11.

Header code	Header type
0	Hop-by-hop options header
43	Routing header
44	Fragment header
51	Authentication header
52	Encapsulating security payload header
60	Destination options header

TABLE 8.3 Extension headers

LARGE PACKET

IPv6 allows a payload size of more than 64 K by using an extension header. The use of payload size greater than 64 K has been promoted mainly by people who work on supercomputers. Figure 8.13 shows the format of the extension header for a packet with a jumbo payload (length that is greater than 65,535 bytes). The next header field identifies the type of header immediately following this header. The value 194 defines a jumbo payload option. The payload length in the basic header must be set to 0. The option length field (opt len) specifies the size of the jumbo payload length field in bytes. Finally, the jumbo payload length field specifies the payload size. With 32 bits the payload size can be as long as 4,294,967,295 bytes.

Fragmentation

As noted previously, IPv6 allows only the source host to perform fragmentation. Intermediate routers are forbidden to fragment any packet. If the packet length is greater than the MTU of the network this packet is to be forwarded to, an intermediate router must discard the packet and send an ICMP error message back to the source. A source can find the minimum MTU along the path from the source to the destination by performing a "path MTU discovery" procedure. An advantage of doing fragmentation at the source only is that routers can process packets faster, which is important in a high-speed environment. A disadvantage is that the path between a source and a destination must remain reasonably static so that the path MTU discovery does not give outdated information. If a source wants to fragment a packet, the source will include a fragment extension header (shown in Figure 8.14) for each fragment of the packet.

The fragment offset M (more fragment) and identification fields have the same purposes as they have in IPv4 except the identification is now extended to 32 bits. Bits 8 to 15 and bits 29 and 30 are reserved for future use.

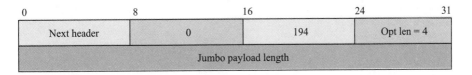

FIGURE 8.13 Extension header for jumbo packet

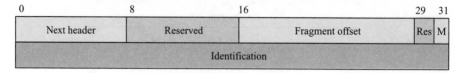

FIGURE 8.14 Fragment extension header

Source Routing

Like IPv4, IPv6 allows the source host to specify the sequence of routers to be visited by a packet to reach the destination. This option is defined by a routing extension header, which is shown in Figure 8.15. The header length specifies the length of the routing extension header in units of 64 bits, not including the first 64 bits. Currently, only type 0 is specified. The segment left field identifies the number of route segments remaining before the destination is reached. Maximum legal value is 23. Initially, this value will be set to the total number of route segments from the source to the destination. Each router decrements this value by 1 until the packet reaches the destination. Each bit in the strict/loose bit mask indicates whether the next destination address must be followed strictly (if the bit is set to 1) or loosely (if the bit is set to 0).

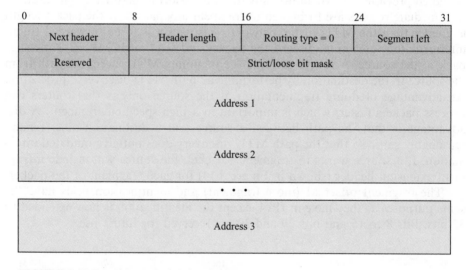

FIGURE 8.15 Routing extension header

8.4   USER DATAGRAM PROTOCOL

Two transport layer protocols, TCP and UDP, build on the best-effort service provided by IP to support a wide range of applications. In this section we discuss the details of UDP.

The **User Datagram Protocol (UDP)** is an unreliable, connectionless transport layer protocol. It is a very simple protocol that provides only two additional services beyond IP: demultiplexing and error checking on data. Recall that IP knows how to deliver packets to a host, but does not know how to deliver them to the specific application in the host. UDP adds a mechanism that distinguishes among multiple applications in the host. Recall also that IP checks only the integrity of its header. UDP can optionally check the integrity of the entire UDP datagram. Applications that use UDP include Trivial File Transfer Protocol, DNS, SNMP, and Real-Time Protocol (RTP).

The format of the UDP datagram is shown in Figure 8.16. The destination port allows the UDP module to demultiplex datagrams to the correct application in a given host. The source port identifies the particular application in the source host to receive replies. The UDP length field indicates the number of bytes in the UDP datagram (including header and data).

The UDP checksum field detects errors in the datagram, and its use is optional. If a source host does not want to compute the checksum, the checksum field should contain all 0s so that the destination host knows that the checksum has not been computed. What if the source host does compute the checksum and finds that the result is 0? The answer is that if a host computes the checksum whose result is 0, it will set the checksum field to all 1s. This is another representation of zero in 1s complement. The checksum computation procedure is similar to that in computing IP checksum except for two new twists. First, if the length of the datagram is not a multiple of 16 bits, the datagram will be padded out with 0s to make it a multiple of 16 bits. In doing so, the actual UDP datagram is not modified. The pad is used only in the checksum computation and is not transmitted. Second, UDP adds a **pseudoheader** (shown in Figure 8.17) to the beginning of the datagram when performing the checksum computation. The pseudoheader is also created by the source and destination hosts only during the checksum computation and is not transmitted. The pseudoheader is to ensure

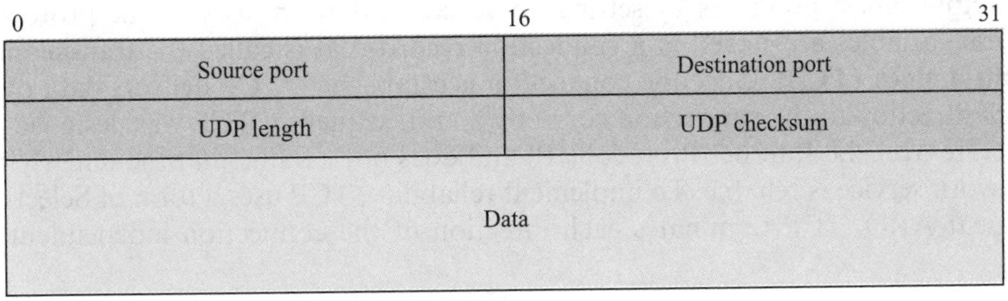

0	16	31
Source port	Destination port	
UDP length	UDP checksum	
Data		

FIGURE 8.16 UDP datagram

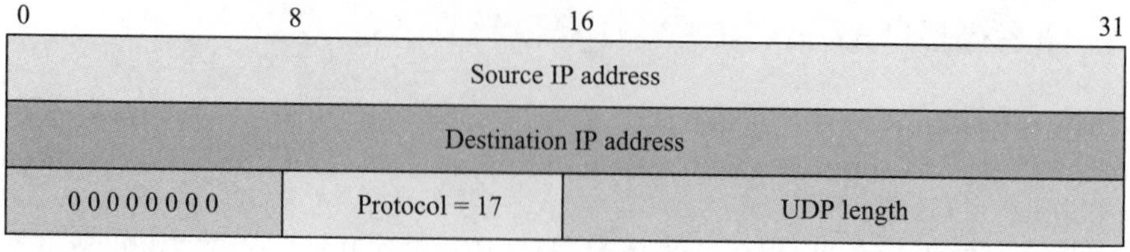

0	8	16	31
Source IP address			
Destination IP address			
00000000	Protocol = 17	UDP length	

FIGURE 8.17 UDP pseudoheader

that the datagram has indeed reached the correct destination host and port. Finally, if a datagram is found to be corrupted, it is simply discarded and the source UDP entity is not notified.

8.5   TRANSMISSION CONTROL PROTOCOL

TCP and IP are the workhorses in the Internet. In this section we first discuss how TCP provides reliable, connection-oriented stream service over IP. To do so, TCP implements a version of Selective Repeat ARQ. In addition, TCP implements congestion control through an algorithm that identifies congestion through packet loss and that controls the rate at which information enters the network through a congestion window.

8.5.1   TCP Reliable Stream Service

The **Transmission Control Protocol (TCP)** provides a logical full-duplex (two-way) connection between two application layer processes across a datagram network. TCP provides these application processes with a connection-oriented, reliable, in-sequence, byte-stream service. TCP also provides flow control that allows receivers to control the rate at which the sender transmits information so that buffers do not overflow. TCP can also support multiple application processes in the same end system.[6]

Before data transfer can begin, TCP establishes a connection between the two application processes by setting up variables that are used in the protocol. These variables are stored in a connection record that is called the **transmission control block (TCB)**. Once the connection is established, TCP delivers data over each direction in the connection correctly and in sequence. TCP was designed to operate over the Internet Protocol (IP) and does not assume that the underlying network service is reliable. To implement reliability, TCP uses a form of Selective Repeat ARQ. TCP terminates each direction of the connection independently,

---

[6]TCP also implements congestion control, but this function is discussed in Chapter 7.

allowing data to continue flowing in one direction after the other direction has been closed.

TCP does not preserve message boundaries and treats the data it gets from the application layer as a byte stream. Thus when a source sends a 1000-byte message in a single chunk (one write), the destination may receive the message in two chunks of 500 bytes each (two reads), in three chunks of 400 bytes, 300 bytes and 300 bytes (three reads), or in any other combination. In other words, TCP may split or combine the application information in the way it finds most appropriate for the underlying network.

In this section we first discuss how TCP provides reliable stream service. We then discuss details of the TCP header and protocol.

8.5.2 TCP Operation

We now consider the adaptation functions that TCP uses to provide a connection-oriented, reliable, stream service. We are interested in delivering the user information so that it is error free, without duplication, and in the same order that it was produced by the sender. We assume that the user information consists of a stream of bytes as shown in Figure 8.18. For example, in the transfer of a long file the sender is viewed as inserting a byte stream into the transmitter's send buffer. The task of TCP is to ensure the transfer of the byte stream to the receiver and *the orderly delivery* of the stream to the destination application.

TCP was designed to deliver a connection-oriented service in an internet environment, which itself offers connectionless packet transfer service, so different packets can traverse a different path from the same source to the same destination and can therefore arrive out of order. Therefore, in the internet old messages from previous connections may arrive at a receiver, thus potentially complicating the task of eliminating duplicate messages. TCP deals with this problem by using long (32 bit) sequence numbers and by establishing randomly selected initial sequence numbers during connection setup. At any given time the receiver is accepting sequence numbers from a much smaller window, so the likelihood of accepting a very old message is very low. In addition, TCP enforces

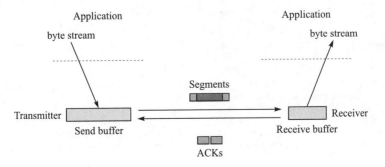

FIGURE 8.18 TCP preview

a time-out period at the end of each connection to allow the network to clear old segments from the network.

TCP uses a sliding-window mechanism, as shown in Figure 8.19. The send window contains three pointers:

- S_{last} points to the oldest byte that has not yet been acknowledged.
- S_{recent} points to the last byte that has been transmitted but not yet acknowledged.
- $S_{last} + W_S - 1$ indicates the highest numbered byte that the transmitter is willing to accept from its application.

Note that W_S no longer specifies the maximum allowable number of outstanding transmitted bytes. A different parameter, discussed below, specifies this value.

The receiver maintains a receive window that contains three pointers:

- R_{last} points to the oldest byte that has not been read by the destination application.
- R_{next} points to the location of the lowest numbered byte that has not yet been received correctly, that is, the next byte it expects to receive.
- R_{new} points to the location of the highest numbered byte that has been received correctly.

Note that R_{new} can be greater than R_{next} because the receiver will accept out-of-sequence, error-free bytes. The receiver can buffer at most W_R bytes at any given time, so $R_{last} + W_R - 1$ is the maximum numbered byte that the receiver is prepared to accept. Note that we are assuming that the user application does not necessarily read the received bytes as soon as they are available.

The transmitter arranges a consecutive string of bytes into a PDU that is called a **segment**. The segment contains a header with address information that

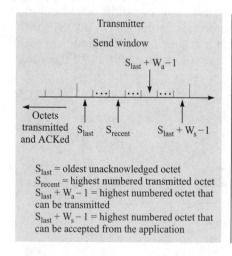
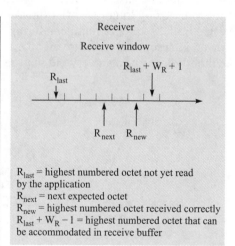

FIGURE 8.19 TCP end-to-end flow control

enables the network to direct the segment to its destination application process. The segment also contains a **sequence number** which corresponds to the number of the first byte in the string that is being transmitted. Note that this differs significantly from conventional ARQ. The transmitter decides to transmit a segment when the number of bytes in the send buffer exceeds some specified threshold or when a timer that is set periodically expires. The sending application can also use a **push command** that forces the transmitter to send a segment.

When a segment arrives, the receiver performs an error check to detect transmission errors. If the segment is error free and is not a duplicate segment, then the bytes are inserted into the appropriate locations in the receive buffer if the bytes fall within the receive window. Note that the receiver will accept out-of-order but error-free segments. If the received segment contains the byte corresponding to R_{next}, then the R_{next} pointer is moved forward to the location of the next byte that has not yet been received. An acknowledgment with the sequence number R_{next} is sent in a segment that is transmitted in the reverse direction. R_{next} acknowledges the correct receipt of all bytes up to $R_{next} - 1$. This acknowledgment, when received by the transmitter, enables the transmitter to update its parameter S_{last} to R_{next}, thus moving the send window forward. Later in the section we give an example of data transfer in TCP.

TCP separates the flow control function from the acknowledgment function. The flow control function is implemented through an **advertised window** field in the segment header. Segments that travel in the reverse direction contain the advertised window size that informs the transmitter of the number of buffers currently available at the receiver. The advertised window size is given by

$$W_A = W_R - (R_{new} - R_{last})$$

The transmitter is obliged to keep the number of outstanding bytes below the advertised window size, that is

$$S_{recent} - S_{last} \leq W_A$$

To see how the flow control takes effect, suppose that the application at the receiver's side stops reading the bytes from the buffer. Then R_{new} will increase while R_{last} remains fixed, thus leading to a smaller advertised window size. Eventually the receive window will be exhausted when $R_{new} - R_{last} = W_R$, and the advertised window size will be zero. This condition will cause the transmitter to stop sending. The transmitter can continue accepting bytes from its application until its buffer contains W_S bytes. At this point the transmitter blocks its application from inserting any more bytes into the buffer.

Finally, we consider the retransmission procedure that is used in TCP. The transmitter sets a timer each time a segment is transmitted. If the timer expires before any of the bytes in the segment are acknowledged, then the segment is retransmitted. The internet environment in which TCP must operate implies that the delay incurred by segments in traversing the network can vary widely from connection to connection and even during the course of a single connection. For this reason TCP uses an adaptive technique for setting the retransmission timeout value. The **round-trip time** t_{RTT} is estimated continuously, using measure-

---

MAXIMUM SEGMENT LIFETIME, REINCARNATION, AND THE Y2K BUG

The Y2K bug is the result of an ambiguity that arises from the limited precision used in early computer programs to specify the calendar year. The use of only two decimal digits to represent a year results in an ambiguity between the year 1900 and the year 2000. Consequently, unanticipated actions may be taken by these programs when the year 2000 is reached. This same problem is faced by millions of TCP processes every second of every day.

A TCP connection is identified by the source and destination port numbers and by the IP address of the source and destination machines. During its lifetime, the TCP connection will send some number of segments using the 32 bit byte sequence numbering. Each segment is encapsulated in an IP packet and sent into an internet. It is possible for an IP packet to get trapped in a loop inside the network, typically while the routing tables adapt to a link or router failure. Such a packet is called a *lost* or *wandering duplicate*. In the meantime TCP at the sending side times out and sends a retransmission of the segment that arrives promptly using the new route. If the wandering duplicate subsequently arrives at the same connection, then the segment will be recognized as a duplicate and rejected. (This scenario assumes that segments are not being sent so fast that the sequence numbers have not already wrapped around.)

It is also possible for the TCP connection to end, even while one of its wandering duplicates is still in the network. Suppose that a new TCP connection is set up between the same two machines and with the same port numbers. The new TCP connection is called an *incarnation* of the previous connection. TCP needs to protect the new connection so that duplicates from previous connections are prevented from being accepted and interpreted by the new connection. For example, the duplicate could be the command to terminate a connection. To deal with these and other problems, we show in the next section that every implementation of TCP assumes a certain value for the **maximum segment lifetime (MSL)**, which is the maximum time that an IP packet can live in an internet. We discuss how the MSL is derived when we discuss IP.

---

ments of the time τ_n that elapses from the instant a segment is transmitted until the moment the corresponding acknowledgment is received

$$t_{RTT}(new) = \alpha t_{RTT}(old) + (1 - \alpha)\tau_n$$

where α is a number between 0 and 1. A typical value of α is 7/8. The time-out value t_{out} is chosen to take into account not only t_{RTT} but also the variability in the estimates for the round-trip time. This variability is given by the standard deviation σ_{RTT} of the round-trip time. The time-out value is then

$$t_{out} = t_{RTT} + k\sigma_{RTT}$$

where k is some suitable constant. Thus if the estimates for round-trip times are highly variable, then the standard deviation will be large and a large time-out value will be used. On the other hand, if the round-trip times are nearly constant, the standard deviation will be close to zero, and the time-out value will be slightly larger than the mean round-trip time. The standard deviation involves estimating the average of the squared deviation, $(\tau_n - t_{RTT})^2$, and taking the square root. In practice, the average of the absolute deviation $|\tau_n - t_{RTT}|$ is simpler to estimate and is used instead:

$$d_{RTT}(new) = \beta d_{RTT}(old) + (1 - \beta)|\tau_n - t_{RRT}|$$

A typical value is $\beta = 1/4$. The time-out value that has been found to work well is then

$$t_{out} = t_{RTT} + 4d_{RTT}$$

8.5.3 TCP Protocol

We now discuss the structure of TCP segments and the setting up of a TCP connection, the data transfer phase, and the closing of the connection. Detailed information about TCP can be found in RFC 793 and RFC 1122.

TCP SEGMENT

Figure 8.20 shows the format of the TCP segment. The header consists of a 20-byte fixed part plus a variable-size options field.

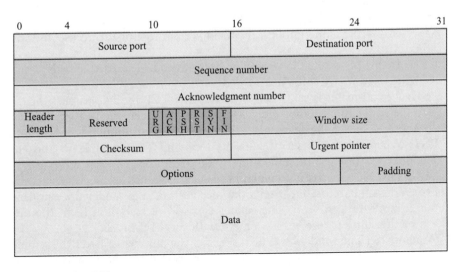

FIGURE 8.20 TCP segment

The description of each field in the TCP segment is given below. The term *sender* refers to the host that sends the segment, and *receiver* refers to the host that receives the segment.

Source port and **destination port:** The source and destination ports identify the sending and receiving applications, respectively. Recall from section 2.3.1 that the pair of ports and IP addresses identify a process-to-process connection.

Sequence number: The 32-bit sequence number field identifies the position of the first data byte of this segment in the sender's byte stream during data transfer (when SYN bit is not set). The sequence number wraps back to 0 after $2^{32} - 1$. Note that TCP identifies the sequence number for each byte (rather than for each segment). For example, if the value of the sequence number is 100 and the data area contains five bytes, then the next time this TCP module sends a segment, the sequence number will be 105. If the SYN bit is set to 1 (during connection establishment), the sequence number indicates the **initial sequence number (ISN)** to be used in the sender's byte stream. The sequence number of the first byte of data for this byte stream will be ISN + 1. It is important to note that a TCP connection is full duplex so that each end point independently maintains its own sequence number.

Acknowledgment number: This field identifies the sequence number of the next data byte that the sender expects to receive if the ACK bit is set. This field also indicates that the sender has successfully received all data up to but not including this value. If the ACK bit is not set (during connection establishment), this field is meaningless. Once a connection is established, the ACK bit must be set.

Header length: This field specifies the length of the TCP header in 32-bit words. This information allows the receiver to know the beginning of the data area because the options field is variable length.

Reserved: As the name implies, this field is reserved for future use and must be set to 0.

URG: If this bit is set, the urgent pointer is valid (discussed shortly).

ACK: If this bit is set, the acknowledgment number is valid.

PSH: When this bit is set, it tells the receiving TCP module to pass the data to the application immediately. Otherwise, the receiving TCP module may choose to buffer the segment until enough data accumulates in its buffer.

RST: When this bit is set, it tells the receiving TCP module to abort the connection because of some abnormal condition.

SYN: This bit requests a connection (discussed later).

FIN: When this bit is set, it tells the receiver that the sender does not have any more data to send. The sender can still receive data from the other direction until it receives a segment with the FIN bit set.

Window size: The window size field specifies the number of bytes the sender is willing to accept. This field can be used to control the flow of data and congestion.

Checksum: This field detects errors on the TCP segment. The procedure is discussed below.

Urgent pointer: When the URG bit is set, the value in the urgent pointer field added to that in the sequence number field points to the last byte of the "urgent data" (data that needs immediate delivery). However, the first byte of the urgent data is never explicitly defined. Because the receiver's TCP module passes data to the application in sequence, any data in the receiver's buffer up to the last byte of the urgent data may be considered urgent.

Options: The options field may be used to provide other functions that are not covered by the header. If the length of the options field is not a multiple of 32 bits, extra padding bits will be added. The most important option is used by the sender to indicate the **maximum segment size (MSS)** it can accept. This option is specified during connection setup. Two other options that are negotiated during connection setup are intended to deal with situations that involve large delay-bandwidth products. The **window scale** option allows the use of a larger advertised window size. The window can be scaled upward by a factor of up to 2^{14}. Normally the maximum window size is $2^{16} - 1 = 65,535$. With scaling the maximum advertised window size is $65,535 \times 2^{14} = 1,073,725,440$ bytes. The **timestamp** option is intended for high-speed connections where the sequence numbers may wrap around during the lifetime of the connection. The timestamp option allows the sender to include a timestamp in every segment. This timestamp can also be used in the RTT calculation.

TCP CHECKSUM

The purpose of the TCP checksum field is to detect errors. The checksum computation procedure is similar to that used to compute an IP checksum (discussed in Chapter 3) except for two features. First, if the length of the segment is not a multiple of 16 bits, the segment will be padded with zeros to make it a multiple of 16 bits. In doing so, the TCP length field is not modified. Second, a **pseudoheader** (shown in Figure 8.21) is added to the beginning of the segment when performing the checksum computation. The pseudoheader is created by the source and destination hosts during the checksum computation and is not transmitted. This mechanism ensures the receiver that the segment has indeed reached the correct destination host and port and that the protocol type is TCP (which is

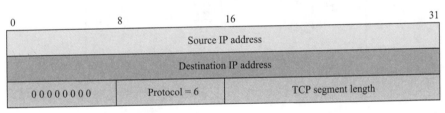

FIGURE 8.21 TCP pseudoheader

assigned the value 6). At the receiver the IP address information in the IP packet that contained the segment is used in the checksum calculation.

CONNECTION ESTABLISHMENT

Before any host can send data, a connection must be established. TCP establishes the connection using a **three-way handshake** procedure shown in Figure 8.22. The handshakes are described in the following steps:

1. Host A sends a connection request to host B by setting the SYN bit. Host A also registers its initial sequence number to use (Seq_no = x).
2. Host B acknowledges the request by setting the ACK bit and indicating the next data byte to receive (Ack_no = x + 1). The "plus one" is needed because the SYN bit consumes one sequence number. At the same time, host B also sends a request by setting the SYN bit and registering its initial sequence number to use (Seq_no = y).
3. Host A acknowledges the request from B by setting the ACK bit and confirming the next data byte to receive (Ack_no = y + 1). Note that the sequence number is set to x + 1. On receipt at B the connection is established.

If during a connection establishment phase, one of the hosts decides to refuse a connection request, it will send a reset segment by setting the RST bit. Each SYN message can specify options such as maximum segment size, window scaling, and timestamps.

Because TCP segments can be delayed, lost, and duplicated, the initial sequence number should be different each time a host requests a connection.[7]

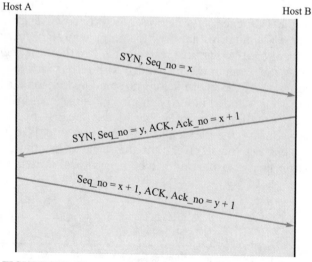

FIGURE 8.22 Three-way handshake

---

[7]The specification recommends incrementing the initial sequence number by one every 4 microseconds.

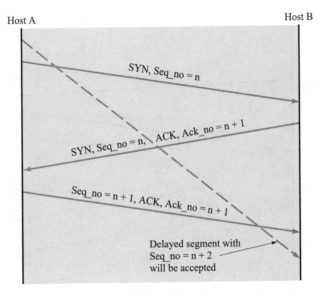

FIGURE 8.23 If a host always uses the same initial sequence number, old segments cannot be distinguished from the current ones

The three-way handshake procedure ensures that both endpoints agree on their initial sequence numbers. To see why the initial sequence number must be different, consider a case in which a host can always use the same initial sequence number, say, n, as shown in Figure 8.23. After a connection is established, a delayed segment from the previous connection arrives. Host B accepts this segment, since the sequence number turns out to be legal. If a segment from the current connection arrives later, it will be rejected by host B, thinking that the segment is a duplicate. Thus host B cannot distinguish a delayed segment from the new one. The result can be devastating if the delayed segment says, for example, "Transfer 1 million dollars from my account." You should verify that if the initial sequence number is always unique, the delayed segment is very unlikely to possess a legal sequence number and thus can be detected and discarded.

Example—A Client Server Application

Let us revisit the connection establishment process depicted in Figure 8.22 in the context of a client/server application. Let host A denote the client and let host B denote the server. Figure 8.24 shows that the server (host B) must first carry out a *passive open* to indicate to TCP that it is willing to accept connections. When using Berkeley sockets, a passive open is performed by the calls `socket`, `bind`, `listen`, and `accept`. When a client, say, from host A, wishes to initiate a session, the client performs an *active open*. This step involves making a `socket` call that creates the socket on the client side and then a `connect` call that initiates the TCP connection. This action causes the host A TCP module to initiate the

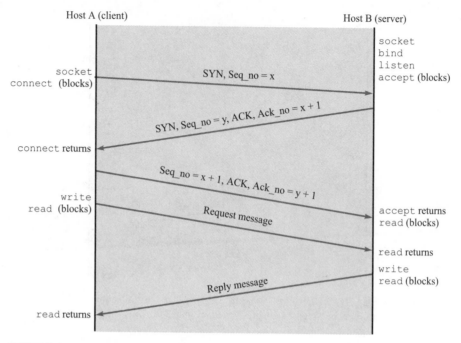

Host A (client)   Host B (server)

FIGURE 8.24 Client server application process actions and TCP

three-way handshake shown in Figure 8.22. When the server (host B) receives the first SYN, it returns a segment with an ACK and its own SYN. When the client (host A) receives this segment, connect returns and the client sends an ACK. Upon receiving this ACK, accept returns in the server, and the server is ready to read data. Client A then issues a write call to send a request message. Upon receipt of this segment by the TCP module in host B, read returns, and the request message is passed to the server. Subsequently, the server sends a reply message. The data transfer is discussed next.

DATA TRANSFER

To provide a reliable delivery service to applications, TCP uses the Selective Repeat ARQ protocol with positive acknowledgment implemented by a sliding-window mechanism. The difference here is that the window slides on a byte basis instead of on a packet basis. TCP can also apply flow control over a connection by dynamically advertising the window size. Flow control is the process of regulating the traffic between two points and is used to prevent the sender from overwhelming the receiver with too much data.

Figure 8.25 illustrates an example of how a TCP entity can exert flow control. Suppose that at time t_0, the TCP module in host B advertised a window of size 2048 and expected the next byte received to have a sequence number 2000. The advertised window allows host A to transmit up to 2048 bytes of unacknow-

Host A   Host B

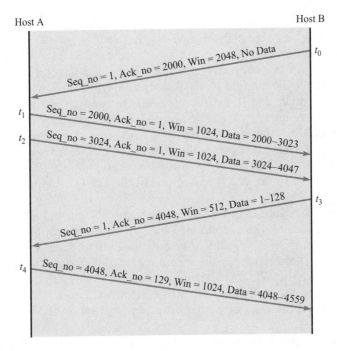

FIGURE 8.25 TCP window flow control

ledged data. At time t_1, host A has only 1024 bytes to transmit, so it transmits all the data starting with the sequence number 2000. The TCP entity also advertises a window of size 1024 bytes to host B, and the next byte is expected to have a sequence number of 1. When the segment arrives, host B chooses to delay the acknowledgment in the hope that the acknowledgment can ride freely with the data. Meanwhile at time t_2, A has another 1024 bytes of data and transmits it. After the transmission, A's sending window closes completely. It is not allowed to transmit any more data until an acknowledgment comes back.

At time t_3, host B has 128 bytes of data to transmit; it also wants to acknowledge the first two segments of data from host A. Host B can simply **piggyback** the acknowledgment (by specifying the acknowledgment number to be 4048) to the data segment. Also at this time, host B finds out that it can allocate only 512 bytes of receive buffer space for this connection because other connections are also competing for the precious memory. So it shrinks the advertised window from 2048 bytes to 512 bytes. When host A receives the segment, A changes its sending window to 512 bytes. If at time t_4, host A has 2048 bytes of data to transmit then it will transmit only 512 bytes. We see that window advertisement dynamically controls the flow of data from the sender to the receiver and prevents the receiver's buffer from being overrun. Another use of window advertisement is to control congestion inside the network. This subject was covered in Chapter 7.

The previous discussion shows how TCP can delay transmission so that the acknowledgment can be piggybacked to the data segment. Another use of delayed transmission is to reduce bandwidth waste. Consider a login session in which a user types one character at a time. When a character arrives from the application, the TCP module sends the segment with one byte of data to the other end. The other end (login server) needs to send an acknowledgment and then an echo character back to the client. Finally, the client needs to send an acknowledgment of the echo character. Thus one character generates four exchanges of IP packets between the client and the server with the following lengths: 41 bytes, 40 bytes, 41 bytes, and 40 bytes (assuming IP and TCP header are 20 bytes each). In the WAN environment, this waste of bandwidth is usually not justified.

A solution to reduce the waste was proposed by Nagle and is called the **Nagle algorithm**. The idea works as follows. When an interactive application wants to send a character, the TCP module transmits the data and waits for the acknowledgement from the receiver. In the meantime, if the application generates more characters before the acknowledgment arrives, TCP will not transmit the characters but buffer them instead. When the acknowledgment eventually arrives, TCP transmits all the characters that have been waiting in the buffer in a single segment.

In the LAN environment where delay is relatively small and bandwidth is plentiful, the acknowledgment usually comes back before another character arrives from the application. Thus the Nagle algorithm is essentially disabled. In the WAN environment where acknowledgments can be delayed unpredictably, the algorithm is self-adjusting. When delay is small, implying that the network is lightly loaded, only a few characters are buffered before an acknowledgment arrives. In this case TCP has the luxury of transmitting short segments. However, when delay is high, indicating that the network is congested, many more characters will be buffered. Here, TCP has to transmit longer segments and less frequently. In some cases the Nagle algorithm needs to be disabled to ensure the interactivity of an application even at the cost of transmission efficiency.

Another problem that wastes network bandwidth occurs when the sender has a large volume of data to transmit and the receiver can only deplete its receive buffer a few bytes at a time. Sooner or later the receive buffer becomes full. When the receiving application reads a few bytes from the receive buffer, the receiving TCP sends a small advertisement window to the sender, which quickly transmits a small segment and fills the receive buffer again. This process goes on and on with many small segments being transmitted by the sender for a single application message. This problem is called the **silly window syndrome**. It can be avoided by having the receiver not advertise the window until the window size is at least as large as half of the receive buffer size, or the maximum segment size. The sender side can cooperate by refraining from transmitting small segments.

Example—MSS and Bandwidth Efficiency

The value of the maximum segment size (MSS) affects the efficiency with which a given transmisson link is used. To discuss the effect of MSS on bandwidth efficiency consider Figure 8.26. Each TCP segment consists of a 20-byte TCP header followed by the block of data. The segment in turn is encapsulated in an IP packet that includes 20 additional bytes of header. The MSS for a connection is the largest block of data a segment is allowed to contain. TCP provides 16 bits to specify the MSS option, so the maximum block of data that can be carried in a segment is 65,495 bytes; that is, 65,535 minus 20 bytes for the IP header and 20 bytes for the TCP header.

In TCP the default MSS is 536 bytes, and the corresponding IP packet is then 576 bytes long. In this case 40 bytes out of every 576 bytes, that is, 7 percent are overhead. If instead, the MSS were 65,495 bytes, then the overhead would only be 0.06 percent. We will see, however, that various networks impose limits on the size of the blocks they can handle. For example, Ethernet limits the MSS to 1460.

Example—Sequence Number Wraparound and Timestamps

The original TCP specification assumed a maximum segment lifetime (MSL) of 2 minutes. Let's see how long it takes to wrap around the sequence number space using current high-speed transmission lines. The 32-bit sequence number wraps around when $232 = 4,294,967,296$ bytes have been sent. Because TCP uses Selective Repeat ARQ, the maximum allowable window size is 2^{31} bytes. In a T-1 line the time required is $(2^{32} \times 8)/(1.544 \times 10^6) = 6$ hours. In a T-3 line (45 Mbps), the wraparound will occur in 12 minutes. In an OC-48 line (2.4 Gbps), the wraparound will occur in 14 seconds! Clearly, sequence number wraparound becomes an issue for TCP operating over very high speed links.

When the timestamp option is in use, the sending TCP inserts a 32-bit timestamp into a 4-byte field in the header of each segment it transmits. The receiving TCP echoes the 32-bit timestamp it receives by inserting the timestamp into a 4-byte field in the header of each ACK segment. By combining the 32-bit

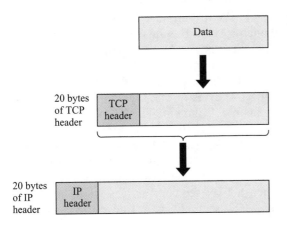

FIGURE 8.26 Header overhead

timestamp with the 32-bit sequence number we obtain what amounts to a 64-bit sequence number to deal with the wraparound problem. To be effective, the timestamp clock should tick forward at least once every 2^{31} bytes sent. This requirement places a lower bound on the clock frequency. The clock should also not be too fast; it should not complete a cycle in less than one MSL period. Clock frequencies in the range of one tick every 1 ms to 1 second meet these bound requirements. For example, a 1 ms clock period works for transmission rates up to 8 terabps, and the timestamp clock wraps its sign bit in 25 days [RFC 1323].

Example—Delay-Bandwidth Product and Advertised Window Size

Consider a link that has a round-trip time of 100 ms. The number of bytes in such a link at a T-1 speed is $1.544 \times 10^6 \times .100/8 = 19,300$ bytes. For a T-3 line (45 Mbps), the number of bytes is 562,500 bytes, and for an OC-48 line it increases to 3 megabytes. Suppose that a *single* TCP process needs to keep these links fully occupied transmitting its information. Its advertised window must then be at least as large as the RTT × bandwidth product. We saw at the beginning of this section that the normal maximum advertised window size is only 65,535, which is not enough for the T-3 and OC-48 lines. The window scale option, however, allows a window size of up to $65,535 \times 2^{14} = 1$ gigabyte, which is enough to handle these cases.

TCP CONNECTION TERMINATION

TCP provides for a **graceful close** that involves the independent termination of each direction of the connection. A termination is initiated when an application tells TCP that it has no more data to send. The TCP entity completes transmission of its data and, upon receiving acknowledgment from the receiver, issues a segment with the FIN bit set. Upon receiving a FIN segment, a TCP entity informs its application that the other entity has terminated its transmission of data. For example, in Figure 8.27 the TCP entity in host A terminates its transmission first by issuing a FIN segment. Host B sends an ACK segment to acknowledge receipt of the FIN segment from A. Note that the FIN segment uses one byte, so the ACK is 5087 in the example.

After B receives the FIN segment, the direction of the flow from B to A is still open. In Figure 8.27 host B sends 150 bytes in one segment, followed by a FIN segment. Host A then sends an acknowledgment. The TCP in host A then enters the **TIME_WAIT state** and starts the TIME_WAIT timer with an initial value set to twice the *maximum segment lifetime* (**2MSL**). The only valid segment that can arrive while host A is in the TIME_WAIT state is a retransmission of the FIN segment from host B (if host A's ACK was lost, and host B's retransmission time-out has expired). If such a FIN segment arrives while host A is the TIME_WAIT state, then the ACK segment is retransmitted and the

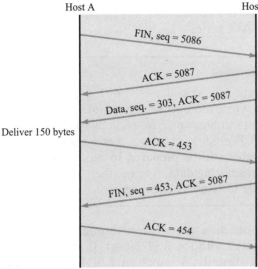

FIGURE 8.27 TCP graceful close

TIME_WAIT timer is restarted at 2MSL. When the TIME_WAIT timer expires, host A closes the connection and then deletes the record of the connection.

The TIME_WAIT state serves a second purpose. The MSL is the maximum time that a segment can live inside the network before it is discarded. The TIME_WAIT state protects future incarnations of the connection from delayed segments. The TIME_WAIT forces TCP to wait at least two MSLs before setting up an incarnation of the old connection. The first MSL accounts for the maximum time a segment in one direction can remain in the network, and the second MSL allows for the maximum time a reply in the other direction can be in the

---

SAYING GOODBYE IS HARD TO DO

Under normal conditions the TCP closing procedure consists of one or more transmissions of a final ACK from host A until a final ACK successfully reaches host B. Host B then closes its TCP connection. Subsequently, host A closes its TCP connection after the expiry of the TIME_WAIT state. It is possible, however, that all retransmitted FIN segments from host B can get lost in the network so that host A closes its connection before host B receives the final ACK. Let's see why this situation is unavoidable. Suppose we insisted that host A not close its connection until it knew that host B had received the final ACK. How is host A supposed to know this? By getting an ACK from B acknowledging A's final ACK! But this scenario is the same problem we started with, all over again, with hosts A and B exchanging roles. Thus the sensible thing to do is for host B to retransmit its FIN some number of times and then to quit without receiving the ACK from A.

network. Thus all segments from the old connection will be cleared from the network at the end of the TIME_WAIT state.

Example—Client/Server Application Revisited

Let us revisit the connection termination process depicted in Figure 8.27 in the context of a client/server application. Once more let host A denote the client and let host B denote the server. The client initiates an *active close* by issuing a `close` call informing the server that the TCP module in host A has no more data to send. The TCP module in host A must still be prepared to receive data from server B. When the server is done transmitting data, it issues the `close` call. This action causes the host B TCP module to issue its FIN segment.

TCP provides for an abrupt connection termination through reset (RST) segments. An RST segment is a segment with the RST bit set. If an application decides to terminate the connection abruptly, it issues an ABORT command, which causes TCP to discard any data that is queued for transmission and to send an RST segment. The TCP that receives the RST segment then notifies its application process that the connection has been terminated.

RST segments are also sent when a TCP module receives a segment that is not appropriate. For example, an RST segment is sent when a connection request arrives for an application process that is not listening on the given port. Another example involves half-open connections that arise from error conditions. For example, host A in one side of the connection may crash. The TCP module in host B, unaware of the crash, may then send a segment. Upon receiving this segment, host A sends an RST segment, thus informing host B that the connection has been terminated.

TCP STATE TRANSITION DIAGRAM

A TCP connection goes through a series of states during its lifetime. Figure 8.28 shows the state transition diagram. Each state transition is indicated by an arrow, and the associated label indicates associated events and actions. Connection establishment begins in the CLOSED state and proceeds to the ESTABLISHED state. Connection termination goes from the ESTABLISHED state to the CLOSED state. The normal transitions for a client are indicated by thick solid lines, and the normal transitions for a server are denoted by dashed lines. Thus when a client does an active open, it goes from the CLOSED state, to SYN_SENT, and then to ESTABLISHED. The server carrying out a passive open goes from the CLOSED state, to LISTEN, SYN_RCVD, and then to ESTABLISHED.

The client normally initiates the termination of the connection by sending a FIN. The associated state trajectory goes from the ESTABLISHED state, to FIN_WAIT_1 while it waits for an ACK, to FIN_WAIT_2 while it waits for the other side's FIN, and then to TIME_WAIT after it sends the final ACK. When the TIME_WAIT 2MSL period expires, the connection is closed and the

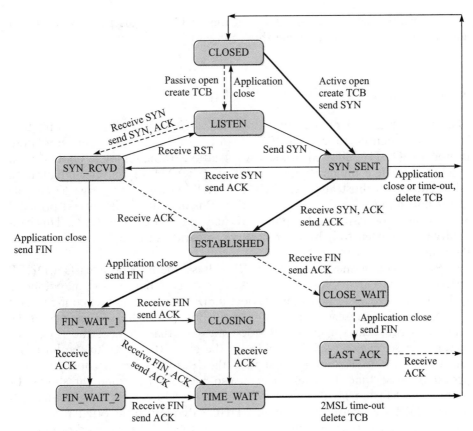

FIGURE 8.28 TCP state transition diagram
Note: The thick solid line is the normal state trajectory for a client; the dashed line is the normal state trajectory for a server

transmission control block that stores all the TCP connection variables is deleted. Note that the state transition diagram does not show all error conditions that may arise, especially in relation to the TIME_WAIT state. The server normally goes from the ESTABLISHED state to the CLOSE_WAIT state after it receives a FIN, to the LAST_ACK when it sends its FIN, and finally to CLOSE when it receives the final ACK. We explore other possible state trajectories in the problem section at the end of the chapter.

8.6 DHCP AND MOBILE IP

A host requires three elements to connect to the Internet: an IP address; a subnet mask; and the address of a nearby router. Each time a user moves or relocates, these elements must be reconfigured. In this section we discuss protocols that have been developed to automate this configuration process.

8.6.1   Dynamic Host Configuration Protocol

The Dynamic Host Configuration Protocol (DHCP) automatically configures hosts that connect to a TCP/IP network. An earlier protocol, Bootstrap Protocol (BOOTP), allowed diskless workstations to be remotely booted up in a network. DHCP builds on the capability of BOOTP to deliver configuration information to a host and uses BOOTP's well-known UDP ports: 67 for the server port and 68 for the client port. DHCP is in wide use because it provides a mechanism for assigning temporary IP network addresses to hosts. This capability is used extensively by Internet service providers to maximize the usage of their limited IP address space.

When a host wishes to obtain an IP address, the host broadcasts a DHCP Discover message in its physical network. The server in the network may respond with a DHCP Offer message that provides an IP address and other configuration information. Several servers may reply to the host, so the host selects one of the offers and broadcasts a DHCP Request message that includes the ID of the server. The selected server then allocates the given IP addres to the host and sends a DHCP ACK message assigning the IP address to the host for some period of lease time T. The message also includes two time thresholds T1 (usually = .5T) and T2 (usually .875T). When time T1 expires the host attempts to extend the lease period by sending a DHCP Request to the original server. If the host gets the corresponding ACK message, it will also receive new values for T, T1, and T2. If the host does not receive an ACK by time T2, then it broadcasts a DHCP Request to any server on the network. If the host does not receive an ACK by time T, then it must relinquish the IP address and begin the DHCP process from scratch.

8.6.2   Mobile IP

Mobile networking is a subject that is becoming increasingly important as portable devices such as personal digital assistants (PDAs) and notebook computers are becoming more powerful and less expensive, coupled with people's need to be connected whenever and wherever they are. The link between the portable device and the fixed communication network can be wireless or wired. If a wireless link is used, the device can utilize a radio or infrared channel. Of these two alternatives, radio channels can traverse longer distance without the line-of-sight requirement, but introduce electromagnetic interference and are often subject to federal regulations (e.g., Federal Communication Commission, or FCC). Infrared channels are often used in shorter distances. A wireless connection enables a user to maintain its communication session as it roams from one area to another, providing a very powerful communication paradigm. In this section we look at a simple IP solution for mobile computers.

Mobile IP allows portable devices called **mobile hosts (MHs)** to roam from one area to another while maintaining the communication sessions. One requirement in mobile IP is that a legacy host communicating with an MH and the intermediate routers should not be modified. This requirement implies that an MH must continuously use its permanent IP address even as it roams to another area. Otherwise, existing sessions will stop working and new sessions should be restarted when an MH moves to another area. The basic mobile IP solution is sketched in Figure 8.29.

The mobile IP routing operates as follows:

- When a **correspondent host (CH)** wants to send a packet to an MH, the CH transmits the standard IP packet with its address as the source IP address and the MH's address as the destination IP address. This packet will be intercepted by the mobile host's router called the **home agent (HA)**, which keeps track of the current location of the MH. The HA manages all MHs in its **home network** that use the same address prefix. If the MH is located in the home network, the HA simply forwards the packet to its home network.
- When an MH moves to a **foreign network**, the MH obtains a **care-of address** from the **foreign agent (FA)** and registers the new address with its HA. The care-of address reflects the MH's current location and is typically the address of the FA. Once the HA knows the care-of address of the MH, the HA can forward the registration packet to the MH via the FA.

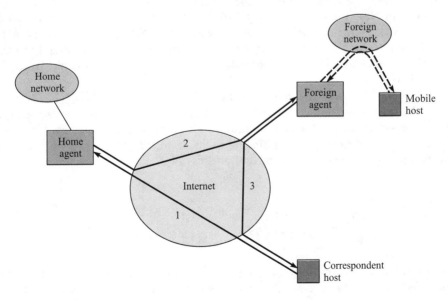

FIGURE 8.29 Routing for mobile hosts

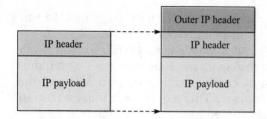

FIGURE 8.30 IP to IP encapsulation

Unfortunately, the HA cannot directly send packets to the MH in a foreign network in a conventional way (i.e., by using the care-of address as the destination address of the IP packet, the packet final destination will the FA rather than the MH). The solution is provided by a **tunneling** mechanism that essentially provides two destination addresses—the destination of the other end of the tunnel (i.e., the FA) and the final destination (i.e., the MH). The IP packet tunneled by the HA is encapsulated with an outer IP header (see Figure 8.30) containing the HA's address as the source IP address and the care-of address as the destination IP address. When the FA receives the packet, the FA decapsulates the packet that produces the original IP packet with the correspondent host's address as the source IP address and the MH's address as the destination IP address. The FA can then deliver the packet to the MH.

Packets transmitted by the MH to the correspondent host typically use a normal IP packet format with the MH's address as the source IP address and the correspondent host's address as the destination IP address. These packets follow the default route.

Observe that the route traveled by the packet from the CH to the MH is typically longer than that from the MH to the CH. For example, it may happen that a CH in New York sends a packet to the HA in Seattle that is tunneled back to New York, since the MH is visiting New York! Several proposals exist to improve the routing mechanism so that the CH may send packets directly to the care-of address endpoint in subsequent exchanges. One solution is illustrated in Figure 8.31.

When the HA receives a packet from a CH destined to an MH (1), it tunnels the packet to the current care-of address (2a), as before. However, it also sends a **binding message** back to the CH containing the current care-of address (2b). The CH can save this message in its **binding cache** so that future packets to the MH can be directly tunneled to the care-of address (4).

8.7   INTERNET ROUTING PROTOCOLS

The global Internet topology can be viewed as a collection of autonomous systems. An **autonomous system (AS)** is loosely defined as a set of routers or networks that are technically administered by a single organization such as a

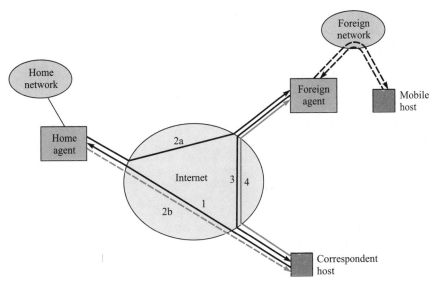

FIGURE 8.31 Route optimization for mobile IP

corporate network, a campus network, or an ISP network. There are no restrictions that an AS should run a single routing protocol within the AS. The only important requirement is that to the outside world, an AS should present a consistent picture of which ASs are reachable through it.

There are three categories of ASs:

1. *Stub AS* has only a single connection to the outside world. Stub AS is also called single-homed AS.
2. *Multihomed AS* has multiple connections to the outside world but refuses to carry **transit traffic** (traffic that originates and terminates in the outside world). Multihomed AS carries only **local traffic** (traffic that originates or terminates in that AS).
3. *Transit AS* has multiple connections to the outside world and can carry transit and local traffic.

For the purpose of AS identification, an AS needs to be assigned with a globally unique AS number (ASN) that is represented by a 16-bit integer and thus is limited to about 65,000 numbers. We show later how ASNs are used in exterior routing. Care must be taken not to exhaust the AS space. Currently, there are about 3500 registered ASs in the Internet. Fortunately, a stub AS, which is the most common type, does not need an ASN, since the stub AS prefixes are placed at the provider's routing table. On the other hand, a transit AS needs an ASN. At present, an organization may request an ASN from the American Registry for Internet Numbers (ARIN) in North America.

Routing protocols in the Internet are arranged in a hierarchy that involves two types of protocols: **Interior Gateway Protocol (IGP)** and **Exterior Gateway**

Protocol (EGP). IGP is used for routers to communicate within an AS and relies on IP addresses to construct paths. EGP is used for routers to communicate among different ASs and relies on AS numbers to construct AS paths. In this section we cover two popular IGPs: Routing Information Protocol and Open Shortest Path First. We will discuss the current de facto standard for EGP, which is Border Gateway Protocol (BGP) version 4 (BGP-4). As an analogy, IGP can thought of as providing a map of a county detailing how to reach each building (host/router), while EGP provides a map of a country, connecting each county (AS).

8.7.1 Routing Information Protocol

Routing Information Protocol (RIP) is based on a program distributed in BSD[8] UNIX called `routed`[9] and uses the distance-vector algorithm discussed in Chapter 7. RIP runs on top of UDP via well-known UDP port number 520. The metric used in the computation of shortest paths is typically configured to be the number of hops. The maximum number of hops is limited to 15 because RIP is intended for use in local area environments where the diameter of the network is usually quite small. The cost value of 16 is reserved to represent infinity. This small number helps to combat the count-to-infinity problem.

A router implementing RIP sends an update message to its neighbors every 30 seconds nominally. To deal with changes in topology such as a link failure, a router expects to receive an update message from each of its neighbors within 180 seconds in the worst case. The reason for choosing a value greater than 30 seconds is that RIP uses UDP, which is an unreliable protocol. Thus some update messages may get lost and never reach the neighbors. If the router does not receive the update message from neighbor X within this limit, it assumes that the direct link to X has failed and sets the corresponding minimum cost to 16 (infinity). If the router later receives a valid minimum cost to X from another neighbor, the router will replace infinity with the new minimum cost.

RIP uses split horizon with poisoned reverse to reduce routing loops.[10] Convergence is speeded up by requiring a router to implement triggered updates. However, a router may want to randomly delay the triggered update messages to avoid loading the network excessively.

The RIP message format is shown in Figure 8.32. The message consists of a command field, a version field, 16 bits that must be set to zero, and a variable number of routing information messages called RIP entries (up to 25 such entries). Each RIP entry is 20 byes long and consists of an address family identifier, an IP address, a metric, and some fields that must be set to 0. Table 8.4 lists the purpose of each field.

---

[8]BSD stands for Berkeley Software Distribution.

[9]Pronounced "route dee" for route daemon.

[10]Split horizon with poisoned reverse was discussed in section 7.5.1.

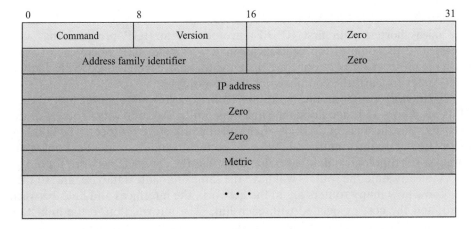

FIGURE 8.32 RIP message format

Although its simplicity is clearly an advantage, RIP also has some limitations, including limited metric use and slow convergence. With the use of hop counts and a small range of value (1 to 15) specified in the metric, the protocol cannot take advantage of network load conditions. Furthermore, the protocol cannot differentiate between high-bandwidth links and low ones. Although the split horizon helps speed up convergence, the protocol may perform poorly under certain types of failures.

RIP-2 allows the RIP packet to carry more information (e.g., subnet mask, next hop, and routing domain). RIP-2 also provides a simple authentication procedure. Unlike RIP-1, RIP-2 can be used with CIDR. The complete specification is documented in RFC 2453.

Field	Description
Command	The command field specifies the purpose of this message. Two values are currently defined: a value of 1 requests the other system to send its routing information, and a value of 2 indicates a response containing the routing information in the sender's routing table.
Version	This field contains the protocol version. RIP-1 set this field to 1, and RIP-2 sets this to 2.
Address family identifier	This field is used to identify the type of address. Currently, only IP address is defined, and the value is 2 for IP.
IP address	This field indicates the address of the destination, which can be a network or host address.
Metric	This field specifies the cost (number of hops) to the destination, which can range from 1 to 15. A value of 16 indicates that the destination is unreachable.

TABLE 8.4 RIP fields

8.7.2 OPEN SHORTEST PATH FIRST

The **open shortest path first (OSPF)** protocol is an IGP protocol that was intended to fix some of the deficiencies in RIP. Unlike RIP where each router learns from its neighbors only the distance to each destination, OSPF enables each router to learn the complete network topology.

Each OSPF router monitors the cost (called the **link state**) of the link to each of its neighbors and then floods the link-state information to other routers in the network. For this reason OSPF is often called a link-state protocol. The flooding of the link-state information allows each router to build an identical **link-state database** (or topological database) that describes the complete network topology.

At steady state the routers will have the same link-state database, and so they will know how many routers are in the network, the interfaces and links between them, and the cost associated with each link. The information in the link-state database allows a router to build the shortest-path tree with the router as the root. The computation of the shortest paths is usually performed by Dijkstra's algorithm, although other routing algorithms can be equally applied. Because the link-state information provides richer information than does distance-vector information, OSPF typically converges faster than RIP when a failure occurs in the network.

Some of the features of OSPF include

- Calculation of multiple routes to a given destination, one for each IP type of service.[11] This functionality provides an added flexibility that is not available in RIP.
- Support for variable-length subnetting by including the subnet mask in the routing message.
- A more flexible link cost that can range from 1 to 65,535. The cost can be based on any criteria.
- Distribution of traffic (load balancing) over multiple paths that have equal cost to the destination. Equal-cost multipath is a simple form of traffic engineering.
- Authentication schemes to ensure that routers are exchanging information with trusted neighbors.
- Multicast rather than broadcast of its messages to reduce the load on systems that do not understand OSPF.
- Use of a *designated router* (and a backup designated router) on multiaccess networks (where routers can talk directly to each other) to reduce the number of OSPF messages that are exchanged. The designated router is elected by a Hello protocol.

To improve scalability, OSPF introduces a two-level hierarchy that allows an AS to be partitioned into several groups called **areas**, that are interconnected by a central **backbone area** as shown in Figure 8.33. An area is identified by a 32-bit number known as the area ID. Continuing our previous analogy, an area can be

---

[11] The type of service field was discussed in section 8.2.1.

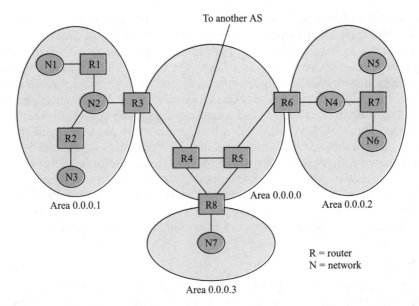

FIGURE 8.33 OSPF areas

thought of as a city or town within a country (AS). The backbone area is identified with area ID 0.0.0.0. The topology of an area is hidden from the rest of the AS in the sense that each router in an area only knows the complete topology inside the area. This approach limits the flooding traffic to an area, which makes the protocol more scalable. The information from other areas is summarized by **area border routers (ABRs)** that have connections to multiple areas. The concept of areas allows OSPF to provide a two-level hierarchy where different areas can exchange packets through the backbone area.

Four types of routers are defined in OSPF. An *internal router* is a router with all its links connected to the networks within the same area. An *area border router* is a router that has its links connected to more than one area. A *backbone router* is a router that has its links connected to the backbone. Finally, an *autonomous system boundary router (ASBR)* is a router that has its links connected to another autonomous system. ASBRs learn about routes outside the AS through an exterior gateway protocol such as BGP. In Figure 8.33 routers 1, 2, and 7 are internal routers. Routers 3, 6, and 8 are area border routers. Routers 3, 4, 5, 6, and 8 are backbone routers. Router 4 is an ASBR.

Two routers are said to be *neighbors* if they have an interface to a common network. A **Hello protocol** allows neighbors to be discovered automatically. Neighbor routers are said to become *adjacent* when they synchronize their topology databases through the exchange of link-state information. Neighbors on point-to-point links become adjacent. Neighbors on multiaccess networks become adjacent to designated routers as explained below. The use of designated routers reduces the size of the topological database and the network traffic generated by OSPF.

A multiaccess network is simply a set of routers that can communicate directly with each other (Think of a multiaccess network as a clique of friends.) In a broadcast multiaccess network, the routers communicate with each other by using a broadcast network such as a LAN. On the other hand, in nonbroadcast multiaccess networks, the routers communicate through a nonbroadcast network, for example, a switched-packet network such as ATM and frame relay. OSPF uses a *designated router* (and a backup designated router) on multiaccess networks. (Think of the designated router as the most popular member of the clique.) The number of OSPF messages that are exchanged is reduced by having the designated router participate in the routing algorithm on behalf of the entire multiaccess network; that is, the designated router generates the link advertisements that list the routers that are attached to its multiaccess network. (The most popular member will promptly give the rest of the clique the lowdown on the network state.) The designated router is elected by the Hello protocol.

OSPF OPERATION

The OSPF protocol runs directly over IP, using IP protocol 89. The OSPF header format is shown in Figure 8.34. Its fields are identified in Table 8.5. Each OSPF packet consists of an OSPF header followed by the body of a particular packet type. There are five types of OSPF packets: hello, database description, link-state request, link-state update, and link-state ACK. OSPF packets are sent to the multicast address 224.0.0.5, which is recognized as the AllSPFRouters address on point-to-point links and on broadcast multiaccess networks. OSPF packets need to be sent to specific IP addresses in nonbroadcast multiaccess networks.

The OSPF operation consists of the following stages.

1. Neighbors are discovered through the sending of Hello messages and designated routers are elected in multiaccess networks.
2. Adjacencies are established and link-state databases are synchronized.
3. Link-state advertisements (LSAs) are exchanged by adjacent routers to allow topological databases to be maintained and to advertise interarea and interAS

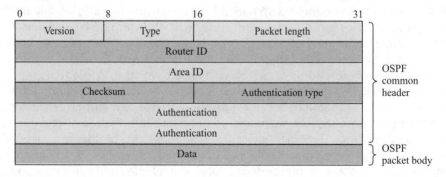

FIGURE 8.34 OSPF common header

Field	Description
Version	This field specifies the protocol version. The most current version is 2.
Type	The type field specifies the type of OSPF packet. The following types are defined: hello, database description, link-state request, link-state update, link-state acknowledgments.
Packet Length	This field specifies the length of OSPF packet in bytes, including the OSPF header
Router ID	This field identifies the sending router. This field is typically set to the IP address of one of its interfaces.
Area ID	This field identifies the area this packet belongs to. The area ID of 0.0.0.0 is reserved for backbone.
Checksum	The checksum field is used to detect errors in the packet.
Authentication type and authentication	The combination of these fields can be used to authenticate OSPF packets.

TABLE 8.5 OSPF header fields

routes. The routers use the information in the database to generate routing tables.

In the following discussion we indicate how the various OSPF packet types are used during these stages.

Stage 1: Hello Packets

OSPF sends Hello packets (type 1) to its neighbors periodically to discover, establish, and maintain neighbor relationships. Hello packets are transmitted to each interface periodically, typically every 10 seconds. The format of the body of the Hello packet is shown in Figure 8.35, its fields are identified in Table 8.6. Each router broadcasts a Hello packet periodically onto its network. When a router receives a Hello packet, it replies with a Hello packet containing

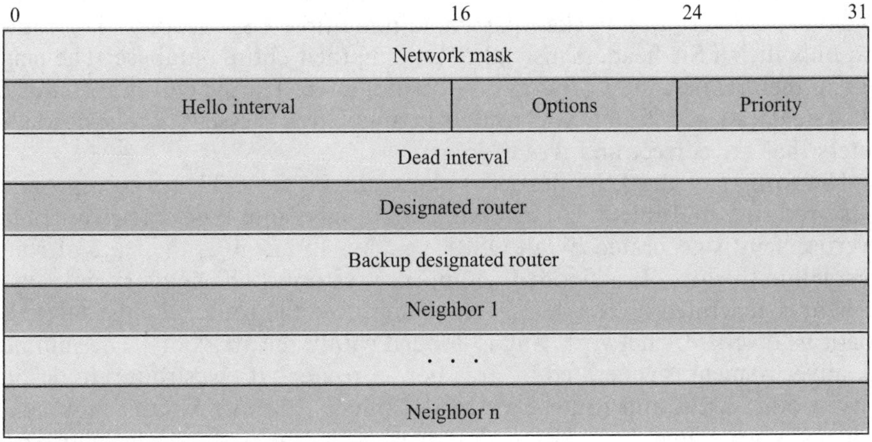

FIGURE 8.35 OSPF Hello packet format

Field	Description
Network mask	The network mask is associated with the interface the packet is sent on.
Hello interval	This field specifies the number of seconds between Hello packets.
Options	Optional capabilities that are supported by the router.
Priority	This field specifies the priority of the router. It is used to elect the (backup) designated router. If set to 0, this router is ineligible to be a (backup) designated router.
Dead interval	This field specifies the number of seconds before declaring a nonresponding neighbor down.
Designated router	This field identifies the designated router for this network. This field is set to 0.0.0.0. if there is no designated router.
Backup designated router	Same as designated router.
Neighbor	This field gives the router ID of each neighbor from whom Hello packets have recently been received.

TABLE 8.6 Hello packet fields

the router ID of each neighbor it has seen. When a router receives a Hello packet containing its router ID in one of the neighbor fields, the router is assured that communication to the sender is bidirectional. Designated routers are elected in each multiaccess network after neighbor discovery. The election is based on the priority and ID fields.

Stage 2: Establishing adjacencies and synchronizing databases

OSPF involves establishing "adjacencies" between a subset of the routers in AS. Only routers that establish adjacencies participate in the operation of OSPF. Once two neighboring routers establish the connectivity between them, they exchange database description packets (type 2) to synchronize their link-state databases. One router acts as a master and the other as a slave. Multiple packets may be used to describe the link-state database. The format of the database description packet is shown in Figure 8.36. Table 8.7 lists the fields. Note that the database description packet may contain multiple LSA headers. The routers send only their LSA headers instead of sending their entire database. The neighbor can then request the LSAs that it does not have. The synchronization of the link-state databases of all OSPF routers in an an area is essential to synthesizing routers that are correct and free of loops.

The format of the LSA header is shown in Figure 8.37. Its corresponding fields are listed in Table 8.8. There are several link-state types. The router link advertisement is generated by all OSPF routers, and it gives the state of router links within the area. It is flooded within the area only. The network link advertisement is generated by the designated router, lists the routers connected to the broadcast or NBMA network, and is flooded within the area only. The summary link advertisement is generated by area border routers; it gives routes to destinations in other areas and routes to ASBRs. Finally, the AS external link advertisement is generated by ASBRs, describes routes to destinations outside the OSPF network, and is flooded in all areas in the OSPF network.

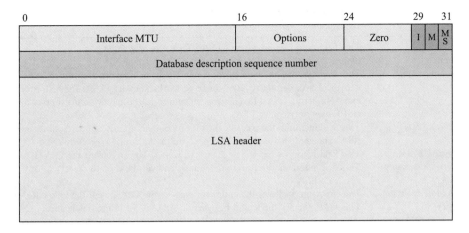

FIGURE 8.36 OSFP database description packet

Field	Description
Interface MTU	This field specifies the maximum transmission unit of the associated interface.
Options	Optional capabilities that are supported by the router.
I bit	The Init bit is set to 1 if this packet is the first packet in the sequence of database description packets.
M bit	The More bit is set to 1 if more database description packets follow.
MS bit	The Master/Slave bit indicates that the router is the master. Otherwise, it is the slave.
Database description sequence number	This field identifies the packet number sequentially so that the receiver can detect a missing packet.
LSA header	The link state advertisement (LSA) header describes the state of the router or network. Each LSA header contains enough information to uniquely identify an entry in the LSA (type, ID, and advertising router).

TABLE 8.7 OSPF database description packet fields

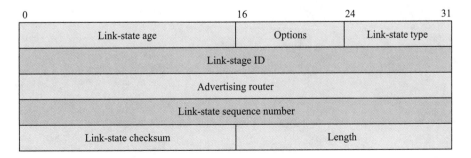

FIGURE 8.37 LSA header

Field	Description
Link-state age	This field describes how long ago the LSA was originated.
Options	Optional capabilities that are supported by the router.
Link-state type	This field specifies the type of the LSA. The possible types are: router LSAs, network LSAs, summary LSAs for IP networks, summary LSAs for ASB routers, and AS-external LSAs. In a network of routers connected by point-to-point links, OSPF uses only router LSAs.
Link-state ID	This field identifies the piece of the routing domain that is being described by the LSA. The contents of this field depend on the link-state type.
Advertising router	This field identifies the router ID of the router that originated the LSA.
Link-state sequence number	This field numbers the LSAs sequentially and can be used to detect old or duplicate LSAs.
Link-state checksum	The checksum includes the entire contents of the LSA except the link-state age.
Length	This field specifies the length in bytes of the LSA including this header.

TABLE 8.8 LSA header fields

Stage 3: Propagation of link-state information and building of routing tables

When a router wants to update parts of its link-state database, it sends a link-state request packet (type 3) to its neighbor. The format of this message is shown in Figure 8.38. Each LSA request is specified by the link-state type, link-state ID, and the advertising router. These three fields are repeated for each link.

In response to a link-state request or when a router finds that its link state has changed, the router will send the new link-state information, using the link-state update message (type 4). The contents of the link-state update message are composed of LSAs, as shown in Figure 8.39.

OSPF uses reliable flooding to ensure that LSAs are updated correctly. Suppose that a router's local state has changed, so the router wishes to update its LSA. The router then issues a link-state update packet that invokes the reliable flooding procedure. Upon receiving such a packet, a neighbor router examines the LSAs in the update. The neighbor installs each LSA in the update

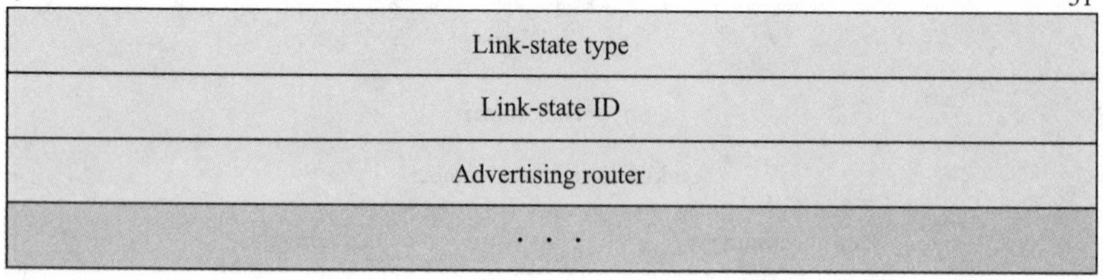

FIGURE 8.38 OSPF link-state request packet

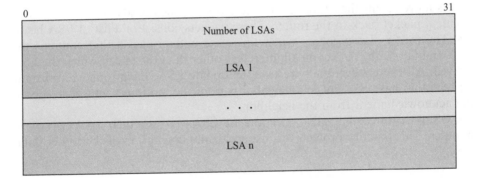

0   31

| Number of LSAs |
| LSA 1 |
| . . . |
| LSA n |

FIGURE 8.39 OSPF link-state update packet

ROUTING HIERARCHY AND INTERNET SCALABILITY

The Internet is a vast collection of networks that are logically linked by a globally unique address space and that provide communications using the TCP/IP protocol suite. Routing protocols are responsible for determining connectivity in the Internet and for generating the routing tables that direct packets to their destinations. In principle, the routing protocols must provide connectivity between every pair of routers in the Internet. This requirement poses a huge scalability challenge with the explosive growth of the Internet. The Autonomous System structure introduces a two-level hierarchy that decomposes the problem of determining Internet connectivity into two parts: routing within an AS (intradomain routing) and routing between ASs (interdomain routing).

At the lower level in the hierarchy, intradomain routing is handled by interior gateway protocols that identify optimal paths within an AS. However, ASs may vary greatly in size; an AS can consist of a campus network such as a university or a large transit network such as a national ISP. The problem of scale arises again in a large AS because the IGP must deal with all routers in the AS. To deal with routing in a large AS. OSPF introduces another level of hierarchy within the AS through the notion of areas. The IGP must then deal only with the routers inside an area.

At the higher level, the introduction of hierarchy also results in the smaller problem of determining connectivity between ASs. Exterior gateway protocols such as BGP address this problem. BGP allows interdomain routers to advertise information about how to reach various networks in the Internet. Furthermore we will see that CIDR addressing allows BGP routers to advertise aggregated paths that reduce the amount of global routing information that needs to be exchanged.

that is more recent than those in its database and then sends an LSA acknowl-edgment packet back to the router. These packets consist of a list of LSA head-ers. The router also prepares a new link-state update packet that contains the LSA and floods the packet on all interfaces other than the one in which the LSA arrived. All routers eventually receive the update LSA. Each router retransmits an LSA that has been sent to a neighbor periodically until receiving correspond-ing acknowledgment from the neighbor.

OSPF requires that all originators of LSAs refresh their LSAs every 30 minutes. This practice protects against accidental corruption of the router data-base.

8.7.3   Border Gateway Protocol

The purpose of an exterior gateway protocol is to enable two different ASs to exchange routing information so that IP traffic can flow across the AS border. Because each AS has its own administrative control, the focus of EGPs is more on policy issues (regarding the type of information that can cross a border) than on path optimality. The **Border Gateway Protocol (BGP)** is an interAS (or inter-domain) routing protocol that is used to exchange network reachability informa-tion among BGP routers (also called BGP speakers).

Each BGP speaker establishes a TCP connection with one or more BGP speakers. In each TCP connection, a pair of BGP speakers exchange BGP mes-sages to set up a BGP session, to exchange information about new routes that have become active or about old routes that have become inactive, and to report error conditions that lead to a termination of the TCP connection. A BGP speaker will advertise a route only if it is actively using the route to reach a particular CIDR prefix. The BGP advertisement also provides attributes about the path associated with the particular prefix.

With BGP the network reachability information that is exchanged contains a sequence of ASs that packets must traverse to reach a destination network, or group of networks reachable through a certain prefix. The information exchanged among BGP speakers allows a router to construct a graph of AS connectivity (see Figure 8.40) whereby routing loops can be pruned and some routing policy at the AS level can be applied.

BGP is a path vector protocol in the sense that BGP advertises the sequence of AS numbers to the destination. An example of a path vector is route 10.10.1.0/24 is reachable via AS1, AS2, AS6, and AS7. Path vector information can easily be used to prevent a routing loop.When a router receives a path advertisement, the router ensures that its AS number does not appear in the path. If it does appear, the router can simply not use that path.

BGP can enforce policy by affecting the selection of different paths to a destination and by controlling the redistribution of routing information. For example, if an AS is to refuse to carry traffic to another AS, the first AS can enforce a policy prohibiting this action.

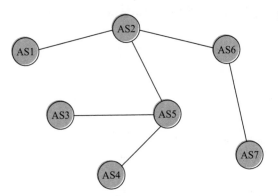

FIGURE 8.40 A graph of ASs

Two speakers exchanging information on a connection are called peers or neighbors. BGP peers run over TCP on port 179 to exchange messages so that reliability is ensured. Using TCP to provide reliable service simplifies the protocol significantly, since complicated timer management is avoided. Using TCP also allows the protocol to update its routing table partially with high confidence.

Initially, BGP peers exchange the entire BGP routing table. Only incremental updates are sent subsequently, instead of periodic refreshes such as in OSPF or RIP. The incremental update approach is advantageous in terms of bandwidth usage and processing overhead. Keepalive messages are sent periodically to ensure that the connection between the BGP peers is alive. The messages are typically sent every 30 seconds and, being only 19 bytes long, should consume minimal bandwidth. Notification messages are sent in response to errors or some exceptions.

Although BGP peers are primarily intended for communications between different ASs, they can also be used within an AS. BGP connections inside an AS are called **internal BGP (iBPG)**, and BGP connections between different ASs are called **external BGP (eBGP)**, as shown in Figure 8.41. Although eBGP peers are directly connected at the link layer, iBGP peers may not be.[12] The purpose of

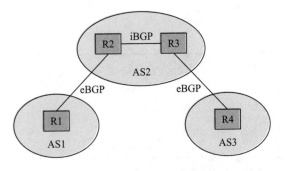

FIGURE 8.41 Internal and external BGP

---

[12]Cisco's implementation allows for eBGP peers to be indirectly connected. This scheme is called eBGP multihop.

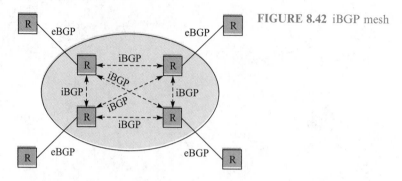

FIGURE 8.42 iBGP mesh

iBGP is to ensure that network reachability information is consistent among the BGP routers in the same AS. For example, if the next-hop external route from R2 to AS1 is via R1, then the next-hop external route from R3 to AS1 is also via R1. A router can easily determine whether to run iBGP or eBGP by comparing the ASN of the other router with its ASN. All iBGP peers must be fully meshed logically so that all eBGP routes are consistent within an AS. As shown in Figure 8.42, an iBGP mesh may create too many TCP sessions. iBGP mesh can be reduced through methods called *confederation* and *route reflection*.

Large ISPs use path aggregation in their BGP advertisements to other ASs. An ISP can use CIDR to aggregate the addresses of many customers into a single advertisement, and thus reduce the amount of information required to provide routing to the customers. The ISP obtains the aggregate by filtering the set of paths so that only aggregates are advertised and more specific paths are not advertised.

BGP MESSAGES

All BGP messages begin with a fixed-size header that identifies the message type. Figure 8.43 shows the BGP message header format.

A description of each field follows.

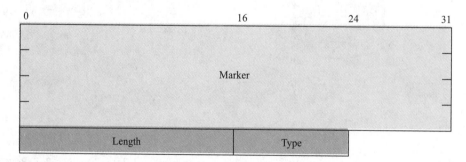

FIGURE 8.43 BGP header format

Marker: The marker field authenticates incoming BGP messages or detects loss of synchronization between a pair of BGP peers. If the message is an OPEN message, the marker field can be predicted based on some authentication mechanism used, which is likely to be Message-Digest Algorithm version 5 (MD-5). If the OPEN message carries no authentication, the marker field must be set to all 1s.

Length: This field indicates the total length of the message in octets, including the BGP header. The value of the length field must be between 19 and 4096.

Type: This field indicates the type of the message. BGP defines four message types:

1. OPEN
2. UPDATE
3. NOTIFICATION
4. KEEPALIVE

Each message type is discussed in the following four sections.

OPEN Message

The first message sent by a BGP router after the transport connection is established is the OPEN message. Figure 8.44 shows the format of the OPEN message.

A description of each field follows.

Version: This field indicates the protocol version number of the message. The current BGP version number is 4.

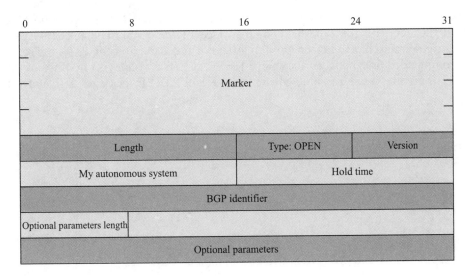

FIGURE 8.44 BGP OPEN message

My autonomous system: This field indicates the AS number of the sending router.

Hold time: This field is used by the sender to propose the number of seconds between the transmission of successive KEEPALIVE messages. The receiving router calculates the value of the Hold Time by using the smaller of its configured hold time and the hold time received in the OPEN message. A hold time value of zero indicates that KEEPALIVE messages will not be exchanged at all.

BGP identifier: This field identifies the sending BGP router. The value is determined by one of the IP local interface addresses of the BGP router and is used for all BGP peers regardless of the interface used to transmit the BPG messages. Cisco's implementation uses the highest IP address in the router or the highest "loopback" address (the IP address of the virtual software interface).

Optional parameters length: This field indicates the total length of the optional parameters field in octets. This field is set to zero if there is no optional parameter.

Optional parameters: This field contains a list of optional parameters, encoded in TLV (that is, parameter type, parameter length, parameter value) structure. Currently, only authentication information (defined as parameter type 1) is defined. The parameter value contains an authentication code field (one octet) followed by an authentication data field, which is variable length. The authentication code specifies the type of authentication mechanism used, the form and meaning of the authentication data, and the algorithm used for computing the value of the marker.

KEEPALIVE Message

BGP speakers continuously monitor the reachability of the peers by exchanging the KEEPALIVE messages periodically. The KEEPALIVE message is just the BGP header with the type field set to 4 (see Figure 8.45). The KEEPALIVE messages are exchanged often enough as to not cause the hold timer to expire. A recommended time between successive KEEPALIVE messages is one-third of the hold time interval. This value ensures that KEEPALIVE messages arrive at the

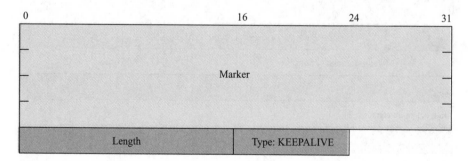

FIGURE 8.45 KEEPALIVE message format

receiving router almost always before the hold timer expires even if the transmission delay of a TCP is variable. If the hold time is zero, then KEEPALIVE messages will not be sent.

NOTIFICATION Message

When a BGP speaker detects an error or an exception, the speaker sends a NOTIFICATION message and then closes the TCP connection. Figure 8.46 shows the NOTIFICATION message format.

The error code indicates the type of error condition, while the error subcode provides more specific information about the nature of the error. The data field describes the reason for the notification, whose length can be deduced from the message length. The error codes and error subcodes are defined in Table 8.9.

UPDATE Message

After the connection is established, BGP peers exchange routing information by using the UPDATE messages. The UPDATE messages are used to construct a graph of AS connectivity. UPDATE messages may contain three pieces of information: unfeasible routes, path attributes, and network layer reachability information (NLRI). Figure 8.47 shows the big picture of the body of the UPDATE message. An UPDATE message can advertise a single route and/or withdraw a list of routes.

The withdrawn routes field provides a list of IP address prefixes for the routes that need to be withdrawn from BGP routing tables. The unfeasible routes length field (two octets) indicates the total length of the withdrawn routes field in octets. An UPDATE message can withdraw multiple unfeasible routes from service. If a BGP router does not intend to withdraw any routes, then the unfeasible routes length field is set to zero and the withdrawn routes field is not present.

A BGP router uses NLRI, the total path attributes length, and the path attributes to advertise a route. The NLRI field contains a list of IP address

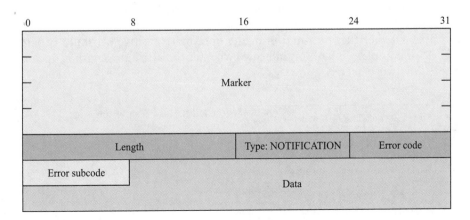

FIGURE 8.46 NOTIFICATION message format

Code	Description (with subcodes where present)		
1	Message header error		
	1 Connection Not Synchronized	3	Bad Message Type
	2 Bad Message Length		
2	OPEN message error		
	1 Unsupported Version Number	4	Unsupported Optional Parameter
	2 Bad Peer AS	5	Authentication Failure
	3 Bad BGP Identifier	6	Unacceptable Hold Time
3	UPDATE message error		
	1 Malformed Attribute List	7	AS Routing Loop
	2 Unrecognized Well-known Attribute	8	Invalid NEXT_HOP Attribute
	3 Missing Well-known Attribute	9	Optional Attribute Error
	4 Attribute Flags Error	10	Invalid Attribute Error
	5 Attribute Length Error	11	Malformed AS_PATH
	6 Invalid Origin Attribute		
4	Hold timer expired		
5	Finite state machine error		
6	Cease		

TABLE 8.9 BGP error codes and subcodes

prefixes that can be reached by the route. The length of the NLRI field is not encoded explicitly, but is deduced from the message length, total path attribute length, and unfeasible routes length. Path attributes describe the characteristics of the route and are used to affect routing behavior. If the NLRI field is present, the total path attribute length field indicates the total length of the path attributes field in octets. Otherwise, this field is set to zero, meaning that no route is being advertised.

An UPDATE message has a variable-length sequence of path attributes. Each path attribute is a triple < Attribute Type, Attribute Length, Attribute Value >, as shown in Figure 8.48. Figure 8.49 displays the format of the attribute type, consisting of attribute flags (one octet) and attribute type code (one octet). The O bit (higher-order bit) indicates whether the attribute is optional (O = 1) or well-known (required). The T bit indicates whether the attribute is transitive or nontransitive (local). Well-known attributes are always transitive. The P bit indicates whether the information in the optional transitive attribute is partial (P = 1) or complete. The E bit indicates whether the attribute length is one octet (E = 0) or two octets.

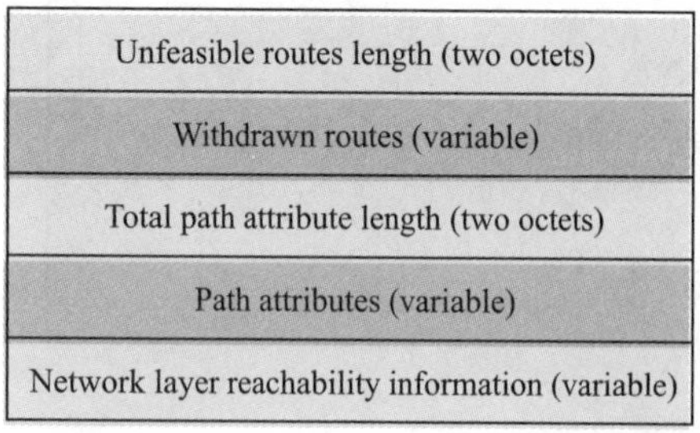

Unfeasible routes length (two octets)
Withdrawn routes (variable)
Total path attribute length (two octets)
Path attributes (variable)
Network layer reachability information (variable)

FIGURE 8.47 UPDATE message

Attribute type	Attribute length	Attribute value

FIGURE 8.48 Format of each attribute

Some explanation on path attributes is in order. There are four categories of path attributes: well-known mandatory, well-known discretionary, optional transitive, and optional nontransitive. Well-known attributes must be recognized by all BGP speakers, and well-known mandatory attributes must be sent in any UPDATE message. On the other hand, optional attributes may not be recognized by a BGP speaker. Paths with unrecognized transitive optional attributes should be accepted and passed to other BGP peers with the P bit set to 1. Paths with unrecognized nontransitive optional attributes must be silently rejected.

The attribute type code field contains the attribute code. The attribute codes that are defined in the standards follow.

ORIGIN (type code 1): The ORIGIN attribute is a well-known mandatory attribute that defines the origin of the NLRI. The attribute value (one octet) can have one of three values. A value of 0 (for IGP) indicates that the NLRI is interior to the originating AS, for example, when IGP routes are redistributed to EGP. A value of 1 (for EGP) indicates that NLRI is learned via BGP. A value of 2 (for INCOMPLETE) indicates that NLRI is learned by some other means. This usually occurs when a static route is distributed to BGP and the origin of the route will be incomplete.

AS_PATH (type code 2): The AS_PATH attribute is a well-known mandatory attribute that lists the sequence of ASs that the route has traversed to reach the destination. When a BGP speaker originates the route, the speaker adds its own AS number when sending the route to its external peers. When a BGP speaker propagates a route that it has learned from another BGP speaker, the propagating speaker prepends its own AS number to the AS_PATH list. BGP uses the AS_PATH attribute to detect a potential loop. If the AS number of the BGP speaker is already contained in the list of AS numbers that the route has already been through, then the route should be rejected. When a customer uses a private AS number (64512-65535), the provider must make sure to strip the private AS number from the AS_PATH list, since these numbers are not globally unique.

NEXT_HOP (type code 3): The NEXT_HOP attribute is a well-known mandatory attribute that defines the IP address of the border router that should be used as the next hop to the destinations listed in the NLRI. Figure 8.50 clarifies the meaning of next hop. For eBGP the next hop is the IP address of the BGP peer. For example, R4 advertises 10.1.2.0/24 to R3 with a next hop of 10.10.1.2. For iBGP the next hop advertised by eBGP should be carried into iBGP. For example, R3 should advertise 10.1.2.0/24

O	T	P	E	0	Attribute type code

FIGURE 8.49 Attribute type

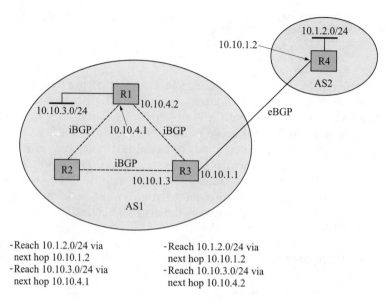

-Reach 10.1.2.0/24 via
  next hop 10.10.1.2
-Reach 10.10.3.0/24 via
  next hop 10.10.4.1

-Reach 10.1.2.0/24 via
  next hop 10.10.1.2
-Reach 10.10.3.0/24 via
  next hop 10.10.4.2

FIGURE 8.50 BGP NEXT_HOP

to R2 with a next hop of 10.10.1.2 instead of 10.10.2.2. To work properly, R2 should be able to reach 10.10.1.2 via IGP.

MULTI_EXIT_DISC (type code 4): The MULTI_EXIT_DISC (MED) attribute is an optional nontransitive **attribute** that is used to discriminate among multiple entry/exit points to a neighboring AS and give a hint to the neighboring AS about the preferred path. The MED field is encoded in a four-octet unsigned integer. An exit or entry point with lower MED value should be preferred. The value can be derived from the IGP metric. The MED attribute received is never propagated to other ASs (i.e., it is nontransitive). Figure 8.51 illustrates the use of the MED attribute. In this example, R2 and R3 advertise the same route (10.1.1.0/24) to R1. R2 uses a MED value of 100, while R3 uses a MED value of 200. It is a hint that AS2 prefers AS1 to use the route via R2 for this case. The usage of the MED attribute becomes complicated when another AS advertises the same route, since the IGP metrics used by different ASs can be different. In such a case comparing a MED value used by one AS with another MED value used by another AS makes no sense.

LOCAL_PREF (type code 5): The LOCAL_PREF attribute is a well-known discretionary attribute that informs other BGP speakers within the same AS of its degree of preference for an advertised route. The LOCAL_PREF attribute is exchanged only among the iBGP peers and should not be exchanged by the eBGP peers. The attribute that indicates the degree of preference for the exit points is encoded in a four-octet unsigned integer. A higher value indicates a higher preference.

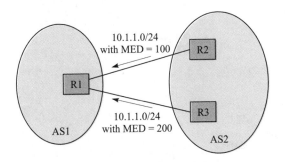

FIGURE 8.51 Example of the MED attribute

ATOMIC_AGGREGATE (type code 6): The ATOMIC_AGGREGATE attribute is a well-known discretionary attribute that informs other BGP speakers that it selected a less specific route without selecting a more specific one that is included in it. The attribute length for ATOMIC_AGGREGATE is zero (no Attribute value). A BGP speaker that receives a route with the ATOMIC_AGGREGATE attribute should not remove the attribute when propagating it to other speakers. Also, the BGP speaker should not make the route more specific when advertising this route to other BGP speakers.

AGGREGATOR (type code 7): The AGGREGATOR attribute is an optional transitive attribute that specifies the last AS number that formed the aggregate route (encoded in two octets) followed by the IP address of the BGP speaker that formed the aggregate route (encoded in four octets). Thus the attribute is six octets long.

8.8 MULTICAST ROUTING

The routing mechanisms discussed so far assume that a given source transmits its packets to a single destination. For some applications such as teleconferencing, a source may want to send packets to multiple destinations simultaneously. This requirement calls for another type of routing called **multicast routing** as illustrated in Figure 8.52. In the figure, source S wants to transmit to destinations with multicast group G1. Although the source can send each copy of the packet separately to each destination by using conventional unicast routing, a more efficient method would be to minimize the number of copies. For example, when router 1 receives a packet from the source, router 1 copies the packet to routers 2 and 5 simultaneously. Upon receipt of these packets, router 2 forwards the packet to its local network, and router 5 copies the packet to routers 7 and 8. The packet will eventually be received by each intended destination.

With multicast routing, each packet is transmitted once per link. If unicast routing is used instead, the link between the source and router 1 will carry four copies for each packet transmitted. In general, the bandwidth saving with multicast routing becomes more substantial as the number of destinations increases.

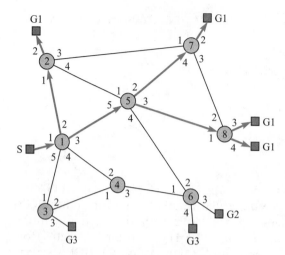

FIGURE 8.52 Multicast tree rooted at source

There are many ways to generate a multicast tree. One approach that is used in **multicast backbone (MBONE)** is called **reverse-path multicasting**. MBONE is basically an overlay packet network on the Internet supporting routing of IP multicast packets (with Class D addresses).

8.8.1   Reverse-Path Broadcasting

The easiest way to understand reverse-path multicasting is by first considering a simpler approach called **reverse-path broadcasting (RPB)**. RBP uses the fact that the set of shortest paths to a node forms a tree that spans the network. See Figure 7.29. The operation of RBP is very simple:

- Upon receipt of a multicast packet, a router records the source address of the packet and the port the packet arrives on.
- If the shortest path from the router back to the source is through the port the packet arrived on, the router forwards the packet to all ports except the one the packet arrived on; otherwise, the router drops the packet. The port over which the router expects to receive multicast packets from a given source is referred to as the **parent** port.

The advantages of RPB are that routing loops are automatically suppressed and each packet is forwarded by a router exactly once. The basic assumption underlying the RPB algorithm is that the shortest path from the source to a given router should be the same as the shortest path from the router to the source. This assumption requires each link to be symmetric (each direction has the same cost). If links are not symmetric, then the router must compute the shortest path from the source to the router. This step is possible only if link-state protocols are used.

To see how RPB works for the network shown in Figure 8.53, assume that source S transmits a multicast packet with group G1. The parent ports for this multicast group are identifed in bold numbers, as shown in Figure 8.53. First

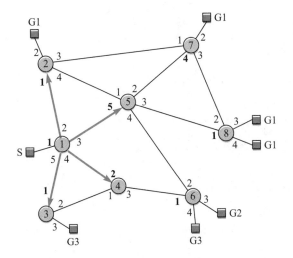

FIGURE 8.53 Router 1 forwards the packet

router 1 receives a packet on port 1 from source S. Because the packet comes from the parent port, the router forwards the packet to all other ports.

Assume that these packets reach routers 2, 3, 4, and 5 some time later. Router 2 computes the shortest path from itself to S and finds that the parent port is port 1. It then forwards the packet through all other ports. Similarly, routers 3, 4, and 5 find that the packet arrives on the parent ports. As a result, each router forwards the packet to its other ports, as shown in Figure 8.54. Note that the bidirectional link indicates that each router forwards the packet to the other.

Next assume that the packets arrive at routers 2 through 8. Because router 2 receives the packet from port 4, which is not the parent port, router 2 drops the packet. Routers 3, 4, and 5 also drop the packet for the same reason. Router 6,

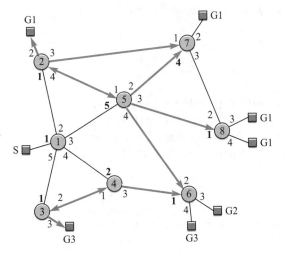

FIGURE 8.54 Routes 2, 3, 4, and 5 forward the packets

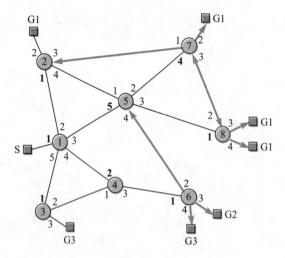

FIGURE 8.55 Routers 6, 7, and 8 forward the packets

however, finds that the parent port is port 1, so router 6 forwards the packet coming from router 4 to its other ports. The packet that came from router 5 to router 6 is dropped. Routers 7 and 8 forward the packet that came from router 5. Figure 8.55 illustrates the process pictorially. The reader should verify that the packet will no longer be propagated after this point.

Note that although the hosts connected to routers 3 and 6 belong to different multicast groups, the packets for group G1 are still forwarded by these routers. Typically, these hosts are attached to the router via a shared medium LAN. Therefore, unnecessary bandwidth is wasted regardless of the multicast group. This problem can be solved by having the router truncate its transmission to the local network if none of the hosts attached to the network belong to the multicast group. This refinement is called **truncated reverse-path broadcasting (TRPB)**. Note that TRPB performs truncation only at the "leaf" routers (routers at the edge of the network). The next section describes a protocol that allows a router to determine which hosts belong to which multicast groups.

8.8.2   Internet Group Management Protocol

The **Internet Group Management Protocol (IGMP)** allows a host to signal its multicast group membership to its attached router. IGMP runs directly on IP using protocol type 2. However, IGMP is usually considered as part of IP. The IGMP message format is very simple, as can be seen from Figure 8.56.

A description of each field follows.

Version: This field identifies the version number.

Type: This field identifies the message type. There are two message types: Type 1 indicates a query message sent by a router, and type 2 indicates a report sent by a host.

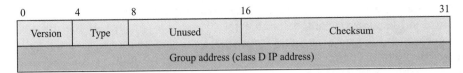

Version	Type	Unused	Checksum
Group address (class D IP address)			

FIGURE 8.56 IGMP message format

Unused: This field must be set to zero.

Checksum: This field contains a checksum for all eight bytes of the IGMP message.

Group Address: This address is the class D IPv4 address. This field is set to zero in a query message, and is set to a valid group address in the response.

When a host wants to join a new multicast group, the host sends an IGMP report specifying the group address to join. The host needs to issue only one IGMP report, even though multiple applications are joining the same group. The host does not issue a report if it wants to leave the group.

A multicast router periodically issues an IGMP query message to check whether there are hosts belonging to multicast groups. The IGMP message is sent with the IP destination address set to the **all-hosts** multicast address (i.e., 224.0.0.1). When a host receives a query, it must respond with a report for each group it is joining. By exchanging queries and reports, a router would know which multicast groups are associated with a given port so that the router would forward an incoming multicast packet only to the ports that have hosts belonging to the group. Note that the router does not have to know how many hosts belong to a particular group. It only has to know that there is at least one host for a particular group.

To make sure that multiple hosts do not send a report at the same time when receiving a query that would lead to a collision, the scheduled transmission time should be randomized. Because the router only has to know that at least one host belongs to a particular group, efficiency can be improved by having a host monitor if another host sends the same report earlier than its scheduled transmission. If another host has already sent the same report, the first host cancels its report.

8.8.3 Reverse-Path Multicasting

As section 8.8.2 just explained IGMP allows hosts to indicate their group members. We now continue with selective multicast routing called **reverse-path multicasting (RPM)**, which is an enhancement of TRPB. Unlike TRPB, RPM forwards a multicast packet only to a router that will lead to a leaf router with group members.

Each router forwards the first packet for a given (source, group) according to TRPB. All leaf routers receive at least the first multicast packet. If a leaf router does not find a member for this group on any of its ports, the leaf router will send a **prune message** to its upstream router instructing the upstream router not to forward subsequent packets belonging to this (source, group). If the upstream router receives the prune messages from all of its downstream neighbors, it will in turn generate a prune message to its further upstream router.

Referring to an earlier example in Figure 8.52, router 3 would send a prune message to routers 1 and 4. When router 1 receives the prune message, router 1 stops forwarding the subsequent multicast packets to port 5. Similarly, router 6 would send a prune message to routes 4 and 5. When router 4 realizes that it has received the prune messages from all of its downstream neighbors, router 4 sends a prune message to router 1, which subsequently stops forwarding the packets to router 4. The reader should verify that the multicast packets eventually follow the paths shown in Figure 8.52.

Each prune information is recorded in a router for a certain lifetime. When the prune timer expires, the router purges the information and starts forwarding the multicast packets to its downstream neighbors.

A host may later decide to join a multicast group after a prune message has been sent by its leaf router. In this case the leaf router would send a **graft message** to its upstream router to cancel its earlier prune message. Upon receiving the graft message, the first upstream router will forward subsequent multicast packets to this leaf router. If the first upstream router has also sent a prune message earlier, the first upstream router will send a graft message to the second upstream router. Eventually, a router in the multicast tree will reactivate its affected port so that the multicast packets will travel downstream toward the host. Figure 8.57 shows the grafting message flow when a host attached to router 6 wants to join the group. Subsequently, router 1 will forward the multicast packets to router 4, which will forward the multicast packets to router 6. Eventually, the multicast packets arrive at the host.

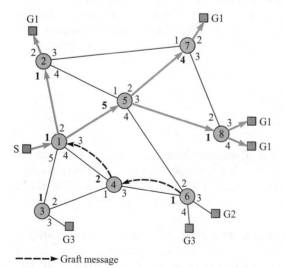

FIGURE 8.57 Grafting to cancel pruning

8.8.4   Distance-Vector Multicast Routing Protocol

Distance-Vector Multicast Routing Protocol (DVMRP) is the multicast routing algorithm that is used in the MBONE routers. DVMRP is based on a combination of RIP and RPM. One main difference between RIP and DVMRP is that RIP determines the next hop toward the destination, whereas DVMRP determines the previous hop toward the source. DVMRP uses a tunneling mechanism to traverse networks that do not support multicasting. Tunnels are manually configured between two multicast router (mrouters).

One of the severe shortcomings of DVMRP is its inability to operate in a large internetwork. The first reason for this shortcoming is that the maximum number of hops is limited to 15. Second, it is not desirable to flood the entire Internet's leaf routers with the first multicast packet for a particular (source, group). Periodic refloodings due to the soft state mechanism associated with the prune information compound the problem. Also, the flat nature of the routing domain make it unsuitable for a large network.

SUMMARY

We began this chapter with an end-to-end view of the flow of application layer messages through the IP layer in the end systems and across an IP internetwork. We then examined the details of how Internet Protocol enables communications across a collection of networks. We paid particular attention to the hierarchical structure of IP addresses and explained their role in ensuring scalability of the Internet. The role of address prefixes and the use of masks was explained. We also examined the interplay between IP addresses, routing tables, and router operation. We discussed how the IP layer must perform segmentation and reassembly in order to provide the higher layers with independence from the underlying network technologies. We also introduced Internet Protocol version 6 and identified its advantages relative to IPv4 in terms of address space and streamlined header design.

The Internet offers two basic communication services that operate on top of IP: TCP reliable stream service and UDP datagram service. We examined the simple operation of UDP, and we then focussed on the details of how TCP provides reliable stream service and flow control. We also discussed the various issues that underlie the design of TCP: the rationale for using the three-way handshake; the separation of acknowledgments and window size; the difficulties in closing a connection, and the consequences of sequence number space limitations as well as of round-trip time variability.

We introduced DHCP and mobile IP and explained their role in extending IP service to mobile and transitory users.

The later sections of the chapter were concerned with IP routing. We introduced the Autonomous System structure of the Internet and explained how the routing problem is then partitioned into intradomain and interdomain components. We introduced RIP and OSPF for intradomain routing. We identified the advantages of OSPF and explained its operation in detail. We discussed how intradomain and inter-domain routing involve different concerns, e.g. optimal path routing versus policy routing. We also explained how BGP4 provides interdomain routing. Finally, we showed how the hierarchy of the Autonomous System structure and the hierarchy inherent in CIDR addressing work together to provide an Internet that can scale to enormous size.

In the last section we provided an introduction to multicast routing.

CHECKLIST OF IMPORTANT TERMS

2MSL time-out
Address Resolution Protocol (ARP)
advertised window
all hosts
area
area border router
autonomous system (AS)
backbone area
Bellman-Ford algorithm
binding cache
binding message
Border Gateway Protocol (BGP)
care-of address
classless interdomain routing (CIDR)
correspondent host (CH)
deflection routing
Dijkstra's algorithm
Distance-Vector Multicast Routing
    Protocol (DVMRP)
dotted-decimal notation
Exterior Gateway Protocol (EGP)
external BGP (eBGP)
flooding
foreign address (FA)
foreign network
fragment
graceful close
graft message
home agent (HA)
home network
host ID

initial sequence number (ISN)
Interior Gateway Protocol (IGP)
internal BGP (iBGP)
Internet Control Message Protocol
    (ICMP)
Internet Group Management
    Protocol (IGMP)
IPv6
link state
link-state database
local traffic
lost/wandering duplicate
Manhattan street network
maximum segment lifetime (MSL)
maximum segment size (MSS)
maximum transmission unit (MTU)
mobile host (MH)
multicast backbone (MBONE)
multicast routing
Nagle algorithm
network diameter
network ID
network routing
open shortest path first (OSPF)
parent port
piggyback
prune message
pseudoheader
push command
Reverse Address Resolution Protocol
    (RARP)

reverse-path broadcasting (RPB)
reverse-path multicasting (RPM)
round-trip time
Routing Information Protocol (RIP)
segment
sequence number
shortest-path algorithm
silly window syndrome
source routing
subnetting
supernet
three-way handshake

timestamp
TIME_WAIT state
time-to-live (TTL) field
transit traffic
transmission control block (TCB)
Transport Control Protocol (TCP)
truncated reverse-path broadcasting
  (TRPB)
tunneling
User Datagram Protocol (UDP)
window scale

FURTHER READING

Braden, R., "TCP Extensions for High Performance: An Update," Internet Draft, June 1993.

Comer, D. E., *Internetworking with TCP/IP, Volume 1*, 3rd ed., Prentice Hall, Englewood Cliffs, New Jersey, 1995.

Degermark, M., A. Brodnik, S. Carlsson, and S. Pink, "Small Forwarding Tables for Fast Routing Lookups," in *Designing and Building Gigabit and Terabit Internet Routers*, ACM SigComm 98 tutorial proceedings, September 1998.

Huitema, C., *IPv6, The New Internet Protocol*, Prentice Hall PTR, Upper Saddle River, New Jersey, 1996.

Huitema, C., *Routing in the Internet*, Prentice Hall, Englewood Cliffs, New Jersey, 1995.

Kosiur, D., *IP Multicasting: The Complete Guide to Interactive Corporate Networks*, Wiley, New York, 1998.

Moy, J. T., *OPSF: Anatomy of an Internet Routing Protocol*, Addison-Wesley, Reading, Massachusetts, 1998.

Murhammer, M. W., O. Atakan, S. Bretz, L. R. Pugh, K. Suzuki, and D. H. Wood, *TCP/IP Tutorial and Technical Overview*, Prentice Hall PTR, Upper Saddle River, New Jersey, 1998.

Perlman, R., *Interconnections: Bridges and Routers*, Addison-Wesley, Reading, Massachusetts, 1992.

Stallings, W., "IPv6: The New Internet Protocol," *IEEE Communications*, Vol. 34, No. 7, pp. 96–108, July 1996.

Stevens, W. R., *TCP/IP Illustrated, Volume 1*, Addison-Wesley, Reading, Massachusetts, 1994.

Stevens, W. R., *UNIX Network Programming, Volume 1: Networking APIs: Sockets and XTI*, 2nd ed., Prentice Hall, Englewood Cliffs, New Jersey, 1998.

Stewart, J., *BGP4: Inter-Domain Routing in the Internet*, Addison-Wesley, Reading, Massachusetts, 1999.

Waldvogel, M., G. Varghese, J. Turner, and B. Plattner, "Scalable High Speed IP Routing Lookups," in *Designing and Building Gigabit and Terabit Internet Routers*, ACM SigComm 98 tutorial proceedings, September 1998.

RFC 793, J. Postel (ed.), "Transmission Control Protocol: DARPA Internet Program Protocol Specification," September 1981.

RFC 1058, C. Hedrick, "Routing Information Protocol," June 1988.

RFC 1122, R. Braden (ed.), "Requirements for Internet Hosts—Communication Layers," October 1989.

RFC 1323, V. Jacobsen, R. Braden, and D. Borman, "TCP Extension for High Performance," May 1992.

RFC 1771, Y. Rekhter and T. Li, "A Border Gateway Protocol 4 (BGP-4)," May 1995.

RFC 2002, C. Perkins, "IP Mobility Support," October 1996.

RFC 2113, D. Katz, "IP Router Alert Option," February 1997.

RFC 2328, J. Moy, "OSPF Version 2," April 1998.

RFC 2453, G. Malkin, "RIP Version 2," November 1998.

PROBLEMS

1. Identify the address class of the following IP addresses: 200.58.20.165; 128.167.23.20; 16.196.128.50; 150.156.10.10; 250.10.24.96.

2. Identify the range of IPv4 addresses spanned by Class A, Class B, and Class C.

3. A university has 150 LANs with 100 hosts in each LAN.
   a. Suppose the university has one Class B address. Design an appropriate subnet addressing scheme.
   b. Design an appropriate CIDR addressing scheme.

4. A small organization has a Class C address for seven networks each with 24 hosts. What is an appropriate subnet mask?

5. A packet with IP address 150.100.12.55 arrives at router R1 in Figure 8.7. Explain how the packet is delivered to the appropriate host.

6. In Figure 8.7 assign a physical layer address 1, 2, ... to each physical interface starting from the top row, moving right to left, and then moving down. Suppose H4 sends an IP packet to H1. Show the sequence of IP packets and Ethernet frames exchanged to accomplish this transfer.

7. ARP is used to find the MAC address that corresponds to an IP address; RARP is used to find the IP address that corresponds to a MAC address. True or false?

8. Perform CIDR aggregation on the following/24 IP addresses: 128.56.24.0/24; 128.56.25.0/24; 128.56.26.0/24; 128.56.27.0/24.

9. Perform CIDR aggregation on the following /24 IP addresses: 200.96.86.0/24; 200.96.87.0/24; 200.96.88.0/24; 200.96.89.0/24.

10. The following are estimates of the population of major regions of the world: Africa 900 million; South America 500 million; North America 400 million; East Asia 1500 million; South and Central Asia 2200 million; Russia 200 million; Europe 500 million.

a. Suppose each region is to be assigned 100 IP addresses per person. Is this possible? If not, how many addresses can be assigned per person.

b. Design an appropriate CIDR scheme to provide the addressing in part (a).

11. Suppose four major ISPs were to emerge with points of presence in every major region of the world. How should a CIDR scheme treat these ISPs in relation to addressing for each major region?

12. Suppose that a Teledesic-like satellite network (Chapter 4) is based on IP routing. Design a CIDR scheme assuming each cell has 100,000 hosts.

13. Look up the netstat command in the manual for your system. Find and try the command to display the routing table in your host.

14. Suppose a router receives an IP packet containing 600 data bytes and has to forward the packet to a network with maximum transmission unit of 200 bytes. Assume that IP header is 20 bytes long. Show the fragments that the router creates and specify the relevant values in each fragment header (i.e., total length, fragment offset, and more bit).

15. Design an algorithm for reassembling fragments of an IP packet at the destination IP.

16. Does it make sense to do reassembly at intermediate routers? Explain.

17. Abbreviate the following IPv6 addresses:
    a. 0000:0000:0F53:6382:AB00:67DB:BB27:7332
    b. 0000:0000:0000:0000:0000:0000:004D:ABCD
    c. 0000:0000:0000:AF36:7328:0000:87AA:0398
    d. 2819:00AF:0000:0000:0000:0035:0CB2:B271

18. What is the efficiency of IPv6 packets that carry 10 ms of 64 kbps voice? Repeat if an IPv6 packets carries 1 frame of 4 Mbps MPEG2 video, assuming 30 frames/second.

19. Why does IPv6 allow fragmentation at the source only?

20. Assuming the population estimates in problem 10, how many IP addresses does IPv6 provide per capita?

21. Suppose that IPv6 is used over a noisy wireless link. What is the effect of not having header error checking?

22. Explain how the use of hierarchy enhances scalability in the following aspects of Internet:
    a. Domain name system
    b. IP addressing
    c. OSPF routing
    d. Interdomain routing

23. The TCP in station A sends a SYN segment with ISN = 1000 and MSS = 1000 to station B. Station B replies with a SYN segment with ISN = 5000 and MSS = 500. Suppose station A has 10,000 bytes to transfer to B. Assume the link between stations A and B

is 8 Mbps and the distance between them is 200 m. Neglect the header overheads to keep the arithmetic simple. Station B has 3000 bytes of buffer available to receive data from A. Sketch the sequence of segment exchanges, including the parameter values in the segment headers, and the state as a function of time at the two stations under the following situations:

a. Station A sends its first data segment at $t = 0$. Station B has no data to send and sends an ACK segment every other frame.

b. Station A sends its first data segment at $t = 0$. Station B has 6000 bytes to send and sends its first data segment at $t = 2$ ms.

24. Suppose that the TCP in station A sends information to the TCP in station B over a two-hop path. The data link in the first hop operates at a speed of 8 Mbps, and the data link in the second hop operates at a speed of 400 kbps. Station B has a 3 kilobyte buffer to receive information from A, and the application at station B reads information from the receive buffer at a rate of 800 kbps. The TCP in station A sends a SYN segment with ISN = 1000 and MSS = 1000 to station B. Station B replies with a SYN segment with ISN = 5000 and MSS = 500. Suppose station A has 10,000 bytes to transfer to B. Neglect the header overheads to keep the arithmetic simple. Sketch the sequence of segment exchanges, including the parameter values in the segment headers, and the state as a function of time at the two stations. Show the contents of the buffers in the intermediate switch as well as at the source and destination stations.

25. Suppose that the delays experienced by segments traversing the network are equally likely to be any value in the interval [50 ms, 75 ms].

a. Find the mean and standard deviation of the delay.

b. Most computer languages have a function for generating uniformly distributed random variables. Use this function in a short program to generate random times in the above interval. Also, calculate t_{RTT} and d_{RTT} and compare to part (a).

26. Suppose that the advertised window is 1 Mbyte long. If a sequence number is selected at random from the entire sequence number space, what is the probability that the sequence number falls inside the advertised window?

27. Explain the relationship between advertised window size, RTT, delay-bandwidth product, and the maximum achievable throughput in TCP.

a. Plot the maximum achievable throughput versus delay-bandwidth product for an advertised window size of 65,535 bytes.

b. In the preceding plot include the maximum achievable throughput when the window size is scaled up by a factor of 2^K, where $K = 4, 8, 12$.

c. Place the following scenarios in the plot obtained in part (b): Ethernet with 1 Gbps and distance 100 meters; 2.4 Gbps and distance of 6000 km; satellite link with speed of 45 Mbps and RTT of 500 ms; 40 Gbps link with distance of 6000 km.

28. Consider the three-way handshake in TCP connection setup.

a. Suppose that an old SYN segment from station A arrives at station B, requesting a TCP connection. Explain how the three-way handshake procedure ensures that the connection is rejected.

b. Now suppose that an old SYN segment from station A arrives at station B, followed a bit later by an old ACK segment from A to a SYN segment from B. Is this connection request also rejected?

29. Suppose that the initial sequence number (ISN) for a TCP connection is selected by taking the 32 low-order bits from a local clock.
   a. Plot the ISN versus time assuming that the clock ticks forward once every $1/R_c$ seconds. Extend the plot so that the sequence numbers wrap around.
   b. To prevent old segments from disrupting a new connection, we forbid sequence numbers that fall in the range corresponding to 2MSL seconds prior to their use as an ISN. Show the range of forbidden sequence numbers versus time in the plot from part (a).
   c. Suppose that the transmitter sends bytes at an average rate $R > R_c$. Use the plot from part (b) to show what goes wrong.
   d. Now suppose that the connection is long-lived and that bytes are transmitted at a rate R that is much lower than R_c. Use the plot from part (b) to show what goes wrong. What can the transmitter do when it sees that this problem is about to happen?

30. Suppose that during the TCP connection closing procedure, a machine that is in the TIME_WAIT state crashes, reboots within MSL seconds and immediately attempts to reestablish the connection, using the same port numbers. Give an example that shows that delayed segments from the previous connections can cause problems. For this reason RFC 793 requires that for MSL seconds after rebooting TCP is not allowed to establish new connections.

31. Are there any problems if the server in a TCP connection initiates an active close?

32. A fast typist can do 100 words a minute, and each word has an average of six characters. Demonstrate Nagle's algorithm by showing the sequence of TCP segment exchanges between a client, with input from our fast typist, and a server. Indicate how many characters are contained in each segment sent from the client. Consider the following two cases:
   a. The client and server are in the same LAN and the RTT is 20 ms.
   b. The client and server are connected across a WAN and the RTT is 100 ms.

33. *Simultaneous Open.* The TCP state transition diagram allows for the case where the two stations issue a SYN segment at nearly the same time. Draw the sequence of segment exchanges and use Figure 8.28 to show the sequence of states that are followed by the two stations in this case.

34. *Simultaneous Close.* The TCP state transition diagram allows for the case where the two stations issue a FIN segment at nearly the same time. Draw the sequence of segment exchanges and use Figure 8.28 to show the sequence of states that are followed by the two stations in this case.

35. Suppose an ISP has 1000 customers and that at any time during the busiest hour of the day the probability that a particular user requires service is 20. The ISP uses DHCP. Is a Class C address enough to make the probability less than 1 percent that no IP address is available when a customer places a request?

36. Compare mobile IP with the procedures used by cellular telephone networks (Chapter 4) to handle roaming users.
    a. Which cellular network components provide the functions of the home and foreign agent?
    b. Is the handling of mobility affected by whether the transfer service is connectionless or connection oriented?

37. Consider a user that can be in several places (home networks) at different times. Suppose that the home networks of a user contain registration servers where users send updates of their location at a given time.
    a. Explain how a client process in a given end system can find the location of a given user to establish a connection, for example, Internet telephone, at a given point in time.
    b. Suppose that proxy servers are available, and their function is to redirect location requests to another server that has more precise location information about the callee. For example, a university might have such a server, which redirects requests for prof@university.edu to departmental servers. Explain how a location request for engineer@home.com might be redirected to a.prof@ece.university.edu.

38. What is the maximum width of an RIP network?

39. Let's consider the bandwidth consumption of the RIP protocol.
    a. Estimate the number of messages exchanged per unit time by RIP.
    b. Estimate the size of the messages exchanged as a function of the size of the RIP network.
    c. Estimate the bandwidth consumption of an RIP network.

40. RIP runs over UDP, OSPF runs over IP, and BGP runs over TCP. Compare the merits of operating a routing protocol over TCP, UDP, IP.

41. Compare RIP and OSPF with respect to convergence time and the number messages exchanged under several trigger conditions, that is, link failure, node failure, and link coming up.

42. Consider the OSPF protocol.
    a. Explain how OSPF operates in an autonomous system that has no defined areas.
    b. Explain how the notion of area reduces the amount of routing traffic exchanged.
    c. Is the notion of area related to subnetting? Explain. What happens if all addresses in an area have the same prefix.

43. Assume that there are N routers in the network and that every router has m neighbors.
    a. Estimate the amount of memory required to store the information used by the distance-vector routing.
    b. Estimate the amount of money required to store the information by the link-state algorithm.

44. Suppose a network uses distance-vector routing. What happens if the router sends a distance vector with all 0s?

45. Suppose a network uses link-state routing. Explain what happens if:
    a. The router fails to claim a link that is attached to it.
    b. The router claims to have a link that does not exist.

46. Consider a broadcast network that has n OSPF routers.
    a. Estimate the number of database exchanges required to synchronize routing databases.
    b. What is the number of database exchanges after a designated router is introduced into the network?
    c. Why is the backup designated router introduced? What is the resulting number of database exchanges?

47. Suppose n OSPF routers are connected to a nonbroadcast multiaccess network, for example, ATM.
    a. How many virtual circuits are necessary to provide the required full connectivity?
    b. Does OSPF function correctly if a virtual circuit fails?
    c. Is the number of required virtual circuits reduced if point-to-multipoint virtual circuits are available?

48. The figure below shows seven routers connected with links that have the indicated costs. Use the Hello protocol to show how the routers develop the same topology database for the network.

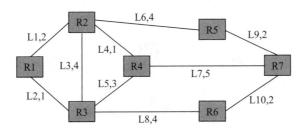

49. Consider the exchange of Hello messages in OSPF.
    a. Estimate the number of Hello messages exchanged per unit time.
    b. Estimate the size of the Hello messages.
    c. Estimate the bandwidth consumed by Hello messages.

50. Consider the notion of adjacency in OSPF.
    a. Explain why it is essential that all adjacent routers be synchronized.
    b. Explain why it is sufficient that all adjacent routers be synchronized; that is, it is not necessary that all pairs of routers be synchronized.

51. Consider the robustness of OSPF.
    a. Explain how the LSA checksum provides robustness in the OSPF protocol.
    b. An OSPF router increments the LS Age each time the router inserts the LSA into a link-state update packet. Explain how this step protects against an LSA that is caught in a loop.
    c. OSPF defines a minimum LS update interval of 5 seconds. Explain why.

52. Assume that for OSPF updates occur every 30 minutes, an update packet can carry three LSAs, and each LSA is 36 bytes long. Estimate the bandwidth used in advertising one LSA.

53. Identify elements where OSPF and BGP are similar and elements where they differ. Explain the reasons for similarity and difference.

54. Discuss the OSPF alternate routing capability for the following cases:
    a. Traffic engineering, that is, the control of traffic flows in the network.
    b. QoS routing, that is, the identification of paths that meet certain QoS requirements.
    c. Cost-sensitive routing, that is, the identification of paths that meet certain price constraints.
    d. Differential security routing, that is, the identification of paths that provide different levels of security.

55. Consider the following arrangement of autonomous systems and BGP routers.

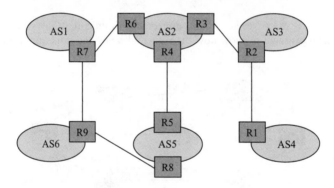

    a. Suppose that a certain network prefix belongs to AS4. Over which router pairs will the route to the given network be advertised?
    b. Now suppose the link between R1 and R2 fails. Explain how a loop among AS1, AS2, AS6, and AS5 is avoided.
    c. Suppose that R9 is configured to prefer AS1 as transit and AS6 is configured to prefer AS1 as transit. Explain how BGP handles this situation.

56. Why does BGP not exchange routing information periodically as RIP does?

57. Consider the network shown in Figure 8.52. Suppose that a source connected to router 7 wishes to send information to multicast group G3.
    a. Find the set of paths that are obtained from reverse-path broadcasting.
    b. Repeat for truncated reverse-path broadcasting.
    c. Repeat for reverse-path multicasting.

58. Discuss the operation of the reverse-path multicasting in the following two cases:
    a. The membership in the multicast group in the network is dense.
    b. The membership in the multicast group in the network is sparse.

CHAPTER 9

ATM Networks

In Chapter 7 we saw that asynchronous transfer mode (ATM) was developed to combine the attributes of time-division circuit-switched networks and packet-switched networks. We also saw how ATM provides the capability of providing Quality-of-Service support in a connection-oriented packet network. In this chapter we present the details of ATM network architecture.

The chapter is organized as follows:

1. *Why ATM?* We first provide a historical context and explain the motivation for the development of ATM networks.
2. *BISDN reference model.* We examine the BISDN reference model that forms the basis for ATM, and we explain the role of the user and control planes.
3. *ATM layer.* We examine the "network layer" of the ATM architecture, and we explain the operation of the ATM protocol. Quality of service (QoS) and the ATM network service categories and the associated traffic management mechanisms are introduced.
4. *ATM adaptation layer.* We introduce the various types of ATM adaptation layer protocols that have been developed to support applications over ATM connections.
5. *ATM signaling.* We provide an introduction to ATM addressing and to ATM signaling standards.
5. *PNNI routing.* We briefly describe a dynamic routing protocol for ATM networks, called PNNI.

In the next chapter we show how IP and ATM networks can be made to work together. We also show how IP networks are evolving to provide QoS support.

9.1   WHY ATM?

The concept of ATM networks emerged from standardization activities directed at the development of *Integrated Services Digital Networks (ISDNs)*. In the 1970s the trend toward an all-digital (circuit switched) telephone network was clearly established and the need to (eventually) extend digital connectivity to the end user was recognized. It was also apparent that data applications (e.g., computer communications and facsimile) and other nonvoice applications (e.g., videoconferencing) would need to be accommodated by future networks. It was also clear that circuit switching would not be suitable for bursty data traffic and that packet switching would have to be provided. The ISDN standards were the first effort at addressing these needs.

The recommendations adopted by the CCITT (now the telecommunications branch of the International Telecommunications Union) in 1984 defined an *ISDN* as a network that provides *end-to-end digital connectivity* to support a *wide range of services* to users through a limited set of *standard user-network interfaces*. The basic rate interface consisted of two constant bit rate 64 kbps B channels and a 16 kbps D channel. The primary rate interface provided for either 23 B channels and a 64 kbps channel, primarily in North America, or for 30 B channels and a 64 kbps channel elsewhere. The ISDN recommendations provided for the establishment of voice and data connections. The recommendations focused exclusively on the interface between the user and the network and did not address the internal organization of the network. Indeed once inside the network, voice and data traffic would typically be directed to separate circuit-switched and packet-switched networks. Thus the network operator was still burdened with the complex task of operating multiple dissimilar networks.

It soon became clear that higher rate interfaces would be required to handle applications such as the interconnection of high-speed local area network (LANs) as well as to transfer high-quality digital television. The initial discussions on *broadband ISDN* (BISDN) focused on defining additional interfaces along the lines of established rates in the telephone digital multiplexing hierarchy. However, eventually the discussions led to a radically different approach, known as ATM, that attempts to handle steady stream traffic, bursty data traffic, and everything in between. ATM involves converting all traffic that flows in the network into 53-byte blocks called *cells*. As shown in Figure 9.1, each cell has 48 bytes of *payload* and a 5-byte *header* that allows the network to forward each cell to its destination.

The connection-oriented cell-switching and multiplexing principles underlying ATM were already discussed in Chapter 7. Here we reiterate the anticipated advantages of ATM networks.

1. The network infrastructure and its management is simplified by using a single transfer mode for the network; indeed, extensive bandwidth management capabilities have been built into the ATM architecture.

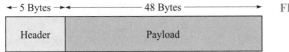

FIGURE 9.1 The ATM cell

2. Unlike shared media networks, ATM is not limited by speed or distance; the switched nature of ATM allows it to operate over LANs as well as global backbone networks at speeds ranging from a few Mbps to several Gbps.
3. The QoS attributes of ATM allow it to carry voice, data, and video, thus making ATM suitable for an integrated services network.

The ATM standardization process has taken place under the auspices of the ITU-T in concert with national and regional bodies such as ANSI in the United States and ETSI in Europe. The development of industry implementation agreements has been mainly driven by the ATM Forum.

9.2   BISDN REFERENCE MODEL

The BISDN reference model is shown in Figure 9.2. The model contains three planes: the user plane, the control plane, and the management plane.The *user plane* is concerned with the transfer of user data including flow control and error recovery. The *control plane* deals with the signaling required to set up, manage, and release connections. The *management plane* is split into a layer management plane that is concerned with the management of network resources and a plane

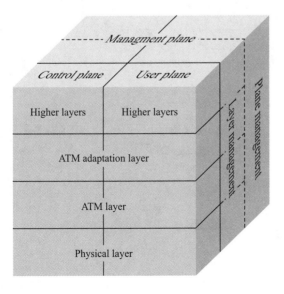

FIGURE 9.2 The broadband ISDN reference model

management plane that deals with the coordination of the other planes. We focus on the user and control planes.

The user plane has three basic layers that together provide support for user applications: the ATM adaptation layer, the ATM layer, and the physical layer. The **ATM adaptation layer (AAL)** is responsible for providing different applications with the appropriate support, much as the transport layer does in the OSI reference model. Several AAL types have been defined for different classes of user traffic. The AAL is also responsible for the conversion of the higher-layer service data units (SDUs) into 48-byte blocks that can be carried inside ATM cells. Figure 9.3 shows how the information generated by voice, data, and video applications are taken by AALs and converted into sequences of cells that can be transported by the ATM network. The AAL entity on the receiver side is responsible for reassembling and delivering the information in a manner that is consistent with the requirements of the given application. Note that the AAL entities reside in the terminal equipment, and hence they communicate on an end-to-end basis across the ATM network as shown in Figure 9.4.

The **ATM layer** is concerned solely with the sequenced transfer of ATM cells in connections set up across the network. The ATM layer accepts 48-byte blocks of information from the AAL and adds a 5-byte header to form the ATM cell. The header contains a label that identifies the connection and that is used by a switch to determine the next hop in the path as well as the type of priority/ scheduling that the cell is to receive.

ATM can provide different QoS to different connections. This requires that a *service contract* be negotiated between the user and the network when the connection is set up. The user is required to describe its traffic and the required QoS when it requests a connection. If the network accepts the request, a contract is

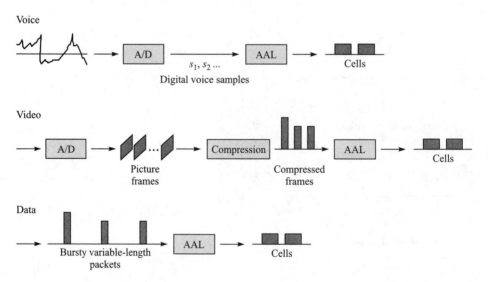

FIGURE 9.3 The AAL converts user information into cells

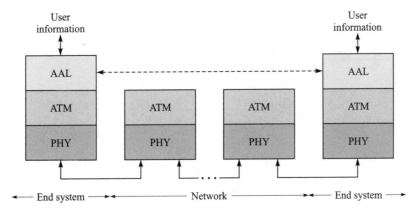

FIGURE 9.4 User plane layers

established that guarantees the QoS as long as the user complies with its traffic description. Queue priority and scheduling mechanisms implemented in ATM switches provide the capability of delivering QoS. To deliver on its QoS commitments, the ATM network uses policing mechanisms to monitor user compliance with the connection contract and may discard cells not found in compliance.

In terms of number of users involved, ATM supports two types of connections: point to point and point to multipoint. Point-to-point connections can be unidirectional or bidirectional. In the latter case different QoS requirements can be negotiated for each direction. Point-to-multipoint connections are always unidirectional.

In terms of duration, ATM provides permanent virtual connections (PVCs) and switched virtual connections (SVCs). PVCs act as "permanent" leased lines between user sites. PVCs are typically provisioned "manually" by an operator. SVCs are set up and released on demand by the end user.

SVCs are set up through signaling procedures. Initially the source user must interact with the network through a **user-network interface (UNI)** (see Figure 9.5). The connection request must propagate across the network and eventually involve an interaction at the destination UNI. Within a network, switches must interact across the **network-network interface (NNI)** to exchange information. Switches that belong to different public networks communicate across a **broadband intercarrier interface (B-ICI)**. The source and destination end systems as well as all switches along the path across the network are eventually involved in the allocation of resources to meet the QoS requirements of a connection.

Signaling can be viewed as an application in which end systems and switches exchange higher-level messages that establish connections across the network. The role of the **control plane** is to support signaling and network control applications. The control plane has the same three basic layers as the user plane. A *signaling AAL* has been defined for the control plane to provide for the reliable exchange of messages between ATM systems. Higher-layer protocols have been defined for use over the UNI, for the NNI, and for the B-ICI.

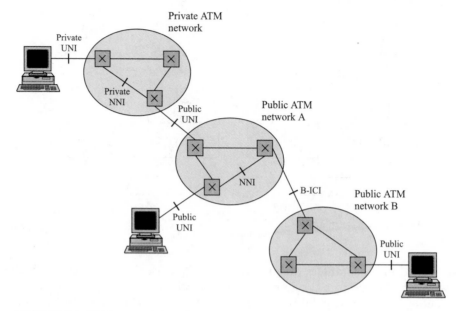

FIGURE 9.5 ATM network interfaces

The *physical layer* is divided into two sublayers as shown in Figure 9.6. The *physical medium dependent sublayer* is the lower of the two layers and is concerned with details of the transmission of bits over the specific medium, such as line coding, timing recovery, pulse shape, as well as connectors. The *transmission convergence sublayer* establishes and maintains the boundaries of the ATM cells in the bit stream; generates and verifies header checksums; inserts and removes "idle" ATM cells when cells are not available for transmission; and, of course, converts ATM cells into a format appropriate for transmission in the given physical medium.

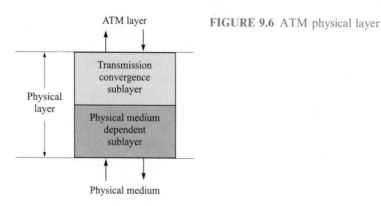

FIGURE 9.6 ATM physical layer

A large number of physical layers have been defined to provide for support for ATM in a wide range of network scenarios, for example, LAN/WAN and private/public scenarios. The approach in the ATM Forum has been to use and adapt existing physical layer standards as much as possible. In addition to SONET/SDH, physical layers have been defined for 1.5 Mbps (DS-1), 2.0 Mbps (E-1), 45 Mbps (DS-3), 34.4 Mbps (E3), 139 Mbps (E-4), 100 Mbps (FDDI), and 155 Mbps (Fiber Channel). Other physical layers will be defined as needed.

9.3 ATM LAYER

The ATM layer is concerned with the sequenced transfer of cells of information across connections established through the network. In this section we examine the operation of the ATM layer in detail. We begin with a description of the ATM cell header. We then discuss how fields in the ATM header are used to identify network connections. This section is followed by a description of the types of ATM network service categories and the traffic management mechanisms required to provide these services. A later section deals with ATM addressing and ATM signaling.

9.3.1 ATM Cell Header

Different ATM cell headers have been defined for use in the UNI and in the NNI. The UNI is the interface point between ATM end users and a private or public ATM switch, or between a private ATM switch and a public carrier ATM network, as shown in Figure 9.5. The NNI is the interface between two nodes (switches) in the same ATM network.

Figure 9.7 shows the 5-byte cell header for the UNI. We first briefly describe the functions of the various fields. We then elaborate on their role in ATM networks.

Generic flow control: The GFC field is 4 bits long and was intended to provide flow control and shared medium access to several terminals at the UNI. It is currently undefined and is set to zero. The GFC field has significance only at the UNI and is not carried end to end across the network. The UNI and NNI cell headers differ in that the GFC field does not appear in the NNI cell header; instead the VPI field is augmented to 12 bits.

Virtual path identifier: The VPI field is 8 bits long, so it allows the definition of up to $2^8 = 256$ virtual paths in a given UNI link. Recall from Figure 7.31 that each virtual path consists of a bundle of virtual channels that are switched as a unit over the sequence of network nodes that correspond to the path.

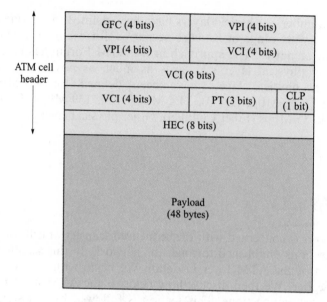

ATM cell header

GFC (4 bits)	VPI (4 bits)	
VPI (4 bits)	VCI (4 bits)	
VCI (8 bits)		
VCI (4 bits)	PT (3 bits)	CLP (1 bit)
HEC (8 bits)		

Payload
(48 bytes)

FIGURE 9.7 ATM cell header format

Virtual channel identifier: The VCI field is 16 bits long, so it allows the definition of up to $2^{16} = 65,536$ virtual channels per virtual path. The VIP/VCI field is the *local* identifier for a given connection in a given link, and the value of the field changes at every switch.

Payload type: The 3-bit payload type field allows eight types of ATM payloads as shown in Table 9.1. The most significant bit is used to distinguish between data cells ($b_3 = 0$) and operations, administration, and maintenance (OAM) cells ($b_3 = 1$).

For data cells ($b_3 = 0$), the second bit serves as the explicit forward congestion indication (EFCI), which is set by switches to indicate congestion and is used by the congestion control mechanism for the available bit rate (ABR) service defined below.

Payload type identifier	Meaning
000	User data cell, congestion not experienced, SDU type = 0 (that is, beginning or continuation of SAR-SDU in AAL5)
001	User data cell, congestion not experienced, SDU type = 1 (that is, end of SAR-SDU in AAL5)
010	User data cell, congestion experienced, SDU type = 0
011	User data cell, congestion experienced, SDU type = 1
100	OAM F5 segment associated cell
101	OAM F5 end-to-end associated cell
110	Resource management (RM) cell (used in traffic management)
111	Reserved for future use

TABLE 9.1. ATM payload types

For data cells ($b_3 = 0$), the least significant bit (b_1) is carried transparently across the network. We show below that $b_1 = 1$ is used by AAL type 5 (AAL5) to signal that a cell carries the end of a SDU.

The payload field (110) defines resource management cells that are used in traffic management.

Cell loss priority: The CLP bit establishes two levels of priorities for ATM cells. A cell that has CLP = 0 is to be treated with higher priority than a cell with CLP = 1 during periods of congestion. In particular, CLP = 1 cells should be discarded before CLP = 0 cells. The CLP bit can be set by terminals to indicate less important traffic or may be set by the network to indicate lower-priority QoS flows or cells that have violated their traffic contract.

Header error control: An 8-bit CRC checksum, using the generator polynomial described in Table 3.8, is calculated over the first four bytes of the header. This code can correct all single errors and detect all double errors in the header. The checksum provides protection against misdelivery of cells from errors that may occur in transit. Two modes are defined. In *detection* mode cells with inconsistent checksums are discarded. In *correction* mode single bit errors are corrected. Correction mode is suitable only in media where single errors predominate over multibit errors. The HEC needs to be recomputed at every switch, since the VPI/VCI value changes at every hop.[1]

9.3.2   Virtual Connections

In Chapter 7 we describe how ATM uses virtual path and virtual channel identifiers in the cell headers to identify a connection across a network. These locally defined identifiers are used to forward cells that arrive at a switch to the appropriate output port. At each switch the VPI/VCI identifier are used to access tables that specify the output port and the VPI/VCI identifier that is to be used in the next hop. In this manner the chain of identifiers define a connection across the network.

The VPI/VCI format allows ATM to switch traffic at two levels. In VP switching, entire bundles of VCs arriving at a given input port and identified by a given VPI are transferred to the same output port. The switch does not look at the VCI value. Prior to transfer along the next hop, the VPI value is mapped into the value that is defined for the next hop. In VP switching, however, the VCI value is not changed. The ability to handle bundles of VCs at a time is very useful to the network operator in facilitating the management of network resources and in simplifying routing topologies.

---

[1]The HEC may also be used for cell delineation.

As indicated above, ATM networks provide two basic types of connections. *Permanent virtual connection* (PVCs) are long-term connections that are typically used by network operators to provision bandwidth between endpoints in an ATM network. *Switched virtual connections* (SVCs) are shorter-term connections that are established in response to customer requests. In SVCs the table entries are established during the call setup procedure that precedes the transfer of ATM cells in a connection.

9.3.3   QoS Parameters

A central objective of ATM is to provide QoS guarantees in the transfer of cell streams across the network. In ATM the QoS provided by the network is specified in terms of the values of several end-to-end, cell-level parameters. A total of six QoS performance parameters have been specified.

The following three QoS network performance parameters are defined in ATM standards. These parameters are not negotiated at the time of connection setup and are indicators of the intrinsic performance of a given network.

Cell error ratio: The *CER* of a connection is the ratio of the number of cells that are delivered with one or more bit errors during a transmission to the total number of transmitted cells. The CER is dependent on the underlying physical medium. The CER calculation excludes blocks of cells that are severely errored (defined below).

Cell misinsertion rate: The *CMR* is the average number of cells/second that are delivered mistakenly to a given connection destination (that is, that originated from the wrong source). The CMR depends primarily on the rate at which undetected header errors result in misdelivered cells. The CMR calculation excludes blocks of cells that are severely errored (defined below). Note that CMR is a rate in cells/second rather than a ratio, since the mechanism that produces misinsertion is independent of the number of cells produced in a connection.

Severely-errored cell block ratio: A severely-errored cell block event occurs when more than M cells are lost, in error, or misdelivered in a given received block of N cells, where M and N are defined by the network provider. The severely-errored cell block ratio (SECBR) is the ratio of severely-errored cell blocks to total number of transmitted cell blocks in a connection. The SECBR is determined by the properties of the error mechanisms of the transmission medium, by buffer overflows, and by operational effects of the underlying transmission system such as losses of information that occur when active transmission links are replaced by backup transmission links in response to faults.

The following three QoS parameters may be negotiated between the user and the network during connection setup.

Cell loss ratio: The *CLR* for a connection is the ratio of the number of lost cells to total number of transmitted cells. The CLR value is negotiated between the user and the network during call setup and specifies the CLR objective for the given connection. It is specified as an order of magnitude in the range of 10^{-1} to 10^{-15}. It can also be left unspecified. The CLR objective can apply to either the CLP = 0 (conforming) cell flow or to the CLP = 0 + 1 (all cells) in the cell flow. The degree to which CLR can be negotiated depends on the sophistication of the buffer-allocation strategies that are available in a given network.

Cell transfer delay: The *CTD* is the time that elapses from the instant when a cell enters the network at the source UNI to the instant when it exists at the destination UNI. The CTD includes propagation delay, processing delays, and queueing delays at multiplexers and switches. In general, different cells in a connection experience different values of delays, so the CTD is specified by a probability density function as shown in Figure 9.8.[2] The standards provide for the negotiation of the "maximum" CTD. As shown in Figure 9.8, the *maximum CTD* is defined as the value D_{max} for which the fraction $1 - \alpha$ of all cells have CTD less than D_{max}, where α is some appropriately small value. For example, the requested value of CLR places an upper bound on the value of α. Queue-scheduling algorithms in the ATM switches can be used to control the CTD experienced by cells in a given connection.

Cell delay variation: The *CDV* measures the variablity of the total delay encountered by cells in a connection. The CDV excludes the fixed component D_0 of the CTD that is experienced by all cells in a connection, for example, the propagation delay and fixed processing delays (see Figure 9.8). Current standards provide for the negotiation of the peak-to-peak

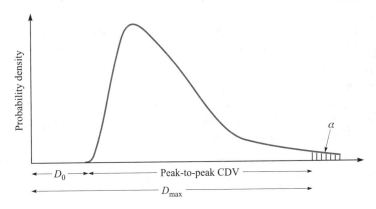

FIGURE 9.8 Probability density function of cell transfer delay

---

[2]The probability that the delay value falls in an interval is the area under the probability density function for that interval. See [Leon-Garcia 1994].

CDV, which is simply the difference between the maximum CTD D_{max} and the fixed delay component D_0. Note that network switches have only limited control over the spread (variance) of CTD values, and consequently the range of CDV values that can be negotiated for a connection is also limited.

9.3.4 Traffic Descriptors

The capability of a network to provide given levels of QoS to a connection depends on the manner in which the connection produces cells for transmission, that is, at a constant smooth rate or in a highly bursty fashion. This capability also depends on the amount of network resources, that is, bandwidth and buffers, that the network allocates to the connection. Therefore, the connection contract between the user and the network must specify the manner in which the source will produce cells. For this purpose standards have specified a number of *source traffic descriptor* parameters. To be enforceable, policing algorithms must be available to monitor the traffic produced by a source to determine whether it conforms to the connection contract.

The following source traffic parameters have been defined to specify this pattern of demand for transmission.

Peak cell rate: The *PCR* specifies the rate in cells/second that a source is never allowed to exceed. The minimum allowable interval between cells is given by T = 1/PCR.

Sustainable cell rate: The *SCR* is the average cell rate, in cells/second, produced by the source over a long time interval.

Maximum burst size: The *MBS*, in a number of cells, specifies the maximum number of consecutive cells that may be transmitted by a source at the peak cell rate (PCR).

Minimum cell rate: The *MCR* is the minimum average cell rate, in cells/second, that the source is always allowed to send.

Even if a source produces cells at exactly the PCR rate, subsequent ATM cell multiplexing and physical layer processing (e.g., insertion of a cell into a bit stream) can produce certain variability about the PCR rate. The policing mechanism must take into account this unavoidable variability. The following interface performance parameter has been defined for this purpose.

Cell delay variation tolerance: The *CDVT* specifies the level of cell delay variation that must be tolerated in a given connection.

9.3.5 ATM Service Categories

ATM connections with arbitrary traffic flow properties and arbitrary QoS are possible by selecting values for the traffic descriptor and the negotiable QoS parameters. In practice there are several clearly identifiable classes of traffic in terms of traffic properties and network QoS requirements. The ATM Forum has defined five *ATM service categories* as shown in Table 9.2. The first two categories apply to connections that are real time in the sense of having stringent delay and timing requirements.

Constant bit rate: The *CBR* ATM service category is intended for traffic with rigorous timing requirements, such as voice, circuit emulation, and certain types of video, that require a constant cell transmission rate for the entire duration of a connection. The traffic rate is specified by the PCR. The QoS is specified by the CTD and CDV, as well as by the CLR.

Real-time variable bit rate: The *rt-VBR* ATM service category is intended for variable-bit-traffic, such as certain types of video, with rigorous timing requirements. The traffic is described by the PCR, SCR, and MBS. The QoS is specified by the CLR, CTD, and CDV.

Three categories of non-real-time connections have been defined.

Non-real-time variable-bit-rate: The *nrt-VBR* ATM service category addresses bursty sources, such as data transfer, that do not have rigorous timing requirements. The traffic is described by the PCR, SCR, and MBS. The QoS is specified by the CLR, and no delay requirements are specified.

Available bit rate: The *ABR* ATM service category is intended for sources that can dynamically adapt the rate at which they transmit cells in response to feedback from the network. This service allows the sources

Attribute	ATM layer service category				
	CBR	rt-VBR	nrt-VBR	UBR	ABR
Traffic parameters					
PCR and CDVT[4,5]	Specified	Specified	Specified	Specified[2]	Specified[3]
SCR, MBS, CDVT[4,5]	n/a	Specified	Specified	n/a	n/a
MCR[4]	n/a	n/a	n/a	n/a	Specified
QoS parameters					
peak-to-peak CDVT	Specified	Specified	Unspecified	Unspecified	Unspecified
maxCTD	Specified	Specified	Unspecified	Unspecified	Unspecified
CLR[4]	Specified	Specified	Specified	Unspecified	See note 1
Other attributes					
Feedback	Unspecified	Unspecified	Unspecified	Unspecified	Specified[6]

Notes: [1]CLR is low for sources that adjust cell flow in response to control information. Whether a quantitative value for CLR is specified is network specific.

[2]May not be subject to connection admission control and usage parameter control procedures.

[3]Represents the maximum rate at which the ABR source may ever send data. The actual rate is subject to the control information.

[4]These parameters are either explicitly or implicitly specified for PVCs or SVCs.

[5]CDVT refers to the cell delay variation tolerance. CDTV is not signaled. In general, CDVT need not have a unique value for a connection. Different values may apply at each interface along the path of a connection.

[6]See discussion on ABR in Chapter 7.

TABLE 9.2. ATM service category attributes [ATM April 1996]

to exploit the bandwidth that is *available* in the network at a given point in time. The traffic is specified by a PCR and MCR, which can be zero. The sources adjust the rate at which they transmit into the network by implementing a congestion control algorithm that dictates their response to resource management cells that explicitly provide rate flow information. Connections that adapt their traffic in response to the network feedback can expect a low CLR as well as a "fair" share of the available bandwidth.

Unspecified bit rate: The *UBR* ATM service category does not provide *any* QoS guarantees. The PCR may or may not be specified. This service is appropriate for noncritical applications that can tolerate or readily adjust to the loss of cells.

In ATM the QoS guarantees are provided on a per connection basis; that is, every connection can expect that its QoS requirements will be met. Since ATM involves the handling of flows of cells from many connections that necessarily interact at multiplexing points, it is worth considering the nature of the QoS guarantees and the corresponding resource allocation strategies for the different ATM service categories.

For CBR connections the source is free to transmit at the negotiated PCR at any time for any duration. It follows then that the network must allocate sufficient bandwidth to allow the source to transmit continuously at the PCR. The scheduling/priority discipline in the multiplexer must ensure that such bandwidth is regularly available to the connection so that the CDV requirement is also met. The steady flow assumed for CBR sources implies that only limited interaction occurs between different CBR flows. In essence, the individual CBR flows act as if they were in separate, isolated transmission links.

For rt-VBR connections the source transmission rate is expected to vary dynamically around the SCR and below the PCR. Therefore, it is to the benefit of the network operator to statistically multiplex these flows to improve the actual utilization of the bandwidth. However, the mixing of rt-VBR flows must be done in a way that maintains some degree of isolation between flows. In particular it is essential to meet the delay and CLR requirements of the connections.

The situation for nrt-VBR connections is similar to that of rt-VBR connections. Again it is in the interest of the network operator to statistically multiplex nrt-VBR sources. In this case the degree of multiplexing is limited only by the commitment to provide conforming flows with the negotiated CLR.

UBR connections, with their lack of any QoS guarantees, provide an interesting contrast to the preceding service categories. When the traffic levels in the network are low, UBR connections may experience performance that is as good as that of the service categories with QoS guarantees. Only as network traffic levels increase, do the QoS guarantees become noticeable. From the point of view of the network operator, a low tariff for UBR service can be used to stimulate demand for bandwidth when the utilization of network sources is low. This approach is useful to a broad range of users when traffic levels are low; however, it is increasingly less useful as traffic levels increase and the network performance experienced becomes less consistent. The ABR service category fills a niche in this context. UBR connections receive no guarantees as network traffic levels vary. ABR connections, on the other hand, are assured of a low level of CLR as long as they conform by responding to the network feedback information. The option to negotiate a nonzero MCR also provides ABR connections with additional assurance of service.

9.3.6   Traffic Contracts, Connection Admission Control, and Traffic Management

Traffic management refers to the set of control functions that together ensure that connections receive the appropriate level of service. To provide QoS guarantees to each connection, the network must allocate an appropriate set of resources to each new connection. In particular the network must ensure that new VCs are assigned to links that have sufficient available bandwidth and to ports that have sufficient buffers to handle the new as well existing connections at the committed levels of QoS. The queue scheduling algorithms presented in section 7.6 can be used to ensure that the cells in specific connections receive appropriate levels of performance in terms of delay and loss.

Connection admission control (CAC) is a network function that determines whether a request for a new connection should be accepted or rejected. If accepted, the user and network are said to enter into a traffic contract. The contract for a connection includes the ATM service category, the traffic descriptors, and the QoS requirements that are introduced in section 9.3.5 and shown in Table 9.2. The CAC procedure takes the proposed traffic descriptors and QoS requirements and determines whether sufficient resources are available along a route from the source to the destination to support the new connection as well as already established connections. Specific CAC algorithms are not specified by the standards bodies and are selected by each network operator.

The QoS guarantees are valid only if the user traffic conforms to the connection contract. Usage parameter control (UPC) is the process of enforcing the traffic agreement at the UNI. Each connection contract includes a cell conformance definition that specifies how a cell can be policed, that is, determined to be either conforming or nonconforming to the connection contract. The generic cell rate algorithm (GCRA) is equivalent to the leaky-bucket algorithm described in section 7.7 and can be used to determine whether a cell conforms to the specified PCR and CDVT. Cells that are found to be not conforming are "tagged" by having their cell loss priority (CLP) bit in the header set to 1 at the access to the network. When congestion occurs in the network, cells with CLP = 1 are discarded first. Thus nonconforming cells are more likely to be discarded. The GCRA algorithm can be modified so that cell conformance to the PCR, CDVT, as well as the SCR and MBS can be checked.

A connection is said to be compliant with its contract if the proportion of nonconforming cells does not exceed a threshold specified by the network operator. As long as a connection remains compliant, the network will provide the

QoS specified in the contract. However, if a connection becomes noncompliant, the network may then cease providing the contracted QoS.

Traffic shaping is a mechanism that allows sources to ensure that their traffic conforms to the connection contract. In traffic shaping, a leaky bucket is used to identify nonconforming cells that are then buffered and delayed so that all cells entering the network are conforming. The token bucket mechanism discussed in Chapter 7 is a method for doing traffic shaping.

Congestion can still occur inside the network even if all cells that enter the network conform to their connection contract. Such congestion will occur when many cells from different connections temporarily coincide. The purpose of *congestion control* is to detect the onset of congestion and to activate mechanisms to minimize the impact and duration of congestion. ATM networks employ two types of congestion control. *Selective cell discarding* involves discarding CLP = 1 cells during periods of congestion. The onset of congestion may be defined by a threshold on the contents of buffers in switches and multiplexers. Discarding low-priority cells helps meet the QoS commitments that have been made to the high-priority cells.

A second class of congestion control involves sending *explicit congestion feedback* from the network to the sources. This type of control is appropriate for ABR connections, which by definition involve sources that can adapt their input rate to network feedback. The rate-based flow control algorithm described in Chapter 7 has been recommended for this type of congestion control.

9.4   ATM ADAPTATION LAYER

An application that operates across an ATM network has a choice of the five ATM connection service categories shown in Table 9.2. Every application involves the transfer of one or more blocks or of a stream of information across the network. The ATM service categories provide for the sequenced transfer of cells across the network with a certain delay or loss performance. At the very least a conversion is required from the application data blocks to ATM cells at the source and a conversion back to the application blocks at the destination. One purpose of the ATM adaptation layer is to provide for the mapping between application data blocks to cells.

Applications naturally specify their QoS requirements in terms of their data blocks, not in terms of ATM cells. It is also possible that the service provided by the ATM layer does not meet the requirements of the application. For example, the ATM layer does not provide reliable stream service by itself, since some cell losses can occur. Another purpose of the ATM adaptation layer, then, is to enhance the service provided by the ATM layer to the level required by the application. It should be noted that multiple higher layers may operate over the AAL, for example, HTTP over TCP over AAL. To be more precise, we emphasize that the function of the AAL is to provide support for the layer directly above it. Thus if the layer above the AAL is TCP, then the AAL need not be concerned with providing reliable stream service. On the other hand, the AAL may be called on to provide reliable stream service when such service is not available in the higher layers, as, for example, in signaling applications.

Different applications require a different combination of functions in an AAL. For example, circuit emulation applications require that information be transferred as if the underlying ATM connection were a dedicated digital transmission line. Real-time voice and certain video applications present similar requirements. On the other hand, frame relay applications require the non-real-time, connection-oriented transfer of a sequence of frames between two end systems. In yet another example, IP routers require the connectionless transfer of a packet to another router, using the ATM network as a "data link." In some cases the IP packets carry payloads that are controlled by TCP entities at the end systems. Each of these examples impose different requirements on the AAL.

There have been several attempts to categorize applications into a small set of classes and to design AALs to meet the requirements of each class. These efforts have not met with success, and no clear correspondence exists between application classes, AALs, and ATM connection service categories. Our approach here is to discuss the formats and services of the AALs that have been developed to date. We then discuss what combinations of applications, AALs, and ATM service categories make sense. It should be noted that users are also free to use proprietary (nonstandard) AALs.

The AAL is divided into two sublayers as shown in Figure 9.9. The purpose of the **segmentation and reassembly (SAR)** sublayer is to segment the PDUs of the higher layer into blocks that are suitable for insertion into the ATM cell payloads at the source and to reassemble the higher-layer PDUs from the sequence of received ATM cell payloads at the destination. The **convergence sublayer (CS)** is divided into a **common part (CPCS)** and a **service-specific part (SSCS)**. The CPCS deals with packet framing and error-detection functions that all AAL users require. The SSCS provides functions that depend on the requirements of specific classes of AAL users. Consequently, each AAL usually has a specific SAR and CPCS sublayer and several optional SSCS sublayers.

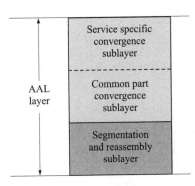

FIGURE 9.9 AAL sublayers

9.4.1   AAL1

The ATM adaptation layer type 1 (AAL1) supports services that require the transfer of information at a constant rate. Examples of this type of service are a single 64 kbps PCM voice call, sub-T-1 connections that consist of n × 64 kbps streams, T-1/E-1 and other digital circuits from the telephone hierarchy, and constant bit-rate digital video. The AAL PDU structure contains fields that enable clock recovery and sequence numbering. It also contains an option for the transfer of the internal (frame) structure within a continuous bit stream.

The generic AAL1 process is shown in Figure 9.10. The convergence sub-layer function takes the user data stream, optionally inserts a 1-byte pointer to provide structure information, and produces 47-byte CS PDUs, which it then passes with three-bit sequence numbering to the segmentation and reassembly sublayer. Thus the CS PDU normally contains either 47 bytes or 46 bytes of user information, depending on whether a pointer is inserted. Note that the CS PDU need not be completely filled with 47 bytes of user information. In low-bit-rate applications with low-delay requirements, for example, a single 64 kbps voice call, the CS PDU may be required to pass 47-byte blocks that are only partially filled with user information.

Figure 9.11 shows the AAL1 PDUs. The SAR sublayer attaches a 1-byte header to each CS PDU as shown in Figure 9.11a. The first four bits constitute the *sequence number (SN)* field. The first bit in the SN field is the *convergence sublayer indicator (CSI)* and is followed by the 3-bit sequence number that can be used for the detection and monitoring of loss and misinsertion of SAR payloads, and hence cells. In even-numbered cells the CSI bit may be used to indicate the existence of a CS sublayer pointer to the destination, as shown in Figure 9.11b. In odd-numbered cells the CSI bit may also optionally be used to convey timing information from the source to the destination. The last four bits of the header contain *sequence number protection (SNP)* check bits that provide error-

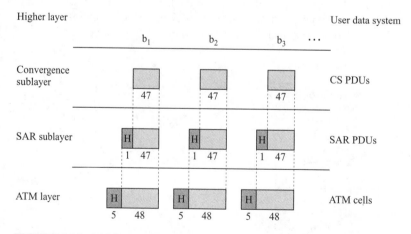

FIGURE 9.10 AAL1 process

(a) SAR PDU header

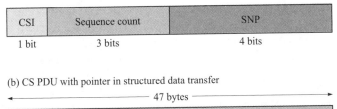

FIGURE 9.11 AAL1 PDUs

detection and correction capability for the SAR header. The four check bits are computed by using a Hamming (7,4) code with an additional overall parity check bit. This makes it possible to correct all single-bit and to detect all double-bit error patterns.

As an example of the use of AAL1 for unstructured information transfer, consider the transfer of a T-1 connection. Recall that the T-1 stream consists of a frame consisting of one framing bit followed by 24 bytes at a repetition rate of 8 kHz. In unstructured AAL1 transfer the bits from the T-1 stream are grouped into blocks of 47 bytes and passed to the SAR. Note that the bytes in the CS PDU will not be aligned to the bytes in the T-1 stream, since each frame has an odd number of bits. Now suppose that the T-1 connection uses *structured data transfer*. The T-1 stream is now viewed as a sequence of 24-byte frames (the framing bit is ignored). The bytes in the T-1 stream are mapped directly into the bytes in the CS PDUs. By prior agreement between the source and destination AAL entities, every so many cells, eight, for example, the first byte of the CS PDU contains a *pointer* that can be used to determine the beginning of the next frame. Because pointers can be inserted only in even-numbered cells, the beginning of the frame can be anywhere in the payloads of the current or the next cell. The periodic insertion of pointers provides protection against loss of synchronization that could result from cell losses.

The convergence sublayer at the destination can provide a number of services that are useful to applications requiring a constant transfer rate. The odd-numbered CSI bits can be used to convey a residual timestamp that provides the relative timing between the local clock and a common reference clock. The destination uses these residual timestamps to reconstruct the source clock and replay the received data stream at the correct frequency. This synchronous residual timestamp method is described in section 5.5. It should be noted that other methods for clock recovery, such as adaptive buffer, do not require the use of CSI bits. These methods are also discussed in Chapter 5.

To deliver information to the user at a fixed rate, the convergence sublayer at the destination can carry out timing recovery and then use a playout technique to absorb the cell delay variation in the received sequence. The convergence sublayer can also use the sequence numbers to detect lost or misinserted cells. The capability to request retransmissions is not provided, so the CS can provide only an indication to the user that a loss or misinsertion has occurred.

The convergence sublayer can also implement either of two forward-error correction (FEC) techniques to correct for cell errors and losses. For applications that have a low-delay requirement, the first FEC technique operates on groups of 15 cells and adds sufficient check bits to form 16 cells. The technique can correct one lost cell per group of 16 and can also correct certain other error patterns. The additional FEC delay incurred is then 15 cells. The second FEC technique uses a variation of the interleaving techniques discussed in Chapter 3 to arrange the CS-PDUs of 124 cells as columns in an array. Four additional columns of check bits are added to provide the capability to correct up to four cell losses as well as certain other error patterns. The technique involves incurring a delay of 124 cells at the source and at the destination, so it is generally appropriate only for non-CTD-sensitive traffic, for example, video streaming.

9.4.2   AAL2

AAL type 2 (AAL2) was originally intended to provide support for applications that generate information at a bit rate that varies dynamically with time and that also has end-to-end timing requirements. The prime example of such an application is video that when compressed produces a bit stream that varies widely depending on the degree of detail and the degree of motion in a scene. The development of an AAL for this type of traffic was never completed and the ITU subsequently began work on the development of an AAL, also designated type 2, for a different class of applications. We will discuss this latter AAL in this section.

The new AAL2 is intended for the bandwidth-efficient transfer of low-bit-rate, short-packet traffic that has a low-delay requirement. In effect the AAL2 adds a third level of multiplexing to the VP/VC hierarchy of ATM so that two or more low-bit-rate users can share the same ATM connection. An example where this functionality is required arises in the transfer of compressed voice information from a base station in a digital cellular system to its mobile telephone switching office as shown in Figure 9.12. The low-bit-rate digital streams for individual voice calls need to be transferred in a timely fashion to the switching office where the actual telephone switching takes place. The low delay and low bit rate imply that cell payloads would be only partially filled if each call had its own VC. AAL2 multiplexes the streams from multiple calls to provide both low delay and high utilization of cell payloads.

Figure 9.13 shows the operation of the AAL2. The AAL2 layer is divided into the common part sublayer (CPCS) and the service-specific convergence sublayer (SSCS). In this discussion we focus on the CPCS that transfers CPCS SDUs from a CPCS user at the source to a CPCS user at the destination. The CPCS provides nonassured operation; that is, SDUs may be delivered incorrectly or not at all, as a result of cell losses. The CPCS can multiplex SDUs from

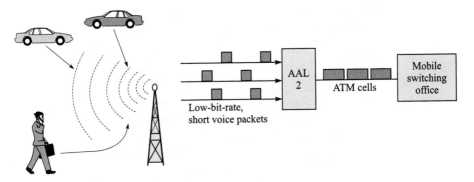

FIGURE 9.12 Application scenario for AAL2

multiple SSCS users. These users can be of different type and involve different SSCS functions.

User layer packets are transferred to the AAL2 layer as shown in Figure 9.13. These packets can vary in size, since users can be of different type. The maximum allowable packet size is 64 bytes. Assuming that the SSCS is not present, a three-byte header is added to each packet to form a CPCS packet. As shown in Figure 9.14a, the first byte is the channel identifier (CID) that identifies each user. These AAL channels are bidirectional, and the same CID is used for both directions. The six higher-order bits of the next byte are a length indicator that specifies one less than the number of bytes in the CPCS-packet payload. The remaining two bits of the second byte in the header specify the

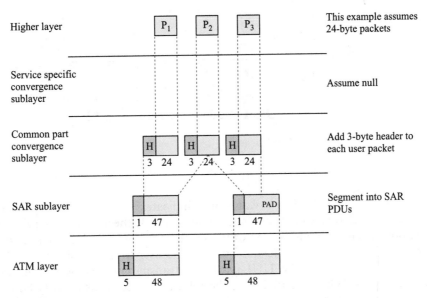

FIGURE 9.13 AAL2 process

(a) CPS packet structure

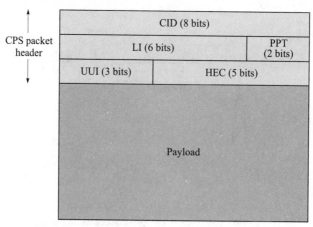

CPS packet header

(b) ATM SDU

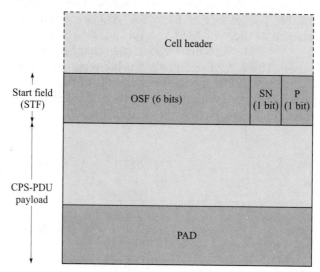

Start field (STF)

CPS-PDU payload

FIGURE 9.14 CPS packets

packet payload type (PPT). A value of 3 indicates that the CPCS packet is serving an OAM function. When the value is not 3, the packet is serving the application-specific functions, for example, the transfer of voice. The higher-order three bits of the third header byte are the user-to-user indication (UUI). When the PPT is not 3 the UUI is carried transparently between the SSCS protocol entities. When the PPT is 3, the UUI is carried transparently between the AAL layer management entities. The final five bits of the header are check bits that are used to detect errors. The encoding uses the generator polynomial $g(x) = x^5 + x^2 + 1$.

As shown in Figure 9.13, the CPCS packets are concatenated one after another prior to segmentation into 48-byte ATM SDUs. Each ATM SDU consists of a 1-byte start field and 47 bytes of CPCS packet bytes or padding (see Figure 9.14b). The start field provides the offset from the end of the STF to the start of the first CPCS packet in the SDU or, in the absence of such, the start of the PAD field. The maximum CPCS packet size is 64 bytes, so a CPCS packet may overlap one or two ATM cell boundaries. To meet the low-delay requirements, it is possible for the payload in the last cell to be only partially filled as shown, for example, in the second cell in Figure 9.13.

9.4.3   AAL3/4

In the early ATM standardization efforts, AAL type 3 (AAL3) was intended for applications that generate bursts of data that need to be transferred in connection-oriented fashion with low loss, but with no delay requirement. AAL type 4 (AAL4) was similarly intended for connectionless transfer of such data. By convention *all* connectionless packets at the UNI use the *same* VPI/VCI number, so a multiplexing ID was introduced in AAL4 to distinguish different packets. The efforts to develop AAL3 and AAL4 were later combined to produce AAL3/4, which can be used for either connection-oriented or connectionless transfer. The distinguishing feature of AAL3/4 is that it allows long messages from multiple users to be simultaneously multiplexed and interleaved in the same ATM VC.[3]

AAL3/4 can operate in two modes: message mode and stream mode. In message mode the AAL accepts a single user message for segmentation into ATM payloads, and the destination delivers the message. In stream mode one or more user PDUs, each as small as a single byte, are accepted at a time by the AAL and are subsequently delivered to the destination without an indication of the boundaries between the original PDUs. Both modes allow for assured or nonassured operation. In assured operation an end-to-end protocol is implemented in the SSCS to allow for the error-free delivery of messages. In nonassured operation, messages may be delivered in error or not at all.

The AAL3/4 process in message mode is shown in Figure 9.15. The user information is first passed to the SSCS and then to the CPCS, which adds fill bytes to make the CPCS payload a multiple of four bytes (32 bits). The CPCS PDU is formed by adding four bytes of header and four bytes of trailer. The CPCS PDU is passed to the SAR sublayer, which produces SAR PDUs with 4 bytes of overhead and 44 bytes of payload. If necessary, the last SAR PDU is

---

[3]See Goralski for a discussion on the merging of AAL3 and AAL4.

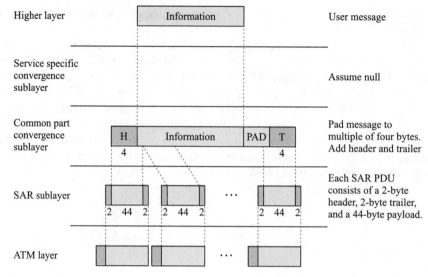

FIGURE 9.15 AAL3/4 process

padded to fill the payload. Finally the 48-byte SAR PDUs are passed to the ATM layer.

The CPCS PDU has a header, followed by the CPCS-PDU payload, possibly with padding, and finally a trailer. The CPCS-PDU payload can have a length of 1 to 65,535 bytes. As shown in Figure 9.16a, the header begins with a one-byte common part indicator (CPI) field that specifies how subsequent fields are to be interpreted. Only CPI = 0 has been defined to indicate that the BAsize and Length fields are to be interpreted in units of bytes. The one-byte beginning tag (Btag) field and the end tag (Etag) field are set to the same value at the

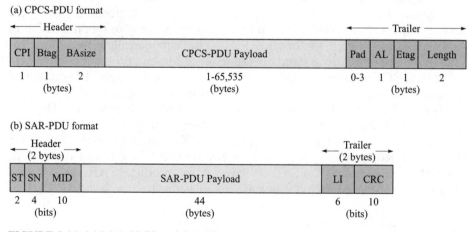

FIGURE 9.16 AAL3/4 CPCS and SAR formats

source and changed for each successive CPCS PDU. This practice allows the destination to detect when the incorrect header and trailer have been associated. The two-byte buffer allocation size indication (BAsize) field informs the destination of the maximum buffer size required to receive the current CPCS PDU. The padding field contains zero to three pad bytes to make the CPCS PDU a multiple of four bytes. This approach ensures that the trailer part will be aligned to a 32-bit boundary, making it easier to process the trailer. The alignment field consists of a byte of zeros to make the trailer four bytes long. The two-byte length field indicates the length of the payload.

Figure 9.16b shows that the SAR PDU contains a 2-byte header, 44 bytes of payload, and a 2-byte trailer. The first two bits in the header are the segment type: the value 10 indicates that the PDU contains the beginning of a message (BOM), 00 denotes continuation of message (COM), 01 indicates end of message (EOM), and 11 indicates a single-segment message (SSM). The next four bits provide the sequence number (SN) of the SAR PDU within the same CPCS PDU. The sequence numbers are used to ensure correct sequencing of cells when the CPCS PDU is rebuilt at the destination. The remaining 10 bits in the header are for the **multiplexing identifier**, also called **message identifier**, **(MID)**. The MID allows the SAR sublayer to multiplex the messages of up to 2^{10} AAL users on a single ATM VC. All SAR PDUs of the same CPCS PDU have the same MID. The six-bit length indicator (LI) in the trailer specifies the size of the payload. Except for the last cell, all cells for a given CPCS PDU are full, so LI = 44. The last cell can have LI from 4 to 44. The 10-bit CRC provides for the detection of errors that may occur anywhere in the PDU.

Figure 9.17 elaborates on how multiplexing is done in AAL3/4. When multiple users share the same VC, the messages from each user are sent to a *different*

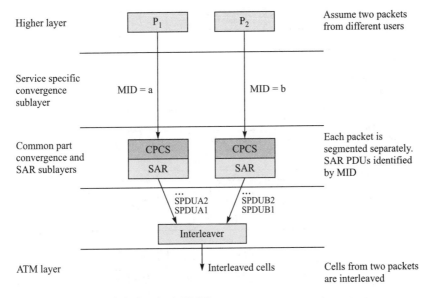

FIGURE 9.17 Multiplexing in AAL3/4

instance of the AAL. Each such AAL will produce one or more SAR PDUs that are then intermixed and interleaved prior to being passed to the ATM layer. Thus the SAR PDUs from different users arrive intermixed at the destination. The MID allows the messages from each user to be reassembled. Note that the MID can be viewed as setting up a very short term connection within the VC. Each such connection is for the duration of a single packet transfer and is delimited by the BOM and EOM cells. The MID feature of AAL3/4 was developed to provide compatibility with the IEEE 802.6 Metropolitan Area Standard, which provides connectionless LAN interconnection service.

A problem with AAL3/4 is that it is heavy in terms of overhead. Each message has at least eight bytes added at the CSCP sublayer, and subsequently each ATM cell payload includes four additional bytes of overhead. The 10-bit CRC and the 4-bit sequence numbering also may not provide enough protection. These factors led to the development of AAL5.

9.4.4 AAL5

AAL type 5 (AAL5) provides an efficient alternative to AAL3/4. AAL5 forgoes the multiplexing capability of AAL3/4 but does support message and stream modes, as well as assured and nonassured delivery.

Figure 9.18 shows the operation of AAL5. A user PDU is accepted by the AAL layer and is processed by the SSCS if necessary. The SSCS then passes a block of data to the CPCS, which attaches 0 to 47 bytes of padding and an 8-byte

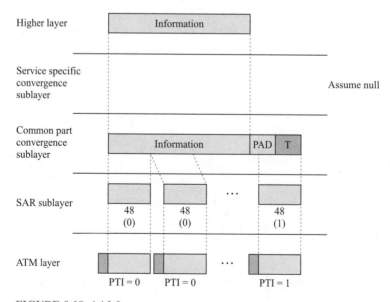

FIGURE 9.18 AAL5 process

Information	Pad	UU	CPI	Length	CRC
0–65,535 (bytes)	0–47	1	1	2 (bytes)	4

FIGURE 9.19 AAL5 PDU

trailer to produce a CPCS PDU that is a multiple of 48 bytes. The maximum CPCS-PDU payload is 65,535 bytes.

As shown in Figure 9.19, the trailer contains one byte of UU that is passed transparently between the end-system entities, one byte of CPI that aligns the trailer to eight bytes, a two-byte LI that specifies the number of bytes of user data in the CPCS-PDU payload, and a four-byte CRC check to detect errors in the PDU. The SAR sublayer segments the CPCS PDU into 48-byte payloads that are passed to the ATM layer. The SAR also directs the ATM layer to set the PTI field in the header of the last cell of a CPCS PDU. This step allows the boundary between groups of cells corresponding to different messages to be distinguished at the destination. Note that unlike AAL3/4, AAL5 can have only one packet at a time in a VC because the cells from different packets cannot be intermixed.

AAL5 is much more efficient than AAL3/4 is, as AAL5 does not add any overhead in the SAR sublayer. AAL5 does not include sequence numbers for the SAR PDUs and instead relies on its more powerful CRC checksum to detect lost, misinserted, or out-of-sequence cells. AAL5 is by far the most widely implemented AAL.

9.4.5 Signaling AAL

The signaling AAL (SAAL) has been standardized as the AAL in the control plane. The SAAL provides reliable transport for the signaling messages that are exchanged among end systems and switches to set up ATM VCs.

The SAAL is divided into a common part and a service-specific part as shown in Figure 9.20. The service-specific part in turn is divided into a service-specific connection-oriented protocol (SSCOP) and a service-specific coordination function (SSCF). The SSCF supports the signaling applications above it by mapping the services they require into the services provided by the SSCOP. SSCF sublayers have been developed for UNI and NNI.

The SSCOP is a peer-to-peer protocol that provides for the reliable transfer of messages. It provides for the ordered delivery of messages in either assured or unassured mode. In assured mode SSCOP uses a form of Selective Repeat ARQ for error recovery. The ARQ protocol is suitable for use in situations that have large delay-bandwidth products, for example, satellite channels and ATM links. Thus the protocol uses large window sizes to provide sequence numbers for the transmitted packets. To achieve bandwidth efficiency, selective retransmission is used. SSCOP makes use of the service provided by the convergence sublayer and

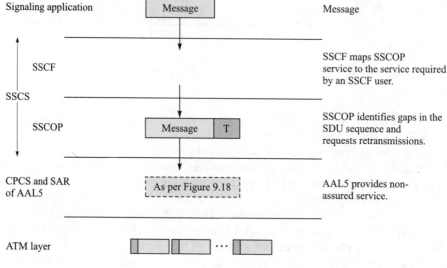

FIGURE 9.20 SAAL process

SAR sublayers of AAL5 as shown in Figure 9.20. The destination AAL5 layer passes SDUs up to the SSCOP layer only if their checksum is correct. The SSCOP buffers all such SDUs and looks for gaps in the SDU sequence. Because ATM is connection-oriented, the SSCOP knows that the SDUs corresponding to these gaps have errors or have been lost. The sender periodically polls the receiver to find the status of the receive window. Based on the response from the receiver, the transmitter then selectively retransmits the appropriate SDUs. This approach allows SSCOP to ensure that SDUs are retransmitted and that messages are delivered in the correct order and without errors.

Figure 9.21 shows the structure of the SSCOP PDU. A trailer is added, consisting of zero to three bytes of padding, a two-bit pad length indicator, a two-bit reserved (unassigned) field, and a four-bit PDU type field to specify the type of message in the payload. The payload types include sequenced data messages as well as the poll and other messages used by the SSCOP protocol. A 24-bit sequence number is provided for each CPCS PDU. Note that the cell error and loss detection capabilities are provided by the 32-bit CRC that is used in the AAL5 layers, as per Figure 9.19.

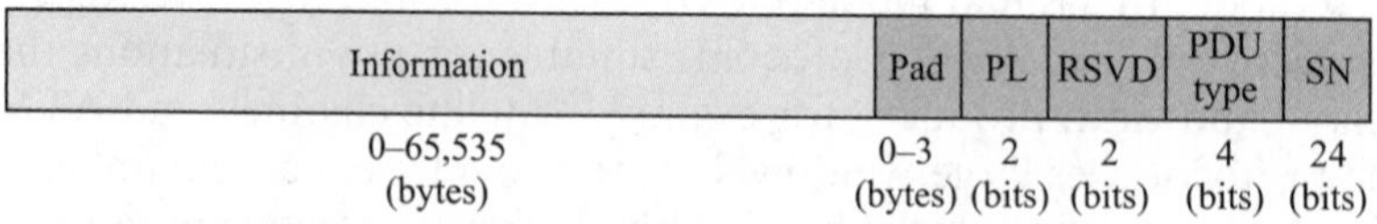

Information	Pad	PL	RSVD	PDU type	SN
0–65,535 (bytes)	0–3 (bytes)	2 (bits)	2 (bits)	4 (bits)	24 (bits)

FIGURE 9.21 SSCOP PDU

9.4.6 Applications, AALs, and ATM Service Categories

We saw in the previous section that AALs can be called upon to support a wide range of applications, from emulating a digital transmission line to transferring packet streams of various types. Table 9.3 lists features that characterize the requirements of various types of applications. Table 9.4 summarizes the capabilities of the various AALs.

The application/higher layer that operates over the AAL may require that information be transferred in the form of a stream or as discrete messages. The transfer may involve a constant rate of transfer or may vary with time. The application may be satisfied with a nonassured delivery of information, where there may be occasional losses or misdeliveries, or it may require high levels of assurance on the correct delivery of all information. In the case of nonassured delivery, there may be various degrees of tolerance to errors in the delivered information. Some applications may be tolerant to relatively high delays and significant levels of delay variation, whereas other applications may require tight tolerance in the delay and the jitter. The ability to multiplex streams/packets from different users is an important requirement in certain settings. Finally, the demand for efficiency in the use of the cell payload will depend on the cost of bandwidth in a given setting.

Various combinations of the features in Table 9.3 appear in different applications. A simple example such as voice can require different combinations of features in different contexts. Voice-based applications are almost always stream oriented. In many situations these applications are error tolerant. However, the degree of error tolerance is different when the stream is carrying voice than when it is carrying modem signals that carry data or fax. Furthermore, as the bit rate of voice signals decreases with compression, the degree of error tolerance also decreases. If silence suppression is used, the stream becomes variable bit rate rather than constant bit rate. Telephone conversations between humans requires real-time transfer with low delay and jitter. But voice mail and voice response applications are much more tolerant of delay. Finally, multiplexing may be important in voice applications where bandwidth costs are significant, but not relevant where bandwidth is cheap as in LANs.

Feature	Application Requirements	
Transfer granularity	Stream	Message
Bit rate	Constant	Variable
Reliability	Nonassured	Assured
Accuracy	Error tolerant	Error intolerant
Delay sensitivity	Delay/jitter sensitive	Delay/jitter insensitive
Multiplexing	Single user	Multiple users
Payload efficiency	Bandwidth inexpensive	Bandwidth expensive

TABLE 9.3. Features that characterize application requirements

Sublayer	Feature	AAL1	AAL2	AAL3/4	AAL5	SAAL
SSCS	Forward error control	Optional	Optional	Optional	Optional	No
	Error detection and retransmission	No	No	Optional	Optional	SSCOP
	Timing recovery	Optional	Optional	No	Optional	No
CPCS	Multiplexing	No	8-bit CID	10-bit MID	No	No
	Framing structure	Yes	No	No	No	No
	Message delimiting	No	Yes	Yes	PTI	PTI
	Advance buffer allocation	No	No	Yes	No	No
	User-to-user indication	No	3 bits	No	1 byte	No
	Overhead	0	3 bytes	8 bytes	8 bytes	4 bytes
	padding	0	0	4 bytes	0–44 bytes	0–47 bytes
	Checksum	No	No	No	32 bit	32 bit
	Sequence numbers	No	No	No	No	24 bit
SAR	Payload/overhead	46–47 bytes	47 bytes	44 bytes	48 bytes	48 bytes
	Overhead	1–2 bytes	1 byte	4 bytes	0	0
	Checksum	No	No	10 bits	No	No
	Timing information	Optional	No	No	No	No
	Sequence numbers	3 bit	1 bit	4 bit	No	No

TABLE 9.4. Capabilities of the AAL types

Given the diversity of requirements in different voice applications, it is not surprising that many of the AALs have been adapted for use with different voice applications. A simple AAL1 has been adapted for the transfer of individual 64 kbps voice calls. Other versions of AAL1 with structured data transfer (SDT) are intended for handling multiples of 64 kbps calls. In addition, AAL2 provides the capability to multiplex several low-bit-rate voice calls. AAL5 has also been adapted to carry voice traffic. The rationale here was that AAL5 was required for signaling, so the cost of adding AAL1 could be avoided by operating voice over AAL5.

The choice of which ATM service category to use below the AAL depends on the performance requirements the AAL is committed to deliver to the service above it. It also depends on the manner in which the user level passes SDUs to the AAL because this feature influences the manner in which the cell traffic is generated. Thus the CBR and the rt-VBR service categories are suitable for applications with a real-time requirement, for example, voice over AAL5 over CBR. However, under certain network loading conditions, the other service categories may provide adequate performance. For example, under low network loads voice over AAL5 over UBR may give adequate performance. Policing may also be required to ensure that the cell stream produced by the AAL conforms to the connection contract.

Data transfer applications require different combinations of features in the AAL. These applications tend to be message oriented and variable in bit rate. Best-effort (nonassured) service is usually sufficient, but certain applications can require assured service. Most data transfer applications are relatively insensitive to delays, but it is possible to conceive of monitoring/control applications that require very low delay. As indicated above, AAL5 with nonassured service is the most widely deployed AAL. AAL5 with SSCOP has also been used to provide assured service in the user plane.

Video applications represent a third broad class of applications that include relatively low-bit-rate videoconferencing, for example, n × 64 kbps, to high-quality video distribution and video on demand. Delay plays an important part in applications that involve real-time interactivity. Cell-delay variation plays an important role in almost all video applications because the receiver must compensate for the CDV and present the data to the video decoder at very nearly constant rate. This stringent CDV requirement arises from the manner in which traditional analog television signals deal with the three color components. Very small errors in synchronization among the color components have a dramatic impact on the picture quality.

As a concrete example, consider the transport of constant-bit-rate MPEG2 video in a video-on-demand application. An issue in the design of such a system is whether the timing recovery should be done by the AAL, by the MPEG2 systems layer,[4] or possibly by a layer in between. Another issue involves the degree of error detection and correction required and the use of error concealment techniques. Efficiency in the use of cell payload is also a concern. The recommendation developed by the ATM Forum addressed the issues as follows. AAL1 was not selected for a number of reasons. The SRTS capability of AAL1 could not be used because many end systems do not have access to the common network clock required by the method. The FEC interleaving technique was deemed to introduce too much delay and to be too costly. On the other hand, AAL5 is less costly because of its much wider deployment. It was also found that AAL5 over CBR ATM service could carry MPEG2 transport packets (of 188 bytes each) with sufficiently low jitter that timing recovery could be done in the MPEG2 systems layer. The recommendation does not require that the CBR ATM stream coming out of the video server be shaped to conform to the negotiated CBR parameters.

9.5 ATM SIGNALING

The utility of a network is directly related to the capability to connect dynamically to any number of destinations. Signaling provides the means for dynamically setting up and releasing switched virtual connections in an ATM network. The establishment of an end-to-end connection involves the exchange of signaling messages across a number of interfaces, for example, user-network interface (UNI), network-network interface (NNI), and broadband intercarrier interface (B-ICI). Signaling standards are required for each of these interfaces. In this section we focus on the signaling standards for UNI and NNI.

---

[4]MPEG video coding is discussed in Chapter 12.

The establishment of dynamic connections requires the ability to identify endpoints that are attached in the network. This function is provided by network addresses, the topic of the next section.

9.5.1 ATM Addressing

ATM uses two basic types of addresses: telephony-oriented E-164 addresses intended for use in public networks and ATM end system addresses (AESAs) intended for use in private networks. E-164 telephone numbers can be variable in length and have a maximum length of 15 digits. For example in the United States and Canada, 11-digit numbers are used: 1-NPA-NXX-ABCD, for example, 1-416-978-4764. The first 1 is the ITU assigned country code; the next three digits, NPA, are the area code; the following three digits, NXX, are the office code; and the final four digits, ABCD, are the subscriber number. Telephone numbers for other countries begin with a different country code and then follow a different format.

AESAs are based on the ISO Network Service Access Point (NSAP) format that consists of 20-byte addresses. The NSAP format has a hierarchical structure as shown in Figure 9.22. Each address has two parts.

- The *initial domain part (IDP)* identifies the administrative authority that is responsible for allocating the addresses that follow.

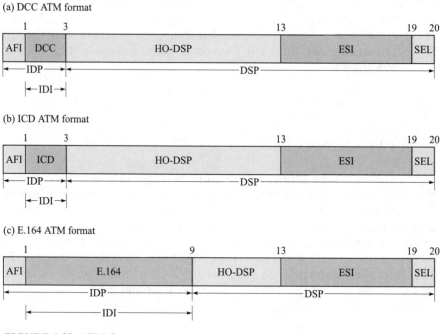

FIGURE 9.22 ATM formats

* The *domain-specific part (DSP)* contains the address allocated by the given authority.

The IDP itself consists of two parts: a one-byte authority and format identifier (AFI) identifies which structure is to follow and the initial domain identifier (IDI) specifies the authority that allocates the DSP that follows.

Figure 9.22 shows the format of the three initial types of AESAs: data country code (DCC) with AFI = 39_{HEX}; international code designator (ICD) with AFI = 47_{HEX}; and E.164 (contained within the AESA format) with AFI = 45_{HEX}. From Figure 9.22a, it can be seen that the IDI in the DCC format is 2 bytes long, followed by a 10 byte "higher order domain specific part (HO-DSP)." For example, IDI = 840_{HEX} identifies the United States, for which ANSI administers the DCC addresses. The format used by ANSI is as follows: the first three bytes identify the organization that uses these addresses, the next byte is used for other purposes, and the final six bytes are for the organization to assign. The ICD addresses are administered by the British Standards Institute. The ICD format, shown in Figure 9.22b, places a four-digit code in the IDI bytes to identify an organization, which is then allowed to administer the next 10 bytes. The six-byte end system identifier (ESI) identifies an end system and usually consists of a MAC address. The one-byte selector (SEL) can be used by the end system to identify higher-layer protocol entities that are to receive the traffic. The AESA format for E-164 addresses is shown in Figure 9.22c. The IDI consists of eight bytes that can hold the 15 digits of the E-164 address.

E-164 "native" addresses are supported in the public UNI and in the B-ICI. AESAs are supported in the private UNI and NNI, as well as in the public UNI. The ATM Forum and other bodies are working on the interworking of these public and private addresses to provide end-to-end connectivity.

9.5.2   UNI Signaling

The ATM signaling standards are based on the standards developed for telephone networks. We saw in Chapter 4 that telephone networks use two signaling standards: ISDN signaling (Q.931) is used in the exchange of call setup messages at the UNI; the ISUP protocol of Signaling System #7 is used to establish a connection from a source switch to a destination switch within the network. ATM signaling has developed along similar lines with signaling procedures developed for the UNI, the NNI, and the B-ICI. In this section we consider UNI signaling.

ITU-T recommendation Q.2931, derived from Q.931, specifies B-ISDN signaling at the *ATM UNI*. ATM Forum UNI signaling 4.0 is based on Q.2931. A number of messages have been defined for use in the setup and release of connections. ATM connections involve many more parameters than narrowband ISDN involves, so the signaling messages carry special fields, called *information elements (IEs)*, that describe the user requests. These signaling messages are transferred across the UNI using the services of the SAAL layer in the control plane. Recall that the SAAL provide reliable message transfer using the SSCOP protocol that operates over AAL5. ITU-T recommendation Q.2130 specifies the SSCF that is used between the Q.2931 signaling application and SSCOP. The signaling cells that are produced by AAL5 use the default virtual channel identifier by VPI = 0 and VCI = 5.

Table 9.5 shows the capabilities provided by UNI 4.0. These capabilities are categorized as being applicable to end-system or switch equipment and as being mandatory (M) or optional (O). Point-to-point as well as point-to-multipoint calls are supported. Point-to-point ABR connections are supported. Signaling of individual QoS parameters and negotiation of traffic parameters are also supported. The leaf-initiated join capability allows an end system to join a point-to-multipoint connection with or without the intervention of the root. Group addressing allows a group of end systems to be identified. Anycast capability allows the setting up of a connection to an end system that is part of an ATM group. The reader is referred to ATM Forum UNI 4.0 signaling specification for details on these capabilities.

Table 9.6 shows a few of the messages used by UNI 4.0 and Q.2931 and their significance when sent by the host or the network. Each signaling message con-

Number	Capability	Terminal equipment	Switching system
1	Point-to-point calls	M	M
2	Point-to-multipoint calls	O	M
3	Signaling of individual QoS parameters	O	M
4	Leaf-initiated join	M	M
5	ATM anycast	O	O
6	ABR signaling for point-to-point calls	O	(1)
7	Generic identifier transport	O	O
8	Virtual UNIs	O	O
9	Switched virtual path (VP) service	O	O
10	Proxy signaling	O	O
11	Frame discard	O	O (2)
12	Traffic parameter negotiation	O	O
13	Supplementary services	–	–
13.1	Direct dialing in (DDI)	O	O
13.2	Multiple subscriber number (MSN)	O	O
13.3	Calling line identification presentation (CLIP)	O	O
13.4	Calling line identification restriction (CLIR)	O	O
13.5	Connected identification presentation (COLP)	O	O
13.6	Connected line identification restriction (COLR)	O	O
13.7	Subaddressing (SUB)	O	(3)
13.8	User-user signaling (UUS)	O	O

Notes: [1]This capability is optional for public networks/switching systems and is mandatory for private networks/switching systems.

[2]Transport of the frame discard indication is mandatory.

[3]This capability is mandatory for network/switching systems (public and private) that support only native E.164 address formats.

TABLE 9.5. Capabilities of UNI 4.0

Message	Meaning (when sent by host)	Meaning (when sent by network)
SETUP	Requests that a call be established	Indicates an incoming call
CALL PROCEEDING	Acknowledges the incoming call	Indicates the call request will be attempted
CONNECT	Indicates acceptance of the call	Indicates the call was accepted
CONNECT ACK	Acknowledges acceptance of the call	Acknowledges making the call
RELEASE	Requests that the call be terminated	Terminates the call
RELEASE COMPLETE	Acknowledges releasing the call	Acknowledges releasing the call

TABLE 9.6. Signaling messages involved in connection setup

tains a *call reference* that serves as a local identifier for the connection at the UNI. Each message also contains a number of information elements. These include obvious parameters such as calling and called party numbers, AAL parameters, ATM traffic descriptor and QoS parameters, and connection identifier. Many other mandatory and optional parameters are defined.

The signaling procedures specify the sequence of message exchanges to establish and release connections. They also address the handling of many error conditions that can arise. In Figure 9.23, we consider the simple case of the establishment of a point-to-point virtual connection using Q.2931.

1. Host A sends a SETUP message on VPI/VCI = 0/5 identifying the destination (host B) and other parameters specifying details of the requested connection.

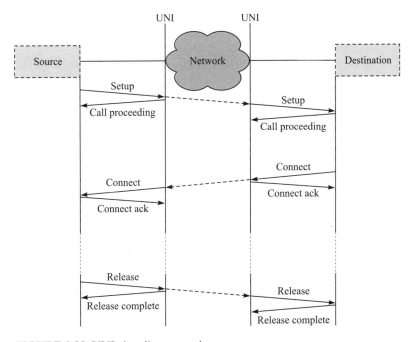

FIGURE 9.23 UNI signaling example

2. The first switch analyzes the contents of the SETUP message to see whether it can handle the requested connection. If the switch can handle the request, the network returns a CALL PROCEEDING message to the host containing the VPI/VCI for the first link. It also forwards the SETUP message across the network to the destination.
3. Upon arrival of the SETUP message, the destination sends a CALL PROCEEDING message.
4. If the destination accepts the call, it sends a CONNECT message that is forwarded across the network back to host A. The CONNECT messages trigger CONNECT ACKNOWLEDGE messages from the network and eventually from the source.
5. The connection is now established, and the source and destination can exchange cells in the bidirectional VC that has been established.
6. Either party can subsequently initiate the termination of the call by issuing a RELEASE message. This step will trigger RELEASE COMPLETE messages from the network and from the other party.

A point-to-multipoint connection is established as follows. The root of the connection begins by establishing a connection to the first destination (leaf) by using the above procedure. It then issues ADD PARTY messages that attach additional destinations (leaves) to the connection. Note that point-to-multipoint connections are unidirectional, so cells can flow only from the root to the leaves.

9.5.3 PNNI Signaling

The ATM Forum has developed the PNNI specification for use between private ATM switches (private network node interface) and between groups of private ATM switches (private network-to-network interface) as shown in Figure 9.24. The PNNI specification includes two types of protocols.

1. A routing protocol that provides for the selection of routes that can meet QoS requirements (this routing protocol is discussed in the next section).

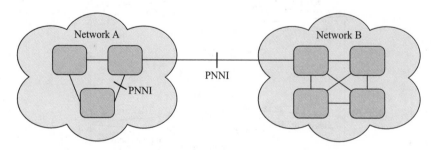

FIGURE 9.24 PNNI contexts

2. A complementary signaling protocol for the exchange of messages between switches and between private networks. In this section we consider the signaling protocol.

The PNNI signaling protocol provides for the establishment and release of point-to-point as well as point-to-multipoint connections. The protocol is based on UNI 4.0 with extensions to provide support for source routing, for crankback (a feature of the routing protocol), and for alternate routing of connection requests in the case of connection setup failure. UNI signaling is asymmetric in that it involves a user and a network. PNNI modifies UNI signaling to make it symmetric. It also includes modifications in the information elements to carry routing information.

PNNI uses source routing where the first switch selects the route to the destination. In Figure 9.25 the source host requests a connection to host B by sending a SETUP message, using UNI signaling. The first switch carries out the connection admission control (CAC) function and returns a CALL PROCEEDING message if it can handle the connection request. The first switch maintains and uses a topology database to calculate a route to the destination that can meet the requirements of the connection contract.[5] The route consists of a vector of switches that are to be traversed. The SETUP message propagates

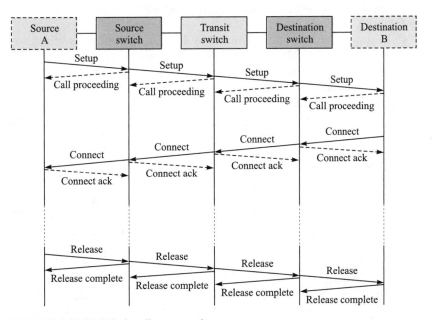

FIGURE 9.25 PNNI signaling example

---

[5]The PNNI routing protocol includes procedures for distributing the link-state information required to determine the routes that can meet specific QoS requirements.

across the network, using the source route. Each switch along the path performs CAC and forwards the SETUP message along the next hop if it can handle the connection request. It also issues a CALL PROCEEDING message to the preceding switch along the route. If the destination accepts the call, a connect message is returned across the network to the source. Connection release proceeds in similar fashion as shown in Figure 9.25.

The PNNI routing protocol introduces hierarchy in the ATM network that provides a switch with detailed routing information in its immediate vicinity and only summary information about distant destinations. The signaling protocol is somewhat more complicated than the preceding example because of this hierarchical feature.

9.6   PNNI ROUTING

Routing in ATM networks is a subject that has not been investigated as much as other ATM areas such as congestion control and switching. The most visible result on ATM routing comes from the ATM Forum standards work on the **private network-to-network interface (PNNI[6])**. In this section, we briefly look at the main concepts behind PNNI routing techniques. Unlike Internet routing, which is divided into intradomain and interdomain routing protocols, PNNI works for both cases.

PNNI adopts the link-state philosophy in that each node would know the topology of the network. However, PNNI adds a new twist to make the routing protocol scalable so that it can work well for small networks as well for large networks with thousands of nodes. This goal is achieved by constructing a **routing hierarchy**, as illustrated in Figure 9.26.

As shown in the figure, a *peer group* (PG) is a collection of nodes (physical or logical) where each maintain an identical view of the group. For example, peer group A.1 consists of three nodes A.1.1, A.1.2, and A.1.3. A PG is abstractly represented in a higher-level routing hierarchy as a *logical group node* (LGN). For example, LGN B represents PG B in the next hierarchy. Note that the definition of PG and LGN is recursive. At the lowest level a PG is a collection of physical switches connected by physical links. At higher levels a PG is a collection of LGNs connected by logical links. Each PG contains a *peer group leader* (PGL) that actually executes the functions of the logical group nodes for that PG. A PGL summarizes the topological information within the PG and injects this information into the higher-order group. A PGL also passes down summarized topological information to its PG. The advantage of using this hierarchical structure is that each switch maintains only a partial view of the entire network topology, thereby reducing the amount of routing information kept at each switch. For example, from the point of view of switch A.1.1, the

---

[6]PNNI also stands for private network node interface.

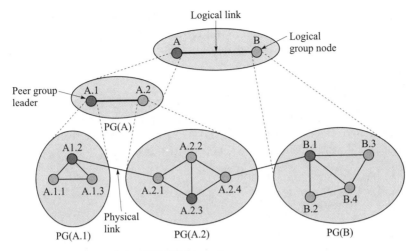

FIGURE 9.26 Example of PNNI hierarchy

topology of the network is shown in Figure 9.27, which is much simpler than that of the entire topology.

PNNI uses source routing to set up connections. First, the source node specifies the entire path across its peer group, which is described by a **designated transit list (DTL)**. Suppose that a station attached to switch A.1.1 requests a connection setup to another station attached to switch B.3. After the source station requests a connection setup, switch A.1.1 chooses the path to be (A.1.1, A.1.2, A.2, B). In this case three DTLs organized in a stack will be built by A.1.1 in the call setup:

DTL: [A.1.1, A.1.2] pointer-2
DTL: [A.1, A.2] pointer-1
DTL: [A, B] pointer-1

The current transit pointer specifies which node in the list is currently being visited at that level, except at the top TDL where the transit pointer specifies the node to be visited next. Thus the top pointer points to A.1.2 and the other pointers point to A.1 and A respectively.

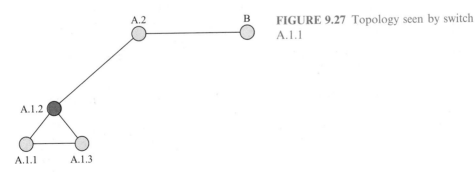

FIGURE 9.27 Topology seen by switch A.1.1

When A.1.2 receives the call setup message, the switch realizes that the entry in the top of the stack is exhausted. After removing the top DTL, A.1.2 finds that the next entry is A.2, which is also its immediate neighbor. The DTLs become

DTL: [A.1, A.2] pointer-2
DTL: [A, B] pointer-1

When A.2.1 receives the call setup message, the switch finds that the target has been reached, since A.2.1 is in A.2. So A.2.1 builds a route to B, say, through A.2.3 and A.2.4, and pushes a new DTL onto the stack:

DTL: [A.2.1, A.2.3, A.2.4] pointer-2
DTL: [A.1, A.2] pointer-2
DTL: [A, B] pointer-1

When A.2.3 receives the call setup message, the switch advances the current transit pointer and forwards the message to A.2.4. When A.2.4 receives the call setup message, it finds that the targets at the top two DTLs have been reached. So A.2.4 removes the top two DTLs and forwards the message with the following DTL to its neighbor:

DTL: [A, B] pointer-2

When B.1 receives the call setup message, B.1 finds that the curernt DTL has been reached. B.1 builds a new DTL, giving

DTL: [B.1, B.3] pointer-2
DTL: [A, B] pointer-2

When the message reaches its destination, B.3 determines that it is a DTL terminator, since all DTLs are at the end and B.3 is the lowest-level node.

It is possible that a call setup will be blocked when the requested resources at a switch are not available. PNNI provides **crankback** and **alternate routing**. When a call is blocked at a particular DTL, it is cranked back to the creator of the DTL.

A switch determines the path based on the connection's traffic descriptor, QoS requirements, and the resources stored in its database. Because CAC is not standardized, PNNI uses a **generic connection admission control (GCAC)** to select a path that is likely to satisfy the connection's end-to-end traffic and QoS requirements. GCAC, however, only predicts the most likely path that can satisfy the connection traffic, since the information known by the switch may be outdated when the actual call request is made at other switches along the path.

GCAC requires the following parameters:

- Available cell rate (ACR): A measure of available bandwidth on the link.
- Cell rate margin (CRM): A measure of the difference between the aggregate allocated bandwidth and sustained rate of the existing connections.
- Variance factor (VF): A relative measure of the CRM normalized by the variance of the aggregate rate.

When a new connection with peak cell rate PCR and sustained cell rate SCR requests a connection setup, the GCAC algorithm examines each link along the path and performs the following decision:

if PCR ≤ ACR
          include the link
else if SCR > ACR
          exclude the link
else if [ACR − SCR][ACR − SCR + 2 * CRM] ≥ VF * SCR (PCR − SCR)
          include the link
else
          exclude the link

The VF is CRM^2/VAR, and the variance is given by

$$VAR = \sum_i SCR(i)[PCR(i) - SCR(i)]$$

After the path has been selected, each switch along the path eventually has to perform its own CAC to decide the acceptance/rejection of the connection request. Note that GCAC uses VF only in the case SCR < ACR < PCR.

ATM IS DEAD, LONG LIVE ATM!

The development of ATM standards was a massive effort that attempted to develop an entire future network architecture from the ground up. BISDN was viewed as a future multiservice network that could encompass LANs and WANs in a common framework. Alas, this was not to be.

The ATM connection-oriented networking paradigm emerged at the same time as the explosion of the World Wide Web. The Web is built on HTTP and the TCP/IP protocol suite which is decidedly connectionless in its network orientation. ATM to the user is not what was needed and so the vision of an end-to-end universal ATM network perished.

ATM technology does have advantages such as facilitating high-speed switching and enabling the management of traffic flows in the network. For this reason ATM has found wide application in carrier backbone networks and, to a lesser extent, in campus networks that require prioritization of traffic flows.

The long-term future of ATM in an all-IP world is uncertain. In Chapter 10 we examine the issues in operating IP over ATM. We discuss a number of initiatives that incorporate features of ATM in IP. In effect, it is possible that ATM may disappear over the long run, but that many of its innovations in high-speed switching, traffic management, and QoS will survive in an IP-networking framework.

SUMMARY

In this chapter we examined the architecture of ATM networks. We discussed the BISDN reference model and the role of the control plane in setting up ATM connections. We then examined the structure of the ATM header and its relationship to virtual paths and virtual connections. The connection setup involves the establishment of a traffic contract between the user and the network that commits the user to a certain pattern of cell transmission requests and the network to a certain level of QoS support. We examined the traffic management mechanisms that allow ATM to provide QoS guarantees across a network. The various types of ATM service categories to meet various information transfer requirements were introduced.

We discussed the role of the ATM adaptation layer in providing support for a wide range of user applications. This support included segmentation and reassembly, reliable transmission, timing recovery and synchronization, and multiplexing.

Last, we introduced PNNI signaling and its associated routing protocol which are used to set up connections with QoS guarantees across ATM networks.

CHECKLIST OF IMPORTANT TERMS

alternate routing
ATM adaptation layer (AAL)
ATM layer
broadband intercarrier interface
  (B-ICI)
common part (CSCP)
control plane
convergence sublayer (CS)
crankback
designated transit list (DTL)
generic connection admission control
  (GCAC)

message identifier (MID)
multiplexing identifier
network-network interface (NNI)
private network-to-network interface
  (PNNI)
routing hierarchy
segmentation and reassembly (SAR)
service specific part (SSCS)
user-network interface (UNI)

FURTHER READING

ATM Forum Technical Committee, "ATM User-Network Interface (UNI) Signaling Specification, Version 4.0," af-sig-0061.000, July 1996.

ATM Forum Technical Committee, "Private Network-Network Interface Specification, Version 1.0," af-pnni-0055.000, March 1996.

ATM Forum Technical Committee, "Traffic Management Specification, Version 4.0," af-tm-0056.000, April 1996.

Black, U., *ATM: Foundation for Broadband Networks*, Prentice Hall, Englewood Cliffs, New Jersey, 1995.

Dutton, H. J. R. and P. Lenhard, *Asynchronous Transfer Mode (ATM)*, Prentice Hall PTR, Upper Saddle River, New Jersey, 1995.

Helgert, H. J., *Integrated Services Digital Networks: Architectures, Protocols, Standards*, Addison-Wesley, Reading, Massachusetts, 1991.

Ibe, O. C., *Essentials of ATM Networks and Services*, Addison-Wesley, Reading, Massachusetts, 1997. Excellent concise introduction to ATM.

ITU draft, I.363.2, "B-ISDN ATM Adaptation Layer Type 2 Specification," November 1996.

Leon-Garcia, A., *Probability and Random Processes for Electrical Engineering*, Addison-Wesley, Reading, Massachusetts, 1994.

McDysan, D. E. and D. L. Spohn, *ATM: Theory and Application*, McGraw-Hill, New York, 1995.

Prycker, M. de, *Asynchronous Transfer Mode: Solution for Broadband ISDN*, Prentice Hall, Englewood Cliffs, New Jersey, 1995.

PROBLEMS

1. Suppose that instead of ATM, BISDN had adopted a transfer mode that would provide constant-bit-rate connections with bit rates given by integer multiples of 64 kbps connections. Comment on the multiplexing and switching procedures that would be required to provide this transfer mode. Can you give some reasons why BISDN did not adopt this transfer mode?

2. a. Compare the bandwidth management capabilities provided by ATM virtual paths to the capabilities provided by SONET networks.

   b. Can ATM virtual paths be adapted to provide the fault tolerance capabilities of SONET rings? Explain your answer.

3. In Chapter 6 we saw that the performance of MACs for LANs can depend strongly on the delay-bandwidth product of the LAN. Consider the use of ATM in a LAN. Does the performance depend on the delay-bandwidth product? If yes, explain how; if no explain why not? Does the same conclusion apply to any LAN that involves the switching packets, rather than the use of a broadcast medium?

4. Does the performance of ATM depend strongly on the delay-bandwidth product of a wide-area network? Specifically, consider the performance of ATM congestion control in a network with large delay-bandwidth product. Does the size of the buffers in the ATM switches influence the severity of the problem? Give an order of magnitude for the amount of buffering required in the switches.

5. Compare a conventional TDM leased line with an ATM PVC from the user's point of view and from the network operator's point of view. Which features of PVCs make them attractive from both points of view?

6. An inverse multiplexer is a device that takes as input a high-speed digital stream and divides it into several lower-speed streams that are then transmitted over parallel transmission lines to the same destination. Suppose that a very high speed ATM stream is to be sent over an inverse multiplexer. Explain which requirements must be met so that the inverse demultiplexer produces the original ATM stream.

7. Suppose an ATM switch has 32 input ports and 32 output ports.
   a. Theoretically, how many connections can the switch support?
   b. What are the table lookup requirements for supporting a large number of connections? Do they limit the practical size on the number of connections that can actually be supported?
   c. (Optional) Do a Web search on content addressable memories (CAMs) and explain how these can help in addressing the table-lookup problem.

8. Explain how the header error checksum can be used to synchronize to the boundary of a sequence of contiguous ATM cells. What is the probability that an arbitrary five-octet block will satisfy the header error checksum? Assume bits are equally likely to be 0 or 1. What is the probability that two random five-octet blocks that correspond to two consecutive headers pass the error check?

9. We saw that ATM provides a GFC field in the UNI ATM header. Suppose that several terminals share a medium to access an ATM network.
   a. Explain why flow control may be required to regulate the access of traffic from the terminals into the network. Explain how the GFC field can be used to do this task?
   b. Explain how the GFC field can be used as a subaddress to provide point-to-multipoint access to an ATM network. Does this usage conflict with the flow control requirement?

10. The purpose of the header error control (HEC) field is to protect against errors in the header that may result in the misaddressing and misdelivery of cells. The CRC used in the ATM header can correct all single errors and can detect (but not correct) all double errors that occur in the header. Some, but not all, multiple errors in excess of two can also be detected.
   a. Suppose that bit errors occur at random and that the bit error rate is p. Find the probability that the header contains no errors, a single error, a double error, more than two errors. Evaluate these probabilities for $p = 10^{-3}$, 10^{-6}, 10^{-9}.
   b. Relate the calculations in part (a) to the CMR experienced in such an ATM network.
   c. In practice the bit errors may not always be random, and so to protect against bursts of errors the following adaptive procedure may be used.
      Normally the receiver is in the "correction" mode. If a header is error free, the receiver stays in this mode; if the receiver detects a single error, the receiver corrects the error and changes to "detection" mode; if the receiver detects a multiple error, the receiver changes to HEC error detection mode. If the receiver is in "detection" mode and one or more errors are detected in the header, then the cell is discarded and the receiver stays in the detection mode; if the header is error free, then the receiver changes to the "detection" mode.
      Explain why this procedure protects against bursts of errors. If the bit errors are independent, what is the probability that two consecutive headers contain single errors and hence that a cell is discarded unnecessarily?

11. What is the difference between CER and CLR? Why is one negotiated during connection setup and the other is not?

12. Why does the calculation of CER and CMR exclude blocks that are counted in the SECBR?

13. Explain how weighted fair queueing scheduling can be used to affect the CLR and CTD experienced by cells in an ATM connection.

14. Explain the effect of a single cell loss from a long packet in a situation that uses an end-to-end ARQ retransmission protocol. Can you think of strategy to deal with such losses? Can you think of a way to ameliorate the impact of packet retransmission?

15. Consider a sequence of ATM cells carrying PCM voice from a single speaker.
    a. What are the appropriate traffic descriptors for this sequence of cells, and what is an appropriate leaky bucket for policing this stream?
    b. Suppose that an ATM connection is to carry the cell streams for M speakers. What are appropriate traffic descriptors for the resulting aggregate stream, and how can it be policed?

16. Consider a sequence of ATM cells carrying PCM voice from a single speaker, but suppose that silence suppression is used.
    a. What are the appropriate traffic descriptors for this sequence of cells, and what is an appropriate leaky bucket(s) arrangement for policing this stream? Which situation leads to nonconforming cells?
    b. Suppose that an ATM connection is to carry the cell streams for M speakers. What are appropriate traffic descriptors for the resulting aggregate stream, and how can it be policed?

17. Suppose that constant-length packets (of size equal to M cells) arrive at a source to be carried by an ATM connection and that such packets are separated by exponential random times T. What are the appropriate traffic descriptors for this sequence of cells, and what is an appropriate leaky bucket(s) arrangement for policing this stream? Which situation leads to nonconforming cells?

18. Explain why each specific set of traffic descriptors and QoS parameters were selected for each of the ATM service categories.

19. Suppose that IP packets use AAL5 prior to transmission over an ATM connection. Explain the transfer-delay properties of the IP packets if the ATM connection is of the following type: CBR, rt-VBR, nrt-VBR, ABR, or UBR.

20. Proponents of ATM argue that VBR connections provide a means of attaining multiplexing gains while providing QoS. Proponents of IP argue that connectionless IP routing can provide much higher multiplexing gains. Can you think of arguments to support each claim. Are these claims conflicting, or can they both be correct?

21. a. Consider a link that carries connections of individual voice calls using PCM. What information is required to perform call admission control on the link?

b. Now suppose that the link carries connections of individual voice calls using PCM but with silence suppression. What information is required to do call admission control?

22. Suppose that an ATM traffic stream contains cells of two priorities, that is, high-priority cells with CLP = 0 in the headers and low-priority cells with CLP = 1.
   a. Suppose we wish to police the peak cell rate of the CLP = 0 traffic to p_0 as well as the peak rate of the combined CLP = 0 and CLP = 1 traffic to P_{0+1}. Give an arrangement of two leaky buckets to do this policing. Nonconforming cells are dropped.
   b. Compare the following policing schemes: (1) police CLP = 0 traffic to peak rate p_0 and police CLP = 1 traffic to peak rate p_1, (2) police the combined CLP = 0 and CLP = 1 traffic to peak rate $p_0 + p_1$. Which approach is more flexible?
   c. Repeat part (a) if nonconforming CLP = 0 cells that do not conform to the p_0 are tagged by changing the CLP bit to 1. Cells that do not conform to p_{0+1} are dropped.

23. Suppose that an ATM traffic stream contains cells of two priorities, that is, high-priority cells with CLP = 0 in the headers and low-priority cells with CLP = 1.
   a. Suppose we wish to police the sustainable cell rate of the CLP = 0 traffic to SCR_0 and BT as well as the peak rate of the combined CLP = 0 and CLP = 1 traffic to p_{0+1}. Give an arrangement of two leaky buckets to do this policing. Nonconforming cells are dropped.
   b. Repeat part (a) if cells that do not conform to SCR_0 and BT are tagged by changing the CLP bit to 1. Cells that do not conform to p_{0+1} are dropped.

24. Suppose that an ATM traffic stream contains cells of two priorities, that is, high-priority cells with CLP = 0 in the headers and low-priority cells with CLP = 1. Suppose we wish to police the peak cell rate and the sustainable cell rate of the combined CLP = 0 and CLP = 1 traffic. Give an arrangement of two leaky buckets to do this policing. Nonconforming cells are dropped.

25. Explain how weighted fair queueing might be used to combine the five ATM service categories onto a single ATM transmission link. How are the different service categories affected as congestion on the link increases?

26. a. Discuss what is involved in calculating the end-to-end CLR, CTD, and CDV.
   b. Compare the following two approaches to allocating the end-to-end QoS to per link QoS: equal allocation to each link; unequal allocation to various links. Which is more flexible? Which is more complex?

27. Suppose that an application uses the reliable stream service of TCP that in turns uses IP over ATM over AAL5.
   a. Compare the performance seen by the application if the AAL5 uses CBR connection; nrt-VBR connection; ABR connection; UBR connection.
   b. Discuss the effect on the peformance seen by the application if the ATM connection encounters congestion somewhere in the network.

28. Suppose that an ATM connection carries voice over AAL1. Suppose that the packetization delay is to be kept below 10 ms.
   a. Calculate the percentage of overhead if the voice is encoded using PCM.

b. Calculate the percentage of overhead if the voice is encoded using a 12 kbps speech-encoding scheme from cellular telephony.

29. Explain how the three-bit sequence number in the AAL1 header can be used to deal with lost cells and with misinserted cells.

30. How much delay is introduced by the two interleaving techniques that can be used in AAL1?

31. a. How many low-bit-rate calls can be supported by AAL2 in a single ATM connection?
    b. Estimate the bit rate of the connection if an AAL2 carries the maximum number of calls carrying voice at 12 kbps.
    c. What is the percentage of overhead in part (b)?

32. Compare the overhead of AAL3/4 with that of AAL5 for a 64K byte packet.

33. Discuss the purpose of all the error checking that is carried out at the end systems and in the network for an ATM connection that carries cells produced by AAL3/4. Repeat for AAL5.

34. Suppose that in Figure 9.17 packets from A and B arrive simultaneously and each produces 10 cells. Show the sequence of SPDUs produced including segment type, sequence number, and multiplexing ID.

35. Consider SSCOP, the AAL protocol for signaling. Discuss the operation of the Selective Repeat ARQ procedure to recover from cell losses. In particular, discuss how the protocol differs from the Selective Repeat ARQ protocol introduced in Chapter 5.

36. Can the SSCOP AAL protocol be modified to provide the same reliable stream service that is provided by TCP? If yes, explain how; if no, explain why not.

37. The "cells in frames" proposal attempts to implement ATM on workstations attached to a switched Ethernet LAN. The workstation implements the ATM protocol stack to produce cells, but the cells are transmitted using Ethernet frames as follows. The payload of the Ethernet frame consists of a four-byte CIF header, followed by a single ATM header, and up to 31 ATM cell payloads.
    a. Find the percentage of overhead of this approach and compare it to standard ATM.
    b. The workstation implements ATM signaling, and the NIC driver is modified to handle several queues to provide QoS. Discuss the changes required in the Ethernet switch so that it can connect directly to an ATM switch.

38. Compare the size of the address spaces provided by E-164 addressing, AESA addressing, IPv4 addressing, IPV6 addressing, and IEEE 802 MAC addressing.

39. Can IP addresses be used in ATM? Explain why or why not?

40. Identify the components that contribute to the end-to-end delay experienced in setting up an ATM connection using PNNI.

41. Describe the sequence of DTLs that are used in setting up a connection from A.1.3 to A.2.2 in Figure 9.26. Repeat for a connection from B.4 to A.1.2.

42. a. Discuss the differences and similarities between PNNI and OSPF.
    b. Can PNNI be modified to provide QoS routing in the Internet? Explain why or why not?

43. Compare the hierarchical features of the combination of BGP4 and OSPF with the features of PNNI.

44. Which aspects of the ATM network architecture depend on the fixed-length nature of ATM cells? What happens if ATM cells are allowed to be variable in length?

45. Which aspects of ATM traffic management change if ATM connections must belong to one of a limited number of classes and if QoS is guaranteed not to individual connections but the class as a whole? Can VPs play a role in providing QoS to these classes?

46. Explain how the ATM architecture facilitates the creation of multiple virtual networks that coexist over the same physical ATM network infrastructure but that can be operated as if they were separate independent networks. Explain how such virtual networks can be created and terminated on demand.

CHAPTER 10

Advanced Network Architectures

In the previous two chapters we examined two key network architectures: (1) IPv4 and its next version based on IPv6 and (2) ATM. IP was designed to provide internetworking functions capable of running over a variety of network technologies. ATM was designed to provide end-to-end Quality-of-Service (QoS) support for a wide range of services. In this chapter we consider proposed enhancements to IP that provide QoS, as well as the interworking of IP and ATM.

We introduce several advanced networking technologies that have been specified in the ATM Forum as well as by the IETF. Some of these technologies will gradually be incorporated into emerging networks, thus helping define the next generation of networks. The chapter is organized as follows:

1. *IP forwarding architectures.* We introduce a forwarding taxonomy for IP networks that is useful in understanding approaches to obtaining high performance. The classification is based on destination-based forwarding and switched forwarding. The switched-forwarding category is further classified into an overlay model and a peer model.
2. *Overlay model.* We present several variations of the overlay model, namely, classical IP over ATM, LAN emulation, Next-Hop Resolution Protocol, and multiprotocol over ATM.
3. *MPLS.* Out of possible variations of the peer model, we present multiprotocol label switching (MPLS), which constitutes the most promising example.
4. *Integrated services for the Internet.* We introduce the integrated services model for the Internet and describe two services that have been standardized: guaranteed service and controlled-load service.
5. *RSVP.* We present the Resource Reservation Protocol (RSVP) which is used to set up resource reservations along the intended path of a packet flow.

6. *Differentiated services.* We describe the differentiated services model that has been introduced by IETF as a scalable approach to provide levels of QoS support in the Internet.

10.1 IP FORWARDING ARCHITECTURES

The Internet has been growing at an exponential rate in terms of the number of hosts and domains, and traffic demand. This high growth in traffic demand has been stressing existing infrastructures in the core networks. To maintain their network performance and avoid congestion collapse, service providers have been constantly upgrading their backbone links, typically with ATM technology. When transmission links are upgraded, routers and switches may have to be upgraded as well to keep up with increasing link speeds. This situation has led to changes in router architecture to extend the achievable performance. A number of IP forwarding solutions have resulted, and they can be categorized as shown in Figure 10.1.

Category 1 retains the same forwarding paradigm as the conventional router architecture but solves the potential bottlenecks in traditional routers by modifying the internal architecture. For example, one change is the replacement of the bus backplane in the data paths with a switch backplane, thus providing simultaneous packet forwarding. In addition, scalability is improved by placing an IP lookup engine at each interface. This approach is incorporated in the design of gigabit routers.

Category 2 simplifies the lookup process by using short, fixed-length labels rather than long, variable-length IP prefixes. A typical way to simplify this process is to run IP over ATM, which uses the VCI and VPI in the lookup process. Because label lookup uses direct indexing, it can be easily performed in hardware. For example, label lookup in ATM simply uses the incoming VPI/ VCI value as a direct index to a connection table entry to determine the outgoing VPI/VCI, the output port, and other relevant information. ATM label lookup can be easily done in one cell time. On the other hand, IP address lookup has

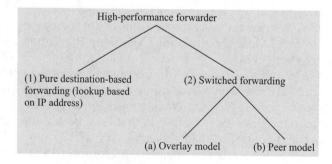

FIGURE 10.1 IP forwarding taxonomy

typically required implementation in software or firmware. Category 2 can be further classified into the *overlay model* and the *peer model*.

In the **overlay model**, ATM switches are not aware of IP addresses and IP routing protocols. This model overlays an IP network onto an ATM network, essentially creating two network infrastructures with two addressing schemes and two routing protocols. Each end system uses both IP and ATM addresses that are uncoupled. Thus an address resolution protocol is required to map from one address to another. One advantage of this model is that the ATM infrastructure can be developed independently of the IP infrastructure. Examples of this model are classical IP over ATM and multiprotocol over ATM.

The **peer model** uses the existing IP addresses (or algorithmically derived ATM addresses) to identify end systems and uses IP routing protocols to set up ATM connections. One advantage of the peer model is that it does not require an address resolution protocol to interwork routeable address spaces and thus simplifies address administration. A node typically has an integrated ATM switching and IP routing function, and the node can be viewed as a "peer" to other routers. The peer model maintains one network infrastructure. The best example of this model is multiprotocol label switching.

10.2 OVERLAY MODEL

IP is the dominant internetworking layer, while ATM is perceived as an economical switching solution for high-speed backbone networks. For this reason, there has been much interest in overlaying IP internetworking protocol on top of ATM. In this section we look at three IP-over-ATM approaches.

10.2.1 Classical IP Over ATM

The **classical IP over ATM (CLIP)** model [RFC 2255] is an IETF specification whereby IP treats ATM as another subnetwork to which IP hosts and routers are attached. In the CLIP model multiple IP subnetworks are typically overlaid on top of an ATM network. The part of an ATM network that belongs to the same IP subnetwork is called a **logical IP subnetwork (LIS)**, as shown in Figure 10.2. All members (IP end systems) in the same LIS must use the same IP address prefix (e.g., the same network number and subnet number). Two members in the same LIS communicate directly via an ATM virtual channel connection (VCC).

Each LIS operates and communicates independently of other LISs on the same ATM network. Communications to hosts outside the LIS must be provided via an IP router that is connected to the LIS. Therefore, members that belong to different LISs must communicate through router(s).

Suppose a host (host S) wants to use CLIP to send packets to another host (host D). When host S sends the first packet to host D in the same LIS, host S

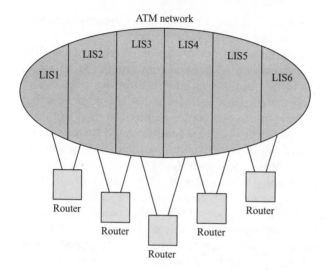

ATM network

FIGURE 10.2 Classical IP over ATM model

knows only the IP address of host D. To set up a VCC, host S needs to know the ATM address of host D. How does host S resolve the ATM address of host D from the IP address?

The solution is provided by implementing an *ATM Address Resolution Protocol* (ATM ARP) server on each LIS. The ATM address of the ATM ARP server is configured at each host. When a host boots up, it registers its IP and ATM addresses to the ATM ARP server on the same LIS. When a host wants to resolve the ATM address of another host from the IP address, the first host asks the ATM ARP server for the corresponding ATM address. After the host receives an ATM ARP reply from the ATM ARP server, the host can establish a VCC to the destination host and send packets over the VCC. This process involves fragmenting the IP packet into ATM cells at the source host and the reassembly of the packet at the destination host.

What happens if the destination host belongs to another LIS? In this situation the source host simply establishes a VCC to the router connected to the same LIS. The router examines the IP packet, determines the next-hop router, establishes a VCC, and forwards the packet along to the next router. The process is continued until the router of the LIS of the destination is reached, and the packet is then delivered to the destination host.

In CLIP, IP packets sent from the source host to the destination host in a different LIS must undergo routing through the LIS router, even if it is possible to establish a direct VCC between the two IP members over the ATM network. This requirement precludes the establishment of a VCC with a specific QoS between end systems.

CLIP allows a permanent virtual connection (PVC) to be established between hosts of a LIS. In this case the connection is preestablished manually between two hosts. What one host needs to find is the IP address of the other end. The host uses an inverse ATMARP (InATMARP) to find the IP address of the other end.

10.2.2 LANE

LAN emulation (LANE) is an ATM Forum specification intended to accelerate the deployment of ATM in the enterprise network. Typically, a host runs an internetwork layer protocol such as IP over a "legacy" LAN such as Ethernet or token ring. LANE enables any software that runs on a legacy LAN to also run on an ATM network without any modification. LANE works by presenting the network layer with an interface that is identical to that of legacy LANs. Figure 10.3 illustrates the changes in the lower layers such that the interface from the device driver to the network layer (e.g., NDIS[1]) remains unchanged. LANE maintains the same interface between the network layer and the data link layer, so in effect an ATM network can be made to appear like an Ethernet or token ring LAN to the higher layer software. This behavior also enables LANE to support other network layer protocols such as IPX and AppleTalk. In contrast, the CLIP model supports IP only.

An **emulated LAN (ELAN)** consists of the following components (see Figure 10.4):

- A set of **LAN emulation clients (LECs)**
- **LAN emulation server (LES)**
- **Broadcast and unknown server (BUS)**
- **LAN emulation configuration server (LECS)**

A LEC resides in the end system (e.g., host, server, bridge, etc.) and performs data forwarding, address resolution, and control functions. Each LEC is identified by a unique ATM address. A LES responds to LEC address resolution requests by resolving MAC addresses to ATM addresses. A BUS handles broadcast, multicast and initial (i.e., before a VCC is established) traffic in a given ELAN. One main purpose of the LAN Emulation Configuration Server (LECS) is to assign LECs to the corresponding ELANs (i.e., associate a LEC to the correct LES).

During the registration phase, each LEC notifies the LES of its ATM and MAC addresses. When a LEC (say, LEC1) wants to send a frame to another

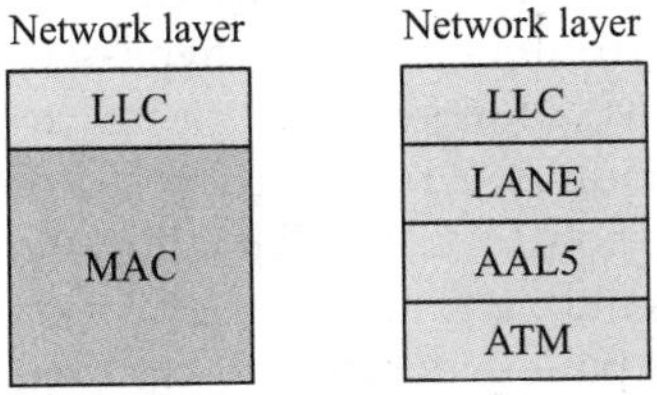

Network layer	Network layer
LLC	LLC
	LANE
MAC	AAL5
	ATM

FIGURE 10.3 Legacy LAN and LANE protocol stacks

[1]NDIS stands for network driver interface specification. Microsoft networking protocols interact with network card drivers by using NDIS. NDIS operates at the logical link control sublayer of the data link layer. NDIS allows the binding of multiple NDIS-compliant NIC cards with one protocol stack, multiple protocols with a single NIC card, or multiple protocols with multiple NICs.

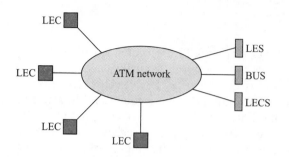

FIGURE 10.4 LANE configuration

LEC (say, LEC2), LEC1 first checks if it knows the ATM address of LEC2. If LEC1 does not know the ATM address, then it sends an LE_ARP request to the LES. In the meantime, LEC1 sends subsequent frames via the BUS. If there was an earlier registration from LEC2, the LES can resolve LEC2's MAC address to the ATM address. After receiving the LE_ARP reply, LEC1 will cache the ATM address of LEC2 and set up a VCC to LEC2. From then on frames from LEC1 to LEC2 are transmitted through the VCC. The cache is aged out so that inactive VCCs are eventually purged.

LANE has several shortcomings. LANE, by virtue of operating at the MAC layer, is susceptible to broadcast storms. The requirement that LANE hide the details of the underlying ATM network from the network layer also implies that the QoS attributes of ATM cannot be made available to the network layer protocols. LANE is defined to operate over UBR and ABR connections, which more closely match the service provided by LAN MAC protocols.

10.2.3   NHRP

LANE enables a station to resolve an ATM address from a MAC address. The **Next-Hop Resolution Protocol (NHRP)** enables a station connected to an ATM network to resolve an ATM address from an IP address. NHRP allows a host to determine the ATM address of another host or of an egress router from the ATM network.

The main objective of NHRP is to find the most efficient shortcut path through the ATM network so that intermediate routers can be bypassed. Recall that the CLIP model resolves only the ATM address that belongs to the same LIS. In other words, the CLIP model requires a router to perform packet forwarding between two different LISs. In contrast, NHRP allows a shortcut to traverse multiple LISs, making it more suitable for larger networks. Figure 10.5 illustrates the key difference between the path generated by classical IP (default path or routed path) and the path generated by NHRP (shortcut path or cut-through path).

NHRP is based on a client/server architecture. An NHRP cloud contains entities called **next-hop clients (NHCs)**, which are responsible for initiating NHRP resolution request packets, and **next-hop servers (NHSs)**, which are

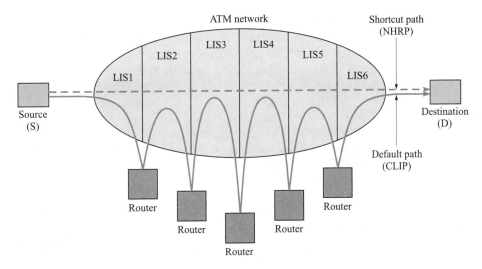

FIGURE 10.5 NHRP and CLIP compared

responsible for answering NHRP resolution requests by means of NHRP replies. Both NHC and NHS maintain an address resolution cache. An NHC in NHRP basically replaces an ATMARP client in the classical IP model, and an NHS basically replaces an ATMARP server.

The NHRP protocol works as follows. When node S (which contains an NHC) wants to transmit data to the destination node D, S may want to send the data through a VCC. If S does not know the ATM address of D, station S first sends an NHRP resolution request along the routed path.[2] The NHRP resolution request packet contains node D's IP address, node S's IP address, and node S's ATM address.

Each router along the routed path has an NHS. When an NHRP resolution request packet arrives at a router, the NHS determines if the router is the "serving NHS" for node D. A serving NHS for node D has one of its router's interfaces connected to the same LIS as node D. If the NHS is not serving node D, it will reinitiate an NHRP resolution request packet to the next NHS along the path to the destination. The process continues until the NHRP resolution request reaches the NHS that is the serving node D. This NHS then resolves node D's ATM address and sends a positive NHRP resolution reply packet back to node S. The NHRP resolution reply packet contains node D's ATM address. If a negative reply is received, S should cache this information so it will not retrigger a resolution request packet. By caching the negative reply, station S knows that data packet must follow the routed path.

A transit NHS that relays the NHRP resolution reply may cache the IP-to-ATM address binding information. When a subsequent NHRP resolution

---

[2]That is, the path using routers between LISs.

request arrives at this transit NHS, it may reply using the cached information. However, such a reply must be identified as *nonauthoritative*. Only the serving NHS can respond with an NHRP *authoritative* resolution reply. In general, nonauthoritative reply speeds up the address resolution process. However, this service comes at the expense of increasing the cache size requirement at the NHS. Also, when the IP-to-ATM address binding at the destination changes, a transit NHS will respond with a wrong resolution reply.

NHS may optionally support address aggregation as follows. The NHS can return not only the address of the destination that is sought but also the subnet mask associated with that address. This practice allows intermediate NHSs that cache this information to reply to requests for all IP addresses with the same prefix. Timers are used to remove stale information from the caches.

While waiting for the NHRP resolution reply, S has three options: (1) drop the data packet, (2) retain the packet until the reply comes back, or (3) forward the packet along the routed path toward node D. The NHRP protocol recommends the third option in order to reduce latency.

10.2.4   MPOA

Multiprotocol over ATM (MPOA) was designed by the ATM Forum to provide internetworking service such as IP, IPX, and AppleTalk over an ATM network. MPOA basically integrates LANE and NHRP. MPOA uses LANE to establish layer 2 connections (i.e., within the same LIS) and NHRP to establish layer 3 connections (i.e., across different LISs). Thus MPOA enables direct ATM connections between MPOA devices that can then use ATM's QoS capabilities.

Figure 10.6 shows an example of an MPOA network configuration with a number of MPOA edge devices (EDs) (the figure shows only two EDs for simplicity). The ED forwards packets between legacy systems and the ATM network. The ED contains an **MPOA client (MPC)** that is used primarily to source and sink shortcut VCCs. An MPOA network may also contain MPOA-capable devices that are directly connected to the ATM network and that include an MPC.

Each router contains an **MPOA server (MPS)** that relies on an NHRP server (NHS) to perform address resolution between IP addresses and ATM addresses.[3] MPC and MPS communicate via an emulated LAN (ELAN).

Suppose that Host1 has data to send to Host2 and that no VCC between MPC1 and MPC2 exists yet. The packets would first follow the path through MPC1 (ingress MPC), through each router along the default path (see Figure 10.6), through MPC2 (egress MPC), and finally to Host2. This process is not an efficient way to transfer data in an ATM network, since each packet has to go through a router when the packet goes to a different subnet.

---

[3]Alternatively, an MPS can reside in a stand-alone ATM-attached route server.

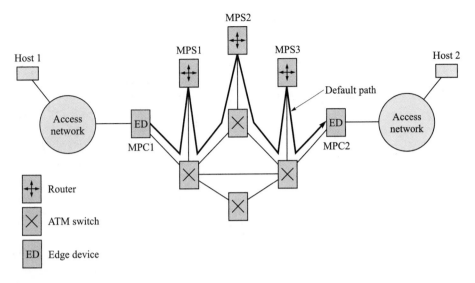

FIGURE 10.6 Example of network topology

When the ingress MPC detects a long-lived flow (i.e., x frames in a given time interval are observed), that MPC will try to establish a shortcut VCC. To do so, the ingress MPC first has to know the ATM address of the egress MPC. If the ingress MPC does not know the ATM address, that MPC would have to make a resolution request all the way to the egress MPC as shown in Figure 10.7. First, the ingress MPC sends an MPOA resolution request to the ingress MPS (step 1). Because the ingress MPS and the egress MPC do not belong to the same subnet, the ingress MPS sends an NHRP resolution request to the next hop (step 2),

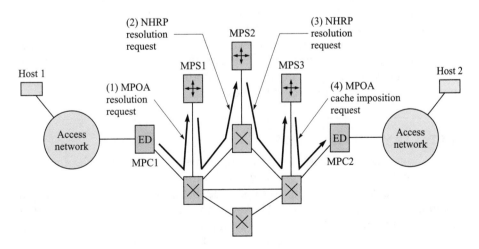

FIGURE 10.7 Request process

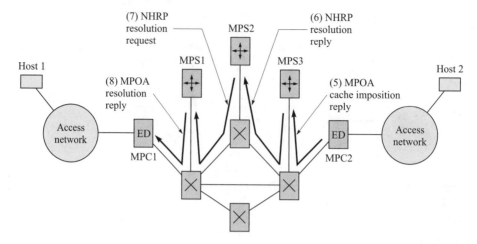

FIGURE 10.8 Reply process

which forwards the request to egress MPS (step 3). Recognizing that the egress MPC belongs to the same subnet, the egress MPS now asks the egress MPC whether it has enough resources to support the connection through an MPOA cache imposition request (step 4).

If the egress MPC has enough resources to accept the connection, that MPC sends an MPOA cache imposition reply to the egress MPS that contains the ATM address of the egress MPC (step 5), as shown in Figure 10.8. The egress MPS then translates the MPOA reply to NHRP resolution reply and forwards it to MPS2 (step 6), which forwards it to the ingress MPS (step 7). Finally, the ingress MPS sends an MPOA resolution reply to the ingress MPC (step 8), containing the ATM address of the egress MPC.

Once the ingress MPC knows the ATM address of the egress MPC, the ingress MPC sets up a VCC and transfers the subsequent packets through the VCC, as shown in Figure 10.9. If there is no more data for that flow, the VCC will eventually be terminated through aging.

The persistent setups and tear-downs of VCCs (due to new flows being detected and existing flows becoming inactive) tend to stress the call processing function in ATM switches. Because of this limitation, MPOA does not scale well in a core network that supports many flows and edge devices.

IS ATM LAYER 2 OR LAYER 3?

Throughout this text we have used the OSI convention that identifies the layer 3 (network layer) as the layer that involves routing and switching from source to destination. ATM was designed to provide this capability, and when used in this manner, ATM is in layer 3. However, in the next section we show that MPLS uses the information from IP routing to assign labels and set tables in ATM and other switches. In this context, ATM is in layer 2 (link layer).

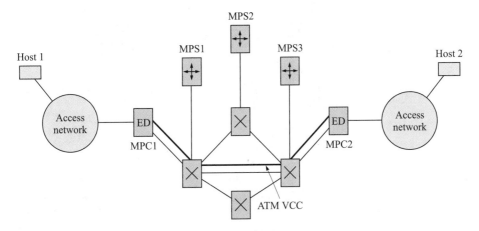

FIGURE 10.9 Data transfer through shortcut VCC

10.3 MPLS

ATM introduced the use of label switching to enable fast forwarding of cells across a network. Label switching provides a low-cost hardware implementation, scalability to very high speeds, and flexibility in the management of traffic flows. For these reasons, IP over ATM networks provided the bandwidth in the network backbone that was needed to meet the growth in Internet traffic in the late 1990s. However, the overlay model of IP-over-ATM has the disadvantage that two network infrastructures need to be managed, each with its own addressing, routing, and management concerns. This is unnecessary in a network that carries mostly IP traffic. Consequently, several peer model approaches to integrating IP and ATM were proposed in the 1990s [Davie 1998]. In the peer model, IP routing and addressing set up ATM flows and only a single network infrastructure needs to be managed.

The IETF established the **MultiProtocol Label Switching** (MPLS) working group to standardize a label-switching paradigm that integrates layer 2 switching with layer 3 routing.[4] The device that integrates routing and switching functions is called a **Label-Switching Router (LSR)**. A key feature of MPLS is the separation of the control and forwarding components in an LSR as shown in Figure 10.10. MPLS assumes that LSRs build and maintain forwarding tables based on information that is exchanged with other routers (through OSPF or BGP), and on control information that may be received through label distribution and setup protocols. The separation of control and forwarding components allows each component to be developed and modified independently. We will show that this separation gives network providers great flexibility in defining the services they offer.

---

[4]The discussion in this section is based on "work in progress" of the IETF MPLS working group.

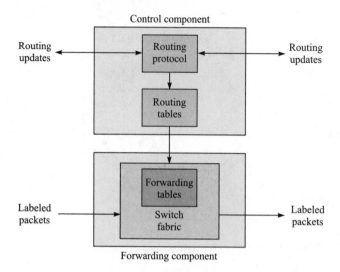

FIGURE 10.10 MPLS separates control and forwarding components

MPLS is initially focused on the IPv4 and IPv6 layer 3 protocols, but may be extended later to include IPX, DECnet, AppleTalk, and so on. Additionally, MPLS is to operate with multiple layer 2 technologies such as ATM, frame relay, PPP and Ethernet. Label switching is expected to improve the forwarding performance through a simplified lookup process, improve scalability through "label stacking" and "merging," and provide traffic engineering via efficient explicit routing. We explain these anticipated improvements in the following sections.

10.3.1 Fundamentals of Labels

A router performs a forwarding process via a "longest prefix match" (i.e., matching the longest prefix of an IP destination address to the entries in a routing table). In MPLS the forwarding process can be simplified by indexing directly into a forwarding table. This behavior is made possible by using a *short fixed-length label*, rather than an IP destination address. A **label** is a form of shorthand for the packet header that simplifies the forwarding decision a node would make for the packet. Unlike an IP address that identifies a specific host or router, a label identifies a virtual circuit between two neighboring LSRs, and the meaning of a label is significant only between the two neighboring LSRs. In the case of ATM, the label is encoded in the VCI and/or VPI fields. A sequence of LSRs that is to be followed by a packet is called a **label-switched path (LSP)**. Thus an LSP is analogous to a virtual channel connection in ATM, except that LSP is unidirectional. The direct benefits derived from the simplified forwarding process by means of label lookup are improved forwarding performance and lower cost.

The label-based forwarding will make use of information in the label and possibly other header information such as a time-to-live field or a class-of-service field. In some cases the label will be provided by existing layer 2 headers (e.g., VPI/VCI in ATM), and in other cases the label will be provided by an MPLS header. MPLS is intended to operate over various layer 2 media, so there will be different types of MPLS headers.

Figure 10.11a shows that the VPI/VCI field in an ATM cell can be used as a label. For layer 2 protocols that do not support a label, such as PPP or Ethernet, an MPLS header is inserted between the layer 2 and IP headers as shown in Figure 10.11b. The 32-bit MPLS header contains a 20-bit label field, a 3-bit class-of-service field, a 1-bit hierarchical stack field, and an 8-bit time-to-live field.

A group of packets that are forwarded in the same manner are said to belong to the same **forwarding equivalence class (FEC)**. An FEC may have many granularities. At one end of the spectrum, an FEC could be associated with a particular application for a particular source and destination host pair. At the other end, an FEC could be associated with an egress LSR (the LSR that handles traffic as it leaves an MPLS domain). We show that this latter extreme involves the aggregation of many traffic flows into an inverted tree rooted at the egress LSR. These different levels of granularity allow MPLS to be used in a wide range of situations.

Label assignment refers to the process of allocating a label and binding a label to an FEC. Label assignment can be driven by control traffic (*topology-driven label assignment* or *request-driven label assignment*) or by data traffic (*traffic-driven label assignment*). In topology-driven label assignment, labels are assigned in response to normal processing of routing protocol control traffic. For example, as an LSR processes OSPF or BGP updates, it makes changes to entries in the routing table and triggers assignments of labels to those entries. Topology-driven label assignment minimizes the call processing power required to set up and tear down LSPs and eliminates the setup latency.

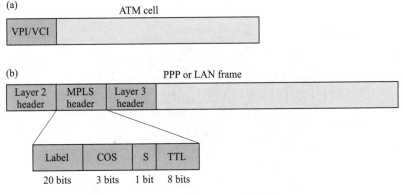

FIGURE 10.11 MPLS labels in ATM and PPP/LAN

In request-driven label assignment, labels are assigned in response to the processing of request control traffic. For example, as RSVP messages are processed, the LSR can make changes to entries in its forwarding tables and assign labels to the entries.

In traffic-driven label assignment, the assignment and distribution of labels are triggered when an LSR identifies patterns in the traffic that justify the use of labels. Traffic-driven label assignment has the advantage of avoiding full-meshed LSPs by establishing the LSPs on demand only when there is traffic to send. However, a latency is incurred between the appearance of a flow and the assignment of a label to the flow. Moreover, traffic-driven label assignment tends to consume high call-processing power to set up and tear down LSPs.

10.3.2 Label Stack

Labels are attached and removed from packets as follows. When a packet enters a particular domain, the ingress LSR creates a new label for the packet by performing a "label push." Subsequent LSRs in the domain only swap the incoming label to the outgoing label. When the packet leaves the domain, the egress LSR performs a "label pop."

It is possible for a packet to have a stack of m labels (or m levels). In a given domain the label at the top of the stack (depth 1) is the only one that determines the forwarding decision. A packet with an empty stack (depth 0) is called an unlabeled packet. An LSR forwards unlabeled packets at the IP layer.

Figure 10.12 illustrates the use of a label stack. Suppose LSR A receives an unlabeled packet from a normal layer 3 router destined for LSR G. First, A pushes on a label from B. B replaces the incoming label with the label that is agreed upon by F. B also pushes on a label from C. C replaces the incoming label with the label from D. Similarly, D replaces the label with the label from F. F pops the label stack and replaces the remaining label with the label from G. G pops the label stack and forwards the unlabeled packet by using normal layer 3 forwarding. As far as MPLS domain 1 is concerned, the LSP is $<A, B, F, G>$,

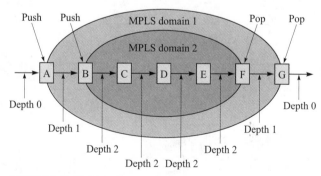

FIGURE 10.12 MPLS tunneling

while as far as MPLS domain 2 is concerned, the LSP is < B, C, D, E, F >. The actual physical path taken by the packet is < A, B, C, D, E, F, G >.

Label stacking has a number of uses. For example, this feature is useful when two labels are used for IGP and BGP.[5] The BGP label is used to forward packets from one BGP speaker to another BGP speaker, while the IGP label is used to forward packets within an autonomous system. The use of MPLS allows the routers internal to a transit routing domain operate without knowledge of the BGP routing information.

10.3.3   VC Merging

Topology-driven label assignment generally requires full-meshed LSPs to be established among the edge LSRs. This requirement puts heavy demand on the usage of labels at each LSR. The number of labels used can be significantly reduced if "merging" is implemented. In such a situation, if multiple incoming streams at a given LSR are going to the same egress LSR, the incoming labels for the incoming streams will be swapped to the same outgoing label. From the point of view of an egress LSR, the merging of streams can be viewed as creating a multipoint-to-point connection rooted at the egress LSR. Note that conventional IP datagram forwarding merges packet flows into just such a multipoint-to-point tree.

In the nonmerging case a multipoint-to-point connection will be emulated as a multiple point-to-point connections. However, connection table entries are wasted, since the table requires $O(n^2)$ entries to support n edge LSRs. Thus the nonmerging approach does not scale well and can be used only for small networks.

In the merging case incoming labels intended for the same egress LSR are translated to the same outgoing label, as illustrated in Figure 10.13. This method reduces the connection table entries essentially from $O(n^2)$ to $O(n)$. This property gives MPLS a great degree of scalability and makes it appropriate for very large domains.

When ATM is used, incoming VCs may be merged to the same outgoing VC. The consequence of "VC merging" is that cells intended for the same egress LSR become indistinguishable at the output of a switch, as they carry the same label. Therefore, cells belonging to different packets for the same egress LSR cannot be interleaved because the receiver will not be able to reassemble the packets. If cells within the same packet are not to be interleaved with others intended for the same egress LSR, then the cells should first be reassembled by using the "end of packet" bit accommodated in AAL5 before they can be merged. Figure 10.14 shows how VC merging reshuffles the output streams to ensure that cells never

---

[5]Interior Gateway Protocol (IGP) and Border Gateway Protocol (BGP) are discussed in Chapter 8.

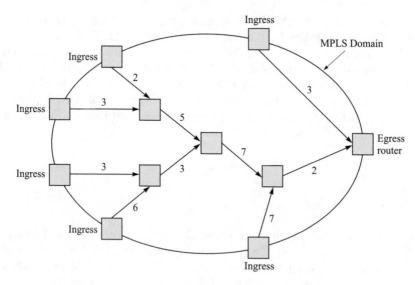

FIGURE 10.13 Labels for the same egress LSR are merged at each LSR

interleave. Packet reassembly for VC merging requires an additional amount of buffering. It turns out that VC merging incurs only a minimal buffer overhead.

10.3.4 Label Distribution Protocol

Label Distribution Protocol (LDP) is the protocol used to distribute label bindings. Label bindings between two LSRs can be distributed by an upstream LSR or a downstream LSR. LDP dictates that a downstream LSR distributes label bindings to an upstream LSR. Figure 10.15 illustrates the label distribution process. When LSR 1 (the upstream LSR) detects that LSR 2 (the downstream LSR) is its next hop for FEC = 10.5/16, LSR 1 sends a label request message to

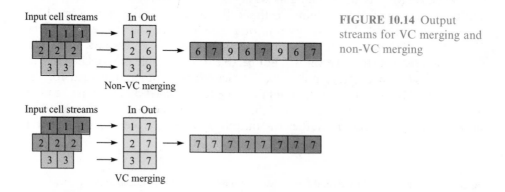

FIGURE 10.14 Output streams for VC merging and non-VC merging

FIGURE 10.15 Label distribution

LSR 2. Upon receiving the message, LSR 2 responds with a label binding message that specifies the label-to-FEC binding.

In this example, called the *downstream-on-demand* mode, LSR 2 distributes a label binding in response to an explicit request from LSR 1. Another mode, called the *downstream (unsolicited)* mode, allows LSR 2 to distribute a label binding even if it has not been explicitly requested.

LDP allows LSP setups to be initiated in two ways. In *ordered control* an LSR distributes a label-to-FEC binding only if the LSR is the egress LSR for that FEC or the LSR has already received a label binding for that FEC from its next hop. In *independent control* each LSR independently binds a label to an FEC and distributes the binding to its peer.

10.3.5   Explicit Routing for Traffic Engineering

Typically, the route taken by a packet in the Internet is determined hop by hop by each router according to the shortest-path objective. In explicit routing the route taken by a packet is determined by a single node, usually the ingress LSR. MPLS facilitates explicit routing, since the sequence of LSRs to be followed need not be carried in the packet header as in conventional datagram networks.

One useful application of explicit routing is traffic engineering that is intended to maximize resource utilization in the network. Figure 10.16 illustrates how resources may be inefficiently utilized by using hop-by-hop routing. Node 3 assumes that the shortest path to node 8 is via node 4. Thus node 3 forwards traffic from nodes 1 and 2 that are destined to node 8 through node 4. This action may cause some links to be congested, while others are lightly loaded.

Figure 10.17 shows how node 2 may decide to establish the explicitly routed LSP through nodes 3, 5, 7, and 8 if it determines that link 4–6 is already heavily utilized. We see that explicit routing allows traffic to be mapped to the physical topology of the network in a more flexible way than hop-by-hop routing.

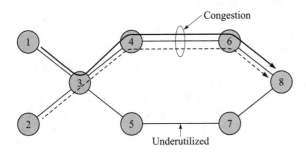

FIGURE 10.16 Inefficient use of resources caused by hop-by-hop routing

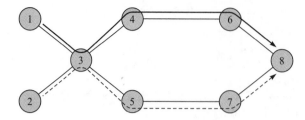

FIGURE 10.17 Traffic engineering using explicit routing

Traffic engineering can be done adaptively according to the state of the network. In one example multiple explicitly routed LSPs are set up between ingress and egress LSRs. Based on the states of the LSPs, the ingress LSR performs load balancing to use the network resources efficiently.

10.3.6   MPLS Support for Virtual Networks

MPLS allows service providers to configure LSPs to create **virtual networks** that support particular classes of traffic flows. For example, in a later section we discuss differentiated services IP which define expedited forwarding service and assured service, in addition to best effort service. In principle an ISP could

---

FROM TELEGRAMS TO CIRCUITS TO DATAGRAMS TO VIRTUAL CIRCUITS TO...

In Chapter 1 we saw that the telegraph network used message switching, a special case of datagram packet switching. The invention of analog voice transmission led to a telephone network based on the switching of circuits. The proliferation of computers in the 1980s triggered extensive research and development into packet-switching technologies of two basic forms: datagram and virtual circuit. The simplicity and robustness of IP led to the growth of the datagram-based Internet. ATM based on virtual-circuit packet switching introduced low-cost, high-performance switching, and in MPLS, we see the use of a virtual-circuit approach as a means of optimizing the core IP network.

We have witnessed the evolution of communication networks from packet switching to circuit switching and then to packet switching again. We have also witnessed the tension between virtual-circuit and datagram packet switching and interestingly their apparent coalescence. Thus we see that the balance of the prevailing technologies becomes manifest in network architectures that can favor one extreme or the other or even a blend of these.

Vinton Cerf made the remark, "Today: you go through a circuit switch to get to a packet switch. Tomorrow: you go through a packet switch to get to a circuit switch." Hmmm . . . What about the day after tomorrow?

configure a separate LSP to carry each class of traffic between each pair of edge LSRs. A more practical solution is to merge LSPs of the same traffic class to obtain multipoint-to-point flows that are rooted at an egress LSR. The LSRs serving each of these flows would be configured to provide the desired levels of performance to each traffic class.

MPLS can also be used to create **Virtual Private Networks (VPNs)**. A VPN provides wide-area-connectivity to a large multilocation organization. MPLS can provide connectivity between VPN sites through LSPs that are dedicated to the given VPN. The LSPs can be used to exchange routing information between the various VPN sites, transparently to other users of the MPLS network, thus giving the appearance of a dedicated wide area network. VPNs involve many security issues related to ensuring privacy in the public or shared portion of a network. These issues are discussed in Chapter 11.

10.4 INTEGRATED SERVICES IN THE INTERNET

Traditionally, the Internet has provided best-effort service to every user regardless of its requirements. Because every user receives the same level of service, congestion in the network often results in serious degradation for applications that require some minimum amount of bandwidth to function properly. As the Internet becomes universally available, there is also interest in providing real-time service delivery to applications such as IP telephony. Thus an interest has developed in having the Internet provide some degree of QoS.

To provide different QoS commitments, the IETF developed the **integrated services model** that requires resources such as bandwidth and buffers to be explicitly reserved for a given data flow to ensure that the application receives its requested QoS. The model requires the use of *packet classifiers* to identify flows that are to receive a certain level of service as shown in Figure 10.18. It also requires the use of *packet schedulers* to handle the forwarding of different packet flows in a manner that ensures that QoS commitments are met. *Admission control* is also required to determine whether a router has the necessary resources to accept a new flow. Thus the integrated services model is analogous to the ATM model where admission control coupled with policing are used to provide QoS to individual applications.

The Resource Reservation Protocol (RSVP), which is discussed later in the chapter, is used by the integrated services model to provide the reservation messages required to set up a flow with a requested QoS across the network. RSVP is used to inform each router of the requested QoS, and if the flow is found admissible, each router in turn adjusts its packet classifier and scheduler to handle the given packet flow.

A *flow descriptor* is used to describe the traffic and QoS requirements of a flow. The flow descriptor consists of two parts: a *filter specification (filterspec)* and a *flow specification (flowspec)*. The filterspec provides the information

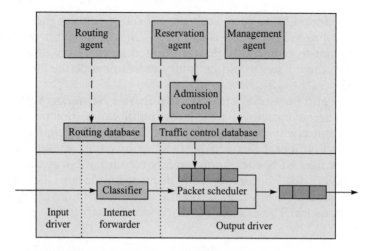

FIGURE 10.18 Router model in integrated services IP

required by the packet classifier to identify the packets that belong to the flow. The flowspec consists of a *traffic specification (Tspec)* and a *service request specification (Rspec)*. The Tspec specifies the traffic behavior of the flow in terms of a token bucket as discussed in Chapter 7. The Rspec specifies the requested QoS in terms of bandwidth, packet delay, or packet loss.

The integrated services model introduces two new services: guaranteed service and controlled-load service.

10.4.1 Guaranteed Service

The **guaranteed service** in the Internet can be used for applications that require real-time service delivery. For this type of application, data that is delivered to the application after a certain time is generally considered worthless. Thus guaranteed service has been designed to provide a firm bound on the end-to-end packet delay for a flow. How does guaranteed service provide a firm delay bound?

Recall that from Chapter 7, if a flow is shaped by a (b, r) token bucket and is guaranteed to receive at least R bits/second, then the delay experienced by the flow will be bounded by b/R with a fluid flow model, assuming that $R > r$. This delay bound has to be adjusted by error terms accounting for the deviation from the fluid flow model in the actual router or switch.

To support guaranteed service, each router must know the traffic characteristics of the flow and the desired service. Based on this information, the router uses admission control to determine whether a new flow should be accepted. Once a new flow is accepted, the router should police the flow to ensure compliance with the promised traffic characteristics.

10.4.2   Controlled-Load Service

The **controlled-load service** is intended for adaptive applications that can tolerate some delay but that are sensitive to traffic overload conditions. These applications typically perform satisfactorily when the network is lightly loaded but degrade significantly when the network is heavily loaded. Thus the controlled-load service was designed to provide approximately the same service as the best-effort service in a lightly loaded network regardless of the actual network condition. The above interpretation is deliberately imprecise for a reason. Unlike the guaranteed service that specifies a quantitative guarantee, the controlled-load service is qualitative in the sense that no target values on delay or loss are specified. However, an application requesting a controlled-load service can expect low queueing delay and low packet loss, which is a typical behavior of a statistical multiplexer that is not congested. Because of these loose definitions of delay and loss, the controlled-load service requires less implementation complexity than the guaranteed service requires. For example, the controlled-load service does not require the router to implement the weighted fair queueing algorithm.

As in the guaranteed service, an application requesting a controlled-load service has to provide the network with the token bucket specification of its flow. The network uses admission control and policing to ensure that enough resources are available for the flow. Flows that conform to the token bucket specification should be served with low delay and low loss. Flows that are nonconforming should be treated as best-effort service.

10.5   RSVP

The resource **ReSerVation Protocol (RSVP)** was designed as an IP signaling protocol for the integrated services model. RSVP can be used by a host to request a specific QoS resource for a particular flow and by a router to provide the requested QoS along the path(s) by setting up appropriate states.[6]

Because IP traditionally did not have any signaling protocol, the RSVP designers had the liberty of constructing the protocol from scratch. RSVP has the following features:

- Performs resource reservations for unicast and multicast (multipoint-to-multipoint) applications, adapting dynamically to changing group membership and changing routes.

---

[6]RSVP can be extended for use in other situations. For example RSVP has been proposed to reserve resources and install state related to forwarding in MPLS [RFC 2430].

- Requests resource in one direction from a sender to a receiver (i.e., a simplex resource reservation). Bidirectional resource reservation requires both end systems to initiate separate reservations.
- Requires the receiver to initiate and maintain the resource reservation.
- Maintains soft state at each intermediate router: A resource reservation at a router is maintained for a limited time only, and so the sender must periodically refresh its reservation.
- Does not require each router to be RSVP capable. Non-RSVP-capable routers use a best-effort delivery technique.
- Provides different reservation styles so that requests may be merged in several ways according to the applications.
- Supports both IPv4 and IPv6.

To enable resource reservations an RSVP process (or daemon) in each node has to interact with other modules, as shown in Figure 10.19. If the node is a host, then the application requiring a QoS delivery service first has to make a request to an RSVP process that in turn passes RSVP messages from one node to another. Each RSVP process passes control to its two local control modules: policy control and admission control. The policy control determines whether the application is allowed to make the reservation. Relevant issues to be determined include authentication, accounting, and access control. The admission control determines whether the node has sufficient resources to satisfy the requested QoS. If both tests succeed, parameters are set in the classifier and packet scheduler to exercise the reservation. If one of the tests fails at any node, an error notification is returned to the originated application.

In RSVP parlance a **session** is defined to be a data flow identified by its destination. Specifically, an RSVP session is defined by its destination IP address, IP protocol number, and optionally destination port. The destination IP address can be unicast or multicast. The optional destination port may be specified by a

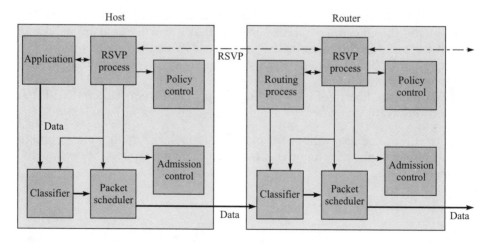

FIGURE 10.19 RSVP architecture

TCP or UDP port number. When the destination address is multicast, it is not necessary to include the destination port, since different applications typically use different multicast addresses. For multicast transmission, there may be multiple senders and multiple destinations in a group. For unicast transmission, there may be multiple senders but one destination.

An RSVP reservation request consists of a **flowspec** and a **filterspec**. The flowspec is used to set parameters in the node's packet scheduler. Generally, the parameters include a service class, an Rspec (R for reserve) that defines the requested QoS, and a Tspec (T for traffic) that describes the sender's traffic characteristics. The filterspec specifies the set of packets that can use the reservation in a given session and is used to set parameters in the packet classifier. The set of packets is typically defined in terms of sender IP address and sender port.

10.5.1 Receiver-Initiated Reservation

RSVP adopts the receiver-initiated reservation principle, meaning that the receiver rather than the sender initiates the resource reservation. This principle is similar in spirit to many multicast routing algorithms where each receiver joins and leaves the multicast group independently without affecting other receivers in the group. The main motivation for adopting this principle is that RSVP is primarily designed to support multiparty conferencing with heterogeneous receivers. In this environment the receiver actually knows how much bandwidth it needs. If the sender were to make the reservation request, then the sender must obtain the bandwidth requirement from each receiver. This process may cause an implosion problem for large multicast groups.

One problem with the receiver-initiated reservation is that the receiver does not directly know the path taken by data packets. RSVP solves this by introducing **Path** messages that originate from the sender and travel along the unicast/multicast routes toward the receiver(s). The main purposes of the Path message are to store the "path state" in each node along the path and to carry information regarding the sender's traffic characteristics and the end-to-end path properties. The path state includes the unicast IP address of the previous RSVP-capable node. The Path message contains the following information:

- **Phop**: The address of the previous-hop RSVP-capable node that forwards the Path message.
- **Sender template**: The sender IP address and optionally the sender port.
- **Sender Tspec**: The sender's traffic characteristics.
- **Adspec**: Information used to advertise the end-to-end path to receivers. The contents of the Adspec may be updated by each router along the path.

Upon receiving the Path message, the receiver sends Resv messages in a unicast fashion toward the sender along the reverse path that the data packets use. A Resv message carries reservation requests to the routers along the path. Figure 10.20 illustrates the traces of Path and Resv messages. When sender S has data to send to receiver Rx, S sends Path messages periodically toward Rx

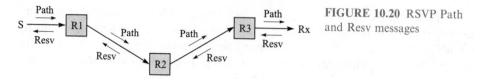

FIGURE 10.20 RSVP Path and Resv messages

along the path determined by the routing protocol. The nodes that receive the Path message record the state of the path. When Rx receives a Path message, Rx can begin installing the reservation by sending a Resv message along the reverse path. The IP destination address of a Resv message is the unicast address of a previous-hop node, obtained from the path state. RSVP messages are sent as "raw" IP datagrams with protocol 46. However, it is also possible to encapsulate RSVP messages as UDP messages for hosts that don't support the raw I/O capability.

10.5.2 Reservation Merging

When there are multiple receivers, the resource is not reserved for each receiver but is shared up to the point where the paths to different receivers diverge. From the receiver's point of view, RSVP merges the reservation requests from different receivers at the point where multiple requests converge. When a reservation request propagates upstream toward the sender, it stops at the point where there is already an existing reservation that is equal to or greater than that being requested. The new reservation request is merged with the existing reservation and is not forwarded further. This practice may reduce the amount of RSVP traffic appreciably when there are many receivers in the multicast tree. Figure 10.21 illustrates the reservation merging example. Here the reservation requests from Rx1 and Rx2 are merged at R3, which are in turn merged at R2 with the request coming from Rx3.

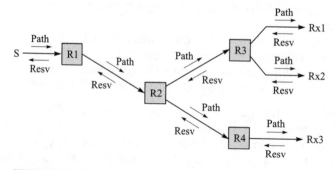

FIGURE 10.21 Merging reservations

10.5.3 Reservation Styles

Three reservation styles are defined in RSVP: wildcard filter, fixed filter, and shared explicit. Figure 10.22 shows a router configuration somewhere in a distribution tree for the purpose of distinguishing reservation styles. Three senders and three receivers are attached to the router. It is assumed that packets from S1, S2, and S3 are forwarded to both output interfaces.

The wildcard-filter (WF) style creates a single reservation shared by all senders. This style can be thought of as a shared pipe whose resource is the largest of the resource requests from all receivers, independent of the number of senders. The WF-style reservation request is symbolically represented by WF(*{Q}), where the asterisk denotes wildcard sender selection and Q denotes the flowspec. WF style is suitable for applications that are unlikely to have multiple senders transmitting simultaneously. Such applications include audio conferencing.

Figure 10.23 shows the WF-style reservation. For simplicity, the flowspec is assumed to be of one-dimensional quantity in multiples of some base resource quantity B. Interface (c) receives a request with a flowspec of 4B from a downstream node, and interface (d) receives two requests with flowspects of 3B and 2B. The requests coming from interface (d) are merged into one flowspec of 3B so that this interface can support the maximum requirement. When forwarded by the input interfaces, the requests from interfaces (c) and (d) are further merged into a flowspec of 4B.

The fixed-filter (FF) style creates a distinct reservation for each sender. Symbolically, this reservation request can be represented by FF(S1{Q1}, S2{Q2}, . . .), where Si is the selected sender and Qi is the resource request for sender i. The total reservation on a link for a given session is the sum of all Qi's.

Figure 10.24 shows the FF-style reservation. Interface (c) receives a request with a flowspec of 4B for sender S1 and 5B for sender S2 and forwards FF(S1{4B}) to (a) and FF(S2{5B}) to (b). Note that the FF-style reservation is shared by all destinations. Interface (d) receives two requests: FF(S1{3B}, S3{B}) and FF(S1{B}). Interface (d) then reserves 3B for S1 and B for S3 and forwards FF(S1{3B}) to (a) and FF(S3{B}) to (b). Interface (a) then merges the two requests received from (c) and (d) and forwards FF(S1{4B}) upstream. Interface (b) packs the two requests from (c) and (d) and forwards FF(S2{5B}, S3{B}) upstream.

The shared-explicit (SE) style creates a single reservation shared by a set of explicit senders. Symbolically, this reservation request can be represented by SE(S1, S2 . . . {Q}), where Si is the selected sender and Q is the flowspec.

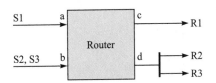

FIGURE 10.22 Example for different reservation styles

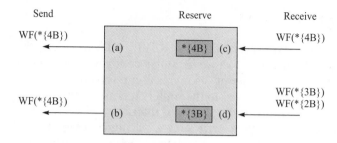

FIGURE 10.23 Wildcard-filter reservation example

Figure 10.25 shows the SE-style reservation. When reservation requests are merged, the resulting filterspec is the union of the original filterspecs and the resulting flowspec is the largest flowspec.

10.5.4   Soft State

The reservation states that are maintained by RSVP at each node are refreshed periodically by using Path and Resv messages. When a state is not refreshed within a certain time-out, the state is deleted. The type of state that is maintained by a timer is called *soft state* as opposed to hard state where the establishment and teardown of a state are explicitly controlled by signaling messages.

Because RSVP messages are delivered as IP datagrams with no reliability requirement, occasional losses can be tolerated as long as at least one of the K consecutive messages gets through. Currently, the default value of K is 3. Refresh messages are transmitted once every R seconds, where the default value is 30 seconds. To avoid periodic message synchronization, the actual refresh period for each message should be randomized, say, using a uniform distribution in the range of $[0.5R, 1.5R]$. Each Path and Resv message carries a TIME_VALUES object containing the refresh period R. From this value, a local node can determine its state lifetime L, which is $L \geq 1.5 * K * R$.[7]

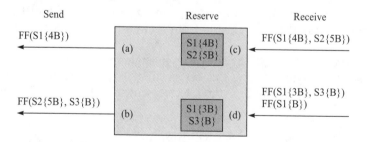

FIGURE 10.24 Fixed-filter reservation example

---

[7]The RSVP specification instead gives this formula: $L \geq (K + 0.5) * 1.5 * R$, which the author thinks is not correct.

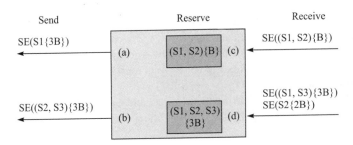

FIGURE 10.25 Shared-explicit reservation example

There is a trade-off between the refresh traffic overhead period and recovery time. The amount of refresh traffic overhead is *(Path_size + Resv_size)/R* bits/ second. A short refresh period increases the refresh traffic overhead, while a long refresh period lengthens the recovery time.

Although a time-out will eventually occur in a path or reservation state if the corresponding refresh message is absent, RSVP can speed up state removals by utilizing teardown messages. There are two types of teardown messages: **PathTear** and **ResvTear**. A PathTear message travels from the point of origin, following the same paths as the Path messages toward the receivers, deleting the path state and dependent reservation state along the paths. A ResvTear message travels from the point of origin toward the upstream senders, deleting the reservation state along the way.

10.5.5 RSVP Message Format

Each RSVP message consists of a common header and a body consisting of a variable number of objects that depend on the message type. The objects in the message provide the information necessary to make resource reservations. The format of the common header is shown in Figure 10.26.

The current protocol version is 1, and no flag bits are currently defined. The seven message types are Path, Resv, PathErr, ResvErr, PathTear, ResvTear, and ResvConf.

The RSVP checksum uses the 1s complement algorithm common in TCP/IP checksum computation. The Send_TTL represents the IP TTL value with which the message was sent. It can be used to detect a non-RSVP hop by comparing the Send_TTL value in the RSVP common header and the TTL value in the IP

0	4	8	16		31
Version	Flags	Message type	RSVP checksum		
Send_TTL		Reserved	RSVP length		

FIGURE 10.26 Format of common header

header. The RSVP length field indicates the total length of the RSVP message in octets, including the common header.

The format of the objects that follow the common header is shown in Figure 10.27. The length field indicates the total object length in octets. The length must be a multiple of four.

The Class-Num field identifies the object class. The C-Type value identifies the subclass of the object. The following object classes are defined:

NULL: A NULL object is ignored by the receiver.

SESSION: A SESSION object specifies the session for the other objects that follow. It is indicated by the IP destination address, IP protocol number, and destination port number. The session object is required in every message.

RSVP_HOP: This object carries the IP address of the RSVP-capable router that sent this message. For downstream messages (e.g., Path from source to receiver), this object represents the previous hop; for upstream messages (e.g., Resv from receiver to source), this object represents the next hop.

TIME_VALUES: This object contains the value of the refresh period R.

STYLE: This object defines the reservation style information that is not in the flowspec or filterspec objects. This object is required in the Resv message.

FLOWSPEC: This object defines the desired QoS in a Resv message.

FILTER-SPEC: This object defines the set of packets that receive the desired QoS in a Resv message.

SENDER_TEMPLATE: This object provides the IP address of the sender in a Path message.

SENDER_TSPEC: This object defines the sender's traffic characteristics in a Path message.

ADSPEC: This object carries end-to-end path information (OPWA[8]) in a Path message.

ERROR_SPEC: This object specifies errors in PathErr and ResvErr, or errors in a confirmation in ResvConf.

POLICY_DATA: This object carries policy information that enables the policy module in a node to determine whether a request is allowed or not.

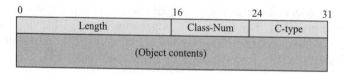

FIGURE 10.27 Format of each object

---

[8]OPWA stands for one pass with advertising. It refers to a reservation model in which downstream messages gather advertisement information that the receiver(s) can use to learn about the end-to-end service.

INTEGRITY: This object carries cryptographic and authentication information that is used to verify the contents of an RSVP message.

SCOPE: This object provides an explicit list of senders that are to receive this message. The object may be used in Resv, ResvErr, or ResvTear messages.

RESV_CONFIRM: This object carries the receiver IP address that is to receive the conformation.

RSVP messages are built from a common header followed by a number of objects. For example, the format of a *Path message* is given as follows:

```
<Path message>::=<Common Header>[<INTEGRITY>]
              <SESSION><RSVP_HOP>
              <TIME_VALUES>
              [<POLICY_DATA>...]
              [<sender descriptor>]
<sender descriptor>::=<SENDER_TEMPLATE><SENDER_TSPEC>[<ADSPEC>]
```

Another important example is the Resv message, which is given as follows:

```
<Resv message>::=<Common Header>[<INTEGRITY>]
              <SESSION><RSVP_HOP>
              <TIME_VALUES>
              [<RESV_CONFIRM>][<SCOPE>]
              [<POLICY_DATA>...]
              <STYLE><flow descriptor list>
```

The flow descriptor list depends on the reservation styles. For Wildcard Filter (WF) style, the list is

```
<flow descriptor list>::=<WF flow descriptor>
  <WF flow descriptor>::=<FLOWSPEC>
```

For Fixed FILTER (FF) style, the list is given by

```
<flow descriptor list>::=<FLOWSPEC><FILTER_SPEC>|
              <flow descriptor list><FF flow descriptor>
  <FF flow descriptor>::=[<FLOWSPEC>]<FILTER_SPEC>
```

For Shared Explicit (SE) style, the flow descriptor list is given by

```
  <flow descriptor>::=<SE flow descriptor>
  <SE flow descriptor>::=<FLOWSPEC><filter spec list>
  <filter spec list>::=<FILTER_SPEC>|
                <filter spec list><FILTER_SPEC>
```

10.6   DIFFERENTIATED SERVICES

The integrated services model was a first step toward providing QoS in the Internet. However, the integrated services model requires a router to keep a flow-specific state for each flow that the router is maintaining. This requirement raises some concerns. First, the amount of state information increases proportionally with the number of flows. Thus routers may need huge storage spaces and demand processing power. Second, the integrated services model may make the routers much more complex, since they need to implement the RSVP protocol, admission control, packet classifier, and sophisticated packet scheduling algorithms. Because of the scalability and complexity issues associated with the integrated services model, IETF has introduced another service model called the **differentiated services (DS) model**, which is intended to be simpler and more scalable. Scalability is achieved in two ways. First, per flow service is replaced with per aggregate service. Second, complex processing is moved from the core of a network to the edge.

Unlike the integrated services model that requires an application to make a resource reservation for each flow, the DS model aggregates the entire customer's requirement for QoS. A customer or organization wishing to receive differentiated services must first have a **service level agreement (SLA)** with its service provider. An SLA is a service contract between a customer and a service provider that specifies the forwarding service that the customer will receive. An SLA includes a **traffic conditioning agreement (TCA)** that gives detailed service parameters such as service level, traffic profile, marking, and shaping. An SLA can be static or dynamic. Static SLAs are negotiated on a relatively long-term basis (e.g., weekly or monthly) between the humans representing the customer and the provider. Dynamic SLAs change more frequently, and thus must use a protocol such as a "bandwidth broker" to effect SLA changes.

If a customer wishes to receive different service levels for different packets, it needs to mark its packets by assigning specific values in the type-of-service (TOS) field (renamed to the DS field) in accordance with the requested service treatments. Packet marking at the customer premises may be done at a host or at a customer's access router as shown in Figure 10.28. In the DS model, different

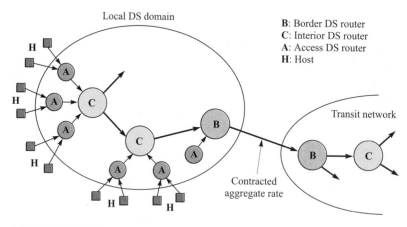

FIGURE 10.28 Router types in differentiated services

values in the DS field correspond to different packet-forwarding treatments at each router, called the **per hop behaviors (PHBs)**. Unlike the integrated services model whereby the router reserves some resources for each flow, the router in the DS model allocates resources on an aggregate basis for each PHB.

To ensure that the traffic entering the service provider's network conforms to the rules specified in the TCA, the service provider performs traffic classification and traffic conditioning at its ingress router. Traffic classification is performed by a **traffic classifier** that directs packets into various outputs based on the content of some portion of the packet header. The content may come from the DS field. It may also come from multiple fields consisting of IP source address, IP destination address, protocol ID number, source port number, and destination port number. Traffic conditioning is performed by a **traffic conditioner** that consists of metering, marking, shaping, and dropping.

10.6.1 DS Field

A DS-capable node uses a classifier to select packets based on the value of the DS field and uses buffer management and scheduling mechanisms to deliver the specific PHB based on the selection result. The value of the DS field is set at network boundaries.

The DS field is intended to supersede the existing definitions of the IPv4 TOS octet and the IPv6 traffic class octet. As shown in Figure 10.29, six bits of the DS field are used as a *differentiated services codepoint* (DSCP) to indicate the PHB a packet should experience at each node. The other two bits in the octet are currently unused (CU) and should be ignored by a DS node.

A DS node uses the entire six-bit DSCP field as a table index to a particular packet handling mechanism. Although implementations are encouraged to use the recommended DSCP-to-PHB mappings, network operators may choose a different DSCP for a given PHB. In such a case the operator may need to re-mark the codepoint at the administrative boundary.

To maintain some form of backward compatibility with current practice, certain codepoints are reserved. In particular, a default PHB codepoint of 000000 remains for use for conventional best-effort traffic, and the codepoint 11x000 is for network control traffic. If a packet is received with an unrecognized codepoint, the packet will be forwarded as if it were marked with the default PHB codepoint.

FIGURE 10.29 The structure of the DS field

10.6.2   Per Hop Behaviors

The standard best-effort treatment that routers perform when forwarding traffic is defined as the default (DE) PHB. To provide additional classes of service, the IETF has defined additional PHBs. So far two additional PHBs have been defined: the expedited forwarding PHB and the assured forwarding PHB.

Expedited Forwarding PHB (EF PHB) provides a low-loss, low-latency, low-jitter, assured-bandwidth, end-to-end service through DS domains. To the connection endpoints, the service obtained by using EF PHB is equivalent to a "virtual leased line." Such service is often called a "premium service."

To ensure very low latency and assured bandwidth, the aggregate arrival rate of packets with EF PHB at every node should be less than the aggregate minimum allowed departure rate. Thus every DS node must be configured with a minimum departure rate for EF PHB independent of other traffic. In addition, the aggregate arrival rate must be shaped and policed so that it is always less than the minimum configured departure rate. The observant reader would notice that the service obtained by using EF PHB is essentially equivalent to CBR service in ATM.

Packets that are marked as EF PHB are encoded with codepoint 101110. When EF packets enter a DS node, they will be placed in a queue that is expected to be short and served quickly so that EF traffic will maintain significantly lower levels of latency, packet loss, and jitter. Several types of queue-scheduling mechanisms such as priority queue and weighted fair queue may be adopted to implement the EF PHB. Note that the DS model specifies the PHB but not the mechanism to provide the PHB.

Assured Forwarding PHB (AF PHB) delivers the aggregate traffic from a particular customer with high assurance (i.e., high probability of the traffic being delivered to the destinations) as long as the aggregate traffic does not exceed the traffic profile (e.g., the subscribed information rate). The customer, however, is allowed to send its traffic beyond the traffic profile with the caveat that the excess traffic may not be given high assurance. Unlike EF PHB, AF PHB is not intended for low-latency, low-jitter applications.

Four independent AF classes in the AF PHB group have been defined to offer several levels of forwarding assurances. Within each AF class, packets are assigned to one of three possible drop-precedence values. If there is congestion in a node, the drop precedence of a packet determines the relative importance of the packet within the AF class. A DS node must maintain the sequence of IP packets of the same microflow belonging to the same AF class regardless of the drop-precedence values.

An IP packet associated with AF class i and drop precedence j is marked with AF codepoint AFij. The recommended values of the AFij codepoint are as follows: AF11 = 001010,   AF12 = 001100,   AF13 = 001110,   AF21 = 010010, AF22 = 010100,   AF23 = 010110,   AF31 = 011010,   AF32 = 011100,   AF33 = 011110, AF41 = 100010, AF42 = 100100, AF43 = 100110. Within each AF class, the codepoint AFx1 yields lower loss probability than the codepoint AFx2, which in turn yields lower loss probability than the codepoint AFx3.

One service that the AF PHB group can implement is the olympic service, which consists of three service classes: bronze, silver, and gold. The objective is to assign packets to these three classes so that packets in the gold class experience lighter load than packets in the silver class and in the bronze class. The bronze, silver, and gold service classes can be mapped to the AF classes 1, 2, and 3, respectively. Packets within each class may be further mapped to one of the three drop-precedence values.

RED AND RIO BUFFER MANAGEMENT

The different packet drop precedence levels in differentiated services IP indicate that packets are to receive different priorities in accessing buffers during periods of congestion. In Chapter 7 we present buffer-access techniques based on queue-size allocation and queue-threshold admission policies. Random Early Detection (RED) is a different type of buffer management technique that can be used with packet flows that are generated by sources that can adapt to congestion. The prime example of such flows are packets generated by TCP sources.

RED drops some packets at random when the average queue length exceeds a given minimum threshold [Floyd 1993]. The dropped packet notifies the corresponding source that congestion is imminent in the network and that the source should reduce the rate at which it transmits packets. The probability of dropping an arriving packet increases in proportion to the growth in average queue size. Ideally enough sources are throttled back that the average queue size stops growing. Of course all packets are dropped if the buffer becomes full. RED provides a subtle form of service differentiation. Well-behaved sources reduce their rate, so the rate at which their packets are discarded is reduced. Sources that do not respond to dropped packets will then constitute a larger fraction of the packet arrivals, and hence they will lose packets at a higher rate. The RED algorithm is described in detail in problem 54 of Chapter 7.

[Clark and Fang 1998] proposed an extension of RED to provide different levels of drop precedence for two classes of traffic. The algorithm is called RED with IN/OUT or RIO for short. Packets are classified as being inside (IN) or outside (OUT) depending on whether they conform to some profile. Two average queue lengths are maintained: Q_{IN} the average number of IN packets, and Q_T the average number of total (IN + OUT) packets in the buffer. IN packets are dropped according to a RED algorithm that is based on Q_{IN} and corresponding packet dropping thresholds. OUT packets are dropped according to Q_T and corresponding thresholds. The thresholds and packet dropping probabilities are selected so that OUT packets are dropped more aggressively than IN packets. Consequently as congestion sets in, OUT packets will be more likely to be dropped. RIO and RIO-like algorithms can therefore be used to provide different levels of packet drop precedence.

10.6.3 Traffic Conditioner

Traffic conditioning is an essential function of a DS node. This section describes the elements that can be combined to provide the traffic conditioning. These elements include meters, markers, shapers, and droppers. Figure 10.30 shows the block diagram of a traffic conditioner.

A border router uses a classifier to identify the service class that should be given to the traffic. Once the traffic is classified, it is typically submitted to a meter. The **meter** measures the traffic to check conformance to a traffic profile, such as the rate and burst size, and provides inputs to other elements. There are various types of meters. The most common one is the token bucket meter that can be used to check conformance against peak rate, average rate, maximum burst size, and other traffic parameters.

A **marker** sets the DSCP in a packet header. Markers may perform marking on unmarked packets (DSCP = 000000) or may re-mark previously marked packets. Traffic that is deemed to be nonconforming may be re-marked to a lower service level. Marking can also be provided by the service provider as a value-added service. For example, the service provider may mark the customer's traffic based on a certain multifield classification.

A **shaper** delays packets so that they are compliant with the traffic profile. Ingress routers may shape the incoming traffic that is deemed noncompliant to protect the DS domain. Egress routers may shape the outgoing traffic before it is forwarded to a different provider's network. This type of shaper ensures that the traffic will conform to the policing action in the subsequent network.

A **dropper** discards traffic that violates its traffic profile. A dropper may be thought of as a shaper with zero buffer size.

10.6.4 Bandwidth Broker

Sections 10.6.2 and 10.6.3 introduce several service classes and mechanisms for providing differentiated services. The problem remains of allocating and controlling the bandwidth within a DS domain so that the objectives of an organization are met. One possible approach is to have users individually decide which

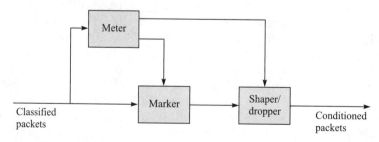

FIGURE 10.30 The functional diagram of a traffic conditioner

service to use, but this approach is unlikely to achieve the desired objectives. Another approach is to have an agent for each domain, called a **bandwidth broker**, that tracks the current allocation of traffic to various services and handles new requests for service according to organizational policies and the current state of traffic allocation [Nichols 1997].

A bandwidth broker manages the bandwidth in a domain in several ways. It is responsible not only for the allocation of traffic to the various service classes within the domain but also for the setting up of packet classifiers and meters in the edge routers. The bandwidth broker maintains a policy database that specifies which users are allowed to request which services at what time. The bandwidth broker first authenticates each requester and then decides whether there is sufficient bandwidth to meet the particular service request.

The bandwidth broker also maintains bilateral agreements with bandwidth brokers in neighboring domains. For flows that request service to a destination in a different domain, the bandwidth broker checks to see that the requested flow conforms to the prearranged allocation through the appropriate next-hop domain. The bandwidth broker then informs the appropriate neighboring bandwidth broker of the new rate allocation and configures the appropriate border router in Figure 10.28 to handle the new flow. The neighboring bandwidth broker then configures its border router to handle the allocated packet flow. This bilateral agreement approach is viewed as a viable means of reaching agreement on the exchange of DS traffic across multiple domains.

It is anticipated that, at least initially, static preallocation of bandwidth in bilateral agreements will predominate. Nevertheless, bandwidth brokers are viewed as capable of handling the range of possible operating points up to dynamic per flow setup of bilateral agreements.

SUMMARY

In this chapter we have examined various network architectures that were designed to improve the performance and value of the current Internet infrastructure. We described the overlay model that overlays an IP network over an ATM network. This approach involves two network infrastructures each with its own addressing scheme and routing protocols. In particular we discussed classical IP over ATM, LAN emulation, Next-Hop Resolution Protocol, and multiprotocol over ATM. Address resolution plays an essential role in these approaches.

We then considered the peer model in which IP routing and addressing are combined with layer 2 forwarding. In particular we introduced multiprotocol label switching, which combines network layer routing with layer 2 forwarding to produce a system that combines low-cost, very high scalability, and flexibility in the routing paradigms that can be supported.

We examined both the integrated services model that provides per flow QoS in the Internet environment and the associated signaling protocol called RSVP. Finally, we examined a new service model called the differentiated services model; it provides a simpler form of QoS than the integrated services model in a more scalable solution.

CHECKLIST OF IMPORTANT TERMS

assured forwarding PHB (AF PHB)
bandwidth broker
broadcast and unknown server (BUS)
classical IP over ATM (CLIP)
controlled-load service
differentiated services (DS) model
dropper
expedited forwarding PHB (EF PHB)
filterspec
flowspec
forwarding equivalence class (FEC)
guaranteed service
integrated services model
label
Label Distribution Protocol (LDP)
label switching router (LSR)
label switched path (LSP)
LAN emulation (LANE)
LAN emulation client (LEC)
LAN emulation configuration server
    (LECS)
LANE emulation server (LES)
logical IP subnetwork (LIS)
marker

meter
MPOA client (MPC)
MPOA server (MPS)
multiprotocol label switching
    (MPLS)
multiprotocol over ATM (MPOA)
next-hop client (NHC)
Next-Hop Resolution Protocol
    (NHRP)
next-hop server (NHS)
overlay model
path
PathTear
peer model
per hop behavior (PHB)
ReSerVation Protocol (RSVP)
ResvTear
service level agreement (SLA)
session
shaper
traffic classifier
traffic conditioner
traffic conditioning agreement (TCA)

FURTHER READING

Alles, A., "ATM Internetworking," Cisco white paper, May 1995.
ATM Forum Technical Committee, "LAN Emulation over ATM Version 2 Specification," July 1997.
ATM Forum Technical Committee, "Multi-Protocol over ATM Version 1 Specification," July 1997.

Callon, R. et al., "A Framework for Multiprotocol Label Switching," work in progress, IETF draft, June 1999.

Clark, D. and W. Fang, "Explicit Allocation of Best Effort Packet Delivery Service," *IEEE Transactions on Networking*, August 1998.

Davie, B., P. Doolan, and Y. Rekhter, *Switching in IP Networks*, Morgan Kaufmann, San Francisco 1998.

Floyd, S. and V. Jacobson, "Random Early Detection Gateways for Congestion Avoidance," *IEEE Transactions on Networking*, August 1993.

Ibe, O., *Essentials of ATM Networks and Services*, Addison-Wesley, Reading, Massachusetts, 1997.

Nichols, K., V. Jacobson, L. Zhang, "A Two-Bit Differentiated Services Architecture for the Internet," work in progress, IETF draft, November 1997.

Rosen, E. et al., "Multiprotocol Label Switching Architecture," work in progress, IETF draft, April 1999.

Widjaja, I. and A. I. Elwalid, "Performance Issues in VC-Merge Capable Switches for Multiprotocol Label Switching," *IEEE Journal in Selected Area on Communications*, Vol. 17, No. 6, June 1999, pp. 1178–1189.

RFC 2205, R. Braden et al. "Resource Reservation Protocol (RSVP)," September 1997.

RFC 2211, J. Wroclawski, "Specification of a Controlled-Load Network Element Service," September 1987.

RFC 2212, S. Shenker et al. "Specification of Guaranteed Quality of Service," September 1997.

RF 2255, M. Laubach and J. Halpern, "Classical IP and ARP over ATM," April 1998.

RFC 2332, J. Luciani et al., "NBMA Next Hop Resolution Protocol (NHRP)," April 1998.

RFC 2430, T. Li and Y. Rekhter, "A Provider Architecture for Differentiated Services and Traffic Engineering (PASTE)," October 1998.

RFC 2475, S. Blake et al., "An Architecture for Differentiated Services," December 1998.

PROBLEMS

1. If a LIS has N members, how many VCCs are required by the LIS to support full connectivity?

2. Suppose a department installs an ATM switch to interconnect a number of workstations and that classical IP over ATM is to be used. Explain how communications can be provided to workstations attached to an existing Ethernet LAN.

3. Explain how classical IP over ATM can be used to connect islands of ATM networks. Do these networks have to be confined to a local area?

4. Discuss whether the ATM connections between the LIS routers should be permanent or not.

5. Suppose a department installs an ATM switch to interconnect a number of workstations and that LANE is to be used. Explain how communications can be provided to workstations attached to an existing Ethernet LAN.

6. Compare a LAN that uses LANE and ATM NIC cards with a LAN that uses Fast Ethernet or Gigabit Ethernet only.

7. Consider classical IP over ATM and LANE.
   a. Compare the address resolution procedures in classical IP over ATM and LANE. Explain the reasons for the differences.
   b. Explain why LANE needs a BUS, but CLIP does not.

8. Assume that there are N stations (LECs) in an emulated LAN. Suppose that each station, on average, maintains X ARP entries and that the station sends an LE_ARP request every T seconds on average. Compute the approximate number of ARP messages that the LES has to process in this environment and discuss the scalability of LANE.

9. Suppose that an enterprise has a wide area ATM network. Discuss whether LANE can be used to create virtual LANs that make LANs in different locations appear to belong to the same LAN.

10. Compare the placement of the LANE server functions in a workstation or in an ATM switch in terms of the impact on performance and reliability.

11. Explain why the originating LEC sends its frames to the BUS while waiting for a response to an LE_ARP request.

12. Which factors affect the MTU size in LANE?

13. A host waiting for an NHRP resolution reply may (1) drop the data packet, (2) retain the packet until the reply comes back, or (3) forward the packet along the routed path toward the destination. Discuss the advantages and disadvantages of each option.

14. In NHRP the ATM shortcut can provide a direct link between two routers that are not adjacent. Consider the interplay between the IP routing algorithm and the ATM routing algorithm. Is it possible for this interplay to result in a stable routing loop?

15. Discuss what type of ATM QoS capability is possible with NHRP?

16. Suppose the hosts in Figure 10.6 are on Ethernet LANs.
    a. Trace the sequence of PDUs that are produced when Host1 sends an IP packet to Host2 in the case where no shortcut has been established between the MPCs.
    b. Trace the sequence of PDUs that are produced when Host1 sends an IP packet to Host2 in the case where a shortcut has already been established between the MPCs.

17. MPOA uses flow detection to set up virtual circuits, whereas MPLS also relies on routing topology to establish virtual circuits. Discuss the advantages and disadvantages of the two approaches.

18. Consider explicit routing in MPLS.
    a. Can explicit routing in MPLS be used to select a route that meets the QoS requirements of a particular flow, for example, guaranteed bandwidth and delay? Explain.

b. Give an example at a fine level of granularity that can make use of the QoS explicit routing in part (a). Give another example at a coarser level of granularity.

c. Compare the case where the first LSR computes an explicit path to the case where all LSRs participate in the selection of the path.

d. What information does the first LSR require to select an explicit path?

19. In an MPLS domain, rank the following three flows in terms of their level of aggregation: (a) all packets destined to the same host, (b) all packets with the same egress router, (c) all packets with the same CIDR address.

20. Consider MPLS for unicast traffic. Explain what information is used to specify labels when labels are determined according to (a) host pairs, (b) network pairs, (c) destination network, (d) egress router, (e) next-hop AS, (f) destination AS.

21. Explain why it may be useful for the MPLS header to carry a time-to-live value.

22. Compare the overall scalability of the following three cases:
    a. Layer 3 forwarding only: each router performs a longest-prefix match to determine the next forwarding hop.
    b. Layer 3 forwarding and some Layer 2 MPLS forwarding.
    c. Layer 2 MPLS forwarding only.

23. Are MPLS label switching and packet filtering at firewalls compatible? Explain.

24. Discuss which factors determine the level of computational load for label assignment and distribution for the following three cases: (a) topology-driven label assignment, (b) request-driven label assignment, (c) traffic-driven label assignement.

25. Compare MPLS label stacks and IP-over-IP encapsulation tunnels in terms of their traffic engineering capabilities.

26. Compare the following two approaches to traffic engineering:
    a. Centralized computation: Network usage information is processed by a central site and a set of paths are generated for use by LSRs.
    b. Egress computation: Each egress node computes path from various ingress nodes to itself.

27. RSVP signaling is very different from ATM signaling. Discuss the differences and list the advantages and disadvantages of each protocol.

28. Can an MPLS label correspond to all packets that match a particular integrated services model filter specification? If yes, explain how such a label may be used. If no, explain why not.

29. Consider the merging of reservations in RSVP. Suppose that a single sender and 2^n receivers are reached through a path that consists of a depth n binary tree that has router located at each node. What is the total number of reservation request messages generated by the receivers?

30. Suppose RSVP is used to set up a virtual private network to interconnect n users across an Internet. Discuss the appropriateness of the three RSVP reservation styles for this application.

31. Select and justify a choice of RSVP reservation style that is appropriate for the following situations:
   a. A multiuser application where a video stream from each participant must be delivered to every other participant.
   b. An audioconferencing application that is self-limiting in that more than two participants are unlikely to speak simultaneously.
   c. A multiuser application in which each participant must receive explicit permission before transmitting information.

32. What information needs to be carried by Resv messages in RSVP to set up guaranteed IP service with a specific bandwidth and a given delay-bound requirement? How does a router process this information?

33. Discuss the interplay between a QoS routing algorithm, RSVP, and integrated services IP.

34. Can RSVP be used to set up explicit routes in MPLS? Explain.

35. Explain how the soft-state feature of RSVP allows it to adapt to failures in the network.

36. Discuss whether RSVP provides a scalable means to set up connections to a session that consists of the broadcasting of a given TV program.

37. Explain what happens if RSVP is used to set up small-bandwidth flows on a large-bandwidth link? How well does RSVP scale in this situation?

38. Discuss possible scheduling mechanisms, for example, head-of-line priority and weighted fair queueing, that you could use to implement EF PHB, AF PHB group, and DE PHB. Discuss the trade-offs between performance and complexity.

39. Can MPLS make use of DS information to determine the forwarding experienced by different traffic flows? If yes, explain how. If no, explain why not.

40. What is an appropriate traffic conditioner for EF PHB packets? AF PHB packets?

41. Suppose that two RSVP-capable users are connected to an intranet that in turn connects them through an Internet backbone that provides differentiated services IP. Suppose that the intranets provide integrated services IP. Explain the functions that must be implemented in the intranet routers to provide controlled-load service and guaranteed service.

42. Compare integrated services IP and differentiated services IP in terms of their ability to protect themselves against theft of service.

43. Explain why shaping may be required at the egress node of a differentiated services domain.

44. Suppose that an MPLS network consisting of N routers supports C DS traffic classes. A trunk is defined as traffic from a single traffic class that is aggregated into an LSP.

    a. What is the maximum number of trunks if a trunk is defined for every pair of ingress router and egress router?

    b. Suppose that the trunks in part (a) are merged if they have the same traffic class and the same egress routers. What is the maximum number of multipoint-to-point trees rooted in an egress router?

    c. Can you think of other ways of merging trunks?

CHAPTER 11

Security Protocols

To provide certain services, some communication protocols need to process the information they transmit and receive. For example, protocols that provide reliable communication service encode the transmitted information to detect when transmission errors have occurred so that they can initiate corrective action. Another example is a security protocol that provides a secure communication service that prevents eavesdroppers from reading or altering the contents of messages and prevents imposters from impersonating legitimate users. In this chapter we discuss the following aspects of security:

1. *Security and cryptographic algorithms*. We introduce the threats that can arise in a network context, and we identify various types of security requirements. We introduce secret key and public key cryptography and cryptographic checksums, and we explain how they meet the above security requirements.
2. *Security protocols*. We develop protocols that provide security services across insecure networks. We also introduce protocols for establishing a security association and for managing keys. We relate these security protocols to the standard security protocols developed for the IP layer—IP Security (IPSec)—and for the transport layer—Secure Sockets Layer (SSL) and Transport Layer Security (TLS).
3. *Cryptographic algorithms*. This optional section describes the Data Encryption Standard (DES) and the Rivest, Shamir, and Adleman (RSA) encryption algorithm.

11.1   SECURITY AND CRYPTOGRAPHIC ALGORITHMS

Public communication networks traditionally have not been secure in the sense of providing high levels of security for the information that is transmitted. As these networks are increasingly used for commercial transactions, the need to provide security becomes critical. We showed earlier that many transactions take the form of a client/server interaction. Figure 11.1 shows several threats that can arise in a network setting:

- Information transmitted over the network is not secure and can be observed and recorded by eavesdroppers. This information can be replayed in attempts to access the server.
- Imposters can attempt to gain unauthorized access to a server, for example, a bank account or a database of personal records.
- An attacker can also flood a server with requests, overloading the server resources and resulting in a *denial of service* to legitimate clients.
- An imposter can impersonate a legitimate server and gain sensitive information from a client, for example, a bank account number and associated user password.

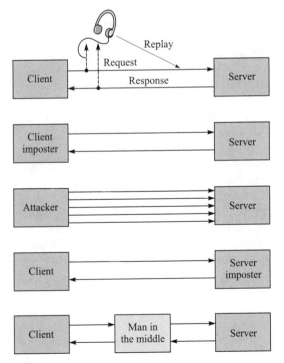

FIGURE 11.1 Network security threats

- An imposter manages to place itself as the *man in the middle*, convincing the server that it is the legitimate client and the legitimate client that it is the legitimate server.

These threats give rise to one or more of the following security requirements for information that is transmitted over a network:

- Privacy or confidentiality: The information should be readable only by the intended recipient.
- Integrity: The recipient can confirm that a message has not been altered during transmission.
- Authentication: It is possible to verify that the sender or receiver is who he or she claims to be.
- Nonrepudiation: The sender cannot deny having sent a given message.[1]

These requirements are not new; they take place whenever people interact. Consequently, people have developed various means to provide these security needs. For example, the privacy and integrity of important and sensitive information is maintained by keeping it under lock and key. In sensitive situations people are required to identify themselves with official documentation. Contracts are signed so that the parties cannot repudiate an agreement. In this and subsequent sections we are interested in developing analogous techniques that can be used to provide security across a network.

The need for security in communications is in fact also not new. This need has existed in military communications for thousands of years. It should not be surprising then that the approaches developed by the military form the basis for providing security in modern networks. In particular, the possession of secrets, such as passwords and keys, and the use of encryption form the basis for network security.

One feature that is new in the threats faced in computer networks is the *speed* with which break-in attempts can be made from a *distance* by using a network. Because the threats are implemented on computers, very high attempt rates are possible. Countermeasures include a wide range of elements from firewalls and network security protocols to security practices within an organization. In this chapter we focus on network protocols that provide security services. We emphasize that the overall security of a system depends on many factors, only some of which are addressed by network protocols.

In the following section we will present an overview of basic cryptographic techniques and show how they can be used to meet the security requirements. We then discuss security in the context of peer-to-peer protocols. Finally, we show how these protocols are used in Internet security standards.

---

[1]On the other hand, some situations require *repudiation*, which enables a participant to plausibly deny that it was involved in a given exchange.

11.1.1   Applications of Cryptography to Security

The science and art of manipulating messages to make them secure is called **cryptography**. An original message to be transformed is called the **plaintext**, and the resulting message after the transformation is called the **ciphertext**. The process of converting the plaintext into ciphertext is called **encryption**. The reverse process is called **decryption**. The algorithm used for encryption and decryption is often called a **cipher**. Typically, encryption and decryption require the use of a secret **key**. The objective is to design an encryption technique so that it would be very difficult if not impossible for an unauthorized party to understand the contents of the ciphertext. A user can recover the original message only by decrypting the ciphertext using the secret key.

For example, **substitution ciphers** are a common technique for altering messages in games and puzzles. Each letter of the alphabet is mapped into another letter. The ciphertext is obtained by applying the substitution defined by the mapping to the plaintext. For example, consider the following substitution:

```
a b c d e f g h i j k l m n o p q r s t u v w x y z
z y x w v u t s r q p o n m l k j i h g f e d c b a
```

where each letter in the first row is mapped into the corresponding letter in the row below. The message "hvxfirgb" is easy to decode if you know the "key," which is the permutation that is applied.

Transposition ciphers are another type of encryption scheme. Here the order in which the letters of the message appear is altered. For example, the letters may be written into an array in one order and read out in a different order. If the receiver knows the appropriate manner in which the reading and writing is done, then it can decipher the message. Substitution and transposition techniques are easily broken. For example, if enough ciphertext is available, such as through eavesdropping, then the frequency of letters, pairs of letters, and so forth can be used to identify the encryption scheme.

Modern encryption algorithms depend on the use of mathematical problems that are easy to solve when a key is known and that become extremely difficult to solve without the key. There are several types of encryption algorithms, and we discuss these next.

SECRET KEY CRYPTOGRAPHY

Figure 11.2 depicts a **secret key cryptographic** system where a sender converts the plaintext P into ciphertext $C = E_K(P)$ before transmitting the original message over an insecure channel. The sender uses a *secret* key K for the encryption. When the receiver receives the ciphertext C, the receiver recovers the plaintext by performing decryption $D_K(C)$, using the same key K. It is the sharing of a secret, that is, the key, that enables the transmitter and receiver to communicate. Symbolically, we can write $P = D_K(E_K(P))$. Secret key cryptography is also referred to as *symmetric key cryptography*.

The selection of the cryptographic method must meet several requirements. First of all, the method should be easy to implement, and it should be deployable

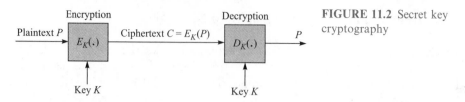

FIGURE 11.2 Secret key cryptography

on large scale. This suggests adopting a standard mathematical algorithm that can be readily implemented. On the other hand, the algorithm must provide security to all of its users. It is here that the use of a key plays an essential role. The key must uniquely specify a particular variation of the algorithm that will produce secure ciphertext. Indeed the best algorithms should prevent an attacker from deriving the key even when a large sample of the plaintext and corresponding ciphertext is known. In general, the number of possible keys must be very large. Otherwise, a brute force approach of trying all possible keys may be successful.

Clearly, secret key cryptography addresses the *privacy* requirement. A message that needs to be kept confidential is encrypted prior to transmission, and any eavesdropper that manages to gain access to the ciphertext will be unable to access the contents of the plaintext message. The Data Encryption Standard (DES) is a well-known example of a secret key system and is discussed in a later section.

A traditional method of *authentication* involves demonstrating possession of a secret. For example, in a military setting a messenger might be confirmed to be authentic if he or she can produce the correct answer to the specific question. A similar procedure can be used over a network, using secret key cryptography. Suppose that a transmitter wants to communicate with a receiver as shown in Figure 11.3 and that the receiver and transmitter share a secret key. The transmitter sends a message identifying itself. The receiver replies with a message that contains a random number r. This action is called a **challenge**. The transmitter sends a **response** with an encrypted version of the random number. The receiver applies the shared key to decrypt the number. If the decrypted number is r, then the receiver knows that it is communicating with the given transmitter. If the transmitter also wishes to authenticate the receiver, the transmitter can then issue a challenge by sending its own random number. It is extremely important that the random numbers used in each challenge be different; otherwise, an eaves-

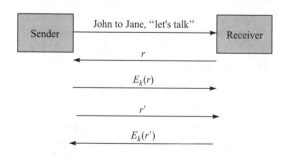

FIGURE 11.3 Secret key authentication

dropper, such as the one in Figure 11.1, could maintain a table of challenges and responses and eventually manage to be authenticated. The term **nonce**, derived from *number once*, describes the desired random numbers.

Integrity refers to the ability to ascertain that a message has not been altered during transmission. One approach is to encrypt the entire message, since the modified ciphertext, when decrypted, will produce gibberish. However, in general the message itself may not have been encrypted prior to transmission. The next section discusses how integrity can be provided in this case.

CRYPTOGRAPHIC CHECKSUMS AND HASHES

The usual approach to providing integrity is to transmit a *cryptographic checksum* or *hash* along with the unencrypted message. The transmitter and receiver share a secret key that allows them to calculate the checksum that consists of a fixed number of bits. To ascertain integrity, the receiver calculates the checksum of the received message and compares it to the received checksum. If the checksums agree, the message is accepted. Note that this procedure corresponds exactly to that used in error detection in Chapter 3.

A cryptographic checksum must be designed so that it is *one way* in that it is extremely difficult to find a message that produced a given checksum. Furthermore, given a message, finding another message that would produce the same checksum should also be extremely difficult. In general the checksum is much shorter than the transmitted message. However, the cryptographic checksum cannot be too short. For example, suppose that we are dealing with messages that are 1000 bits long and that we are using a checksum that is 128 bits long. There are 2^{1000} possible messages and "only" 2^{128} possible checksums, so on average $2^{1000}/2^{128} = 2^{872}$ messages produce the same checksum. This calculation does not mean that it is easy to find a message that corresponds to a given checksum. Suppose we start going through all 2^{1000} possible messages, computing their checksum, and comparing to the desired checksum. The fraction of messages with the desired checksum is $1/2^{128}$, so on average it will take us 2^{128} tries to get to the first message that matches the desired checksum. If it takes 1 microsecond to generate and check a message, then the average time required to find one that matches the checksum is 10^{25} years.

The **message digest 5 (MD5) algorithm** is an example of a hash algorithm. The MD5 algorithm begins by taking a message of arbitrary length and padding it into a multiple of 512 bits. A buffer of 128 bits is then initialized to a given value. At each step the algorithm modifies the content of the buffer according to the next 512-bit block. When the process is completed, the buffer holds the 128-bit "hash" code. The MD5 algorithm itself does not require a key.

The **keyed MD5**, which combines a secret key with the MD5 algorithm, is widely used to produce a cryptographic checksum. First the message is padded to a multiple of 512 bits. The secret key is also padded to 512 bits and attached to the front and back of the padded message. The MD5 algorithm then computes the hash code [Kaufman et al., 1995, p. 120]. Note that in addition to the key, an ID and other information could be appended. This technique would also allow the receiver to authenticate that the authorized sender sent the information. A hash function that depends on a secret key and on a message is called a **message authentication code**.

The **secure hash algorithm 1 (SHA-1)** is another example of a hash function. SHA-1 was developed for use with the Digital Signature Standard. SHA-1 produces an 160-bit hash and is considered more secure than MD5. A keyed SHA-1 hash is produced in the same manner as a keyed MD5 hash.

A general method for improving the strength of a given hash function is to use the **hashed message authentication code (HMAC)** method. Using MD5 as an example, HMAC works as follows. First, the shared secret is padded with zeros to 512 bits. The result is XORed with ipad, which consists of 64 repetitions of 00110110. Second, the message is padded to a multiple of 512 bits. Third, the concatenation of the blocks in the first two steps is applied to the MD5 algorithm to obtain a 128-bit hash. The hash is padded to 512 bits. Fourth, the shared secret is padded with zeros to 512 bits, and the result is XORed with opad, which consists of 64 repetitions of 01011010. Fifth, the blocks in the previous two steps are applied to the MD5 algorithm to produce the final 128-bit hash. The general HMAC procedure involves adjusting the block size (512 bits for MD5) and the hash size (128 bits for MD5) to the particular hash function. For example, SHA-1 works with a block size of 512 and a hash size of 160 bits.

PUBLIC KEY CRYPTOGRAPHY

Unlike secret key cryptography, keys are not shared between senders and receivers in **public key cryptography** (sometimes also referred to as *asymmetric cryptography*). Public key cryptography was invented in 1975 by Diffie and Hellman. It relies on two different keys, a public key and a private key. A sender encrypts the plaintext by using a public key, and a receiver decrypts the ciphertext by using a private key, as illustrated in Figure 11.4. Symbolically, a public key cryptographic system can be expressed as $P = D_{K2}(E_{K1}(P))$, where $K1$ is the public key and $K2$ is the private key. In some systems the encryption and decryption process can be applied in the reverse order such as $P = E_{K1}(D_{K2}(P))$. One important requirement for public key cryptography is that it must not be possible to determine $K2$ from $K1$. In general the public key is small, and the private key is large. The best-known example of public key cryptography is the one developed by Rivest, Shamir, and Adleman, known as **RSA**.[2]

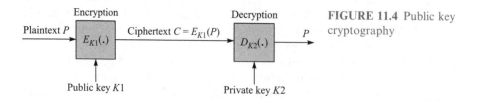

FIGURE 11.4 Public key cryptography

---

[2]RSA is described in section 11.3.2.

Public key cryptography provides for *privacy* as follows. The transmitter uses the public key to encrypt its message P and then transmits the corresponding ciphertext to the receiver. Only the holder of the private key can decrypt the message; therefore, the privacy of the message is assured. Similarly, the messages from the receiver to the transmitter are kept private by encrypting them with the transmitter's public key. Note that *integrity* is not assured, since an intruder can intercept the message from the transmitter and insert a new message using the public key. This problem can be addressed by having the transmitter encrypt the message with its *private* key and transmit $D'_{K2}(P)$. No intruder can successfully alter this ciphertext, and the receiver can decrypt it by using the public key, $P = E'_{K1}(D'_{K2}(P))$.

Public key cryptography can also be used for *authentication* as shown in Figure 11.5. Here the transmitter begins by identifying itself. The receiver picks a nonce r, encrypts it by using the transmitter's public key, and issues a challenge. The transmitter uses its private key to determine the nonce and responds with the nonce r.

Public key cryptography can also be used to produce a **digital signature**. To sign a message the transmitter first produces a noncryptographic checksum or hash of the message. The transmitter then encrypts the checksum or hash using its private key to produce the signature. No one else can create such a signature. The transmitter then sends the message and the signature to the receiver. The receiver confirms the signature as follows. First the receiver applies the public key encryption algorithm to the signature to obtain a checksum. The receiver then computes the checksum directly from the message. If the two checksums agree, then only the given transmitter could have issued the message. Note that the digital signature confirms that the transmitter produced the message and that the message has not been altered.

COMPARISON OF SECRET KEY AND PUBLIC KEY CRYPTOGRAPHIC SYSTEMS

In terms of capabilities public key systems are more powerful than secret key systems. In addition to providing for integrity, authentication, and privacy, public key systems also provide for digital signatures. Unfortunately, public key cryptography has a big drawback in that it is much slower than secret key cryptography. For this reason, public key cryptography is usually used only during the setup of a session to establish a so-called *session key*. The session key is then used in a secret key system for encrypting messages for the duration of the session.

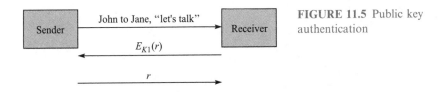

FIGURE 11.5 Public key authentication

Example—Pretty Good Privacy

Pretty Good Privacy (PGP) is a secure e-mail protocol developed by Phillip Zimmerman. PGP became notorious for being freely available in the Internet. The protocol uses public key cryptography and so it provides privacy, integrity, authentication, and digital signatures. PGP can also be used to provide privacy and integrity for stored files.

11.1.2   Key Distribution

In principle, secret key systems require every pair of users to share a separate key. Consequently, the number of keys can grow as the square of the number of users, making these systems infeasible for large-scale use. This problem can be addressed through the introduction of a **key distribution center (KDC)** as shown in Figure 11.6. Every user has a shared secret key with the KDC. If user A wants to communicate with user B, user A contacts the KDC to request a key for use with user B. The KDC authenticates user A, selects a key K_{AB}, and encrypts it by using its shared keys with A and B to produce $E_{KA}(K_{AB})$ and $E_{KB}(K_{AB})$. The KDC sends both versions of the encrypted key to A. Finally, user A can contact user B and provide a *ticket* in the form of $E_{KB}(K_{AB})$ that allows them to communicate securely.

Example—Kerberos Authentication Service

Kerberos is an authentication service designed to allow clients to access servers in a secure manner over a network. Kerberos uses a KDC that shares a secret key with every client. When a user logs on to a workstation, he supplies a name and password that is used to derive the user's secret key. The workstation is authenticated by the KDC. The KDC returns a session key encrypted with the user's secret key. The KDC also returns a ticket-granting ticket (TGT) that contains the session key and other information encrypted with the KDC's own secret key.

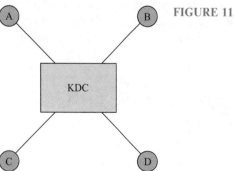

FIGURE 11.6 Key distribution

Each time the client wishes to access a particular server, it sends a request to the KDC along with the TGT and the server's name. The KDC decrypts the TGT to recover the session key. The KDC then returns a ticket to the client that allows access to the desired server.

Public key systems require only one pair of keys per user, but they still face the problem of how the public keys are to be distributed. Because an imposter can advertise certain public keys as belonging to other parties, the public keys must be certified somehow. One approach to addressing this problem is to establish a **certification authority (CA)**. The function of a CA is to issue certificates that consist of a *signed* message, stating the name of a given user, his or her public key, a serial number identifying the certificate, and an expiration date. The certificates can be stored anywhere they can be conveniently accessed through a directory service. Each user is initially configured to have the public key of the CA. To communicate with user B, user A contacts a server to obtain a certificate for B. The signature in the certificate authenticates the message and its integrity.

Note that the public key of certain users will inevitably be revoked, but retrieving a certificate after it has been issued is not easy. For this reason, a *certificate revocation list (CRL)* must also be issued periodically by the CA. The CRL lists the serial numbers of certificates that are no longer valid. Thus

---

X.509 AUTHENTICATION SERVICE

The ITU.500 series recommendations is intended to facilitate the interconnection of systems to provide *directory services*. A server or distributed set of servers together hold information that constitutes a *directory*. This information is used to facilitate communications between applications, people, and communication devices. An example of information provided by a directory is the mapping of user names to network addresses. The X.509 recommendation provides a framework for the provision of *authentication service* by a directory to its users. The recommendation specifies the form of authentication information held by the directory, describes how authentication information may be obtained from the directory, states the assumptions about how authentication information is formed and placed in the directory, defines ways in which applications may use this authentication information to perform authentication, and describes how other security services may be supported by authentication. An important feature of X.509 is that the directory can hold user certificates, which can be freely communicated within the directory system and obtained by users of the directory. The certificates can contain a user's public key that has been signed by a certificate authority. Many protocols including IP Security, SSL, and TLS described later in the chapter use the X.509 certificate format.

we conclude that each time a user wishes to communicate with another, the user must retrieve not only a certificate for the given user but also a current CRL.

KEY GENERATION: DIFFIE-HELLMAN EXCHANGE

An alternative to key distribution using KDCs or CAs is to have the transmitter and receiver create a *secret shared key* by using a series of exchanges over a public network. Diffie and Hellman showed how this can be done. The procedure assumes that the transmitter and receiver have agreed on the use of a large prime number p, that is, for example, about 1000 bits long, and a *generator* number g that is less than p. The transmitter picks a random number x and calculates $T = g^x$ modulo p. Similarly the receiver picks a random number y and calculates $R = g^y$ modulo p. The transmitter sends T to the receiver, and the receiver sends R to the transmitter. At this point the transmitter and receiver both have T and R, so they can compute the following numbers:

- The transmitter calculates R^x modulo $p = (g^y)^x$ modulo $p = g^{xy}$ modulo $p = K$.
- The receiver calculates T^y modulo $p = (g^x)^y$ modulo $p = g^{xy}$ modulo $p = K$.

Therefore, both transmitter and receiver arrive at the same number K, as shown in Figure 11.7. An eavesdropper would have p, g, T, and R available, but neither x nor y. To obtain these values, the eavesdropper would need to be able to compute discrete logarithms, that is, $x = \log_g(T)$ and $y = \log_g(R)$. It turns out that this computation is exceedingly difficult to do for large numbers. Thus the transmitter and receiver jointly develop a shared secret K, which they can use in subsequent security operations.

The Diffie-Hellman exchange in Figure 11.7 has weaknesses. The required exponentials need many multiplications for large prime number p, so the algorithm is computationally intensive. This feature makes the system susceptible to a flooding attack, which could produce a heavy computational burden on a machine and result in denial of service to legitimate clients. In the next section we show how the use of "cookies" can thwart this type of attack.

The Diffie-Hellman exchange is also susceptible to a man-in-the-middle attack. Suppose that the intruder is able to intercept T and R as shown in Figure 11.8. The intruder keeps R and sends $R' = g^{y'}$ *modulo* p to the transmitter. At this point the transmitter and the intruder have established a secret key K_1 based on x and y'. Similarly, the intruder can establish a shared secret K_2 with the intended receiver based on x' and y. From now on the intruder is privy to all communications between the transmitter and receiver, and the transmitter and receiver can do nothing to detect the intruder. In the next section we show how

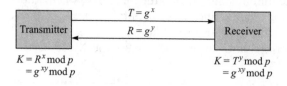

FIGURE 11.7 Diffie-Hellman exchange

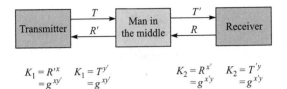

$$K_1 = R'^x \quad K_1 = T^{y'} \qquad K_2 = R^{x'} \quad K_2 = T'^y$$
$$\quad = g^{xy'} \quad \quad = g^{xy'} \qquad \quad = g^{x'y} \quad \quad = g^{x'y}$$

FIGURE 11.8 Man-in-the-middle attack

the intruder can be prevented from inserting itself into the middle through the authentication of the transmitter and receiver and their keys T and R.

11.2 SECURITY PROTOCOLS

A security protocol refers to a set of rules governing the interaction between peer processes to provide a certain type of security service. The protocol specifies the messages that are to be exchanged, the type of processing that is to be implemented, and the actions that are to be taken when certain events occur. In this section we describe the basic protocols that have been developed for the Internet.

11.2.1 Application Scenarios

Figure 11.9 shows a number of scenarios that may require secure communication services. In part (a) two host computers communicate across the public Internet. The hosts are exposed to all the security threats that were outlined earlier, such as eavesdropping, server and client imposters, and denial-of-service attacks. The two hosts can protect some of the information transmitted by encrypting the packets that are exchanged between them. Note, however, that the routers in the Internet need to read the IP header so the header cannot be encrypted. This introduces an element of insecurity in that the source and destination addresses reveal the existence of the hosts, their addresses, and the fact that they are communicating with each other. A patient eavesdropper can also analyze the timing and frequency of packet exchanges and the length of the messages to gain some insights into the nature of the interaction. Note that the hosts in Figure 11.9a can choose which layer to apply security to. Instead of encrypting the payloads of IP packets, the hosts might encrypt the payloads of TCP segments, or only certain application layer PDUs.

In part (b) two host computers on local networks have gateways between their local network and the public Internet. These gateways could be **firewalls** whose role is to enforce a security policy between the secure internal networks and the Internet. A firewall is implemented in a computer or a router, and its role is to control external access to internal information and services. A simple firewall system may involve *packet filtering* carried out by a router to enforce certain rules. Various fields in arriving packets are examined to determine whether they

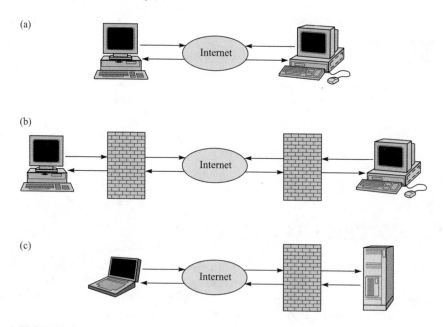

FIGURE 11.9 Scenarios requiring secure communication services

should be allowed to pass or be discarded. These fields can include source and destination IP addresses and TCP/UDP port numbers, ICMP message types, and fields inside the IP and TCP payloads. Secure communications between host A and host B in Figure 11.9b can be provided by establishing a secure *tunnel* between the two gateways. The packets that are exchanged between the hosts can then be completely encrypted, so the internal addresses of the packets are secure as well. Part (c) shows a situation that is typical when a mobile host attempts to access its home office across the public Internet. The mobile host and the home gateway must have procedures in place to allow the mobile host access while barring intruders.

Firewall systems can include elements that operate at layers above the network layer. For instance, packet filtering cannot understand the contents, for example, application-level commands, in the payload. For this reason, *application-level gateways* or *proxies* have been developed to enforce security policy at the level of specific services. An application-level gateway is interposed between the actual client and the actual server: The gateway monitors and filters the messages and relays them from client to server and from server to client. The gateway can be configured so that users need to provide authentication information and so that only certain features of the service are made available. Application level gateways for Telnet, FTP, and HTTP are in common use.

The *circuit-level gateway* is another type of firewall system. This gateway restricts the TCP connections that are allowed across it. End-to-end connections are not allowed and instead must be broken into a connection from the client to the gateway and a connection from the gateway to the client. Once a user has been authenticated, the gateway relays segments between the two end systems.

11.2.2   Types of Service

In this section we are concerned with three communication security requirements:

1. Integrity: The recipient of the information from the network should be able to confirm that the message has not been altered during transmission.
2. Authentication: The recipient should be able to ascertain that this sender is who he or she claims to be.
3. Privacy: The information should be readable only by the intended recipient.

In applications that require security, the first two requirements are usually essential. The third requirement involves encryption of information that can entail significant additional processing and hence is not always justified.

First we consider the case of two peer processes that have already established a **security association** that specifies the type of processing to be carried out by the processes, including the type of cryptographic algorithms and the shared secret keys. Let us consider how a communication service can provide integrity, authentication, and privacy.

INTEGRITY AND AUTHENTICATION SERVICE

We saw in section 11.1 that integrity can be provided through the use of a *cryptographic checksum* algorithm that takes a concatenation of the shared secret key and the message and produces a shorter fixed-length message using a one-way function. The algorithm for producing the checksum or hash is public, but without the secret key it is extremely difficult to modify the message and still produce the right checksum. On the other hand, the receiver can easily verify that the received information has not been altered. It simply concatenates the shared secret key with the received message to recalculate the checksum. If the recalculated and the received checksums are the same, then the receiver can be very confident that the message was not altered. If they do not agree, then the receiver can be certain that something is wrong. Note that the checksum calculation can be extended to cover any information whose integrity needs to be verified. In particular, if the cryptographic checksum includes the shared secret key, the sender's identity, and the message, then the identity of the sender is also authenticated.

Figure 11.10a shows a typical packet structure for providing integrity and authentication service. An **authentication header** is sandwiched in between the normal packet header and its payload (SDU).

The normal header includes a field set to indicate the presence of the authentication header. The authentication header itself contains a field that identifies the security association, a field for a sequence number, and a field for the cryptographic checksum. Ideally, the cryptographic checksum should cover the entire

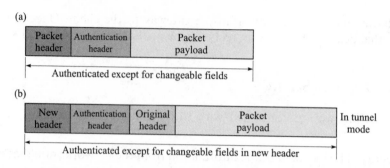

FIGURE 11.10 Packet structure for authentication and integrity service

packet. However, some fields in the packet header are changed while the packet traverses the network, and so these need to be excluded. The cryptographic checksum algorithm is applied to the packet in Figure 11.10a with the values in the changeable fields and the cryptographic checksum field set to zero. The resulting checksum is then inserted in the authentication header, and the packet is transmitted.

To verify integrity and authentication, the receiver recalculates the cryptographic checksum based on the received packet and the shared secret key, with the appropriate fields set to zero. If the recalculated checksum and the received checksum do not agree, the packet is discarded and the event is recorded in an audit log.

The authentication header can have a *sequence number* whose purpose is to provide protection against **replay attacks**. A typical ploy by eavesdroppers is to replay previously recorded sequences of packets to gain access to a system. When a security association is established, the sequence number is set to zero. Each time a new packet is sent, it receives a new sequence number. The transmitter has the task of ensuring that a sequence number is never reused. Instead when the number space is exhausted, the security association will be closed and a new one will be established. A receiver is prepared to accept a packet only *once*, so a replay attack will not work. Because packets can arrive out of order, at any given time the receiver maintains a sliding window of sequence numbers. Packets with sequence numbers to the left of the window are rejected. Packets inside the window are checked against a list to see whether they are new, and if so, their cryptographic checksum is checked. Packets to the right of the window are treated as new, and their cryptographic checksum is checked.

A **tunnel** can be established to provide security between two gateways only as is shown in Figure 11.11. The tunnel is established by encapsulating a packet inside another packet as shown in Figure 11.10b. Here the header in the inner packet has the address of the destination host, and the header in the outer packet has the address of the gateway. In tunnel mode the authentication header can protect the *entire* inner packet and the unchangeable part of the outer packet.

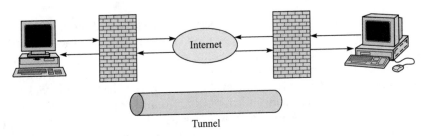

Tunnel

FIGURE 11.11 Tunnel between two firewall systems

PRIVACY SERVICE

The integrity and authentication service do not prevent eavesdroppers from reading the information in each packet. Encryption is required to provide privacy in the transmission of packet information. Again we suppose that a security association is already in place, so the transmitter and receiver have already agreed on the cryptographic algorithms and shared secret keys that are to be used.

Figure 11.12a shows a typical packet structure where an **encryption header** is inserted after the normal header. The normal header contains a field set to indicate the presence of an encryption header. The encryption header contains fields to identify the security association and to provide a sequence number. This header is followed by an encrypted version of the payload and possibly by initialization, padding, and other control information. Note that the packet header and the encryption header are *not* encrypted, so privacy is provided only by the upper-layer protocols encapsulated in the payload. The receiver takes each packet and decrypts the payload. If the payload portion has been

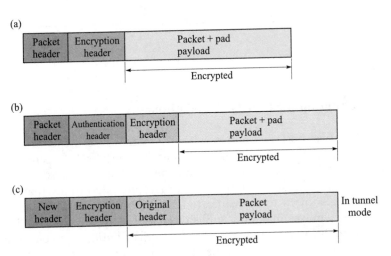

FIGURE 11.12 Packet structure for privacy service

altered during transmission, then the decrypted message will differ greatly from the original message. Such changes may be detectable by the higher-layer protocol if the decrypted message does not make sense.

Figure 11.12b shows a packet structure that can be used to provide privacy for the payload and integrity and authentication for the overall packet. The packet header contains a field indicating that an authentication header follows. The authentication header in turn contains a field indicating that an encryption header follows. The payload is encrypted first, and then a cryptographic check-sum over the entire packet is calculated with some of the fields set to zero as before. This packet structure also provides the antireplay capability.

These packet structures allow eavesdroppers to read the information in the packet header. A stronger form of privacy is obtained by using encryption in a *tunnel mode* as shown in Figure 11.12c. The inner packet header carries the addresses of the ultimate source and destinations, and the outer packet header may contain other addresses, for example, those of a security gateway. All information afer the encryption header is encrypted and hence not readable by an eavesdropper.

There are many options for combining the application of authentication and encryption headers in normal and tunnel modes. We explore these in the problems.

Example—Virtual Private Networks

A virtual private network (VPN) has traditionally referred to a dedicated set of network resources (leased lines) that provide wide area connectivity for different sites of a large company. The low cost of Internet communications has made the establishment of VPNs across the Internet very attractive. The key requirement here is to create private communication across a public insecure medium. For this reason, VPNs over the Internet are generally based on the establishment of secure tunnels.

11.2.3   Setting up a Security Association

Suppose that two host computers wish to establish secure communications across a public network. The computers must first establish a security associa-tion. This step requires them to agree on the combination of security services they wish to use and on the type of cryptographic algorithms that are to be utilized. They also need to authenticate that the other host is who it claims to be. Finally, they must somehow establish a shared secret key that can be used in the subsequent cryptographic processing. Key distribution is a central problem in meeting these requirements and therefore to enabling widespread use of secur-ity services.

One approach is to "manually" distribute keys to the appropriate machines ahead of time. This approach, however, is not very scalable and not suitable for highly dynamic environments. As discussed in section 11.1.2, another approach is to use either key distribution centers to obtain secret keys or certification authorities that issue digitally signed certificates that provide the public keys of various users. A very appealing alternative to these approaches is a procedure that allows the two hosts to establish a security association and a common secret key *independently* of other hosts or servers. In this section we describe the **Internet key exchange (IKE)** protocol that has been developed to provide such a procedure.

We saw that the Diffie-Hellman exchange allows two hosts to establish a common secret key through an exchange of information over an insecure network. We saw however that the exchange was susceptible to a man-in-the-middle attack where an intruder can manage to insert itself transparently between the two hosts. In addition, we saw that the exchange made the hosts susceptible to denial of service attacks. The IKE protocol addresses these problems by adding an authentication step after the Diffie-Hellman exchange that can detect the presence of a man-in-the-middle and by using "cookies" to thwart denial-of-service attacks.

Figure 11.13 shows a series of message exchanges between two hosts to establish a security association. The exchange makes use of a digital signature for authentication and features the use of a pair of cookies generated by the hosts to identify the security association and to prevent flooding attacks. The cookie generation must be fast and must depend on the source and destination address, the date and time, and a local secret. A cryptographic hash algorithm such as MD5 can be used for this purpose. To protect themselves against flooding attacks, the hosts always check for the validity of the cookies in arriving packets first.

Note the following about the exchange of messages:

- The initiator first generates a unique pseudorandom number, of say 64 bits, which is called the initiator's cookie C_i. The initiator associates this cookie

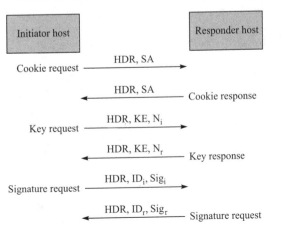

FIGURE 11.13 Establishing a security association

value with the expected address of the responder. The initiator host sends a *cookie request* message to the responder host requesting a security association. The header (HDR) contains the initiator's cookie. The *security association (SA) field* in the message offers a set of choices regarding encryption algorithm, hash algorithm, authentication method, and information about the parameters to be used in the Diffie-Hellman exchange.

- The responder checks to see whether the initiator's cookie is not already in use by the source address in the packet header. If not, the responder generates its cookie C_r. The responder associates this cookie value with the expected address of the initiator. The responder replies with a *cookie response* message in which it selects one of the offered choices in the initiator's SA field. The header includes both cookies, C_r and C_i.

- Upon receiving the response, the initiator first checks the address and initiator cookie in the arriving packet against its list. From now on the initiator will identify the security association by the pair (C_i, C_r). At this point it records the association as "unauthenticated." Next the initiator sends a *key request* message including its public Diffie-Hellman value $T = g^x$ modulo p and a nonce N_i.

- The responder host first checks the responder cookie in the arriving message. If the cookie is not valid, the message is ignored. If the cookie is valid, the security association will henceforth be identified by the pair (C_i, C_r). At this point the association is recorded as "unauthenticated." The responder sends a *key response* message with its public value $R = g^y$ modulo p and a nonce N_r.

- After this exchange, both the initiator and responder hosts have the secret constant $K = g^{xy}$ modulo p. Both parties now compute a secret string of bits SKEYID known only to them; for example, they might obtain a hash of the concatenation of the two nonces N_i and N_r and K. SKEYID might be 128 bits long.

- The initiator now prepares a signature stating what it knows: namely, SKEYID, T, R, C_i, C_r; the contents of the SA field; and the initiator identification. For example, the initiator can obtain a hash of the concatenation of the binary representations of all this information. The initiator identification and the signature are then encrypted with an algorithm specified in the security association by using a key derived from K, and (C_i, C_r); which are known to the initiator and responder. The initiator sends this information in a *signature request* message.

- The responder decrypts the message from the initiator. The responder then recalculates the hash of its version of the shared information, namely, SKEYID, T, R, C_i, C_r; the contents of the SA field; and the initiator identification. If the recalculated hash agrees with the received hash, then the initiator and the Diffie-Hellman public values have been authenticated. The security association and keys are recorded as authenticated. A man in the middle would have been detected at this point, since it could not have knowledge of SKEYID, which was derived from K.

- The responder now prepares its signature stating what it knows: namely, SKEYID, R, T, C_i, C_r; the contents of the SA field; and the responder identification. The responder identification and the signature are then encrypted, and the *signature response* message is sent to the initiator.

• The initiator recalculates the hash of its version of the shared information to authenticate the responder as well as the values of the Diffie-Hellman public values. At this point the security association is established. The security association and keys are recorded as authenticated.

Once the security association is established, the two hosts can derive additional shared secret keys from the cookie values and K. The availability of shared secret key information also allows the hosts to quickly establish additional security associations as needed.

11.2.4   IPSec

The goal of IPSec is to provide a set of facilities that support security services such as authentication, integrity, confidentiality, and access control at the IP layer. IPSec also provides a key management protocol to provide automatic key distribution techniques. The security service can be provided between a pair of communication nodes, where the node can be a host or a gateway (router or firewall).

IPSec uses two protocols to provide traffic security: **authentication header** and **encapsulating security payload**. These protocols may be applied alone or together to provide a specific security service. Each protocol can operate in either **transport mode** or **tunnel mode**. In the transport mode the protocols provide security service to the upper-layer protocols. In the tunnel mode the protocols provide security service to the tunneled packets.

AUTHENTICATION HEADER

Authentication and integrity of an IP packet can be provided by an authentication header (AH). The location of the AH is after other headers that are examined at each hop and before any other headers that are not examined at an intermediate hop. With IPv4, the AH immediately follows the IPv4 header, as shown in Figure 11.14.[3] The protocol value in the IP header is set to 51 to identify the presence of an AH. The authentication information is calculated over the entire IP packet, including the IP header, except over those fields that change in transit (i.e., mutable fields such as TTL, header checksum, and fragment offset) that are set to zero in the calculation.

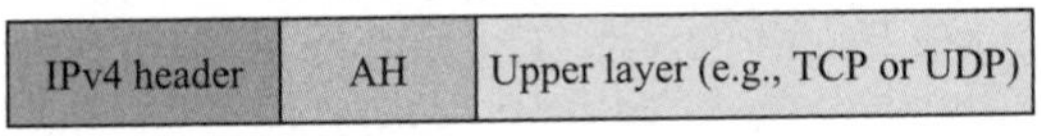

| IPv4 header | AH | Upper layer (e.g., TCP or UDP) |

FIGURE 11.14 Example with IPv4

---

[3]In the tunnel mode the AH is located immediately before the inner IP header.

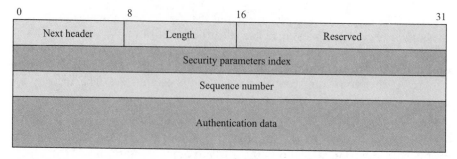

FIGURE 11.15 Format of authentication header

The format of the AH is shown in Figure 11.15. The next header field is used to identify the next payload after the AH. In other words, it has the same purpose as the IP protocol number in IPv4. The length field indicates the length of the authentication data in multiples of four octets. The security parameters index (SPI) identifies the security association for this packet. The value of the sequence number field is incremented by one for each packet sent. Its purpose is to protect against replay attacks. The result of the authentication algorithm is placed into the authentication data field. AH implementations must support HMAC with MD5 and HMAC with SHA-1 for the authentication algorithms.

The AH is incorporated in IPv6. The AH appears after the hop-by-hop, routing, and fragmentation extension headers. The destination options extension header can appear either before or after the AH. The protocol header that precedes the AH contains the value 51 in the next header field.

ENCAPSULATING SECURITY PAYLOAD

The encapsulating security payload (ESP) provides confidentiality, authentication, and data integrity. An ESP can be applied alone or in combination with an AH.

The ESP consists of a header and a trailer, as shown in Figure 11.16. The authenticated coverage starts from the SPI field until the next header field, and the encrypted coverage starts from the payload data field until the next header field. The protocol number immediately preceding the ESP header is 50. The SPI and sequence number have the same meanings as before. The payload data is variable-length data whose contents are specified in the next header field. The padding field is optional; its purpose is to conform to a particular encryption algorithm that requires the plaintext to be a multiple of some number of octets. The pad length field indicates the length of the padding field in octets. If the padding field is not present, then the pad length is zero. The authentication data field is optional, and its purpose is to provide authentication service.

The ESP is incorporated in IPv6 and appears after the hop-by-hop, routing, and fragmentation extension headers. The destination options extension header can appear either before or after the ESP header. The protocol header that precedes the ESP header contains the value 50 in the next header field.

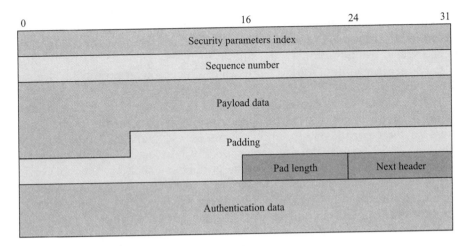

FIGURE 11.16 Format of ESP

11.2.5 Secure Sockets Layer and Transport Layer Security

The Secure Sockets Layer (SSL) protocol was developed by Netscape Communications to provide secure HTTP connections. SSL operates on top of a reliable stream service such as TCP and provides a secure connection for applications such as HTTP, as well as FTP, Telnet, and so on. SSL version 3.0 [Freier et al., 1996] was submitted to the IETF for standardization and, after some changes, led to the Transport Layer Security (TLS version 1.0) protocol in RFC 2246. We cover the TLS protocol as described in [RFC 2246].

As shown in the Figure 11.17, the TLS protocol consists of protocols that operate at two layers: the TLS Record protocol and the TLS Handshake protocol, along with the Change Cipher Spec protocol and the Alert protocol.

The TLS Record protocol provides a secure connection with the attributes of privacy and reliability. The connection provides *privacy* through the use of symmetric (secret key) encryption. The specific encryption algorithms can be selected from a wide range of standard algorithms. The secret keys that are used in the symmetric encryption algorithms are generated uniquely for each connection and are derived from a secret that is negotiated by another protocol, for example, the TLS Handshake protocol. The TLS Record protocol can operate without

Handshake protocol	Change cipher spec protocol	Alert protocol	HTTP protocol
TLS record protocol			
TCP			
IP			

FIGURE 11.17 TLS in the TCP/IP protocol stack

encryption. The connection provides *reliability* through the use of a keyed message authentication code (MAC).[4] The specific hash function for a connection can be selected from a set of standard hash functions. The TLS Record protocol typically operates without a MAC only while a higher-layer protocol is negotiating the security parameters.

The TLS Handshake protocol, along with the Change Cipher Spec protocol and the Alert protocol is used to negotiate and instantiate the security parameters for the record layers, to authenticate the users, and to report error conditions. The TSL Handshake protocol is used by a server and a client to establish a **session**. The client and server negotiate parameters such as protocol version, encryption algorithm, and method for generating shared secrets. They can authenticate their identity by using asymmetric (public key) encryption with an algorithm that can be selected from a set of supported schemes. Typically only the server is authenticated to protect the user from a man-in-the-middle attack. The client and server also negotiate a shared secret that is kept secure from eavesdroppers. The negotiation is made reliable so that its parameters cannot be modified by an attacker without being detected. Once a client and server have established a session, they can set up multiple secure connections, using the parameters that have been established for the session.

THE TLS HANDSHAKE PROTOCOL

The client and server use the Handshake protocol to negotiate a session that is specified by the following parameters [RFC 2246, p. 23]:

- *Session identifier*: An arbitrary byte sequence chosen by the server to identify an active or resumable session state.
- *Peer certificate*: An X509.v3 certificate of the peer. This element of the state may be null.
- *Compression method*: The algorithm used to compress data prior to encryption.
- *Cipher spec*: Specifies the bulk data encryption algorithm (such as null or DES) and a MAC algorithm (such as MD5 or SHA). It also defines cryptographic attributes such as hash size.
- *Master secret*: A 48-byte secret shared between the client and the server.
- *Is resumable*: A flag indicating whether the session can be used to initiate new connections.

The TLS Record layer uses these session parameters to create its security parameters. The resumption feature allows many connections to be instantiated using the same session.

The TLS handshake process is shown in Figure 11.18.

Step 1: The client and server exchange hello messages to negotiate algorithms, exchange random values, and initiate or resume the session.

---

[4]Readers please note: MAC in this chapter refers to message authentication code, not to medium access control.

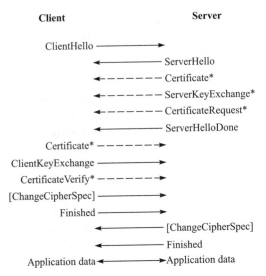

Client	Server
ClientHello ───────────▶	
◀─────────── ServerHello	
◀─ ─ ─ ─ ─ ─ Certificate*	
◀─ ─ ─ ─ ─ ─ ServerKeyExchange*	
◀─ ─ ─ ─ ─ ─ CertificateRequest*	
◀─────────── ServerHelloDone	
Certificate* ─ ─ ─ ─ ─ ─▶	
ClientKeyExchange ───────────▶	
CertificateVerify* ─ ─ ─ ─ ─ ─▶	
[ChangeCipherSpec] ───────────▶	
Finished ───────────▶	
◀─────────── [ChangeCipherSpec]	
◀─────────── Finished	
Application data ◀────────▶ Application data	

FIGURE 11.18 The TLS handshake process

Notes: * indicates optional or situation-dependent messages that are not always sent. The ChangeCipherSpec is not aTLS handshake message.

In Figure 11.18 the client sends a ClientHello message to request a connection. The ClientHello message includes the client version of the TLS protocol; the current time and date in standard UNIX 32-bit format and a 28-byte random value; an optional session ID (if not empty, this value identifies the session whose security parameters the client wishes to reuse, this field empty if no session ID available or if new security parameters are sought); a CipherSuite list that contains the combinations of key exchange, bulk encryption, and MAC algorithms that are supported by the client; and a list of compression algorithms supported by the client.

The server examines the contents of the ClientHello message. The server sends a ServerHello message if it finds an acceptable set of algorithms in the lists proposed by the client. If the server cannot find a match, it sends a handshake failure alert message and closes the connection. The ServerHello message contains the following parameters: the server version, a random value that is different from and independent of the value sent by the client, a session ID, a CipherSuite selected from the list proposed by the client, and a compression algorithm from the list proposed by the client.

Step 2: The client and server exchange cryptographic parameters to allow them to agree on a premaster secret. If necessary, they exchange certificates and cryptographic information to authenticate each other. They then generate a master secret from the premaster secret and exchange random values.

In Figure 11.18, after the ServerHello message the server may also send the following three messages. The Certificate message is sent if the server needs to be authenticated. The message generally includes an X509.v3 certificate that contains a key for the corresponding key exchange method. The ServerKeyExchange is sent immediately after the Certificate message and is required when the server certificate message does not contain enough data to allow the client to exchange a premaster secret, as for example in a Diffie-Hellman exchange. The

CertificateRequest message is sent by the server if the client is required to be authenticated. Finally, the server sends the ServerHelloDone message, indicating that it is done sending messages to support the key exchange. The server then waits for the client's response.

The client examines the messages from the server and prepares appropriate responses. If required to, the client sends a certificate message with a suitable certificate. If it has no suitable certificate, the client may reply with a certificate message containing no certificate, but the server may then respond with a fatal handshake failure alert message that results in a termination of the connection. The client may follow the certificate message with a ClientKeyExchange message that provides information to set the premaster secret. The CertificateVerify message is sent by a client after it has sent a certificate message. The purpose of the message is to explicitly verify the client certificate. The client prepares a digital signature of the sequence of messages that have been exchanged from the client hello up to but not including this message. The client uses its private key to prepare the signature. This step allows the server to verify that the client owns the private key for the client certificate.

Step 3. The client and server provide their record layer with the security parameters. The client and server verify that their peer has calculated the same security parameters and that the handshake occurred without tampering by an attacker.

The ChangeCipherSpec message is part of the Change Cipher protocol that signals changes in the ciphering strategy. The protocol consists of a single message that is encrypted and compressed according to the current connection state. When sent by the client, the ChangeCipherSpec message notifies the server that subsequent records will be protected under a new CipherSpec and keys. After the client sends the ChangeCipherSpec message in Figure 11.18, the server copies its pending CipherSpec into the current CipherSpec. The client follows immediately with a finished message, which is prepared using the new CipherSpec algorithms. The finished message allows the server to verify that the key exchange and authentications have been successful.

The server responds with ChangeCipherSpec and finished messages of its own. Once the client and server have validated the finished messages they receive, they finally can begin to exchange information over the connection.

THE TLS RECORD PROTOCOL

The TLS Record protocol is used for the encapsulation of higher-level protocols. The protocol provides privacy through the use of symmetric encryption and provides reliability through the use of a message authentication code. The operation of the TLS Record protocol is determined by the TLS connection state, which specifies the compression, encryption, and MAC algorithms; the MAC secret; the bulk encryption keys; and the initialization vectors for the connection in the read (receive) and write (send) directions. The initial state always specifies that no encryption, compression, or MAC be used. The TLS Handshake protocol sets the security parameters for pending read and write states, and the ChangeCipherSpec protocol makes the pending states current.

The connection state also includes the following elements: the state of the compression algorithm, the current state of the encryption algorithm, the MAC secret for the connection, and the record sequence number that is initially 0 and may not exceed $2^{64} - 1$. The connection state needs to be updated each time a record is processed.

The sender in the TLS Record protocol is responsible for taking messages from the application layer, fragmenting them into manageable blocks, optionally compressing them, applying a Message Authentication Code applying encryption, and transmitting the results. The receiver is responsible for decryption, verification, decompression, and message reassembly and delivery to the application layer.

BACKWARD COMPATIBILITY WITH SSL

The TLS v1.0 protocol is based on the SSL 3.0 but is sufficiently different to not interoperate. However, TLS 1.0 does incorporate a feature that allows a TSL implementation to back down to SSL 1.0. When a TLS client wishes to negotiate with a SSL 3.0 server, the client sends a ClientHello message using the SSL 3.0 record format and client structure but sends {3,1} for the version field to indicate that it supports TLS 1.0. A server that supports only SSL 3.0 will respond with an SSL 3.0 ServerHello message. A server that supports TLS will respond with a TLS ServerHello. The negotiations will then proceed as appropriate.

A TLS server that wishes to handle SSL 3.0 clients should accept SSL 3.0 ClientHello messages and respond with SSL 3.0 ServerHello messages if the ClientHello message has version field {3,0}, indicating that the client does not support TLS.

For a discussion of the differences between TLS 1.0 and SSL 3.0, see [RFC 2246].

11.3   CRYPTOGRAPHIC ALGORITHMS

We have so far intentionally avoided discussing the details of particular cryptographic algorithms. This approach enabled us to emphasize the fact that security procedures are independent of specific cryptographic algorithms as long as certain requirements are met. In this section we discuss two specific cryptographic algorithms that are used in current standards. The reader is referred to [Kaufman et al., 1995], [Schneier 1996], and [Stallings 1999] for more detailed discussions of cryptographic algorithms.

11.3.1   DES

Data Encryption Standard (DES) was developed by IBM in early 1970s and adopted by the National Bureau of Standards, now the National Institute of

Standards and Technology (NIST), in 1977. DES is now the most widely used shared key cryptographic system. DES can be implemented much more efficiently in hardware than in software.

In the encryption process DES first divides the original message into blocks of 64 bits. Each block of 6-bit plaintext is separately encrypted into a block of 64-bit ciphertext. DES uses a 56-bit secret key. The choice of the key length has been a controversial subject. It is generally agreed that 56 bits are too small to be secure. The DES encryption algorithm, which has 19 steps, is outlined in Figure 11.19. The decryption basically runs the algorithm in reverse order.

Each step in the DES algorithm takes a 64-bit input from the preceding step and produces a 64-bit output for the next step. The first step performs an initial permutation of 64-bit plaintext that is independent of the key. The last step performs a final permutation that is the inverse of the initial permutation. It is generally believed that the initial and final permutations do not make the algorithm more secure. However, they may add some value when multiple DES encryptions are being performed with multiple keys.

The next-to-last stage swaps the 32 bits on the left with the 32 bits on the right. Each of the remaining 16 iterations performs the same function but uses a different key. Specifically, the key at each iteration is generated from the key at the preceding iteration as follows. First a 56-bit permutation is applied to the

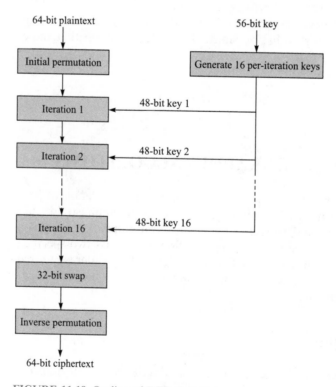

FIGURE 11.19 Outline of DES algorithm

key. Then the result is partitioned into two 28-bit blocks, each of which is independently rotated left by some number of bits. The combined result undergoes another permutation. Finally, a subset of 48 bits is used for the key at the given iteration.

The operation at each iteration is shown in Figure 11.20. The 64-bit input is divided into two equal portions denoted by L_{i-1} and R_{i-1}. The output generates two 32-bit blocks denoted by L_i and R_i. The left part of the output is simply equal to the right part of the input. The right part of the output is derived from the bitwise XOR of the left part of the input and a function of the right part of the input and the key at the given iteration.

The preceding algorithm simply breaks a long message into 64-bit blocks, each of which is independently encrypted using the same key. In this scheme DES is said to be operating in the **electronic codebook** (ECB) mode. This mode may not be secure when the structure of the message is known to the attacker.

Deficiency in the ECB mode can be removed by introducing dependency among the blocks. A simple way to introduce the dependency is to XOR the current plaintext block with the preceding ciphertext block. Such a scheme is shown in Figure 11.21 and is called **cipher block chaining** (CBC). The first plaintext block is XORed with a given *initialization vector (IV)* that can be transmitted to the receiver in ciphertext for maximum security. CBC is generally the preferred approach for messages longer than 64 bits.

As alluded earlier, using a 56-bit key makes DES vulnerable to brute-force attack. One well-known improvement is called **triple DES**, which actually uses two keys, extending the overall key length to 112 bits. Choosing two keys instead of three keys reduces the overhead significantly without comprising the security for commercial purposes. Triple DES performs the following encryption algorithm:

$$C = E_{K1}(D_{K2}(E_{K1}(P)))$$

and the following decryption algorithm:

$$P = D_{K1}(E_{K2}(D_{K1}(C)))$$

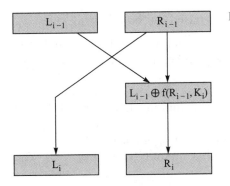

FIGURE 11.20 Each iteration in DES

(a) Encryption

FIGURE 11.21 Cipher block chaining

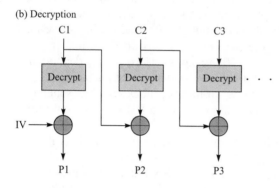

(b) Decryption

Note that the encryption algorithm uses an encryption-decryption-encryption (EDE) sequence rather than triple encryption EEE. This practice allows triple DES to talk to a single-key DES by making K1 and K2 the same, since $E_{K1}(D_{K1}(E_{K1}(P))) = E_{K1}(P)$.

11.3.2   RSA

The RSA algorithm (named after its inventors, Rivest, Shamir, and Adleman) is a widely accepted scheme for public key cryptography. It utilizes modular arithmetic and factorization of large numbers.

The public and private keys are generated based on the following rules:

1. Choose two large prime numbers p and q such that the product is equal to n. The plaintext P that is represented by a number must be less than n. In practice, n is a few hundred bits long.
2. Find a number e that is relatively prime to $(p-1)^*(q-1)$. Two numbers are said to be relatively prime if they have no common factors except 1. The public key consists of $\{e, n\}$.
3. Find a number d such that $d^*e = 1 \bmod ((p-1)^*(q-1))$. In other words, d and e are multiplicative inverses of each other modulo $((p-1)^*(q-1))$. The private key consists of $\{d, n\}$.

The RSA algorithm is based on the fact that if n, p, q, d, and e satisfy properties 1 to 3 above, then for any integer $P < n$ the following key property holds:

$$P^{de}(\bmod\ n) = P(\bmod\ n)$$

The RSA algorithm uses binary keys that are several hundred bits long, typically 512 bits. RSA takes a binary block of plaintext of length smaller than the key length and produces a ciphertext that is the same length of the key. Suppose that P is an integer that corresponds to a block of plaintext. RSA encrypts P as follows:

$$C = P^e(\bmod\ n)$$

The above calculation will yield an integer between 0 and n, and hence will require the same number of bits as the key.

To decrypt the ciphertext C, the RSA algorithm raises C to the power e and reduces the result modulo n:

$$C^d(\bmod\ n) = (P^e)^d(\bmod\ n) = P^{de}(\bmod\ n) = P(\bmod\ n) = P.$$

Thus we see that the aforementioned key property ensures that the plaintext can be recovered by the owner of the private key.

Why is the RSA algorithm secure? We know the public key, n and d, so can't we just determine the other factor e? It turns out that factoring large integers n is very computationally intensive using currently available techniques.

Example—Using RSA

Let us consider the following simple example. Suppose we have chosen $p = 5$, and $q = 11$. Then $n = 55$, and $(p - 1)(q - 1) = 40$. Next find a number e that is relatively prime to 40, say, 7. The multiplicative inverse of 7 modulo 40 yields $d = 23$. From number theory we can find such a d only if e is relatively prime to $((p - 1)(q - 1))$. Now the public key is $\{7, 55\}$, and the private key is $\{23, 55\}$.

Suppose a message "RSA" is to be protected. For simplicity the message is represented numerically as 18, 19, 1 and uses three plaintexts: $P_1 = 18$, $P_2 = 19$, and $P_3 = 1$. The resulting ciphertexts are

$$C_1 = 18^7 \bmod 55 = 17$$
$$C_2 = 19^7 \bmod 55 = 24$$
$$C_3 = 1^7 \bmod 55 = 1$$

The reader may verify that the decryption produces

$$17^{23} \bmod 55 = 18$$
$$24^{23} \bmod 55 = 19$$
$$1^{23} \bmod 55 = 1$$

How do we calculate modular arithmetic involving large numbers such as 17^{23} mod 55? Fortunately, we can simplify the computation by using the following property:

$$(a^{*}b) \bmod n = ((a \bmod n)^{*}(b \bmod n)) \bmod n$$

As an example, we can write the first decryption as

$$17^{23} \bmod 55 = 17^{16+4+2+1} \bmod 55 = (17^{16}\ 17^{4}\ 17^{2}\ 17) \bmod 55$$

Now 17^{2} mod 55 = 14. Then we continue with 17^{4} mod 55 = $(17^{2})^{2}$ mod 55 = 14^{2} mod 55 = 4046 mod 55 = 31. Similarly, 17^{8} mod 55 = $(17^{4})^{2}$ mod 55 = 31^{2} mod 55 = 961 mod 55 = 26, and 17^{16} mod 55 = $(17^{8})^{2}$ mod 55 = 26^{2} mod 55 = 676 mod 55 = 16.

Finally, we can write

$$17^{23} \bmod 55 = (17^{16}\ 17^{4}\ 17^{2}\ 17) \bmod 55 = (16 \times 31 \times 14 \times 17) \bmod 55 = 18$$

which is the original plaintext P_{1}.

SUMMARY

Security protocols are required to deal with various threats that arise in communication across a network. These threats include replay attacks, denial of service attacks, and various approaches to impersonating legitimate clients or servers. We saw that security protocols build on cryptographic algorithms to provide privacy, integrity, authentication, and non-repudiation services.

Secret key cryptography requires the sharing of the same secret key by two users. Secret key algorithms have the advantage that they involve modest computation. However the need for clients and servers to manage a different key for each association is too complicated. We saw that key distribution centers can help deal with the problem of managing keys in a secret key system.

Public key cryptography simplifies the problem of key management by requiring each user to have a private key that is kept secret and a public key that can be distributed to all other users. Public key cryptography nevertheless requires certification authorities that can vouch that a certain public key corresponds to a given user. Public key algorithms tend to be more computationally intensive than secret key algorithms. Consequently security protocols tend to use public key techniques to establish a master key at the beginning of a session, which is then used to derive a session key that can be used with a secret key method. The classic Diffie-Hellman exchange is used to establish a joint secret across a network. We saw how recent security protocol standards address weaknesses in the Diffie-Hellman exchange through the use of cookies and authentication.

Security protocols can be used at various layers of the protocol stack. The IP Security standards provide for extensions to IPv4 and IPv6 headers so that authentication, privacy, and integrity service can be provided across a public internet. Virtual private networks can be created using the tunneling capabilities of IPSec. We examined the Secure Sockets Layer and Transport Layer Security protocols which can provide secure connections over TCP, for application layer protocols such as HTTP. We also mentioned the important examples of PGP and Kerberos authentication service which demonstrate how security can be applied at the application layer or above.

Finally, we showed that security protocols can operate over various cryptographic algorithms as long as certain general requirements are met. We provided only a brief discussion of two important cryptographic algorithms, DES and RSA. The subject of cryptographic algorithms is currently an area of intense research. The student is referred to several excellent texts on the subject that are listed in the references.

CHECKLIST OF IMPORTANT TERMS

authentication header (AH)
certification authority (CA)
challenge
cipher
cipher block chaining (CBC)
ciphertext
cryptograph
Data Encryption Standard (DES)
decryption
digital signature
electronic codebook (ECB)
Encapsulating Security Payload
  (ESP)
encryption
encryption header
firewall
hashed message authentication code
  (HMAC)
internet key exchange (IKE)
key
key distribution center (KDC)

keyed MD5
message authentication code (MAC)
message digest 5 (MD5) algorithm
nonce
plaintext
public key cryptography
replay attack
response
Rivest, Shamir, and Adleman (RSA)
  algorithm
secret key cryptography
secure hash algorithm-1 (SHA-1)
security association
session
substitution cipher
transport mode
transposition cipher
triple DES
tunnel
tunnel mode

FURTHER READING

Chapman, D. B. and E. D. Zwicky, *Building Internet Firewalls*, O'Reilly, Cambridge, 1995.

Freier, A. O., P. Karlton, and P. C. Kocher, "The SSL Protocol, Version 3.0," Transport Layer Security Working Group, November 18, 1996.

ISO/IEC 9594-8, "IT-OSI-The Directory: Authentication Framework," 1997.

Kaufman, C., R. Perlman, and M. Speciner, *Network Security: Private Communication in a Public World*, Prentice Hall, Upper Saddle River, New Jersey, 1995.

Khare, R. and S. Agranat, "Upgrading to TLS within HTTP/1.1," Network Working Group, June 22, 1999.

Schneier, B., *Applied Cryptography*, John Wiley & Sons, 1996.

Stallings, W., *Cryptography and Network Security: Principles and Practice*, Prentice Hall, Upper Saddle River, New Jersey, 1999.

Wagner, D. and B. Schneier, "Analysis of the SSL 3.0 Protocol," November 19, 1996, available at http://www.counterpane.com/ssl.html.

Yeager, N. J. and R. E. McGrath, *Web Server Technology: The Advanced Guide for World Wide Web Information Providers*, Morgan Kaufmann, San Francisco, 1996.

RFC 2246, T. Dierks and C. Allen, "The TLS Protocol Version 1.0," January 1999.

RFC 2401, S. Kent and R. Atkinson, "Security Architecture for the Internet Protocol," November 1998.

RFC 2402, S. Kent and R. Atkinson, "IP Authentication Header," November 1998.

RFC 2406, S. Kent and R. Atkinson, "IP Encapsulating Security Payload (ESP)," November 1998.

RFC 2408, D. Maughan, M. Schertler, M. Schneider, and J. Turner, "Internet Security Association and Key Management Protocol (ISAKMP)," November 1998.

RFC 2409, D. Harkins and D. Carrel, "The Internet Key Exchange," November 1998.

RFC 2412, H. K. Orman, "The OAKLEY Key Determination Protocol," November 1998.

PROBLEMS

1. Suppose that a certain information source produces symbols from a ternary alphabet {A, B, C}. Suppose also that the source produces As three times as frequently as Bs and Bs twice as frequently as Cs. Can you break the following ciphertext?
   a. CBA
   b. CBAACCBACBBCCCCBACC

2. Suppose that a certain information source produces symbols from a ternary alphabet {A, B, C}. Suppose that it is known that As always occur in pairs. Suppose that it is also known that an encryption scheme takes blocks of four symbols and permutes their order. Can you break the following ciphertext: BABACAABACACBAAB?

3. In authentication using a secret key, consider the case where the transmitter is the intruder.
   a. What happens if the transmitter initiates the challenge?
   b. How secure is the system if the receiver changes the challenge value once every minute.

c. How secure is the system if the receiver selects the challenge value from a set of 128 choices?

4. Consider a system where a user is authenticated based on an ID and password that are supplied by the transmitter in plaintext. Does it make any difference if the password and ID are encrypted? If yes, explain why? If no, how would you improve the system?

5. Compare the level of security provided by a server that stores a table of IDs and associated passwords as follows:
   a. Name and password stored unencrypted.
   b. Name and password stored in encrypted form.
   c. Hash of name and password stored.

6. Explain why the processsing required to provide privacy service is more complex than the processing required for authentication and for integrity.

7. Consider the following hashing algorithm. A binary block of length M is divided into subblocks of length 128 bits, and the last block is padded with zeros to a length of 128. The hash consists of the XOR of the resulting 128-bit vectors. Explain why this hashing algorithm is not appropriate for cryptographic purposes.

8. The *birthday problem* is a standard problem in probability textbooks, and it is related to the complexity of breaking a good hashing function.
   a. Suppose there are n students in a class. What is the probability that at least two students have the same birthdate? Hint: Find the probability that all students have a different birthdate.
   b. In part (a) for what value of n is the probability that at least two students have the same birthdate approximately $\frac{1}{2}$?
   c. Suppose we use a 64-bit hash function. If we obtain the hash outputs for n distinct messages, what is the probability that no two hashes are the same? For what value of n the probability of finding at least one matching pair $\frac{1}{2}$?
   d. Concoct a digital signature scenario in which the ability to find a pair of messages that lead to the same hash output would be useful to an unscrupulous person.

9. Consider the operation of the key distribution center: the KDC sends $E_{KA}(K_{AB})$ and E_{KB} (K_{AB}) to host A; Host A in turn sends the "ticket" $E_{KB}(K_{AB})$ to host B. Explain why the term *ticket* is appropriate in this situation.

10. Suppose a directory that stores certificates is broken into, and the certificates in the directory are replaced by bogus certificates? Explain why users will still be able to identify these certificates as bogus.

11. Suppose that KDC A serves one community of users and KDC B serves a different community of users. Suppose KDC A and KDC B establish a shared key K_{AB}. Develop a method to enable a user α from KDC A to obtain a shared key with a user β from KDC B. Note that KDCs A and B must participate in the process of establishing the shared key $K_{\alpha\beta}$.

12. Consider a highly simplified Diffie-Hellman exchange in which $p = 29$ and $g = 5$. Suppose that user A chooses the random number $x = 3$ and user B chooses the number $y = 7$. Find the shared secret K.

13. Consider a firewall consisting of a packet-filtering router.
   a. Explain how this firewall could be set up to allow in only HTTP requests to a specific server.
   b. Suppose an outside intruder attempts to send packets with forged internal addresses, which presumably are allowed access to certain systems. Explain how a packet-filtering router can detect these forged packets.
   c. Explain how packets from a given remote host can be kept out.
   d. Explain how inbound mail to specific mail servers is allowed in.
   e. Explain how inbound Telnet service can be blocked.

14. The ICMP "Host Unreachable" error message is sent by a router to the original sender when the router receives a datagram that it cannot deliver or forward. Suppose that the router in question is a packet-filtering router. Discuss the pros and cons of having the router return an ICMP message from the following viewpoints: processor load on the router, traffic load in the network, susceptibility to denial-of-service attacks, and disclosure of filtering operation to intruders.

15. Compare the following two approaches to operation of a packet-filtering router: discard everything that is not expressly permitted versus forward everything that is not expressly prohibited.
   a. Which policy is more conservative?
   b. Which policy is more visible to the users?
   c. Which policy is easier to implement?

16. Identify the fields in the IPv4 header that cannot be covered by the authentication header. What can be done to protect these fields?

17. Explain the benefits the authentication header in IPSec brings to the operation of a packet-filtering router.

18. Explain why a packet-filtering router should reassemble a packet that has been fragmented in the network and check its authentication header, instead of forwarding the fragments to the destination.

19. a. Explain how the use of cookies thwarts a denial-of-service attack in the Diffie-Hellman exchange.
   b. Explain how authentication thwarts the man-in-the-middle attack.

20. Consider Figure 11.9a where two hosts communicate directly over an internet. Explain the rationale for the following uses of AH and ESP modes in IPSec. Show the resulting IP packet formats.
   a. Transport mode: AH alone.
   b. Transport mode: ESP alone.
   c. Transport mode: AH applied after ESP.
   d. Tunnel mode: AH alone.
   e. Tunnel mode: ESP alone.

21. Explain how IPSec can be applied in the exchange of routing information, for example, OSPF.

22. Consider the arrangement in Figure 11.9b where two internal networks are protected by firewalls. Explain the rationale for establishing a tunnel between gateways using AH and having the end hosts use ESP in transport mode.

23. Consider the arrangement in Figure 11.9c where a mobile host communicates with an internal host across a firewall. Explain the rationale for establishing an AH tunnel between the remote host and the firewall and ESP in transport mode between the two hosts.

24. When is IPSec appropriate? When is SSL/TLS appropriate?

25. Suggest some SSL/TLS situations in which it is appropriate to authenticate the client.

26. HTTP over SSL uses URLs that begin with https: and use TCP port 443, whereas HTTP alone uses URLs that begin with http: and use TCP port 80. Explain how a secure and unsecured mode using the same port number 80 may be possible by introducing an upgrade mechanism in HTTP.

27. Suppose DES is used to encrypt a sequence of plaintext blocks $P_1, P_2, P_3, \ldots, P_i, \ldots, P_N$ into the corresponding ciphertext blocks $C_1, C_2, C_3, \ldots, C_i \ldots, C_N$.
   a. If block C_i is corrupted during transmission, which block(s) will not be decrypted successfully using the ECB mode?
   b. Repeat (a) using the CBC mode.

28. Using the RSA algorithm, encrypt the following:
   a. $p = 3, q = 11, e = 7, P = 12$
   b. $p = 7, q = 11, e = 17, P = 25$
   c. Find the corresponding ds for (a) and (b) and decrypt the ciphertexts.

CHAPTER 12

Multimedia Information and Networking

In Chapter 1 we discussed the trend toward communication networks that can handle a wide range of information types including multimedia. This chapter has two purposes. The first purpose is to give a more detailed presentation of the formats, the bit rates, and other properties of important types of information including text, speech, audio, facsimile, image and video. In the first three sections we show that the representation and audio and video information can require higher bit rates and lower delays than have been traditional in packet networks. Advances such as those discussed in Chapter 10 will provide packet networks with the responsiveness to meet these requirements. The second purpose of the chapter is to introduce the protocols that are available for the transport of multimedia information over packet networks. We introduce the Real-Time Transport Protocol (RTP), which has been developed for the transfer of real-time information over packet networks. We also introduce two approaches to session control protocols that will play a key role in the introduction of telephone service in the Internet.

This chapter is organized as follows:

1. *Lossless data compression.* We present techniques that reduce the number of bits required to represent a file of information under the condition that the original file can be recovered in its exact form. We discuss how existing standards apply lossless data compression to text and other computer files, as well as to facsimile.
2. *Digital representation of analog signals.* We discuss how existing lossy data compression standards covert images, voice, audio, and video to binary form.
3. *Techniques for increasing compression.* We present techniques that reduce the number of bits required to represent a block or stream of information under the condition that the original file can be recovered approximately but within

some level of fidelity. In particular we introduce the MPEG compression standards for audio and video.

4. *The Real-Time Transport Protocol.* We introduce RTP, which operates over UDP and provides the framework for the transport of multimedia over Internet.

5. *Session control protocols.* We introduce the Session Initiation Protocol, which has been designed to work with RTP. We also introduce the ITU-T H.323 recommendation for packet-based multimedia communication systems. We also briefly discuss the interworking of traditional telephony and corresponding systems in IP networks.

Note that sections 12.4 and 12.5 do not require the first three sections as prerequisite material.

12.1 LOSSLESS DATA COMPRESSION

In lossless data compression we are given a block of digital information, and we are interested in obtaining an efficient digital, usually binary, representation. Examples of such a block of information are any computer files generated by applications, such as word processors, spreadsheets, and drawing programs. Lossless data compression techniques are also used by modems and facsimile machines to reduce the time to transmit files and by operating systems and utility programs to reduce the amount of disk space required to store files.

We will view a block of information as a sequence of symbols from some alphabet as shown in Figure 12.1. The objective of lossless data compression is to map the original information sequence into a string of binary digits so that (1) the average number of bits/digital symbol is small and (2) the original digital information sequence can be recovered exactly from encoded binary stream. We will assess the performance of any given lossless data compression code by the average number of encoded bits/symbol.

12.1.1 Huffman Codes

The Morse code uses the basic principle for achieving efficient representations of information: frequently occurring symbols should be assigned short codewords, and infrequent symbols should be assigned longer codewords. In this manner the average number of bits/symbol is reduced. In 1954 Huffman invented an

FIGURE 12.1 Lossless data compression

algorithm for identifying the code that is optimal in the sense of minimizing the average number of bits/symbol. A **Huffman code** segments the original sequence of symbols into a sequence of fixed-length blocks. Each block is assigned a variable-length binary stream called a codeword. The codewords are selected so that no codeword is a prefix of another **codeword**. The set of all codewords then form the terminal nodes of a binary tree such as the one shown in Figure 12.2.

Before discussing the Huffman procedure for obtaining the best code, we will discuss a simple example to demonstrate the encoding and decoding procedures. For simplicity we will consider encoding individual symbols, that is, blocks of length 1. This example assumes that our information source generates a stream of symbols from the alphabet {a,b,c,d,e} and that the symbols occur with respective probabilities {1/4, 1/4, 1/4, 1/8, 1/8}. Figure 12.2 shows a binary code that assigns two-bit and three-bit codewords to the symbols. The figure also shows the binary string that results from encoding the sequence of symbols aedbbad. ... The binary tree code in this figure shows how the codewords correspond to the five terminal nodes of the tree.

The tree in Figure 12.2 can be used to demonstrate the general decoding procedure. At the beginning of the decoding, we begin at the top node of the tree. Each encoded bit determines which branch of the tree is followed. The sequence of encoded bits then traces a path down the tree until a terminal node is reached. None of the intermediate nodes prior to this terminal node can correspond to a codeword by the design of the code. Therefore, as soon as a terminal node is reached, the corresponding symbol can be output. The decoder then returns to the top node of the tree and repeats this procedure. This technique, while efficiently compacting the number of binary bits, increases the effect of transmission errors. For example, if the 00 for *a* incurs an error giving 11, the result will be not only a change in the letter but also an error in the alignment of sets of data bits with letters.

Let $\ell(s)$ be the length of the codeword assigned to symbol s. The performance of the code is given by the average number of encoded bits/symbol, which is given by

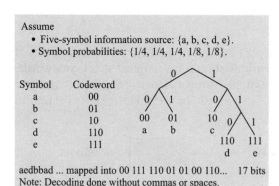

Assume
• Five-symbol information source: {a, b, c, d, e}.
• Symbol probabilities: {1/4, 1/4, 1/4, 1/8, 1/8}.

Symbol	Codeword
a	00
b	01
c	10
d	110
e	111

aedbbad ... mapped into 00 111 110 01 01 00 110... 17 bits
Note: Decoding done without commas or spaces.

FIGURE 12.2 Huffman coding example

$$E[\ell] = \ell(a)P[a] + \ell(b)P[b] + \ell(c)P[c] + \ell(d)P[d] + \ell(e)P[e]$$
$$= 2(.25) + 2(.25) + 2(.25) + 3(.125) + 3(.125)$$
$$= 2.25\text{bits/symbol.}$$

The simplest code to use for this information source would assign codewords of equal length to each symbol. Since the number of codewords is five, 3-bit codewords would be required. The performance of such a code is three bits/symbol. Thus, for example, a file consisting of 10,000 symbols would produce 30,000 bits using three-bit codewords and an average of 22,500 bits using the Huffman code.

The extent to which compression can be achieved depends on the probabilities of the various symbols. For example, suppose that a source produces symbols with equal probability. Clearly the best way to code the symbols is to give them codewords of equal length to the extent possible. In particular, if the source produces symbols from an alphabet of size 2^m, then the best code simply assigns an m-bit codeword to each symbol. No other code can produce further compression. The Shannon entropy, which is introduced in the next section, specifies the best possible compression performance for a given information source.

The design of the Huffman code requires knowledge of the probabilities of the various symbols. In certain applications these probabilities are known ahead of time, so the same Huffman code can be used repeatedly. If the probabilities are not known, codes that can adapt to the symbol statistics "on the fly" are preferred. Later in this section we present the Lempel-Ziv adaptive code.

◆ HUFFMAN CODE ALGORITHM AND THE SHANNON ENTROPY

We are now ready to discuss the Huffman algorithm. We assume that the probabilities of the symbols are known. The algorithm starts by identifying the two symbols with the smallest probabilities. It can be shown that the best code will connect the terminal nodes of these two symbols to the same intermediate node. These two symbols are now combined into a new symbol whose probability is equal to the sum of the two probabilities. You can imagine the two original symbols as being placed inside an envelope that now represents the combined symbol. In effect we have produced a new alphabet in which the two symbols have been replaced by the combined symbol. The size of the new alphabet has now been reduced by one symbol. We can again apply the procedure of identifying the two symbols with the smallest probabilities from the reduced alphabet and combining them into a new symbol. Each time two symbols are combined, we connect the associated nodes to form part of a tree. The procedure is repeated until only two symbols remain. These are combined to form the root node of the tree.

The Huffman procedure is demonstrated in Figure 12.3a where we consider a source with five symbols {a, b, c, d, e} with respective probabilities {.50, .20, .15, .10, .05}. The first step of the Huffman algorithm combines symbols d and e to form a new symbol, which we will denote by (de) and which has combined probability .15. The terminal node for symbols d and e are combined into the intermediate node denoted by 1 in Figure 12.3a. The new alphabet now consists

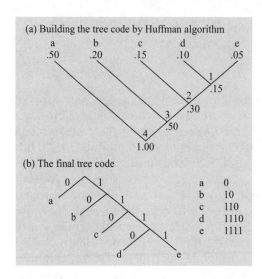

(a) Building the tree code by Huffman algorithm

(b) The final tree code

a	0
b	10
c	110
d	1110
e	1111

FIGURE 12.3 Building a Huffman tree code

of the symbols {a, b, c, (de)} with probabilities {.50, .20, .15, .15}. The second step of the Huffman algorithm combines the symbols c and (de) into the new symbol (c(de)) with combined probability of .3. These are combined into intermediate node 2 in Figure 12.3a. The third step of the algorithm combines symbols b and (c(de)) into (b(c(de))). The final step combines the two remaining symbols a and (b(c(de))) to form the root node of the tree.

In Figure 12.3b we have rearranged the tree code obtained from the Huffman algorithm and obtained the codewords leading to the terminal nodes. This is done by starting at the root node and assigning a 0 to each left branch and a 1 to each right branch. The resulting code is shown in Figure 12.3b. Let $\ell(s)$ be the length of the codeword assigned to symbol s. The performance of the code is given by the average number of encoded bits/symbol, which is given by

$$E[\ell] = \ell(a)P[a] + \ell(b)P[b] + \ell(c)P[c] + \ell(d)P[d] + \ell(e)P[e]$$
$$= 1(.50) + 2(.20) + 3(.15) + 4(.10) + 4(.05)$$
$$= 1.95 \text{ bits/symbol}$$

The simplest code that could have been used for this information source would assign three-bit codewords to each symbol. Thus we see that the Huffman code has resulted in a representation more efficient than the simple code.

We indicated above that in general Huffman coding deals with blocks of n symbols. To obtain the Huffman codes where n is greater than 1, we arrange all possible n-tuples of symbols and their corresponding probabilities. We then apply the Huffman algorithm to this superalphabet to obtain a binary tree code. For example, the $n = 2$ code in Figure 12.3 would consider the alphabet consisting of the 25 pairs of symbols {aa, ab, ac, ad, ae, ba, bb, ..., ea, eb, ec, ed, ee}. In general, performance of the code in terms of bits/symbol will improve as the block length is increased. This of course, is accompanied by an increase in complexity.

In 1948 Shannon addressed the question of determining the best performance attainable in terms of encoded bits/symbol for any code of any block length. He found that the best performance was given by the entropy function, which depends on the probabilities of sequences of symbols. Suppose that the symbols are from the alphabet $\{1, 2, 3, \ldots, K\}$ with respective probabilities $\{P[1], P[2], P[3], \ldots, P[K]\}$. If the sequence of symbols are statistically independent, the **entropy** is given by

$$H = -\sum_{k=1}^{K} P[k] \log_2 P[k]$$
$$= -P[1] \log_2 P[1] - P[2] \log_2 P[2] \ldots - P[K] \log_2 P[K]$$

Shannon proved that no code can attain an average number of bits/symbol smaller than the entropy. In the preceding expression the logarithms are taken to the base 2, which can be obtained as follows:

$$\log_2 x = \ln x / \ln 2$$

where $\ln x$ is the natural logarithm.

For the example in Figure 12.3, the entropy is given by

$$H = (-.50 \ln .50 - .20 \ln .20 - .15 \ln .15 - .10 \ln .10 - .05 \ln .05)/\ln 2$$
$$= 1.923$$

This result indicates that the code obtained in Figure 12.3 is sufficiently close to the best attainable performance and that coding of larger block lengths is unnecessary. If the performance of the code had differed significantly from the entropy, then larger block lengths could be considered.

As an additional example consider the code in Figure 12.2. It can be easily shown that Huffman algorithm will produce this code. The average number of bits/symbol for this code is 2.25 bits. Furthermore, the entropy of this code is also 2.25 bits/symbol. In other words, this code attains the best possible performance, and no increase in block length will yield any improvements.

The entropy formula can also be used as a guideline for obtaining good codes. For example, suppose that a source has a K symbol alphabet and that symbols occur with equal probability, that is, $1/K$. The entropy is then given by

$$H = -\sum_{k=1}^{K} P[k] \log_2 P[k] = -\sum_{k=1}^{K} \frac{1}{K} \log_2 \frac{1}{K} = \log_2 K$$

In the special case where $K = 2^m$, we have $H = \log_2 2^m = m$ bits per symbol. Note that we can assign each symbol an m-bit codeword and achieve the entropy. Thus in this case this simple code achieves the best possible performance. This result makes intuitive sense, since the fact that the symbols are equiprobable suggests that they should have the same length. More generally, by comparing the expression for the entropy and the expression for the average number of bits/symbol, we see that the length of the kth symbol can be identified with the term $-\log_2 1/P[k]$. This suggests that a good code will assign to a symbol that has probability $P[k]$ a binary codeword of length $-\log_2 1/P[k]$. For example, if a given symbol has probability $1/2$, then it should be assigned a one-bit codeword.

12.1.2 Run-Length Codes

In many applications one symbol occurs much more frequently than all other symbols, as shown in Figure 12.4. The sequence of symbols produced by such information sources consist of many consecutive occurrences of the frequent symbol, henceforth referred to as **runs**, separated by occurrences of the other symbols. For example, the files corresponding to certain types of documents will contain long strings of blank characters. Facsimile information provides another example of where the scanning process produces very long strings of white dots separated by short strings of black dots. Run-length coding is a very effective means of achieving lossless data compression for these types of information sources. Instead of breaking the sequence of symbols into fixed-length blocks and assigning variable-length codewords to these blocks, a **run-length code** parses the sequence into variable-length strings consisting of consecutive occurrences of the common symbol and the following other symbol. The codeword for each run consists of a binary string to specify the length of the run, followed by a string that specifies the terminating symbol. Now if very short binary codewords are used to specify the length of very long runs, then it is clear that huge compression factors are being achieved. In the remainder of this section we focus on the run-length coding of binary sources that are typified by facsimile.

Figure 12.5 describes the simplest form of run-length coding for binary sources where 0s (white dots) are much more frequent than 1s (black dots). The sequence of binary symbols produced by the information source is parsed into runs shown in the left column. The encoder carries out this parsing simply by counting the number of 0s between 1s and producing an m-bit binary code-word that specifies the number of 0s. When two consecutive 1s occur, we say that a run of length 0 has occurred. The figure shows that the maximum complete run that can be encoded has length $2^m - 2$. Thus when $2^m - 1$ consecutive 0s are observed, the encoder must terminate its count. The code in Figure 12.5 has the encoder output a codeword consisting of all 1s that tells the decoder to output $2^m - 1$ consecutive 0s; the encoder resets its counter to 0 and starts a new run.

Figure 12.6 shows a string of 137 consecutive binary symbols that are encoded into four-bit codewords, and then decoded to obtain original sequence.

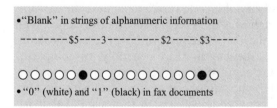

• "Blank" in strings of alphanumeric information

-------- \$5----3--------- \$2----- \$3-----

○○○○○●○○○○○○○○○○○●○
• "0" (white) and "1" (black) in fax documents

FIGURE 12.4 Run-length coding—Introduction

Inputs: ■□□□□□□□□□□□■□□□□□□□■□□□■□□□■□□□□□□

Run	Length	Codeword	Codeword ($m = 4$)
1	0	00...0	0000
01	1	00...1	0001
001	2	00...10	0010
0001	3	00...11	0011
00001	4	.	.
000001	5	.	.
0000001	6	.	.
.	.	.	.
.	.	.	.
000...01	$2^m - 2$	11...10	1110
000...00	run $> 2^m - 2$	11...11	1111

$\longleftarrow m \longrightarrow$

FIGURE 12.5 Run-length coding—Example 1

In the example the 137 original binary symbols are compacted into four bits. As before, the performance of the lossless data compression coding scheme is given by the average number of encoded bits per source symbol, which is given by $m/E[R]$ where m is the number of bits in the codeword and $E[R]$ is the average number of symbols encoded in each run. For example, suppose that the probability of a zero is 0.96 and that $m = 4$. A simple numerical calculation shows that the average run length is 11.45. The compression ratio is then $E[R]/m = 11.45/4 = 2.86$.

Another way of viewing the previous run-length coding procedure is to consider it as a two-step process. The first step maps the original symbol sequence into a sequence of lengths. The second step carries out the binary encoding of this sequence of lengths. The lengths themselves can be viewed as a sequence of symbols that are produced by an information source. In Figure 12.5 we use a fixed-length binary codeword to encode the lengths. It is clear then that in general we may be able to improve performance by using a Huffman code specifically designed for the probability of occurrence of these new symbols, that

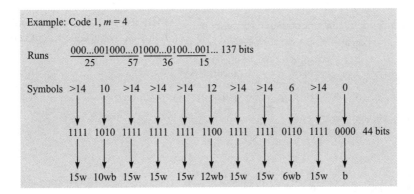

FIGURE 12.6 Run-length encoding and decoding

is, of the lengths. Again we will use a simple example to illustrate how we can improve the performance of run-length coding.

Figure 12.7 shows a run-length coding scheme that uses variable-length codewords to specify the lengths. As before, the sequence of binary symbols is parsed into runs of consecutive 0s followed by a 1. In this scheme the maximum complete run length is $2^m - 1$. When $2^m - 1$ consecutive 0s are encountered, the run is truncated as before. Suppose that the probability of this occurrence is approximately 1/2. Then according to our discussion of Huffman codes, it makes sense to assign this pattern a one-bit codeword. The code in Figure 12.7 assigns the codeword consisting of a single 0 to this pattern. All other runs are assigned a fixed-length codeword consisting of 1 followed by the binary representation of the length of the run. The code in Figure 12.7 attempts to achieve a higher compaction than that in Figure 12.5 by assigning a one-bit codeword to the longest word.

Figure 12.8 considers the same sequence of binary symbols as shown in Figure 12.6. The new code now encodes the original 137 binary symbols into 26 bits. The new coding scheme we have just presented maps a variable number of information symbols into variable-length binary codewords. The performance of this scheme is now given by $E[\ell]/E[R]$ where $E[\ell]$ is the average number of encoded bits per run and $E[R]$ is the average number of symbols encoded in each run. Suppose again that the probability of a zero is 0.96. A numerical calculation then gives $E[\ell] = 2.92$ and $E[R] = 12$, for a compression ratio of $12/2.92 = 4.11$.

FACSIMILE CODING STANDARDS

Run-length coding forms the basis for the coding standards that have been developed for facsimile transmission. In facsimile a black and white image is scanned and converted into a rectangular array of dots called **pixels**. Each pixel corresponds to a measurement made at a given point in the document, and each pixel is assigned a value 0 (for white) or 1 (for black) according to the measured intensity. The International Telecommunications Union (ITU) standard defines several options for encoding facsimile images. The standards

Inputs: ■□□□□□□□□□□□■□□□□□□□■□□□■□□□■□□□□□□

Run	Length	Codeword	Codeword ($m = 4$)
1	0	10...0	10000
01	1	10...1	10001
001	2	10...10	10010
0001	3	10...11	10011
00001	4		
000001	5	·	·
0000001	6	·	·
·	·	·	·
·	·	·	·
·	·	·	·
000...01	$2^m - 1$	11...11	11111
000...00	run $> 2^m - 1$	0	0

$$\overleftrightarrow{m + 1}$$

FIGURE 12.7 Run-length coding followed by variable-length coding

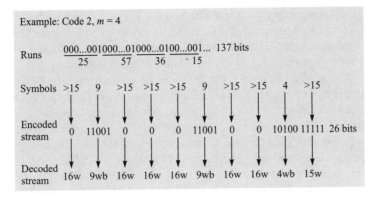

FIGURE 12.8 Variable-to-variable run-length coding using Huffman coding

assume a scanning resolution of 200 dots/inch in the horizontal direction and 100 dots/inch in the vertical direction. For an 8.5-×-11-inch standard North American document, the scanning process will produce $8.5 \times 200 \times 11 \times 100 = 1,870,000$ pixels. Because each pixel is represented by one bit, each page corresponds to 235 kilobytes where a byte consists of eight bits. Using a modem that transmits at a speed of 10,000 bits per second, we see that the transmission of each page will require about 3 minutes. Clearly lossless data compression is desirable in this application.

The ITU considered a set of eight documents, listed in Figure 12.9, in defining the facsimile coding standards. These documents are a business letter, a

Documents scanned at 200×100 pixels/square inch				
CCITT	G-II	G-III	G-IV	Time for G-IV (9.6 kbps)
1. Business letter	256K	17K	10K	8.33 sec
2. Circuit diagram	256K	15K	5.4K	4.50 sec
3. Invoice	256K	31K	14K	11.67 sec
4. Dense text	256K	54K	35K	29.17 sec
5. Technical paper	256K	32K	16K	13.33 sec
6. Graph	256K	23K	8.3K	6.92 sec
7. Dense Japanese text	256K	53.5K	34.6K	28.83 sec
8. Handwriting and simple graphics	256K	26K	10K	8.33 sec
Average	256K	31.4K	16.6K	
Compression ratio	1	8.2	15.4	
Max comp ratio	17.1	47.4		
Note: G-IV intended for documents scanned at 400×100 pixels/square inch and ISDN transmission at 64 kilobits/second.				

FIGURE 12.9 Compression results for ITU coding standards

simple hand-drawn circuit diagram, an invoice form, a page with dense text, a page from a technical paper, a graph, a page of dense Japanese text, and a page with simple graphics and handwriting. After scanning and prior to compression, each document required 256 kilobytes. (The ITU was considering A4 standard-size paper that is slightly larger than the 8.5-×-11-inch North American standard.) Four groups of standards are defined for facsimile by the ITU. The Group I and Group II standards are analog in nature. The Group II column in Figure 12.9 corresponds to a scanned but uncompressed version of the Group II standard. Groups III and IV use lossless data compression techniques.

The Group III standard defined by ITU uses a so-called one-dimensional modified Huffman coding technique that combines run-length coding and Huffman coding, as shown in Figure 12.10a. The ITU standards committee measured the statistics of runs of white symbols followed by runs of black symbols for the combined set of test documents. In addition to the symbols specifying the run lengths, special characters such as end-of-line and Escape need to be considered. Using this information, it is then possible to apply the Huffman algorithm to obtain a code for encoding the run length. However, because of the long runs that can occur, the resulting Huffman code can be quite complex. As a result, the ITU committee developed a modified, but simpler to implement, Huffman code to encode the run lengths. Because the statistics of the white runs and black runs differ, the Group III standard uses different Huffman codes to encode the white runs and black runs.

The Group III column in Figure 12.9 shows the number of kilobytes for each test document after compression using Group III encoding scheme. The business letter and the circuit diagram, which contained a large number of white pixels, are compressed by factors of 15 or so. However, the documents with denser text are only compressed by factors of 5 or so. Overall the compression ratio for the entire set of documents is about 8. To the extent that these documents are typical, compression ratios of this order can be expected. Figure 12.9 shows

(a) Huffman code is applied to white runs and black runs

FIGURE 12.10 Facsimile coding standards

(b) Encode differences between consecutive lines

the transmission times for the various documents using the CCITT V.29 fax modem speed of 9.6 kbps. The later standard V.17 allows for a speed of 14.4 kbps.

The Group IV standard exploits the correlation between adjacent scanned lines. In black-and-white documents, black areas form connected regions so that if a given pixel is black in a given scanned line, the corresponding pixel in the next line is very likely to also be black. By taking the modulo 2 sum of pixels from adjacent lines, we will obtain 1s only at the pixel locations where the lines differ as shown in Figure 12.10b. Only a few 1s can be expected, since adjacent lines are highly correlated. In effect, this processing has increased the number and the length of the white runs leading to higher compression performance. The Group IV standard developed by the ITU uses a modified version of this approach in which Huffman codes encode the position and number of pixel changes relative to the previous scanned line. In Figure 12.9 we see that the Group IV standard achieves a compression ratio approximately double that of Group III. Because a scanned line is encoded relative to a previous line, an error in transmission can cause an impairment in all subsequent lines. For this reason the ITU standard recommends that the one-dimensional technique be applied periodically and that the number of lines encoded using the two-dimensional technique be limited to some predefined number.

The scanning density of the original ITU standards has long been superseded by laser and inkjet printing technology. The ITU has modified the standard to allow for resolutions of 200, 300, and 400 pixels per inch. As the scanning resolution is increased, we can expect that the lengths of the horizontal runs will increase and that the correlation between adjacent vertical lines will increase. Hence it is reasonable to expect to attain significantly higher compression ratios than those shown in Figure 12.9.

12.1.3 Adaptive Codes

The techniques so far require that we know the probability of occurrence of the information symbols. In certain applications these probabilities can be estimated once, and it is reasonable to assume that they will remain fixed. Examples include facsimile and certain types of documents. On the other hand, the statistics associated with certain information sources either are not known ahead of time or vary with time. For this reason it is desirable to develop **adaptive lossless data compression** codes that can achieve compaction in these types of situations.

The most successful adaptive lossless data compression techniques were developed by Ziv and Lempel in the late 1970s. In run-length coding, the information stream is parsed into black runs and white runs, and the runs are encoded using a variable-length code. The Lempel-Ziv approach generalizes the principle behind run-length coding and parses the information stream into *strings of symbols* that occur frequently and encodes them using a variable-length code. In the Lempel-Ziv algorithm the encoder identifies repeated patterns of symbols and encodes in the following manner. Whenever a pattern is repeated in the

"All tall We all are tall. All small We all are small."

Can be mapped into

"All_ta[2, 3]We_ [6, 4]are[4, 5]._ [1, 4]sm[6, 15][31, 5]."

FIGURE 12.11 Lossless data compression using the Lempel-Ziv algorithm

information sequence, the pattern is replaced by a pointer to the first occurrence of the pattern and a value that indicates the length of the pattern. The algorithm is adaptive in that the frequently occurring patterns are automatically identified by the encoding process as more and more of the symbol sequence is processed.

In Figure 12.11, we give an example of the algorithm for a rhyme from Dr. Seus [1963]. Most beginning readers have high redundancy, since that is one of the main characteristics that make the reader easy to read. The first six characters, All_ta, are left unchanged, since they do not contain a repeated pattern. The next three symbols, ll_, are seen to have occurred before, so the marker [2,3] is used to indicate the decoder should refer back to the second character and reproduce the next three characters. The next three characters, We_, are left unchanged, but the following four characters, all_, are seen to have occurred starting with character 6. The next three characters, are, are left unchanged, but the next five characters, _tall, are seen to have occurred starting with character 4. The next two characters, ._, are left unchanged, and the following four characters, All_, are seen to have occurred starting with character 1. The next two characters, sm, are unchanged, and we then hit the jackpot by finding that the next 15 characters, all_We_all_are_, occurred starting with character 6. Finally, the last word is seen to have occurred starting with character 31. The final period completes the encoding. In this example the original sequence consisted of 53 ASCII characters, and the compressed sequence consists of 29 ASCII characters, where we assume that each number in the pointer takes one ASCII character. Isn't this fun? In the problem section you get a chance to decode another one of our favorite Dr. Seuss rhymes.

ERROR CONTROL AND LOSSLESS DATA COMPRESSION

Lossless data compression is achieved by removing a redundancy in the sequence of symbols used to represent a given set of information. The reduced redundancy implies that the resulting compressed set is more vulnerable to error. As shown for the case of two-dimensional facsimile coding, individual errors can propagate and cause significant impairments in the recovered image. Similarly, errors introduced in the encoded sequence that results from Huffman coding can result in complete loss of the information. The effect of errors on information that has been compressed can be reduced through the use of error-correction techniques and through the insertion of synchronization information that can limit error propagation.

The above version of the Lempel-Ziv algorithm requires the encoder to search backward through the entire file to identify repeated patterns. The algorithm can be modified by restricting the search to a window that extends from the most recent symbol to some predetermined number of prior symbols. The pointer no longer refers to the first occurrence of the pattern but rather to the most recent occurrence of the pattern. Another possible modification of the Lempel-Ziv algorithm involves limiting the search to patterns contained in some dictionary. The encoder builds this dictionary as frequently occurring patterns are identified.

Adaptive lossless data compression algorithms are used widely in information transmission and storage applications. In transmission applications, lossless data compression can reduce the number of bits that need to be transmitted and hence the total transmission time. In storage applications lossless data compression techniques may reduce the number of bits required to store files and hence increase the apparent capacity of a given storage device. Three types of lossless data compression techniques have been implemented in conjunction with modems. The MMP5 protocol uses a combination of run-length coding and adaptive Huffman coding. The MMP7 protocol exploits the statistical dependence between characters by applying context-dependent Huffman codes. The V.42bis ITU modem standard use the Lempel-Ziv algorithm. The Lempel-Ziv algorithm has also been implemented in storage applications such as the UNIX COMPRESS command. The Lempel-Ziv algorithm yields compression ratios ranging from 2–6 in file-compression applications.

12.2 DIGITAL REPRESENTATION OF ANALOG SIGNALS

We now consider the digital representation of analog information. Analog information is characterized by the fact that its exact representation requires an infinite number of bits. Examples of analog information include speech and audio signals that consist of the continuous variation of amplitude versus time. Another example of analog information is image information that consists of the variation of intensity over a plane. Video and motion pictures are yet another example of analog information where the variation of intensity is now over space and time. All of these signals can assume a continuum of values over time and/or space and consequently require infinite precision in their representation. All of these signals are also important in human communications and are increasingly being incorporated into a variety of multimedia applications.

Lossy data compression involves the problem of representing information within some level of approximation. In general, data compression is required to represent analog signals using only a finite number of bits. Therefore, a figure of merit for a lossy data compression technique is the quality, fidelity, or degree of precision that is achieved for a given number of representation bits. We present a number of techniques for achieving lossy data compression. These include quantization, predictive coding, and transform coding. We also discuss how these techniques are used in standards for coding speech, audio, image, and video signals.

12.2.1 Properties of Analog Signals

Many signals that are found in nature can be represented as the sum of sinusoidal signals. For example, many speech sounds consist of the sum of a sinusoidal wave at some fundamental frequency and its harmonics. These analog signals have the form:

$$x(t) = \sum a_k \cos(2\pi k f_0 t + \phi_k).$$

For example, Figure 12.22 (shown later in the chapter) contains the periodic waveform for the sound "ae" as in cat.

As another example, suppose we are transmitting binary information at a rate of 8 kilobits/second. A binary 1 is transmitted by sending a rectangular pulse of amplitude 1 and of duration 0.125 milliseconds, and a 0 by sending a pulse of amplitude -1. Figure 12.12a shows the signal that results if we repeatedly send the octet 10101010, and Figure 12.12b shows the signal that results when we send 11110000. Note that the signals result in square waves that repeat at rates of 4 kHz and 1 kHz, respectively. By using Fourier series analysis (see Appendix 3B), we can show that the first signal is given by

$$x_1(t) = (4/\pi)\{\sin(2\pi(4000)t) + (1/3)\sin(2\pi(12000)t + (1/5)\sin(2\pi(20000)t + \ldots\}$$

(a)

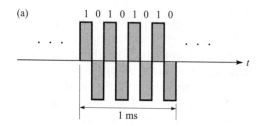

1 0 1 0 1 0 1 0

1 ms

FIGURE 12.12 Signals corresponding to repeated octet patterns

(b)

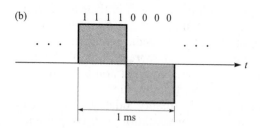

1 1 1 1 0 0 0 0

1 ms

and has frequency components at the odd multiples (harmonics) of 4 kHz. Similarly, we can show that the second signal has harmonics at odd multiples of 1 kHz and is given by

$$x_2(t) = (4/\pi)\{\sin(2\pi(1000)t) + (1/3)\sin(2\pi(3000)t + (1/5)\sin(2\pi(5000)t + \ldots\}$$

Figure 12.13a and Figure 12.13b show the "spectrum" for the signals $x_1(t)$ and $x_2(t)$, respectively. The **spectrum** gives the magnitude of the amplitudes of the sinusoidal components of a signal. It can be seen that the first signal has significant components over a much broader range of frequencies than the second signal; that is, $x_1(t)$ has a larger bandwidth than $x_2(t)$. In general the bandwidth is an indicator of how fast a signal varies with time. Signals that vary quickly have a larger bandwidth. For example, in Figure 12.13 $x_1(t)$ varies four times faster than $x_2(t)$ and thus has the larger bandwidth.

Not all signals are periodic. For example, Figure 12.20 (shown later in the chapter) shows the sample waveform for the sentence "The speech signal level varies with time." It can be seen that indeed the level varies with time. Furthermore if we "zoom in" on the parts of the signal that contain ringing, we would find that the signal is periodic during those intervals. The long-term average spectrum of the voice signal ends up looking like that shown in Figure 12.14.[1]

(a) Frequency components for 10101010

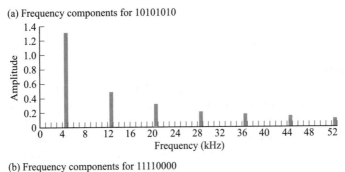

(b) Frequency components for 11110000

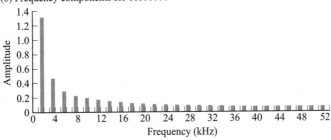

FIGURE 12.13 (a) Frequency components for pattern 10101010. (b) Frequency components for pattern 11110000

---

[1] For examples of long-term spectrum measurements of speech, see [Jayant and Noll 1984, p. 40].

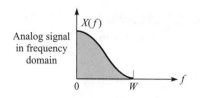

Analog signal
in frequency
domain

FIGURE 12.14 Analog waveform $x(t)$ with lowpass spectrum

Sample the signal every T seconds

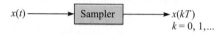

We define the **bandwidth of an analog signal** as the range of frequencies at which the signal contains nonnegligible power, that is, nonnegligible a_k. There are many ways of defining the bandwidth of a signal. For example, the 99 percent bandwidth is defined as the frequency range required to contain 99 percent of the power of the original signal. Usually the appropriate choice of bandwidth of a signal depends on the application. For example, the human ear can detect signals in the range 20 Hz to 20 kHz. In telephone communications frequencies from 200 Hz to 3.5 kHz are sufficient for speech communications. However, this range of frequencies is clearly inadequate for music that contains significant information content at frequencies higher than 3.5 kHz.

12.2.2   Analog-to-Digital Conversion

Suppose we have an analog waveform, a time function $x(t)$, that has a spectrum with bandwidth W Hz as shown in Figure 12.14. To convert the signal to digital form, we begin by taking instantaneous samples of the signal every T seconds to obtain $x(nT)$ for integer values n (see Figure 12.15). Because the signal varies continuously in time, we obtain a sequence of real numbers that for now we assume have an infinite level of precision. Intuitively, we know that if the samples are taken frequently enough relative to the rate at which the signal varies, we can recover a good approximation of the signal from the samples, for example, by drawing a straight line between the sample points. Thus the sampling process replaces the continuous function of time by a sequence of real-valued numbers. The very surprising result, due to Nyquist, is that we can recover the original signal $x(t)$ precisely as long as the sampling rate is higher than some minimum value! The Nyquist sampling theorem states that if the sampling rate is $1/T >$

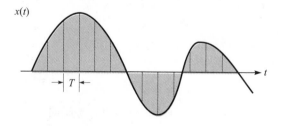

FIGURE 12.15 Sampling of an analog signal

$2W$ samples/second, then $x(t)$ can be recovered from the sequence of sample values $\{x(nT)\}$. We refer to $2W$ as the **Nyquist sampling rate**. The reconstruction of $x(t)$ is carried out by interpolating the samples $x(nT)$.

As an example consider the voice signal in the telephone system that has a nominal bandwidth of 4 kHz. The Nyquist sampling rate then requires that the voice signal be sampled at a rate of 8000 samples/second. For high-quality audio signals the bandwidth is 22 kHz, leading to a sampling rate of 44,000 samples/ second. As a final example consider an analog television. The analog TV signal has a bandwidth of 4 MHz, leading to a sampling rate of 8,000,000 samples/ second.

◆ NYQUIST SAMPLING THEOREM

We now explain how the Nyquist sampling theorem comes about. Let $x(kT)$ be the sequence of samples that result from the sampling of the analog signal $x(t)$. Consider a sequence of very narrow pulses $\delta(t - kT)$ that are spaced T seconds apart and whose amplitudes are modulated by the sample values $x(kT)$ as shown in Figure 12.16.

$$y(t) = \sum_k x(kT)\delta(t - kT)$$

Signal theory enables us to show that $y(t)$ has the spectrum in Figure 12.16, where the spectrum of $x(t)$ is given by its Fourier transform:

$$X(f) = \int_{-\infty}^{\infty} x(t)e^{-j2\pi ft}\,dt = \int_{-\infty}^{\infty} x(t)\cos 2\pi ft\,dt + j\int_{-\infty}^{\infty} x(t)\sin 2\pi ft\,dt$$

The spectrum $X(f)$ is defined for positive and negative frequencies and is complex valued. The spectrum of the sampled signal is

$$Y(f) = \sum_k X\left(f - \frac{k}{T}\right)$$

(a) Original

(b) $\frac{1}{T} > 2W$

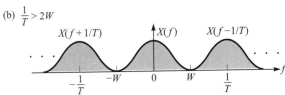

(c) $\frac{1}{T} < 2W$

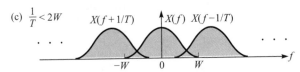

FIGURE 12.16 Spectrum of sampled signal

Nyquist's results depend on having these repeated versions of the spectrum be sufficiently apart. If the sampling rate $1/T$ is greater than $2W$, then the translated versions of $X(f)$ will be nonoverlapping. When a signal is applied to an ideal lowpass filter, the spectrum of the output signal consists of the portion of the spectrum of the input signal that falls in the range zero to W. Therefore, if we apply a lowpass filter to $y(t)$ as shown in Figure 12.16, then we will recover the original exact spectrum $X(f)$ and hence $x(t)$. We conclude that the analog signal $x(t)$ can be recovered exactly from the sequence of its sample values as long as the sampling rate is $2W$ samples/second.

Now consider the case where the sampling rate $1/T$ is less than $2W$. The repeated version of $X(f)$ now overlap, and we cannot recover $x(t)$ precisely. If we were to apply $y(t)$ to a lowpass filter in this case, the output would include an **aliasing error** that results from the additional energy that was introduced by the tails of the adjacent signals. In practice, signals are not strictly bandlimited, and so measures must be taken to control aliasing errors. In particular, signals are frequently passed through a lowpass filter prior to sampling to ensure that their energy is confined to the bandwidth that is assumed in the sampling rate.

12.2.3   Digital Transmission of Analog Signals

Figure 12.17 shows the standard arrangement in the handling of the analog information by digital transmission (and storage) systems. The signal produced by an analog source $x(t)$, which we assume is limited to W Hz, is sampled at the Nyquist rate, producing a sequence of samples at a rate of $2W$ samples/second. These samples have infinite precision, so they are next input into a quantizer that produces an approximation within a specified accuracy. The level of accuracy determines the number of bits m that the quantizer uses to specify the approx-

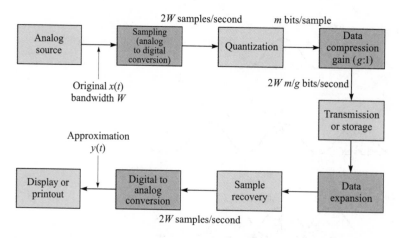

FIGURE 12.17 Digital transmission of analog signal

imation. A lossless data compression technique may be applied next to produce an efficient representation of the quantizer output sequence. The bit rate out of the quantizer is $2Wm$ bits/second, since samples occur at a rate of $2W$ samples/second and each sample requires m bits.

Suppose the lossless data compression scheme reduces the bit rate by a factor of g. The final representation of the analog source is $2Wm/g$ bits/second. At this point we have obtained a digital representation of the original analog signal within a specified accuracy or quality. This digital representation can be stored or transmitted any number of times without additional distortion as long as no errors are introduced into the digital representation.

The approximation to the original signal $x(t)$ is recovered by the mirror process shown in Figure 12.17. A data expansion step recovers the original quantizer outputs from the data compressed representation. Next a sample recovery step takes the quantizer outputs and produces the approximations to the original samples. Finally a digital-to-analog conversion system uses these samples to modulate a sequence of narrow pulses and applies them to a lowpass filter to produce an analog signal that approximates the original signal within the prescribed accuracy. Note that this process applies equally to the storage or transmission of analog information.

Consider now the operation of the quantizer. Its task is to take sample values $x(kT)$ and produce an approximation $y(kT)$ than can be specified using a fixed number of bits. In general, quantizers have a certain number, say, $M = 2^m$, of approximation values that are used to represent the quantizer inputs. For each input $x(kT)$ the closest approximation point is found, and the index of the approximation point is specified using m bits. The decoder on the receiver side is assumed to have the set of approximation values so the decoder can recover the values from the indices. The design of a quantizer requires knowledge about the range of values that are assumed by the signal $x(t)$. The set of approximation values is selected to cover this range. For example, suppose $x(t)$ assumes the values in the range $-V$ to V. Then the set of approximation values should be selected to cover only this range. Selecting approximation values outside this range is unnecessary and will lead to inefficient representations. Note that as we increase m, we increase the number of intervals that cover the range $-V$ to V. Consequently, the intervals become smaller and the approximations become more accurate. We next quantify the trade-off between accuracy and the bit rate $2Wm$.

Figure 12.18 shows the simplest type of quantizer, the **uniform quantizer**, in which the range of the amplitudes of the signal are covered by equally spaced approximation values. The range $-V$ to V is divided into 2^m intervals of equal length Δ. When the input $x(kT)$ falls in a given interval, then its approximation value $y(kT)$ is the midpoint of the interval. The output of the quantizer is simply the m bits that specify the interval.

In general, the approximation value is not equal to the original signal value, so an error is introduced in the quantization process. The value of the **quantization error** $e(kT)$ is given by

$$e(kT) = y(kT) - x(kT)$$

Uniform quantizer

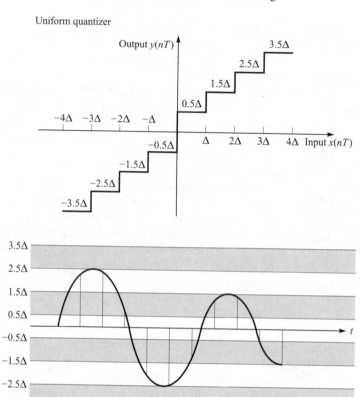

$x(t)$ and the corresponding quantizer approximations

FIGURE 12.18 A uniform quantizer

Figure 12.19 shows the value of the quantization error as the function of the quantizer input. It can be seen that the error takes on values between $-\Delta/2$ and $\Delta/2$. When the interval length Δ is small, then the quantization error values are small and the quantizer can be viewed as simply adding "noise" to the original signal. For this reason the figure of merit used to assess the quality of the approximation produced by a quantizer is the **signal-to-noise ratio (SNR)**:

$$\text{SNR} = \frac{\text{average signal power}}{\text{average noise power}}$$

The standard deviation σ_x is a measure of the spread of the signal values about the mean, which we are assuming is zero. On the other hand, V is the maximum value that the quantizer assumes can be taken on by the signal. Frequently, selecting V so that it corresponds to the maximum value of the signal is too inefficient. Instead V is selected so that the probability that a sample $x(kT)$ exceeds V is negligible. This practice typically leads to ratios of approximately $V/\sigma_x \approx 4$.

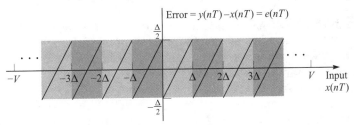

If the number of levels M is large, then the error is approximately uniformly distributed between $(-\Delta/2, \Delta/2)$.

Mean square error:

$$\sigma_e^2 = \int_{-\frac{\Delta}{2}}^{\frac{\Delta}{2}} x^2 \frac{1}{\Delta}\ dx = \frac{\Delta^2}{12}$$

FIGURE 12.19 Quantizer error

SNR is usually stated in decibels. In the next section we show that the SNR is given by

$$\text{SNR dB} = 10\log_{10}\sigma_x^2/\sigma_e^2 = 6m + 10\log_{10} 3\sigma_x^2/V^2$$
$$\approx 6m - 7.27\text{dB for } V/\sigma^2 = 4$$

The preceding equation states that each additional bit used in the quantizer will increase the SNR by 6 dB. This result makes intuitive sense, since each additional bit doubles the number of intervals, and so for a given range $-V$ to V, the intervals are reduced in half. The average magnitude of the quantization error is also reduced in half, and the average quantization error power is reduced by a quarter. This result agrees with $10\log_{10} 4 = 6$ dB. We have derived this result for the case of uniform quantizers. More general quantizers can be defined in which the intervals are not of the same length. The SNR for these quantizers can be shown to also have the form of the preceding equations where the only difference is in the constant that is added to $6m$ [Jayant and Noll 1984].

We noted before that the human ear is sensitive to frequencies up to 22 kHz. For audio signals such as music, a high-quality representation involves sampling at a much higher rate. The Nyquist sampling rate for $W = 22$ kHz is 44,000 samples/second. The high-quality audio also requires finer granularity in the quantizers. Typically 16 or more bits are used per sample. For a stereo signal we therefore obtain the following bit rate:

$$44,000\ \frac{\text{samples}}{\text{second}} \times 16\ \frac{\text{bits}}{\text{sample}} \times 2 \text{ channels} = 1.4\text{Mbps}$$

We see that high-quality audio signals can require much higher rates than are required for more basic signals such as those of telephony speech. The bit rate for audio is increased even further in modern surround-sound systems. For example,

PCM

The standard for the digital representation of voice signals in telephone networks is given by the **pulse code modulation (PCM)** format. In PCM the voice signal is filtered to obtain a lowpass signal that is limited to $W = 4\,\text{kHz}$. The resulting signal is sampled at the Nyquist rate of $2W = 8\,\text{kHz}$. Each sample is then applied to an $m = 8$ bit quantizer. The number of levels in the quantizer is therefore 256. The type of quantizers used in telephone systems are **nonuniform quantizers**. A technique called **companding** is used so that the size of the intervals increases with the magnitude of the signal x. The SNR formula is given by

$$\text{SNR dB} = 6m - 10 \text{ for PCM speech}$$

Because $m = 8$, we see that the SNR is $38\,\text{dB}$. Note that an SNR of 1 percent corresponds to $40\,\text{dB}$. In the backbone of modern digital telephone systems, voice signals are carried using the *log-PCM* format, which uses a logarithmic scale to determine the quantization intervals. The ISDN standard extends the use of the PCM format all the way to the telephone at the user's end.

the Digital Audio Compression (AC-3) that is part of the U.S. ATSC high-definition television standard involves five channels (left, right, center, left-surround, right-surround) plus a low-frequency enhancement channel for the 3 Hz to 100 Hz band.

The derivation for the performance of the uniform quantizers assumes that we know the exact dynamic range $-V$ to V in which the signal values will fall. Frequently the system does not have sufficient control over signal levels to ensure that this is the case. For example, the distance a user holds a microphone from his or her mouth will affect the signal level. Thus if a mismatch occurs between the signal level for which the system has been designed and the actual signal level, then the performance of the quantizer will be affected. Suppose, for example, that the quantizer is designed for the dynamic range $-V$ to V and the actual signal level is in the range $-V/2$ to $V/2$. Then in effect we are using only half the signal levels, and it is easy to show that the SNR will be 6 dB less than if the signal occupied the full dynamic range. Conversely, if the actual signal level exceeds the dynamic range of the quantizer, then the larger-magnitude samples will all be mapped into the extreme approximation values of the quantizer. This type of distortion is called **clipping** and in the cassette recording systems that you are surely familiar with is indicated by a red light in the signal-level meter.

Figure 12.20 shows the speech waveform for the following sentence: The speech signal level varies with time. The waveform is about 3 seconds long and was sampled at a rate of 44 kHz, so it consists of approximately 130,000 samples. Large variations in signal levels can be observed.

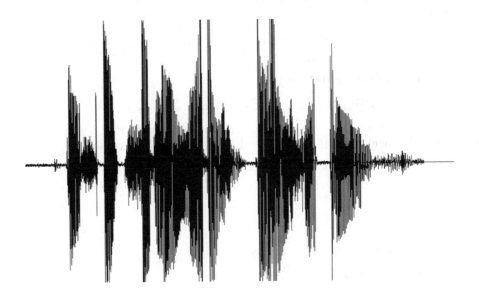

FIGURE 12.20 Variation of signal level for speech

One way of dealing with variations in signal level is to actually measure the signal levels for a given time interval and to multiply the signal values by a constant that maps the values into the range for which the quantizer has been designed. The constant is then transmitted along with the quantizer values. These types of quantizers are called **adaptive quantizers**. A passive way of dealing with variations in the signal level is to use nonuniform quantizers where the quantizer intervals are roughly proportional to the signal level. The companders used for telephone speech are an example of this.

◆12.2.4   SNR Performance of Quantizers

We now derive the SNR performance of a uniform quantizer. When the number of levels M is large, then the error values are approximately uniformly distributed in the interval $(-\Delta/2, \Delta/2)$. The power in the error signal is then given by

$$\sigma_e^2 = \int_{-\frac{\Delta}{2}}^{\frac{\Delta}{2}} x^2 \frac{1}{\Delta} \, dx = \frac{\Delta^2}{12}$$

Let σ_x^2 be the average power of the signal $x(t)$. Then the SNR is given by

$$SNR = \frac{\sigma_x^2}{\Delta^2/12}$$

From the definition of the quantizer, we have that $\Delta = 2V/M$ and that $M = 2^m$; therefore

$$SNR = \frac{\sigma_x^2}{\Delta^2/12} = \frac{12\sigma_x^2}{4V^2/M^2} = 3\left(\frac{\sigma_x}{V}\right)^2 M^2 = 3\left(\frac{\sigma_x}{V}\right)^2 2^{2m}$$

The SNR, stated in decibels, is then

$$\text{SNR dB} = 10\log_{10}\sigma_x^2/\sigma_e^2 = 6m + 10\log_{10}3\sigma_x^2/V^2$$

As indicated before, V is selected so that the probability that a sample $x(kT)$ exceeds V is negligible. If we assume that $V/\sigma_x \approx 4$, then we obtain

$$\text{SNR dB} \approx 6m - 7.27\text{dB for } V/\sigma^2 = 4$$

12.3   TECHNIQUES FOR INCREASING COMPRESSION

The quantizers that were introduced in section 12.2 simply convert the samples into digital form. They do not deal with statistical dependencies between samples. For example, in sections where the signal is changing slowly, consecutive samples will tend to have values that are close to each other, as shown in Figure 12.21. In another example, the signal is approximately periodic, and longer-term dependencies between sample values can be observed. Figure 12.22 shows the waveform for the sound corresponding to the English vowel sound "ae", as in the word *cat*. The long-term dependency between sample values is evident. This periodic property is common to the so-called voiced sounds that include the vowel sounds as well as many consonant sounds such as "n," "l," and "r."

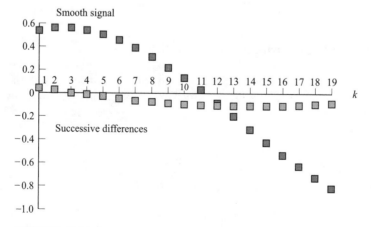

FIGURE 12.21 A smooth signal and its successive differences

FIGURE 12.22 Sample waveform of "ae" sound as in cat

Figure 12.23 gives an example of an image where, in smooth portions of the picture, neighboring samples tend to have the same values. Long-term dependencies can also be observed when images contain periodic patterns. The compression techniques discussed in this section attempt to exploit the redundancies to attain greater compression and hence to provide more efficient representations.

Three basic compression techniques exploit the redundancies in signals. **Predictive coding** techniques, also called **differential coding** techniques, attempt to predict a sample value in terms of previous sample values. In Figure 12.21 the sequence of differences occupies a smaller dynamic range and consequently can be coded with greater accuracy by a quantizer. Predictive coding techniques are used extensively in the coding of speech signals. A second type of technique involves transforming the sequence of sample values into another domain that yields a signal that is more amenable to lossless data compression. These types of techniques are called transform coding techniques. Transform coding is used

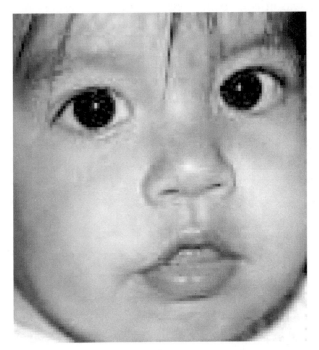

FIGURE 12.23 Image sample with smooth portions

extensively in image and video coding. Subband coding is an important special case of transform coding. A third type of technique is to design quantizers that deal not with individual samples, but instead map blocks of samples into approximation points that have been designed to represent blocks of samples. This third technique, which is called **vector quantization**, is usually used in combination with other techniques. In the remainder of this section we discuss the first two types of techniques.

12.3.1 Predictive Coding

Figure 12.24 shows the block diagram of a **differential PCM (DPCM)** system, which is the simplest type of predictive coding. The next sample $x(n)$ is predicted by linear combination of N previous outputs of the system:

$$\hat{x}(n) = h_1 y(n-1) + h_2 y(n-2) + \ldots + h_N y(n-N)$$

In the simplest case we would use a first order predictor where $\hat{x}(n)$ is predicted by $h_1 y(n-1)$. In general the prediction becomes more accurate as more terms are used in the predictor.

The difference between the sample value and the predicted value, which is called the **prediction error**, is applied to the quantizer, and the output is transmitted to the decoder. Note that the prediction is not in terms of the N previous samples $x(n-1), x(n-2), \ldots x(n-N)$, but rather in terms of the N previous outputs of the encoder. The reason is that the previous sample values are not available at the decoder. By using the above predictor, both the encoder and

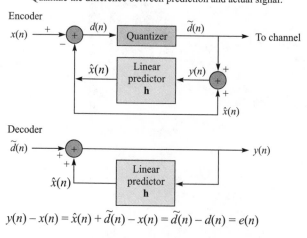

Quantize the difference between prediction and actual signal:

$$y(n) - x(n) = \hat{x}(n) + \tilde{d}(n) - x(n) = \tilde{d}(n) - d(n) = e(n)$$

The end-to-end error is only the error introduced by the quantizer!

FIGURE 12.24 Differential PCM coding

decoder will generate the same sequence of prediction values. It can be shown that the SNR of a DPCM system is given by

$$SNR_{DPCM} = SNR_{PCM} + 10\log_{10} G_p$$

This equation has the following interesting interpretation. A PCM in isolation has the SNR performance given by SNR_{PCM}. The introduction of the predictor reduces the power of the input to the quantizer from σ_x^2 to σ_d^2. This reduction factor, given by $G_p = \sigma_x^2/\sigma_d^2$, is directly translated into an improvement in SNR.

The design of a DPCM system involves finding the set of optimum prediction coefficients $h_1, h_2, \ldots, h_N$ that maximize the prediction gain. This process involves computing the statistics of a long sequences of speech samples from a variety of speakers and then carrying out certain matrix computations that ultimately yield the best predictor. See [Jayant and Noll 1984] for details. Typical values of G_p range from 6 to 10 dB. These gains can be used to reduce the bit rate by between one and two bits/sample. For example, the quality of a 64 kbps PCM, which uses eight bits/sample, can be obtained by using six bits/sample, giving a bit rate of six bits/sample × 8000 samples/second = 48 kbps.

In the section on PCM it was indicated that the dynamic range of speech signals can vary considerably. The correlation properties of speech signals also vary with speaker and also according to the sound. The ITU has standardized a 32 kbps speech-coding method that enhances the preceding DPCM technique by using an adaptive quantizer as well as by computing the prediction coefficients on the fly in attempt to maximize the instantaneous coding gain. This algorithm is called **adaptive DPCM (ADPCM)**. The encoded speech using this algorithm is of comparable subjective quality as that produced by eight-bit PCM coding. ADPCM is simple to implement and is used extensively in private network applications and in voice-mail systems.

Predictive coding techniques can be refined even further by analyzing speech samples so that the prediction coefficients are calculated essentially on a per sound basis. For example, for sounds such as the one in Figure 12.22 long-term predictors are used to predict the waveform from one period to the next. A further improvement involves replacing the coding of the prediction error with very low bit rate approximation functions. The class of **linear predictive coders (LPC)** uses this approach. Several variations of LPC coders have been selected for use in digital cellular telephone systems, for example, quadrature code-excited linear prediction (QCELP) coding at a bit rate of 14.4 kbps for use with the Qualcomm CDMA system, vector sum-excited linear prediction (VSELP) coding at a bit rate of 8 kbps for use with TDMA, and regular pulse-excited long term prediction (RPELTP) coding at a bit rate of 13 kbps for use in the GSM system.

In this section, we have focused on the application of predictive coding methods to speech signals. It should be noted, however, that the techniques are applicable to any signals where significant redundancy occurs between signal samples and are also used in image and video coding.

◆SNR PERFORMANCE OF DPCM

We are interested in obtaining an expression for SNR performance of the DPCM system, which is given by the ratio of the average signal power and the average power of the end-to-end error introduced by the system. Let $d(n)$ be the prediction error

$$d(n) = x(n) - \hat{x}(n)$$

and let $\tilde{d}(n)$ be the output of the quantizer for the input $d(n)$. The end-to-end error of the system is given by $y(n) - x(n)$. By noting that $y(n) = \hat{x}(n) + \tilde{d}(n)$, we then have

$$y(n) - x(n) = \hat{x}(n) + \tilde{d}(n) - x(n) = \tilde{d}(n) - d(n) = e(n)$$

This equation shows that the end-to-end error in the system is given solely by the error introduced by the quantizer. Thus the SNR of the system is given by

$$SNR = \frac{\sigma_x^2}{\sigma_e^2} = \frac{\sigma_x^2}{\sigma_d^2} \frac{\sigma_d^2}{\sigma_e^2} = G_p \frac{\sigma_d^2}{\sigma_e^2}$$

where

$$G_p = \frac{\sigma_x^2}{\sigma_d^2}$$

and the first equality is the definition of the SNR and the second equality is obtained by multiplying and dividing by σ_d^2. The term G_p is called the **prediction gain**, and the term σ_d^2/σ_e^2 is simply the performance of the quantizer in isolation because it is the ratio of the power of input to the quantizer and the power of the error introduced by the quantizer. In decibels the SNR equation becomes

$$SNR_{DPCM} = 10 \log_{10} \frac{\sigma_d^2}{\sigma_e^2} + 10 \log_{10} G_p = SNR_{PCM} + 10 \log_{10} G_p.$$

12.3.2   Transform Coding

Transform coding takes the sequence of signal sample values and converts them via a transformation into another sequence. In some instances the resulting sequence is more easily compressed or compacted. In other cases the transformed sequence allows the encoder to produce the quantization noise in a less perceivable form. Examples of the transformations that can be used are discrete Fourier and cosine transforms which involve taking time-domain signals into frequency-domain signals. A recently introduced example involves the use of wavelet transforms. In the next sections we consider subband coding and discrete cosine transform (DCT) coding and their application to speech, audio, image, and video signals.

12.3.3   Subband Coding

When the number of quantization levels is not small, the quantization error signal has a flat spectrum. That is, its energy is evenly distributed over the range of frequencies occupied by the signal. Most information signals have a spectrum that is nonuniform. Figure 12.25a shows the spectrum of a typical signal and the associated quantization noise. In this case the relative signal power to noise power differs according to frequency. In certain applications, such as audio, the ear is sensitive to these frequency-dependent SNRs. This situation suggests that systems could be designed to distribute the quantization error in a less perceivable form.

The subband coding technique was developed to do so. In **subband coding** the signal is decomposed into the sum of K component signals that are obtained by applying $x(t)$ to a bank of filters that are nonoverlapping in frequency. This procedure is shown in Figure 12.25b. Each component signal is quantized using a different quantizer. The number of levels in the quantizer determines the power of the quantization noise. Therefore, the encoder can control the amount of quantization noise introduced in each component through the number of bits assigned to the quantizer. The encoder typically performs the bit allocation based on the instantaneous power in the different subbands.

The MPEG standard for digital audio uses subband coding for the compression of audio.[2] Audio signals are sampled at a 16-bit resolution at sampling rates of 16 kHz, 22.05 kHz, 24 kHz, 32 kHz, 44.1 kHz, or 48 kHz. The sequence of time samples of the audio signal are filtered and divided into 32 disjoint component signals. The audio signal is also transformed into the frequency domain

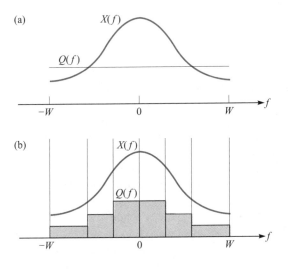

FIGURE 12.25 Subband coding of a typical signal

---

[2]MPEG video compression is discussed later in section 12.3.7.

WHAT IS MP3?

MP3 stands for MPEG layer 3 audio compression. The layer 3 compression option can reduce the bit rate of an audio signal by a factor of 12 with very low loss in sound quality. MP3 software for compression and playback can be downloaded from various websites. High-quality audio MP3 files can be created from music CDs, so MP3 has become very popular for "recording" selections from CDs in personal computers. The MP3 phenomenon has led to problems with the illegal distribution of copyrighted music material over the Internet. The recording industry is working on the development of secure digital recording methods.

using a fast Fourier transform to assess the frequency composition of the signal. The subband signals are quantized according to the bit allocation specified by the encoder, which takes into account the instantaneous powers in the different frequency bands as well as their relative degree of perceptual importance. The **MPEG audio compression** standard has adjustable compression ratios that can deliver sound ranging from high-fidelity CD quality to telephone voice quality. MPEG audio compression can operate at three layers, from 1 to 3 in order of increasing complexity. A decoder of a given layer is required to decode compressed signals using all lower layers. The compressed signals have bit rates that range between 32 kbps to 384 kbps. The algorithm has a mode that allows it to compress stereo audio signals, taking into account the dependencies between the two channels.

12.3.4   Discrete Cosine Transform Coding

Figure 12.26a shows samples of the signal that is smooth in the sense that it varies slowly in time. The figure also shows the corresponding **discrete cosine transform (DCT) coding** of the sequence. The smooth signal is primarily lowpass in nature so that the DCT of the signal is concentrated at low frequencies. If, instead of coding the signal in the time domain, we encode the signal in the frequency domain, we see that the frequency domain values will contain many zeros that can be encoded efficiently through the use of Huffman and run-length coding. In this section we refrain from the equations associated with the various transforms and focus on the qualitative behavior. Appropriate references are included at the end of the chapter.

The effectiveness of the DCT becomes more evident when the two-dimensional DCT is applied to images. In these applications the images are sampled to obtain a rectangular array of pixel values, as shown in Figure 12.27. This array is divided into square blocks consisting of 8×8 pixels. Figure 12.26b shows the values of the pixels obtained from a certain image. Note that in this case the time variable is replaced by the two space variables. The DCT then produces a block

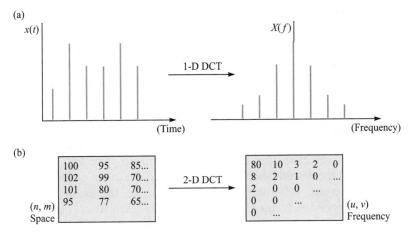

FIGURE 12.26 One-dimensional and two-dimensional DCTs

of values defined over two spatial frequency variables. The DCT of this 8×8 block is also shown in Figure 12.26b. Because the original block consisted of the smooth section of the image the DCT has nonzero values concentrated at the low frequencies that correspond to the upper-left corner of the transformed block. It can also be seen that all other pixel values are zero. It is clear that the transformed block can be more readily compressed.

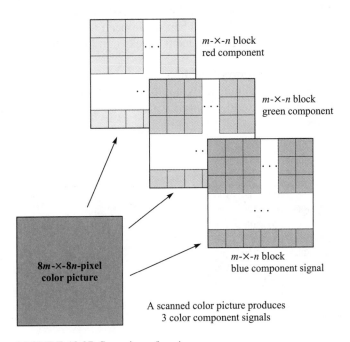

FIGURE 12.27 Scanning of an image

12.3.5   The JPEG Image-Coding Standard

The **Joint Photograph Expert Group (JPEG)** image-coding standard uses the DCT in the following way. The original image is segmented into blocks of 8 × 8 pixel values. The DCT is then applied to each of these blocks. The coefficients in each DCT block are then quantized. Typically, a different quantizer is used for different frequency coordinates because of the differences in perceptual importance. The 8 × 8 array of pixels is converted into a block of 64 consecutive values through the zigzag scanning process shown in Figure 12.28. It can be seen that only the first few values in this block will be nonzero. The first component in each block, called the **dc component**, corresponds to the mean of the original pixel values and is encoded separately. The remainder of the components, the **ac components**, are encoded using Huffman codes designed to encode the nonzero values and run-length codes to deal with the long runs of zeros. Consecutive blocks tend to have similar means, so the dc components of consecutive blocks are encoded using DPCM. This technique can reduce the number of bits required to represent an image by compression ratios of 5 to 30. Figure 12.29 summarizes the JPEG image-coding standard.

The choice of 8-×-8-pixel blocks in the JPEG coding algorithm was primarily determined by the size of block for which the DCT algorithm could be implemented in VLSI. By transforming larger blocks of information, it is possible to capture the redundancies among larger groups of samples and hence achieve larger compression ratios. Another issue in JPEG coding is artifacts that result from the block-oriented coding. At very low coding bit rates, the boundaries between the blocks within the image become visible. The "blockiness" in the picture can become quite annoying. There have been a number of proposals for using blocks that overlap so that adjacent blocks can be matched at the boundaries.

Consider now the question of dealing with color images. Modern video display systems produce color images by combining variable amounts of the three primary colors: red, green, and blue (RGB). Any color can be represented as a

In image and video coding, the picture array is divided into 8-×-8-pixel blocks that are coded separately.

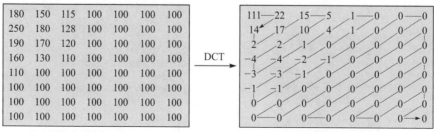

8 × 8 block of 8-bit pixel values   Quantized DCT coefficients

FIGURE 12.28 Zigzag scanning process

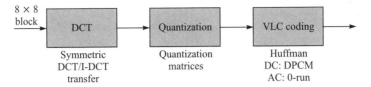

8 × 8 block → DCT → Quantization → VLC coding →

Symmetric DCT/I-DCT transfer   Quantization matrices   Huffman DC: DPCM AC: 0-run

FIGURE 12.29 JPEG image-coding standard

linear combination of these three colors. A color video camera or scanner produces an **RGB representation** which is a three-dimensional signal consisting of the three color components. Thus the compression of color images involves the compression of three component images, one for each primary color. For example, "full color" display systems use eight bits to represent each color component, giving a total of 24 bits/pixel and a capability of representing $2^{24} = 16,777,216$ possible colors. An SVGA system has a resolution of 800×600 pixels/screen, and so it requires 1.44 megabytes of memory to store one screen. Similarly, an XGA system has a resolution of 1024×768 pixels/screen, and so it requires $1024 \times 768 \times 3 = 2.4$ megabytes of memory. If you play with the resolution control in your personal computer, you will be given a number of choices for the resolution and the total number of colors that the available memory can accommodate.

To compress a color image, the above DCT coding algorithm can be applied separately to each component image. A more effective compression method involves first applying a transformation to the RGB representation of the image. The R, G, and B components for a pixel are converted into a luminance component, Y, and two chrominance components, I and Q, by the following equations:

$$x_Y = 0.30x_R + 0.59x_G + 0.11x_B$$
$$x_I = 0.60x_R - 0.28x_G - 0.32x_B$$
$$x_Q = 0.21x_R - 0.52x_G + 0.31x_B$$

This transformation was developed as part of the design of color television systems that required compatibility with black-and-white television. The luminance component provides the information required to produce the image in a black-and-white television. The two chrominance components provide the additional information required to produce a color image. One of the observations made in the design of color television was that the human eye is much more sensitive to the luminance signal than to the chrominance signals because the luminance signal provides information about edges and transitions in images. An analogy can be made to the coloring books that you surely grew up with; the black lines on a page that define the basic image correspond to the luminance signal, and the coloring on the page corresponds to the chrominance components. Image compression systems make use of this difference in sensitivity by processing the luminance signal at the original resolution and processing the chrominance signals at lower resolutions.

A commonly used format involves having the chrominance components represented at one-quarter of the resolution of the luminance signal. For example, one frame of a television signal consists of 480 lines with 640 pixels/line. This frame will produce 4800 8-×-8-pixel blocks for the luminance signal for one frame. The chrominance signals are originally sampled at this resolution but blocks of 2×2 pixels are collapsed into a single pixel by taking an average of the four pixel values. As a result, each chrominance signal will produce 1200 pixel blocks for each frame. The DCT compression algorithm can then be applied to the blocks produced by the three component signals and the resulting representation can be transmitted or stored. Note that this approach halves the total number of blocks that need to be processed from $3 \times 4800 = 14{,}400$ blocks to $4800 + 2 \times 1200 = 7200$ blocks. The decoder then recovers the luminance frame and the two chrominance frames. The latter are expanded to the original resolution by creating a 2×2 set of pixels from each pixel value.

As an example, consider a scanned image of a $8'' \times 10''$ color picture at a resolution of 1200 pixels/inch. Inexpensive scanners with this resolution are now available. The number of pixels produced in the scanning process is then

$$8 \times 1200 \times 10 \times 1200 = 115.2 \times 10^6 \text{ pixels}$$

Assuming that 24 bits are used to represent each pixel value, the uncompressed scanned picture will yield a file consisting of 350 megabytes! If the JPEG algorithm can produce a high-quality reproduction with a compression ratio of 10 for this particular image, then the file size is reduced to 35 megabytes. If in addition we use one-quarter resolution for the chrominance components, the file size can then be reduced to about 17.5 megabytes.

12.3.6 Compression of Video Signals

Television and motion pictures are based on the principle that a rapid sequence of pictures can give the appearance of motion. Consider, for example, a booklet of pictures that change incrementally. When the pages of the booklet are flicked in rapid succession, the sequence of pictures merge to produce animation. This principle is the basis for motion pictures where a series of photographs are projected onto a screen at a rate of 24 picture frames/second. Figure 12.30 shows how the principle is incorporated into television.

The television signal consists of a one-dimensional function of time that is used to produce a sequence of images. The television signal specifies the intensity that is to be drawn in a horizontal line across the television screen. **Synchronization** signals control the tracing of consecutive lines down the screen as well as the return to the top of the screen after each picture frame. The television system in North America and Japan uses 525 lines/frame and displays 30 frames/second. Systems in Europe used 625 lines/frame and a display rate of

Information = M bits/pixel × (W × H) pixels/frame × F frames/second

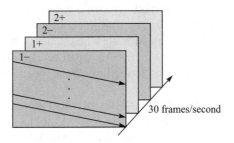

FIGURE 12.30 The TV/video signal

25 frames/second. At these rates some flicker is still noticeable, but instead of increasing the frame rate, a technique called **interlacing** is used. A frame is divided into two fields that consist of the odd and even lines, respectively. Odd and even fields are displayed in alternation, resulting in a display rate of 60 fields/second. The odd and even fields draw lines in nonoverlapping portions of the picture. The display screen consists of material that maintain each line for a sufficient period of time to produce the appearance of a higher frame rate. Modern technology now makes it possible to implement progressive or non-interlaced video systems that use higher frame rates.

From Figure 12.30 we can calculate the uncompressed bit error rate required to represent a standard television signal. In the figure we assume that each frame consists of 720 lines by 480 pixels/line, as specified by a industry standard for digital television. If we use eight bits to represent each color component of a given pixel, then the bit rate for the uncompressed television signal is 24 × 720 × 480 × 30 = 248 megabits/second. As another example, consider one of the high-definition television standards that has a resolution of 1920 × 1080 pixels/frame at 60 frames/second. The uncompressed bit rate for this HDTV signal is 3 gigabits/second. Clearly, effective compression algorithms are desirable for these signals.

In the discussion relating to JPEG, we presented a DCT coding method as a means for exploiting the spatial redundancies in a picture. In **intraframe video coding** each picture is encoded independently of all other pictures. The motion JPEG system involves applying the JPEG algorithm to each frame of a video sequence. For the 720 × 480 television signal described above, the JPEG algorithm can produce digital video with high quality at bit rates of about 20 megabits/second.

The temporal nature of video implies that video signals also contain temporal redundancy. **Interframe video coding** methods exploit this redundancy to achieve higher compression ratios. Video signals can be viewed as consisting of scenes in which the setting is fixed and the time is continuous. In scenes with little or no motion, the consecutive frames differ in very small ways. In this case the notion of predictive coding suggests that we encode the differences between consecutive frames. The pixels in these different frames will predominantly consist of zeros and cover a narrow dynamic range. From our results on the performance of quantizers, we know that such signals will be highly compressible. However, in scenes that contain a significant amount of motion the difference between successive frames will fail to capture the temporal relation between them. Motion compensation algorithms have been developed to address this problem.

Figure 12.31 shows the operation of the video compensation method. As before, frames are segmented into 8-x-8-pixel blocks. The coding of given frame is performed relative to the previous frame, which is stored in a buffer. For each 8×8 block of pixels, a search is carried out to identify the 8×8 block in the preceding frame that best matches the block that is to be encoded. A **motion vector** is calculated specifying the relative location of the encoded block to the block in the preceding frame. The DCT is applied to the difference between the two blocks, and the result is quantized and encoded as in the JPEG method. The resulting compressed information and motion vector allow the decoder to reconstruct the approximation of the original signal.

Motion compensation is the most time-intensive part in current video compression algorithms. The region that is searched to identify the best fit to a block is typically limited to a region in the vicinity of the block. Note that the algorithm that is used to conduct the search does not need to be standardized, since the decoder needs to know the motion vectors but not how they were found. Another consideration in selecting the motion compensation algorithm is whether the encoding has to be done in real time. In a studio production situation where the uncompressed video sequence has been prerecorded and stored, the encoder can use complex motion compensation algorithms to produce efficient and high-quality compressed versions of the signal. In real-time applications, such as broadcasting and videoconferencing, the system has much less time to carry out motion compensation, so simpler, less effective algorithms are used.

Figure 12.32 shows the block diagram for the ITU H.261 video encoding standard for videoconferencing applications. This standard provides for various

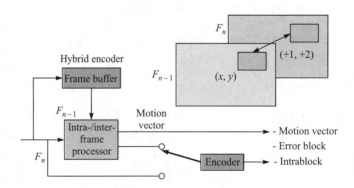

FIGURE 12.31 Video motion compensation

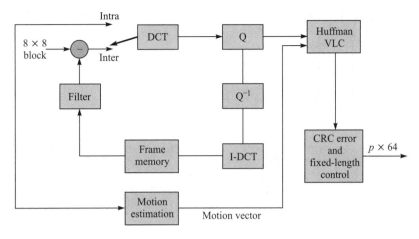

FIGURE 12.32 The H.261 videoconferencing standard

coding options that can result in bit rates of the form $p \times 64$ kbps. Recall that 64 kbps is the standard bit rate required to handle a single voice signal within the telephone network. The $p = 1$ and $p = 2$ versions can be accommodated in the basic rate interface of ISDN. The $p = 24$ version occupies a T-1 telephone line. The H.261 system is defined for frames of size 360×240 pixels and 180×120 pixels, which are 1/4 and 1/16 the resolution of the 720×480 system discussed earlier. The encoder uses a combination of intraframe DCT coding and interframe DCT coding with motion compensation. Reasonable picture quality is provided by these systems when the scenes contain little motion. However, the encoding delay is significant, so the time that elapses from when a person moves to when the motion is seen on the screen is noticeable. This delay makes the interaction between the participants a bit awkward.

12.3.7   The MPEG Video Coding Standards

Video on demand and related applications require that the encoded video signal accommodate capabilities associated with VCR controls. These include the ability to fast forward, reverse, and access a specific frame in a video sequence. The **Motion Picture Expert Group (MPEG)** standards have been developed to meet these needs. The MPEG-1 and MPEG-2 standards use three types of encoding that produce three types of frames as shown in Figure 12.33. **I-frames** are encoded with intraframe coding. Since their decoding is independent of all other frames, I-frames facilitate fast-forward, reverse, and random-access capabilities. I-frames have the lowest compression, but they can be encoded and decoded faster than the other frame types. **P-frames** (predictive) are encoded by using motion compensation relative to the most recent I- or P-frame. P-frames have better compression levels than I-frames. **B-frames** (bidirectional) use motion

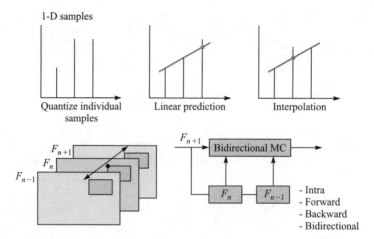

FIGURE 12.33 Prediction and interpolation in MPEG

compensation relative to both the preceding and the following I- or P-frames. The motion vector for a given block can refer to either or both these two frames. B-frames achieve the highest compression of the three frame types, but they also take the longest time to encode. In addition, they cause significant delay at the encoder, since they require the availability of the following I- or P-frame. In the MPEG standards the motion compensation algorithm attempts to find the best match for macroblocks that consist of 16-x-16-pixel blocks of luminance symbols. The motion vector that results from this process is then used for the associated 8-x-8-pixel chrominance signal blocks.

The encoded frames in MPEG are arranged in **groups of pictures** as shown in Figure 12.34. I-frames are inserted into the sequence at regular intervals to provide VCR capabilities as well as to limit the propagation of errors associated with predictive coding. The remainder of the frames consist of P-frames and B-frames.

The MPEG-1 standard was developed to produce VCR-quality digital video at about 1.2 Mbps. This standard makes storage possible in CD-ROM and

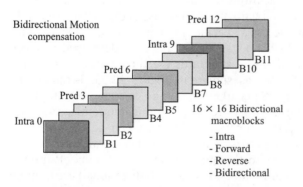

FIGURE 12.34 MPEG group of picture structure

transmission possible over T-1 digital telephone lines. The MPEG-2 standard is much broader in scope. It can accommodate frame sizes of 352×240 pixels, 720×480 pixels, 1440×1152 pixels, and 1920×1080 pixels. The MPEG-2 standard defines a system of profiles where each consists of a set of coding tools and parameters. For example, the MPEG-2 main profile uses three types of frames, namely, I-, B-, and P-frames. It focuses on the 720-x-480-pixel resolution that corresponds to conventional television. Very good quality video is attainable using MPEG-2 at bit rates in the range of 4 to 6 Mbps. Another profile accommodates simpler implementation by avoiding the use of B-frames.

Other profiles address scalability options where users with differing requirements wish to access the same video information, as shown in Figure 12.35. In **SNR scalability** different users require the same image sizes but have different bit rates available. The encoded video then needs to be represented as two streams: one at a basic quality that needs the lower of the bit-rate requirements; the other that provides information that allows the higher bit-rate user to enhance the quality of the recovered signal. SNR scalability is also referred to as layered coding because it can provide the basis for maintaining basic-quality transmission in the presence of errors. This level of performance is achieved by having the base stream transmitted with higher priority or greater error protection than the enhancement stream. In **spatial scalability** different users wish to access the same information at different resolutions. Again, the encoded information is divided into two streams: one that provides the basic resolution and the other that provides the enhancement required to obtain the higher resolution. The aim of spatial scalability in MPEG-2 is to support multiresolution applications that involve backward compatibility, for example, between MPEG-2 and MPEG-1 or H.261, as well as compatibility between standard-format television and HDTV.

The MPEG-2 profiles also consider HDTV. As discussed in Chapter 3, conventional television uses an aspect ratio of 4:3, giving a squarish picture. HDTV

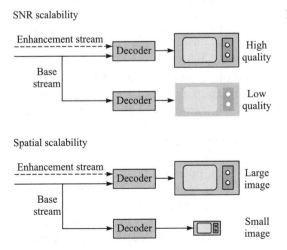

SNR scalability

FIGURE 12.35 Scalable video coding

Spatial scalability

uses an aspect ratio of 16:9, giving a picture that is closer to that of motion pictures. MPEG-2 allows for two larger resolutions to accommodate HDTV: 1440×1152 and 1920×1080. The higher resolution and frame rates of these signals result in huge bit rates for the uncompressed signal. MPEG-2 coding can produce high-quality HDTV compressed signals at bit rates in the range 19 to 38 Mbps. The Advanced Television System Committee (ATSC) has developed an MPEG-2–based coding standard for HDTV systems in the United States for use over terrestrial broadcast and cable systems over conventional analog MHz bandwidth channels.

The MPEG standards group has already begun work on an MPEG-4 coding standard for use in *very low bit rate* applications such as those encountered in wireless networks. It is clear that the requirement for multimedia communications will soon extend to wireless networks. The problems are that bandwidth is scarce and communication is unreliable in wireless networks. So the challenge here is how to send images, for example, remote telemetry or security, over these types of channels.

12.3.8   MPEG Multiplexing

The systems part of the MPEG-2 standard deals with the multiplexing and demultiplexing of streams of audio, video, and control information. MPEG-2 has the ability to carry multiple elementary streams that need not have a common time base. Figure 12.36 shows the case of an audio and a video stream that have a common time base. The outputs of the encoders are packetized, and timestamps derived from a common clock are inserted into each packet. This process results in *packetized elementary streams* (PES). Each PES packet contains information that identifies the stream; specifies the packet length, and gives a PES priority that can be used with layered coding, as well as presentation and decoding timestamps and packet transmission rate information. As shown in the figure, two multiplexing options are supported: program stream and transport stream.

MPEG-2 program stream results from the multiplexing of PES streams that have a common time base into a single stream. The program stream is intended for error-free environments, and so it uses packets that are variable length and relatively long. The transport stream multiplexes one or more PES streams with one or more time bases. The transport stream is intended for situations where packet loss may occur, for example, network transmission, and so it uses fixed-length 188-byte packets that facilitate recovery from the losses. The transport stream packets contain program clock reference timestamps that can be used for clock recovery.

For a transport stream the decoder demultiplexes the stream and removes the transport stream and PES headers to reconstruct the compressed video and audio elementary streams. These streams are placed in a buffer and retrieved by the video and audio decoders at the appropriate time. Two timing processes are in play here. First the output from the video and audio decoders must synchro-

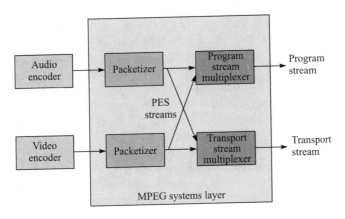

FIGURE 12.36 MPEG-2 multiplexing

nize to have "lip synch." This process requires the synchronization of the two media streams and hence the compensation for any misalignments in the relative timing of packets that may have occurred during transmission. The synchronization is done through the use of presentation timestamps and decoding timestamps. The second timing process is the recovery of the encoder clock. The decoder must reconstruct the timing of the clock that was used at the encoder. The encoder inserts program clock reference timestamps. Chapter 5 covers this type of timing recovery.

12.4 THE REAL-TIME TRANSPORT PROTOCOL

Traditional real-time communications have taken place over circuit-switched networks that can provide low transfer delay of a steady stream of information. Telephone calls and low-bit-rate videoconferencing are the primary examples of this type of communications. Packet networks were developed primarily for the transport of data information that did not have such stringent timing requirements. Advances in compression algorithms and computer processing power combined with improvements in transmission bandwidth and packet-switching capability are making it possible to support real-time communications over packet networks. Advances in computer processing in particular make it feasible for a wide range of media types to coexist in multimedia applications. Real-time packet communications make it possible to exploit the ubiquity of the Internet. Packet communications also bring capabilities such as multicasting that are not easily provided by circuit-switched networks.

   Real-time packet communications, however, must deal with impairments inherent in packet networks, which include packet delay and jitter, packet loss, and delivery of out-of-sequence or duplicate packets. The **Real-Time Transport Protocol (RTP)** [RFC 1889] provides end-to-end transport functions for applications that require real-time transmission, such as audio or video over unicast or multicast packet network services. RTP services include payload type identification, sequence numbering, timestamping, and delivery monitoring. RTP normally runs on top of UDP, but it can also run over other suitable networks or transport protocols. RTP does not provide any means to ensure the timely delivery of information or Quality-of-Service guarantees. For these RTP must depend on the services of the underlying network layer. RTP instead provides the mechanisms for dealing with impairments such as jitter and loss, as well as for timing recovery and intermedia synchronization.

The **RTP Control Protocol (RTCP)** is a companion protocol for monitoring the quality of service observed at a receiver and for conveying this and other information about participants to the sender. This capability is especially useful in situations where the sender can adapt its algorithm to the prevailing network conditions, for example, available bandwidth or network delay/jitter.

RTP is intentionally not a complete protocol and is intended to be sufficiently flexible so that it can be incorporated into the application processing, instead of being implemented as a separate layer. The use of RTP in a particular application therefore requires one or more companion documents. A *profile* specification document defines attributes and/or modifications and extensions to RTP for a class of applications, for example, audio and video. A *payload format* document in turn defines how a particular audio or video encoding is to be carried over RTP, for example, MPEG-2 video or ADPCM audio. [RFC 1890] specifies an initial set of payload types. [RFC 2250] describes a packetization scheme for MPEG video and audio streams.

The RTP/RTCP protocols are in wide use in the Internet supporting audio- and video-streaming applications as well as Internet telephony and other real-time applications.

12.4.1   RTP Scenarios and Terminology

Let's introduce some terminology, using first an audio conference example and then an audiovisual conference example.

An *RTP session* is an association among a group of participants communicating via RTP. Suppose for now that the RTP session involves an audio conference. The chair of the conference makes arrangements to obtain an IP multicast group address and a pair of consecutive UDP port numbers that identify the RTP session. The first port number (even) is for RTP audio and the other port number (odd) is for the corresponding RTCP stream. The address and port information is distributed (by some means beyond the scope of RTP) to the participants.

Once the audio conference is underway, each participant can send fixed-duration blocks of audio information as payloads in an RTP PDU, which in turn is incorporated in a UDP datagram. The RTP header specifies the type of audio encoding. It also includes sequence and timestamp information that can be used to deal with packet loss and reordering and to reconstruct the encoder clock. Each source of a stream of RTP packets is identified by a 32-bit *synchronization source (SSRC)* ID that is carried in the RTP header. All packets for a given SSRC use the same timing and sequence number space to allow receivers to regroup and resynchronize the given packet sequence.

Each RTP periodically multicasts a receiver report on the RTCP port. The report provides an indication of how well the RTP packets are being received. The report also identifies who is participating in the conference. Upon leaving the conference, each site transmits an RTCP BYE packet.

Now consider the case of an audiovisual conference. Each media is transmitted using a *separate* RTP session, that is, using separate multicast and UDP port pair addresses. The audio and video RTP sessions are treated as completely separate except that their association is indicated by a unique name that is carried in their RTCP packets. This practice allows the synchronized playback of the audio and video streams.

There are a number of reasons for using a separate RTP session for each media stream. Different Quality-of-Service or discard treatments can be provided for each stream, for example, under congestion, video packets are discarded and audio packets kept. Some users may choose or be capable of receiving only audio and not video. In the case of layered coding, different terminals may choose to receive a different set of layered video streams either because of their processing capability or because of the available bandwidth in their direction.

RTP allows for the use of devices called *mixers*. A mixer is an intermediate system that receives RTP packers from one or more sources, possibly changes the data format, and combines the packet in some manner and then forwards new RTP packets. For example, a mixer may combine several audio streams into a single stream. The mixer usually needs to make timing adjustments among the streams, so it generates new timing for the packet sequence it produces. All the packets thus generated will have the mixer SSRC identified as their synchronization source. The mixer inserts in each RTP packet header a *contributing source (CSRC)* list of the sources that contributed to the combined stream.

RTP also allows for devices called *translators* that relay RTP packets with the SSRC left intact. Translators include devices that convert formats without mixing, replicators from multicast to unicast, and application-level filters in firewalls.

12.4.2. RTP Packet Format

Figure 12.37 shows the packet header format for RTP. The first three rows (12 bytes) are found in every packet. The fourth row (CSRC) is used when a mixer has handled the information in the payload.

The RTP packet fields are used as follows:

Version (V): This two-bit field identifies the version of RTP. The current version is 2.

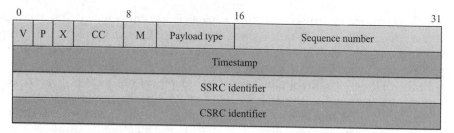

FIGURE 12.37 RTP packet header format

Padding (P): This one-bit field indicates that the packet contains one or more additional padding bytes that are not part of the payload. The last byte of the padding contains a count of how many padding bytes are to be ignored, including itself.

Extension (X): When the one-bit extension field is set, the fixed header must be followed by exactly one header extension.

CSRC count (CC): This four-bit number specifies the number of CSRC identifiers that follow the fixed header.

Marker (M): This one-bit field is defined by a profile and is intended to mark significant events such as frame boundaries in the packet stream.

Payload type (PT): This seven-bit field identifies the format of the RTP payload and determines its interpretation by the application.

Sequence number: This 16-bit field is incremented by one each time an RTP packet is sent. The number can be used by the receiver to detect packet loss and to recover packet sequence. The initial value is selected at random.

Timestamp: This 32-bit number specifies the sampling instant of the first byte in the RTP data packet. The sampling instant must be derived from a clock that increments monotonically and linearly in time, so that the number can be used for synchronization and jitter calculations. The initial value is selected at random.

SSRC: The randomly chosen number is used to distinguish synchronization sources within the same RTP session. It indicates either where the data was combined or the source of the data if there is only one source.

CSRC list: This list of 0 to 15 thirty-two bit items specifies the contributing sources for the payload contained in the packet. The number of identifiers is given by the CC field.

12.4.3   RTP Control Protocol (RTCP)

RTCP involves the periodic transmission of control packets to all participants in the session. The primary function of RTCP is to provide feedback on the quality of the data distribution, which can then be used for control of adaptive encodings or to diagnose faults in the distribution. This feedback information is sent in the form of RTCP sender and receiver reports.

RTCP also carries a persistent transport-level identifier called the Canonical name (CNAME) that is used to keep track of each participant. For example, CNAME can be given by user@host or by a host in single-user systems. CNAME is also used to associate multiple RTP sessions, for example, to synchronize audio and video in related RTP sessions to achieve lip synch.

Because all participants are required to send RTCP packets, a mechanism has been developed to control the rate at which RTCP packets are transmitted. Each participant is able to independently determine the number of participants from the RTCP packets it receives. This information is used to adjust the RTCP transmission intervals.

RTCP can optionally provide minimal session control information such as participant identification. A higher-level session control protocol is needed to provide all the support required by a given application. Such a protocol is beyond the scope of RTCP.

RTCP defines several types of packets to carry different types of control information:

Sender report (SR): The SR distributes transmission and reception statistics from active senders.

Receiver report (RR): The RR distributes reception statistics from participants that are not active senders.

Source description (SDES): SDES provides source description items such as CNAME, e-mail, name, phone number, location and application tool/version.

BYE: This message indicates the end of participation by the sender.

APP: This packet provides application-specific functions that are defined in profile specifications.

The SR packets provide a sender and several reception reports. The sender report information includes (1) a "wall clock" time given by a network time protocol (NTP) timestamp that is seconds elapsed since 0 hour January 1, 1900 and (2) the same time instant as the NTP timestamp but using the clock used to produce the RTP timestamps. This correspondence can then be used for intra- and intermedia synchronization. The sender report also gives the number of RTP packets transmitted as well as the total number of payload bytes transmitted by the sender since starting transmission. Each reception report provides the following statistics on a single synchronization source: fraction of RTP data packets lost since the preceding SR or RR packet was sent, cumulative number of RTP data packets lost since beginning reception, extended highest sequence number received, interarrival jitter, last SR timestamp, and delay since last SR.

The RR packets are the same as SR packets except that a sender report is not included.

12.5   SESSION CONTROL PROTOCOLS

The RTP Control Protocol (RTCP) was designed to provide minimal control functionality. In particular RTCP does not provide explicit membership control and session setup. The intent is that a separate session control protocol would provide this functionality. In this section we introduce several protocols and recommendations that can provide this functionality.

12.5.1   Session Initiation Protocol

The **Session Initiation Protocol (SIP)** [RFC 2543] is an application layer control protocol that can be used to establish, modify, and terminate multimedia sessions or calls with one or more participants. These multimedia sessions can be an Internet telephony call, a multimedia videoconference, a distance-learning session, or multimedia distribution. The participants in the session can be people or various types of media devices, for example, media servers. SIP provides support for user mobility by proxying and request redirection to a user's current location. SIP is designed to be independent of lower-layer transport protocols.[3]

SIP is a text-based client/server protocol. A *transaction* consists of the issuing of a request by a client and the returning of one or more responses by one or more servers. Basic signaling functions are implemented with one or more transactions. As in the case of HTTP, each SIP request invokes a *method* in a server. SIP provides six methods: INVITE, ACK, OPTIONS, BYE, CANCEL, and REGISTER. INVITE is the most basic method and is used to initiate calls.

A SIP system has two components: user agents and network servers. The *user agent* is software in an end system that acts on behalf of a human user. The user agent has two parts: a protocol client, called *user agent client* (UAC), initiates a call; a protocol server, called *user agent server* (UAS), answers a call. Together the UAC and AUS allow for peer-to-peer operation using a client/server protocol.

The function of the network servers is to do the call routing to establish a call, that is, to find the desired user in the network. The network servers can be of two types: proxy and redirect. A *proxy server* receives a request, determines which server to send it to, and then forwards the request. Several servers may be traversed as a request flows from UAC to UAS. The eventual response traverses the same set of servers but in the reverse direction. SIP allows a proxy server to *fork* a request and forward it simultaneously to several next-hop servers. Each of these branches can issue a response, so SIP provides rules for merging the returning responses. A *redirect server* does not forward a request and instead returns a message to the client with the address of the appropriate next-hop server.

---

[3]In this section we follow the development in [Schulzrinne et al. 1998, 1999].

To establish a call, an INVITE request is sent to the UAS of the desired user. In general, the IP address or host name of the desired user is not known. As a result, this information must be obtained from a name such as an e-mail address, telephone number, or other identifier. The UAC sends this name to an appropriate network server, which in turn may proxy or direct the call to other servers until a server is found that knows the IP address of the desired user. The *response* to an INVITE request from the UAS contains a reach address that the UAC can use to send further transactions *directly* to the UAS. Consequently, the SIP network servers do not need to maintain call state.

The INVITE request contains addresses for the caller and callee, subject of the call, call priority, call routing requests, caller preferences for user location, and desired features of the response. The body of the request contains a description of the medial content for the session. **The Session Description Protocol** (SDP) format can be used provide this description. SDP provides information about the number and types of media streams in a session such as the destination addresses for each stream, the sending and receiving UDP ports, and the payload types. Other formats also provide this information, for example, H.245 capability descriptors. The response to the INVITE provides the information about the media content for the callee.

The REGISTER methods sends location information to a SIP server. For example, a user can send REGISTER to help a server map an incoming address into an outgoing address that can reach the user or a proxy that knows how to reach the user. A typical example involves a user sending a REGISTER message to its usual SIP server, giving it a temporary forwarding address.

BYE terminates a connection between two users. OPTIONS requests information about the capabilities of a callee without setting up a call. ACK ensures reliability in the message exchanges and CANCEL terminates a pending request.

INTERNET TELEPHONY AND IP TELEPHONY

The terms *Internet telephony* and *IP telephony* describe two different approaches to providing telephone service over IP that are different in a subtle but fundamental way. Internet telephony describes telephone service that uses the classical Internet approach, where the service control is in the end user's system, for example, the PC. The SIP protocol provides the means for providing the end systems with telephone service control. This approach is characterized by smart and powerful end systems.

IP telephony describes telephone service over IP in which service control is provided by intelligence inside the network. This approach follows the traditional telephone network setting where the terminal, for example, is a very low cost, low-functionality device. The actual setting up of a connection across the network is done by switches attached to the terminal equipment.

The interworking of telephone service that spans the Internet and the telephone network is addressed by the introduction of gateways, as provided by H.323 and related approaches.

12.5.2   H.323 Multimedia Communications Systems

ITU-T recommendation H.323 consists of a set of standards to provide support for real-time multimedia communications on LANs and packet networks that do not provide QoS guarantees. H.323 evolved out of the H.320 videoconferencing standards for ISDN. H.323 terminals and equipment can carry voice, video, data, or any combination of these. H.323 addresses call control, multimedia management, bandwidth management, and interfaces to other networks. In particular it provides a means for interworking telephone-based and IP-based conferencing.

As shown in Figure 12.38, an H.323 network involves several components. In addition to H.323 terminals, the network involves gateways, gatekeepers, and multipoint control units.

The gateways provide interworking between H.323 terminals in the packet network and other terminal types, for example, telephone sets in a conventional telephone network. The gateway is responsible for the translation between audio and video codec formats. For example, a speech signal may be encoded using ITU-T G.729 8 kbps compression with RTP framing in the packet network and 64 kbps PCM in a telephone network. The gateway is responsible for the translation between the two media formats.

The gateway is also responsible for the mapping of signaling messages from the packet side of the network to other networks. In particular, in telephony applications the gateway performs call setup between the packet network and the public telephone network. In particular, the gateways terminate H.323 signaling on the packet network side and usually ISDN signaling on the telephone network side. Part of the call setup involves establishing a path for the call across the gateway.

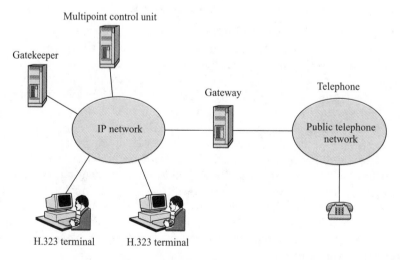

FIGURE 12.38 Components of H.323 network and a telephone network

The gatekeepers are responsible for call control for calls within an H.323 network. Gatekeepers grant permission or deny requests for connections. They manage the bandwidth that can be used by calls. Gatekeepers perform name-to-address translation, and they direct calls to appropriate gateways when necessary.

H.323 terminals in a multipoint conference can send audio and video directly to other terminals by using multicasting. They can also use multipoint control units that can combine incoming audio streams and video streams and transmit the resulting streams to all terminals.

Figure 12.39 shows the scope of an H.323 terminal. H.225 specifies the call control procedures that are used for setting up H.323 calls in the H.323 network. The procedures use a subset of the Q.931 messages that are used in conventional ISUP signaling.[4] Terminals are addressed using either IP addresses or names (e-mail address, telephone number) that can be mapped to an IP address. H.225 also stipulates that RTP/RTCP is to be used in the packetization of the audio and video streams.

The H.245 control channel is a reliable channel (operating over TCP) that carries the control messages to set up logical channels, including the exchange of transmit and receive capabilities. The RAS control deals with registration, admission control, bandwidth management between end points, and gatekeepers.

H.323 specifies audio and video codes that are supported. All H.323 terminals are required to support the G.711 voice standard for log-PCM voice compression. Additional audio codecs are also specified. Video is optional in H.323 terminals. QCIF H.261 video is the baseline video mode.

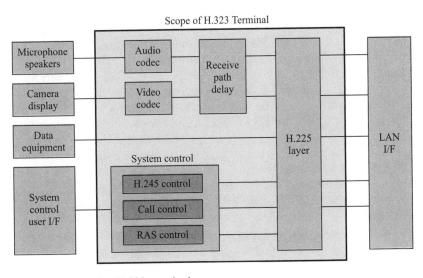

FIGURE 12.39 An H.323 terminal

---

[4]ISUP signaling is discussed in Chapter 4.

12.5.3   Media Gateway Control Protocols

The H.323 recommendation assumes relatively powerful end systems attached to the packet network. The signaling and processing requirements are too complex for simple terminal equipment such as telephones. This restriction has led to proposals for approaches that allow simple terminal equipment to connect to the Internet and provide telephone service. The approaches have two aspects. The first aspect involves the introduction of a *residential gateway* that is interposed between a telephone and the Internet and provides the required processing capability. The second aspect involves the partitioning of the functions of the H.323 gateway into two parts. The *media gateway* is placed between the Internet and the telephone network. The role of the gateway is to carry out media format conversion. The call control function is provided by *call agents* that are placed in the Internet. Residential gateways interact with call agents to set up a telephone call. Call agents in turn interact with *SS7 gateways* that allow the agents to interact with the telephone signaling system. The call agents use a *media gateway control* protocol to control the setup of connections across the media gateways.[5]

CONTROL AT THE EDGE OR INSIDE THE NETWORK?

The Internet and the traditional telephone network are fundamentally different in their approach to service control. The Internet has always placed control in the end system and made minimal assumptions about the network—in fact, the network or networks. The telephone network has always placed service control in the switches inside the network. The two approaches lead to fundamentally different philosophies with respect to the introduction of new services. The traditional Internet approach, by keeping control in the end system, allows any new application that operates over TCP/IP to be readily introduced into the network without the permission of the operator. The traditional telephone approach, by keeping control inside the network, allows the network operator to decide which services to deploy in the network. The experience with the World Wide Web has demonstrated that the Internet approach leads to a much faster rate of service introduction because the Internet empowers a much larger community to define and introduce services. Nevertheless, it is possible that, as the Internet service industry matures and a few dominant operators emerge, there will be a temptation to control the services that are available to the end users by progressively moving service control back inside the network. We can only hope that network architectures will emerge that foster competition and continue to allow end users to drive the introduction of new services.

---

[5]The media gateway control protocols are work in progress, so the reader is referred to the IETF Web site (http://www.ietf.org) for the current status of these standards.

SUMMARY

Multimedia networking involves the transfer of a variety of information types that may be integrated in a single application. In this chapter we presented the basic techniques that are used to obtain efficient digital representations for various types of information media. We took a close look at the format, bit rates, and other properties of different types of information, such as speech, audio, facsimile, image, and video. We saw that audio and video information can require higher bit rates than those typically provided by packet networks. We also presented the important compression standards from the classic logarithmic-PCM for telephone voice to the current MPEG2 and MP3 standards for video and audio.

Advances in packet networks will allow them to carry real-time audio and video. We introduced the RTP protocol, which provides the framework for the transport of multimedia over the Internet. We also introduced protocols for initiating and controlling multimedia sessions, SIP and H.323. Finally we highlighted the philosophical differences between service control in the end system and service control in the network.

CHECKLIST OF IMPORTANT TERMS

ac component
adaptive lossless data compression
adaptive DPCM (ADPCM)
adaptive quantizer
◆aliasing error
B-frame
bandwidth of an analog signal
clipping
codeword
companding
dc component
delay
differential coding
differential PCM (DPCM)
digital information
discrete cosine transform (DCT)
◆entropy
groups of pictures
Huffman code
I-frame
interframe video coding
interlacing
intraframe video coding
Joint Photograph Expert Group
 (JPEG)
linear predictive coders (LPC)
Motion Picture Expert Group
 (MPEG)
motion vector
nonuniform quantizers
Nyquist sampling rate
out of order
P-frame
pixel
prediction error
prediction gain
predictive coding
pulse code modulation (PCM)
quantization error
RGB representation
Real-Time Transport Protocol (RTP)
RTP Control Protocol (RTCP)

run   spectrum
run-length code   subband coding
Session Initiation Protocol (SIP)   synchronization
signal-to-noise ratio (SNR)   transform coding
SNR scalability   uniform quantizer
spatial scalability   vector quantization

FURTHER READING

Arango, M., A. Dugan, I. Elliott, C. Huitema, and S. Pickett "Media Gateway Control Protocol (MGCP)," draft-huitema-megaco-mgcp-v0r1-05.txt, February 21, 1999.

Gibson, J. D., T. Berger, T. Lookabaugh, D. Lindbergh, and R. L. Baker, *Digital Compression for Multimedia: Principles and Standards*, Morgan Kaufmann, San Francisco, 1998.

Huitema, C., J. Cameron, P. Mouchtaris, and D. Smyk, "An Architecture for Residential Internet Telephony Service," *IEEE Network*, Vol. 13, No. 3, 1999, pp. 50–56.

Jayant, N. S. and P. Noll, *Digital Coding of Waveforms*, Prentice Hall, Englewood Cliffs, New Jersey, 1984.

Schulzrinne, H., S. L. Casner, R. Frederick, and V. Jacobson, "RTP: A Transport Protocol for Real-Time Applications," draft-ietf-rtp-new-04.ps, June 25, 1999.

Schulzrinne, H. and J. Rosenberg, "The IETF Internet Telephony Architecture and Protocols," *IEEE Network*, Vol. 13, No. 3, 1999, pp. 18–23.

Schulzrinne, H. and J. Rosenberg, "A Comparison of SIP and H.323 for Internet Telephony," *Proceedings International Workshop on Network and Operating System Support for Digital Audio and Video (NOSSDAV)*, Cambridge, England, July 1998.

Schulzrinne, H. and J. Rosenberg, "The Session Initiation Protocol: Providing Advanced Telephony Services across the Internet," *Bell Labs Technical Journal*, Vol. 3, October–December 1998, pp. 144–160.

Seuss, Dr. [pseudo.], *Hop on Pop*, Random House, New York, 1963.

T. M. Denton Consultants, "Netheads vs Bellheads: Research into Emerging Policy Issues in the Development and Deployment of Internet Protocols," available at http://www.tmdenton.com/netheads3.htm.

RFC 1889, H. Schulzrinne, S. Casner, R. Frederick, and V. Jacobson, " A Transport Protocol for Real-Time Applications," January 1996.

RFC 1890, H. Schulzrinne, "RTP Profile for Audio and Video Conferences with Minimal Control," January 1996.

RFC 2250, D. Hoffman, G. Fernando, V. Goyal, and M. Civanlar, "RTP Payload Format for MPEG1/MPEG2 Video," January 1998.

RFC 2543, M. Handley, H. Schulzrinne, E. Schooler, and J. Rosenberg, "SIP: Session Initiation Protocol," March 1999.

PROBLEMS

1. a. Give three examples of applications in which the information must be represented in a lossless way.

b. Give three examples of applications in which information can be represented in a lossy manner.

2. The probabilities for the letters in the English alphabet are given in Table 1.1. The space between letters constitutes one-sixth of the characters in a page.
   a. Design a Huffman code for the letters of the English alphabet and the space character.
   b. What is the average number of bits/symbol?
   c. Compare the answer of part (b) to the entropy.

3. Suppose an information source generates symbols from the alphabet {a,b,c} and suppose that the three symbols are equiprobable.
   a. Find the Huffman code for the case where the information is encoded one symbol at a time?
   b. Find the Huffman codes when the symbols are encoded in blocks of length 2 and in blocks of length 3.
   c. Compare the answers in parts (a) and (b) with the entropy of the information source.

4. Consider a binary information source that "stutters" in that even-numbered symbols are repetitions of the previous odd numbered symbols. The odd-numbered symbols are independent of other symbols and take on the values 0 and 1 with equal probabilities.
   a. Design a Huffman code for encoding pairs of symbols where the first component of the pair is an odd-numbered symbol and the second component is the following even-numbered symbol.
   b. Now design a Huffman code for encoding pairs of symbols where the first component of the pair is an even-numbered symbol and the second component is the following odd-numbered symbol.
   c. Compare the performance of the two codes. What is the entropy of this information source?

5. The MNP5 lossless data compression algorithm used in various modem products uses a combination of run-length coding and adaptive variable-length coding. The algorithm is designed to encode binary octets, that is, groups of eight bits. The algorithm keeps a running count of the frequency of occurrence of the 256 possible octets and has a table listing the octets from the most frequent to the least frequent. The two most frequent octets are given the codewords 0000 and 0001. The remaining 254 octets are assigned a codeword that consists of a three-bit prefix that specifies the number of remaining bits and a suffix that specifies the rank of the codeword within a group:

most frequent octet	000	0
second most frequent octet	000	1
third most frequent octet	001	0
fourth most frequent octet	001	1
fifth most frequent octet	010	00
sixth most frequent octet	010	01
seventh most frequent octet	010	10
eighth most frequent octet	010	11
ninth most frequent octet	011	000
... ...		
16th most frequent octet	011	111

17th most frequent octet	100	0000
... ...		
32nd most frequent octet	100	1111
... ...		
129th most frequent octet	111	0000000
... ...		
256th most frequent octet	111	1111111

a. Suppose that a certain source generates the 256 octets with respective probabilities $c(1/2)^i$, $i = 1, 2, \ldots, 256$, where $c \approx 1$. What is the average number of bits/octet? What is the entropy of this source?

b. Suppose that the 256 octets are equiprobable? What is the average number of bits/octet? Is coding useful in this case?

c. For which set of probabilities is this the Huffman code? What is the entropy for such a source?

6. a. Use the $m = 3$ and $m = 5$ run-length codes to encode the binary string used in Figure 12.5.

b. Repeat part (a) using the $m = 3$ and $m = 5$ codes from Figure 12.6.

7. Consider a binary source that produces information bits b_i independent of each other and where $P[b_i = 0] = p$ and $P[b_i = 1] = 1 - p$. Let l be the number of consecutive 0s before a 1 occurs.

a. Show that $P[l = i] = (1 - p)p^i$, for $i = 0, 1, 2, \ldots$
b. Show that $P[l > L] = p^{L+1}$.
c. Find the performance of the run-length code in Figure 12.5 for $m = 4$ when $p^{16} = 1/2$.
d. Find the performance of the run-length code in Figure 12.7 for $m = 4$ when $p^{16} = 1/2$ and compare it to the result in part (c).

8. a. Use the Lempel-Ziv algorithm, as described in the chapter, to encode the string of characters in part (c) and part (d) of problem 7.

b. Repeat (a) where the algorithm refers to the most recent occurrence of the repeated pattern.

9. Decode the following Lempel-Ziv encoded passage from Dr. Seuss and find the compression ratio:

I_do_not_like_them_in_a_box._[1,19]with[22,3]f[26,4][1,24]house[28,28]m[86,25][16,2]r[13,2]or [14,2][144,4][1,19]anyw[153,20]gree[81,2]eggs[177,3]d[84,2]am[28,21]S[218,2]-I-[218,3]

10. Apply the Lempel-Ziv algorithm to the binary string in Figure 12.5.

11. A way of visualizing the Nyquist theorem is in terms or periodic sampling of the second hand of a clock that makes one revolution around the clock every 60 seconds. The Nyquist sampling rate here should correspond to two samples per cycle, that is, sampling should be done at least every 30 seconds.

a. Suppose we begin sampling when the second hand is at 12 o'clock and that we sample the clock every 15 seconds. Draw the sequence of observations that result. Does the second hand appear to move forward?

b. Now suppose we sample every 30 seconds. Does the second hand appear to move forward or backward? What if we sample every 29 seconds?

c. Explain why a sinusoid should be sampled at a little more than twice its frequency.

d. Now suppose that we sample every 45 seconds. What is the sequence of observations of the second hand?

e. Motion pictures are made by taking a photograph 24 times a second. Use part (c) to explain why car wheels in movies often appear to spin backward while the cars are moving forward!

12. Researchers are currently developing "software radios" that will be able to demodulate and decode any radio signal regardless of format or standard. The basic idea in software radio is to immediately convert the transmitted radio signal into digital form so that digital signal processing software can be used to do the particular required processing. Suppose that a software radio is to demodulate FM radio and television. What sampling rate is required in the A/D conversion? The transmission bandwidth of FM radio is 200 kHz, and the transmission bandwidth of television is 6 MHz.

13. An AM radio signal has the form $x(t) = m(t)\cos(2\pi f_c t)$, where $m(t)$ is a lowpass signal with bandwidth W Hz. Suppose that $x(t)$ is sampled at a rate of $2W$ samples/second. Sketch the spectrum of the sampled sequence. Under which conditions can $m(t)$ be recovered from the sampled sequence?

14. A black-and-white image consists of a variation in intensity over the plane.

a. By using an analogy to time signals, explain spatial frequency in the horizontal direction; in the vertical spatial direction. Hint: Consider bands of alternating black and white bands. Do you think there is a Nyquist sampling theorem for images?

b. Now consider a circle and select a large even number N of equally spaced points around the perimeter of the circle. Draw a line from each point to the center and color alternating regions black and white. What are the spatial frequencies in the vicinity of the center of the circle? Will aliasing occur if the colored disk is sampled?

15. A high-quality speech signal has a bandwidth of 8 kHz.

a. Suppose that the speech signal is to be quantized and then transmitted over a 28.8 kbps modem. What is the SNR of the received speech signal?

b. Suppose that instead a 64 kbps modem is used? What is the SNR of the received speech signal?

c. What modem speed is needed if we require an SNR of 40 dB?

16. An analog television signal is a lowpass signal with a bandwidth of 4 MHz. What bit rate is required if we quantize the signal and require an SNR of 60 dB?

17. An audio digitizing utility in a PC samples an input signal at a rate of 44 KHz and 16 bits/sample. How big a file is required to record 20 seconds?

18. Suppose that a signal has amplitudes uniformly distributed between $-V$ and V.

a. What is the SNR for a uniform quantizer that is designed specifically for this source?

b. Suppose that the quantizer design underestimates the dynamic range by a factor of 2; that is, the actual dynamic range is $-2V$ to $2V$. Plot the quantization error versus signal amplitude for this case. What is the SNR of the quantizer?

19. A telephone office line card is designed to handle modem signals of the form $x(t) = A\cos(2\pi f_c t + \phi(t))$. These signals are to be digitized to yield an SNR of 40 dB using a uniform quantizer. Due to variations in the length of lines and other factors, the value of A varies by up to a factor of 100.

   a. How many levels must the quantizer have to produce the desired SNR?

   b. Explain how an adaptive quantizer might be used to address this problem?

20. The basic idea in companding is to obtain robustness with respect of variations in signal level by using small quantizer intervals for small signal values and larger intervals for larger signal values. Consider an eight-level quantizer in which the inner four intervals are Δ wide and the outer four intervals are 2Δ wide. Suppose the quantizer covers the range -1 to 1. Find the SNR if the input signal is uniformly distributed between $-V$ and V for $1/2 < V < 1$. Compare to the SNR of a uniform quantizer.

21. Suppose that a speech signal is A/D and D/A converted four times in traversing a telephone network that contains analog and digital switches. What is the SNR of the speech signal after the fourth D/A conversion?

22. What are the consequences of transmission errors in PCM? What are the consequences in DPCM?

23. a. Use the features in the sample waveforms shown in Figures 12.20 and 12.22 to explain how the ADPCM algorithm achieves compression.

   b. What additional information, relative to DPCM, does the ADPCM encoder have to transmit to the decoder?

24. Suppose that a sound such as that shown in Figure 12.22 has a duration of 30 ms. Suppose that you compute a Fourier series to produce a periodic function to approximate the signal.

   a. Estimate the number of bits required to specify the Fourier series.

   b. Now suppose you take the difference between the periodic function produced by the Fourier series and the actual function. How would you encode the difference signal to produce an improved quality signal? Give a rough estimate of the additional number of bits required?

25. If you have a microphone and a utility such as Sound Blaster Wave Studio in your PC that allows you to sample speech waveforms, describe the properties of the following sounds: "e" (as in *beet*), "i" (as in *bit*), "m" (as in *moon*), "s" (as in *sit*), and "h" (as in *hit*). How do the first three waveforms differ from the latter two?

26. In Figure 12.25a is the subband coding system better off not transmitting the outer band where the noise level exceeds the signal level?

27. Current AM transmission signals occupy a band of 10 kHz. Suppose that we wish to transmit stereo CD-quality audio over an AM frequency band.

   a. What is the minimum size of a modem signal constellation required to achieve this using uniform quantization?

   b. What is the minimum size of signal constellation if MP3 is used instead?

28. FM radio signals use a transmission bandwidth of 200 kHz. Can MPEG-1 or MPEG-2 video be transmitted over one of these channels? Explain.

29. An analog TV channel occupies a bandwidth of 6 MHz. How many MPEG-1 channels can be carried over such a channel if a digital transmission scheme is used with four constellation points?

30. (*The Globe and Mail*, Jan. 25, 1995) The CD-video standard proposed by Sony/Phillips provides for movies of up to 135 minutes with a total storage of 3.7 Gbytes. Can the JPEG or MPEG coding standards meet these bit rate requirements?

31. How many minutes of music can be stored in the CD system of problem 30, using uniform quantization? MP3 coding?

32. An XGA graphics display has a resolution of 1024×768 pixels. How many bytes are required to store a screenful of data if the number of colors is 256? 65,536? 16,777,216?

33. A scanner can handle color images up to a maximum size of 8.5 inches by 11 inches at resolutions up to 4800 pixels/inch. What file size is generated by the maximum-size image sampled at 24 bits/pixel? What range of file size will result if JPEG compression is used?

34. In JPEG the DCT coefficients are expressed as numbers in the range 0 to 255. The DCT coefficients (prior to quantization) corresponding to an 8×8 block of pixels is given below:

$$
\begin{array}{cccccccc}
148 & 92 & 54 & 20 & 6 & 2 & 2 & 0 \\
86 & 72 & 45 & 16 & 8 & 1 & 0 & 0 \\
56 & 48 & 32 & 10 & 7 & 2 & 0 & 0 \\
20 & 14 & 8 & 4 & 2 & 0 & 0 & 0 \\
4 & 3 & 2 & 1 & 0 & 0 & 0 & 0 \\
2 & 1 & 1 & 0 & 0 & 0 & 0 & 0 \\
1 & 1 & 1 & 0 & 0 & 0 & 0 & 0 \\
0 & 0 & 0 & 0 & 0 & 0 & 0 & 0
\end{array}
$$

In JPEG a DCT coefficient is quantized by dividing it by a weight 2^k and rounding the quotient, where $k = 0, 1, \ldots, 6$ is an integer selected according to the relative importance of the DCT coefficient. Coefficients corresponding to the lower frequencies have small values of k, and those corresponding to higher frequencies have higher values of k. The 8×8 quantization table specifies the weights that are to be used.

a. Explain how the larger values of k correspond to coarser quantizers.

b. From the following matrix of quantization table, find the resulting block of quantized DCT coefficients.

1	1	2	4	8	16	32	64
1	1	2	4	8	16	32	64
2	2	2	4	8	16	32	64
4	4	4	4	8	16	32	64
8	8	8	8	8	16	32	64
16	16	16	16	16	16	32	64
32	32	32	32	32	32	32	64
64	64	64	64	64	64	64	64

c. Find the one-dimensional sequence that results after zigzag scanning.

35. a. What impact does the speed of a CD-ROM drive have on the applications it can support?
    b. How does speed interact with total storage capacity?

36. A DS-3 digital transmission system has a bit rate of 45 Mbps and is the first level of high-speed transmission available to users.
    a. How many PCM calls can be accommodated in one DS-3 line?
    b. How many MPEG-1 or MPEG-2 television channels can be similarly accommodated?

37. What minimum delay does the MPEG decoder have to introduce to decode B frames? What impact does this factor have on the storage required at the receiver?

38. Suppose a video signal is to be broadcast over a network to a larger number of receivers and that the receivers have displays with resolutions of 720×480, 360×240, and 180×120. Explain how the scalability options of MPEG coding may be used in this multicasting application.

39. A video server system is designed to handle 100 simultaneous video streams. Explain the tasks that need to be carried out in servicing one stream. What is the aggregate input/output requirements of the system?

40. Should RTP be a transport layer protocol? Explain why or why not.

41. Can RTP be used to provide reliable real-time communication? If yes, explain how. If no, explain why not.

42. Can RTP operate over AAL5? Explain.

43. Suppose that the marker bit in the RTP header is used to indicate the beginning of a talkspurt. Explain the role of the timestamps and sequence numbers at the receiver.

44. Discuss the relative usefulness of RTCP for unicast and multicast connections.

45. Suppose a router acts as a firewall and filters packets. Can the router identify RTP packets? Can the router identify RTP packets from different connections?

46. Suppose you want to implement an IP telephony system.

a. Look up the RTP payload format for G.711. Discuss the transmission efficiency for the resulting system.
b. Repeat for G.729.

47. Explain the relevance to scalability of the SIP protocol not having to maintain state. Contrast this with the case of traditional telephone networks.

48. Consider whether and how SIP can be used to provide the following services:
   a. Call display: the number and/or name of the calling party is listed on a screen before the call is answered.
   b. Call waiting: a special sound is heard when the called party is on the line and another user is trying to reach the called party.
   c. Call answer: if the called party is busy or after the phone rings a prescribed number of times, the network gives the caller the option of leaving a voice message.
   d. Three-way calling: allows a user to talk with two other people at the same time.

49. A professor has three locations at his university: his regular professorial office, his lab, and the administrative office for a research center he operates.
   a. Explain how the *fork* capability of SIP may be used to direct an Internet phone call to the professor.
   b. Suppose that the professor spends Tuesdays working at a small startup, where he also has an Internet phone. Explain how a REGISTER message can be used to direct calls from the professor's office in the university to the startup.
   c. Suppose that on Wednesday, the professor forgets to change the forwarding instructions from the day before. Can a call placed to him at the startup reach him at the office? at the lab?

50. A recent proposal involves replacing the H.323 gateway with a trunking gateway that is responsible solely for media format conversion. Signaling to establish a call is handled by a call agent that talks to the client on the packet network side, to an SS7 gateway to the telephone network, and to the trunking gateway.
   a. Explain how a telephone call is set up between an Internet terminal and a telephone set.
   b. Does this system have greater scalability than an H.323 gateway has?
   c. Can the call agents be used to move telephone network intelligent network functionality to the Internet?

Trends in Network Architectures

We have reached the end of our journey only to find that the story is not all told. Networks are evolving at a rapid pace, and many changes are yet to come. We hope we have succeeded at least partially in teaching the student the fundamentals that are lasting and that provide the means to read the future in the present. We end with a brief discussion of current trends in network architecture.

In Chapter 1 we noted how network architecture is influenced by the cost of bandwidth and the cost of processing. The cost of processing and bandwidth have both been steadily declining. Let us examine the impact of these costs on network architecture.

The decrease in the cost of processing has resulted in greater protocol processing capability. Faster QoS-capable switches and routers result from hardware-based packet classifiers, prefix matching, schedulers, and interconnection packet-forwarding fabrics. Improved speed and capability in software-based functions also result from lower cost and faster microprocessors. Notwithstanding these advances, the scalability problems inherent in networking continue to pose a challenge with the tremendous rate of growth in the Internet.

The cost of bandwidth in the form of optical transmission has also been steadily decreasing, but with the introduction of dense wavelength-division multiplexing (WDM) systems a major discontinuity will occur. A single fiber will soon carry 160 wavelengths at 10 Gbps/wavelength. When one considers that a duct can contain several hundred optical fibers, it is clear that routes carrying petabits/second (10^{15} bps) will soon become available. This abundant available bandwidth will surely have a dramatic effect on the architecture of the core network, surely in a way that addresses existing scalability challenges.

WDM with associated add-drop multiplexers provide a circuit-based capability for creating virtual networks that can be reconfigured to meet the demand at any given point in time. This WDM infrastructure provides network-protocol flexibility in that the digital stream carried in each wavelength can carry any type of network protocol. The overall network capacity will be increased further with the availability of wavelength-switching nodes that can switch information from one incoming wavelength into a different outgoing wavelength; this technology will allow a greater degree of wavelength reuse.

Terabit routers are already under development to connect to the optical pipes that will become available with dense WDM. Router companies are already claiming designs that can handle links at speeds of 10 Gbps, and even 40 Gbps. Efforts are also under way to standardize "light" framing procedures for encapsulating IP packets directly onto optical links. Researchers are already trying to develop all-optical packet switching. One can speculate on whether a form of all-optical label switching will be developed in the longer term that can enable MPLS-type capabilities in optical networks. Especially desirable would be the scalability that results from frame merging in inverted tree paths leading from ingress routers to egress routers. The combination of advances in optical core networks and in the routers that connect to them implies that the core of the network will become quite transparent to the end systems.

Traffic from an array of access networks will be aggregated in community and metropolitan networks and fed into the terabit routers that connect to the core network. Access networks will be quite diverse, ranging from conventional telephone and LAN access to newer forms of access such as DSL and cable modem. Various new types of access networks will also appear: home LANs using power-utility wiring and various new wireless standards, fixed wireless-access networks, and wireless IP networks that will evolve out of existing cellular networks. An intriguing possibility is the deployment of community fiber networks as an essential infrastructure, like the water supply. Such systems would enable very high-speed, Ethernet-based access from the residence. Such deployment also underscores the growing gap between the haves and the have-nots in society.

The purpose of networks is to support applications, so it is not surprising that the action will be at the edge of the network where the clients and the servers are located. Applications will include multimedia such as various forms of videostreaming; real-time services such as Internet telephony and associated mobility services; and of course, Web-based services and various forms of electronic commerce. We saw in Chapter 12 that the control of network-based services depends on a variety of servers. Policy servers will be required to control the network as well the services provided over the network. In addition to name servers, location servers will be required to enable communications with mobile users. We saw in Chapter 11 that security servers will be required to manage keys and associated certificates. Bandwidth brokers will be required to control access to QoS services as well as to manage bandwidth resources within a domain and across interdomain boundaries. Access servers and firewalls will be needed to control access to private networks and other resources.

A concern is that the network infrastructure cannot evolve fast enough and may prevent the deployment of the new protocols that will support and enable new applications. This concern is one of the motivations for the new area of research into programmable networks. The purpose of programmable networks is to provide programmable network nodes that can simultaneously support several processing environments that provide different end users with different views of the network and its protocols. Such an infrastructure not only enables the rapid introduction of new protocols and services but also allows the coexistence of different generations of protocols.

Programmable networks also allow the introduction of application layer processing inside the network. This approach facilitates certain types of applications such as caching, multicasting, and some forms of multimedia that are best implemented inside the network. Other types of applications, for example, firewalls, are by necessity already interposed in the network and will also be facilitated by programmable networks.

This discussion of network trends is not intended to be exhaustive. For example, we have not touched upon the important issues of business competition, regulation, and public policy. Much of the story of networks is yet to unfold.

Delay and Loss Performance

A key feature of communication networks is the sharing of resources such as transmission bandwidth, storage, and processing capacity. Because the demand for these resources is unscheduled, the situation can arise where resources are not available when a user places a request. This situation typically leads to a delay or loss in service. In this appendix we develop some simple but fundamental models to quantify the delay and loss performance. The appendix is organized as follows:

1. *Little's formula* relates the average occupancy in the system to the average time spent in the system. This formula is extremely powerful in obtaining average delay performance of complex systems.
2. *Basic queueing model* for a multiplexer allows us to account for arrival rate, message length, transmission capacity, buffer size, and performance measures such as delay and loss.
3. *M/M/1* model provides a simple, basic multiplexer model that allows us to explore trade-offs among the essential system parameters.
4. *M/G/1* model provides a more precise description of service times and message lengths.
5. *Erlang B blocking formula* quantifies blocking performance in loss systems.

A.1 DELAY ANALYSIS AND LITTLE'S FORMULA

Figure A.1 shows a basic model for a delay/loss system. Customers arrive to the system according to some arrival pattern. These customers can be connection requests, individual messages, packets, or cells. The system can be an individual transmission line, a multiplexer, a switch, or even an entire network. The custo-

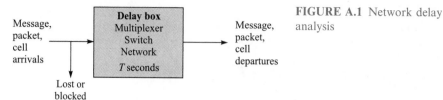

FIGURE A.1 Network delay analysis

mer spends some time T in the system. After this time the customer departs the system. It is possible that under certain conditions the system is in a blocking state, for example, due to lack of resources. Customers that arrive at the system when it is in this state are blocked or lost. We are interested in the following performance measures:

- Time spent in the system: T.
- Number of customers in the system: $N(t)$.
- Fraction of arriving customers that are lost or blocked: P_b.
- Average number of messages/second that pass through the system: throughput.

Customers generally arrive at the system in a random manner, and the time that they spend in the system is also random. In this section we use elementary probability to assess the preceding performance measures.

A.1.1   Arrival Rates and Traffic Load Definitions

We begin by introducing several key system variables and some of their averages. Let $A(t)$ be the number of arrivals at the system in the interval from time 0 to time t. Let $B(t)$ be the number of blocked customers and let $D(t)$ be the number of customer departures in the same interval. The number of customers in the system at time t is then given by

$$N(t) = A(t) - D(t) - B(t)$$

because the number that have entered the system up to time t is $A(t) - B(t)$ and because $D(t)$ of these customers have departed by time t. Note that we are assuming that the system was empty at $t = 0$. The long-term **arrival rate** at the system is given by

$$\lambda = \lim_{t \to \infty} \frac{A(t)}{t} \text{customers/second}$$

The **throughput** of the system is equal to the long-term departure rate, which is given by

$$\text{throughput} = \lim_{t \to \infty} \frac{D(t)}{t} \text{customers/second}$$

The **average number in the system** is given by

$$E[N] = \lim_{t \to \infty} \frac{1}{t} \int_0^t N(t')dt' \text{ customers}$$

The **fraction of blocked customers** is then

$$P_b = \lim_{t \to \infty} \frac{B(t)}{A(t)}$$

Figure A.2 shows a typical sample function $A(t)$, the number of arrivals at the system. We assume that we begin counting customers at time $t = 0$. The first customer arrives at time τ_1, and so $A(t)$ goes from 0 to 1 at this time instant. The second arrival is τ_2 seconds later. Similarly the nth customer arrival is at time $\tau_1 + \tau_2 + \ldots + \tau_n$, where τ_i is the time between the arrival of the $i - 1$ and the ith customer. The arrival rate up to the time when the nth customer arrives is then given by $n/(\tau_1 + \tau_2 + \ldots + \tau_n)$ customers/second. Therefore, the long-term arrival rate is given by

$$\lambda = \lim_{n \to \infty} \frac{n}{\tau_1 + \tau_2 + \ldots + \tau_n} = \lim_{n \to \infty} \frac{1}{(\tau_1 + \tau_2 + \ldots + \tau_n)/n} = \frac{1}{E[\tau]}$$

In the preceding expression we assume that all of the interarrival times are statistically independent and have the same probability distribution and that their average or expected value is given by $E[\tau]$. *Thus the average arrival rate is given by the reciprocal of the average interarrival time.*

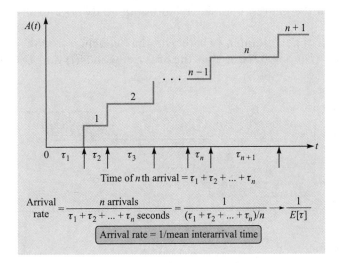

FIGURE A.2 Arrivals at a system as a sample function

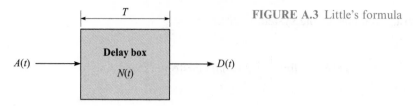

FIGURE A.3 Little's formula

A.1.2   Little's Formula

Next we will develop **Little's formula**, which relates the average time spent in the system $E[T]$ to the arrival rate λ and the average number of customers in the system $E[N]$ by the following formula:

$$E[N] = \lambda E[T]$$

We will assume, as shown in Figure A.3, that the system does not block any customers. The number in the system $N(t)$ varies according to $A(t) - D(t)$.

Suppose we plot $A(t)$ and $D(t)$ in the same graph as shown in Figure A.4. $A(t)$ increases by 1 each time a customer arrives, and $D(t)$ increases by 1 each time a customer departs. The number of customers in the system $N(t)$ is given by the difference between $A(t)$ and $D(t)$. The number of departures can never be greater than the number of arrivals, and so $D(t)$ lags behind $A(t)$ as shown in the figure. Assume that customers are served in first-in, first-out (FIFO) fashion. Then the time T_1 spent by the first customer in the system is the time that elapses between the instant when $A(t)$ goes from 0 to 1 to the instant when $D(t)$ goes from 0 to 1. Note that T_1 is also the area of the rectangle defined by these two time instants in the figure. A similar relationship holds for all subsequent times $T_2, T_3, \ldots$

Consider a time instant t_0 where $D(t)$ has caught up with $A(t)$; that is, $N(t_0) - A(t_0) - D(t_0) = 0$. Note that the area between $A(t)$ and $D(t)$ is given by

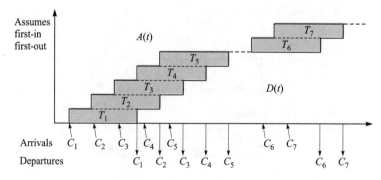

FIGURE A.4 Arrivals and departures in a FIFO system

the sum of the times T_0 spent in the system by the first $A(t_0)$ customers. The time average of the number of customers in the system up to time t_0 is then

$$\frac{1}{t_0}\int_0^{t_0} N(t')dt' = \frac{1}{t_0}\sum_{j=1}^{A(t_0)} T_j$$

If we multiply and divide the preceding expression by $A(t_0)$, we obtain

$$\frac{1}{t_0}\int_0^{t_0} N(t')dt' = \frac{A(t_0)}{t_0}\left\{\frac{1}{A(t_0)}\sum_{j=1}^{A(t_0)} T_j\right\}$$

This equation states that, up to time t_0, the average number of customers in the system is given by the product of the average arrival rate $A(t_0)/t_0$ and the arithmetic average of the times spent in the system by the first $A(t_0)$ customers. Little's formula follows if we assume that

$$E[T] = \lim_{A(t_0)\to\infty}\left\{\frac{1}{A(t_0)}\sum_{j=1}^{A(t_0)} T_j\right\}$$

It can be shown that Little's formula is valid even if customers are not served in order of arrival [Bertsekas 1987].

Now consider a system in which customers can be blocked. The above derivation then applies if we replace $A(t)$ by $A(t) - B(t)$, the actual number of customers who enter the system. The actual arrival rate into a system with blocking is $\lambda(1 - P_b)$, since P_b is the fraction of arrivals that are blocked. It then follows that Little's formula for a system with blocking is

$$E[N] = \lambda(1 - P_b)E[T]$$

In the preceding derivation we did not specify what constitutes a "system," so Little's formula can be applied in many different situations. Thus we can apply Little's formula to an individual transmission line, to a multiplexer, to a switch, or even to a network.

We now show the power of Little's formula by finding the average delay that is experienced by a packet in traversing a packet-switching network. Figure A.5 shows an entire packet-switching network that consists of interconnected packet switches. We assume that when a packet arrives at a packet switch the packet is routed instantaneously and placed in a multiplexer to await transmission on an outgoing line. Thus each packet switch can be viewed as consisting of a set of multiplexers. We begin by applying Little's formula to the network as a whole. Let N_{net} be the total number of packets in the network, let T_{net} be the time spent by the packet in the network, and let λ_{net} be the total packet arrival rate to the network, Little's formula then states that

$$E[N_{net}] = \lambda_{net}E[T_{net}]$$

This formula implies that the average delay experienced by packets in traversing the network is

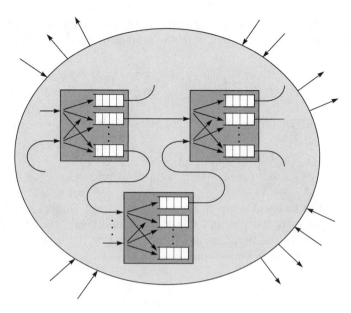

FIGURE A.5 Packet-switching network delay

$$E[T_{net}] = E[N_{net}]/\lambda_{net}$$

We can refine the preceding equation by applying Little's formula to each individual multiplexer. For the mth multiplexer Little's formula gives

$$E[N_m] = \lambda_m E[T_m]$$

where λ_m is the packet arrival rate at the multiplexer and $E[T_m]$ is the average time spent by a packet in the multiplexer. The total number of packets in the network N_{net} is equal to the sum of the packets in all the multiplexers:

$$E[N_{net}] = \sum_m E[N_m] = \sum_m \lambda_m E[T_m]$$

By combining the preceding three equations, we obtain an expression for the total delay experienced by a packet in traversing the entire network:

$$E[T_{net}] = E[N_{net}]/\lambda_{net} = \frac{1}{\lambda_{net}} \sum_m \lambda_m E[T_m]$$

Thus the network delay depends on the overall arrival rates in the network, the arrival rate to individual multiplexers, and the delay in each multiplexer. The arrival rate at each multiplexer is determined by the routing algorithm. The delay in a multiplexer depends on the arrival rate and on the rate at which the associated transmission line can transmit packets. Thus the preceding formula succinctly incorporates the effect of routing as well as the effect of the capacities of the transmission lines in the network. For this reason the preceding expression is frequently used in the design and management of packet-switching networks. To obtain $E[T_m]$, it is necessary to analyze the delay performance of each multiplexer. This is our next topic.

A.2 BASIC QUEUEING MODELS

The pioneering work by Erlang on the traffic engineering of telephone systems led to the development of several fundamental models for the analysis of resource-sharing systems. In a typical application customers demand resources at random times and use the resources for variable durations. When all the resources are in use, arriving customers form a line or "queue" to wait for resources to become available. *Queueing theory* deals with the analysis of these types of systems.

A.2.1   Arrival Processes

Figure A.6 shows the basic elements of a queueing system. Customers arrive at the system with interarrival times $\tau_1, \tau_2, \ldots, \tau_n$. We will assume that the interarrival times are independent random variables with the same distribution. The results for the arrival process developed in Figure A.2 then hold. In particular, the *arrival rate to the system* is given by

$$\lambda = \frac{1}{E[\tau]} \text{customers/second}$$

Several special cases of arrival processes are of interest. We say that arrivals are *deterministic* when the interarrival times are all equal to the same constant value. We say that the arrival times are *exponential* if the interarrival times are exponential random variables with mean $E[\tau] = 1/\lambda$:

$$P[\tau > t] = e^{-t/E[\tau]} = e^{-\lambda t} \text{ for } t > 0$$

The case of exponential interarrival times is of particular interest because it leads to tractable analytical results. It can be shown that when the interarrival

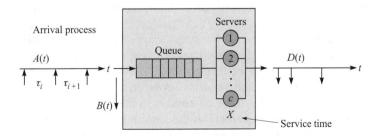

FIGURE A.6 Queueing model

times are exponential, then the number of arrivals $A(t)$ in an interval of length t is given by a Poisson random variable with mean $E[A(t)] = \lambda t$:

$$P[A(t) = k] = \frac{(\lambda t)^k}{k!} e^{-\lambda t} \text{ for } k = 0, 1, \dots$$

For this reason, the case of exponential interarrival times is also called the *Poisson arrival process*.

A.2.2   Service Times

Resources are denoted by "servers" because their function is to serve customer requests. The time required to service a customer is called the **service time** and is denoted by X. In our discussion the server is typically a transmission line and the service time can be the time required to transmit a message or the duration of a telephone call. The maximum rate at which a server can process customers is attained when the server is continuously busy. When this is the case, the average time between customer departures is equal to the average service time. The processing capacity of a single server is given by the maximum throughput or departure rate. From the discussion leading to the arrival rate formula, clearly the processing capacity is given by

$$\mu = \frac{1}{E[X]} \text{customers/second}$$

The processing capacity μ can be likened to the maximum flow that can be sustained over a pipe. The number of servers c in a queueing system can be greater than one. The total processing capacity of a queueing system is then given by $c\mu$ customers/second.

An ideal queueing system is one where customers arrive at equal intervals and in which they require a constant service time. As long as the service time is less than the interarrival time, each customer arrives at an available server and there is no waiting time. In general, however, the interarrival time and the service times are random. The combination of a long service time followed by a short interarrival time can then lead to a situation in which the server is not available for an arriving customer. For this reason, in many applications a queue is provided so that a customer can wait for an available server, as shown in Figure A.6. When a server becomes available the next customer to receive service is selected according to the service discipline. Possible service disciplines are FIFO; last-in, first-out (LIFO); service according to priority class; and random order of service. We usually assume FIFO service disciplines.

The *maximum* number of customers allowed in a queueing system is denoted by K. Note that K includes both the customers in queue and those in service. We denote the total number of customers in the system by $N(t)$, the number in queue by $N_q(t)$, and the number in service by $N_s(t)$. When the system is full, that is, $N(t) = K$, then new customers arrivals are blocked or lost.

A.2.3   Queueing System Classification

Queueing systems are classified by a notation that specifies the following characteristics:

- Customer arrival pattern.
- Service time distribution.
- Number of servers.
- Maximum number in the system.

For example, in Figure A.7 the queueing system $M/M/1/K$ corresponds to a queueing system in which the interarrival times are exponentially distributed (M)[1]; the service times are exponentially distributed (M); there is a single server (1); and at most K customers are allowed in the system. The $M/M/1/K$ model was used to illustrate the typical delay and loss performance of a data multiplexer in Chapter 5. If there is no maximum limit in the number of customers allowed in the system, the parameter K is left unspecified. Thus the $M/M/1$ system is identical to the above system except that it has no maximum limit on the number of customers allowed in the system.

The $M/G/1$ is another example of a queueing system where the arrivals are exponential, the service times have a *general* distribution, there is a single server, and there is no limit on the customers allowed in the system. Similarly, the $M/D/1$ system has constant, that is, *deterministic*, service times.

Figure A.8 shows the parameters that are used in analyzing a queueing system. The total time that a customer spends in the system is denoted by T, which consists of the time spent waiting in queue W plus the time spent in service X. When a system has blocking, P_b denotes the fraction of customers that are blocked. Therefore, the actual arrival rate into the system is given by $\lambda(1 - P_b)$. This value is the arrival rate that should be used when applying Little's formula.

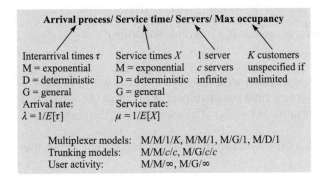

FIGURE A.7 Queueing model classification

---

[1]The notation M is used for the exponential distribution because it leads to a Markov process model.

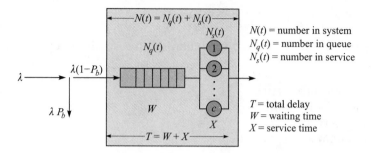

FIGURE A.8 Queueing system variables

Thus the average number in the system and the average delay in the system are related by

$$E[N] = \lambda(1 - P_b)E[T]$$

If we apply Little's formula where the "system" is just the queue, then the average number of customers in queue and the average waiting time are related by

$$E[N_q] = \lambda(1 - P_b)E[W]$$

Finally, if the system is defined as a set of servers, then the average number of customers in service and the average service time are related by

$$E[N_s] = \lambda(1 - P_b)E[X]$$

The preceding three equations are very useful in relating average occupancy and average delay performance. Typically it is relatively simple to obtain the averages associated with occupancies, that is, $N(t)$, $N_q(t)$ and $N_s(t)$.

Finally, we revisit some of the terms introduced earlier in the appendix. The *traffic load* or offered load is the rate at which "work" arrives at the system:

$$a = \lambda \text{ customers/second} \times E[X] \text{ seconds/customer}$$
$$= \lambda/\mu \text{ Erlangs.}$$

The *carried load* is the average rate at which the system does work. It is given by the product of the average service time per customer, $E[X]$, and the actual rate at which customers enter the system, $\lambda(1 - P_b)$. Thus we see that the carried load is given by $a(1 - P_b)$.

The *utilization* ρ is defined as the average fraction of servers that are in use:

$$\rho = \frac{E[N_s]}{c} = \frac{\lambda}{c\mu}(1 - P_b)$$

Note that when the system has a single server, then the utilization ρ is also equal to the proportion of time that the server is in use.

A.3   M/M/1: A BASIC MULTIPLEXER MODEL

In this section we develop the M/M/1/K queueing system, shown in Figure A.9, as a basic model for a multiplexer. The interarrival times τ in this system have mean $E[\tau] = 1/\lambda$ as an exponential distribution. Let $A(t)$ be the number of arrivals in the interval 0 to t; then as indicated above $A(t)$ has a Poisson distribution.

The average packet length is $E[L]$ bits per packet, and the transmission line has a speed of R bits/second. So the average packet transmission time is $E[X] = E[L]/R$ seconds. This transmission line is modeled by a single server that can process packets at a maximum rate of $\mu = R/E[L]$ packets/second. We assume that the packet transmission time X has an exponential distribution:

$$P[X > t] = e^{-t/E[X]} = e^{-\mu t} \text{ for } t > 0$$

We also assume that the interarrival times and packet lengths are independent of each other. We will first assume that at most K packets are allowed in the system. We later consider the case were K is infinite.

In terms of long-term flows, packets arrive at this system at a rate of λ packets/second, and the maximum rate at which packets can depart is μ packets/second. If $\lambda > \mu$, then the system will necessarily lose packets because the system is incapable of handling the arrival rate λ. If $\lambda < \mu$, then on the average the system can handle the rate λ, but it will occasionally lose packets because of temporary surges in arrivals or long consecutive service times. We will now develop a model that allows us to quantify these effects.

Consider what events can happen in the next Δt seconds. In terms of arrivals there can be 0, 1, or >1 arrivals. Similarly there can be 0, 1, or >1 departures. It can be shown that if the interarrival times are exponential, then

$$P[1 \text{ arrival in } \Delta t] = \lambda \Delta t + o(\Delta t)$$

where $o(\Delta t)$ denotes terms that are negligible relative to Δt, as $\Delta t \to 0$.[2] Thus the probability of a single arrival is proportion to λ. Similarly, it can also be shown that probability of no arrivals in Δt seconds is given by

$$P[0 \text{ arrival in } \Delta t] = 1 - \lambda \Delta t + o(\Delta t)$$

The preceding two equations imply that only two events are possible as Δt becomes very small: one arrival or no arrival. Since the service times also have an

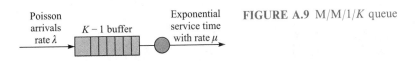

Poisson arrivals rate λ $K-1$ buffer   Exponential service time with rate μ **FIGURE A.9** M/M/1/K queue

---

[2]In particular, a function $g(x)$ is $o(x)$ if $g(x)/x \to 0$ as $x \to 0$; that is, $g(x)$ goes to 0 faster than x does.

exponential distribution, it can be shown that a customer in service will depart in the next Δt seconds with probability

$$P[1 \text{ departure in } \Delta t] = \mu \Delta t + o(\Delta t)$$

and that the probability that the customer will continue its service after an additional Δt seconds is

$$P[0 \text{ departure in } \Delta t] = 1 - \mu \Delta t + o(\Delta t)$$

A.3.1 M/M/1 Steady State Probabilities and the Notion of Stability

We can determine the probability of changes in the number of customers in the system by considering the various possible combinations of arrivals and departures:

$$P[0 \text{ arrival \& 0 departure in } \Delta t] = \{1 - \mu \Delta t + o(\Delta t)\}\{1 - \lambda \Delta t + o(\Delta t)\}$$
$$= 1 - (\lambda + \mu)\Delta t + o(\Delta t)$$

The preceding equation gives the probability that the number in the system is still $n > 0$ after Δt seconds.

$$P[1 \text{ arrival \& 0 departure in } \Delta t] = \{\lambda \Delta t + o(\Delta t)\}\{1 - \mu \Delta t + o(\Delta t)\}$$
$$= \lambda \Delta t + o(\Delta t)$$

The preceding equation gives the probability that the number in the system increases by 1 in Δt seconds.

$$P[\text{no arrival \& 1 departure in } \Delta t] = \{1 - \lambda \Delta t + o(\Delta t)\}\{\mu \Delta t + o(\Delta t)\}$$
$$= \mu \Delta t + o(\Delta t).$$

The preceding equation gives the probability that the number in the system decreases by 1 in Δt seconds. Note that the preceding equations imply that $N(t)$ always changes by single arrivals or single departures.

Figure A.10 shows the state transition diagram for $N(t)$, the number in the system. $N(t)$ increases by 1 in the next Δt seconds with probability $\lambda \Delta t$ and decreases by 1 in the next Δt seconds with probability $\mu \Delta t$. Note that every transition n to $n+1$ cannot recur until the reverse transition $n+1$ to n occurs.

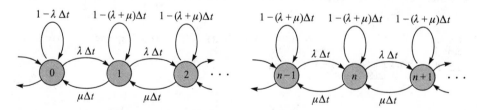

FIGURE A.10 State transition diagram

Therefore, if the system is stable, that is, if it does not grow steadily to infinity, then the long-term transition rate from n to $n+1$ must equal the long-term transition rate from $n+1$ to n.

Let p_n be the probability that n customers are in the system; then p_n is also the proportion of time that the system is in state n. Therefore, $p_n \lambda \Delta t$ is the transition from state n to state $n+1$. Similarly, $p_{n+1} \mu \Delta t$ is the transition rate from $n+1$ to n. This discussion implies that the two transition rates must be equal. Therefore

$$p_{n+1} \mu \Delta t = p_n \lambda \Delta t$$

This implies that

$$p_{n+1} = (\lambda/\mu) p_n \quad n = 0, 1, \ldots, K$$

Repeated applications of the preceding recursion imply that

$$p_{n+1} = (\lambda/\mu)^{n+1} p_0 = p^n p_0 \quad n = 0, 1, \ldots, K$$

To find p_0, we use the fact that the probabilities must add up to 1:

$$1 = p_0 + p_1 + p_2 + \ldots + p_K = p_0\{1 + \rho + \rho^2 + \rho^3 + \ldots \rho_K\}$$
$$= p_0 \frac{1 - \rho^{K+1}}{1 - \rho}$$

which implies that

$$p_0 = \frac{1 - \rho}{1 - \rho^{K+1}}$$

Finally we obtain the probabilities for the number of customers in the system:

$$P[N(t) = n] = p_n = \frac{(1 - \rho)\rho^n}{1 - \rho^{K+1}}, \text{ for } n = 0, 1, \ldots, K$$

The probability of blocking or loss in the M/M/1/K system is given by $P_{loss} = p_K$, which is the proportion of time that the system is full.

Consider what happens to the state probabilities as the load ρ is varied. For ρ less than 1, which corresponds to $\lambda < \mu$, the probabilities decrease exponentially as n increases; thus the number in the system tends to cluster around $n = 0$. In particular, adding more buffers is beneficial when $\lambda < \mu$, since the result is a reduction in loss probability. When $\rho = 1$, the normalization condition implies that all the states are equally probable; that is, $p_n = 1(K + 1)$. Once ρ is greater than 1, the probabilities actually increase with n and tend to cluster toward $n = K$; that is, the system tends to be full, as expected. Note that adding buffers when $\lambda > \mu$ is counterproductive, since the system will fill up the additional buffers. This result illustrates a key point in networking: The arrival rate should not be allowed to exceed the maximum capacity of a system for extended periods of time. The role of *congestion control* procedures inside the network is to deal with this problem.

The average number of customers in the system $E[N]$ is given by

$$E[N] = \sum_{n=0}^{K} np_n = \sum_{n=0}^{K} n\frac{(1-\rho)\rho^n}{1-\rho^{K+1}} = \frac{\rho}{1-\rho} - \frac{(K+1)\rho^{K+1}}{1-\rho^{K+1}}$$

The preceding equation is valid for ρ not equal to 1. When $\rho = 1$, $E[N] = K/2$. By applying Little's formula, we obtain the average delay in an M/M/1/K system:

$$E[T] = \frac{E[N]}{\lambda(1-P_K)}$$

Now consider the M/M/1 system that has $K = \infty$. The state transition diagram for this system is the same as in Figure A.10 except that the states can assume all nonzero integer values. The probabilities are still related by

$$p_{n+1} = (\lambda/\mu)^{n+1}p_0 = \rho^n p_0 \quad n = 0, 1, \ldots$$

where $\rho = \lambda/\mu$. The normalization condition is now

$$1 = p_0 + p_1 + p_2 + \ldots = p_0\{1 + \rho + \rho^2 + \rho^3 + \ldots\}$$
$$= p_0\frac{1}{1-\rho}$$

Note that the preceding power series converges only if $\rho < 1$, which corresponds to $\lambda < \mu$. This result agrees with our intuition that an infinite-buffer system will be stable only if the arrival rate is less than the maximum departure rate. If not, the number in the system would grow without bound. We therefore find that the state probabilities are now

$$P\{N(t) = n\} = p_n = (1-\rho)\rho^n \quad n = 0, 1, \ldots \rho < 1$$

The average number in the system and the average delay are then given by

$$E[N] = \sum_{n=0}^{\infty} n(1-\rho)\rho^n = \frac{\rho}{(1-\rho)}$$

$$E[T] = \frac{E[N]}{\lambda} = \frac{1/\mu}{(1-\rho)}$$

The average waiting time is obtained as follows:

$$E[W] = E[T] - E[X] = \frac{(1/\mu)\rho}{(1-\rho)}.$$

These equations show that the average delay and the average waiting time grow without bound as ρ approaches 1. Thus we see that when arrivals or service times are random, perfect scheduling is not possible, so the system cannot be operated at $\lambda = \mu$.

A.3.2   Effect of Scale on Performance

The expressions for the M/M/1 system allow us to demonstrate the typical behavior of queueing systems as they are increased in scale. Consider a set of m separate M/M/1 systems, as shown in Figure A.11. Each system has an arrival rate of λ customers/second and a processing rate of μ customers/second. Now suppose that it is possible to combine the customer streams into a single stream with arrival rate $m\lambda$ customers/second. Also suppose that the processing capacities are combined into a single processor with rate $m\mu$ customers/second. The mean delay in the separate systems is given by

$$E[T_{separate}] = \frac{1/\mu}{(1 - \rho)}$$

The combined system has an arrival rate $\lambda' = m\lambda$ and a processing rate $\mu' = m\mu$; therefore, its utilization is $\rho' = \lambda'/\mu' = \rho$. Therefore, the mean delay in the combined system is

$$E[T_{combined}] = \frac{1/\mu'}{(1 - \rho)} = \frac{1/m\mu}{(1 - \rho)} = \frac{1}{m} E[T_{separate}]$$

Thus we see that the combined system has a total delay of $1/m$ of the separate systems.

The improved performance of the combined system arises from improved global usage of the processors. In the separate systems some of the queues may be empty while others are not. Consequently, some processors can be idle, even though there is work to be done in the system. In the combined system the processor will stay busy as long as customers are waiting to be served.

A.3.3   Average Packet Delay in a Network

In the beginning of this section, we used Little's formula to obtain an expression for the average delay experienced by a packet traversing a network. As shown in Figure A.5 the packet-switching network was modeled as many interconnected multiplexers. The average delay experienced by a packet in traversing the network is

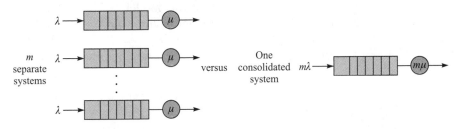

FIGURE A.11 Multiplexing gain and effect of scale

$$E[T_{net}] = \frac{1}{\lambda_{net}} \sum_m \lambda_m E[T_m]$$

where $E[T_m]$ is the average delay experienced in each multiplexer. To apply the results from our M/M/1 analysis, we need to make several assumptions. The most important assumption is that the service times that a packet experiences at different multiplexers are independent of each other. In fact, this assumption is untrue, since the service time is proportional to the packet length, and a packet has the same length as it traverses the network. Nevertheless, it has been found that this *independence assumption* can be used in larger networks.

If we model each multiplexer by an M/M/1 queue, we can then use the corresponding expression for the average delay:

$$E[T_m] = (1/\mu)(1 - \rho_m), \quad \text{where } \rho_m = \lambda_m/\mu$$

where λ_m is the packet arrival rate at the mth multiplexer. The average packet delay in the network is then

$$E[T_m] = \sum_m \frac{1}{\lambda_{net}} \left(\frac{\rho_m}{1 - \rho_m} \right)$$

This simple expression was first derived by [Kleinrock 1964]. It can be used as the basis for selecting the transmission speeds in a packet-switching network. It can also be used to synthesize routing algorithms that distribute the flows of packets over the network so as to keep the overall packet delay either minimum or within some range.

A.4   THE M/G/1 MODEL

The M/M/1 models derived in the previous sections are extremely useful in obtaining a quick insight into the trade-offs between the basic queueing systems parameters. The M/G/1 queueing model allows us to consider a more general class of resource sharing systems. In particular, the service time can have any distribution and is not restricted to be exponential. The derivation of the M/G/1 results is beyond the scope of our discussion; we simply present the results and apply them to certain multiplexer problems. The reader is referred to [Leon-Garcia 1994] for a more detailed discussion of this model.

As shown in Figure A.12, the M/G/1 model assumes Poisson arrivals with rate λ, service times X with a general distribution $F_x(x) = P[X \leq x]$ that has a mean $E[X]$ and a variance $VAR[X]$, a single server, and unlimited buffer space.

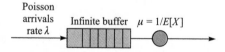

Poisson arrivals rate λ Infinite buffer $\mu = 1/E[X]$

FIGURE A.12 M/G/1 queueing system

The mean waiting time in an M/G/1 queueing system is given by

$$E[W] = \frac{\lambda E[X^2]}{2(1-\rho)}$$

Using the fact that $E[X^2] = VAR[X] + E[X]^2$, we obtain

$$E[W] = \frac{\lambda(VAR[X] + E[X]^2)}{2(1-\rho)} = \frac{\rho(1 + C_X^2)}{2(1-\rho)} E[X]$$

where the coefficient of variation of the service time is given by $C_X^2 = VAR[X]/E[X]^2$.

The mean delay of the system is obtained by adding the mean service time to $E[W]$:

$$E[T] = E[W] + E[X]$$

The mean number in the system $E[N]$ and the mean number in queue $E[N_q]$ can then be found from Little's formula.

A.4.1 Service Time Variability and Delay

The mean waiting time in an M/G/1 system increases with the coefficient of variation of the service time. Figure A.13 shows the coefficient of variation for several service time distributions. The exponential service time has a coefficient of variation of 1 and serves as a basis for comparison. The constant service time has zero variance and hence has a coefficient of variation of zero. Consequently, its mean waiting time is one-half that of an M/M/1 system. The greater randomness in the service times of the M/M/1 system results in a larger average delay in the M/M/1 system.

Figure A.13 also shows two other types of service time distributions. The Erlang distribution has a coefficient of variation between 0 and 1, and the hyperexponential distribution has a coefficient of variation greater than 1. These two distributions can be used to model various degrees of randomness in the service time relative to the M/M/1 system.

In Chapter 6 we used the M/G/1 formula in assessing various types of schemes for sharing broadcast channels.

	M/D/1	M/Er/1	M/M/1	M/H/1
Interarrivals	Constant	Erlang	Exponential	Hyperexponential
C_X^2	0	<1	1	>1
$E[W]/E[W_{M/M/1}]$	1/2	1/2<, <1	1	>1

FIGURE A.13 Comparison of mean waiting times in M/G/1 systems

A.4.2   Priority Queueing Systems

The M/G/1 model can be generalized to the case where customers can belong to one of K priority classes. When a customer arrives at the system, the customer joins the queue of its priority class. Each time a customer service is completed, the next customer to be served is selected from the head of the line of the highest priority nonempty queue. We assume that once a customer begins service, it cannot be preempted by subsequent arrivals of higher-priority customers.

We will assume that the arrival at each priority class is Poisson with rate λ_k and that the average service time of a class k customer is $E[X_k]$, so the load offered by class k is $\rho_k = \lambda_k E[X_k]$.

It can be shown that if the total load is less than 1; that is

$$\rho = \rho_1 + \rho_2 + \ldots + \rho_K < 1$$

then the average waiting time for a type k customer is given by

$$E[W_k] = \frac{\lambda E[X^2]}{(1 - \rho_1 - \rho_2 - \ldots - \rho_{k-1})(1 - \rho_1 - \rho_2 - \ldots - \rho_k)}$$

where

$$E[X^2] = \frac{\lambda_1}{\lambda} E[X_1^2] + \frac{\lambda_2}{\lambda} E[X_2^2] + \ldots + \frac{\lambda_K}{\lambda} E[X_K^2]$$

The mean delay is found by adding $E[X_k]$ to the corresponding waiting time. The interesting result in the preceding expression is in the terms in the denominator that indicate at what load a given class saturates. For example, the highest priority class has average waiting time

$$E[W_1] = \frac{\lambda E[X^2]}{(1 - \rho_1)}$$

so it saturates as ρ_1 approaches 1. Thus the saturation point of class 1 is determined only by its own load. On the other hand, the waiting time for class 2 is given by

$$E[W_2] = \frac{\lambda E[X^2]}{(1 - \rho_1)(1 - \rho_1 - \rho_2)}$$

The class 2 queue will saturate when $\rho_1 + \rho_2$ approaches 1. Thus the class 2 queue saturation point is affected by the class 1 load. Similarly, the class k queue saturation point depends on the sum of the loads of the classes of priority up to k. This result was used in Chapter 7 in the discussion of priority queueing disciplines in packet schedulers.

A.4.3   Vacation Models and Multiplexer Performance

The M/G/1 model with vacations arises in the following way. Consider an M/G/1 system in which the server goes on vacation (becomes unavailable) whenever it empties the queue. If upon returning from vacation, the server finds that the system is still empty, the server takes another vacation, and so on until it finds customers in the system. Suppose that vacation times are independent of each other and of the other variables in the system. If we let V be the vacation time, then the average waiting time in this system is

$$E[W] = \frac{\lambda E[X^2]}{2(1-\rho)} + \frac{E[V^2]}{2E[V]}$$

The M/G/1 vacation model is very useful in evaluating the performance of various multiplexing and medium access control systems. As a simple example consider an ATM multiplexer in which cells arrive according to a Poisson process and where cell transmission times are constrained to begin at integer multiples of the cell time. When the multiplexer empties the cell queue, then we can imagine that the multiplexer goes away on vacation for one cell time, just as modeled by the M/G/1 vacation model. If the cell transmission time is the constant X, then a vacation time is also $V = X$. Therefore, the average waiting time in this ATM multiplexer is

$$E[W] = \frac{\lambda X^2}{2(1-\rho)} + \frac{E[X^2]}{2E[X]} = \frac{\rho X}{2(1-\rho)} + \frac{X}{2}$$

The first term is the average waiting time in an ordinary M/G/1 system. The second term is the average time that elapses from the arrival instant of a random customer arrival to the beginning of the next cell transmission time.

A.5   ERLANG B FORMULA: M/M/c/c System

In Figure A.14 we show the M/M/c/c queueing model that can be used to model a system that handles trunk connection requests from many users. We assume trunk requests with exponential interarrival times with rate λ requests/second. Each trunk is viewed as a server, and the connection time is viewed as the service time X. Thus each trunk or server has a service rate $\mu = 1/E[X]$. We assume that connection requests are blocked if all the trunks are busy.

The state of the preceding system is given by $N(t)$, the number of trunks in use. Each new connection increases $N(t)$ by 1, and each connection release decreases $N(t)$ by 1. The state of the system then takes on the values $0, 1, \ldots, c$. The probability of a connection request in the next Δt seconds is given by $\lambda \Delta t$. The M/M/c/c queueing model differs from the M/M/1 queueing model in terms of the departure rate. If $N(t) = n$ servers are busy, then each server will complete its service in the next Δt seconds with probability $\mu \Delta t$. The

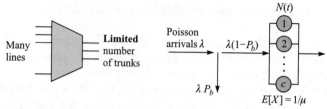

- Blocked calls are cleared from the system; no waiting allowed.
- Performance parameter: P_b = fraction of arrivals that are blocked.
- $P_b = P[N(t) = c] = B(c, a)$ where $a = \lambda/\mu$.
- $B(c, a)$ is the Erlang B formula, which is valid for **any** service time distribution.

$$B(c, a) = \frac{\dfrac{a^c}{c!}}{\displaystyle\sum_{j=0}^{c} \frac{a^j}{j!}}$$

FIGURE A.14 M/M/c/c and the Erlang B formula

probability that one of the connections completes its service and that the other $n - 1$ continue their service in the next Δt seconds is given by

$$n(\mu\Delta t)(1 - \mu\Delta t)^{n-1} \approx n\mu\Delta t$$

The probability that two connections will complete their service is proportional to $(\mu\Delta t)^2$, which is negligible relative to Δt. Therefore, we find that the departure rate when $N(t) = n$ is $n\mu$.

Figure A.15 shows the state transition diagram for the M/M/c/c system. Proceeding as in the M/M/1/K analysis, we have

$$p_{n+1}(n + 1)\mu\Delta t = p_n \lambda \Delta t$$

This result implies that

$$p_{n+1} = \frac{\lambda}{(n + 1)\mu} p_n \quad n = 0, 1, \ldots, c$$

Repeated applications of the preceding recursion imply that

$$p_{n+1} = \frac{(\lambda/\mu)^{n+1}}{(n + 1)!} p_0 \quad n = 0, 1, \ldots, c$$

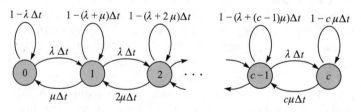

FIGURE A.15 State transition diagram for M/M/c/c

Let the offered load be denoted by $a = \lambda/\mu$. To find p_0, we use the fact that the probabilities must add up to 1:

$$1 = p_0 + p_1 + p_2 + \ldots + p_c = p_0 \sum_{n=0}^{c} \frac{a^n}{n!}$$

Finally, we obtain the probabilities for the number of customers in the system:

$$P[N(t) = n] = p_n = \frac{a^n}{\displaystyle\sum_{k=0}^{c} \frac{a^k}{k!}}, \text{ for } n = 0, 1, \ldots, c$$

The probability of blocking in the M/M/c/c system is given by $P_b = p_c$, which is the proportion of time that the system is full. This result leads to the Erlang B formula $B(c, a)$ that was used in Chapter 4.

Finally, we note that the Erlang B formula also applies to the M/G/c/c system; that is, the service time distribution need not be exponential.

FURTHER READING

Bertsekas, D. and R. Gallager, *Data Networks*, Prentice-Hall, Englewood Cliffs, New Jersey, 1987.

Kleinrock, L., *Communication Nets: Stochastic Message Flow and Delay*, McGraw-Hill, New York, 1964.

Leon-Garcia, A., *Probability and Random Process for Electrical Engineering*, Addison-Wesley, Reading, Massachusetts, 1994.

Network Management

All networks, whether large or small, benefit from some form of management. Network management involves configuring, monitoring, and possibly reconfiguring components in a network with the goal of providing optimal performance, minimal downtime, proper security, and flexibility. This type of management is generally accomplished by using a *network management system*, which contains a software bundle designed to improve the overall performance and reliability of a system. In a small network, network management systems might be used to identify users who present security hazards or to find misconfigured systems. In a large network, network management systems might also be used to improve network performance and track resource usage for the purpose of accounting or charging.

As different types of networks have evolved over the years, so have different approaches to network management. The most common computer network management system currently implemented is the Simple Network Management Protocol (SNMP), which was originally intended to be a short-term solution to the network management issue. Alternatively, there is an OSI-based network management system called Common Management Information Protocol (CMIP). The OSI system was to be developed as the long-term solution. However, the ease with which SNMP was first implemented, as well as certain differences that emerged during the development of the OSI version, resulted in a continued separate development of SNMP.

The components that make up a network often come from many different vendors, and the operating systems that run the different systems in a network are often different. It is therefore important that a network management system be based on standards so that interoperability is also ensured. In this appendix we examine the concepts underlying network management in general and then present the most common approach that has been implemented for communication networks, namely, SNMP. Specifically we address the following topics:

1. *Network management overview.* We present the functions that can be performed by a network management system, and we describe the components that make up a network management system.
2. *SNMP.* We discuss the standards behind the most common network management protocol in use, the IETF-developed SNMP.
3. *Structure of Management Information.* We describe the Structure of Management Information (SMI), which defines the rules for describing management information.
4. *Management Information Base.* We describe the collection of objects, called the Management Information Base (MIB), that are managed by SNMP.
5. *Remote Network Monitoring.* We briefly discuss remote monitoring (RMON), which offers extensive network diagnostic, planning, and performance information.

B.1   NETWORK MANAGEMENT OVERVIEW

Network management involves monitoring and controlling a networking system so that it operates as intended. It also provides a means to configure the system while still meeting or exceeding design specifications. Note that management may be performed by a human, by an automated component, or both.

The functions performed by a network management system can be categorized into the following five areas:

Fault management refers to the detection, isolation, and resolution of network problems. Because a fault can cause part of a network to malfunction or even cause the whole network to fail completely, fault management provides a means for improving the reliability of the network. Examples include detecting a fault in a transmission link or network component, reconfiguring the network during the fault to maintain service level, and restoring the network when the fault is repaired.

Configuration management refers to the process of initially configuring a network and then adjusting it in response to changing network requirements. This function is perhaps the most important area of network management because improper configuration may cause the network to work suboptimally or to not work at all. An example is the configuration of various parameters on a network interface.

Accounting management involves tracking the usage of network resources. For example, one might monitor user load to determine how to better allocate resources. Alternatively, one might examine the type of traffic or the level of traffic that passes through a particular port. Accounting management also includes activities such as password administration and charging.

Performance management involves monitoring network utilization, end-to-end response time, and other performance measures at various points in a network. The results of the monitoring can be used to improve the performance of the network. Examples include tracking Ethernet utilization on all switched interfaces and reconfiguring new switched interfaces if performance is deemed to be below some specific level.

Security management refers to the process of making the network secure. This process, of course, involves managing the security services that pertain to access control, authentication, confidentiality, integrity, and non-repudiation. Security is discussed in Chapter 11.

Although each specific network management architecture is based on different premises, certain concepts are common to most approaches to network management. A network contains a number of *managed devices* such as routers, bridges, switches, and hosts. Network management essentially involves monitoring and/or altering the configuration of such devices. An **agent** is a part of a network management system that resides in a managed device. The agent's tasks are to provide *management information* about the managed device and to accept instructions for configuring the device. It is also possible for an agent to not reside in the managed device; such an agent is called a *proxy agent.*

A **network management station** provides a text or graphical view of the entire network (or one of its components). This view is provided by way of a management application or manager that resides on the station. The manager exchanges management information with the agent by using a *network management protocol.* A management station allows a human or automated process to reconfigure a network, react to faults, observe performance, monitor security, and track usage—in short to implement the various network management functions required for that network.

More than one network management station can exist in a network. One network might have several stations, each of which provides a different view of the same portion of the network. Alternatively, a network might have several stations, each responsible for a different geographical portion of the network.

Figure B.1 shows a portion of a departmental network to illustrate how the network management concepts might apply. As shown in the figure, each host contains an agent that collects management information pertaining to the host. Similarly, the router also contains its own agent. The manager in the management station can poll a particular agent to obtain specific management information, which for example, can be the number of packet losses in the router. The management station can also alter some values in a managed device.

Note that a network management system may operate in a centralized or distributed manner or include both types of computing. In a centralized system one computer system runs most of the applications required for network management. Similarly one database is located on the central machine. A distributed system may have several peer network management systems running simultaneously with each system managing a specific part of the network. The systems may be distributed geographically. Finally, a hierarchical network mangement system can have a centralized system at the root, with distributed peer systems running as children of the root.

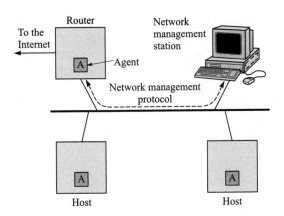

FIGURE B.1 A managed network

B.2 SIMPLE NETWORK MANAGEMENT PROTOCOL (SNMP)

In the early days of the Internet, the Internet Activities Board recognized the need for a management framework by which to manage TCP/IP implementations. The framework consists of three components:

1. A conceptual framework that defines the rules for describing management information, known as the Structure of Management Information (SMI).
2. A virtual database containing information about the managed device known as the Management Information Base (MIB).
3. A protocol for communication between a manager and an agent of a managed device, known as Simple Network Management Protocol (SNMP).

Essentially, the data that is handled by SNMP must follow the rules for objects in the MIB, which in turn are defined according to the SMI. These components are discussed separately below.

SNMP is an application layer protocol that is used to read and write variables in an agent's MIB. The most current version is SNMPv3. For convenience here SNMP refers to all versions of SNMP unless otherwise indicated.

SNMP is based on an asynchronous request-response protocol enhanced with trap-directed polling. The qualifier *asynchronous* refers to the fact that the protocol need not wait for a response before sending other messages. *Trap-directed polling* refers to the fact that a manager polls in response to a trap message being sent by an agent, which occurs when there is an exception or after some measure has reached a certain threshold value. SNMP operates in a connectionless manner with UDP being the preferred transport mode. An SNMP manager sends messages to an agent via UDP destination port 161, while an agent sends trap messages to a manager via UDP destination port 162. The connectionless mode was chosen partly to simplify SNMP's implementation and because connectionless is usually the preferred mode by management applications that need to talk to many agents.

The messages (PDUs) exchanged via SNMP consist of a header and a data part. The header contains a version field, a community name field, and a PDU type field. The community name transmits a cleartext password between a manager and an agent, so this field can serve as a limited form of authentication. The other fields are self-explanatory.

SNMP provdes three ways to access management information.

1. Request/response interaction in which a manager sends a request to an agent and the agent responds to the request. The request is usually to retrieve or modify management information associated with the network device in question. Specific information is requested by a manager, using one of the following requests:
   - `GetRequest-PDU` for requesting information on specific variables.
   - `GetNextRequest-PDU` for requesting the next set of information (usually from a management information table).
   - `GetBulkRequest-PDU` for requesting bulk information retrieval. This request was introduced in SNMPv2 to allow the retrieval of as much information as possible in a packet.
   - `SetRequest-PDU` for creating or modifying management information.
   The agent must always reply using a `Response-PDU`.

2. Request/response interaction in which a manager sends a request to another manager and the latter responds to the request. The request is usually to notify a manager of management information associated with the the manager, using `InformRequest-PDU`.

3. Unconfirmed interaction in which an agent sends an unsolicited `Trap-PDU` to a manager. This request is usually to notify the manager of an exceptional situation that has resulted in changes to the management information associated with the network device.

Table B.1 shows another way of looking at the manager/agent roles and the functions they can perform.

A typical interaction between a manager and agent would proceed as follows. The manager issues some form of get request that contains a unique request-id to match the response with the request, a zero-valued error status/ error index, and one or more variable bindings. The agent issues a response containing the *same* request-id, a zero-valued error status if there is no error, and the *same* variable bindings. Figure B.2 shows a typical interaction for an information request.

If an exception occurs for one or more of the variables, then the particular error status for each relevant variable is returned as well. For example, if the

Role	Generate	Receive

Agent	`Response-PDU, Trap-PDU`	`GetRequest-PDU,`
		`GetNextRequest-PDU,`
		`GetBulkRequest-PDU,`
		`SetRequest-PDU,`
		`InformRequest-PDU`

Manager	`GetRequest-PDU,`
	`GetNextRequest-PDU`
	`GetBulkRequest-PDU,`
	`SetRequest-PDU,`
	`InformRequest-PDU`
		`Response-PDU, Trap-PDU`

TABLE B.1. Request/Response interactions by role

agent does not implement a particular variable, then the `noSuchName` error status is returned. Other error status values are `tooBig`, indicating that the response is too large to send; `badValue`, indicating that the set operation specified an invalid value or syntax; `readOnly`, indicating that a manager attempted to write a read-only variable; and `genErr` for some other errors.

Version 3 of SNMP was formally documented in early 1998 [RFC 2271]. It presents a more complex framework for message exchange, the complexity being required both for extensibility and for security reasons. The security system contains a user-based security model, as well as other security models that may be implemented. This model is intended to protect against the unauthorized modification of information of SNMP messages in transit, masquerading, eavesdropping, or deliberate tampering with the message stream (that is, a deliberate reordering, delay, or replay of messages). It does not protect against denial of service or unauthorized traffic analysis. The model uses the MD5 encryption scheme for verifying user keys, a SHA message digest algorithm (HMAC-SHA-96) to verify message integrity and to verify the user on whose behalf the message was generated, and a CBC-DES symmetric encryption protocol for privacy. See [RFC 2274] for further information on the user-based security model.

FIGURE B.2 Typical request/response interaction

B.3 STRUCTURE OF MANAGEMENT INFORMATION

The **Structure of Management Information (SMI)** defines the rules for describing managed objects. In the SNMP framework managed objects reside in a virtual database called the Management Information Base (MIB). Collections of related objects are defined in MIB modules. The modules are written using a subset of **Abstract Syntax Notation One (ASN.1)**, which describes the data structures in a machine-dependent language. SNMP uses the **Basic Encoding Rule (BER)** to transmit the data structures across the network unambiguously.

Several data types are allowed in SMI. The primitive data types consist of INTEGER, OCTET STRING, NULL, and OBJECT IDENTIFIER. Additional user-defined data types are application specific. Primitive data types are written in uppercase, while user-defined data types start with an uppercase letter but contain at least one character other than an uppercase letter. Table B.2 lists some of the data types permitted in SMI.

An OBJECT IDENTIFIER is represented as a sequence of nonnegative integers where each integer corresponds to a particular node in the tree. This data type provides a means for identifying a managed object and relating its place in the object hierarchy. Figure B.3 shows the object identifier tree as it has been defined for various internet objects. A *label* is a pairing of a text description with an integer for a particular node, also called a subidentifier. The root node is un-labeled. Similarly, the integers that make up an object identifier are separated by periods (.). For example, as shown by the tree, the object identifiers for all Internet objects start with 1.3.6.1.

The internet (1) subtree itself has six subtrees:

Data type	Description
INTEGER	A 32-bit integer
OCTET STRING	A string of zero or more bytes with each byte having a value between 0 to 255
Display STRING	A string of zero or more bytes with each byte being a character from the NVT ASCII set
NULL	A variable with no value
OBJECT IDENTIFIER	An authoritatively defined data type described below
IpAddress	A 32-bit Internet address represented as an octet string of length 4
Counter	A nonnegative integer that increases from 0 to $2^{32} - 1$ and then wraps back to 0
Gauge	A nonnegative integer that can increase or decrease, but which latches at a maximum value
TimeTicks	A nonnegative integer that counts the time in hundredths of a second since some epoch
Opaque	An opaquely encoded data string

TABLE B.2. SMI data types

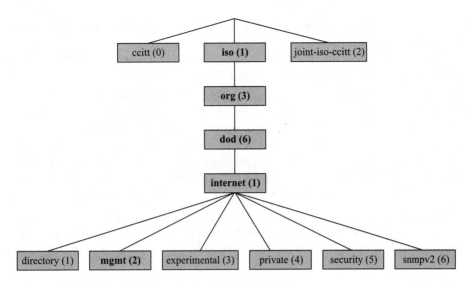

FIGURE B.3 Object identifier tree for internet (1) objects

- The directory (1) subtree is reserved for future use describing how OSI directory may be used in the Internet.
- The mgmt (2) subtree is used to identify "standard" objects that are registered by the Internet Assigned Numbers Authority (IANA).
- The experimental (3) subtree is for objects being used experimentally by working groups of the IETF. If the object becomes a standard, then it must move to the mgmt (2) subtree.
- The private (4) subtree is for objects defined by a single party, usually a vendor. It has a subtree enterprise (1), which allows companies to register their network objects.
- The security (5) subtree is for objects related to security.
- The snmpv2 (6) subtree is reserved for housekeeping purposes for SNMPv2. This subtree includes object information for transport domains, transport proxies, and module identities.

At the time of writing, the mgmt (2) subtree has only one subtree, mib-2 (1), defined. Figure B.4 shows the subtrees for mib-2. For example, according to the

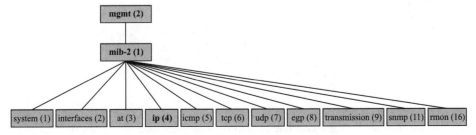

FIGURE B.4 Standard system objects of mib-2 (1) subtree

tree any object definition regarding an IP module would contain the object identifier 1.3.6.1.2.1.4.

Object definitions are generally packaged into **information modules**. Three types of information modules are defined using the SMI:

- MIB modules, which serve to group definitions of interrelated objects.
- Compliance statements for MIB modules. These define a set of requirements that managed nodes must meet with respect to one or more MIB modules.
- Capability statements for agent implementations. These specify the degree to which a managed node is able to implement objects that are defined in a MIB module. Capability statements are often provided by vendors with regard to a particular product and how well it can implement particular MIB modules.

The names of all standard information modules must be unique. Information modules that are developed for a particular company, known as enterprise information modules, must have names that are unlikely to match a standard name.

B.4 MANAGEMENT INFORMATION BASE

The **Management Information Base (MIB)** is a virtual database used to define the functional and operational aspects of network devices. The database should contain an object for each functional aspect of a device that needs to be managed. These objects are usually grouped into different information modules. The information provided by the MIB represents the common view and structure of management capabilities that are shared between the management station and device's agent.

Each definition of a particular object contains the following information about the object: its name, the data type, a human-readable description, the type of access (read/write), and an object identifier. For example, the ip (4) subtree in Figure B.2 refers to an ip group that contains a number of object definitions pertaining to common ip objects. One such object is:

```
ipInHdrErrors   OBJECT-TYPE
    SYNTAX      Counter
    ACCESS      read-only
    STATUS      mandatory
    DESCRIPTION
                "The number of input datagrams discarded due to
                errors in their IP headers, including bad
                checksums, version number mismatch, other format
                errors, time-to-live exceeded, errors discovered in
                processing their IP options, etc."
    ::={ ip 4 }
```

We see that a manager can use this managed information to obtain statistics of the number of IP packets that are discarded because of various errors. Many more objects have been defined in the mib-2 subtree (See RFC 1213 and its recent updates).

B.5   REMOTE NETWORK MONITORING

An additional set of modules, known as Remote Network Monitoring (RMON), was developed in 1995. These are considered to be not only an extension of the mib-2 but also an improvement. In SNMP, managed information that is to be used to monitor a device must be collected by polling. Even if trap-based polling is used, a certain amount of overhead is associated with obtaining the information. RMON uses a technique called *remote management* to obtain monitoring data. In this approach a network monitor (often called a *probe*) collects the data from the device. The probe may stand alone or be embedded within the managed device. Management applications communicate with an RMON agent in the probe by using SNMP; they do not communicate directly with the device itself. This separation from the device makes it easier to share information among multiple management stations. Furthermore, if the management application loses its connection to the RMON agent, the application can usually retrieve the data later, as the RMON agent will continue collecting data (assuming it is able to) even in the absence of a connection to the management application. Because a probe has considerable resources, it can also store various historical statistical information that can later be played back by the network management station. Several switch manufacturers incorporate RMON software in their switches so as to facilitate network management of distributed LANs.

RMON also provides for a higher level of standardization of the information collected. The data collected by mib-2, while standard, is relatively raw and is generally in the form of counters. RMON turns the raw data into a form more suitable for management purposes.

RMON is included as a subtree of mib-2 (rmon (16)), as shown in Figure B.4. Its objects are divided into 10 subtrees based on their function. RMON focuses on network management at layer 2 (data link). An extension of RMON, RMON-2, was proposed to provide network management layer 3 (network) and higher but especially for network layer traffic.

FURTHER READING

International Standard 8824. Information processing systems—Open Systems Interconnection—"Specification of Abstract Syntax Notation One (ASN.1)," International Organization for Standardization, December 1987.

Leinwand, A. and K. Fang, *Network Management: A Practical Perspective*, Addison-Wesley, Reading, Massachusetts, 1993.

Perkins, D. T., *RMON: Remote Monitoring of SNMP-Managed LANs*, Prentice Hall PTR, Upper Saddle River, New Jersey, 1999.

Rose, M., *The Simple Book: An Introduction to Networking Management*, Prentice Hall PTR, Upper Saddle River, New Jersey, 1996. Provides an excellent discussion of SNMP.

Stallings, W., *SNMP, SNMPv2, and CMIP: The Practical Guide to Network-Management Standards*, Addison-Wesley, Reading, Massachusetts, 1993.

RFC 1155, M. Rose and K. McCloghrie, "Structure and Identification of Management Information for TCP/IP-Based Internets," May 1990.

RFC 1157, J. Case, M. Fedor, M. Schoffstall, and J. Davin, "A Simple Network Management Protocol (SNMP)," May 1990.

RFC 1213, K. McCloghrie and M. Rose, "Management Information Base for Network Management of TCP/IP-Based Internets: MIB-II," March 1991.

RFC 1905, J. Case, K. McCloghrie, M. Rose, and S. Waldbusser, "Protocol Operations for Version 2 of the Simple Network Management Protocol (SNMPv2)," January 1996.

RFC 1907, J. Case, K. McCloghrie, M. Rose, and S. Waldbusser, "Management Information Base for Version 2 of the Simple Network Management Protocol (SNMPv2)," January 1996.

RFC 2578, K. McCloghrie, D. Perkins, J. Shoenaelder, eds., "Structure of Management Information for Version 2" (authors of previous version: J. Case, K. McCloghrie, M. Rose, and S. Waldbusser), April 1999.

RFC 2271, D. Harrington, R. Presuhn, and B. Wijnen, "An Architecture for Describing SNMP Management Frameworks," January 1998. This RFC documents SNMPv3 architecture.

RFC 2274, U. Blumenthal and B. Wijnen, "User-Based Security Model (USM) for Version 3 of the Simple Network Management Protocol (SNMPv3)," January 1998.

Index

FSK	frequency shift keying	LECS	LAN emulation configuration server
FTP	File Transfer Protocol	LES	LANE emulation server
GCAC	Generic Connection Admission Control	LIS	logical IP subnetwork
		LLC	logical link control
GCRA	generalized cell rate algorithm	LPC	linear predictive coders
GEO	geostationary earth orbit	LSP	label switched path
GFC	generic flow control	LSR	label switch router
GIF	graphical interchange format	MAC	medium access control
GSM	Global System for Mobile Communications	MAC	message authentication code
		MBONE	multicast backbone
HA	home agent	MBS	maximum burst size
HDLC	High-level Data Link Control	MD5	message digest 5 algorithm
HMAC	hashed message authentication code	MH	mobile host
HEC	header error check	MIB	management information base
HOL	head-of-line priority queueing	MID	message identifier
HTML	hypertext markup language	MPC	MPOA client
HTTP	Hypertext Transfer Protocol	MPEG	Motion Picture Expert Group
iBGP	internal BGP	MPLS	multiprotocol label switching
ICMP	Internet Control Message Protocol	MPOA	multiprotocol over ATM
IEEE	Institute of Electrical and Electronics Engineers	MPS	MPOA server
		MSC	mobile switching office
IETF	Internet Engineering Task Force	MSL	maximum segment lifetime
IGMP	Internet Group Management Protocol	MSS	maximum segment size
IGP	Interior Gateway Protocol	MTSO	mobile telephone switching office
IKE	internet key exchange	MTU	maximum transmission unit
IP	Internet Protocol	NAK	negative acknowledgment frame
IPSec	IP security	NAV	network allocation vector
IS	Interim Standard	NCP	Network Control Protocol
ISDN	Integrated Services Digital Network	NHC	next-hop client
ISN	initial sequence number	NHRP	Next-Hop Resolution Protocol
ISO	International Organization for Standardization	NHS	next-hop server
		NIC	network interface card
ITU	International Telecommunications Union	NNI	network-network interface
		NRM	normal response mode
IXC	interexchange carrier	NRZ	nonreturn-to-zero encoding
JPEG	Joint Photograph Expert Group	NRZ-I	NRZ-Inverted
KDC	key distribution center	NVT	network virtual terminal
LAN	local area network	OC	optical carrier level
LANE	LAN emulation	OSI	open systems interconnection
LAP	link access procedure	OSPF	open shortest path first
LATA	local access transport areas	PAP	Password Authentication Protocol
LCP	Link Control Protocol	PCF	point coordination function
LDP	Label Distribution Protocol	PCM	pulse code modulation
LEC	LAN Emulation Client	PDU	protocol data unit
LEC	local exchange carrier	PHB	per-hop behavior